Operator Theory: Advances and Applications
Vol. 145

Nonlinear Hyperbolic Equations, Spectral Theory, and Wavelet Transformations

A Volume of
Advances in Partial Differential Equations

Sergio Albeverio
Michael Demuth
Elmar Schrohe
Bert-Wolfgang Schulze
Editors

Springer Basel AG

Editors:

Sergio Albeverio
Institut für Angewandte Mathematik
Universität Bonn
53115 Bonn
Germany
e-mail: albeverio@uni-bonn.de

Elmar Schrohe
Institut für Mathematik
Universität Potsdam
14415 Potsdam
Germany
e-mail: schrohe@math.uni-potsdam.de

Michael Demuth
Institut für Mathematik
Technische Universität Clausthal
38678 Clausthal-Zellerfeld
Germany
e-mail: demuth@math.tu-clausthal.de

Bert-Wolfgang Schulze
Institut für Mathematik
Universität Potsdam
14415 Potsdam
Germany
e-mail: schulze@math.uni-potsdam.de

2000 Mathematics Subject Classification 35Lxx, 35P15, 42C40, 47A53

A CIP catalogue record for this book is available from the
Library of Congress, Washington D.C., USA

Bibliographic information published by Die Deutsche Bibliothek
Die Deutsche Bibliothek lists this publication in the Deutsche Nationalbibliografie; detailed bibliographic data is available in the Internet at <http://dnb.ddb.de>.

ISBN 978-3-0348-9429-6 ISBN 978-3-0348-8073-2 (eBook)
DOI 10.1007/978-3-0348-8073-2

Printed on acid-free paper produced from chlorine-free pulp. TCF ∞
Cover design: Heinz Hiltbrunner, Basel

ISBN 978-3-0348-9429-6

9 8 7 6 5 4 3 2 1 www.birkhauser.ch

Contents

Preface

This book is the seventh volume of "Advances in Partial Differential Equations", a series originating from the work of the research group "Partial Differential Equations and Complex Analysis" at the University of Potsdam.

The present volume focuses on recent developments in nonlinear and hyperbolic equations. In the first contribution, P. Popivanov of Sofia studies the singularities of solutions of several classes of nonlinear partial differential equations and systems. He begins with a survey of the known theory on propagation and interaction of singularities and then presents his own results which have applications to the Monge–Ampère equation, to quasi-linear systems arising in fluid mechanics as well as to integro-differential equations for mechanics of media with memory.

There follows an article by F. Hirosawa (Tsukuba) and M. Reissig (Freiberg) on $L_p - L_q$ decay estimates for Klein–Gordon equations with time-dependent coefficients. They explain, in particular, the influence of the relation between the mass term and the wave propagation speed on the estimates.

The third paper is by M. Dreher (Freiberg). He investigates quasi-linear weakly hyperbolic equations. His main topics are the local existence of solutions in Sobolev spaces and C^∞, blow-up criteria, domains of dependence, and C^∞ regularity.

Spectral theory of semibounded selfadjoint operators is the topic of the contribution by A. Noll (Darmstadt). He proves upper and lower bounds for the bottom eigenvalue as well as an upper bound for the second eigenvalue in terms of capacitary estimates.

The Gabor transformation and the wavelet transformation are two distinguished tools in signal processing with a wide range of applications. In their joint article, M. Holschneider (Potsdam) and B. Nazaret (Lyon) construct an interpolating family between them, leading to the construction of anisotropic scales of Banach spaces.

Several papers complete the present volume; they are concerned with subjects as diverse as the Weyl–Hörmander calculus for manifolds, singular fibrations, and index theory.

Bonn, Clausthal, and Potsdam, March 2003

S. Albeverio M. Demuth E. Schrohe B.-W. Schulze

Operator Theory:
Advances and Applications, Vol. 145, 1–94

Nonlinear PDE. Singularities, Propagation, Applications

P.R. Popivanov

To my wife Angela

Preface

This text deals with the singularities of the solutions of several classes of nonlinear partial differential equations and systems. Applications of the results here obtained are given for the Monge–Ampère equation, for quasi-linear systems arising in fluid mechanics, and for some nonlinear integrodifferential equations useful in solid body mechanics in media with memory. In our investigations we have used two different approaches – the classical method of characteristics in the case of systems with one space variable and the machinery of paradifferential operators in the multidimensional case.

Our text comprises three chapters divided in 10 sections. Chapter I consists of §1, §2, §3, §3 being Appendix I. Chapter II consists of §4, §5, §6, §7, §8, §8 being Appendix II. Chapter III is divided into §9 and §10.

Section 1 is a survey on the propagation and interaction of singularities of one-dimensional semilinear hyperbolic systems. The main theorems here were proved by Rauch–Reed, L. Micheli, I. Iordanov and others. The solutions are allowed to have jump discontinuities in the initial data. A detailed study of the propagation of the jump discontinuities is given in this section. The singularities mentioned before propagate along the corresponding characteristics till their intersection (as in the linear case). After their collision new singularities can appear which will propagate along the full set of characteristics starting from the collision point. The newly created singularities are weaker than the initial ones. Often they are called anomalous singularities. Interesting effects can appear when the system possesses two tangential at one point characteristics. Thus we have nonlinear interaction between two singularities – the first one created by the jump discontinuity in the Cauchy data and the second one generated by the weak hyperbolicity.

Both in the strictly hyperbolic and the weakly hyperbolic cases, a necessary and sufficient condition for the existence of anomalous singularities is proved. Moreover, the optimal order of the anomalous singularities is also found.

Section 2 deals with the interaction of three conormal waves for the semilinear wave equation with two space variables. As it was proved by Bony and Melrose–Ritter, in this situation new singularities in comparison with the linear case appear.

We follow here (up to some modifications) the approach of Keller–Ting. The only new feature is the proof of convergence and of pointwise differentiation of the series, giving the solution inside and up to the light cone inscribed in the characteristic pyramid. The proof is elementary, as it involves the Gauss hypergeometric function, avoiding this way the heavy technic of second microlocalization.

In Appendix I we propose blow up results for the solutions of the Cauchy problem for some classes of quasi-linear systems in the plane. This is another type of singularities when the solutions remain bounded (eventually becoming multi-valued after some time) but their gradient blows up in a finite time. Interesting books on the subject are written by Hörmander [31], Li Ta-tsien [48], Alinhac [3]. We prove here a result of ours and we apply it to the nonlinear vibrating string equation.

§4 (Chapter II) deals with the singularities of the solutions of quasi-linear hyperbolic systems in the plane and "in the generic case". To do this we linearize our system via the classical hodograph transformation and we investigate the singularities of this transformation by applying the Whitney theorem for C^∞ normal forms in $\mathbf{R}^2$.

Our main results in this chapter are concentrated in §5–§7. By using the paradifferential approach we prove several theorems for microlocal singularities of the solutions of some classes of fully nonlinear systems. To be more precise, we find sufficient conditions for nonexistence of isolated singularities in a conical neighborhood of the corresponding characteristic point. Applications are given to the Monge–Ampère equation and to some systems arising in fluid mechanics. The theorems here formulated and proved are mainly due to the author.

To complete the results of this chapter and to compare them with the results from §1, Chapter I, we include Appendix II containing, with proofs, several results of Bony on interaction of conormal waves for semilinear PDE.

Chapter III is devoted to the travelling waves for several classes of nonlinear integrodifferential operators arising in solid body mechanics. We construct C^1 smooth solutions of them in the form of travelling waves of tension. Blow up results in the case of a solid body located in a medium with exponential memory are also formulated.

We have tried to give simple and clear proofs of the central results given here. They are illustrated by many examples and 28 figures. That is why we believe that our text could be interesting and useful for young mathematicians beginning their work in the domain of nonlinear partial differential equations and their applications. Probably, there are new results for the specialists in the domain of nonlinear PDE as well.

Acknowledgments are due to Prof. B.-W. Schulze and to Prof. M. Demuth for their expert co-operation. The main part of the theorems in Chapter II and Chapter III were communicated at first at the seminar of Prof. B.-W. Schulze in Potsdam, as well as at the seminars on PDE of Prof. Rodino (Torino), Prof. Parenti (Bologna), Prof. Zanghirati (Ferrara) and at the seminar of PDE in Univ. P. Sabatier – Toulouse.

In preparing the LaTeXversion of the manuscript we had the valuable help of L. Apostolova and P. Marinov. Many, many thanks to them. Acknowledgments are due also to I. Iordanov for support in preparing §1, Chapter I and to P. Roussev for Lemma 2.2 from §2, Chapter I.

Chapter I
Interaction of singularities of the solutions of semilinear hyperbolic equations and systems

1. Singularities of the solutions of one-dimensional semilinear hyperbolic systems

1. Rather complete results have been obtained for the singularities of linear and semilinear hyperbolic systems with one space dimension. The corresponding initial data are very often assumed to have singularities of the type "finite jump in finitely many points", i.e., the jumps of all derivatives are finite and the jump discontinuities exist in a discrete set. The methods used in the proof of these results are relatively elementary and transparent. Unfortunately, this approach works only in the case of one space variable and it seems impossible to generalize it for several space variables.

It is well known that the singularities of the solutions of one space dimensional linear hyperbolic systems generated by the singularities of their initial data propagate along the corresponding characteristics (linear effect). The paper [69] deals with a new, nonlinear effect for a class of semilinear hyperbolic equations in which anomalous singularities in comparison with the linear case appear. The new singularities are created by the interaction of already existing singularities generated by the initial data and propagating along the characteristics. The new-born singularities are weaker than the ones which create them. Moreover, they appear in the cross points of the characteristics carrying out the initial singularities and they propagate along (third) characteristics starting from these cross points. A full investigation of the problem mentioned above in the strictly hyperbolic case is contained in [70].

More precisely, a necessary and sufficient condition for the existence of anomalous singularities is found in [70].

A very important example of the appearance of anomalous singularities for one-dimensional weakly hyperbolic semilinear systems was proposed in [71] and it was studied in detail by L. Micheli in [57]. It concerns the nonlinear interaction between two singularities – the first one created by a jump discontinuity in the initial data and the second one generated by the weak hyperbolicity (linear effect). The optimal order of the anomalous singularity is found in [57]. An essential simplification of the approach of L. Micheli is given in [37]. This way it is possible to find a necessary and sufficient condition for the existence of anomalous singularities (the latter being identical to the condition from [70] for strictly hyperbolic systems). A weakly hyperbolic system having three tangential characteristics is studied similarly in [79]. Propagation of singularities of quasi-linear strictly hyperbolic systems was considered in [56].

2. For the sake of simplicity we shall study systems written in a diagonal form

$$D_i u_i \equiv \partial_t u_i + \lambda_i(x,t)\partial_x u_i = f_i(x,t,u), \quad i = 1,2,\dots,n, \tag{1.1}$$

where $u = (u_1, \dots, u_n)$, $f_i \in C^\infty(\mathbf{R^{2+n}})$, the real-valued function $\lambda_i \in C^\infty(\mathbf{R}^2)$ and

$$\inf_{i \neq j, (x,t) \in \mathbf{R}^2} |\lambda_i(x,t) - \lambda_j(x,t)| \geq \delta > 0. \tag{1.2}$$

Thus the hyperbolic system (1.1) is investigated in the "trapezoidal" domain

$$G_T = \{(x,t) \in \mathbf{R}^2 \ | 0 \leq t \leq T, \ x(t) \leq x \leq X(t)\},$$

where having in mind the local character of our study, we assume that the curves $x = x(t)$, $x = X(t)$ are defined in the whole interval $[0,T]$ and they satisfy the equations

$$\frac{dx(t)}{dt} = \max_{1 \leq k \leq n} \{\lambda_k(x,t)\}, \ x(0) = a,$$
$$\frac{dX(t)}{dt} = \min_{1 \leq k \leq n} \{\lambda_k(X,t)\}, \ X(0) = b.$$

Moreover, we suppose that $x(T) \leq X(T)$.

It is obvious that in the strictly hyperbolic case (1.2) we have $\lambda_1(x,t) < \lambda_2(x,t) < \cdots < \lambda_n(x,t)$ and therefore the above written equations become very simple. The more general form of the equations for $x(t)$ and $X(t)$ will be used in the nonstrictly hyperbolic case.

The curve $\xi = \xi_i(\tau; x,t) \equiv \xi_i(\tau)$ given by

$$\frac{d\xi(\tau)}{d\tau} = \lambda_i(\xi(\tau),\tau), \ \xi(t) = x \tag{1.3}$$

is called an i-characteristic of (1.1) passing through the point (x,t). The arcs $\xi = \xi_i^\pm$, where $\xi_i^\pm \equiv \xi(\tau)$ for $\pm(\tau - t) > 0$, are called forward (ξ^+) and respectively backward i-characteristic through the point (x,t). Without loss of generality we shall assume that if $(x,t) \in G_T$, then all i-characteristics exist globally, i.e., they exist until leaving G_T. Obviously, ξ_i are C^∞ smooth with respect to their arguments.

The domain G_T can be defined geometrically as follows: For each point $(x,t) \in G_T$ the backward characteristics $\xi = \xi_i(\tau; x,t)$, $\tau < t$, $i = 1,2,\dots,n$, are contained in G_T and they cross the lower base $[a,b]$ of G_T.

The Cauchy data for the system (1.1) are given by

$$u_i(x,0) = u_i^0(x), \quad i = 1,2,\dots,n, \tag{1.4}$$
$$u_i^0 \in C^\infty([a,b] \setminus \{x_0^1, \dots, x_0^s\}).$$

Moreover, for each point $x_0^k \in (a,b)$, $1 \leq k \leq s$, there exists a neighborhood $V_k \ni x_0^k$ such that $u_i^0 \in C^h(V_k)$, $h = h(k,i) \geq -1$, $h \in \mathbf{Z}$, the $(h+1)$ derivative of

u_i^0 is discontinuous at the point x_0^k and the corresponding left and right limits of $u_i^{0(p)}$ at x_0^k for all derivatives of order $p \geq h+1$ are finite.

Thus the integer h is the order of the highest derivative of u_i^0 which is continuous at x_0^k. Obviously, $h = -1$ iff the function u_i^0 has a finite jump discontinuity at x_0^k. We shall say that the initial data u_i^0 has a (singularity of the type) finite jump of order $h(i,k)$ at x_0^k and we shall denote this fact by : $u_i^0 \in C_*^h(V_k \setminus \{x_0^k\})$, $C_*^h(V_k \setminus \{x_0^k\}) = C^h(V_k) \cap C_*^\infty(V_k \setminus \{x_0^k\})$. Similar notation will be used in the case of two variables: $C_*^h(V \setminus C)$, where $V \subset \mathbf{R}^2$ is a domain and the curve $C \subset V$. Often we shall say: a function of finite order h at C.

The union S of all outgoing characteristics starting from the points

$$(x_0^1, 0), \ldots, (x_0^s, 0),$$

i.e., the outgoing characteristics starting from their cross points etc., will play a crucial role in studying the Cauchy problem (1.1), (1.2) (1.4). To do this we shall prove at first the existence of a solution having finite order at the arcs of S. In our second step we shall find the sharp order of the singularities (more precisely, of the jump of the solution).

The first problem mentioned above is solved in [70], [57] by replacing the system (1.1) with an integral system equivalent to it. The contraction mapping principle applied to the latter system in the functional class $C_*^N(G_T \setminus S)$ gives the existence result. Another approach was proposed in [36]. It combines the classical theorem for the existence and uniqueness of low regularity solution with the C^∞ regularity of the solution of (1.1), (1.2) (1.4) in $G_T \setminus S$ – proved by L. Micheli. This way the linear and nonlinear effects can be followed easily.

To realize the program mentioned above we shall remind the reader of the classical theorem for existence and uniqueness of the solution of the Cauchy problem for the linear system (1.1), (1.2) (1.4), i.e.,

$$f_i(x,t,u) = \sum_{j=1}^{n} a_{ij}(x,t)u_j + b_i(x,t), \quad i = 1, 2, \ldots, n, \tag{1.5}$$

supposing the Cauchy data $u_i^0 \in C^0[a,b]$.

We shall say that the vector-valued function $(u_1(x,t), \ldots, u_n(x,t))$ defined in G_T is a solution of the problem (1.1), (1.2) (1.4) if it is continuous in G_T, each component u_i is continuously differentiable along the vector field $\partial_t + \lambda_i(x,t)\partial_x$ and the equalities (1.1), (1.4) hold. Similar but more general results are given in the book [74]. The just proposed notion of a solution according to this book is called a solution in a broad sense. In contrast with it the notion of a classical solution for the same problem assumes C^1 differentiability of u in the interior of the domain G_T.

For an arbitrary point $(x,t) \in G_T$ we rewrite (1.1) this way:

$$\frac{d}{d\tau} u_i(\xi_i(\tau; x,t), \tau) = f_i(\xi_i(\tau; x,t), \tau, u(\xi_i(\tau), \tau)), \quad i = 1, 2, \ldots, n \tag{1.6}$$

and therefore into integral form

$$u_i(x,t) = u_i^0(\xi_i(0;x,t)) + \int_0^t f_i(\xi_i(\tau),\tau,u(\xi_i(\tau),\tau))d\tau. \tag{1.7}$$

Obviously, the finding of a continuous solution of the integral system (1.7) is equivalent to the existence of a broad sense solution for (1.1), (1.2) (1.4).

From (1.7) and having in mind (1.5), we define the successive approximations for the linear case:

$$\overset{(s+1)}{u_i}(x,t) = \overset{(0)}{u_i}(x,t) + \int_0^t \sum_{j=1}^n a_{ij}(\xi_i(\tau;x,t),\tau) \overset{(s)}{u_j}(\xi_i(\tau),\tau)d\tau, \tag{1.8}$$

where $\overset{(0)}{u_i}(x,t) = \overset{(0)}{u_i}(\xi_i(0;x,t)) + \int_0^t b_i(\xi_i(\tau),\tau)d\tau$ and $i = 1,2,\ldots,n$.

It is evident that $\overset{(s)}{u}_i \in C^0(G_T)$ and that each component of $\overset{(s)}{u}_i$ possesses a continuous derivative along the corresponding characteristic. That is, $D_i \overset{(s)}{u}_i \equiv (\partial_t + \lambda_i(x,t)\partial_x) \overset{(s)}{u}_i$ is continuous (providing that $\overset{(s-1)}{u_i}$ is continuous).

For a domain $G \subset \mathbf{R}^2$ we introduce some notation. Put

$$\|u\|_G = \sup_{(x,t)\in G} \sqrt{\sum_{i=1}^n u_i^2(x,t)}$$

and denote

$$G_\tau = G_T \cap \{(x,t)|\ t \leq \tau\}, \quad \overset{(s)}{V}(t) = \|\overset{(s+1)}{u} - \overset{(s)}{u}\|_{G_t}, \quad t \in [0,T].$$

One can easily see that

$$\overset{(s)}{V} \leq C \int_0^t \overset{(s-1)}{V}(\tau)d\tau, \tag{1.9}$$

where the constant C is estimated by $\max\limits_{1\leq i,j\leq n,(x,t)\in G_T} |a_{ij}(x,t)|$.

Applying induction we conclude that $\overset{(s)}{V}(t) \leq \dfrac{(Ct)^s}{s!} \overset{(0)}{V}(t)$ and consequently the sequence $\{\overset{(s)}{u}(x,t)\}$ is uniformly convergent to some continuous function $u(x,t)$. Passing to the limit in (1.8) we see that u satisfies the integral system (1.7). The proof of the existence of a broad-sense solution is completed.

We point out that the uniqueness result holds true in the same functional class. In fact, denote by u the difference between two broad-sense solutions of the linear system and put $V(t) = \|u\|_{G_t} \geq 0$. Then we have

$$V(t) \leq C \int_0^t V(\tau)d\tau. \tag{1.10}$$

According to the Gronwall inequality we get $V(t) \equiv 0$ as $V(0) = 0$.

Remark 1.1. In the previous proof we did not use the strict hyperbolicity of the system. Thus the same result for existence and uniqueness of a broad-sense solution is also valid in the weakly hyperbolic case.

Regularity of the broad-sense solution depends on the regularity of the initial data. More precisely, suppose that $u_i^0 \in C^1[a,b]$, $1 \le i \le n$. Then obviously $\overset{(s)}{u} \in C^1$, $\forall s \in \mathbf{N}$. Differentiating (1.8) with respect to x we get

$$\begin{aligned}\partial_x \overset{(s+1)}{u}_i (x,t) &= \partial_x \overset{(0)}{u}_i (x,t) \\ &+ \sum_{j=1}^{n} \int_0^t [\partial_1 a_{ij}(\xi_i(\tau;x,t),\tau)\partial_2\xi_i(\tau;x,t) \overset{(s)}{u}_j (\xi_i(\tau),\tau) \\ &+ u_{ij}(\zeta_i(\tau),\tau)\partial_1 \overset{(s)}{u}_j . \partial_2\xi_i(\tau;x,t)]d\tau,\end{aligned}$$

where ∂_μ, $\mu = 1,2$, stands for the derivative with respect to the μ-variable of the corresponding smooth function. Having in mind the uniform boundedness of a_{ij} and of the derivatives of ξ_i in G_T, we get

$$\begin{aligned}&\|\partial_x \overset{(s+1)}{u} - \partial_x \overset{(s)}{u} \|_{G_t} \\ &\le C_1 \int_0^t \| \overset{(s)}{u} - \overset{(s-1)}{u} \|_{G_\tau} d\tau + C_2 \int_0^t \|\partial_x \overset{(s)}{u} - \partial_x \overset{(s-1)}{u} \|_{G_\tau} d\tau\end{aligned} \tag{1.11}$$

with appropriate constants $C_1, C_2 > 0$.

Denote now $\overset{(s)}{W}(t) = \|\partial_x(\overset{(s)}{u} - \overset{(s-1)}{u})\|_{G_t}$, $\overset{(s)}{V}(t) = \| \overset{(s)}{u} - \overset{(s-1)}{u} \|_{G_t}$ and $C_0 = \max(C, C_1, C_2)$ where the constant C participates in (1.9).

Applying (1.11) we obtain $\overset{(1)}{W}(t) \le C_0 t \overset{(0)}{V}(t) + C_0 t \overset{(0)}{W}(t)$. In a similar way we have $\overset{(2)}{W}(t) \le C_0 \int_0^t \overset{(1)}{V}(\tau)d\tau + C_0 \int_0^t (C_0\tau \overset{(0)}{V}(\tau) + C_0\tau \overset{(0)}{W}(\tau))d\tau$.

But according to (1.9), $\overset{(1)}{V}(\tau) \le \dfrac{C_0\tau}{1!} \overset{(0)}{V}(\tau)$ and thus $\overset{(2)}{W}(t) \le \dfrac{(C_0 t)^2}{2!} \overset{(0)}{V}(t) + \dfrac{(C_0 t)^2}{2!} \overset{(0)}{V}(t) + \dfrac{(C_0 t)^2}{2!} \overset{(0)}{W}(t)$.

Applying induction to (1.11) we find that for each $s \in N$,

$$\overset{(s)}{W}(t) \le s\frac{(C_0 t)^s}{s!} \overset{(0)}{V}(t) + \frac{(C_0 t)^s}{s!} \overset{(0)}{W}(t).$$

Thus the sequence $\{\partial_x \overset{(s)}{u}\}$ turns out to be uniformly convergent, i.e., the broad-sense solution $u(x,t)$ is differentiable along the x-axis and therefore (1.1) implies it to be differentiable in t as well, i.e., $u \in C^1(G_T)$. Similarly, initial data being from C^k, the solution belongs to $C^k(G_T)$, $k \ge 1$, $k \in \mathbf{N}$.

Now we are able to study the singularities of the solution of the linear Cauchy problem (1.1), (1.2) (1.4), (1.5). Let the initial data possess finitely many discontinuity points. For example, if

$$\begin{cases} u_i^0(x) & \in \quad C_*^{-1}([a,b] \setminus \{x_0^1\}), \\ u_j^0 & \in \quad C[a,b], j \neq i, \end{cases}$$

then (see (1.8)) $\overset{(s)}{u_i}$ is discontinuous across the i-characteristic C_i, starting from the point $(x_0^1, 0)$ (see Fig. 1). All the other components $\overset{(s)}{u}_j$, $j \neq i$, of the s-approximation are continuous because of the continuity of their initial data and $\overset{(s)}{u}_i$ participating under the integral sign.

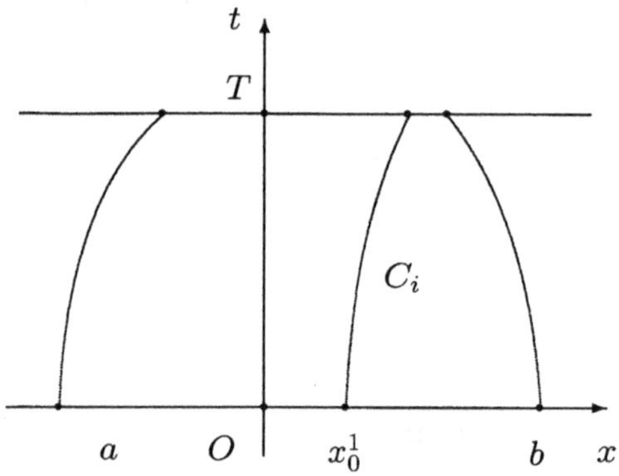

Fig. 1

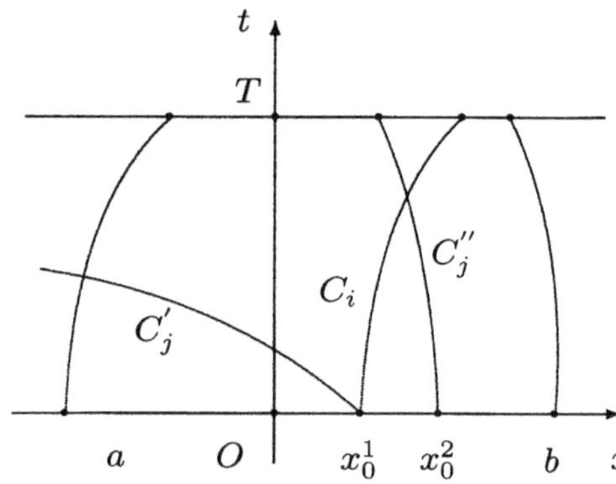

Fig. 2

Certainly, the boundary values of $\overset{(s)}{u}_i$ from both sides of C_i are different, i.e., $\overset{(s)}{u}_i$ is a double-valued function on C_i.

If another initial datum also has a jump, e.g., if $u_j^0(x) \in C_*^{-1}((a,b) \setminus \{x_0^1, x_0^2\})$ (see Fig. 2) for some j, we conclude inductively again that $\overset{(s)}{u}_i$ is discontinuous across C_i , $\overset{(s)}{u}_j$ is discontinuous across C_j', C_j'' and all other components are continuous.

In order to study the Cauchy problem (1.1), (1.4) with discontinuous initial data, the remarks above lead to the following generalization of the notion of broad-sense solution (for short – discontinuous solution): We assume that the vector-valued function $u = (u_1, u_2, \dots, u_n)$ defined in G_T has the properties given below:

(i) The restriction of each component u_k along each k-characteristic is continuous, and a piecewise differentiable function such as the derivative $D_k u_k$ is assumed to possess a finite jump in finitely many points of the chosen characteristic

(ii) There exist at most finitely many k-characteristics C_k, $(k = 1, \dots, n)$ on which the component u_k is not single-valuedly defined, i.e., the limits of u_k at the points of such a characteristic C_k (from both sides of C_k) are different.

Remark 1.2. A typical example of a function satisfying the conditions of that definition is $w_k s(x,t)$ given by the equality

$$w_k(x,t) = v_k(x,t) + \int_0^t f(\xi_k(\tau;x,t),\tau,u(\xi_k(\tau),\tau)d\tau,$$

where each component u_j of $u = (u_1,\ldots,u_n)$ is discontinuous across at most finitely many characteristics of type C_j and the functions v_k are discontinuous across at most finitely many characteristics of type C_k. It is clear that for a fixed characteristic C_k passing through the point (x,t) the discontinuities of the components u_j, $j \neq k$, do not influence the continuity of the integral term. On the other hand, the effect of discontinuity of u_k just across the chosen characteristic C_k (along which we integrate) is the same as the effect of discontinuity of v_k across C_k, namely – the function $w_k(x,t)$ is double-valued on C_k.

Having in mind the definition of discontinuous solution we expect for the case studied in Fig. 2 a solution with the following singularities:

- the component u_i is double-valued on C_i and in $G_T \setminus C_i$ the derivative $D_i u_i$ exists and is continuous except for the points of C_j', C_j'' where it possesses a finite jump across them;
- the component u_j is continuous in each of the three closed domains into which G_T is divided by $C_j' \cup C_j''$, having $D_j u_j$ discontinuous only at the points of C_i;
- the other components of u are continuous at G_T.

So let the initial data for the linear system (1.1) be discontinuous in finitely many points of $[a,b]$. Let S_0 be the union of the respective characteristics starting from these points (i.e., if $u_i^0 \in C^{-1}((a,b)\backslash x_0)$, then the i-characteristic through the point $(x_0,0)$ is included in S_0). The curves of S_0 divide G_T into finitely many closed subsets G_k. It becomes clear from the remarks made above that the sequence $\{\overset{(s)}{u}\}$ of successive approximations is continuous on each of the domains G_k (it is possible for some components of $\overset{(s)}{u}$ to have different boundary values on each of their common boundaries). If considering a fixed component j, then the sequence $\{\overset{(s)}{u}_j\}$ is continuous on each of the closed domains G_m^j obtained through the division of G_T by the characteristics of the type j included in S_0. In fact, each of the domains G_m^j is a union of several subsets G_k, having possibly in common only contour common points. The estimate (1.9) is obtained in a similar way and therefore $\{\overset{(s)}{u}\}$ is uniformly convergent.

The limit function $u(x,t)$ is a discontinuous solution of (1) with the chosen initial data, because:

- $u \in C^0(\bar{G}_k)$, $\forall k \in N$ due to the continuity of the functions from the sequence $\{\overset{(s)}{u}\}$ on each subset $\bar{G}_k$.
- $D_j u_j (= f_j(x,t,u))$ exists in each subdomain G_k and is discontinuous only on the characteristics dividing G_m^j into its component domains G_k.

Now we can generalize the result of regularity proved by the lack of discontinuities. Let us suppose that the initial data belong to the class

$$C^{-1}((a,b) \setminus \{x_0^1, \dots, x_0^k\}) \cap C^0((a,b) \setminus \{y_0^1, \dots, y_0^m\}), \tag{1.12}$$

where some point y_0^r may coincide with some point x_0^l. Let $u_i^0(x)$ be discontinuous at the point x_0. Then, certainly, a C^{-1} singularity of the i-component (i.e., $u_i \in C_*^{-1}(G_T \setminus C_i(x_0,0)))$ propagates along the i-characteristic through $(x_0,0)$ denoted by $C_i(x_0,0)$. It means that a C^0 singularity (even if $u_j^0(x)$ is smooth in a neighborhood of x_0) propagates along any other characteristic $C_j(x_0,0)$, $j \neq i$.

Example 1. For the system

$$(\partial_t + \partial_x)u = 0,$$
$$\partial_t v = u,$$

where $u^0(x) = \begin{cases} 1, & x \le 0 \\ 2, & x > 0 \end{cases}$, $v^0(x) \equiv 0$, we have $v(x,t) = \int_0^t u(x,\tau)d\tau$.

This function is continuous but $\lim\limits_{x\to 0^-} \partial_x v(x,t) = 0$ and $\lim\limits_{x\to 0+} \partial_x v(x,t) = 1$ (see Fig. 3). These calculations are direct because we have $u = 1$ on the left of the characteristic $t = x$ and $u = 2$ on the right of it. The same characteristic bears the singularity of u given by the initial data.

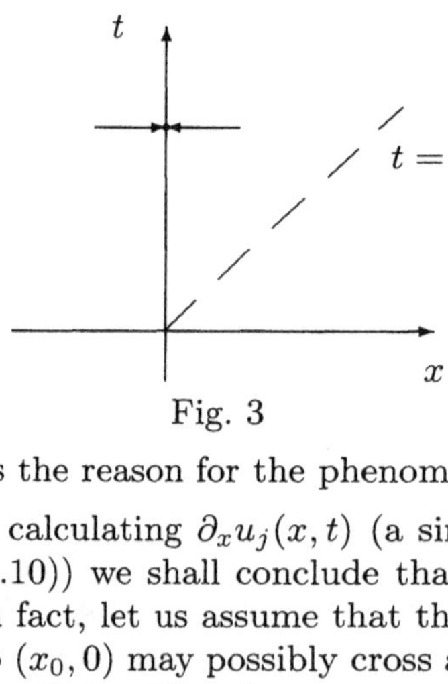

Fig. 3

This example explains the reason for the phenomenon formulated before. In a more general context, in calculating $\partial_x u_j(x,t)$ (a similar calculation for $\partial_x \overset{(s)}{u_j}$ was made after formula (1.10)) we shall conclude that this derivative is discontinuous across $C_j(x_0,0)$. In fact, let us assume that the j-characteristics through the points A and B near to $(x_0,0)$ may possibly cross a single bearing singularity characteristic, namely $C_i(x_0,0)$ (see Fig. 4).

In order to check the continuity of $\partial_x u_j$ at the point D $\in C_j(x_0,0)$, we calculate that derivative at the nearby points (x,t) on the left and on the right of $C_j(x_0,0)$ using the expression

$$u_j(x,t) = u_j^0(\xi_j(0;x,t)) + \int_0^t (\dots)d\tau.$$

For the characteristic $C_j(A)$ the result is quite analogous to the expression mentioned above for $\partial_x \overset{(s+1)}{u_i}$ after formula (1.10). For the characteristic $C_j(B)$ in the

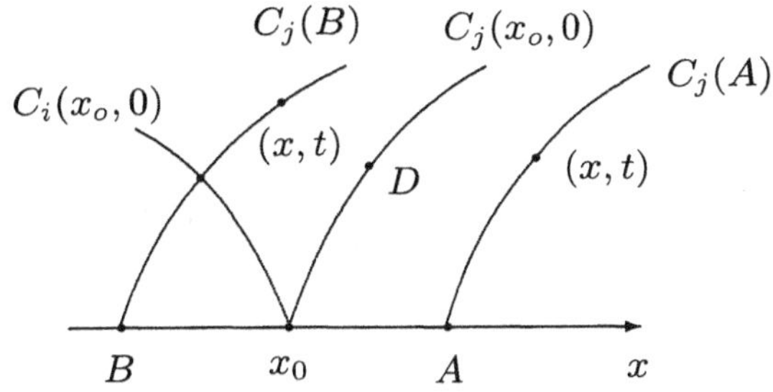

Fig. 4

function under the integral sign there exists a discontinuity (on the characteristic $C_i(x_0,0)$) and therefore we use the representation

$$\int_0^t a_{ji}(\xi_j(\tau,B),\tau)u_i(\xi_j(\tau,B),\tau)d\tau = \int_0^{t_1}(\ldots) + \int_{t_1}^t(\ldots), \tag{1.13}$$

where $t_1 = t_1(x,t)$ is the value of the parameter τ giving the coordinates of the cross point $C_i(x_0,0)\times C_j(B)$. The x derivative of (1.13) is

$$\int_0^t \{\partial_x(a_{ji}.u_i)\}d\tau + [a_{ji}u_i]_{(t_1)}.\partial_x t_1,$$

where $\{\partial_x g\}$ denotes the derivative of g evaluated almost everywhere (i.e., everywhere except for the point of discontinuity of u_i) and $[h]_{(t_1)}$ stands for the jump $h(t_1-0)-h(t_1+0)$.

Having in mind that $\tau = t_1$ is the solution of the equation

$$\xi_j(\tau;B) = \xi_i(\tau;x,t),$$

after differentiation with respect to x we get

$$\partial_x t_1(x,t) = \frac{\partial_x\xi_i(t_1;x,t)}{\lambda_j(\xi_j(t_1;B),t_1)-\lambda_i(\xi_i(t_1;x,t),t_1)}.$$

According to our assumption (1.2) the function $\partial_x t_1$ depends continuously on (x,t).

Remark 1.3. Suppose that the initial data u_k^0, u_m^0, $k\neq m$ have jump discontinuities at the point x_0. Apart from the C^{-1} singularities on the respective characteristics C_k, C_m a propagation of a C^0 singularity along $C_j, j\neq k,m$ is possible (see Fig. 5); but it is also possible that the singularities compensate.

That problem has not yet been studied in detail.

As a conclusion of these remarks we shall note that if some initial datum has a singularity of 0 order at a fixed point (i.e., its first derivative is discontinuous there) then a singularity of the same order shall be propagated by it only along the respective characteristic. Only the second derivatives can be expected to be discontinued across the other characteristics starting from the same point.

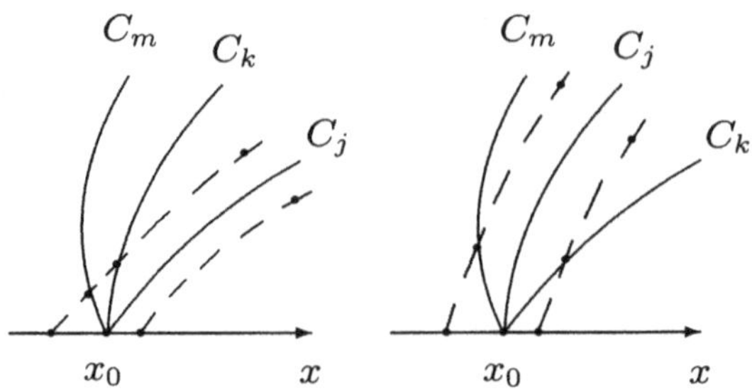

Fig. 5

We shall mark two important properties.

Remark 1.4. We established that in the linear case the singularities appearing from jumps in the initial data propagate along the respective characteristics. It can be easily seen that the singularity across a certain characteristic cannot disappear (nor can a singularity appear across some characteristic if the singularity is lacking in the initial moment). For that purpose, from the integral representation (1.7) written for characteristics near to the considered C_i, we find the following expression for the jump $[u_i](x,t)$ of u_i across C_i:

$$[u_i](x,t) = [u_i^0(x,t)](\xi_i(0;x,t)) + \int_0^t a_{ii}(\xi_i(\tau),\tau)[u_i](\xi_i(\tau),\tau)d\tau.$$

Now the Grownwall inequality implies everything.

Remark 1.5. Except for the singularities described, there will not be new singularities at a later moment of time. We can easily see this even in the cases when the backward characteristic through the studied point crosses more than one bearing singularity characteristic (see Fig. 6).

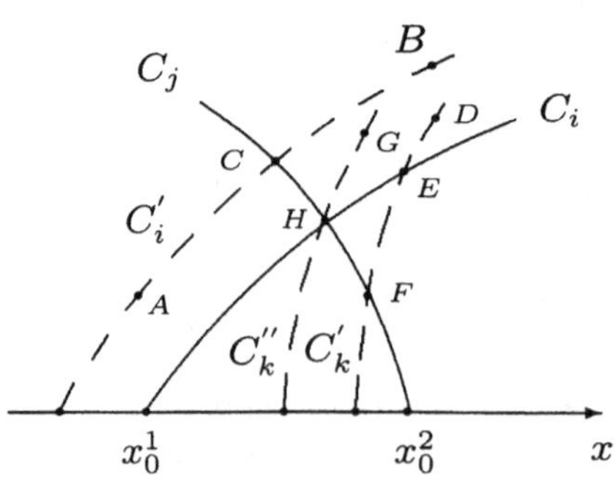

Fig. 6

For that purpose it is sufficient to consider in detail the differentiability of $\overset{(s)}{u}$ in different cases – the appearance of jumps in points of type C, E, F but that bring about no changes in the result. Even when the studied characteristic C_k'' passes through the crossing point $H = C_i \times C_j$ of the two bearing singularities characteristics, new singularities do not appear across it (if lacking in the initial moment). The reason is that in the function under the integral sign there are two different summands discontinuing across the respective characteristics, namely $a_{kj} \overset{(s-1)}{u_j}$ is discontinuous across C_j and $a_{ki} \overset{(s-1)}{u_i}$ across C_i. The lack of a term discontinuous simultaneously across the two intersecting characteristics plays a key role for a typical linear phenomenon: the singularities "meet" at the point H and do not interact, i.e., at the points from $C_k'' \setminus \{H\}$ the solution is smooth.

Although the approach to studying semilinear systems is ostensibly the same, the properties of the solution are principally different. The principal part of the system (1.1) being linear, the definition of "trapezoidal" domain G_T does not alter. The uniqueness theorem of the broad-sense solution, i.e., continuous in G_T and differentiable along the characteristics, may be proved almost the same way. If u and v are two broad-sense solutions with the same initial data, then $w \equiv u - v$ satisfies the conditions:

$$w_i(x,t) = \int_0^t [f_i(\xi_i(\tau),\tau,u(\xi_i(\tau),\tau)) - f_i(\xi_i(\tau),\tau,v(\xi_i(\tau),\tau)]d\tau, i = 1,2,\dots,n,$$

i.e.,

$$w_i(x,t) = \int_0^t \left[\sum_{s=1}^n g_{is}(\tau,u,v).w_s(\xi_i(\tau),\tau)\right] d\tau,$$

where $g_{is} \equiv \int_0^t \frac{\partial f_i}{\partial w_s}(\xi_i(\tau),\tau,v(\xi_i(\tau),\tau) + \lambda(v-u)(\xi_i(\tau),\tau))d\lambda.$

Due to the assumptions made the functions g_{is} are uniformly bounded on G_T, which ensures an inequality of the type (1.10) completing the proof. The uniqueness of the discontinuous solution defined by properties (i), (ii) before Remark 1.2 is proved without modifications. Let $t_1,\dots,t_m$ be the points at which the differentiability of $w = v - u$ along D_i is violated.

Therefore, by using analogous notation,

$$w_i(x,t) = \sum_{k=0}^m \sum_{s=1}^n \int_{t_k}^{t_{k+1}} g_{is}^k(\tau,u,v)w_s(\xi_i(\tau),\tau)d\tau,$$

$$t_0 = 0, \ t_{m+1} \equiv t.$$

From $\max_k |g_{is}^k| \le$ const we have

$$|w_i(x,t)| \le C\sum_k \int_{t_k}^{t_{k+1}} |w|d\tau \le C\int_0^t |w(\xi_i(\tau),\tau)|d\tau,$$

$$C = \text{const} > 0$$

implying the uniqueness.

In contrast with the linear case where the existence theorem is proved in an arbitrary, prescribed "trapezoidal" domain, in the semilinear case we come across a well-known difficulty – the solution exists only locally.

Hence, we shall assume the existence of a broad-sense solution, using the successive approximations only for studying its regularity (as described above).

Remark 1.2 shows that the successive approximations for a semilinear strictly hyperbolic system possess the same properties as the successive approximations in the linear case (with which the existence of a discontinuous solution was achieved with) and naturally we expect analogous results.

Unfortunately, the convergence of the successive approximations is only local (as without special restrictions for the right-hand sides f_i in (1.1) the uniform boundedness of the approximations could not be established). Even if we assume the existence of the solution in some interval $[0,T]$ the convergence under the usual assumptions would be only local. The use of approximations in that case would lead to difficulties, with possible appearance of new singularities from the initial data on the consecutive step that do not exist de facto (see [70], p. 213).

The problem of proving the convergence of the successive approximations to the solution (existing by assumption) can be overcome by replacing them with modified successive approximations (an approach hinted at by E. Horozov).

Let $\bar{u}(x,t)$ be a discontinuous solution of (1.1) existing in G_T. We denote $M = 2\max_{G_T}|\bar{u}(x,t)|$ and then consider the system

$$D_i u_i = \tilde{f}_i(x,t,u), \quad i = 1,2,\ldots,n, \tag{1.14}$$

where the functions $\tilde{f}_i$ are obtained from f_i after multiplying by a cut-off function belonging to $C_0^\infty(\mathbf{R}^{2+n}_{(x,t,u)})$ which is identically equal to 1 on the set $\{(x,t,u)|(x,t)$ belongs to a neighborhood of G_T, $|u| \le M\}$. As $\tilde{f}_i$ are compactly supported smooth functions, the norms of $\tilde{f}_i$, $\frac{\partial \tilde{f}_i}{\partial u}$ are uniformly bounded, the latter meaning the existence of a Lipschitz constant L for the functions $\tilde{f}_i$ with respect to the variable u, independent of the size of u (compare with the proof of the uniqueness theorem, where $\max|g_{is}|$ depends on u, v). On the other hand, the very solution $\bar{u}$ of the initial problem (1.1) is an obvious solution of (1.14). The respective successive approximations defined via the integral form of (1.14) tend uniformly to it over $[0,T]$. To see this we should have in mind that the so-defined successive approximations are uniformly convergent over G_T to a function $u(x,t)$ which is a solution of (1.14). According to the uniqueness theorem $u \equiv \bar{u}$.

As in the linear case we can prove C^1 regularity of the solution under C^1-initial data etc.

One by one we investigate the different cases of finite jump type singularities in the Cauchy data and we get the same results as in the linear case. Concluding these remarks we should point out the apparent fact that de facto the derivative with respect to x of the solution satisfies a linear system but with coefficients depending on the solution.

For the first time a new effect appears when we study a triple intersection of the characteristics, i.e., when the k-characteristic crosses two different, bearing singularities characteristics of the type i, j, $i,j \neq k$ (see Fig. 6 with $G = (x,t)$). The initial function $u_k^0(x)$ is assumed to be smooth in a neighborhood of the point $\xi_i(0;x,t)$. This situation leads to a new effect (nonlinear) found by Rauch–Reed in [69]. It concerns the creation of a singularity for a component having smooth Cauchy data. This is the corresponding example:

$$(\partial_t + \partial_x)u = 0, \quad u^0(x) = \begin{cases} 0, & x \le -1, \\ 1, & x > -1, \end{cases}$$

$$(\partial_t - \partial_x)v = 0, \quad v^0(x) = \begin{cases} 1, & x \le 1, \\ 0, & x > 1, \end{cases}$$

$$\partial_t w = uv, \quad w^0(x) \equiv 0.$$

Simple calculation shows that w is a continuous function but $\partial_x w$ has a finite jump across the part of the characteristic $x = 0$, starting from the cross point $(0, 1)$ of the characteristics bearing the singularities of the components u, v (see Fig. 7).

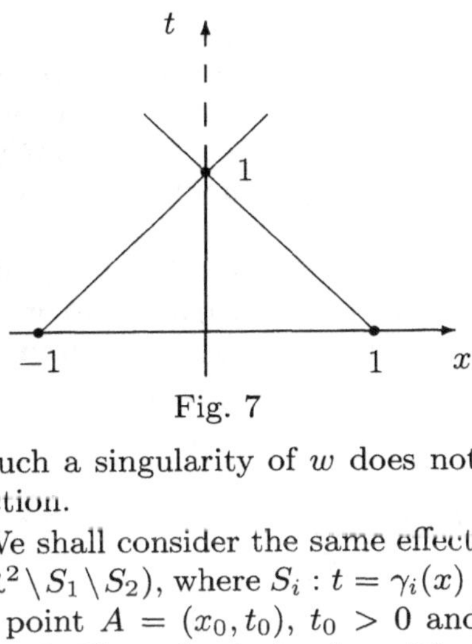

Fig. 7

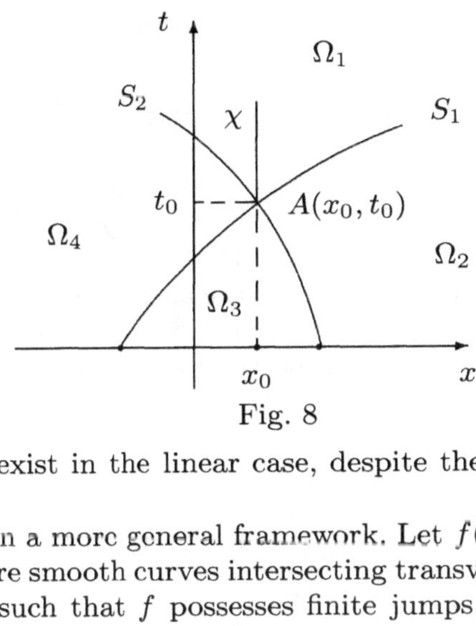

Fig. 8

Such a singularity of w does not exist in the linear case, despite the triple interaction.

We shall consider the same effect in a more general framework. Let $f(x,t) \in C_*^{-1}(\mathbf{R}^2 \setminus S_1 \setminus S_2)$, where $S_i : t = \gamma_i(x)$ are smooth curves intersecting transversally at the point $A = (x_0, t_0)$, $t_0 > 0$ and such that f possesses finite jumps across $S_i, i = 1, 2$. Then the derivative with respect to x of the function

$$u(x,t) = \int_0^t f(x,\tau)d\tau$$

is continuous at the points of the characteristic of $\partial_t : \chi = \{(x,t) | t \ge t_0, x = x_0\}$, O_t being transversal to S_i, if and only if

$$f_1 + f_3 = f_2 + f_4, \tag{1.15}$$

where $f_i = \lim\limits_{(x,t) \to (x_0,t_0), (x,t) \in \Omega_i} f(x,t)$ and the domains Ω_i are the angles between the curves S_1, S_2 clockwise numbered, $\chi \subset \Omega_1$ (see Fig. 8).

The proof of the result mentioned above is given in [36]. Obviously, the same assertion holds when replacing ∂_t by an operator of the type $\partial_t + \alpha(x,t)\partial_x$.

As a consequence of this new effect different cases could appear:

Case I. There exist two systems of characteristics only. Then the underintegral functions participating in the definitions of the successive approximations contain discontinuities across only one system of the characteristics. This case is similar to the linear one.

Case II. The system (1.1) possesses more than two systems of characteristics but the height of the trapezoidal domain G_T is bounded by the first cross point of the characteristics starting from the points $(x_0^s, 0)$, $s = 1, 2, \dots, k$, where the Cauchy data have singularities of finite order. (Such a cross point can exist for $t = T$.) Then the solution $u \in C^\infty(G_T \setminus S_0)$ and S_0 stands for the union of the respective characteristics (moreover, it has across them singularities of finite order). That result combined with a C^∞ regularity result between the characteristics (see Lemma 1.6 below) enable us to investigate the regularity of the solution above the cross point.

Case III. The system is of type $n \times n$, $n \geq 3$ and G_T contains the cross point of bearing singularity characteristics. A central result in this case is the theorem for existence of an anomalous singularity of strictly hyperbolic one space dimensional system [70]. That theorem can be proved in a simple way by applying the necessary and sufficient condition (1.15).

Thus assume that the component u_1 of (1.1) possesses a singularity of finite order n_1 across the 1-characteristic C_1 through the point $(x_1, 0)$ and the component u_2-singularity of finite order n_2 across the 2-characteristic C_2 starting from $(x_2, 0)$ (see Fig. 9).

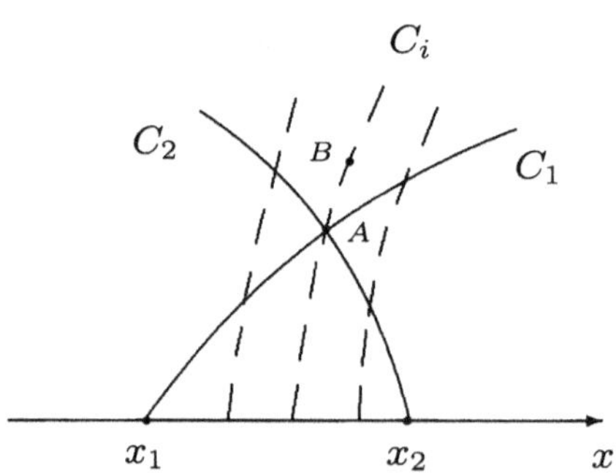

Fig. 9

We shall show that some derivative $\partial_x^s u_i$ of a component u_i, $i \neq 1, 2$, having C^∞ initial datum, is discontinuous across the arc of the i-characteristic C_i located above the intersection point $A = C_1 \times C_2$. We remind the reader of the integral equation satisfied by u_i :

$$u_i(x,t) = u_i^0(\xi_i(0; x, t)) + \int_0^t f_i(\xi_i(\tau; x, t), \tau, u(\xi_i(\tau; x, t), \tau))d\tau.$$

Differentiating $u_i(x,t)$ several times in x we get integral terms and summands containing jumps of the function f_i (or some of its derivatives) across C_1 or C_2. Assuming the discontinuity to arise across only one of these characteristics we can easily see that the expressions mentioned above are continuous functions of (x,t) in a neighborhood of the point B. The successive derivatives of f_i can be written as (we apply the Faa di Bruno formula):

$$\sum_{\gamma+\delta_1+\ldots+\delta_n\le s,\alpha+\beta\le s} \partial^\alpha_{\xi_i}\partial^\beta_u f_i.\prod_{\gamma\le\alpha}\partial^\gamma_x\xi_i(\tau;x,t).\prod_{\delta_1+\ldots+\delta_n\le s-\alpha}\partial^{\delta_j}_x u_j.$$

These summands will contain singularities of the finite jump type if the derivatives $\partial^p_x u_1, p\ge n_1+1$ or $\partial^p_x u_2, p\ge n_2+1$ are to be part of them. According to Case I the derivative $\partial^s_x u_i$ is continuous in a neighborhood of B if that singularity (i.e., finite jump type) is across only one of the two characteristics C_1, C_2. A term containing singularities of the finite jump type across both intersecting characteristics appears for the first time in the expression for $\partial^{n_1+n_2+2}_x(f_i)$. Obviously, this is the derivative

$$\frac{\partial^2 f_i}{\partial u_1\partial u_2}.\partial^{n_1+1}_x u_1.\partial^{n_2+1}_x u_2.$$

Put $\partial_{u_2}\partial_{u_1}f_i|_A = F$. Having in mind that the derivative $\partial_{u_2}\partial_{u_1}f$ is continuous with respect to (x,t) and denoting by U_i+h_i, $h_i\ne 0$, $i=1,2$, respectively by U_i the limits at the point A of $\partial^{n_i+1}_x u_i$ from above and from below the curve C_i, we can apply the condition (1.15) and conclude that an anomalous singularity of order (n_1+n_2+3) arises at A if and only if

$$F[(U_1+h_1)(U_2+h_2)+U_1U_2]\ne F[(U_1+h_1)U_2+U_1(U_2+h_2)],$$

i.e., iff $Fh_1h_2\ne 0$, i.e., iff

$$\frac{\partial^2 f_i}{\partial u_1\partial u_2}|_A\ne 0 \tag{1.16}$$

as $h_i\ne 0$. The inequality (1.16) coincides with the results of [70] for existence of an anomalous singularity.

According to (1.16) the existence of an anomalous singularity in the case of double interaction does not depend on the size of the jumps h_1 and h_2.

Assume now that we have a triple (at least) transversal interaction of characteristics bearing finite order singularities. The previous approach shows that the necessary and sufficient condition for the existence of anomalous singularities depends on the size of the jumps at the interaction point of the respective characteristics.

We shall formulate now an interesting result of L. Micheli [57].

Lemma 1.6. *Under the notation introduced before, consider the domain G_T containing only one point of interaction $A=(x_0,t_0)$ where the crossing of at least two characteristics C_i, $i=1,2,\ldots,k$, bearing singularities of the finite jumps type, takes place.*

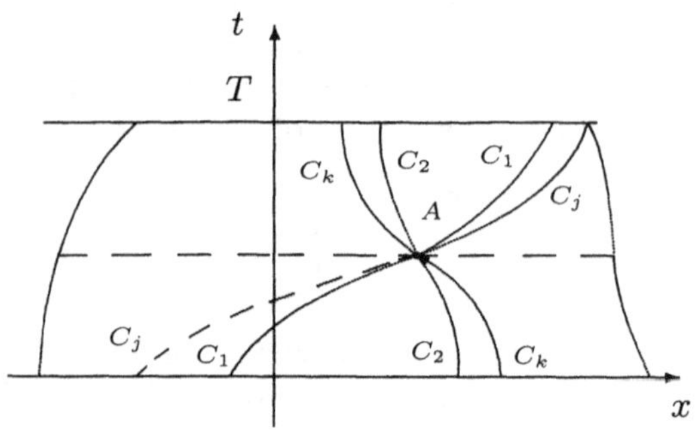

Fig. 10

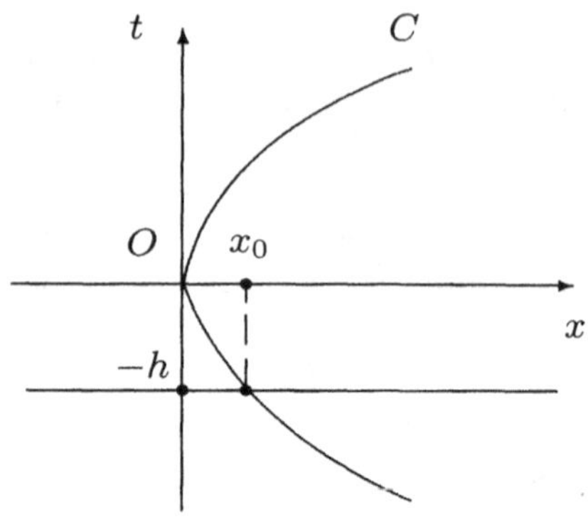

Fig. 11

Let the solution belong to $C_*^\infty(G_{t_0} \setminus \cup C_i)$. Then it belongs to $C_*^\infty(G_T \setminus S)$, where S is the union of all characteristics passing through the point A.

A short sketch of the proof is given further on. Thus, there are no singularities along the punctured characteristic C_j, $j > k$. A singularity could be born at A, propagating along the forward branch of C_j passing through A.

New phenomena appear in weakly hyperbolic systems, e.g., it enables singularities of the finite jump type across one characteristic to give rise to a singularity of the infinite jump type across a characteristic tangent to the original one. The effect is linear and can be illustrated by the following example.

Example 2. ([71]) For the system

$$\begin{aligned} Xu &\equiv (\partial_t + pt^{p-1}\partial_x)u = 0, \\ &\partial_t v = u, \end{aligned}$$

with initial data (see Fig. 11)

$$u|_{t=-h} = u^0(x) = \begin{cases} 0, & x \leq x_0 \\ 1, & x > x_o \end{cases}, v|_{t=-h} = v^0(x) \equiv 0,$$

where p is an even positive integer, it is easy to see that in the "corner" between the X-characteristic $C : x = t^p$ and the positive t semi-axis (to which C is tangent at the origin) the function $v(x,t) \equiv 2x^{\frac{1}{p}}$, while $v \equiv 0$ for $x < 0$. Therefore at the points $\{(0,t),\ t > 0\}$ the derivative $\partial_x v$ has unbounded right limit with the behavior of $x^{\frac{1}{p}-1}$. We shall note that $\partial_t v$ is bounded, i.e., not all the derivatives do blow up.

In a more general situation a similar problem has been studied by L. Micheli. The paper [57] deals with a weakly hyperbolic system of the type:

$$(WH) \begin{cases} X_1 u_1 &= f_1(x,t,u), \\ X_2 u_2 &= f_2(x,t,u), \\ X_i u_i &= f_i(x,t,u),\ i \geq 3, \end{cases}$$

where $X_1 = \partial_t + t^{p-1}\alpha(x,t)\partial_x$, $\alpha(x,t) \geq \alpha_0 > 0, \alpha \in C^\infty$, $p \in \mathbf{N}$ is even, $X_2 = \partial_t$ and the characteristics of the operators $X_i = \partial_t + \lambda_i(x,t)\partial_x$, $\lambda_i \in C^\infty$ are transversal to the X_1 and X_2 characteristics (and to each other). We choose initial data for $t = -h$:

$$(CWH)\ u_1^0(x) \in C^k(\mathbf{R}) \cap C_*^\infty(\mathbf{R} \setminus \{x_0\}),$$

$$u_i^0 \in C^\infty(\mathbf{R}), \quad i \geq 2,$$

where the point x_0 is defined by the condition: the X_1 characteristic through the p. $(x_0, -h)$ is tangent to the t-axis at the origin (see Fig. 11).

For a similar problem the hypothesis for the order of the anomalous singularities ([71]) by the way of analogy with the strictly hyperbolic case is:

$$k + (k+1) + 2 + p - 1,$$

p being the order of contact between the tangent characteristics. [57] achieves quite a better result, namely the order of the anomalous singularity is

$$[k + (k+1) + 2]p.$$

The latter statement is motivated in [37] with an appropriate example but lacks a theorem with a necessary and sufficient condition of the type of the result in [70]. The method of investigation differs from the one used up to now, consisting of successive differentiation along the x-axis of the respective integral equation. While the strictly hyperbolic case always provides for bounded piecewise smooth functions by successive differentiations, Example 2 shows that there should arise unbounded derivatives – Schwartz distributions being non-integrable functions. This difficulty is overcome in [57] by using only appropriate derivatives (as not all the derivatives blow up; see Example 2). In order to make possible the use of integral equations, rather complicated changes of the variables have been done successively.

In [37] that approach is modified and the obtained simplification allows for a proof of an analogue of the condition (1.16) for existence of an anomalous singularity. The new feature of this modification is the systematic use of the same differential system and not of its integral equivalents. This enables the differentiation to be in the sense of distribution theory and afterwards, depending on the properties of the right-hand side of the equation, our coming to the conclusion of the continuity of the solution. Our arguments are based on the following

Lemma 1.7. *(Lemma 1.4 from [37].) Assume that $X = \partial_t + \nu(x,t)\partial_x$, $Y = \partial_t + \mu(x,t)\partial_x$, $\nu, \mu \in C^\infty(\mathbf{R}^2)$, $|\nu - \mu| \geq const > 0$ are two real-valued vector fields. Let $l : t = g(x)$ be a smooth curve transversal to the characteristics of Y. Denote by Ω the curvilinear quadrangle depicted by the arcs of the Y-characteristics passing through the points of l. The symbols Ω_+ (Ω_-) stand for the open parts of Ω located above (below) l. Assume now that $a, w \in C^0(\Omega) \cap C^1(\Omega^-)$, $Xa \in C^0(\Omega)$ and the first-order Schwartz distribution $v = Xw \in D^{1\prime}(\Omega)$ satisfies the linear differential equation $Yv + av = YF$, where $F \in C^0(\Omega)$.*

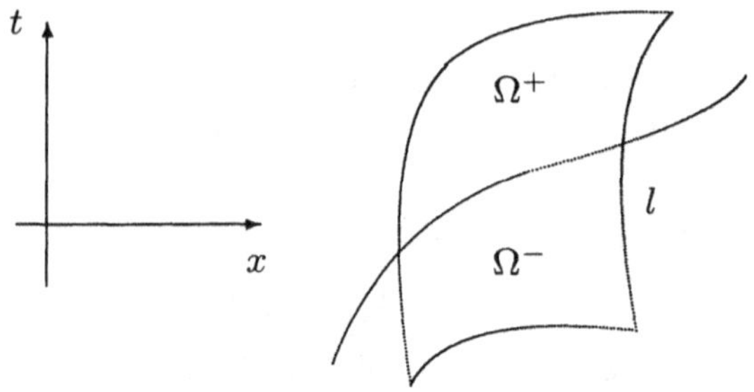

Fig. 12

Then $v \in C^0(\Omega)$ and therefore v turns out to be a classical X-derivative of w (see Fig. 12).

We remind the reader that the first-order distribution is a distribution in Ω represented in the form of a generalized first-order derivative of a continuous function in Ω. The set of such distributions is denoted by $D^{1\prime}(\Omega)$.

To give an application of Lemma 1.7, we shall sketch out the proof of Lemma 1.6, supposing for the sake of simplicity that the solution is continuous in G_T. There exists in G_{t_0} at least one "angle" between two characteristics that the solution belongs to C^∞ (according to Case II considered above). As a starting point we can make use of a part of an "angle" located under $t = t_0$. We shall show at first the way C^∞ regularity is established in the adjacent angles, etc. through the final proof of the lemma. Let C_i be the characteristic through (x_0, t_0) and Ω^-, Ω^+ be two "corner" domains bounded by C_i and the two neighboring characteristics (see Fig. 13). (It does not matter whether these three characteristics are backward or forward.)

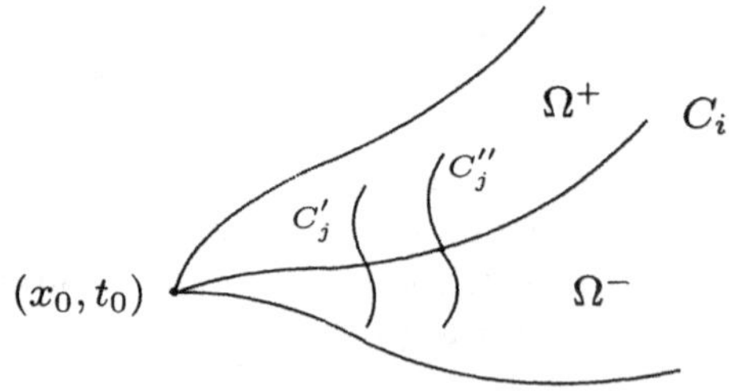

Fig. 13

We assume that the solution $u \in C^\infty$ in Ω^-. Due to the strict hyperbolicity of (1) the j-characteristics C_j, $j \neq i$ are transversal to C_i. Suppose that it is known that for some integer $m \geq 1$ the derivatives $X_i^m u_j$, $j = 1, 2, \ldots, n$, $j \neq i$ are continuous in $\Omega^- \cup \Omega^+$. If we prove that $X_i^{m+1} u_j, j \neq i$ exist and are continuous in Ω^+, then it follows that $u_j \in C^{m+1}(\Omega^+)$. In fact each $(m+1)$order derivative of u_j containing at least one differentiation along X_j is continuous and the vector fields $X_i, X_j, i \neq j$ are transversal.

To show the continuity of the derivative $X_i^{m+1}u_j$ we shall apply the following commutator identity in the sense of distribution theory:

$$X_j X_i^{m+1} u_j = X_i^{m+1} f_j + b X_i^{m+1} u_j + P_m, \tag{1.17}$$

where $b \in C^\infty$ and $P_m \in C^0$, as it is a polynomial of order m of $X_i^\alpha u_k$, $\alpha \leq m$, $k = 1, 2, \ldots, n$. Here $X_i(X_i^m u_j) \in D^{1\prime}$, $X_i(X_i^m)f_j \in D^{1\prime}$.

The result (1.17) coincides with Lemma 1.9. in [37].

On the other hand, according to [69] we can use the chain rule in our situation, obtaining this way that

$$X_i^{m+1} f_j = \sum_{l=1, l\neq i}^{n} \frac{\partial f_j}{\partial u_l} X_i^{m+1} u_l + P'_m, \quad P'_m \in C^0.$$

In fact, $X_i f_j(x,t,u) = (X_i f_j)(x,t,u) + \sum_{k=1}^{n} (\partial_{u_k} f_j) X_i u_k$, $X_i u_i = f_i \in C^0$ etc.

But $X_i = \alpha X_j + \beta X_l$, $l \neq i, j$, where the coefficients $\alpha, \beta \in C^\infty$ and therefore

$$\begin{aligned} \frac{\partial f_j}{\partial u_l} X_i^{m+1} u_l &= \frac{\partial f_j}{\partial u_l}(\alpha X_j + \beta X_l) X_i^m u_l \\ &= A X_j X_i^m u_l + B X_l X_i^m u_l \\ &= X_j(A X_i^m u_l) - X_j(A) X_i^m u_l + \text{a continuous function} \\ &= X_j(F_l) + \text{a continuous function}, \quad F_l = A X_i^m u_l. \end{aligned} \tag{1.18}$$

To obtain (1.18) we have applied (1.17) to $X_l X_i^m u_l$ and the fact that the commutator $[X_i, X_l] = e(X_i - X_l), e \in C^\infty$. Then

$$\begin{aligned} X_l X_i^m u_l &= X_l X_i X_i^{m-1} u_l \\ &= (X_i X_l + [X_l, X_i]) X_i^{m-1} u_l \\ &= X_i X_l X_i^{m-1} u_l + e(X_l - X_i) X_i^{m-1} u_l = \cdots \\ &= X_i^m X_l u_l + d X_i^m u_l + P_{m-1} \\ &= X_i^m f_l + d X_i^m u_l + P_{m-1}, \end{aligned}$$

where $d \in C^\infty$. The functions $X_i^m u_l \in C^0(\Omega)$, $i \neq l$ and $P_{m-1} \in C^0$ according to our assumption, $X_i^m u_l \in C^0(\Omega)$.

Certainly, $A, B \in C^0$. Put $w = X_i^{m+1} u_j$. Then we get from (1.17), (1.18)

$$X_j w - (b + \frac{\partial f_j}{\partial u_j}) w = X_j(F) + \text{a continuous function and } F = \sum_{l=1, l\neq i,j}^{n} F_l \in C^0.$$

One can see that the assumptions of Lemma 1.7 hold for $Y = X_j$, $X = X_i$ and consequently $w \in C^0$. Thus everything is proved.

Remark 1.8. In the weakly hyperbolic case we cannot write down X_j, $j \geq 3$ as a linear combination of the vector fields near the hyperplane $t = 0$. Then we must use a generalization of Lemma 1.7 for 2×2 linear system.

We shall study now the regularity of the solution of the Cauchy problem (CWH) for the weakly hyperbolic system (WH) in a neighborhood of the j-characteristics with $j \geq 3$. Let $C_i^+(C_i^-), i = 1, 2, \ldots, n$, be the outgoing (backward) characteristic through the origin. The discontinuity of $\partial_x^{k+1} u_1$ across C_1^- leads to discontinuities of all components of the solution. One can find the size of these discontinuities at the point $(x, -x^{\frac{1}{p}})$ in an explicit form as a function of x. The estimates of the size of $\partial_x^m u_i$, $i = 1, 2, \ldots, n$, $m \geq k+2$ in the zone about C_1^- can be prolonged in the whole domain located to the right of C_1 and afterwards we have there a direct estimate of the derivatives $X_1^m \partial_x^s u_i$, $i = 1, 2, \ldots, n$ by means of the expression $1 + x^{\frac{1}{p}+k+1-s-\frac{m}{p}}$.

The estimates obtained are also valid in a neighborhood of C_1^+ defined by a line l through $(0,0)$ dividing the zero angle between C_1^+ and C_2^+. In the remaining part of the considered trapezoidal neighborhood R of $(0,0)$ (namely, on the left of $C_1^- \cup l$) we estimate similarly the derivatives $X_2^m \partial_x^s u_i$, $i = 1, 2, \ldots, n$.

A detailed geometrical illustration is given in Fig. 14.

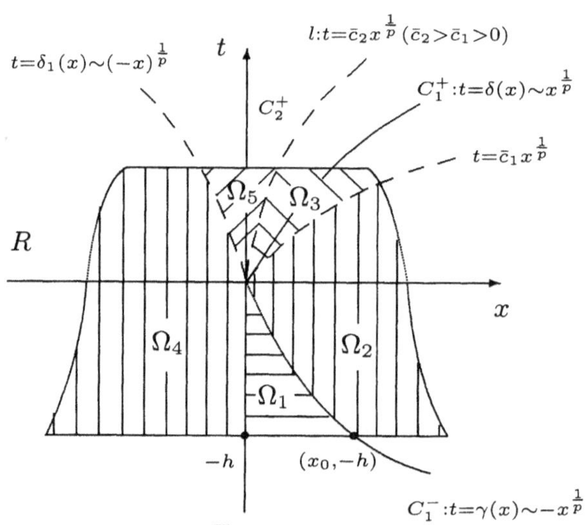

Fig. 14

Hypothesis $H_1(q), q \in \mathbf{N}$. Let Ω_s, $s = 1, \ldots, 5$ be the domains in Fig. 14. The solution u of (WH), (CWH) satisfies:

(i) $u \in C^q(\mathbf{R} \setminus C_1 \setminus C_2^+)$.

(ii) $|X_1^m \partial_x^s u_i| \leq \text{const}(1 + x^{\frac{1}{p}+k+1-s-\frac{m}{p}})$ holds true in $\Omega_1 \cup \Omega_2$ for $i = 1, 2, \ldots, n$ provided $m + s \leq q$ and in the domain Ω_3 if in addition $s \leq k+1$.

(iii) Under the assumption (ii) and $\dfrac{\partial f_2}{\partial u_1}(0, 0, u(0,0)) \neq 0$ there exist constants $c_j(m,s)$ not all equal to 0 such that if $m = \theta p + r$, $r \in \{1, 2, \ldots, p\}$,

$\theta \in \mathbf{N} \cup \{0\}$, then

$$|X_1^m \partial_x^s u_2 - \sum_{j=1}^{m-\theta} c_j t^{jp-r} x^{\frac{1}{p}+k+1-s-\theta-j})|$$
$$\leq \text{const } (1 + x^{\frac{1}{p}+k+1-s-\theta-\frac{m}{p}+\varepsilon}),$$

$\varepsilon > 0$ is valid in the domain Ω_2. A similar inequality holds in Ω_3 and the corresponding most singular part is

$$\sum_{j=1}^{m} \tilde{c}_j t^{jp-m} x^{\frac{1}{p}+k+1-s-j}, \quad \text{with some} \quad \tilde{c}_j \neq 0.$$

Define the differential operators

$$L_1 = X_1, \quad L_2 = \partial_t X_1, \quad L_3 = X_1 \partial_t X_1, \quad \ldots, \quad L_{2k+2} = (\partial_t X_1)^{k+1},$$

$L_r^i = X_i^{r-(2k+2)} L_{2k+2}$ for $r > 2k+2$ and $i = 1$ or $i = 2$.

This is our second hypothesis.

Hypothesis $H_2(q)$**.** The derivatives $L_r^i u_j \in C^0(R \setminus C_1 \setminus C_2^+)$ for $j \geq 3$ and $r \leq q$ and satisfy the inequalities

$$|L_r^i u_j(x,t)| \leq \text{ const}(1 + x^{k+2-\frac{r}{p}}) \text{ if } (x,t) \in \Omega_1 \cup \Omega_2 \cup \Omega_3.$$

Then in [37] the validity of four statements is proved which we shall formulate as lemmas.

Lemma 1.9. *Suppose* $H_1(q)$ *is true. Then* $H_1(q+1)$ *holds for* u_1, u_2.

Lemma 1.10. *The derivatives* $L_r u_j \in C^0(R)$ *for all* $j \geq 3$ *and* $r \leq 2k+2$.

Lemma 1.11. *Let* $q > k+1$. *Then if* $q \leq (2k+3)p - (k+1)$, *the hypothesis* $H_1(q) \Rightarrow H_2(q+k+1)$ *holds.*

Lemma 1.12. $H_1(q)$, $H_2(q+1)$ *imply* $H_1(q+1)$ *for all* u_i, $i = 1, 2, \ldots, n$.

The maximal regularity of the component u_j, $j \geq 3$ near C_j^+ is given by Lemma 1.11.

In order to find a necessary and sufficient condition for the existence of anomalous singularities, we first use condition (iii) of $H_1(q)$. This way the maximal singularity of $X_1^m \partial_x^s u_2$ is known in a neighborhood of C_1^+ (Ω_3). Then an appropriate ODE for a $(2k+3)p+1$-order derivative of $u_j, j \geq 3$ (the derivative being everywhere transversal to C_j) is considered and Lemma 1.7 is applied to it. The sharp estimates enable us to represent the solution of the ODE mentioned above as a sum of two functions. The first one is continuous while the second one has a jump discontinuity when crossing C_j^+ if and only if

$$\frac{\partial^2 f_j}{\partial u_1 \partial u_2}(0, 0, u(0,0)) \neq 0. \tag{1.19}$$

This way we can formulate the following theorem, supposing

$$\frac{\partial f_2}{\partial u_1}(0,0,u(0,0)) \neq 0,$$

$$f_1 \equiv 0,\ u_1^0 = \begin{cases} c_1(x-x_0)^{k+1} + o\left((x-x_0)^{k+1}\right), & x < x_0, x \to x_0 \\ c_2(x-x_0)^{k+1} + o\left((x-x_0)^{k+1}\right), & x \geq x_0, x \to x_0 \end{cases} \quad c_1 \neq c_2,$$

$c_1 \neq 0$.

Theorem 1.13. *The* $[(2k+3)p+1]$ *transversal derivatives of the components* $u_j, j \geq 3$ *of the solution* u *of (HW), (CHW) are discontinuous across* C_j^+ *iff* (1.19) *holds.*

2. New created singularities of the solutions of semilinear hyperbolic equations

2.1. Introduction

It is well known that, after the interaction of several nonlinear waves carrying some initial singularities, new singularities in comparison with the linear case can appear. Rather general results in the case of two space dimensions $N = 2$ were proved by Bony in [10], [11] and by Melrose–Ritter in [54]. To prove the corresponding theorems, Bony developed in the C^∞ category the method of second microlocalization. The proof of Melrose–Ritter is also rather difficult. Our main aim in this paragraph is to prove the appearance of "new-born" singularities after the interaction of the initial travelling waves for $N = 2$ and under very special nonlinear conditions. Moreover, the strength of the new singularities for $t > 0$ will be found. We shall follow here with slight modifications the paper of J. Keller–L. Ting [41]. The only new moment is the proof of the convergence of the corresponding series giving the solution inside and up to the characteristic (light) cone for the wave equation. In our opinion the proof there is simple, clear and could be useful in some other investigations. The main tool in our consideration is the explicit construction of the solution of the semilinear wave equation. To do this we introduce a special nonlinear function $f(u)$ for which the semilinear equation

$$\Box u = f(u),\ \Box = \partial_t^2 - \Delta_x, x \in \mathbf{R}^N, N \geq 2$$

becomes linear inside some simplex $S^N(t)$, $0 < t < t_0$ such that $\cup_{0<t\leq t_0} \partial S^N(t)$ is a sum of $N+1$ characteristic hyperplanes $t - x.n_j = 0$, $1 \leq j \leq N+1$, $|n_j| = 1$, n_j, $x \in \mathbf{R}^N$. After this we use the existence of a solution of the Goursat problem with constant data on the boundary. The theory of the hypergeometric Gauss's function enables us to find the sharp order of Hölder singularity created after the collision of the initial travelling waves.

2.2. Formulation of the problem

We shall confine ourselves to the special case $N = 2$. Thus we consider the equation

$$\Box u = f(u), x \in \mathbf{R}^N, N = 2, u = u(x,t). \tag{2.1}$$

The characteristic hyperplanes of (2.1) are given by

$$t = x.n, n = (n^1, n^2), |n| = 1, N = 2. \tag{2.2}$$

We assume that for $t < 0$ the solution u has three jump discontinuities equal to 1 on the following characteristic hyperplanes:

$$t = x.n_j, |n_j| = 1, 1 \leq j \leq 3. \tag{2.3}$$

The hyperplanes (2.3) pass through the origin. We suppose that for each fixed $t \neq 0$ these hyperlines bound a simplex $S^3(t)$ of finite volume. This simplex shrinks to 0 for $t = 0$ and then it expands for $t > 0$. In the case $N = 2 : S^2(t)$ is an isosceles (equilateral) triangle and this is the geometrical picture:

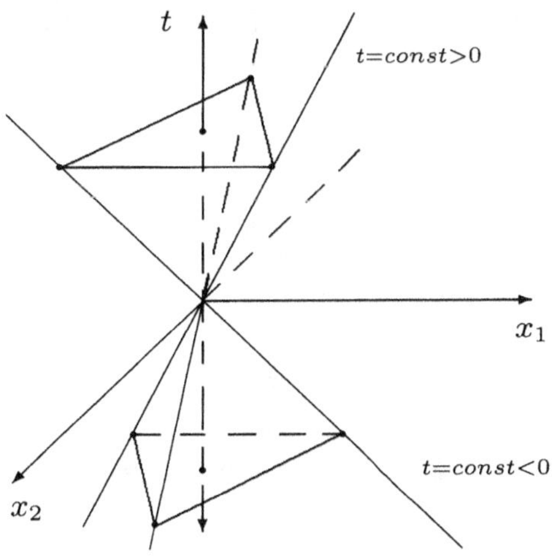

Fig. 15

The unit normal vectors n_j point inside $S^2(t)$ for $t < 0$ and outside $S^2(t)$ for $t > 0$ (see Fig. 16).

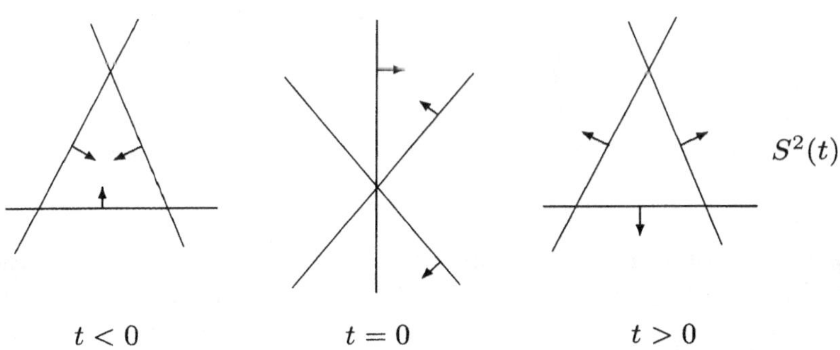

Fig. 16

For $t < 0$ we define $u(x,t)$ by the formula

$$u(x,t) = \sum_{j=1}^{3} H(t - x.n_j), \tag{2.4}$$

where $H(s)$ is the standard Heaviside function:

$$H(s) = \begin{cases} 1, & s > 0, \\ 0, & s < 0. \end{cases} \tag{2.5}$$

Obviously, u is a distribution solution of the homogeneous wave equation for $t < 0$, i.e., $\Box u = 0$, $t < 0$ and u has unit jump discontinuities on the hyperplanes (2.3). Moreover, $u = 0$ in $S^2(t)$, $t < 0$ and u takes values 1,2 outside $S^2(t)$. Geometrically, $N = 2$ implies the validity of the following Fig. 17, where the triangle is $S^2(t)$.

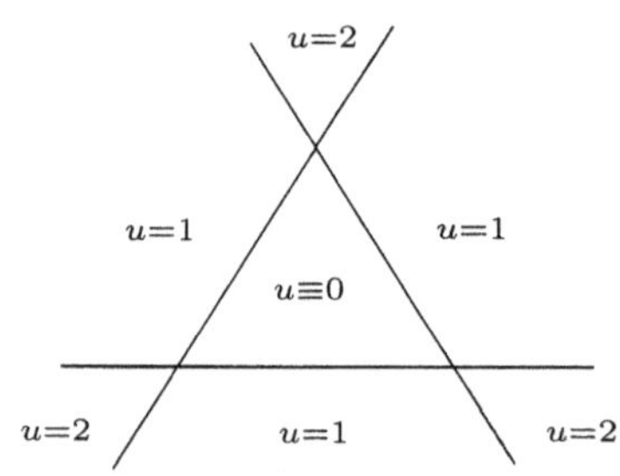

Fig. 17

Suppose now that $t > 0$. Then (2.4) satisfies, in the distribution sense, $\Box u = 0$. This way we conclude that the function (2.4) is a solution of (2.1) for $t < 0$ if

$$f(0) = f(1) = f(2) = 0. \tag{2.6}$$

The piecewise constant function u takes values 1,2 outside $S^2(t)$, $t > 0$, while $u|_{S^2(t)} = 3$, $t > 0$. Certainly, there are two possibilities for $f(u)$: $f(3) = 0$ and $f(3) \neq 0$. In the first case the function u, defined by (2.4) for each t, turns out to be a solution of (2.1) for all (t,x). Consequently, no new singularities in comparison with these in the linear case $\Box u = 0$ appear. So we shall assume that

$$f(3) \neq 0. \tag{2.7}$$

Then (2.4) is a solution of (2.1) outside $S^2(t)$ but (2.4) is not a solution of the equation (2.1) inside $S^2(t)$, $t > 0$. Having in mind that the function u given by (2.4) is equal to 2 on the outer boundary of $S^2(t)$, $t > 0$ and has unit jumps across (2.3), we conclude that on the inner boundary of $S^2(t)$ the function $u \equiv 3$. This is the geometrical illustration:

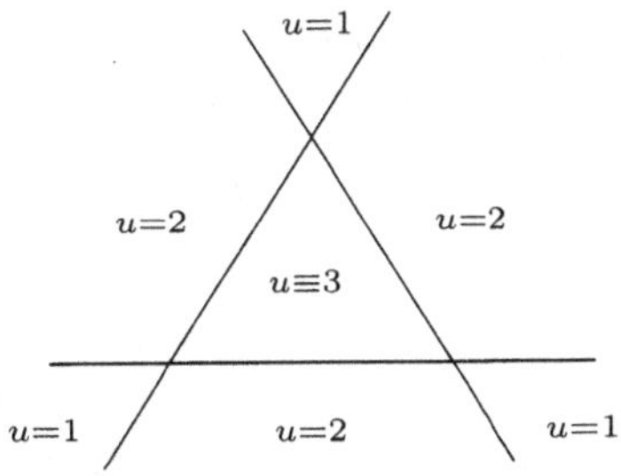

Fig. 18

This way, for the solution u of the equation (2.1) in $S^2(t)$, $t > 0$, we obtain the nonlinear Goursat problem

$$\begin{aligned} \Box u = f(u), \qquad & x \in S^2(t), t > 0, \\ u = 3, \qquad & x \in \partial S^2(t), t > 0. \end{aligned} \tag{2.8}$$

We seek a continuous solution of (2.8) in the domain $\{x \in S^2(t), t \geq 0\}$. This is the main assumption on the smooth function f:

$$\begin{aligned} &\text{There exists } \delta > 0 \text{ and such that } f(u) = f(3) \\ &\qquad \text{for each } u \text{ for which } |u - 3| < \delta. \end{aligned} \tag{2.9}$$

Thus, there exists $t_0 > 0$ with the property $|u - 3| < \delta$ for $x \in S^2(t)$, $0 < t < t_0$. In fact, u is continuous and $u = 3$ on $S^2(t)$. The nonlinear Goursat problem (2.8) is reduced for $0 < t < t_0$ to the linear Goursat problem

$$\begin{aligned} \Box u &= f(3), x \in S^2(t), 0 < t < t_0, \\ u &= 3, x \in \partial S^2(t). \end{aligned} \tag{2.10}$$

We shall prove the existence of a singular solution of (2.10) for $t > 0$, t sufficiently small. The singularities once created for $0 < t < t_0$ will propagate outward along the bicharacteristics for large t.

2.3. Singularities of the solution u of the Goursat problem on the characteristic circle

The smooth solution of the Goursat problem (2.10) will be found in the form:

$$u(x, t) = 3 + \frac{t^2}{2} f(3)[1 - w(x, t)] = 3 + \frac{t^2}{2} f(3) - \frac{t^2}{2} f(3) w. \tag{2.11}$$

Obviously, w will satisfy a simpler Goursat problem:

$$\begin{aligned} \Box(t^2 w) &= 0, x \in S^2(t), t > 0, \\ w &= 1, x \in \partial S^2(t), t > 0. \end{aligned} \tag{2.12}$$

When studying the Goursat problem with data on the characteristic (light) cone it is possible to investigate a much more general problem than (2.12), namely

$$\begin{aligned} \Box(t^2 w) &= 0, x \in S^N(t), t > 0, N \geq 2, \\ w &= 1, x \in \partial S^N(t), t > 0. \end{aligned} \tag{2.13}$$

As it is well known (see [24]) the Goursat problem for the wave equation has a unique solution inside the light cone. For this reason we prefer to concentrate on (2.13) inside the characteristic cone inscribed in the simplex $S^N(t)$ and having a vertex at the origin. Certainly, in this case (2.11) takes the form

$$u(x,t) = N + 1 + \frac{t^2}{2} f(N+1) - \frac{t^2}{2} f(N+1) w, \tag{2.14}$$

$f(0) = f(1) = \cdots = f(N) = 0$, $f(N+1) \neq 0$ and the function u has $N+1$ equal to 1 jump discontinuities on the characteristic hyperplanes $t = x.n_j$, $|n_j| = 1$, $1 \leq j \leq N+1$, $n_j = (n_j^1, \ldots, n_j^N)$. We shall verify that $w(x,t)$ is a homogeneous of order-zero function in (x,t), i.e., w depends on $\frac{x}{t}$, $t > 0$ only. In other words, $w = w(r, \varphi)$, where $r = \frac{|x|}{t}$ is the so-called scaled radius and $\varphi = (\varphi_1, \ldots, \varphi_{N-1})$ are the spherical angles defined by:

$$x_1 = |x| \cos \varphi_1,$$

$$x_2 = |x| sin\varphi_1 \cos \varphi_2,$$

$$\ldots,$$

$$x_N = |x| sin\varphi_1 sin\varphi_2 \ldots sin\varphi_{N-1},$$

$0 \leq \varphi_k \leq \pi$, $1 \leq k \leq N-2$, $-\pi \leq \varphi_{N-1} \leq \pi$.

Then (2.14) becomes

$$u(x,t) = N + 1 + \frac{t^2}{2} f(N+1)[1 - w(\frac{|x|}{t}, \varphi)], \tag{2.15}$$

while (2.13) can be written as:

$$\begin{aligned} &(r^2 - 1) w_{rr} - (2r + \frac{N-1}{r}) w_r + 2w - \frac{1}{r^2} \delta_N w = 0, \\ &r \frac{x}{|x|} \in S^N(1), \\ &w(r, \varphi) = 1, r \frac{x}{|x|} \in \partial S^N(1), \varphi = \frac{x}{|x|}. \end{aligned} \tag{2.16}$$

In the previous formula δ_N stands for the Laplace–Beltrami operator on the unit sphere in $\mathbf{R}^N$, namely

$$\delta_N = \sum_{j=1}^{N-1} \frac{1}{q_j sin^{N-j-1} \varphi_j} \partial_{\varphi_j} (sin^{N-j-1} \varphi_j \partial_{\varphi_j}), q_1 \equiv 1,$$

$$q_j = q_{j-1} sin^2 \varphi_{j-1}, j \geq 2.$$

It is evident that the differential operator in (2.16) is strictly hyperbolic for $r > 1$ and is elliptic for $0 < r < 1$. The equation is parabolic if $r = 1$. Thus, (2.16) is elliptic inside the characteristic cone $\{|x| < t\}$.

This is our plan for a further treatment of the equation (2.16). We shall solve (2.16) in the case $N = 2$ in the hyperbolic region $r > 1$, i.e., outside the unit circle $r = 1$ and we shall study the inward propagation of the waves created by w and starting from $\partial S^2(1)$, where $w = 1$, through the simplex $S^2(1)$ till the unit circle inscribed in $S^2(1)$. This way we shall determine $w(1, \varphi)$ on the unit circle $r = 1$. By solving the Dirichlet problem for the elliptic equation (2.16) in the domain $r < 1$ and with $w(1, \varphi)$ given before, we shall find $w(r, \varphi)$ inside the elliptic region. The hyperbolic-elliptic change of the type of (2.16) at the point $r = 1$ could give rise to some new singularities of w. Our considerations for $r < 1$, i.e., $\{|x| < t\}$ are valid for each dimension $N \geq 2$. This way the Goursat problem with data on the characteristic cone $\{|x| = t\}$ can be investigated too. Because of this reason we give below a complete study of the Goursat problem in arbitrary space dimension $N \geq 2$.

The function $w(r, \varphi)$ is smooth and even analytic for $0 < r < 1$ as the equation (2.16) is elliptic there. We are looking for a solution of the equation (2.16) in the form (separation of variables):

$$w(r, \varphi) = \sum_{n=0}^{\infty} \sum_{k=1}^{k_{n,N}} \alpha_{n,k} w_{n,k}(r) Y_{n,N}^{(k)}(\varphi), \tag{2.17}$$

where $Y_{n,N}^{(k)}$ are the classical spherical harmonics of order n,

$$k_{n,N} = (2n + N - 2)\frac{(n + N - 3)!}{(N - 2)!n!}, k_{n,2} = 2, \qquad Y_{n,2}^{(k)} = \left\{\frac{\sin n\varphi}{\sqrt{\pi}}, \frac{\cos n\varphi}{\sqrt{\pi}}\right\}_{n=0}^{\infty}.$$

As it is known (see for example [76], [78]), the functions $\{Y_{n,N}^{(k)}\}$ form a full orthonormal system (basis) in $L^2(\Sigma)$ and Σ is the unit sphere in $\mathbf{R}^N$. Moreover, $Y_{n,N}^{(k)}$ are eigenfunctions of δ_N, i.e.,

$$\delta_N Y_{n,N}^{(k)} = \lambda_n Y_{n,N}^{(k)}(\varphi), \forall \varphi \in \Sigma, \tag{2.18}$$

$\lambda_n = -n(n + N - 2)$, $n = 0, 1, 2, \ldots$. In the case $N = 2$: $\delta_2 = \partial^2/\partial\varphi_1^2$, $\lambda_n = -n^2$.

Putting (2.17) in (2.16) we obtain

$$(r^2 - 1)w''_{n,k} - \left(2r + \frac{N - 1}{r}\right) w'_{n,k} + \left(2 - \frac{\lambda_n}{r^2}\right) w_{n,k} = 0, \tag{2.19}$$

$$0 < r < 1.$$

Therefore

$$w''_{n,k} - \frac{2r^2 + N - 1}{r(r^2 - 1)} w'_{n,k} + \frac{2r^2 - \lambda_n}{r^2(r^2 - 1)} w_{n,k} = 0, \quad 0 < r < 1. \tag{2.20}$$

The previous equation can be written in the form:

$$w'' + p(r)w' + q(r)w = 0, \tag{2.21}$$

with p, q-meromorphic functions. A point at which p and q are analytic functions is called a normal one. A point is called singular if it is not normal. The points $r = 0, 1$ are the only singular points of (2.20) in the interval $[0, 1]$. We shall assume further on that the point $r = c$ of the equation (2.21) is such that $(r - c)p(r)$ and $(r - c)^2 q(r)$ are analytic near c. Then the point c is said to be regular for the equation (2.21). Suppose that c is a regular point of (2.21). Then (2.21) can be written as well as:

$$(r-c)^2 f(r)w'' + (r-c)g(r)w' + h(r)w = 0 \tag{2.22}$$

and f, g, h are analytic functions near $r = c$. According to [80], [62] the algebraic equation

$$r(r-1)f(c) + rg(c) + h(c) = 0 \tag{2.23}$$

is called an indicial equation of (2.22) at $r = c$.

Let r_1, r_2 be the roots of (2.23) for which $d = r_1 - r_2 \geq 0$. Then each solution of (2.22) near $r = c$ is given by (see [80], [62])

$$w = Aw_1 + Bw_2, A, B = \text{const.}, \tag{2.24}$$

where

$$w_1 = (r-c)^{r_1}\left(1 + \sum_{k=1}^{\infty} c_k (r-c)^k\right), \tag{2.25}$$

the power series being convergent near $r = c$ and

$$w_2 = \begin{cases} (r-c)^{r_2}(1 + \sum_{k=1}^{\infty} c_k^*(r-c)^k), \text{if } d \notin \mathbf{Z} \\ w_1 \log(r-c) + (r-c)^{r_2} \sum_{k=1}^{\infty} d_k (r-c)^k, \text{if } d = 0 \\ \text{const.} w_1 \log(r-c) + (r-c)^{r_2}(-\frac{1}{d} + \sum_{k=1}^{\infty} d_k (r-c)^k), \text{if } d \in \mathbf{N}. \end{cases} \tag{2.26}$$

The power series above are convergent near $r = c$. Our equation (2.20) has the form

$$r^2(r-1)^2(r+1)w''_{n,k} - r(r-1)(2r^2 + N - 1)w'_{n,k} \tag{2.27}$$
$$+(r-1)(2r^2 - \lambda_n)w_{n,k} = 0.$$

Its indicial equation at the regular point $r = 0$ is

$$r(r-1) + (N-1)r + \lambda_n = 0, \tag{2.28}$$

as (2.27) coincides with (2.22) for $f(r) = (r-1)^2(r+1)$, $g(r) = -(r-1)(2r^2+N-1)$, $h(r) = (r-1)(2r^2 - \lambda_n)$. The roots of (2.28) are $r_1 = n$, $r_2 = 2 - n - N$; $d = r_1 - r_2 = N + 2n - 2 > 0$, i.e., $d \in \mathbf{N}$.

According to (2.25), (2.26) we have that:

$$w_1 = r^n\left(1 + \sum_{k=1}^{\infty} c_k r^k\right), \tag{2.29}$$

$$w_2 = c_1 w_1 \log r + r^{2-n-N}\left(-\frac{1}{d} + \sum_{k=1}^{\infty} d_k r^k\right). \tag{2.30}$$

Thus (2.27) possesses only one regular solution near $r = 0$, namely w_1. The indicial equation of (2.27) at the regular point $r = 1$ is

$$2r(r-1) - (N+1)r = 0$$

and its roots are: $r_1 = \frac{N+3}{2}$, $r_2 = 0$, $d = \frac{N+3}{2} > 0$. Applying the asymptotic formulas (2.25), (2.26) we obtain the following solutions:

$$w_3 = (1-r)^{\frac{N+3}{2}} \left(1 + \sum_{k=1}^{\infty} c_k (1-r)^k\right), \tag{2.31}$$

$$w_4 = \begin{cases} 1 + \sum_{k=1}^{\infty} c_k^* (1-r)^k, \text{for } N - \text{even}, \\ \text{const.} w_3 \log(1-r) \\ +(-\frac{1}{d} + \sum_{k=1}^{\infty} d_k (1-r)^k), \text{for } N - \text{odd}. \end{cases} \tag{2.32}$$

We point out that the singularities of the solutions of the equation (2.27) at the regular point $r = 1$ do not depend on the summation index n.

We shall solve (2.19) by an explicit formula. To do this put

$$z = r^2, w_{n,k} = z^{n/2} F(z) \Rightarrow w_{n,k}(r) = r^n F(r^2). \tag{2.33}$$

After this change (2.19) is transformed into:

$$z(z-1)F''(z) - \left[n + \frac{N}{2} - \left(n - \frac{1}{2}\right) z\right] F'(z) + \frac{(n-2)(n-1)}{4} F(z) = 0. \tag{2.34}$$

This is just the hypergeometric equation (see [39], [62], [77])

$$z(z-1)F'' - [c - (a+b+1)z]F' + abF = 0 \tag{2.35}$$

with parameters $a = \frac{n-2}{2}$, $b = \frac{n-1}{2}$, $c = n + \frac{N}{2}$. The hypergeometric function $F(a, b, c, z)$ is the regular solution at $z = 0$ of (2.35). We remind the reader that (see [28], [62], [77]):

$$\begin{aligned} F(a,b,c,z) &= 1 + \frac{ab}{c} z + \frac{a(a+1)b(b+1)}{c(c+1)2!} \\ &+ \frac{a(a+1)(a+2)b(b+1)(b+2)}{c(c+1)(c+2)3!} + \cdots \\ &= \frac{\Gamma(c)}{\Gamma(a)\Gamma(b)} \sum_{j=0}^{\infty} \frac{\Gamma(a+j)\Gamma(b+j)}{\Gamma(c+j)j!} z^j, |z| < 1 \end{aligned} \tag{2.36}$$

and Γ is the famous Euler's gamma function.

The power series (2.36) is absolutely convergent for $|z| < 1$. Moreover, $c - a - b = \frac{N+3}{2} > 0$ implies that (2.36) is absolutely convergent for $|z| = 1$.

Assume that $n = 0$. Then $a = -1$ and according to (2.36) and (2.33) we get that $w_{0,k}(r) = A_0(N + r^2)$. If $n = 1$, then $b = 0$ and therefore $w_{1,k}(r) = A_1 r$. The assumption $n = 2 \Rightarrow a = 0$ and then $w_{2,k} = A_2 r^2$. We observe that $n \geq 3 \Rightarrow a > 0$, $b > 0$, $c > 0$.

Our next step is to investigate the behavior of the derivatives of $w_{n,k}$ near the point $r = 1$. Thus

$$\begin{aligned} \frac{d^p w_{n,k}}{dr^p}(r) &= \frac{d^p(r^n F(r^2))}{dr^p} \qquad (2.37)\\ &= \frac{\Gamma(c)}{\Gamma(a)\Gamma(b)} \sum_{j=0}^{\infty} \frac{\Gamma(a+j)\Gamma(b+j)}{\Gamma(c+j)j!} \\ &\times (n+2j)(n+2j-1)\dots(n+2j-p+1)r^{n+2j-p} \\ &= \sum_{j=0}^{\infty} a_{n+2j-p} r^{n+2j-p}. \end{aligned}$$

The coefficients $a_{n+2j-p} \geq 0$ for $n \geq 3$, $j \geq 1$. According to the Gauss criterion [28] the series $\sum_{j=0}^{\infty} a_j$, $a_j \geq 0$ is convergent (divergent) if the parameter μ in the representation $\frac{a_j}{a_{j+1}} = 1 + \frac{\mu}{j} + O(\frac{1}{j^2})$ is $> 1 (\leq 1)$.

The power series (2.37) is convergent for $r < 1$ and is divergent for $r > 1$ as the d'Alembert criterion shows. On the other hand, $r = 1$ gives:

$$\begin{aligned} \frac{a_{n+2j-p}}{a_{n+2j+2-p}} &= \frac{(1+\frac{1}{j})(1+\frac{c}{j})(1+\frac{2+n-p}{2j})(1+\frac{n-p+1}{2j})}{(1+\frac{a}{j})(1+\frac{b}{j})(1+\frac{2+n}{2j})(1+\frac{n+1}{2j})} \\ &= 1 + \frac{1-p+c-a-b}{j} + O(\frac{1}{j^2}). \end{aligned}$$

In fact, $\frac{1}{1+\frac{\alpha}{j}} = 1 - \frac{\alpha}{j} + \frac{\alpha^2}{1+\frac{\alpha}{j}} \cdot \frac{1}{j^2}$.

Applying Gauss's criterion with $\mu = 1 - p + c - a - b = 1 - p + \frac{N+3}{2}$, we conclude that the power series (2.37) is convergent at $r = 1$ for $p < \frac{N+3}{2}$ and is divergent for $p \geq \frac{N+3}{2}$. This way we have that the regular solutions $w_{n,k}$ of (2.19) at $r = 0$, $n \geq 3$ possess singular derivatives of order $\frac{N+3}{2}$ at the point $r = 1$. Going back to the old variables x, t and combining (2.15), (2.29), (2.30), (2.31), (2.32), (2.37) we come to the following theorem.

Theorem 2.1. *Suppose that $u(x,t)$ satisfies the Goursat problem:*

$$\Box u = f(N+1), x \in S^N(t), t > 0,$$

$$u = N+1, x \in \partial S^N(t), t > 0.$$

Then this is the behavior of $u(x,t)$ within the characteristic sphere $\{|x| = t\}$:

$$u(x,t) = N+1+t^2 f(N+1)[h(\frac{|x|}{t},\varphi) + (1-\frac{|x|}{t})^{\frac{N+3}{2}} g(\frac{|x|}{t},\varphi)], \qquad (2.38)$$

for N even, $|x| < t$, $|x| \to t$,

$$u(x,t) = N+1+t^2 f(N+1)[h(\frac{|x|}{t},\varphi) + (1-\frac{|x|}{t})^{\frac{N+3}{2}} \log(1-\frac{|x|}{t}) g(\frac{|x|}{t},\varphi)], \qquad (2.39)$$

for N odd, $|x| < t$, $|x| \to t$.

The function $h(r,\varphi)$ has $\frac{N+4}{2}$, respectively $\frac{N+3}{2}$ derivatives with respect to r at $r=1$, while $g(r,\varphi)$ is a bounded function.

Our next step is to determine the functions h, g outside the sphere $|x|=t$, i.e., for $|x|>t$. To do this we shall investigate u outside this sphere (circle), i.e., in the "hyperbolic region" inside $S^2(t)$ but outside $\{|x|=t\}$. So we are working again in the case $N=2$.

2.4. Behavior of u outside of the characteristic circle

Then for each fixed $t>0$, $S^2(t)$ is an isosceles (equilateral) triangle (see Fig. 19).

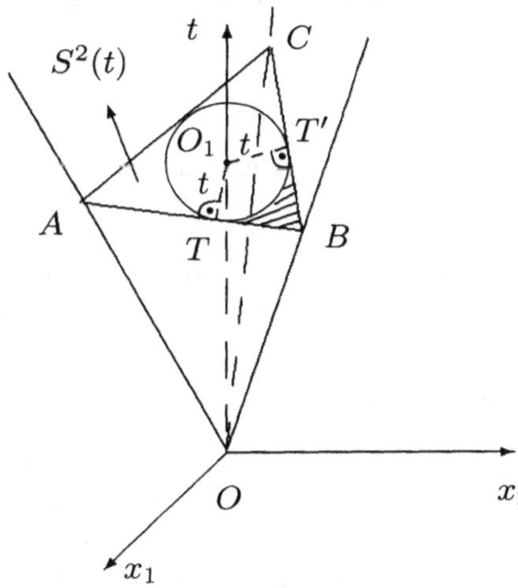

Fig. 19

Thus $S^2(t) = \Delta ABC$ and the characteristic sphere coincides with the inscribed circle in ΔABC with center at O_1 and radius $t>0$. We shall find the solution u of the problem (2.10) in an explicit form at each curvilinear triangle outside the inscribed circle. To fix the ideas we shall only consider the case of the curvilinear triangle TBT' in dashed lines in Fig. 19 (see Fig. 20):

We choose a new orthogonal coordinate system O_1zy and the axis $O_1B = O_1z$ is collinear to the bisectrix O_1B. Denote $\alpha = \angle ABO_1 = \angle CBO_1$ and $\frac{1}{m} = \sin\alpha > 0 \Rightarrow m>1$. Thus $O_1B = tm$ and therefore $B=(tm,0)$ in the new coordinate system. From the triangle O_1QT' we have that $O_1Q = \frac{t}{m}$, $T'Q = \frac{t}{m}\sqrt{m^2-1}$ and consequently $T' = (\frac{t}{m}, -\frac{t}{m}\sqrt{m^2-1})$ in the new coordinate system. Obviously, $T = (\frac{t}{m}, -\frac{t}{m}\sqrt{m^2-1})$. Then the straight lines BT', BT are given by the following equations:

$$BT' : y = \frac{mt-z}{\sqrt{m^2-1}}, \quad BT : y = -\frac{mt-z}{\sqrt{m^2-1}}.$$

The points located inside the angle $\angle TBT'$ satisfy the inequality

$$|y| < (m^2-1)^{-1/2}(mt-z).$$

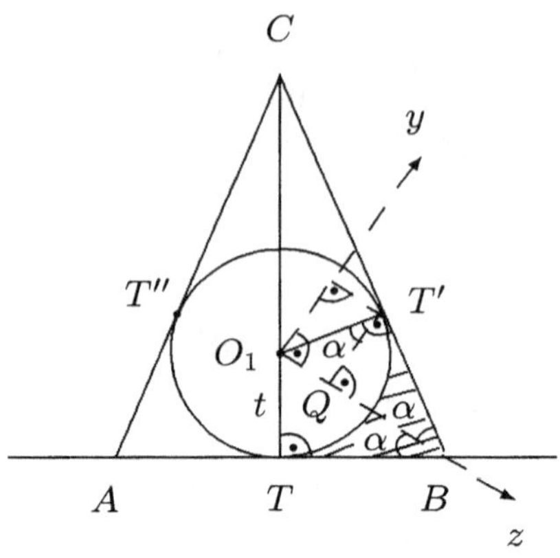

Fig. 20

Denote by v the unit vector collinear to O_1z, $v = (v_1, v_2)$ in the old coordinates and by $l_y = (l_y^1, l_y^2)$ the unit vector collinear to O_1y. Then $z = v.x = v_1x_1 + v_2x_2$, $y = l_y.x = l_y^1x_1 + l_y^2x_2$. As the Laplace operator Δ is invariant under rotations, we rewrite the Goursat problem (2.10) in $\angle TBT'$ in the form:

$$\begin{aligned} u_{tt} - u_{zz} - u_{yy} &= f(3), |y| < (m^2-1)^{-1/2}(mt - z), \\ u &= 3, |y| = (m^2-1)^{-1/2}(mt-z). \end{aligned} \tag{2.40}$$

The previous Goursat problem certainly has a solution. We shall find it by a simple formula. To show this fact we make the following change of the variable: $\tau = (m^2-1)^{-1/2}(mt-z)$, seeking $u = u(\tau, y)$. Then the problem (2.40) becomes

$$\begin{aligned} u_{\tau\tau} - u_{yy} &= f(3), |y| < \tau, \\ u(\tau, y) &= 3, |y| = \tau. \end{aligned} \tag{2.41}$$

This is again a Goursat problem in the plane. After a standard change of variables $\tau_1 = \tau - y$, $y_1 = \tau + y$, $\tau_1 > 0$, $y_1 > 0$, we come to the Goursat problem

$$\begin{aligned} u_{\tau_1 y_1} &= \frac{1}{4}f(3), \tau_1 > 0, y_1 > 0, \\ u(0, y_1) &= u(\tau_1, 0) = 3. \end{aligned} \tag{2.42}$$

Its unique solution is $u(\tau_1, y_1) = 3 + \frac{1}{4}f(3)\tau_1 y_1$ and consequently $u(\tau, y) = 3 + \frac{1}{4}f(3)(\tau^2 - y^2)$, $|y| < \tau$. Going back to the old variables (t, x) we find the solution $u(x,t)$ of the Goursat problem (2.40) in the closed curvilinear triangle BTT':

$$u(x,t) = 3 + \frac{1}{4}f(3)\left[\frac{(mt - v.x)^2}{m^2-1} - y^2\right], \tag{2.43}$$

$$|y| < (m^2-1)^{-1/2}(mt - vx),$$

$y = l_y.x$, l_y being a constant vector. In a similar way we can find u in the curvilinear triangles ATT'', $T''CT'$ outside and on the inscribed circle $|x| = t$. Inside the inscribed circle we are seeking u in the form (2.15), namely

$$u(x,t) = 3 + \frac{t^2}{2} f(3) \left[1 - w\left(\frac{|x|}{t}, \varphi\right)\right]. \tag{2.44}$$

Then

$$\begin{aligned} u|_{|x|=t} &= 3 + \frac{t^2}{2} f(3) \left[1 - w(1,\varphi)\right] \\ &= 3 + \frac{1}{4} f(3)|x|^2 \left(\frac{(m - v.\omega)^2}{m^2 - 1} - (l_y, \omega)^2\right), \\ \omega &= (\cos\varphi, \sin\varphi), -\pi \le \varphi \le \pi. \end{aligned} \tag{2.45}$$

So $w(1,\varphi) = 1 - \frac{1}{2}(\frac{(m - v\omega)^2}{m^2-1} - (l_y, \omega)^2)$.

After this we expand $w(r,\varphi)$ in the Fourier series (2.17) and we repeat in the case $N = 2$ the analysis fulfilled in Section 2.3. Certainly, $\alpha_{n,k}.w_{n,k}$ are the Fourier coefficients of the function $w(1,\varphi)$. Applying Theorem 2.1 we see that the function $u(x,t)$ can be represented inside and up to the characteristic circle $|x| = t$ as:

$$u(x,t) = 3 + t^2 f(3) \left[h\left(\frac{|x|}{t}, \varphi\right) + \left(1 - \frac{|x|}{t}\right)^{5/2} g\left(\frac{|x|}{t}, \varphi\right)\right], \tag{2.46}$$
$$|x| < t, \quad |x| \to t,$$

where $h(r,\varphi)$ has three order derivatives with respect to r for $r \le 1$ and $g(r,\varphi)$ is bounded function.

To be more precise, we shall carefully study the behavior of the series $w(r,\varphi)$ (2.17) for $r < 1$ and with Dirichlet data $w(1,\varphi)$ on the boundary $r = 1$. This way we are dealing with the Goursat type characteristic problem for the wave equation in the light cone $\{|x| < t\}$.

At first we expand $w(1,\varphi)$ in $L_2(\Sigma)$ via the orthonormal system (basis) $Y^{(k)}_{n,N}$, namely:

$$w(1,\varphi) = \sum_{n=0}^{\infty} \sum_{k=1}^{k_{n,N}} p_{n,k} Y^{(k)}_{n,N}(\varphi), p_{n,k} = \left(w(1,\varphi), Y^{(k)}_{n,N}(\varphi)\right)_{L^2(\Sigma)}.$$

Certainly, $w(1,\varphi) \in L_2(\Sigma)$ as it is continuous.

Having in mind (2.17) we obtain that

$$w(r,\varphi) = \sum_{n=0}^{\infty} \sum_{k=1}^{k_{n,N}} \frac{p_{n,k}}{w_{n,k}(1)} w_{n,k}(r) Y^{(k)}_{n,N}(\varphi), \tag{2.47}$$

where

$$w_{n,k}(r) = r^n F\left(\frac{n}{2} - 1, \frac{n}{2} - \frac{1}{2}, n + \frac{N}{2}, r^2\right), 0 \le r \le 1, \tag{2.48}$$

$w_{n,k}(1) = F(\frac{n}{2} - 1, \frac{n}{2} - \frac{1}{2}, n + \frac{N}{2}, 1)$.

Estimates guaranteeing the existence of a solution of the Goursat problem in the light cone with data at $\{|x| = t\}$ are given below.

Lemma 2.2. *Assume that* $F(\frac{n}{2}-1, \frac{n}{2}-\frac{1}{2}, n+\frac{N}{2}, x)$, $0 \le x \le 1$ *is a hypergeometric function,* $n > 2$, $N \ge 2$ *and* $0 \le x \le x_0 < 1$, *with* x_0 *being positive. Then there exists a constant* $C(N, x_0)$, C *independent of* n *and such that*

$$\begin{aligned} 0 &\le F(\frac{n}{2}-1, \frac{n}{2}-\frac{1}{2}, n+\frac{N}{2}, x) \\ &\le C(N, x_0).2^n n^{1/2} \left(\frac{1-\sqrt{1-x_0}}{x_0}\right)^n, \end{aligned}$$
$$0 \le x \le x_0 < 1.$$

Conclusion: $0 \le w_{n,k}(r) \le C(N, x_0) 2^n n^{1/2} \left(\frac{1-\sqrt{1-r_0^2}}{r_0^2}\right)^n$, $0 \le r \le r_0 < 1$, $r_0 > 0$.

Having in mind the fact that if $\lambda(x) = \frac{1-\sqrt{1-x}}{x}$, $0 < x \le 1$, we have: $\lambda(x) > 0$ for $0 < x < 1$, $\lambda'(x) > 0$ for $0 < x < 1$, $\lambda(x)_{x \to 0} \to 1/2$ and therefore $0 < \lambda(x) \le \lambda(x_0) < 1$, $0 \le x \le x_0 < 1$. Thus the series (2.47) is absolutely and uniformly convergent for $0 \le r < 1$. The only new feature in verifying this fact is the estimate for $w_{n,k}(1) = F(\frac{n}{2}-1, \frac{n}{2}-\frac{1}{2}, n+\frac{N}{2}, 1)$ because $|p_{n,k}| \le$ const.

As it is shown in [62]:

$$F(a, b, c, 1) = \frac{\Gamma(c)\Gamma(c-a-b)}{\Gamma(c-a)\Gamma(c-b)} = \frac{\Gamma(n+\frac{N}{2})\Gamma(\frac{N}{2}+\frac{3}{2})}{\Gamma(\frac{n}{2}+\frac{N}{2}+1)\Gamma(\frac{n}{2}+\frac{N}{2}+\frac{1}{2})}$$

and according to the famous Stirling formula $\Gamma(x) \sim e^{-x} x^x (\frac{2\pi}{x})^{-1/2}(1+o(1))$, $x \to \infty$ we conclude that

$$\frac{1}{w_{n,k}(1)} = 2^{-n-N-\frac{1}{2}} n^{\frac{N}{2}+1} \left(1 + O\left(\frac{1}{n}\right)\right), n \to \infty.$$

Proof of Lemma 2.2. The well-known integral representation for the hypergeometric functions [62], $0 \le x < 1$:

$$F(a, b, c, x) = \frac{\Gamma(c)}{\Gamma(b)\Gamma(c-b)} \int_0^1 \frac{t^{b-1}(1-t)^{c-b-1}}{(1-tx)^a} dt,$$
$$c > b > 0,$$

$a = \frac{n}{2}-1$, $b = \frac{n}{2}-\frac{1}{2}$, $c = n+\frac{N}{2}$, $n > 2$ will be used. Put

$$\begin{aligned} J\left(\frac{n}{2}-1, \frac{n}{2}-\frac{1}{2}, n+\frac{N}{2}, x\right) &= \int_0^1 \frac{t^{\frac{n}{2}-\frac{3}{2}}(1-t)^{\frac{n}{2}+\frac{N}{2}-\frac{1}{2}}}{(1-tx)^{n/2-1}} dt \\ &= \int_0^1 t^{-1/2}(1-t)^{\frac{N}{2}+\frac{1}{2}} \left\{\frac{t(1-t)}{1-tx}\right\}^{\frac{n-2}{2}} dt, \quad n > 2. \end{aligned}$$

Consider now the function $\varphi(x;t) = \frac{t(1-t)}{1-tx}$, $\{0 \le x \le x_0 < 1, 0 \le t \le 1\}$. As $\varphi_t' = \frac{1-2t+xt^2}{(1-tx)^2}$ one can easily see that $\varphi_t' = 0 \iff 0 < t_1(x) = \frac{1-\sqrt{1-x}}{x} < 1$

and φ reaches its maximum at $t = t_1(x)$, i.e., $\varphi(x,t) \le \varphi(x, t_1(x)) = (\frac{1-\sqrt{1-x}}{x})^2$, $0 \le t \le 1$.

Therefore,

$$J\left(\frac{n}{2}-1, \frac{n}{2}-\frac{1}{2}, n+\frac{N}{2}, x\right) \le \text{const}(N)\left(\frac{1-\sqrt{1-x}}{x}\right)^{n-2},$$

$$n > 2, 0 \le x \le x_0 < 1.$$

Applying again the Stirling formula we see that

$$\frac{\Gamma(n+\frac{N}{2})}{\Gamma(\frac{n}{2}-\frac{1}{2})\Gamma(\frac{n}{2}+\frac{N}{2}+\frac{1}{2})} \le \text{const}(N) 2^n n^{1/2}$$

which completes the proof of Lemma 2.2.

Our next step is to prove that the series (2.47) is termwise differentiable with respect to (r, φ), $0 \le r < 1$ infinitely many times. There are no problems in differentiation with respect to φ [78].

Having in mind that $\frac{d^p(r^n F(r^2))}{dr^p}$ is a linear combination of terms of the type $2^p r^{p+n} F^{(p)}(r^2), \ldots, n(n-1)\ldots(n-p+1)r^{n-p}F(r^2)$, we see that the termwise differentiability will follow from

Lemma 2.3. *Denote* $F_n(x) = F(\frac{n}{2}-1, \frac{n}{2}-\frac{1}{2}, n+\frac{N}{2}, x)$, $n \ge 3$. *Then for each* $p \in \mathbf{N}$ *there exists a constant* $C(N, p, x_0)$ *such that*

$$\left|\frac{\frac{d^p}{dx^p}F_n(x)}{F_n(1)}\right| \le C(N,p,x_0) n^{\frac{N}{2}+\frac{3}{2}+2p}\left(\frac{1-\sqrt{1-x_0}}{x_0}\right)^n,$$

$$0 \le x \le x_0 < 1.$$

Proof. As it is known from [62]:

$$\frac{d^p F(a,b,c,x)}{dx^p} = (a)_p (b)_p F(a+p, b+p, c+p),$$

$$p \in \mathbf{N}, n \ge 3,$$

$N \ge 2$, where $(a)_p = a(a+1)\ldots(a+p-1) = \frac{\Gamma(a+p)}{\Gamma(a)}$, $(b)_p = \frac{\Gamma(b+p)}{\Gamma(b)}$.

So:

$$\frac{d^p F(a,b,c,x)}{dx^p} = \frac{\Gamma(a+p)\Gamma(c+p)}{\Gamma(a)\Gamma(b)\Gamma(c-b)} \cdot \int_0^1 \frac{t^{b+p-1}(1-t)^{c-b-1}}{(1-tx)^{a+p}} dt.$$

Put

$$\begin{aligned} J(a,b,c,x) &= \int_0^1 \frac{t^{b+p-1}(1-t)^{c-b-1}}{(1-tx)^{a+p}} dt \\ &= \int_0^1 \left(\frac{t(1-t)}{1-tx}\right)^{\frac{n}{2}-1} \cdot \frac{t^{p-1/2}(1-t)^{\frac{N}{2}+\frac{1}{2}}}{(1-tx)^p} dt, \qquad n \ge 3. \end{aligned}$$

As in Lemma 2.2 we obtain

$$\begin{aligned} J &\leq \left(\frac{1-\sqrt{1-x}}{x}\right)^{n-2} \int_0^1 \frac{t^{p-1/2}(1-t)^{\frac{N}{2}+\frac{1}{2}}}{(1-x)^p} dt \\ &\leq \operatorname{const}(N,p)(1-x)^{-p} \left(\frac{1-\sqrt{1-x}}{x}\right)^{n-2}, \quad n \geq 3, \end{aligned}$$

$$0 \leq x \leq x_0 < 1.$$

i.e., $J \leq \operatorname{const}(N, x_0, p)(\frac{1-\sqrt{1-x_0}}{x_0})^n$, $n \geq 3$, $0 \leq x \leq x_0 < 1$.

According to the Stirling formula

$$\frac{\Gamma(n+\frac{N}{2}+p)\Gamma(\frac{n}{2}+p-1)}{\Gamma(\frac{n}{2}-1)\Gamma(\frac{n}{2}-\frac{1}{2})\Gamma(\frac{n}{2}+\frac{N}{2}+\frac{1}{2})} \leq \operatorname{const}(N,p)2^n n^{2p+\frac{1}{2}}$$

and thus Lemma 2.3 is proved.

Obviously, $w(r,\varphi) \in C^\infty(r<1)$, i.e., inside the light cone $|x|<t$ and for arbitrary L_2 data on $\{|x|=t\}$. Even C^∞ smoothness of the boundary data on $\{|x|=t\}$ does not imply smoothness of w up to the boundary. In fact, assume that the sum defining $w(1,\varphi)$ is finite. Then (2.47) is a finite sum too but the functions $w_{n,k}(r)$ have singularity at $r=1$ of order $\frac{N+3}{2}$ (see (2.37)).

As we know $u(x,t)$ is smooth inside and up to the boundary of the three curvilinear triangles and inside the inscribed circle $|x|=t$. Assume now that the three different smooth solutions of the type (2.43) could be extended inside the characteristic circle $|x|=t$. They are certainly unequal in the set $\{|x|<t\}$ because they contain different combinations of the variables (x,t). Moreover, the solution inside $\{|x|<t\}$ is uniquely determined.

Conclusion: The solution u has a singularity from inside on the characteristic circle $|x|=t$. So $g(r,\varphi) \not\equiv 0$ and the strength of this singularity is $(1-\frac{|x|}{t})^{5/2}$, $|x|<t$, $|x| \to t$. There are no other singularities outside and inside $\{|x|=t\}$.

2.5. A microlocal approach to the propagation of singularities. Concluding remarks

Following Beals [6] a simpler proof of the smoothness of the solution of the semi-linear wave equation inside the light cone can be proposed. Thus put

$$\Box u = \frac{\partial^2 u}{\partial x_n^2} - \sum_{j=1}^{n-1} \frac{\partial^2 u}{\partial x_j^2} = f(x,u), \tag{2.49}$$

$$x \in \mathbf{R}^n, f \in C^\infty(\mathbf{R}^{n+1})$$

and suppose that $u \in H^s_{loc}(\mathbf{R}^n)$, $s > \frac{n}{2}$.

The radial vector field M is defined as follows:

$$M = \sum_{j=1}^n x_j \frac{\partial}{\partial x_j}. \tag{2.50}$$

Lemma 2.4. *The real-valued function $u \in H^s_{loc}(\mathbf{R}^n)$ satisfies (2.49) and $M^j u \in H^s_{loc}(\mathbf{R}^n_-)$ for all j. Then $u \in C^\infty(\{x'^2 < x_n^2\} \cap \mathbf{R}^n_+)$, where $\mathbf{R}^n_\pm = \{x \in \mathbf{R}^n : x_n > (<)0\}$ and $x' = (x_1, \dots, x_{n-1})$.*

Proof. Obviously, the commutator $[\Box, M] = 2\Box$. Inductively we shall prove that $M^j u \in H^s_{loc}(\mathbf{R}^n)$, $\forall j \in \mathbf{Z}_+$. The case $j = 0$ is obvious. So assume that $M^t u \in H^s_{loc}(\mathbf{R}^n)$ for some $j \in \mathbf{Z}_+$ and each $0 \le t \le j$ and that $\Box(M^j u) = f_j(x, u, Mu, \dots, M^j u)$, the functions f_j being C^∞ smooth with respect to their arguments. It is well known that $f_j(x, u, Mu, \dots, M^j u) \in H^s_{loc}(\mathbf{R}^n)$. Then

$$\begin{aligned}\Box(M^{j+1}u) &= \Box(M.M^j u) = M\Box(M^j u) + [\Box, M](M^j u)\\ &= Mf_j(x, u, Mu, \dots, M^j u) + 2\Box(M^j u)\\ &= Mf_j(x, u, Mu, \dots, M^j u) + 2f_j(x, u, Mu, \dots, M^j u) \equiv A_{j+1}.\end{aligned}$$

Thus, $\Box(M^{j+1}u) \in H^{s-1}_{loc}(\mathbf{R}^n)$, $M^{j+1} \in H^{s-1}_{loc}(\mathbf{R}^n)$ and $M^{j+1} \in H^s_{loc}(\mathbf{R}^n_-)$.

According to the linear version for propagation of singularities of the wave equation we conclude that $M^{j+1}u \in H^s_{loc}(\mathbf{R}^n)$. Moreover, then the chain rule holds, i.e.,

$$A_{j+1} = f_{j+1}(x, u, Mu, \dots, M^{j+1}u), f_{j+1} \in C^\infty.$$

In fact, $M(f(x, u, v)) = (Mf)(x, u, v) + f_u(x, u, v)Mu + f_v(x, u, v)Mv$, $u, v \in H^s_{loc}$.

Assume now that $(x_0, \xi_0) \in WF(u)$ and $x_0 \neq 0$. Then the linear form $\langle x_0, \xi_0\rangle = 0$. To prove this fact suppose that $\langle x_0, \xi_0\rangle \neq 0$. As $\langle x, \xi\rangle$ is up to a nonvanishing factor the symbol of the operator M and $M^j u \in H^s_{loc}(\mathbf{R}^n)$, $\forall j \in \mathbf{Z}_+$, we see that $u \in H^\infty_{mcl}(x_0, \xi_0)$, i.e., we come to a contradiction. Define now

$$K = \left\{x \in \mathbf{R}^n : x'^2 < x_n^2\right\}, \partial K = \left\{x \in \mathbf{R}^n : x'^2 = x_n^2\right\}.$$

Assume that the point $0 \neq x_0 \in K \cap \mathbf{R}^n_+$. Obviously, $(x_0, \xi_0) \notin WF(u) \Rightarrow u \in H^\infty_{mcl}(x_0, \xi_0)$. Thus consider the case $(x_0, \xi_0) \subset WF(u)$. Then we know that $\langle x_0, \xi_0\rangle = 0$. Having in mind the fact that $K \cap \mathbf{R}^n_+$ is a proper cone with an opening $\frac{\pi}{2}$ we conclude that $\xi_0 \notin \partial K$, i.e., $(\xi^0_n)^2 \neq (\xi'^0)^2$. The wave operator $\Box$ is microelliptic at the point (x_0, ξ_0) and therefore $u \in H^{s+2}_{mcl}(x_0, \xi_0)$ as $\Box u \in H^s_{loc}(\mathbf{R}^n)$. So $u \in H^{s+2}(x_0)$, i.e., $u \in H^{s+2}_{loc}(K \cap \mathbf{R}^n_+)$. Standard bootstrap arguments prove Lemma 2.4.

Definition 2.5. The function $u \in H^s_{loc}$ is said to be conormal with respect to the smooth hyperspace $\Sigma \subset \mathbf{R}^n$ if for each choice of smooth, tangential to Σ vector fields $M_1, \dots, M_j$, $j \in \mathbf{Z}_+$ the following relation holds:

$$M_1 M_2 \dots M_j u \in H^s_{loc}.$$

If the hypersurface Σ is given locally by $\{x_1 = 0\}$, the vector fields M_j are generated by $x_1 \frac{\partial}{\partial x_1}, \frac{\partial}{\partial x_2}, \dots, \frac{\partial}{\partial x_n}$.

Denote by $N^*\Sigma$ the conormal bundle over Σ. Then $WF(u) \subset N^*\Sigma$. In fact, $\nabla\varphi(x)$ is the normal vector at the point $x \in \Sigma$ to the hypersurface $\Sigma = \{\varphi(x) = 0\}$. Suppose that the covector $0 \neq \xi$ is not collinear to $\nabla\varphi(x)$ at x. Then there exists a vector field $M_j \in T(\Sigma)$ such that $\langle \xi, M_j(x) \rangle \neq 0$. The microellipticity of M_j at (x, ξ) implies our inclusion.

In a similar way one can define conormality of the function u with respect to several hypersurfaces (see [11], [13]).

Consider now the hyperplanes $\Sigma_\omega : x_n = \omega.x'$, $x = (x', x_n)$, $|\omega| = 1$. Obviously, they are characteristics for $\Box$ as they are tangential to ∂K. Moreover, the radial vector field $M(x)$, $x \neq 0$ is tangential to ∂K and to the characteristic hyperplanes Σ_ω.

Corollary 2.6. *Let $u \in H^s_{loc}(\mathbf{R}^3)$, $s > \frac{3}{2}$ be a solution of (2.49) with $n = 3$ and u be conormal with respect to $\Sigma = (\Sigma_1, \Sigma_2, \Sigma_3)$ in $\mathbf{R}^3_-$ where Σ is a family of three characteristic hyperplanes intersecting transversally at the origin. Then*

$$u \in C^\infty \left(\mathbf{R}^3 \setminus (\Sigma_1 \cup \Sigma_2 \cup \Sigma_3) \cup \left\{ x_3 = \sqrt{x_1^2 + x_2^2} \right\} \right).$$

Proof. We omit the more delicate part of the proof, namely $u \in C^\infty(\mathbf{R}^3 \setminus (\Sigma_1 \cup \Sigma_2 \cup \Sigma_3) \cup \{\sqrt{x_1^2 + x_2^2} \leq x_3\})$ (see [10]). Having in mind that the radial vector field is tangential to Σ we have according to the definition of conormality that $M^j u \in H^s_{loc}(\mathbf{R}^3_-)$, $\forall j \in \mathbf{Z}_+$. Applying Lemma 2.4 we see that $u \in C^\infty(x_3 > \sqrt{x_1^2 + x_2^2})$.

Thus everything is proved.

3. Appendix I. Blow up of the solutions of the Cauchy problem for quasi-linear hyperbolic systems in the plane

We propose in this Appendix a blow up result for the solutions of the Cauchy problem for some classes of quasi-linear systems in the plane. This is another type of singularities when the solutions remain bounded but their gradients blow up in a finite time. Certainly, there are a lot of papers on the subject but we shall mention only the fundamental papers of P. Lax [45] and F. John [33] as well as the monographs of L. Hörmander [31], Li Ta-tsien [48] and S. Alinhac [1]. Interesting theorems were also proved by Klainerman–Majda [42], Colombini–del Santo [20] and R. Manfrin [52]. Below we apply our result to the nonlinear vibrating string equation and we show that in many cases an explosion of the gradient of its solutions takes place. Our main theorem generalizes some results from [48].

3.1. Statement of the problem and main results

We study here hyperbolic systems of the type:

$$\begin{aligned} \frac{\partial r}{\partial t} + \lambda(r,s)\frac{\partial r}{\partial x} &= 0, \\ \frac{\partial s}{\partial t} + \mu(r,s)\frac{\partial s}{\partial x} &= 0, \ t \geq 0, \\ r|_{t=0} = r_0(x), \ \ s|_{t=0} &= s_0(x), \end{aligned} \tag{3.1}$$

$\lambda(r,s) < \mu(r,s)$, $\lambda, \mu \in C^1$, assuming the Cauchy data $r_0, s_0 \in C^1(R^1)$ to possess sufficiently small C^0 norms. This way blow up of r_x, s_x for a finite time $T > 0$ is possible. We shall consider the following case:

$$(A) \quad \frac{\partial \lambda}{\partial r} \geq 0, \; r_0' \geq 0, \; \frac{\partial \mu}{\partial s} \geq 0.$$

This is our main result.

Theorem 3.1. *Assume that under condition (A) the next condition (B) is fulfilled:*

$$(B) \quad \int^{+\infty} \frac{\partial \mu}{\partial s}(r_0(\beta), s_0(\alpha))\, d\beta = +\infty$$

and the point $\alpha \in R^1$ is such that $s_0'(\alpha) < 0$.

Then s_x blows up in a finite time $T > 0$ supposing the uniform norms of the Cauchy data r_0, s_0 to be sufficiently small.

Without loss of generality we can assume that $\lambda(0,0) < \mu(0,0)$ in the formulation of Theorem 3.1. The integral in (B) is taken from an arbitrary finite β_0 till $+\infty$ as the function $\frac{\partial \mu}{\partial s}(r_0(\beta), s_0(\alpha))$ is integrable on each compact interval $\beta \in [-N, N]$.

Put $\tilde{r} = \varepsilon r_0(x)$, $\tilde{s} = \varepsilon s_0(x)$, $0 < \varepsilon \leq \varepsilon_0 < 1$ and consider the system (3.1) with Cauchy data $\tilde{r}$, $\tilde{s}$. Obviously: $-\infty < r_0(-\infty) \leq r_0(+\infty) < +\infty$.

Corollary 3.2. *Assume that $\frac{\partial \mu}{\partial s}(\varepsilon r_0(+\infty), \varepsilon s_0(\alpha)) > 0$, $\mu_s' \geq 0$, $s_0'(\alpha) < 0$. Then s_x blows up for a finite time if ε_0 is sufficiently small.*

We shall illustrate the case $\frac{\partial \mu}{\partial s}(\varepsilon r_0(+\infty), \varepsilon s_0(\alpha)) = 0$ by some examples.

Corollary 3.3. *Assume that $\mu_s = r^2 \sum_{i,j=0}^n a_{ij} r^i s^j$, $a_{ij} =$ const., $\sum_{i,j=0}^n a_{ij} r^i s^j \geq 0$. Then it follows immediately that $a_{00} \geq 0$ and if $a_{00} = 0$ we have that $a_{10} = a_{01} = 0$. Evidently, $\mu_s'(\varepsilon r_0, \varepsilon s_0) = \varepsilon^2 r_0^2 \sum_{i,j=0}^n a_{ij} \varepsilon^{i+j} r_0^i s_0^j$.*

Suppose that

1) $r_0(+\infty) \neq 0$, $\sum_{i,j=0}^n a_{ij} \varepsilon^{i+j} r_0^i(+\infty) s_0^j(\alpha) > 0$ *for some α. Then s_x has a blow up for a finite time.*
2) $r_0(+\infty) = 0$. *Then condition (B) is reduced to*

$$\int^{+\infty} r_0^2(\beta) d\beta = +\infty$$

if $a_{00} > 0$. Then s_x blows up in finite time.

We shall apply now Theorem 3.1 to the study of the nonlinear vibrating string equation. We remind the reader that the Cauchy problem for this equation is written in the form

$$u_{tt} - (K(u_x))_x = 0, \; t \geq 0, \tag{3.2}$$
$$u|_{t=0} = u_0(x), \;\; u_t|_{t=0} = u_1(x).$$

The functions $K \in C^2(R^1)$, $u_0(x) \in C^2(R^1)$, $u_1 \in C^1(R^1)$ and the uniform norms $||u_1||_{C^0(R^1)}$, $||u_0||_{C^1(R)}$ are supposed to be sufficiently small. Moreover, we assume that

$$K(0) = 0, K^{'} > 0 \text{ everywhere}, \quad K^{''} \leq 0 \text{ everywhere}. \tag{3.3}$$

Define now the functions

$$s_0(x) = \frac{1}{2}\left(u_1(x) + \int_0^{u_0^{'}} \sqrt{K'(\lambda)}d\lambda\right), \tag{3.4}$$

$$r_0(x) = \frac{1}{2}\left(u_1(x) - \int_0^{u_0^{'}} \sqrt{K'(\lambda)}d\lambda\right).$$

Let $v = G(\nu) \in C^2$ be the inverse function of the strictly monotonically increasing function $\nu = \int_0^v \sqrt{K'(\lambda)}d\lambda \in C^2(R^1)$.

We formulate our second result.

Theorem 3.4. *Consider the Cauchy problem* (3.2) *with sufficiently small uniform norms* $||u_1||_{C^0(R^1)}$, $||u_0||_{C^1(R)}$ *and suppose that* $s_0^{'} \geq 0$, $r_0^{'}(\beta) < 0$ *at some point* $\beta \in R^1$ *and*

$$(B^{'}) \quad \int^{+\infty} \frac{K^{''}}{K^{'}}(G(-r_0(\beta) + s_0(\alpha)))d\alpha = -\infty.$$

Then the classical solution $u(t,x)$ *of* (3.2) *will develop a singularity in a finite time.*

Obviously, the condition $(B^{'})$ is rather implicit. To simplify it we assume that

(C) $K^{'} > 0$ and $K^{''} \leq 0$ is not identically zero on any open interval of the real line.

Corollary 3.5. *Consider the Cauchy problem* (3.2) *in* $[0,+\infty) \times R_x$ *and suppose that the condition (C) holds, the function* $s_0(x)$ *given by* (3.4) *is monotonically increasing and* $r_0(x)$ *is not monotonically increasing. Then the classical solution* $u(t,x)$ *of* (3.2) *with sufficiently small norms* $||u_1||_{C^0(R^1)}$, $||u_0||_{C^1(R)}$ *will develop a singularity in a finite time* $T > 0$.

3.2. Proof of Theorem 3.1

We consider the system (3.1). We shall investigate also systems with Cauchy data $\tilde{r} = \varepsilon r_0(x)$, $\tilde{s} = \varepsilon s_0(x)$, $0 < \varepsilon \leq \varepsilon_0$, ε_0 sufficiently small. The system is hyperbolic if $\lambda(r,s) < \mu(r,s)$ everywhere. If we are to work with the small Cauchy data, then we shall assume that $\lambda(r,s)$, $\mu(r,s)$ are defined in a small neighborhood of (0,0). When studying the blow up we can suppose without loss of generality that $\lambda(0,0) = -1$, $\mu(0,0) = +1$. We shall argue by contradiction. Thus let r, s be C^1 solution of (3.1) on $[0,+\infty) \times R_x$.

We introduce the characteristic curves of the solutions r, s respectively:

$$I \quad \begin{cases} \frac{dx_1}{dt}(t,\beta) = \lambda(r_0(\beta), s(t, x_1(t,\beta))), \\ x_1(0,\beta) = \beta \in R^1, \end{cases}$$

(in fact, $r(t, x_1(t,\beta)) \equiv r_0(\beta)$).

$$II \quad \begin{cases} \frac{dx_2}{dt}(t,\alpha) = \mu(r(t,x_2(t,\alpha)), s_0(\alpha)), \\ x_2(0,\alpha) = \alpha \in R^1, \end{cases}$$

(in fact, $s(t, x_2(t, \alpha)) \equiv s_0(\alpha)$).

So we conclude: $||r||_{C^0} = ||r_0||_{C^0}$, $||s||_{C^0} = ||s_0||_{C^0}$.

The curve I is called backward characteristic.

The curve II is called forward characteristic.

If we put $\tilde{r}$ instead of εr_0 and $\tilde{s}$ instead of εs_0 we see that $||\tilde{r}||_{C^0}$, $||\tilde{s}||_{C^0} \leq$ const.ε_0, i.e., λ, μ are well defined for ε_0 sufficiently small. Moreover, if $|\lambda| \leq \varepsilon_0$, $|\mu| \leq \varepsilon_0$, then $-3/2 \leq \lambda(r,s) \leq -1/2$, $1/2 \leq \mu(r,s) \leq 3/2$. This way we can estimate for each $0 < \varepsilon \leq \varepsilon_0$ the functions $x_1(t,\beta)$, $x_2(t,\alpha)$ in the following way:

$$\begin{aligned} \beta - 3/2t &\leq x_1(t,\beta) \leq \beta - 1/2t, \ \forall t \geq 0, \\ \alpha + 1/2t &\leq x_2(t,\alpha) \leq \alpha + 3/2t, \ \forall t \geq 0. \end{aligned} \tag{3.5}$$

Evidently, $x_1(t,\beta)$ is a strictly monotonically decreasing function with respect to $t \geq 0$ for β fixed and $x_2(t,\alpha)$ is a strictly monotonically increasing function with respect to $t \geq 0$ for α fixed. For each β we denote by $\alpha(t,\beta)$ the unique solution of the equation $s(t, x_1(t,\beta)) = s_0(\alpha(t,\beta)))$. Below we propose the geometrical interpretation.

From geometrical reasons it is clear that

$$\alpha(t,\beta) \leq \beta, \ \forall t \geq 0, \ \forall \beta \in R^1,$$

$$\alpha(0,\beta) = \beta, \ \forall \beta \in R^1,$$

$$\alpha(t',\beta) < \alpha(t,\beta) \text{ if } t' > t,$$

i.e., $\frac{\partial \alpha}{\partial t}(t,\beta) \leq 0$. So the smooth function $\alpha(t,\beta)$ is monotonically decreasing with respect to t for each fixed β.

Having in mind (3.5) we have

$$\alpha(t,\beta) < x_1(t,\beta) \leq \beta - 1/2t, \tag{3.6}$$

i.e., $\alpha(t,\beta)_{t\to+\infty} \to -\infty$, β fixed.

The relation (3.6) implies that

$$0 \leq t \leq 2(\beta - \alpha(t,\beta)). \tag{3.7}$$

We shall find now an expression for $\frac{\partial \alpha}{\partial t}(t,\beta)$. Having in mind the fact that

$$x_2(t, \alpha(t,\beta)) = x_1(t,\beta), \tag{3.8}$$

(see Fig. 21) we get:

$$\frac{\partial x_2}{\partial t}(t,\alpha(t,\beta)) + \frac{\partial x_2}{\partial \alpha}(t,\alpha(t,\beta)).\frac{\partial \alpha}{\partial t}(t,\beta) = \frac{\partial x_1}{\partial t}(t,\beta),$$

$$\frac{\partial x_2}{\partial t}(t,\alpha(t,\beta)) = \mu(r_0(\beta(t,\alpha)), s_0(\alpha)),$$

$$\frac{\partial x_1}{\partial t}(t,\beta) = \lambda(r_0(\beta), s_0(\alpha(t,\beta))),$$

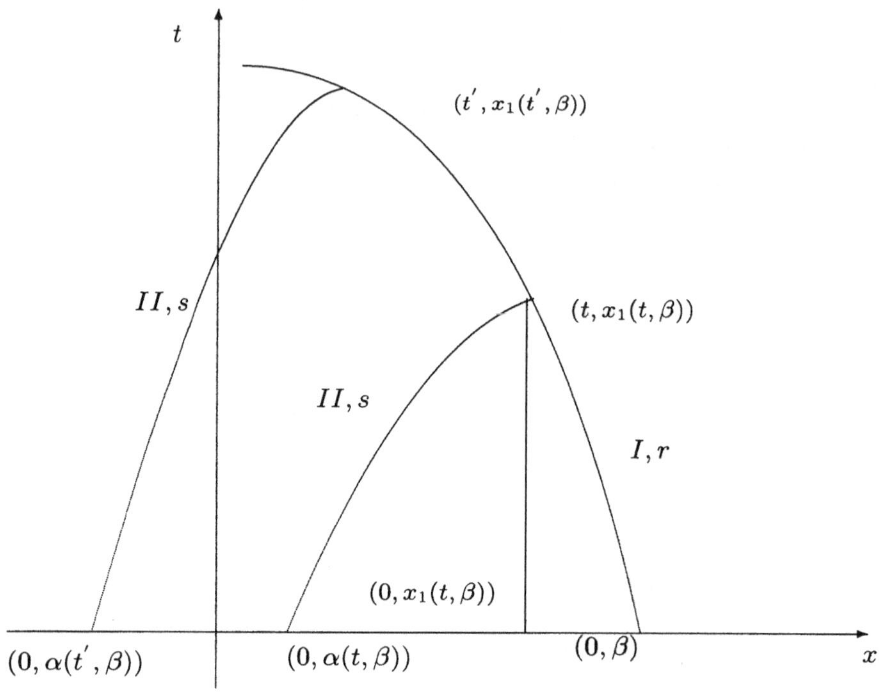

Fig. 21

where $\beta(t, \alpha)$ is the unique solution of the equation

$$r(t, x_2(t, \alpha)) = r_0(\beta(t, \alpha)), \ \forall t \geq 0, \ \forall \alpha \in R^1.$$

Thus $(r_0(\beta), s_0(\alpha(t, \beta))) = (r_0(\beta(t, \alpha)), s_0(\alpha))$ and therefore $\frac{\partial x_2}{\partial \alpha} \frac{\partial \alpha}{\partial t} = \lambda - \mu < 0$. As $\frac{\partial \alpha}{\partial t} \leq 0$ we conclude that $\frac{\partial \alpha}{\partial t} < 0$ and therefore there exists a smooth inverse function $t(\alpha, \beta)$, $\frac{\partial t}{\partial \alpha} < 0$ for each β fixed.

In the same way we investigate the smooth function $\beta(t, \alpha)$, (see Fig. 22).

The equality

$$x_1(t, \beta(t, \alpha)) = x_2(t, \alpha)$$

for each fixed α; $t' > t$ implies $\beta(t', \alpha) > \beta(t, \alpha)$, i.e., $\beta(t, \alpha)$ is monotonically increasing with respect to t ($\frac{\partial \beta}{\partial t} \geq 0$). Moreover,

$$\beta(0, \alpha) = \alpha, \ \forall \alpha \in R^1,$$

$$\beta(t, \alpha) \geq x_2(t, \alpha) \geq \alpha + 1/2t, \ \beta(t, \alpha) \geq \alpha, \forall \alpha.$$

Thus $\beta(t, \alpha)_{t \to +\infty} \to +\infty$, α fixed,

$$0 \leq t \leq 2(\beta(t, \alpha) - \alpha)). \tag{3.9}$$

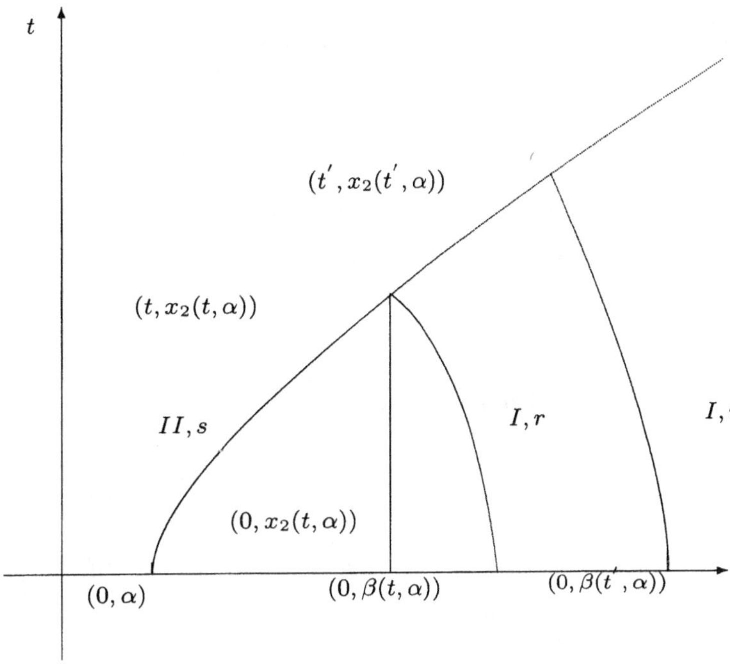

Fig. 22

From the identity $x_1(t, \beta(t, \alpha)) \equiv x_2(t, \alpha)$ we get

$$\frac{\partial x_1}{\partial t} + \frac{\partial x_1}{\partial \beta} \cdot \frac{\partial \beta}{\partial t} = \frac{\partial x_2}{\partial t}.$$

As in the previous identity λ, μ are evaluated at the same point we get that $\frac{\partial \beta}{\partial t} > 0$ for each fixed α, i.e., there exists a smooth inverse function $t(\beta, \alpha)$, $\frac{\partial t}{\partial \beta} > 0$.

As it is known from [48,51]

$$\frac{\partial r}{\partial x}(t, x_1(t, \beta)) = r_0'(\beta) \frac{e^{h_1(r_0(\beta), s_0(\beta)) - h_1(r_0(\beta), s(t, x_1(t, \beta)))}}{A_1(t, \beta)},$$

where

$$\begin{aligned} A_1(t, \beta) &= 1 + r_0'(\beta) \int_0^t \frac{\partial \lambda}{\partial r}(r_0(\beta), s(\tau, x_1(\tau, \beta))e^{h_1(r_0(\beta), s_0(\beta))} \\ &\times \ e^{-h_1(r_0(\beta), s(\tau, x_1(\tau, \beta)))} d\tau, \end{aligned}$$

$$\frac{\partial h_1}{\partial s} = \frac{1}{\lambda - \mu} \frac{\partial \lambda}{\partial s} \Rightarrow h_1(r, s) = \int_0^s \frac{\lambda_s(r, \gamma) d\gamma}{(\lambda - \mu)(r, \gamma)}.$$

So:

$$\frac{\partial r}{\partial x}(t, x_1(t,\beta)) = r_0'(\beta)\frac{e^{h_1(r_0(\beta),s_0(\beta))-h_1(r_0(\beta),s_0(\alpha(t,\beta)))}}{A_1(t,\beta)}, \tag{3.10}$$

$$\begin{aligned} A_1(t,\beta) &= 1 + r_0'(\beta)\int_0^t \frac{\partial \lambda}{\partial r}(r_0(\beta), s_0(\alpha(\tau,\beta))e^{h_1(r_0(\beta),s_0(\beta))} \\ &\times \ e^{-h_1(r_0(\beta),s_0(\alpha(\tau,\beta)))}d\tau. \end{aligned}$$

Remark 3.6. $A_1(t,\beta) = 0$ for some t,β implies $r_0'(\beta) \neq 0$, as $r_0'(\beta) = 0 \Rightarrow A_1(t,\beta) = 1$.

In the same way we have that

$$\frac{\partial s}{\partial x}(t, x_2(t,\alpha)) = s_0'(\alpha)\frac{e^{h_2(r_0(\alpha),s_0(\alpha))-h_2(r_0(\beta(t,\alpha)),s_0(\alpha))}}{A_2(t,\alpha)}, \tag{3.11}$$

$$\frac{\partial h_2}{\partial r} = \frac{1}{\mu-\lambda}\frac{\partial \mu}{\partial r} \Rightarrow h_2(r,s) = \int_0^r \frac{\mu_r(\gamma,s)d\gamma}{(\mu-\lambda)(\gamma,s)},$$

$$\begin{aligned} A_2(t,\alpha) &= 1 + s_0'(\alpha)\int_0^t \frac{\partial \mu}{\partial s}(r_0(\beta(\tau,\alpha)), s_0(\alpha))e^{h_2(r_0(\alpha),s_0(\alpha))} \\ &\times \ e^{-h_2(r_0(\beta(\tau,\alpha)),s_0(\alpha))}d\tau. \end{aligned}$$

Remark 3.7. Put in the expressions containing h_2 $\tilde{r}_0$ instead of εr_0 and $\tilde{s}_0$ instead of εs_0. Having in mind that $||\tilde{r}_0||_{C^0} \le c.\varepsilon_0 < 1$, $c = \text{const.} > 0$, $||\tilde{s}_0||_{C^0} \le c.\varepsilon_0 < 1$ we conclude that one can find a constant $k(\varepsilon_0) > 0$ such that

$$\begin{aligned} \frac{1}{k} &\le \ e^{h_1(\tilde{r}_0(\beta),\tilde{s}_0(\beta))-h_1(\tilde{r}_0(\beta),\tilde{s}_0(\alpha(t,\beta)))} \le k, \\ \frac{1}{k} &\le \ e^{h_2(\tilde{r}_0(\alpha),\tilde{s}_0(\alpha))-h_2(\tilde{r}_0(\beta(t,\alpha)),\tilde{s}_0(\alpha))} \le k, \quad k > 1. \end{aligned} \tag{3.12}$$

We remind the reader that according to our condition (A) $A_1(t,\beta) \ge 1$ and therefore $|r_x| \le$ const., $|r_t| \le$ const. We shall impose conditions on s_0 leading to blow up of s_x for a finite time $T > 0$.

Consider the integral

$$I(t,\alpha) = \int_0^t \frac{\partial \mu}{\partial s}(r_0(\beta(\tau,\alpha)), s_0(\alpha)))\gamma(\beta(\tau,\alpha))d\tau, \tag{3.13}$$

where $\gamma(\beta(\tau,\alpha)) = e^{h_2(r_0(\alpha),s_0(\alpha))-h_2(r_0(\beta(\tau,\alpha)),s_0(\alpha))}$.

For a fixed $\alpha \in R^1$ we make in (3.13) the change of the variable $\beta = \beta(\tau,\alpha)$ $\Rightarrow d\beta = \frac{\partial \beta}{\partial \tau}d\tau \Rightarrow d\tau = \frac{\partial \tau}{\partial \beta}d\beta$ and $\tau(\beta,\alpha)$ is the inverse function of $\beta(\tau,\alpha)$. So we get

$$I = \int_\alpha^{\beta(t,\alpha)} \frac{\partial \mu}{\partial s}(r_0(\beta), s_0(\alpha))\gamma(\beta)\frac{\partial \tau}{\partial \beta}d\beta,$$

as $\beta(0,\alpha) = \alpha$.

We know that $\frac{\partial \beta}{\partial \tau} = \frac{\mu-\lambda}{\frac{\partial x_1}{\partial \beta}(\tau,\beta(\tau,\alpha))} \Rightarrow \frac{\partial \tau}{\partial \beta} = \frac{\frac{\partial x_1}{\partial \beta}(\tau,\beta(\tau,\alpha))}{\mu-\lambda}$.

But $1 \leq \mu - \lambda \leq 3 \Rightarrow$

$$\frac{1}{3}\frac{\partial x_1}{\partial \beta} \leq \frac{\partial \tau}{\partial \beta}(\beta, \alpha) \leq \frac{\partial x_1}{\partial \beta}(\tau, \beta). \tag{3.14}$$

On the other hand, according to [48,51]

$$\frac{\partial x_1}{\partial \delta}(t, \delta) = A_1(t, \delta) e^{h_1(r_0(\delta), s_0(\alpha(t,\delta))) - h_1(r_0(\delta), s_0(\delta))} \tag{3.15}$$

and $x_1(t, \delta)$ is a solution of (I) with δ instead of β.

Therefore $\frac{\partial x_1}{\partial \delta}(t, \delta) \geq \frac{1}{k}$, $\forall t \geq 0$, as $A_1(t, \delta) \geq 1$ and

$$\begin{aligned} I(t, \alpha) &= \int_\alpha^{\beta(t,\alpha)} \frac{\partial \mu}{\partial s}(r_0(\beta), s_0(\alpha)) \gamma(\beta) \frac{\frac{\partial x_1}{\partial \beta}(\tau, \beta)}{\mu - \lambda} d\beta \\ &\geq \frac{1}{3k^2} \int_\alpha^{\beta(t,\alpha)} \frac{\partial \mu}{\partial s}(r_0(\beta), s_0(\alpha)) d\beta. \end{aligned} \tag{3.16}$$

Fix α and let $t \to +\infty \Rightarrow$

$$\underline{\lim}_{t \to +\infty} I(t, \alpha) \geq \frac{1}{3k^2} \int_\alpha^{+\infty} \frac{\partial \mu}{\partial s}(r_0(\beta), s_0(\alpha)) d\beta = +\infty.$$

Having in mind that

$$\frac{\partial}{\partial t} \int_\alpha^{\beta(t,\alpha)} \frac{\partial \mu}{\partial s}(r_0(\beta), s_0(\alpha)) d\beta = \frac{\partial \beta}{\partial t} \cdot \frac{\partial \mu}{\partial s}(r_0(\beta(t, \alpha), s_0(\alpha)) \geq 0,$$

we conclude that the previous integral is monotonically increasing with respect to t.

We see that our condition (B) implies that $I(t, \alpha) \nearrow_{t \to +\infty} +\infty$.

Consider now the continuous function $A_2(t, \alpha)$. We have that $A_2(0, \alpha) = 1$, $A_2(t, \alpha)_{t \to +\infty} \to -\infty$. Having in mind that (3.13) is monotonically increasing we conclude that there exists a unique point $t_0(\alpha) > 0$ such that $A_2(t, \alpha) > 0$, $0 \leq t < t_0$, $A_2(t_0, \alpha) = 0$ and therefore there exists $s_x(t, x_2(t, \alpha))$ for $0 \leq t < t_0$, while $\lim_{t \to t_0 - 0} s_x(t, x_2(t, \alpha)) = -\infty$.

This completes the proof of Theorem 3.1

Remark 3.8. According to F. John, if the system (3.1) is genuinely nonlinear i.e., $\frac{\partial \lambda}{\partial r} > 0$, $\frac{\partial \mu}{\partial s} > 0$, then $r_0' \geq 0$, $s_0' \geq 0$ is the necessary and sufficient condition for existence of global solution r, s. In our case $\frac{\partial \lambda}{\partial r} \geq 0$, $\frac{\partial \mu}{\partial s} \geq 0$. Obviously, if $\frac{\partial \mu}{\partial s}(0,0) > 0$, then the condition (B) is fulfilled and we have a blow up result for s_x.

Remark 3.9. Condition (B) can be rewritten as

$$(B) \quad \int^\infty \frac{\partial \mu}{\partial s}(\varepsilon r_0(\beta), \varepsilon s_0(\alpha)) d\beta = +\infty, 0 < \varepsilon \leq \varepsilon_0$$

for Cauchy data $\tilde{r}$, $\tilde{s}$ and it is a growth condition for $r_0(\beta)$ at $+\infty$. The integral in (B) is taken from arbitrary finite β_0 till $+\infty$ and therefore only the divergence of the integral at $+\infty$ is important.

Fix then $0 < \varepsilon \leq \varepsilon_0$ and $\alpha \in R^1$. Obviously $\frac{\partial\mu}{\partial s}(\varepsilon r_0(\beta), \varepsilon s_0(\alpha)) \to_{\beta\to+\infty}$ $\frac{\partial\mu}{\partial s}(\varepsilon r_0(+\infty), \varepsilon s_0(\alpha))$. If $\frac{\partial\mu}{\partial s}(\varepsilon r_0(+\infty), \varepsilon s_0(\alpha)) > 0$, then (B) is satisfied.

3.3. Proof of Theorem 3.4

We reduce the second-order nonlinear hyperbolic equation (3.2) via the change $v = u_x$, $w = u_t$ to the system

$$\begin{aligned} v_t &= w_x, \\ w_t &= K'(v)v_x. \end{aligned} \tag{3.17}$$

This is a strictly hyperbolic quasi linear homogeneous system with two distinct characteristic roots $\lambda = -\sqrt{K'(v)} < 0$, $\mu = \sqrt{K'(v)} > 0$.

Introduce now the Riemann invariants r, s:

$$2r = w - \int_0^v \sqrt{K'(\lambda)}d\lambda, \qquad 2s = w + \int_0^v \sqrt{K'(\lambda)}d\lambda, \tag{3.18}$$

hence,

$$r + s = w, \qquad s - r = \int_0^v \sqrt{K'(\lambda)}d\lambda, \tag{3.19}$$

and consider the strictly monotonically increasing function $\nu = \int_0^v \sqrt{K'(\lambda)}d\lambda$. As we know $v = G(\nu)$ is its inverse, i.e., (3.19) implies that $v = G(s - r)$. Then the system (3.17) reduces globally and in an equivalent way to the system

$$\begin{aligned} &\frac{\partial s}{\partial t} - k(s-r)\frac{\partial s}{\partial x} = 0, \\ &\frac{\partial r}{\partial t} + k(s-r)\frac{\partial r}{\partial x} = 0, \\ &s|_{t=0} = s_0, r|_{t=0} = r_0, \end{aligned} \tag{3.20}$$

where $k(\nu) = \sqrt{K'(G(\nu))}$, r_0, s_0 are given by (3.4).

Obviously, (3.20) is a system of the type (3.1) with $\lambda(r, s) = -k(s - r) < 0$, $\mu(r, s) = k(r - s) > 0 \Rightarrow \frac{\partial\lambda}{\partial s} = -k'(s - r)$, $\frac{\partial\mu}{\partial r} = -k'(s - r)$. Having in mind that $k'(\nu) = \frac{1}{2}\frac{K''(G(\nu))}{K'(G(\nu))}$ we see that we are in the case (A).

Condition (B) takes the form

$$-\int^{+\infty} k'(s_0(\alpha) - r_0(\beta))d\alpha = +\infty, \tag{3.21}$$

i.e., we obtain the condition (B'). Applying Theorem 3.1 to the system (3.20) we get the desired conclusion for the Cauchy problem (3.2).

Corollary 3.5 is proved in the following way. As $s_0' \geq 0$ and s_0 is bounded, then $s_0(\alpha)_{\alpha\to\infty} \to s_0(+\infty) \neq \infty$. If $k'(s_0(+\infty) - r_0(\beta)) \neq 0$, then the integral (3.21) is divergent. Assuming $k'(s_0(+\infty) - r_0(\beta)) = 0$ we apply the condition (C) and we conclude that near β one can find a point $\bar{\beta}$ and such that $r_0'(\bar{\beta}) < 0$, $k'(s_0(+\infty) - r_0(\bar{\beta})) \neq 0$. Thus everything is proved.

Chapter II
On the regularity of the solutions for several classes of nonlinear systems of partial differential equations and some applications

Introduction

We consider in this chapter singularities of the solutions of several classes of nonlinear partial differential equations (PDE) and systems of PDE arising in gas dynamics and differential geometry. In Section 4 we study quasi-linear hyperbolic systems in the plane via the well-known hodograph transformation. Applying the famous Whitney theorem on the normal forms of smooth mappings from the plane into the plane, we obtain qualitative results on the geometry of the characteristic curves in the generic case. In § 5, 6 a paradifferential technique is used in order to study the propagation of singularities for some classes of nonlinear PDE. Moreover, in many cases we prove nonexistence of isolated singularities for nonlinear systems of PDE. As an application we propose several regularity results to the Meyer flow of a Tricomi gas for nozzle problem, to the Monge–Ampère equation with Gaussian curvature, changing its sign, and to hyperbolic-elliptic quasi-linear systems with characteristics degenerating at a single point. We have given a geometrical illustration of the theorems proved in this chapter. As we mentioned, a (non)linear microlocal analysis approach is used. Thus, the reader is supposed to be familiar with the classical theory of the pseudodifferential operators, paradifferential operators, microlocalized Sobolev spaces H^s_{mcl} etc.([32], [8]).

4. Singularities of the solutions of quasi-linear hyperbolic systems in the plane

1. This paragraph deals with singularities of the solutions of several classes of nonlinear hyperbolic systems in the plane as well as with the regularity of the solutions of some second-order nonlinear partial differential equations of mixed type. In our investigations of quasi-linear hyperbolic systems we use the hodograph transformation in the plane and its normal forms via the famous Whitney theorem on singularities of mappings of the plane into the plane. A very interesting mechanical interpretation of these results was proposed in the book [23].

2. Consider now the following quasi-linear system:

$$\begin{pmatrix} 1 & 0 \\ 0 & 1 \end{pmatrix}\begin{pmatrix} u \\ v \end{pmatrix}_x + \begin{pmatrix} a(u,v) & b(u,v) \\ c(u,v) & d(u,v) \end{pmatrix}\begin{pmatrix} u \\ v \end{pmatrix}_y = 0, \tag{4.1}$$

where a, b, c, d are C^∞ real-valued functions of their arguments u, v, $b(u,v) \neq 0$ everywhere and $\begin{pmatrix} u \\ v \end{pmatrix} \in H^s(\Omega)$, $s > 3$ is a real-valued solution of (4.1), i.e., according to Sobolev's embedding theorem $\begin{pmatrix} u \\ v \end{pmatrix} \in C^{2+\varepsilon}(\Omega)$, $\varepsilon > 0$. Thus we shall study the mapping

$$\Omega \ni (x,y) \to (u(x,y), v(x,y)) \in R^2, \tag{4.2}$$

locally near a fixed point $(x_0, y_0) \in \Omega$. Suppose that $j = \begin{vmatrix} u_x & u_y \\ v_x & v_y \end{vmatrix} \neq 0$ at (x_0, y_0) and put $u(x_0, y_0) = u_0$, $v(x_0, y_0) = v_0$. According to the inverse function theorem there exist two functions $x(u,v) \in C^{2+\varepsilon}$, $y(u,v) \in C^{2+\varepsilon}$ near (u_0, v_0) such that $J = \begin{vmatrix} x_u & x_v \\ y_u & y_v \end{vmatrix}_{(u_0,v_0)} \neq 0$ and the mapping $(u,v) \to (x(u,v), y(u,v))$ is inverse to (4.2) in a neighborhood of (u_0, v_0). Moreover,

$$u_x = jy_v,\ u_y = -jx_v,\ v_x = -jy_u,\ v_y = jx_u.$$

The transformation (4.2) is called a "hodograph transformation" (see [23]).

It is obvious that locally near the point (x_0, y_0) the system (4.1) is equivalent to the next system (near (u_0, v_0)):

$$\begin{pmatrix} b & 0 \\ d & -1 \end{pmatrix} \begin{pmatrix} x \\ y \end{pmatrix}_u + \begin{pmatrix} -a & 1 \\ -c & 0 \end{pmatrix} \begin{pmatrix} x \\ y \end{pmatrix}_v = 0. \tag{4.3}$$

It is interesting to note that the system (4.3) is linear with respect to the unknown functions $x(u,v)$, $y(u,v)$ as a, b, c, d depend only on the arguments (u,v).

Introduce now the characteristics for a linear system of partial differential operators. So, let

$$Pu = \sum_{j=1}^{n} A_j(x) \frac{\partial u}{\partial x_j} + B(x)u = f(x),\ x \in \Omega \subset R^n, \tag{4.4}$$

where A_j, B are smooth $N \times N$ matrices and $u = (u_1, \ldots, u_N)^t$, $f = (f_1, \ldots, f_N)^t$.

Definition 4.1. The smooth nondegenerate surface $S : \Phi(x) = 0$, $x \in \Omega$, $\nabla\Phi(x) \neq 0$ is called characteristic for (4.4) if for each $x \in S$,

$$\det \left(\sum_{j=1}^{n} A_j(x) \frac{\partial \Phi}{\partial x_j}(x) \right) = 0.$$

Definition 4.2. The vector (direction) $\xi \in R^n \setminus 0$ is called characteristic for P if $\det(\sum_{j=1}^{n} A_j(x)\xi_j) = 0$.

It is well known that characteristics are invariant under smooth nondegenerate change of the variables.

Let $A_0(t,x)$, $\det A_0 \neq 0$ be a smooth $N \times N$ matrix.

Definition 4.3. The system

$$A_0 u_t + \sum_{j=1}^{n} A_j(t,x) u_{x_j} + Bu = f$$

is t-strictly hyperbolic if for each $(t,x) \in \tilde{D}$, $\forall \xi \in R^n \setminus 0$, the algebraic equation with respect to τ

$$\det(\tau A_0 + \sum_{j=1}^{n} A_j(t,x)\xi_j) = 0$$

has only real distinct roots $\tau_1(t,x,\xi) < \cdots < \tau_N(t,x,\xi)$ (one can easily see that they are smooth in $(t,x,\xi \neq 0)$).

This way we conclude that (4.1) is strictly hyperbolic with respect to x iff the algebraic equation

$$\xi^2 + (a+d)\xi\eta + \eta^2(ad-bc) = 0$$

has only real distinct roots with respect to ξ, i.e., iff the discriminant

$$\mathcal{D} = (a+d)^2 - 4(ad-bc) > 0. \tag{4.5}$$

We shall continue our considerations for a fixed solution $u(x,y)$, $v(x,y)$ of (4.1) satisfying (4.5). In this case the corresponding characteristic curves depend on the fixed solution (u,v) of (4.1).

The characteristic curves of (4.1) in the (x,y) plane are given by

$$C_+ : \frac{dy_+}{dx} = \xi_+, \qquad C_- : \frac{dy_-}{dx} = \xi_-, \tag{4.6}$$

where $\xi_+ \neq \xi_-$ are the distinct real roots of

$$\lambda^2 - (a+d)\lambda + (ad-bc) = 0, \ \xi_\pm = \frac{a+d \pm \sqrt{\mathcal{D}}}{2},$$

$\xi_\pm = \xi_\pm(u,v)$. Thus through each point $(\tilde{x},\tilde{y}) \in \Omega$ we have passing two transversal characteristics.

The characteristics of the linear system (4.3) are the next one:

$$\frac{du_+}{dv} = \zeta_+(u,v), \qquad \frac{du_-}{dv} = \zeta_-(u,v),$$

where $\zeta_\pm$ are the roots of $b\mu^2 - (d-a)\mu - c = 0$, i.e., $\zeta_\pm(u,v) = \frac{d-a\pm\sqrt{\mathcal{D}}}{2b} = \frac{\xi_\pm - a}{b}$ $\Rightarrow \zeta_+(u,v) \neq \zeta_-(u,v)$, i.e.,

$$\Gamma_+ : b(u,v)\frac{du_+}{dv} = \xi_+ - a, \qquad \Gamma_- : b(u,v)\frac{du_-}{dv} = \xi_- - a, \tag{4.7}$$

are the characteristics of (4.3):

Outside the set where $j = 0$, the characteristics $C_\pm$ are mapped onto the Γ_+ characteristics under the hodograph transformation.

Consider the set of all C^∞ smooth mappings from R^2 into itself equipped with the standard topology. According to Whitney [14] one can find an open and dense set $T \subset C^\infty(R^2, R^2)$ such that for each $x \in R^2$ and $f \in T$ the corresponding germ $\tilde{f}$ is smoothly equivalent to one of the following three germs:

(i) $f : R^2 \to R^2 : (x, z) \to (x, z)$ – regular case.

(ii) $f : R^2 \to R^2 : (x, z) \to (x, z^2) = (x, y)$.

(iii) $f : R^2 \to R^2 : (x, z) \to (x, z^3 - xy) = (x, y)$.

We point out that in the case (ii) $j = 0 \Leftrightarrow z = 0$ and $(x, 0) \to (x, 0)$, while in the case (iii) $j = 0 \Leftrightarrow x = 3z^2$ and $j = 0$ is mapped under f onto $y = \pm\frac{2}{3\sqrt{3}}x^{2/3}$, $x \geq 0$. Thus if we put γ for the set where $j = 0$ and $\tilde{\gamma}$ for its image under the mapping f, we conclude that in the case (ii) both γ and $\tilde{\gamma}$ are smooth nondegenerate curves, while in the case (iii) $\tilde{\gamma}$ has a cusp point at 0. The kernel of the Jacobian of the mapping f, (ii) is of the type $(0, \beta)$, $\beta \neq 0$ and this vector will be called an exceptional direction. The vector $(0, 1)$ is transversal to $z = 0$. We shall concentrate our study on case (ii) as in this situation both γ and $\tilde{\gamma}$ are smooth curves. Moreover, we are interested in the qualitative behavior of the solutions of the system (4.1) in the generic case.

Let $f : u = x,\ v = y^2$ and consider a smooth nondegenerate curve δ in R^2_{xy} passing through the point $(a, 0)$.

There are two different cases to be investigated.

1) δ has a nonexceptional direction at $(a, 0)$.

2) δ has an exceptional direction at $(a, 0)$.

In the case 1) the image $\tilde{\delta}$ of δ under the mapping f is tangential to the axis Ou at the point $(a, 0)$. In the case 2) $\tilde{\delta}$ has a cusp point at $(a, 0)$ but both its branches are nontangential to Ou. For a geometric interpretation see Fig. 23.

Remark 4.4. Consider the analytic curve near 0:

$$\delta : \begin{cases} u = a_1 t^{n_1}, \\ v = b_1 t^{m_1} + b_2 t^{m_2} + \cdots, n_1 \leq m_1 < m_2 < \cdots \in \mathbf{N}, \end{cases}$$

and at least one m_k is not of the type $m_k = ln_1$, $\forall l \in Z$. If n_1, m_1 are even, then the situation (′) appears. If n_1 is even, m_1 - odd and m_1 is not of the type $m_1 = ln_1$, $\forall l \in Z$, then the situation (″) appears.

Going back to our initial system (4.1) we assume that the following condition (A) is fulfilled:

Condition (A): $j = \begin{vmatrix} u_x & u_y \\ v_x & v_y \end{vmatrix} = 0$ along a smooth nondegenerate curve γ in R^2_{xy}, $j \neq 0$ outside γ and its image $\tilde{\gamma}$ under the hodograph transformation (4.2) is also a smooth nondegenerate curve in R^2_{uv}.

Theorem 4.5. *In the generic case and under the condition (A) γ is a C- characteristic of* (4.1) *and its image $\tilde{\gamma}$ is Γ-characteristic.*

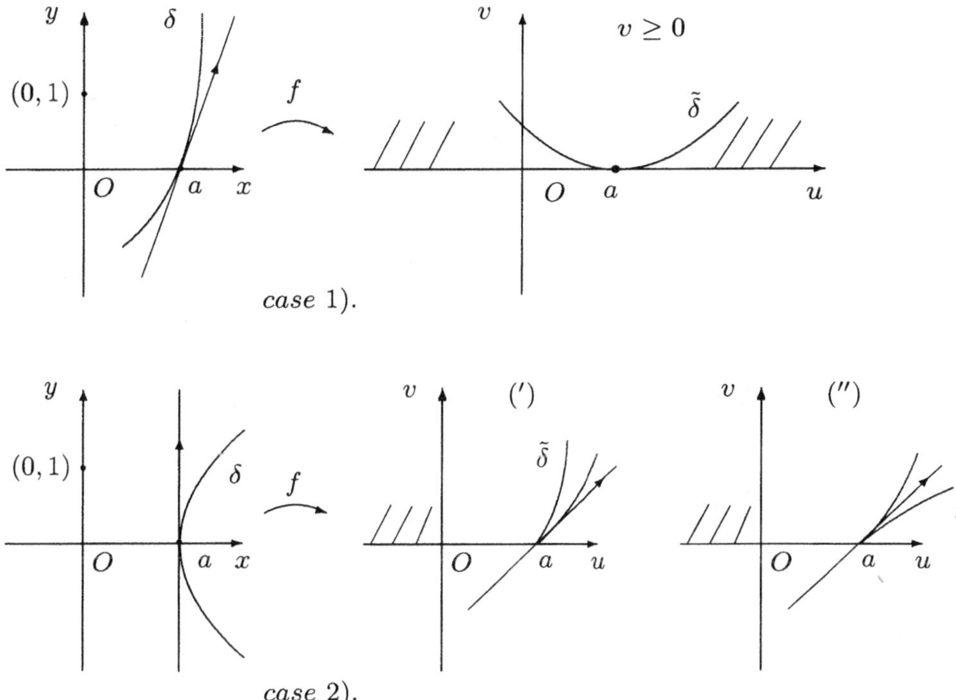

Fig. 23

The C-characteristics of the other type passing through each point of γ are, with exceptional directions, their images run through a section of a Γ-characteristic up to γ in one direction and then backward in the opposite direction. The images of all smooth curves in R^2_{xy} crossing γ in directions different from the exceptional ones are tangential to $\tilde{\gamma}$.

This way the image of a full neighborhood of 0 is a half neighborhood doubly covered. The image domain consists of two sheets which meet along $\tilde{\gamma}$. The curve $\tilde{\gamma}$ is also called an edge of the image.

Proof. Let $P \in \gamma$. As the characteristic curves C_+, C_- passing through P are nontangential each to other, then at least one of them, say C_-, has nonexceptional direction at P (the exceptional direction is transversal to γ). If C_+ has nonexceptional direction too, then the images of C_+, C_- under (4.2) Γ_+, Γ_- would be tangential to $\tilde{\gamma}$ at the point $\tilde{P}$. The last conclusion contradicts (4.7). So, C_+ has an exceptional direction which implies that "in general" Γ_+ has a cusp point at $\tilde{P}$ but its branches are not tangential to $\tilde{\gamma}$. This way we conclude that Γ_+ is a single curve passed two times – forward and backward. As it concerns $\tilde{\gamma}$, it is an envelope of Γ_- characteristics having characteristic directions at each of its points. Therefore, $\tilde{\gamma}$ is a Γ_- characteristic as through any point there is only one curve

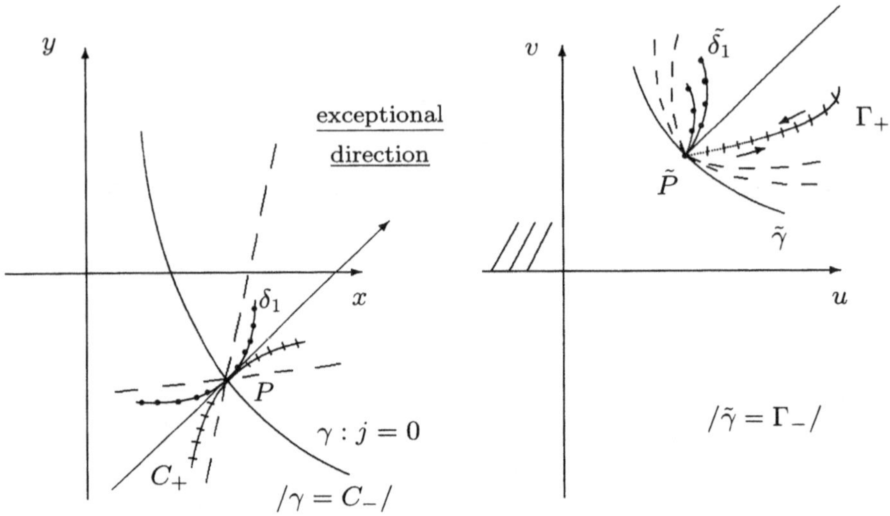

Fig. 24

with this property – the characteristic curve itself. So, γ turns out to be a C_- characteristic.

For the geometric interpretation of the previous result see Fig. 24.

3. We are going to analyse (under several restrictions) the case when double characteristic roots of the equation $\xi^2 + (a+d)\xi\eta + \eta^2(ad - bc) = 0$ appear.

To simplify things we shall study the special system

$$\begin{pmatrix} 1 & 0 \\ 0 & 1 \end{pmatrix} \begin{pmatrix} u \\ v \end{pmatrix}_x + \begin{pmatrix} a(u,v) & b(u,v) \\ c(u,v) & -a(u,v) \end{pmatrix} \begin{pmatrix} u \\ v \end{pmatrix}_y \tag{4.8}$$
$$= f(x,y) \in C^\infty.$$

Define now $r = \begin{pmatrix} a & b \\ c & -a \end{pmatrix}$, $\Delta(u,v) = \det r(u,v)$, $\Delta(u(x,y),v(x,y)) = \Delta(x,y)$, $r = r(u,v) = r(u(x,y),v(x,y))$, where $\begin{pmatrix} u \\ v \end{pmatrix} \in H^s_{loc}(\Omega)$, $s > 3$ are fixed solutions of (4.8).

As the discriminant $\mathcal{D} = -4\Delta$, we have three cases to study:

1) $\Delta(u(x,y),v(x,y)) < 0$ – strictly hyperbolic case, well studied in [8];
2) $\Delta(u(x,y),v(x,y)) > 0$ – elliptic case, well studied in [8];
3) $\Delta(u(x,y),v(x,y)) = 0$.

In the last case the characteristic equation has a double root $\xi = 0$.

Assume now

(a) there exists a point (x_0, y_0) such that $\Delta(x_0,y_0) = 0$ but $\nabla_{x,y}\Delta(x_0,y_0) \neq 0$.

As the case $\partial_x \Delta(x_0, y_0) \neq 0$ is simpler to deal with, we shall concentrate on the more complicated case

(b) $\partial_x \Delta(x_0, y_0) = 0$, $\partial_y \Delta(x_0, y_0) \neq 0$.

So (b) implies that one can find a smooth nondegenerate curve $\gamma \ni (x_0, y_0)$ with the properties $\Delta|_\gamma = 0$, $\Delta \neq 0$ outside γ.

Suppose, moreover

(c) $\partial_{xx} \Delta(x_0, y_0) \neq 0$.

The last assumption means that γ is a strictly convex (concave) curve – parabola type near (x_0, y_0), with vertex at $P = (x_0, y_0)$.

We point out that $\xi = \pm\sqrt{-\Delta}\eta$ and Δ has a sharp first-order zero along γ. The equation of the characteristics passing through the hyperbolic region $\Delta < 0$ is $(dy)^2 + \Delta(dx)^2 = 0$, i.e.,

$\frac{dy_\pm}{dx} = +\sqrt{-\Delta}$. Let $(p_0, q_0) \in \gamma \setminus P$ ($\Rightarrow \partial_x \Delta(p_0, q_0) \neq 0$).

It is trivial to verify that $y_\pm(x) = q_0 + c|x - p_0|^{3/2}$, $c(x, y) \neq 0$, i.c., the characteristics $C_\pm$ starting from the hyperbolic region form cusp points at γ. The characteristic through P degenerates at a single point. This is the corresponding figure.

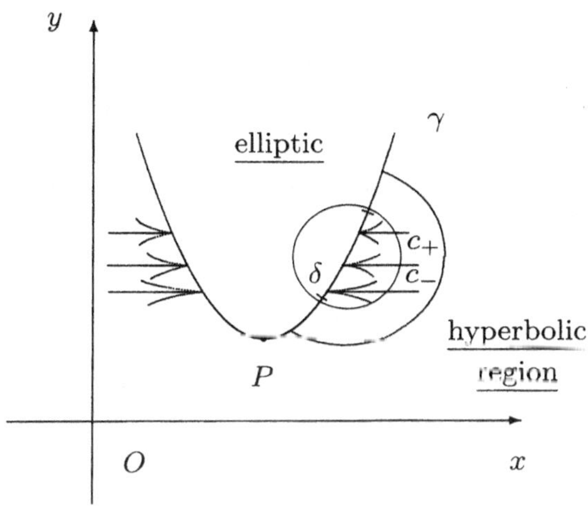

Fig. 25

Obviously, the difficulties in studying the smoothness of the solution of our system is due to the point P.

Applying methods of nonlinear microlocal analysis we can prove the next result.

Theorem 4.6. *Consider a closed arc δ of γ which does not contain P and assume that* $\begin{pmatrix} u \\ v \end{pmatrix} \in C^\infty$ *in a half neighborhood of δ lying in the hyperbolic region. Then* $\begin{pmatrix} u \\ v \end{pmatrix} \in C^\infty$ *in a full neighborhood of δ.*

In the case $\begin{pmatrix} u \\ v \end{pmatrix} \in C^\infty$ *in the hyperbolic region we assume moreover that condition (B) is satisfied:*

(B) if $\Delta_y(x_0, y_0) > 0$, *then* $\Delta_y(x_0, y_0)(s - \frac{1}{2}) > \mu_1(x_0, y_0)$; *if* $\Delta_y(x_0, y_0) < 0$, *then* $|\Delta_y(x_0, y_0)|(s - \frac{1}{2}) > -\mu_2(x_0, y_0)$, *where* $\mu_1(x_0, y_0) > \mu_2(x_0, y_0)$ *are the eigenvalues of the matrix* $\frac{K+K^*}{2}(x_0, y_0)$,

$$\begin{aligned} K(x_0, y_0) &= \left\{ -r_u u_x - r_v v_x - \left(r_u r - r \begin{pmatrix} a_u & a_v \\ c_u & c_v \end{pmatrix} \right) u_y \right. \\ &\quad \left. + \left(r \begin{pmatrix} b_u & b_v \\ -a_u & -a_v \end{pmatrix} - r_v r \right) v_y \right\} (x_0, y_0). \end{aligned}$$

Then $\begin{pmatrix} u \\ v \end{pmatrix} \in C^\infty$ *everywhere.*

The proof of the first part of Theorem 4.6 is a simple consequence of [8]. The proof of the second part will be given in §7.

4. We propose here several examples appearing in gas dynamics and in differential geometry.

The quasi-linear system

$$\begin{pmatrix} s \\ \theta \end{pmatrix}_\Psi = \begin{pmatrix} 0 & 1 \\ s & 0 \end{pmatrix} \begin{pmatrix} s \\ \theta \end{pmatrix}_\Phi \tag{4.9}$$

was studied by L.Bers [7] and it was said that (4.9) expresses some properties of a Meyer flow of a Tricomi gas for the nozzle problem. The symbols s, θ, Ψ and Φ mean the speed, the inclination of velocity, the stream function and the velocity potential respectively. Rewrite (4.9) in the form

$$\begin{pmatrix} u \\ v \end{pmatrix}_x = \begin{pmatrix} 0 & 1 \\ u & 0 \end{pmatrix} \begin{pmatrix} u \\ v \end{pmatrix}_y, \tag{4.10}$$

$$u = u(x, y), v = v(x, y).$$

When investigating the existence of shock waves, Oleinik considered the system

$$\begin{pmatrix} u \\ v \end{pmatrix}_x = \begin{pmatrix} 0 & 1 \\ 1-u & 0 \end{pmatrix} \begin{pmatrix} u \\ v \end{pmatrix}_y \tag{4.11}$$

which is hyperbolic for $u < 1$ and elliptic for $u > 1$. In their paper [49] Tay Ping Liu and Xin studied the system arising in the theory of overcompressive shock waves:

$$\begin{pmatrix} u \\ v \end{pmatrix}_t + \begin{pmatrix} au & b \\ 0 & v \end{pmatrix} \begin{pmatrix} u \\ v \end{pmatrix}_x = 0, \tag{4.12}$$

$$a = \text{const} > 0, b = \text{const} > 0.$$

The study of the following system was proposed by B. Kheifitz

$$(f(\omega))_x + (g(\omega))_y = 0, \tag{4.13}$$

where $f = (f_1, f_2)$, $g = (g_1, g_2)$, $\omega = (u, v)$, i.e.,

$$A(\omega)\partial_x\omega + B(\omega)\partial_y\omega = 0, A(\omega) = df, B(\omega) = dg.$$

The Monge–Ampère equation with Gaussian curvature $K(x, y)$,

$$u_{xx}u_{yy} - u_{xy}^2 = K(x, y)(1 + u_x^2 + u_y^2)^2, \tag{4.14}$$

belongs to the same class of hyperbolic-elliptic systems if for each point (x_0, y_0) for which $K(x_0, y_0) = 0$ we have that $dK(x_0, y_0) \neq 0$.

We shall study all the systems mentioned above from the point of view of microlocal analysis and shall find sufficient conditions for nonexistence of isolated singularities.

To begin with we shall consider the linearization of (4.10), namely

$$\begin{pmatrix} u_1 \\ v_1 \end{pmatrix}_x = \begin{pmatrix} 0 & 1 \\ u & 0 \end{pmatrix} \begin{pmatrix} u_1 \\ v_1 \end{pmatrix}_y + \begin{pmatrix} 0 & 0 \\ u_y & 0 \end{pmatrix} \begin{pmatrix} u_1 \\ v_1 \end{pmatrix} + f, \tag{4.15}$$

i.e., we shall obtain a system with the symbol

$$P_1 = I_2(i\xi) - \begin{pmatrix} 0 & 1 \\ u(x, y) & 0 \end{pmatrix} (i\eta) - \begin{pmatrix} 0 & 1 \\ u_y(x, y) & 0 \end{pmatrix}.$$

For simplicity, we assume $u \in C^2$. So we find the principal symbol p_1^0:

$$-ip_1^0 = \begin{pmatrix} \xi & -\eta \\ -u\eta & \xi \end{pmatrix} \text{ and } -ip_0 = i\begin{pmatrix} 0 & 0 \\ u_y & 0 \end{pmatrix}, I_2 = \begin{pmatrix} 1 & 0 \\ 0 & 1 \end{pmatrix}.$$

Without loss of generality we suppose that our operator has the symbol $-iP_1$, i.e.,

$$p_1^0 + p_0 = \begin{pmatrix} \xi & -\eta \\ -u\eta & \xi \end{pmatrix} + i\begin{pmatrix} 0 & 0 \\ u_y & 0 \end{pmatrix}, \tag{4.16}$$

$$\operatorname{Char} p_1 = \{\det p_1^0 = \xi^2 - u(x, y)\eta^2 = 0\}.$$

The Hamiltonian vector field of $\det p_1^0$ is given by the formula:

$$H_{\det p_1^0} = 2\xi\partial_x - 2\eta u(x, y)\partial_y + u_x(x, y)\eta^2\partial_\xi + u_y(x, y)\eta^2\partial_\eta.$$

According to Bony's nonlinear microlocal theory, if

1) $u(x_0, y_0) < 0$, then $\rho^0 = (x_0, y_0, \xi_0, \eta_0)$ is an elliptic point for each $|\xi^0| + |\eta^0| > 0$ and the microlocal smoothness of $\begin{pmatrix} u \\ v \end{pmatrix}$ at ρ^0 is studied in [8].
2) $u(x_0, y_0) > 0$, $\rho^0 \in \operatorname{Char} p_1$, then ρ^0 is a hyperbolic point and we have propagation of singularities along the integral curve $\mathcal{L}$ of $H_{\det p_1^0}$ passing through the point ρ^0 (see [8]).
3) $u(x_0, y_0) = 0$; then $\xi^0 = 0$, $\eta^0 \neq 0$ if $\rho \in \operatorname{Char} p_1$.

Let us assume that $u_x(x_0, y_0) \neq 0$. Then $H_{\det p_1^0(\rho_0)} = (\eta^0)^2(u_x(x_0, y_0)\partial_\xi + u_y(x_0, y_0)\partial_\eta)$ is not parallel to $\Xi(\rho^0)$, where $\Xi(\rho^0) = \xi_0\partial_\xi + \eta_0\partial_\eta = \eta_0\partial_\eta$ is the radial vector field at the point ρ^0. In that case the singularities again propagate along $\mathcal{L}$.

Suppose now

4) $u_x(x_0, y_0) = 0$ but $u_y(x_0, y_0) \neq 0$. Then $H_{\det p_1^0(\rho^0)} \parallel \Xi(\rho^0)$. This is the case to be studied here.

The case where $u(x_0, y_0) = \nabla_{x,y} u(x_0, y_0) = 0$ is rather complicated because of the appearance of double characteristics and it remains out of our investigations.

Let us now multiply p_1^0 by its cofactor matrix $co\, p_1^0$:

$$co\ p_1^0 . p_1^0 = \begin{pmatrix} \xi & \eta \\ \eta u & \xi \end{pmatrix} \begin{pmatrix} \xi & -\eta \\ -u\eta & \xi \end{pmatrix} = (\xi^2 - u\eta^2) I_2. \tag{4.17}$$

So we have a reduction to the scalar case but our operators (their symbols) are not C^∞ smooth with respect to (x, y). They are only C^1 smooth.

As we shall see, some of the properties of the linearized operator will remain true for the nonlinear operator under consideration.

Let

$$L = -\lambda x\xi - y\eta + d, 0 < \lambda < 1, d = id_1, d_1 \in R^1, \tag{4.18}$$

i.e., $L(x, D) = -\lambda x D_x - y D_y + d = i(\lambda x \partial_x + y\partial_y + d_1)$, as ξ is the symbol of the operator $D_x = -i\partial_x$.

As it is known from [29], we can construct a Schwartz distribution v such that $Lv \in C^\infty$ while it has a fixed singularity along the conic ray (0,1), i.e., $WF(v) = (0, 0, 0, 1) = \rho^0$. In the case $\lambda = 1$ there exists a function v such that $Lv \in C^\infty$, $WF(v) = (0, 0, \Gamma_\varepsilon)$ where $\Gamma_\varepsilon = \{|\varphi - \frac{\pi}{2}| \leq \frac{\varepsilon}{2}, \varepsilon > 0\}$ and φ is the polar angle in the plane $O\xi\eta$.

Definition 4.7. We shall say that $u \in H_{loc}^{s-0}$ iff $u \in H_{loc}^t$, $\forall t < s$ and $u \notin H_{loc}^s$.

One can easily see that there exists a solution v of the equation $Lv = f \in C^\infty$ with the properties $v \in H_{mcl}^{s_0 - 0}(\rho^0)$, $WF(v) = (0, 0, 0, 1)$, $\rho_0 = (0, 0, 0, 1)$, $s_0 = \frac{\lambda+1}{2} - d_1$, $0 < \lambda < 1$. In the case $\lambda = 1$ we have $s_0 = 1 - d_1$ and v has an isolated singularity but in an angle: $(0, 0, \Gamma_\varepsilon)$.

Let us remark that $s_0 = -\mathrm{Im} L_0'$ and L_0' is the subprincipal symbol of L. The definition of L_0' is given below in (5.3).

In all the cases $0 < \lambda \leq 1$ the Hamiltonian vector field $H_{L(\rho_0)} = \Xi(\rho_0)$. More precisely, if $0 < \lambda < 1$, then

$$H_{L(0,0,\xi,\eta)} \parallel \Xi(0, 0, \xi, \eta)$$

iff either $\xi = 0$, $\eta = \pm 1$ or $\xi = \pm 1, \eta = 0$,

$$H_{L(0,0,\xi,\eta)} \parallel \Xi(0, 0, \xi, \eta), \forall (\xi, \eta)\ if\ \lambda = 1.$$

Remark also that the larger $-d_1 > 0$ is, the more regular in C^k spaces the solution v is.

There are no difficulties in proving that there exists a solution w of the semilinear equation with analytic coefficients

$$x\partial_x w + y\partial_y w + 2\mu w = 2f\sqrt{w}, \sqrt{1} = 1, \tag{4.19}$$

and such that

$$w \in H^{s_0-0}_{loc}, s_0 > 2, w \in H^{2s_0-1-0}_{mcl}(0,0,0,1),$$

where $w(0,0) > 0$, the analytic function $f(0,0) \neq 0$, $\mu \in R^1$.

Thus, starting with $s_0 - 0$ a smooth solution, we have an optimal microlocal regularity $2s_0 - \frac{n}{2}$, $s_0 > \frac{n}{2} + 1$, $n = 2$.

Therefore, in the nonlinear case a natural restriction on the microlocal smoothness appears and it is of the type $t \leq 2s - \frac{n}{2}$.

It is worth studying the properties of the operator (4.18) from the point of view of nonexistence of singularities. So consider the operator $P = \sum_{j=1}^n \lambda_j x_j \partial_{x_j}$ in R^n and $L = P + a$, $a = \text{const.} \in R^1$, $\lambda_j > 0$.

Applying the method of characteristics one can show that:

a) we have a unique classical solution $u \in C^1(\Omega)$, $\Omega \ni O$ if $a > 0$ and O is the origin in R^n. Moreover,

$$O \cap \text{ singsupp } Lu = O \cap \text{ singsupp } u\,,\ \forall u \in C^1(\Omega);$$

b) we have uniqueness in $C^1(\Omega)$ modulo an additive constant if $a = 0$ and $f(O) = 0$, $u \in C^1(\Omega) \Rightarrow O \cap$ singsupp $Lu = O \cap$ singsupp u;

c) we have nonuniqueness in $C^1(\Omega)$ if $a < 0$. Moreover, $u \in C^1(\Omega)$ implies that $u(O) = 0$.

We have mentioned above that a solution of $Lv = f \in C^\infty$ with an isolated singularity at a ray exists if $s_0 = \frac{1+\lambda}{2} - a$, $0 < \lambda < 1$, $a > 0$. Thus we get $s_0 < 1$. The function $v \notin C^1(\Omega)$, $\Omega \ni O$, as if $v \in C^1(\Omega)$, $f \in C^\infty(\Omega)$, then $v \in C^\infty(\Omega)$ near O. This is the parametrix R of the operator L in the case a): $Rf = \int_0^1 t^{a-1} f(t^\lambda x) dt$, $t^\lambda x = (t^{\lambda_1} x_1, \dots, t^{\lambda_n} x_n)$.

5. Statement of the problem and formulation of the main results

1. Let Ω be a domain in R^n and suppose that the real-valued function $u \in H^s_{loc}(\Omega)$, $s > \frac{n}{2} + m + 1$, is a solution of the nonlinear system of partial differential equations:

$$P_j(x, u(x), \dots, \partial^\beta u(x), \dots)_{|\beta| \leq m} = 0,\ 1 \leq j \leq N, \tag{5.1}$$

where $P_j(x, u, \dots, u_\beta, \dots)_{|\beta| \leq m}$ are real-valued C^∞ functions of their arguments: $(x, u, \dots, u_\beta, \dots)_{|\beta| \leq m}$, $\partial^\beta = \partial^{\beta_1}_{x_1} \dots \partial^{\beta_n}_{x_n}$ if $\beta = (\beta_1, \dots, \beta_n)$, $u = (u_1, \dots, u_N)$, $\partial^\beta u = (\partial^\beta u_1, \dots, \partial^\beta u_N)$, $u_\beta = (u_{1,\beta}, \dots, u_{N,\beta})$.

The linearization of (5.1) contains the following two symbols:

$$i^m \left[\sum_{|\alpha| = m} \frac{\partial P_j}{\partial u_{k,\alpha}} (x, \partial^\beta u(x))_{|\beta| \leq m} \xi^\alpha \right. \tag{5.2}$$

$$\left. - \ i \sum_{|\alpha| = m-1} \frac{\partial P_j}{\partial u_{k,\alpha}} (x, \partial^\beta u(x))_{|\beta| \leq m} \xi^\alpha \right]^{\downarrow j = 1, \dots, N}_{\rightarrow k = 1, \dots, N} .$$

The corresponding principal symbol is real valued,

$$p_m(x,\xi) = \sum_{|\alpha|=m} \frac{\partial P_j}{\partial u_{k,\alpha}}(x, \partial^\beta u(x))_{|\beta|\le m}\xi^\alpha \in C^{1+\varepsilon}$$

for some $\varepsilon > 0$, while

$$P_{m-1}(x,\xi) = -i \sum_{|\alpha|=m-1} \frac{\partial P_j}{\partial u_{k,\alpha}}(x, \partial^\beta u(x))_{|\beta|\le m}\xi^\alpha \in C^{1+\varepsilon}$$

is purely imaginary. This is the definition of the subprincipal symbol:

$$p'_{m-1}(x,\xi) = p_{m-1}(x,\xi) + \frac{i}{2}\sum_{j=1}^{n} \frac{\partial^2 p_m}{\partial x_j \partial \xi_j}(x,\xi). \tag{5.3}$$

Consider now the point $\rho^0 = (x_0, \xi^0) \in T^*(\Omega)\backslash 0$. It is proved in [8] that det $p_m(\rho^0) \neq 0$ implies that $u \in H^{2s-m-\frac{n}{2}}_{mcl}(\rho^0)$.

So we suppose

(i) det $p_m(\rho^0) = 0$.

Assume that there exists an $N \times N$ square symbol $\tilde{p}_{1-m}(x,\xi)$, positively homogeneous of order $(1-m)$ with respect to ξ and C^∞ smooth for $\xi \neq 0$, $C^{\tilde{\sigma}}$ smooth with respect to x, $\tilde{\sigma} > 1$ and such that in some conical neighborhood of ρ^0 the following identity holds:

(ii) $\tilde{p}_{1-m}.p_m = q_1 I_N$ (I_N is the identity matrix in C^N, q_1 is a first-order real-valued scalar symbol). The notation $ord_\xi q_1 = 1$ means that q_1 is a positively homogeneous symbol of order 1 with respect to ξ.

The matrix $\tilde{p}_{1-m}$ is assumed to be real valued, $q_1 \in C^{\min(1+\varepsilon,\tilde{\sigma})}$ with respect to x. Further on we denote by $\Sigma^m_\rho(\Omega)$ the set of symbols positively homogeneous of order m with respect to ξ, $\xi \neq 0$, C^∞ with respect to $\xi \neq 0$ and belonging to C^ρ with respect to $x \in \Omega$, where $\rho > 0$ is not an integer. The notation $Op(\Sigma^m_\rho)$ stands for the set of corresponding properly supported paradifferential operators (see [8]).

Denote by $\rho^0 = (x_0, \xi^0)$, $|\xi^0| = 1$ such a characteristic point (i.e., det $p_m(\rho^0) = 0$) for which:

(iii) $H_{q_1(\rho^0)} + c\Xi(\rho^0) = 0$ for some constant $c < 0$.

By $H_{q_1(\rho^0)} = \sum_{j=1}^n (\frac{\partial q_1(\rho^0)}{\partial \xi_j}\partial_{x_j} - \frac{\partial q_1(\rho^0)}{\partial x_j}\partial_{\xi_j})$ we denote the Hamiltonian vector field of the scalar function q_1 at ρ^0 and $\Xi(\rho^0) = \sum_{j=1}^n \xi^0_j \partial_{\xi_j}$ is the radial vector field at the point ρ^0. Note that det $p_m(\rho^0) = 0$ implies that $q_1(\rho^0) = 0$.

In [26] Dencker supposed q_1 to be a symbol of real principal type, i.e., $H_{q_1}(\rho^0)$ is not parallel to $\Xi(\rho^0)$.

Unlike him we assume in (iii) that H_{q_1} and Ξ are collinear at $\rho^0 \in \mathrm{Char}P = \{\rho : \det\ p_m(\rho) = 0\}$. The subprincipal symbol of $\tilde{p}_{1-m}$ is given by the formula:

$$\tilde{p}'_{-m} = \frac{i}{2}\sum_{j=1}^{n} \frac{\partial^2 \tilde{p}_{1-m}}{\partial x_j \partial \xi_j}.$$

Introduce now the matrix-valued symbol

$$R' = -\frac{i}{2}\{\tilde{p}_{1-m}, p_m\} + \tilde{p}_{1-m}p'_{m-1} + \tilde{p}'_{-m}p_m$$

with Hölder coefficients in x and $ord_\xi R = 0$. As usual, $\{.,.\}$ denotes the Poisson bracket.

If A is a square $N \times N$ matrix in $\mathbf{C}^N$ and A^* is its adjoint, then $\mathrm{Im}\ A = \frac{A-A^*}{2i}$.

(iv) Suppose that there exists a conical neighborhood $\Gamma \ni \rho^0$ with the next property: $u \in H^t_{mcl}(\Gamma\backslash\rho^0)$, $t < 2s - 1 - \frac{n}{2} - m$, $s - 1 - m - \frac{n}{2} \notin \mathbf{Z}$ and H^t_{mcl} is the microlocalized Sobolev space at $\Gamma\backslash\rho^0$.

This is our main result, proved by the methods of microlocal analysis.

Theorem 5.1. *Assume that the solution $u \in H^s_{loc}(\Omega)$, $s > \frac{n}{2} + m + 1$ of the system* (5.1) *satisfies (i)–(iv) and*

(v) $-cs + \min_{\|z\|=1}(\mathrm{Im}\ R'(\rho^0)z, z) > 0$.

Then $u \in H^t_{mcl}(\Gamma)$.

Thus we conclude that the singularity at ρ^0 and under the condition (v) is not isolated, i.e., it can be removed.

The restriction $t < 2s - \frac{n}{2} - 1 - m$ can be omitted if P is a classical scalar linear differential operator with C^∞ coefficients. So if $Pu \in C^\infty$, $P = p_m + p_{m-1} + \cdots$, $u \in H^s_{loc}$, $-cs + \mathrm{Im}\ p'_{m-1}(\rho^0) > 0$, then $u \in H^s_{mcl}(\Gamma\backslash\rho^0)$ implies that $u \in H^t_{mcl}(\Gamma)$ for each $t > s$. Therefore if $u \in H^s_{loc}$, $-cs + \mathrm{Im}\ p'_{m-1}(\rho^0) > 0$, $Pu \in C^\infty$, $WF(u) \cap (\Gamma\backslash\rho^0) = \emptyset$, then $\rho^0 \notin WF(u)$.

The proof of Theorem 5.1 is reduced to the proof of a theorem from the theory of paradifferential operators. Our assertion is nontrivial if $s < t < 2s - m - 1 - \frac{n}{2}$.

We shall verify Theorem 5.1 applying Th. 3 from [8] with $d = m$, $\rho = s - \varepsilon - \frac{n}{2}$, $0 < \varepsilon \ll 1$, $\sigma = \rho - m$. Thus, there exists a paradifferential operator $\mathcal{P} \in Op(\Sigma^m_\sigma)$, $\sigma > 1$, $\sigma \notin \mathbf{Z}$ with the symbol (5.2) and such that

$$\mathcal{P}u \in H^{s-m+\sigma}_{loc} = H^{s+\sigma-1+(1-m)}_{loc}, u \in H^s_{loc}(\Omega) \subset C^\rho, \rho > m + 1,$$

$\rho \notin \mathbf{Z}$, $u \in H^t_{mcl}(\Gamma\backslash\rho^0)$. As $t < s + \sigma - 1$ we have that $\mathcal{P}u \in H^{t-m+1}_{loc}$. Without loss of generality we assume $\tilde{p}_{1-m} \in \Sigma^{1-m}_\sigma$. So $\mathcal{P}u = f \in H^{t-m+1}_{loc}$, $u \in H^s_{loc}$, $u \in H^t_{mcl}(\Gamma\backslash\rho^0)$ and therefore $\tilde{p}\mathcal{P}u = f_1 \in H^t_{loc}$, $\tilde{p}\mathcal{P} \in Op(\Sigma^1_\sigma)$.

Applying the paradifferential calculus we get:

$$\tilde{p}\mathcal{P} = \tilde{p}_{1-m}p_m + \tilde{p}_{1-m}p_{m-1} + \frac{1}{i}\sum_{j=1}^{n} \frac{\partial \tilde{p}_{1-m}}{\partial \xi_j}\frac{\partial p_m}{\partial x_j} + \tilde{R},$$

where $\tilde{R}$ is a $(-1+\sigma)$ smoothing operator in each Sobolev space (i.e., $\tilde{R}: H^s \to H^{s-1+\sigma}$, $\tilde{R}$ being continuous for each s).

According to (ii)

$$\begin{aligned}\frac{\partial^2}{\partial x_j \partial \xi_j}(\tilde{p}_{1-m} p_m) &= \frac{\partial^2 \tilde{p}_{1-m}}{\partial x_j \partial \xi_j} p_m + \tilde{p}_{1-m} \frac{\partial^2 p_m}{\partial x_j \partial \xi_j} \\ &\quad + \frac{\partial \tilde{p}_{1-m}}{\partial \xi_j} \times \frac{\partial p_m}{\partial x_j} + \frac{\partial \tilde{p}_{1-m}}{\partial x_j} \times \frac{\partial p_m}{\partial \xi_j} \\ &= \frac{\partial^2 q_1}{\partial x_j \partial \xi_j} I_N.\end{aligned}$$

So

$$\tilde{p}\mathcal{P} = \left(q_1 - \frac{i}{2} \sum_{j=1}^{n} \frac{\partial^2 q_1}{\partial x_j \partial \xi_j} \right) I_N - \frac{i}{2}\{\tilde{p}_{1-m}, p_m\} + \tilde{p}_{1-m} p'_{m-1} + \tilde{p}'_{-m} p_m + \tilde{R}.$$

Let us put

$$R = \tilde{p}\mathcal{P} = q_1 I_N - \frac{i}{2} \sum_{j=1}^{n} \frac{\partial^2 q_1}{\partial x_j \partial \xi_j} I_N + R' + \tilde{R}, \tag{5.4}$$

$R \in Op(\Sigma_\sigma^1)$.

Obviously, the subprincipal symbol $R'_0 = R'$.

Using standard bootstrap arguments we can see that Theorem 5.1 is a corollary of the following

Theorem 5.2. *Let the paradifferential operator $R \in Op(\Sigma_\sigma^1)$, $\sigma > 1$, $\sigma = 1 + \frac{1}{l}$, $l \in \mathbf{N}$, $l > 1$ be given by (5.4). Suppose that $u \in H_{comp}^{t-\frac{1}{2l}}(\Omega)$, $Ru \in H_{mcl}^t(\Gamma)$, $u \in H_{mcl}^t(\Gamma \backslash \rho^0)$ and*

$$-ct + \min_{\|z\|=1} (\mathrm{Im}\ R'_0(\rho^0) z, z) > 0. \tag{5.5}$$

Then $u \in H_{mcl}^t(\Gamma)$.

Obviously, $\min_{\|z\|=1}(\mathrm{Im}\ R'_0(\rho^0)z, z)$ is the smallest eigenvalue of the Hermitian matrix $\mathrm{Im}\ R'_0(\rho^0)$ and $\max_{\|z\|=1}(\mathrm{Im}\ R'_0(\rho^0)z, z)$ is its largest one. We propose several remarks below.

1. In many cases we multiply p_m by $\tilde{p}_{m_1} \in \Sigma_\sigma^{m_1}$, $\sigma \notin \mathbf{Z}$ and then $\tilde{p}_{m_1} p_m = q_{m+m_1} I_N$, det $p_m(\rho^0) = 0$. So $q_{m+m_1}(\rho^0) = 0$; $H_{q_{m+m_1}(\rho^0)} + c\Xi(\rho^0) = 0$, $c < 0$.

Replacing (v) by $(v)'$,

$$(v)' \quad -c\left(s - \frac{m + m_1 - 1}{2}\right) + \min_{\|z\|=1} (\mathrm{Im}\ R'_0(\rho^0) z, z) > 0,$$

where

$$R'_0 = -\frac{i}{2}\{\tilde{p}_{m_1}, p_m\} + \tilde{p}_{m_1} p'_{m-1} + \tilde{p}'_{m_1-1} p_m,$$

we obtain the same conclusion that $u \in H^t_{mcl}(\Gamma)$ if $u \in H^s_{loc}$, $s > \frac{n}{2} + m + 1$, $t < 2s - \frac{n}{2} - m - 1$.

More specially, if $\tilde{p}_{m_1} = {}^{co}\, p_m \in \Sigma_\sigma^{m(N-1)}$, i.e., $m_1 = m(N-1)$, det $p_m \in \Sigma_\sigma^{mN}$, we can reformulate our main result for the systems of the type (4.10), (4.11). As usual, ${}^{co}p$ is the matrix of cofactors of the matrix p.

2. If $c > 0$, Theorem 5.1 remains true with $(v)''$ instead of $(v)'$, namely

$$(v)'' \quad c\left(s - \frac{m+m_1-1}{2}\right) - \max_{\|z\|=1}\left(\text{Im } R_0'(\rho^0)z, z\right) > 0.$$

In the scalar case we take $\tilde{p}_{m_1} = 1$ and then (v) is replaced by

$$(v)''' \quad -c\left(s - \frac{m-1}{2}\right) + \text{Im } p_{m-1}'(\rho^0) > 0.$$

Let us consider now the operator (4.18) with $Lu \in C^\infty$, i.e., $m = 1$, $c = -1$, $u \in H^s_{loc}$, $s + \text{Im } L_0' > 0$. The last inequality is equivalent to $s > s_0$.

Then $u \in H^t_{mcl}(\Gamma \backslash \rho^0)$, $\rho^0 = (0,0,0,1)$, $\rho^0 \in \Gamma$ implies $u \in H^t_{mcl}(\Gamma)$, $t > s$.

3. This is a slight generalization of Theorem 5.1 for quasi-linear and semilinear systems. The same conclusion remains true under weaker conditions:

(a) $u \in H^s$, $s > \frac{n}{2} + m + 1$, $t < 2s - 1 - \frac{n}{2} - m$ for fully nonlinear systems,

(b) $s > \frac{n}{2} + m$, $t < 2s - \frac{n}{2} - m$ in the quasi-linear system case,

(c) $s > \frac{n}{2} + m - 1$, $t < 2s - \frac{n}{2} - m + 1$ in the semilinear case.

Certainly, then $u \in H^t_{mcl}(\Gamma)$.

6. Some applications of the previous results

1. Let us consider at first the case 3) for the system (4.10).

Having in mind that the principal symbol $\det p_1^0 = \xi^2 - u(x,y)\eta^2$, we have that if $u(x_0, y_0) = 0$, then $\xi^0 = 0$, $\eta^0 = \pm 1$ if $(x_0, y_0, \xi^0, \eta^0) \in \text{Char} \det p_1$.

Let us consider now the curve $\gamma : u = 0$, $u(x_0, y_0) = 0$. According to the implicit function theorem (say $u_x(x_0, y_0) \neq 0$) we find $x = x(y) \in C^3$, $x_0 = x(y_0)$.

The Hamiltonian system is corresponding to $\det p_1^0$

$$\begin{aligned}
\dot{x} &= 2\xi,\ x(0) = x_0,\\
\dot{y} &= 2u\eta,\ y(0) = y_0,\\
\dot{\xi} &= -u_x\eta^2,\ \xi(0) = 0,\\
\dot{\eta} &= -u_y\eta^2,\ \eta(0) = \pm 1.
\end{aligned}$$

Obviously, $\dot{x}(0)=0$, $\dot{y}(0)=0$,

$$\begin{array}{rcl}\ddot{x}(0) & = & -2u_x(x_0,y_0)\neq 0,\\ \ddot{y} & = & 2(u_x\dot{x}+u_y\dot{y})\eta+2u\dot{\eta}\Rightarrow \ddot{y}(0)=0,\\ \dddot{y} & = & 2(u_{xx}\dot{x}^2+2u_{xy}\dot{x}\dot{y}+u_{yy}\dot{y}^2)\eta\\ & & +2(u_x\ddot{x}+u_y\ddot{y})\eta+2(u_x\dot{x}+u_y\dot{y})\dot{\eta}+2u\ddot{\eta}+2\dot{u}\dot{\eta}\Rightarrow\\ \dddot{y}(0) & = & -4u_x^2(x_0,y_0),\end{array}$$

i.e., the characteristic curve δ in the plane Oxy is given by the equation

$$\delta\left|\begin{array}{l} x(t)=x_0-u_x(x_0,y_0)t^2+O(t^3),\\ y(t)=y_0-\frac{2}{3}u_x^2(x_0,y_0)t^3+O(t^3)\end{array}\right.$$

and has a cusp point at (x_0,y_0) being located in the domain where $u>0$.

Conclusion. Let us consider the system (4.10) in the case 3) ($u_x(x_0,y_0)\neq 0$) and suppose that the solution $u\in H^s(\Omega)$, $s>4$, $\Omega\ni(x_0,y_0)$ is such that $u\in C^\infty(\Omega_+)$, where $\Omega_+=\{(x,y):\ u(x,y)>0\}$. Then $\begin{pmatrix}u\\ v\end{pmatrix}\in C^\infty$ in a full neighborhood of the point (x_0,y_0).

If the curve $\gamma:\ u(x,y)=0,\ u(x_0,y_0)=0,\ \nabla u(x_0,y_0)\neq 0$ has one and only one horizontal tangent passing through the point (x_0,y_0), then $u\in H^s$, $s>4$, $\Omega\ni(x_0,y_0)$, $u\in C^\infty(\Omega_+)$ implies $\begin{pmatrix}u\\ v\end{pmatrix}\in C^\infty(\Omega)$ according to the previous remark and Theorem 5.1. Below we give the details.

We shall apply now Theorem 5.1 to the system (4.10).

Assume that $\begin{pmatrix}u\\ v\end{pmatrix}\in H^s_{loc}$, $s>2$ (i.e., $u\in C^{1+\varepsilon}$, $\varepsilon>0$) and $u(x_0,y_0)=0$, $u_x(x_0,y_0)=0$, $u_y(x_0,y_0)\neq 0$, $\xi^0=0$, $\eta^0=\pm1$. So $c=-sgn\ \eta_0u_y(x_0,y_0)$. Moreover, $R'(\rho^0)=\eta^0iu_y(x_0,y_0)\begin{pmatrix}1&0\\0&-1\end{pmatrix}$. Thus, for $2<s<t<2s-2$ solutions of (4.10) with isolated H^t singularities at $(x_0,y_0;0,\pm1)$ do not exist.

A similar result is true for (4.11) with $u(x_0,y_0)=1$ instead of $u(x_0,y_0)=0$.

2. Let us consider now a slight generalization of (4.14):

$$\left|\begin{array}{cc}u_{xx}&u_{xy}\\ u_{xy}&u_{yy}\end{array}\right|=K(x,y)f(x,y,u,\nabla u),\tag{6.1}$$

where $K(x_0,y_0)=0$, $\nabla K(x_0,y_0)\neq 0$, $K\in C^\infty$, $f\in C^\infty$, $f(x,y,u,p,q)>0$ everywhere, ($p=u_x$, $q=u_y$). Certainly, u,K,f are assumed to be real valued. Let $u\in H^s_{loc}$, $s>4$, (i.e., $u\in C^{3+\varepsilon}(\Omega)$). Unlike [30] we suppose that the curve

$$\gamma=\{(x,y)\in\Omega:K(x,y)=0\}\tag{6.2}$$

is characteristic at the point (x_0,y_0) for equation (6.1) linearized on u.

It is easy to see that at each point $(x,y)\in\gamma$ at least one of the tangential and normal vectors is noncharacteristic for the linearized equation.

Proposition. *Let $u\in C^\infty(\Omega\backslash(x_0,y_0))$ and* (6.2) *be fulfilled. Then $u\in C^\infty(\Omega)$.*

Proof. By a rotation of the coordinates that always leaves the equation (6.1) invariant, we can assume that

$$K(x,y) = y + O(x^2 + y^2),$$

i.e., $(x_0, y_0) = 0$, $\nabla K(0) = (0,1)$, the tangential vector to γ at 0 is $(1,0)$.

Equation (6.1) linearized on u has the following symbol:

$$p = u_{yy}\xi^2 - 2u_{xy}\xi\eta + u_{xx}\eta^2 + iKf_p\xi + iKf_q\eta - Kf_u, \tag{6.3}$$

where

$$f_p = f_p(x, y, u, u_x, u_y).$$

According to (6.3): $u_{xx}(0,0) = 0$, $u_{yy}(0,0) \neq 0$ and therefore due to (6.1) $u_{xy}(0,0) = 0$. The point $\rho^0 = (0,0,\xi_0,\eta_0)$ is noncharacteristic for (6.3) if $\xi_0 \neq 0$. Then $u \in H^{2s-3}_{mcl}(\rho^0)$. So let $\rho^0 = (0,0,0,\eta^0)$, $\eta_0 \neq 0$.

Thus $p_2^0(\rho^0) = 0$, $H_{p_2^0(\rho^0)} = -\eta^{0^2}(u_{xxx}(0,0)\partial_\xi + u_{xxy}(0,0)\partial_\eta)$.

Differentiating (6.1) with respect to x, y we find that: $u_{xxy}(0,0) \neq 0$, $u_{xxx}(0,0) = 0$, i.e., $0 \neq H_{p_2^0(\rho^0)} \parallel \Xi(\rho^0)$, $\eta_0^2 = 1$. The subprincipal symbol $p_2'(\rho^0) = 0$ and therefore the conditions $(v)'''$, $(v)''$ are satisfied. Let Γ be a sufficiently small conic neighborhood of $\rho^0 = (0,0,0,\eta^0)$. Then we know that $u \in H^{2s-3}_{mcl}(\Gamma\backslash\rho^0)$ [1] and therefore $u \in H^{2s-4}_{mcl}(\rho^0)$ which implies $u \in H^{2s-4}(0)$ (i.e., $\exists\varphi \in C_0^\infty$, $\varphi \equiv 1$ near 0, $\varphi u \in H^{2s-4}$). This way we raise the smoothness of u at 0 with $s - 4 > 0$. An iteration of the same procedure leads to $u \in C^\infty$ near $(0,0)$.

Corollary. Consider the equation (6.1) and assume that the punctured curve $\gamma \setminus \{(x_0, y_0)\}$ is noncharacteristic for equation (6.1) linearized on u. Then $u \in C^\infty(\Omega_-)$, $\Omega_- = \{(x,y) : K(x,y) < 0\}$ implies that $u \in C^\infty$ in a full neighborhood of the point (x_0, y_0).

The geometric illustration is given at Fig. 26, where $A = (x_0, y_0)$ and the characteristics in Ω_- are depicted by arrows.

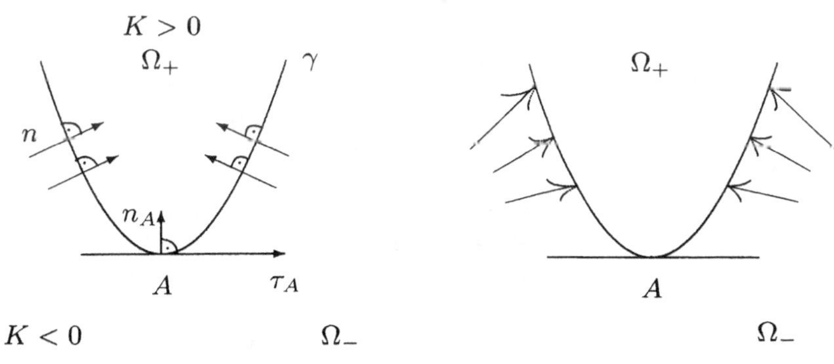

Fig. 26

[1] Here we use the fact that $u \in C^\infty(\Omega\backslash(x_0, y_0))$.

Remark 6.1. To complete the investigation of the Monge–Ampère equation we shall formulate and give a short proof of the result of Hong–Zuily [30]. Thus, we suppose that $u \in C^\infty(\Omega_-)$.

We shall concentrate on the characteristic points of p as the microlocal behavior of the solution near the noncharacteristic points is well known. So we assume that the curve γ is noncharacteristic at the point $(0,0)$ for equation (6.1) linearized on u. This implies that $u_{xx}(0,0) \neq 0$. So two cases must be considered:

$a) u_{yy}(0,0) = 0, \quad b) u_{yy}(0,0) \neq 0.$

Case a) Applying the equation (6.1) we get $u_{xy}(0,0) = 0$. If $\eta^0 \neq 0 \Rightarrow A = (0,0,\xi^0,\eta^0) \notin \operatorname{Char} p$. So we shall study the case $\eta^0 = 0$ only. The equation of the bicharacteristic passing through the point A with $\eta^0 = 0$, $\xi^0 \neq 0$ is as follows:

$$\dot{x} = p_\xi = 2\xi u_{yy} - 2\eta u_{xy}, x(0) = 0,$$

$$\dot{y} = p_\eta = 2u_{xx}\eta - 2u_{xy}\xi, y(0) = 0,$$

$$\dot{\xi} = -p_x, \dot{\eta} = -p_y, \xi(0) = \xi^0 \neq 0, \eta(0) = 0.$$

Thus, $\dot{x}(0) = 0$, $\dot{y}(0) = 0$. On the other hand,

$$\dot{\eta} = -p_y = -u_{yyy}\xi^2 + 2u_{xyy}\xi\eta - u_{xxy}\eta^2$$

implies that $\dot{\eta}(0) = -u_{yyy}(0)\xi_0^2$.

Differentiating (6.1) with respect to y and putting $x = y = 0$ we obtain

$$u_{xx}(0,0)u_{yyy}(0,0) = f(0, u(0), \nabla u(0)) \Rightarrow$$

$u_{yyy}(0,0) = \frac{f(0,u(0),\nabla u(0))}{u_{xx}(0,0)} \neq 0$, i.e., $\dot{\eta}(0) = c\xi_0^2$, $c = \text{const.} \neq 0$.

Having in mind that

$$\ddot{y} = p_{\eta x}\dot{x} + p_{\eta y}\dot{y} + p_{\eta\xi}\dot{\xi} + p_{\eta\eta}\dot{\eta}$$

we obtain $\ddot{y}(0) = 2u_{xx}(0,0)\dot{\eta}(0) = -2f(0,u(0),\nabla u(0))\xi_0^2 = c_1\xi_0^2$, $c_1 < 0$.

In a similar way we get $\ddot{x}(0) = 0$. This is the equation of the characteristic curve β passing through the origin:

$$\beta \left| \begin{array}{ll} x(t) = o(t^2), & t \to 0, \\ y(t) = \frac{c_1}{2}t^2\xi_0^2 + o(t^2). & \end{array} \right.$$

The curve β goes to the hyperbolic region $\Omega_- = \{(x,y) : K(x,y) < 0\}$ as $y(t) + O(x^2(t) + y^2(t)) = \frac{c_1}{2}t^2(1 + O(t^4))$, $t \to 0$. The fact that $u \in C^\infty(\Omega_-)$ and the Bony propagation of singularities theorem [8] imply that $u \in C^\infty$ in a full neighborhood of the origin.

Case b) The equation (6.1) at $(0,0)$ gives us that $u_{xy}(0,0) \neq 0$. One can easily see that $p(0,0,\xi,\eta) = 0 \iff$

$$\eta = \frac{u_{xy}(0)}{u_{xx}(0)}\xi \iff \xi = \frac{u_{xy}(0,0)}{u_{yy}(0,0)}\eta. \tag{6.4}$$

To do this we observe that $u_{xx}(0)u_{yy}(0) = u_{xy}^2(0) \neq 0$. The bicharacteristic, passing through the characteristic point $(0,0,\xi^0,\eta^0)$, ξ^0, η^0 given by (6.4) is uniquely determined. Moreover, its projection on the (x,y) plane is the curve

$$\left| \begin{array}{ll} x(t) = d_1\xi_0^2 t^2 + o(t^2), & d_1 \neq 0, \\ y(t) = d_2\xi_0^2 t^2 + o(t^2), & d_2 < 0. \end{array} \right.$$

The details of this computation are left to the reader. Obviously, the characteristic curves go when $t \neq 0$ to the hyperbolic region Ω_- and therefore $u \in C^\infty$ in a full neighborhood of the origin.

There are three possible pictures of the location of the characteristic curve trough 0 in Ω_-. We propose them below in Fig. 27 and Fig. 28.

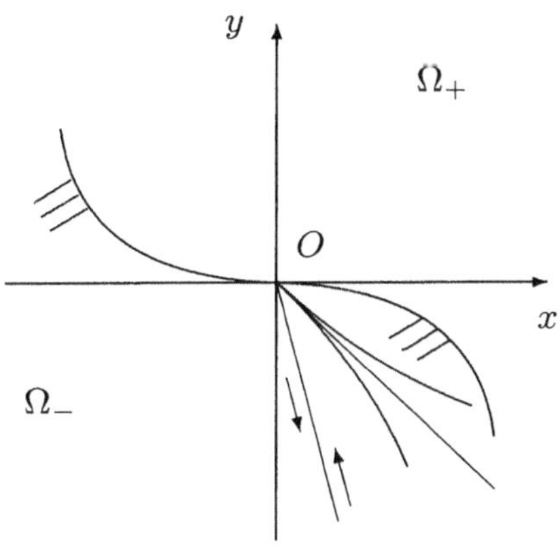

Fig. 27

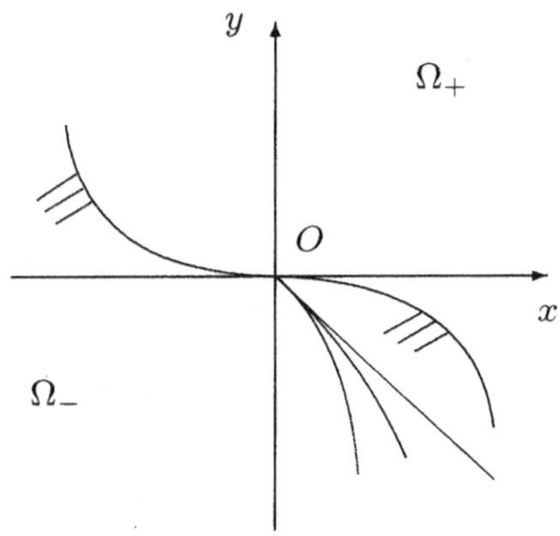

Fig. 28

7. Proof of Theorem 5.2 and Theorem 4.6

1. Let us consider the symbol

$$c_1 = \kappa_1(x)\gamma_1(\xi)(1+\mu^2|\xi|^2)^{-\delta}\,,\ \delta - \text{const} > 0\,,\ \mu \subset (0,1]$$

($\mu \to 0$ further on, i.e., $0 < \mu \ll 1$), $\kappa_1 \in C_0^\infty$, $\kappa_1 \equiv 1$ near x_0, $ord_\xi\gamma_1 = r$, $\gamma_1|\xi|^{-r} \equiv 1$ in a conic neighborhood of ξ^0, $\operatorname{conesupp} c_1 \in \Gamma_1 \subset\subset \Gamma$.

A simple calculation shows that $\{q_1, c_1\} = (1+\mu^2|\xi|^2)^{-\delta}A_r$, where $A_r = \{q_1, \kappa_1\gamma_1\} + 2\delta\kappa_1\gamma_1 \times \sum_{j=1}^n \frac{\partial q_1}{\partial x_j}\mu^2\xi_j(1+\mu^2|\xi|^2)^{-1}$. But $(1+\mu^2|\xi|^2)^{-1}$ is a bounded family in Σ_λ^0, $\forall\lambda > 0$ and $\frac{\mu^2\xi_j}{1+\mu^2|\xi|^2}$ is a bounded family in Σ_λ^{-1}, $\forall\lambda > 0$, $\mu \in (0,1]$. Thus $A_r \in \Sigma_{\sigma-1}^r$ uniformly with respect to μ and $\operatorname{conesupp} A_r \in \Gamma_1 \subset\subset \Gamma$.

Define now the cut-off symbol η, $\eta \equiv 1$ in a conic neighborhood of ρ^0, $\eta \in S_{1,0}^0$, $\operatorname{conesupp}\eta \in \{(x,\xi) : \kappa_1(x)\gamma_1(\xi) \neq 0\}$.

So

$$\begin{aligned}\{q_1, c_1\} &= \eta^2 c_1 \frac{A_r}{\kappa_1\gamma_1} + (1-\eta^2)(1+\mu^2|\xi|^2)^{-\delta} A_r \\ &= \eta^2 c_1 B_0 + (1-\eta^2)(1+\mu^2|\xi|^2)^{-\delta} A_r\end{aligned}$$

and $B_0 \in \Sigma^0_{\sigma-1}$ uniformly with respect to μ, conesupp$B_0 \subset\subset \Gamma$. Thus

$$B_0 = D_0 + 2\delta(1+\mu^2|\xi|^2)^{-1} \sum_{j=1}^{n} \frac{\partial q_1}{\partial x_j} \mu^2 \xi_j$$

and $D_0(\rho) = \frac{\{q_1, \kappa_1\gamma_1\}}{\kappa_1\gamma_1} = \{q_1, \kappa_1\gamma_1\}$, assuming $\rho = (x, \xi)$, $|\xi| = 1$ in a tiny conic neighborhood of ρ^0.

But

$$\begin{aligned}B_0(\rho^0) &= <\nabla_\xi q_1(\rho^0), \nabla_x(\kappa_1\gamma_1(\rho^0)> - <\nabla_x q_1(\rho^0), \nabla_\xi(\kappa_1\gamma_1(\rho^0)> \\ &\quad +2\delta(1+\mu^2)^{-1}\mu^2 <\nabla_x q_1(\rho^0), \xi^0> \\ &= -c <\xi^0, \nabla_\xi \gamma_1(\xi^0)> + 2\delta c \frac{\mu^2}{1+\mu^2} \\ &= -c\left(r - \frac{2\delta\mu^2}{1+\mu^2}\right) \\ &= -c\left(r + O(\mu^2)\right) \quad \mu \to 0.\end{aligned}$$

(Here we used Euler's identity for the homogeneous function γ_1, $ord_\xi \gamma_1 = r$.)

Let us consider the symbol

$$C_\mu^2 = \kappa^2(x)\gamma^2(\xi)|\xi|^{2t}(1+\mu^2|\xi|^2)^{-2},$$

where κ, γ, $ord_\xi \gamma = 0$ have the same properties as κ_1 and γ_1, i.e., $\delta = 2$, $r = 2t$ ($\Rightarrow C_\mu \in \Sigma^{t-2}_\sigma$ for each fixed $\mu > 0$).

This way we get

$$\{q_1, C_\mu^2\} = \eta^2 C_\mu^2 B_0 + (1-\eta^2)(1+\mu^2|\xi|^2)^{-2} A_{2t}, \tag{7.1}$$

where $B_0 \in \Sigma^0_{\sigma-1}$ uniformly with respect to μ, $A_{2t} \in \Sigma^{2t}_{\sigma-1}$ uniformly with respect to μ, conesupp B_0 and conesupp $A_{2t} \subset\subset \Gamma$ and

$$B_0(\rho^0) = -2c(t + O(\mu^2)), \ \mu \to 0. \tag{7.2}$$

We note that $u \in H^t_{mcl}$ (conesupp $(1-\eta^2)A_{2t}$) and that $t > s$ and (v) imply

$$\frac{1}{2} B_0(\rho^0) + \min_{||z||=1} (\mathrm{Im} R_0'(\rho^0) z, z) > 0. \tag{7.3}$$

($C_\mu u = C_\mu I_N u$, $R = q_1 I_N + R_0 + \tilde{R}$.)

The bilinear form $(C_\mu f_1, C_\mu u)$, $f_1 = Ru$ is well defined in H^0 as $C_\mu f_1 \in H^2_{comp}$, $C_\mu u \in H^{2-\frac{1}{2l}}_{comp}$ (conesupp$C_\mu \subset\subset \Gamma$, $\forall \mu > 0$).

So we obtain

$$\operatorname{Im}(C_\mu f_1, C_\mu u) = \operatorname{Im}(C_\mu q_1 I_N u, C_\mu u) + \operatorname{Im}(C_\mu R_0 u, C_\mu u) + K, \tag{7.4}$$

$|K| \leq C||u||^2_{t-\frac{\sigma-1}{2}}$ and the constant C does not depend on μ.

In fact, $C_\mu \in \Sigma^t_\sigma$ uniformly with respect to μ and $\tilde{R}$ is a smoothing operator of order $-(\sigma-1)$, i.e., $|(C_\mu \tilde{R}u, C_\mu u)| \leq ||C_\mu \tilde{R}u||_{\frac{\sigma-1}{2}} ||C_\mu u||_{\frac{1-\sigma}{2}} \leq C||u||^2_{t-\frac{\sigma-1}{2}}$, $\frac{\sigma-1}{2} = \frac{1}{2l}$.

The left-hand side of (7.4) can be easily estimated as $f_1 \in H^t_{mcl}(\Gamma)$, conesupp $C_\mu \in \Gamma$ imply

$$|(C_\mu u, C_\mu f_1)| \leq \varepsilon ||C_\mu u||^2_{0,mcl(\Gamma)} + C(\varepsilon)||f_1||^2_{t,mcl(\Gamma)},$$

$\forall \varepsilon > 0$ and $C(\varepsilon)$ does not depend on μ.

The identity $C_\mu u = \eta C_\mu u + (1-\eta)C_\mu u$, $(\eta - 1)C_\mu \in \Sigma^t_\sigma$ uniformly with respect to μ, the relations $u \in H^t_{mcl}(\Gamma \backslash \rho^0)$ and conesupp η is concentrated near ρ^0, $\eta \equiv 1$ near ρ^0, show that $||C_\mu u||_0 \leq ||\eta C_\mu u||_0 + d_1 ||u||_{t,mcl(\Gamma'')}$, $\rho^0 \notin \bar{\Gamma}''$.

Thus,

$$|(C_\mu f_1, C_\mu u)| \leq \varepsilon ||\eta C_\mu u||^2_0 + C(\varepsilon)||f_1||^2_{t,mcl(\Gamma)} + C_1(\varepsilon)||u||^2_{t,mcl(\Gamma'')}, \quad \forall \varepsilon > 0. \tag{7.5}$$

Evidently, $\operatorname{Im}(C_\mu q_1 I_N u, C_\mu u) = \operatorname{Im}(q_1 C_\mu u, C_\mu u) + \operatorname{Im}([C_\mu, q_1]u, C_\mu u)$. Put $v = C_\mu u$. Then $(q_1 v, v) = (v, q_1^* v)$, $q_1 = q_1 I_N$ and according to [8] the symbol of the L_2 adjoint operator q_1^* of q_1 is given by the formula:

$$q_1^* = q_1 - i \sum_{|\alpha|=1} q^{(\alpha)}_{1(\alpha)} + \tilde{\tilde{R}},$$

$\tilde{\tilde{R}}$ being a smoothing continuous operator of order $-(\sigma-1)$ in the Sobolev spaces. So

$$(q_1 v, v) = (v, q_1 v) + i \sum_{|\alpha|=1} \left(v, q^{(\alpha)}_{1(\alpha)}\right) + O\left(||u||^2_{t-\frac{\sigma-1}{2}}\right)$$

$$\Rightarrow 2\operatorname{Im}(q_1 v, v) = \sum_{|\alpha|=1} \left(v, q^{(\alpha)}_{1(\alpha)} v\right) + O\left(||u||^2_{t-\frac{\sigma-1}{2}}\right).$$

Having in mind the fact that the symbol $q^{(\alpha)}_{1(\alpha)} \in \Sigma^0_{\sigma-1}$ is real valued we get, applying again Theorem 3.3 from [8]:

$$2\operatorname{Im}(q_1 v, v) = \sum_{|\alpha|=1} \left(q^{(\alpha)}_{1(\alpha)} v, v\right) + O\left(||u||^2_{t-\frac{\sigma-1}{2}}\right).$$

Therefore,

$$\operatorname{Im}(q_1 C_\mu u, C_\mu u) = \frac{1}{2} \sum_{|\alpha|=1} \left(q^{(\alpha)}_{1(\alpha)} C_\mu u, C_\mu u\right) + O\left(||u||^2_{t-\frac{1}{2l}}\right), \tag{7.6}$$

and the remainder $O(.)$ is independent of μ.

In a similar way we have that

$$\operatorname{Im}(C_\mu R_0 u, C_\mu u) = \operatorname{Im}(R_0 C_\mu u, C_\mu u) + O\left(||u||^2_{t-\frac{1}{2l}}\right).$$

In fact, $[C_\mu, R_0] \in \Sigma^{t-1}_{\sigma-2}$, for $\sigma > 2$ as $R_0 \in \Sigma^0_{\sigma-1}$ and $[C_\mu, R_0]$ is a $(-t+\sigma-1)$ regularizing operator, uniformly with respect to μ, in the case $1 < \sigma < 2$. Then

$$\mathrm{Im}(R_0 C_\mu u, C_\mu u) = (\mathrm{Im} R_0 C_\mu u, C_\mu u) + O\left(||u||^2_{t-\frac{1}{2l}}\right), \tag{7.7}$$

$O(.)$ is independent of μ, and $\mathrm{Im} R_0 = \frac{R_0 - R_0^*}{2i}$ is a Hermitian selfadjoint matrix.

We know that $[C_\mu, q_1] \in \Sigma^t_{\sigma-1}$ uniformly with respect to μ and it has the principal symbol $\frac{1}{i}\{C_\mu, q_1\}$. So

$$\mathrm{Im}([C_\mu, q_1]u, C_\mu u) = -\mathrm{Re}\left(\{C_\mu, q_1\}u, C_\mu u\right) + O\left(||u||^2_{t-\frac{1}{2l}}\right),$$

i.e.,

$$\mathrm{Im}([C_\mu, q_1]u, C_\mu u) = \frac{1}{2}\mathrm{Re}\left(\{q_1, C_\mu^2\}u, u\right) + O\left(||u||^2_{t-\frac{1}{2l}}\right). \tag{7.8}$$

Combining (7.4)–(7.8) we conclude that

$$\begin{aligned} &\varepsilon||\eta C_\mu u||_0^2 + C(\varepsilon)||f_1||^2_{t,mcl(\Gamma)} + C_1(\varepsilon)||u||^2_{t,mcl(\Gamma'')} \\ &\geq \frac{1}{2}\mathrm{Re}\left(\{q_1, C_\mu^2\}u, u\right) + (\mathrm{Im} R_0' C_\mu u, C_\mu u) + O\left(||u||^2_{t-\frac{1}{2l}}\right). \end{aligned} \tag{7.9}$$

The identity (7.1) shows that

$$\begin{aligned} \mathrm{Re}\left(\{q_1, C_\mu^2\}u, u\right) &= \mathrm{Re}\left(\eta C_\mu B_0 u, \eta C_\mu u\right) + O\left(||u||^2_{t-\frac{1}{2}}\right) \\ &\quad + \mathrm{Re}\,(\underbrace{(1-\eta^2)(1+\mu^2|D|^2)^{-2} A_{2t}}_{J}\, u, u), \end{aligned}$$

where $\mathrm{conesupp}\, A_{2t} \subset \Gamma$, and $A_{2t} \in \Sigma^{2t}_{\sigma-1}$ uniformly with respect to μ.

According to Bony([8], Corollary 3.5 a), b))

$$|(Ju, u)| \leq d_2\left(||u||^2_{t-\frac{\sigma-1}{2}} + ||u||^2_{t,mcl(\Gamma'')}\right). \tag{7.10}$$

Having in mind that

$$(\mathrm{Im} R_0' C_\mu u, C_\mu u) = (\eta^2 \mathrm{Im} R_0' C_\mu u, C_\mu u) + (\underbrace{C_\mu^*(1-\eta^2)\mathrm{Im} R_0' C_\mu}_{J_1}\, u, u), \tag{7.11}$$

$J_1 \in \Sigma^{2t}_{\sigma-1}$ uniformly with respect to μ, and $\mathrm{conesupp}\, J_1 \subset \Gamma$, $J_1 \equiv 0$ near ρ^0, we can draw the conclusion that

$$|(J_1 u, u)| \leq d_3\left(||u||^2_{t-\frac{\sigma-1}{2}} + ||u||^2_{t,mcl(\Gamma'')}\right) \tag{7.12}$$

with d_2, d_3 independent of μ. Obviously,

$$(\eta^2 \mathrm{Im} R_0' C_\mu u, C_\mu u) = (\mathrm{Im} R_0' \eta C_\mu u, \eta C_\mu u) + O(||u||^2_{t-\frac{\sigma-1}{2}})$$

as $[\eta, \mathrm{Im} R_0']$ is a smoothing operator of order $-(\sigma-1)$ and $O(.)$ does not depend on μ.

Thus

$$(\mathrm{Im} R_0' C_\mu u, C_\mu u) = (\mathrm{Im} R_0' \eta C_\mu u, \eta C_\mu u) + O\left(||u||^2_{t-\frac{\sigma-1}{2}} + ||u||^2_{t,mcl(\Gamma'')}\right). \tag{7.13}$$

According to the condition (5.5):
1) $\frac{1}{2}B_0 + \mathrm{Im}R_0^{'} \in \Sigma_{\sigma-1}^0$, and 2) $\frac{1}{2}B_0(\rho^0)I_N + \mathrm{Im}R_0^{'}(\rho^0)$ is a Hermitian and positively definite matrix which implies that $\frac{1}{2}B_0(\rho)I_N + \mathrm{Im}R_0^{'}(\rho)$ is a Hermitian and positively definite matrix near ρ^0.

Combining (7.9)–(7.13) and having in mind that

$$(\eta C_\mu B_0 u, \eta C_\mu u) = (B_0 \eta C_\mu u, \eta C_\mu u) + O\left(||u||^2_{t-\frac{\sigma-1}{2}}\right)$$

we get that $\forall \varepsilon > 0$,

$$\begin{aligned} &\varepsilon||\eta C_\mu u||_0^2 + C(\varepsilon)||f_1||^2_{t,mcl(\Gamma)} + d_4||u||^2_{t,mcl(\Gamma'')} \\ &\geq \mathrm{Re}\left(\left(\frac{1}{2}B_0 + \mathrm{Im}R_0^{'}\right)\eta C_\mu u, \eta C_\mu u\right) + d_5||u||^2_{t-\frac{1}{2l}}. \end{aligned} \tag{7.14}$$

The constants $C(\varepsilon)$, d_4, d_5 in (7.14) are independent of μ.

Taking conesupp $\eta \subset \{\rho : \frac{1}{2}B(\rho) + \mathrm{Im}R_0^{'}(\rho) > 0\}$ we can apply Gårding's inequality for a positive system of paradifferential operators [8] and obtain:

$$\mathrm{Re}\left(\left(\frac{1}{2}B_0 + \mathrm{Im}R_0^{'}\right)(\eta C_\mu u), \eta C_\mu u\right) \geq d_6||\eta C_\mu u||_0^2 + O\left(||\eta C_\mu u||^2_{-\theta}\right) \tag{7.15}$$

for some $\theta = \mathrm{const} > 0$ and $d_6 = \mathrm{const} > 0$; $O(.)$ are independent of μ.

The standard interpolation inequality in Sobolev spaces gives us that

$$||\eta C_\mu u||_0 \leq \mathrm{const},$$

i.e., for sufficiently small $\varepsilon > 0$ we have:

$$u \in H^t_{mcl(\rho^0)} \Rightarrow u \in H^t_{mcl}(\Gamma).$$

Thus everything is proved.

2. To prove Theorem 4.6 we shall use a microlocal approach, being in the framework of Bony's theory of paradifferential operators. The linearization of the semilinear system (4.8) has the following principal symbol:

$$p_1 = \begin{pmatrix} \xi + a\eta & b\eta \\ c\eta & \xi - a\eta \end{pmatrix} \quad \text{and} \quad p_0 = -i\begin{pmatrix} A & B \\ C & D \end{pmatrix}$$

where

$$\begin{aligned} A &= a_u u_y + b_u v_y, \\ B &= a_v u_y + b_v v_y, \\ C &= c_u u_y - a_u v_y, \\ D &= c_v u_y - a_v v_y. \end{aligned}$$

Certainly the coefficients $a_u, \ldots, c_v$ depend on the fixed solution $(u(x,y), v(x,y))$.

Obviously, $\det p_1 = 0 \Leftrightarrow \xi^2 + \eta^2\Delta = 0$, i.e., $\xi = \pm\eta\sqrt{-\Delta}$.

1. In the hyperbolic region $\Delta < 0$ the solution $(u, v) \in C^\infty$.
2. In the elliptic region $\Delta > 0$ we can use Bony's theorem for microlocal raising of smoothness of the solutions of nonlinear PDE [8].
3. Assume that $\bar{\rho} = (\bar{x}, \bar{y}, \bar{\xi}, \bar{\eta})$ is such that $\bar{\xi}^2 + \bar{\eta}^2\Delta = 0$. Then $\Delta = 0 \Rightarrow \bar{\xi} = 0$, $\bar{\eta} = \pm 1$. This way we conclude that if $\tilde{p}_1 = co\, p_1 = \begin{pmatrix} \xi - a\eta & -b\eta \\ -c\eta & \xi + a\eta \end{pmatrix}$, then $\tilde{p}_1 p_1 = q_1 I_2$, where $q_1 = \det p_1$ and moreover, $H_{q_1(\bar{\rho})} = -\Delta_x \partial_\xi - \Delta_y \partial_\eta$ while $\Xi(\bar{\rho}) = \bar{\eta}\partial_\eta$.

So, $H_{q_1(\bar{\rho})} \| \Xi(\bar{\rho})$ iff $\Delta_x = 0$, $\Delta_y \neq 0$. But $\partial/\partial x\, \Delta(u(\bar{x}, \bar{y}), v(\bar{x}, \bar{y})) = 0$, $\Delta(u(\bar{x}, \bar{y}), v(\bar{x}, \bar{y})) = 0 \Leftrightarrow (\bar{x}, \bar{y}) = (x_0, y_0)$ and $P = (x_0, y_0)$ is the vertex of the parabola γ. Therefore each point of $T^*(\gamma \setminus P)$ is either elliptic or q_1 is of real principal type at γ. The projection of the integral curves of H_{q_1} at $R^2_{x,y}$, i.e., the characteristics of (4.8) form cusp points at $\gamma \setminus P$ (Fig. 25).

Applying Bony's theorem for propagation of singularities from [8] we prove the first part of Theorem 4.6 In the case $\rho^0 = (x_0, y_0, 0, \eta)$, $\eta^2 = 1$ we shall use Theorem 5.1 Let us check its requirements having in mind that $(x_0, y_0, \xi_0 \neq 0, \eta_0)$ is an elliptic point for our system. Straightforward computations show that

$$R_0'(x_0, y_0, 0, \eta) = i\eta \left(-r_x - r_y r + r \begin{pmatrix} A & B \\ C & D \end{pmatrix} \right).$$

Putting $K = -r_x - r_y r + r \begin{pmatrix} A & B \\ C & D \end{pmatrix}$ we see that

$$\operatorname{Im} R_0' = \eta \frac{K + K^*}{2}.$$

The expression for K can be rewritten in the following form:

$$\begin{aligned} K(x_0, y_0) \quad = \quad & \left\{ -r_u u_x - r_v v_x - \left(r_u r - r \begin{pmatrix} a_u & a_v \\ c_u & c_v \end{pmatrix} \right) u_y \right. \\ & \left. + \left(r \begin{pmatrix} b_u & b_v \\ -a_u & -a_v \end{pmatrix} - r_v r \right) v_y \right\} (x_0, y_0). \end{aligned}$$

Our condition (B) is equivalent to $(v)'$. As the Hamiltonian vector field passing through the point $(x_0, y_0, 0, \eta)$ is parallel to the radial vector field, the corresponding bicharacteristic to H_{q_1} is a single conic ray and its projection on R^2_{xy} – the characteristic curve through (x_0, y_0) – degenerates at a point. Condition (B) is sufficient for nonexistence of isolated singularities of the solution (u, v) of the system (4.8).

Thus everything is proved.

8. Appendix II. Interaction of conormal waves for semilinear PDE

1. The results on creation of new singularities proposed in §1 of Chapter I relied heavily on the fact that only the case of two variables was considered. Unfortunately, it seems impossible to generalize the proofs from §1 in the multi-dimensional case. It is interesting to note that similar results on creation (interaction) of singularities hold true in this situation, the notion of characteristics being replaced by the notion of characteristic hypersurfaces. To prove these results the machinery of paradifferential operators introduced by Bony is used. Thus, a rather different approach works in the multi-dimensional case. For the sake of completeness we shall formulate here a classical and very interesting result of Bony for interaction of simple waves. A short sketch of the proof will be given too. A refinement of the theorems to be given below was proposed by Alinhac in [1,2]. We shall confine ourselves to Bony's theorems in order to stress the main ideas, to simplify the proofs and to compare the geometrical interpretation of semilinear waves in the 1-dimensional space case with waves propagating in the multi-dimensional case.

2. We shall formulate now Bony's theorem for propagation of nonlinear waves "under the shock". So consider the fully nonlinear partial differential equation:

$$F(x, u, \dots, \partial^\alpha u, \dots)_{|\alpha|\le m} = 0, \tag{8.1}$$

where $F \in C^\infty$ with respect to its arguments and F, u are real-valued functions.

For each sufficiently smooth solution u we can define the linearized on u operator P_u:

$$P_u = \sum_{|\alpha|\le m} \frac{\partial F}{\partial u_\alpha}(x, u, \dots, \partial^\beta u(x), \dots)\partial^\alpha$$

and its principal symbol (up to i^m)

$$p_m(x,\xi) = \sum_{|\alpha|=m} \frac{\partial F}{\partial u_\alpha}(x, u, \dots, \partial^\beta u(x), \dots)\xi^\alpha.$$

As usual, the characteristic set $\mathrm{Char}\, P = \{(x,\xi), \xi \neq 0 : p_m(x,\xi) = 0\}$. The zero-bicharacteristic $\mathcal{L}$ passing through the point $\rho^0 \in \mathrm{Char}\, P$ is defined as the integral curve of the Hamilton vector field: $H_{p_m} = \sum_{j=1}^n (\frac{\partial p_m}{\partial \xi_j}\cdot\frac{\partial}{\partial x_j} - \frac{\partial p_m}{\partial x_j}\cdot\frac{\partial}{\partial \xi_j})$.

(H): Assume now that $u \in H^s(\Omega)$, $s > \frac{n}{2} + m$, $0 \in \Omega$ and that the operator P_u is strictly hyperbolic with respect to the direction t. Put $\Omega^\pm = \Omega \cap \{\pm t > 0\}$. Moreover, we suppose that Ω^+ is contained in the domain of influence of Ω^-, i.e., each backward zero bicharacteristic, starting from a point in Ω^+, meets with Ω^- before leaving Ω.

Theorem 8.1. *(Bony) Suppose that $u \in H^s(\Omega)$, $s > \frac{n}{2} + m$ satisfies* (8.1). *Then*

a) *If $\rho^0 \notin \mathrm{Char}\, P \Rightarrow u \in H^{2s-m-\frac{n}{2}}_{mcl}(\rho^0)$.*

b) *If $s > \frac{n}{2} + m + 2$, $\rho^0 \subset \mathrm{Char}\, P$, $\sigma < 2s - \frac{n}{2} - m - 1$ and $u \in H^\sigma_{mcl}(\rho^0)$, then $u \in H^\sigma_{mcl}(\mathcal{L})$.*

In the semilinear case the assertion b) remains true under the weaker condition $s > \frac{n}{2} + m - 1$, $\sigma < 2s - \frac{n}{2} - m + 1$.

Corollary 8.2. *Under the assumptions b) of the previous theorem, suppose that* (8.1) *is hyperbolic with respect to the variable* x_1 *and* **(H)** *holds. Then* $u|_{x_1<0} \in H^\sigma$, $\sigma > s$ *implies that* $u \in H^\sigma(\Omega)$.

The symbol Σ stands for a smooth variety or for the union of two smooth varieties Σ_i intersecting transversally.

Definition 8.3. The function u belongs to the conormal distribution space $H^s(\Sigma, k)$, $k \in \mathbf{N} \cup \{\infty\}$ if $u \in H^s_{loc}$ and for each integer $0 \le l \le k$ $M_1 M_2 \ldots M_l u \in H^s$, where M_j are arbitrary real smooth vector fields tangential to Σ.

One can see that if $f \in C^\infty$ and $u \in H^s(\Sigma, k)$, $s > \frac{n}{2}$, then $f(u) \in H^s(\Sigma, k)$. Moreover, $u \in H^s(\Sigma, k) \Rightarrow u \in H^{s+k}(\Omega \setminus \Sigma)$. In the special case $\Sigma = \Sigma_1 \cup \Sigma_2$ it is evident that $WF_{s+k}(u) \subset N^*(\Sigma_1) \cup N^*(\Sigma_2) \cup N^*(\Sigma_1 \cap \Sigma_2)$, $N^*(\Sigma_1)$ being the conormal bundle over Σ_1.

Suppose now that $\Sigma_1, \ldots, \Sigma_m$ are smooth hypersurfaces intersecting two-by-two transversally along a variety Γ of codimension 2. Put $\Sigma = \cup_j \Sigma_j$.

Definition 8.4. The function u belongs to the conormal distribution space $H^s(\Sigma, k)$ if $u \in H^s_{loc}$ and for each integer $0 \le l \le k$:

$M_1 \ldots M_l u \in H^s$, where M_j are first-order pseudodifferential operators whose principal symbols vanish on $\cup N^*(\Sigma_i) \cup N^*(\Gamma)$.

It can be proved that if $u \in H^s_{loc}$, $s > \frac{n}{2}$ and $f \in C^\infty$, then $u \in H^s(\Sigma, k)$ $\Rightarrow f(u) \in H^s(\Sigma, k)$.

Theorem 8.5. *(Bony). Consider the semilinear equation*

$$P(x, D)u = F(x, u, \ldots, \partial^\alpha u, \ldots)_{|\alpha| \le m-1} \tag{8.2}$$

and assume that $u \in H^s_{loc}(\Omega)$, $s > \frac{n}{2} + m$ *satisfies* (8.2), *the real-valued function* $F \in C^\infty$ *with respect to its arguments,* P *has real-valued smooth coefficients. Suppose, moreover, that the condition* **(H)** *holds,* Σ *is a characteristic hypersurface of* P *and* $u|_{\Omega^-} \in H^s(\Sigma, k)$. *Then* $u \in H^s(\Sigma, k)$ *in* Ω.

Theorem 8.6. *(Bony). Assume that* $u \in H^s_{loc}(\Omega)$, $s > \frac{n}{2} + m$ *satisfies* (8.2), *the condition* **(H)** *holds and* Σ_1, Σ_2 *are two characteristic hypersurfaces of* P *intersecting transversally at* Γ, *codim* $\Gamma = 2$, $\Gamma \cap \Omega^- = \emptyset$. *Moreover, there are no other characteristic hypersurfaces passing through* Γ. *Then* $u|_{\Omega^-} \in H^s(\Sigma, k)$, $\Sigma = \Sigma_1 \cup \Sigma_2$ *implies that* $u \in H^s(\Sigma, k)$ *in* Ω.

Remark 8.7. Each second-order strictly hyperbolic operator satisfies the geometrical condition imposed on the characteristic hypersurfaces. The interaction between them will not cause anomalous singularities as it did in §1 of Chapter I, i.e., the result is the same in both dimensions.

Theorem 8.8. *(Bony). Suppose that $u \in H^s_{loc}(\Omega)$, $s > \frac{n}{2} + m$ is a solution of* (8.2), *the condition* **(H)** *holds, $\Sigma_{1,2}$ are two characteristic hypersurfaces of P intersecting transversally at Γ, $\Gamma \cap \Omega_- = \emptyset$ and $\Sigma_3, \dots, \Sigma_m$ are characteristic hypersurfaces issuing from Γ and intersecting transversally each other and transversal to $\Gamma_{1,2}$. Assume that in Ω^- the solution $u \in H^{s+k}$ outside $\Sigma_1 \cup \Sigma_2$, $u|_{\Omega^-} \in H^s(\Sigma_i, k)$, $i = 1, 2$. Then we claim that in Ω:*

a) *$u \in H^{s+k}$ outside of $\cup_{i=1}^m \Sigma_i$,*
b) *$u \in H^s(\Sigma_i, k)$ near $\Sigma_i \setminus \Gamma$, $i = 1, 2$,*
c) *$u \in H^s(\Sigma_j, [s + k - t])$ near $\Sigma_j \setminus \Gamma$, $j \geq 3$, $\sigma = \min(s + k, 2s - \frac{n}{2} - m + 1)$.*

Obviously, in case c): $\sigma > s$. Suppose that $k = \infty$. Then $u \in C^\infty$ outside $\Sigma_1 \cup \Sigma_2 \cup \cup_3^m \Sigma_j^+$, $u \in H^{2s - \frac{n}{2} - m + 1}(\Sigma_j^+, \infty)$, $j \geq 3$, where Σ_j^+ is the forward part of Σ_j issuing from Γ and contained in Ω^+. Thus when Σ_1, Σ_2 carrying conormal singularities intersect transversally at Γ, anomalous singularities could appear at Γ. They are of weaker strength in comparison with the initial singularities and propagate along the forward part of new characteristic hyperfaces starting from Γ. The new effect here marked is illustrated in Fig. 7 of §1 (1-space variable case).

Proof of Theorem 8.5. After a local change of variables we can assume that Σ is given by the equation $\varphi \equiv x_1 = 0 \Rightarrow \nabla\varphi = (1, 0, \dots, 0)$. According to the definition of characteristic hypersurface we have $p_m(0, x', \nabla\varphi) = 0$, p_m being the principal symbol of the linear hyperbolic operator P, $x' = (x_2, \dots, x_n)$. Thus the coefficient $a_{m0\dots0}$ in front of ξ_1^m must vanish for $x_1 = 0 \Rightarrow a_{m0\dots0}(x) = x_1 A_1(x)$. So the operator P takes the form

$$P = A_1(x) x_1 \partial_{x_1}^m + \sum_{j=2}^n A_j(x, D)\partial_{x_j} + A_0(x, D) \tag{8.3}$$

and $A_j(x, D)$, $j = 2, \dots, n$ are differential operators of order $m - 1$ with smooth coefficients.

Put $\mathcal{M}$ for the set of all tangential vectors M to Σ. Evidently, $\mathcal{M}$ is generated by the following vector fields: $x_1\partial_{x_1}, \partial_{x_2}, \dots, \partial_{x_n}$.

The well-known commutator identity $[AB, C] = A[B, C] + [A, C]B$ enables us to conclude that

$$[\partial_{x_i}, P] = \sum_{j=2}^n (\partial_{x_i} A_j)\partial_{x_j} + (\partial_{x_i} A_1) x_1 \partial_{x_1}^m + (\partial_{x_i} A_0), \quad i \geq 2, \tag{8.4}$$

$$[x_1\partial_{x_1}, P] = \sum_{j=2}^n [x_1\partial_1, A_1]\partial_{x_j} + [x_1\partial_1, A_0] + (x_1\partial_{x_1} A_1 - (m - 1)A_1) x_1 \partial_{x_1}^m,$$

and $x_1 \partial_{x_1}^m = \partial_{x_1}^{m-1}(x_1\partial_{x_1}) - (m - 1)\partial_{x_1}^{m-1}$.

So we have that

$$[M, P] = \sum_{j=1}^n B_{j,0} M_j + B_0, \tag{8.5}$$

where $B_{j,0}$, B_0 , A_0 are differential operators of order $(m - 1)$.

Inductively one can show that for each multiindex I the following identity holds:

$$[M^I, P] = \sum_{|J|\leq |I|} B_{J,I} M^J \tag{8.6}$$

and $B_{J,I}$ are differential operators of order $(m-1)$.

Applying the chain rule which is valid in this situation we have

$$\partial_{x_j}(F(x,u,\dots,\partial^\beta u)) = \frac{\partial F}{\partial x_j}(x,\dots,\partial^\beta u) + \dots + \frac{\partial F}{\partial u_\beta}.\partial_{x_j}\partial^\beta u,$$

$j \geq 2, |\beta| \leq m-1,$

$$x_1\partial x_1(F(x,u,\dots,\partial^\beta u)) = (x_1\partial_{x_1}F)(x,\dots,\partial^\beta u) + \dots + \frac{\partial F}{\partial u_\beta} x_1\partial^\beta(\partial_{x_1}u).$$

On the other hand, $x_1\partial^\beta(\partial_{x_1}u) = \partial^\beta(x_1\partial_1 u) + C\partial^{\beta-1}\partial_{x_1}u$, $C = \text{const}$. Therefore,

$$M(F(x,u,\dots,\partial^\beta u)) = F_1(x,u,\dots,\partial^\gamma u, x_1\partial_{x_1}u, \nabla\partial_{x_2}u,\dots,\partial^\gamma(x_1\partial_{x_1}u)), \tag{8.7}$$

the function F_1 being C^∞ smooth with respect to its arguments, $|\gamma| \leq m-1$.

There are no difficulties in verifying that

$$M^I(F(x,\dots,\partial^\beta u)) = F_I(x,u,\dots,\partial^\gamma M^{I'}u),$$

where $|\gamma| \leq m-1$, if $|I| \leq l$, then $|I'| \leq l$ and F_I is C^∞ smooth with respect to its arguments.

To fix the idea we shall study the simplest case only.

So $Pu = F \Rightarrow PMu = MF + [P,M]u$.

Put $V_1 = (u, x_1\partial_{x_1}u, \partial_2 u, \dots, \partial_n u)$. According to (8.5) and (8.7) we get

$$PV_1 + BV_1 = F_1(x, V_1, \dots, \partial^\beta V_1)_{|\beta|\leq m-1} \tag{8.8}$$

and B is a linear differential operator of order $(m-1)$.

As $u \in H^s_{loc}(\Omega) \Rightarrow V_1 \in H^{s-1}_{loc}(\Omega)$ and moreover, $V_1 \in H^s(\Omega^-)$. Applying Corollary 8.2 we conclude that $V_1 \in H^s(\Omega)$, i.e., $u \in H^s(\Sigma,1)$. Inductively we get that $u \in H^s(\Sigma,k)$.

Proof of Theorem 8.6. After a local smooth change of variables we can assume that $\Sigma_1 : \varphi_1 \equiv x_1 = 0$, $\Sigma_2 : \varphi_2 \equiv x_2 = 0$. Each hyperplane Σ_3 passing through $\Gamma = \Sigma_1 \cap \Sigma_2$ and different from $\Sigma_{1,2}$ is given by the equation: $\Sigma_3 : \alpha_1 x_1 + \alpha_2 x_2 = 0$, $\alpha_1\alpha_2 \neq 0$. The normal vector to Σ_3 is $(\alpha_1, \alpha_2, 0)$. As we know Σ_3 is noncharacteristic for P. Having in mind the proof of Theorem 8.5 we see that

$$P = x_1 B_1(x)\partial^m_{x_1} + x_2 B_2(x)\partial^m_{x_2} + \sum_{j=3}^n B_j(x,D)\partial_{x_j} + K\partial_{x_1}\partial_{x_2} + B_0.$$

In the previous equality B_j are differential operators of order $(m-1)$ with smooth coefficients, while the differential operator K is of order $(m-2)$. We rewrite P in the form:

$$P = K\partial_{x_1}\partial_{x_2} + \sum_{j=1}^{n} A_j(x,D)Z_j + A_0, \tag{8.9}$$

where $Z_1 = x_1\partial_{x_1}$, $Z_2 = x_2\partial_{x_2}$, $Z_j = \partial_{x_j}$, $j \geq 3$ are generators of the space of all vector fields tangential to Σ.

According to our assumption $p_m(x_1, x_2, x''', \alpha_1, \alpha_2, 0) \neq 0$ on Σ_3, $x''' = (x_3, \dots, x_n)$. Taking $x_1 = x_2 = 0$ we obtain $K(0, x''', \alpha_1, \alpha_2, 0) \neq 0$. Moreover, $\frac{\partial p_m}{\partial \xi_j}(0, 0, x''', \alpha_1, \alpha_2, 0) \neq 0$ for $\alpha_j = 0$, $j = 1, 2$. Thus, $K(0, 0, x''', \alpha_1, \alpha_2, 0) \neq 0$ for $\alpha_1\alpha_2 = 0$. Therefore, K is a microelliptic operator on $N^*(\Gamma) \setminus \{0\}$. Denote by H its inverse pseudodifferential operator of order $2-m$. We have

$$HP = HK\partial_1\partial_2 + \sum_{j=1}^{n} HA_jZ_j + HA_0 \equiv HK\partial_1\partial_2 \tag{8.10}$$

$$+ \sum_{j=1}^{n} R_jZ_j + R_0,$$

R_j being pseudodifferential operators of order 1. Proceeding as before we conclude that

$$\begin{aligned} P(Z_ju) &= Z_j(F(x,u,\dots,\partial^\beta u)) + [P, Z_j]u, \\ [P, Z_j] &= \sum_{j=1}^{n} C_jZ_j + C_0 + D\partial_{x_1}\partial_{x_2}, \end{aligned} \tag{8.11}$$

and C_j are differential operators of order $(m-1)$ while D is a differential operator of order $(m-2)$. Combining (8.10) and the fact that $HK \equiv 1$ in a conical neighborhood of $N^*(\Gamma)$, we have

$$D\partial_{x_1}\partial_{x_2}u = DHPu + \sum_{j=1}^{n} E_jZ_ju + E_0u, \tag{8.12}$$

where E_j are pseudodifferential operators of order $(m-1)$, $ord\, DH = 0$.

Put $V_1 = (u, x_1\partial_{x_1}, x_2\partial_{x_2}, \partial_3, \dots, \partial_n)$. According to (8.7) and (8.11) we obtain the system

$$PV_1 + BV_1 = LG_1(x, V_1, \dots, \partial^\beta V_1), |\beta| \leq m-1 \tag{8.13}$$

in which B is a matrix-valued pseudodifferential operator of order $(m-1)$, L is a matrix-valued pseudodifferential operator of order 0 and G_1 depends C^∞ smoothly on its arguments.

Certainly, $V_1 \in H^{s-1}_{loc}(\Omega)$ and $V_1 \in H^s_{loc}(\Omega^-)$. Fortunately, the result formulated in Corollary 8.2 remains true for systems with diagonal principal symbol of type (8.13) (see [9]). So we have that $V_1 \in H^s(\Omega) \Rightarrow u \in H^s(\Sigma, 1)$ etc.

The proof and the above result for propagation of singularities are with local validity. Theorem 8.8 has a rather technical proof and we omit it.

3. We propose now several concluding remarks. They deal with removable singularities of the solutions of several classes of fully nonlinear systems of partial differential operators of first order. As a paradifferential approach will be used in investigating these systems, our results will be formulated via the properties of the first variation (linearization) of the corresponding system.

So we assume that $F_k(x, u, p)$, $u = (u_1, \dots, u_N)$, $p = (u_{11}, \dots u_{ij}, \dots u_{Nn})$, $1 \le j \le n$, $1 \le i, k \le N$ are real-valued C^∞ functions of their arguments $x \in \Omega$, $u \in \mathbf{R}^N$, $p \in \mathbf{R}^{nN}$ and consider the following fully nonlinear system with real-valued solution u:

$$F_k(x, u, \nabla u) = 0, 1 \le k \le N, x \in \Omega, \tag{8.14}$$

Ω domain in $\mathbf{R}^n$, $u(x) = (u_1(x), \dots, u_N(x)) \in H^s_{loc}(\Omega)$, $s > \frac{n}{2} + 2$, i.e., $u \in C^{2+\varepsilon}(\Omega)$, $\varepsilon > 0$.

Then the linearization (the first variation) of (8.14) on the solution u can be written as

$$\mathcal{P}v = \sum_{j=1}^{n} A_j(x) D_j v - iB(x)v, \xi_j \leftrightarrow D_j = \frac{1}{i}\frac{\partial}{\partial x_j}, \tag{8.15}$$

where the matrix

$$A_j(x) = \left\| \frac{\partial F_k}{\partial u_{ij}}(x, u, \nabla u(x)) \right\|_{1 \le i,k \le N}, j = 1, \dots, n$$

and

$$B(x) = \left\| \frac{\partial F_k}{\partial u_i}(x, u(x), \nabla u(x)) \right\|_{1 \le i,k \le N}.$$

Define now the matrix

$$K(x) = \frac{B + B^*}{2} - \frac{1}{2}\sum_{j=1}^{n} \frac{\partial A_j(x)}{\partial x_j},$$

B^* being the adjoint matrix to B.

As usual the characteristic set of (8.15) is given by

$$\operatorname{Char} p_1 = \left\{ \rho = (x, \xi) \in T^*(\Omega) \setminus 0 : \det \sum_{j=1}^{n} A_j(x)\xi_j = 0 \right\}.$$

In our further investigations we shall suppose that the matrices $A_j(x)$ are symmetric, i.e.,

$$(i) \quad A_j(x) = A_j^*(x), \forall x \in \Omega$$

(i.e., $\frac{\partial F_k}{\partial u_{ij}}(x, u(x), \nabla u(x)) = \frac{\partial F_i}{\partial u_{kj}}(x, u(x), \nabla u(x))$, $\forall x \in \Omega$).

Theorem 8.9. *(see[64]). Consider the nonlinear system* (8.14) *under the assumption (i) and the matrix $A_{j_0}(x)$ being positive (negative) definite in Ω. Suppose, moreover, that there exists a closed set W conical with respect to ξ and having compact base in Ω such that for each characteristic point $\rho^0 \in \text{Char } p_1 \cap \partial W \cap \{x_{j_0} \geq \delta\}$, $\delta = \text{const}$, we have $u \in H^t_{mcl}(\rho^0)$ for some $t < 2s - 2 - \frac{n}{2}$. Then $u \in H^t_{mcl}(\rho^0)$, $\forall \rho^0 \in \text{Char } p_1 \cap W \cap \{x_{j_0} \geq \delta\}$.*

It is well known that if $\rho^0 \notin \text{Char } p_1$, then $u \in H^{2s-1-\frac{n}{2}}(\rho^0)$.

Geometrically we can formulate Theorem 8.9 in a simpler way. If the solution is microlocally H^t smooth in the characteristic points located at the "northern hemisphere" $\partial W \cap \{x_{j_0} \geq \delta\}$ it becomes microlocally H^t smooth in the points of the "northern massive hemisphere" $W \cap \{x_{j_0} \geq \delta\}$. Thus $u \notin H^t_{mcl}(\rho^0)$, ρ^0 being a characteristic point inside $W \cap \{x_{j_0} \geq \delta\}$, implies that there exists at least one characteristic point $\bar{\rho}$ at the northern hemisphere with the property $u \notin H^t_{mcl}(\bar{\rho})$.

The proof of the previous result can be found in [64].

Theorem 8.10. *(see [65]). Consider the system* (8.14), *(i) and suppose that the real-valued solution $u \in H^s_{loc}$, $s > \frac{n}{2}+3$. We shall assume that $\rho^0 = (x^0, \xi^0) \in \text{Char } p_1$, $|\xi^0| = 1$, Γ is a conical neighborhood of ρ^0 and such that $u \in H^t_{mcl}(\Gamma \setminus \rho^0)$, $t < 2s - 2 - \frac{n}{2}$. Then if*

$$(ii)_0 \quad K(x^0) > \gamma Id_N, \gamma = \text{const} > 0$$

$$(iii)_t \quad K(x^0) + t \sum_{k,j=1}^{n} \frac{\partial A_j(x^0)}{\partial x_k} \xi^0_k \xi^0_j > \gamma Id_N$$

are fulfilled we claim that $u \in H^t_{mcl}(\rho^0)$.

Certainly, Id_N denotes the identity matrix on $\mathbf{R}^N$, $K(x^0)$ and the matrix participating in the left-hand side of $(iii)_t$ are symmetric.

Remark 8.11. Consider the same system (8.14) on the n-dimensional torus $\mathbf{T}^n$, i.e., $x \in \mathbf{T}^n$ and suppose that $\frac{\partial F_k}{\partial u_{kj}}(x, u, p) = \frac{\partial F_i}{\partial u_{kj}}(x, u, p)$, $\forall x \in \mathbf{T}^n$ and for each (u, p), s.t. $|u| + |p| < 1$. Let $u \equiv 0$ satisfy (8.14) and $(ii)_0$, $(iii)_t$, $t > 3 + \frac{n}{2}$, are fulfilled for each $x \in \mathbf{T}^n$, $\forall \xi$, $|\xi| = 1$, $u = p = 0$. Then there exists $s_1 > \frac{n}{2} + 3$ with the property: for every $s \in (\frac{n}{2} + 3, s_1)$ one can find in $H^s(\mathbf{T}^n)$ such a neighborhood $W \ni 0$ that the system $F_k(x, u, \nabla u) = f_k$ is solvable in $H^s(\mathbf{T}^n)$ for arbitrary $f = (f_1, \dots, f_N)$ in W. The proof of the assertion just formulated is published in [4]. Therefore, our conditions for removability of the singularities turn out to be a microlocalized variant of the conditions under which global existence results are valid.

Chapter III
Travelling waves for several classes of nonlinear integrodifferential operators

9. Introduction

1. This chapter deals with travelling wave solutions for several classes of nonlinear partial differential equations and nonlinear integrodifferential equations arising in the applications – solid body mechanics in a medium with memory, shallow water equation, etc. In the case of solid body mechanics in a medium with memory, we construct a C^1 solution having the form of a travelling wave of tension. The solvability of the problem under consideration is reduced to the investigation of a nonlinear integral equation. Unfortunately, the contraction mapping principle does not hold in this situation. A nontrivial solution of our problem is found via the construction of successive approximations for an appropriate positive operator acting in the cone of nonnegative monotone functions. For a shallow water equation a weak solution having a corner at its peak is constructed (see [21]).

We shall begin with one-dimensional waves in a medium with memory. Following [50] we shall denote by x a co-ordinate of a point belonging to a solid body, by t – the time variable, by ε – the deformation, by σ – the tension and by ρ – the density. For the sake of simplicity we shall assume that ρ is equal to 1. In a medium with memory there is a link between the tension and the deformation which depends on the history of the process. We shall confine ourselves to the next relation between ε and σ:

$$\varepsilon(t) = \int_{-\infty}^{t} \sqrt{1+K^*}\sqrt{a'(\sigma)}\sqrt{1+K^*}\sqrt{a'(\sigma)}\sigma_t'\, dt. \tag{9.1}$$

In the previous equality K^* is the convolution operator:

$$K^*u(t) = \int_{-\infty}^{t} K(t-\tau)u(\tau)\, d\tau, \tag{9.2}$$

$\sqrt{1+K^*}$ stands for the development of the operator $1+K^*$ into a power series and the integral operator $\sqrt{1+K^*}$ as well as the multiplication operator $a'(\sigma)$ are acting on the function σ_t'.

It is well known from classical mechanics that the next equation holds:

$$\frac{\partial^2\varepsilon}{\partial t^2} - \frac{\partial^2\sigma}{\partial x^2} = 0, \tag{9.3}$$

supposing ε and σ to be smooth functions of (t,x).

Putting (9.1) into (9.3) we conclude that the tension $\sigma(t,x)$ satisfies a rather complicated nonlinear integro-differential equation. According to Theorem 7.1. from [50] the equation (9.3) with ε given by (1) can be sharply factorized into two first order factors describing the propagation of two waves of tension to the left- and to the right-hand side respectively.

Here are the factors:

$$\frac{\partial}{\partial t}\sqrt{1+K^*}\sqrt{a'(\sigma)} \pm \frac{\partial}{\partial x},$$

$$\sqrt{1+K^*}\sqrt{a'(\sigma)}\frac{\partial}{\partial t} \pm \frac{\partial}{\partial x}. \tag{9.4}$$

We shall concentrate our attention to (9.4).

Putting $(\sqrt{1+K^*})^{-1} = 1 - \Phi^*$,

$$\Phi^* u = \int_{-\infty}^{t} \Phi(t-\tau)u(\tau)\,d\tau$$

we see that each smooth solution σ of the nonlinear integro-differential equation

$$\sqrt{a'(\sigma)}\frac{\partial \sigma}{\partial t} \pm (1-\Phi^*)\frac{\partial \sigma}{\partial x} = 0,\ \sigma \in C^2(x \geq 0) \tag{9.5}$$

will satisfy (9.3) with ε given by (9.1).

According to mechanics terminology the function Φ is called "kernel of heredity". Assume that

$$\Phi(t) = ke^{-kt},\ k > 0.$$

So we have that a wave of tension, propagating "to the right-hand side" is given by the nonlinear first-order equation

$$\sqrt{a'(\sigma)}\frac{\partial \sigma}{\partial t} + \frac{\partial \sigma}{\partial x} - k\int_{-\infty}^{t} e^{-k(t-\tau)}\frac{\partial \sigma(\tau,x)}{\partial x}\,d\tau = 0. \tag{9.6}$$

We shall assume, moreover, that

$$a(0)=0,\ a \in C^2,\ a'(\sigma) > 0 \text{ and } \sigma \in C^2(x \geq 0), \tag{9.7}$$

$$\sigma = 0 \text{ for } x \geq 0,\ t \leq 0, \sigma(t,0) = \sigma_0(t) \in C^2(R). \tag{9.8}$$

Obviously, $\sigma_0(t) \equiv 0$ for $t \leq 0$.

We shall construct a classical solution of the mixed problem (9.6), (9.8) and we shall formulate results for global existence in time $t \geq 0$, $x \geq 0$ and for blow up of the corresponding solution. The symbol $||\sigma_0||_{C^0(R^1)}$ stands for the uniform norm of the function σ_0. We suppose further on that $||\sigma_0||_{C^0(R)}, ||\sigma_0'||_{C^0(R)} < \infty$.

This is our first result.

Theorem 9.1. *Consider the mixed problem* (9.6), (9.8). *Then*

(i) *There exists a constant* $C(||\sigma_0(t)||_{C^0})$ *such that if* $\frac{||\sigma_0'||_{C^0}}{k} \times C(||\sigma_0||_{C^0}) < 1$*; then the problem* (9.6), (9.8) *possesses a unique global classical solution* $\sigma \in C^2(x \geq 0, t \geq 0)$.

The constant $C(||\sigma_0||_{C^0})$ *can be estimated in the following way:*

$$C(||\sigma_0||_{C^0}) \leq \max_{|\bar\sigma| \leq ||\sigma_0||_{C^0}} \sqrt{a'(\bar\sigma)}$$

$$\times \frac{1}{\min_{|\bar\sigma| \leq ||\sigma_0||_{C^0}} \sqrt{a'(\bar\sigma)}} \times 1/2 \max_{|\bar\sigma| \leq ||\sigma_0||_{C^0}} \frac{|a''(\bar\sigma)|}{|a'(\bar\sigma)|}.$$

(ii) *σ_t blows up for a finite $X > 0$ if*

a) *one can find a point $\beta_0 > 0$ with the property*

$$1 + \frac{\sigma_0'(\beta_0)}{kw_0(\beta_0)}\left(\sqrt{a'(\sigma_0(\beta_0))} - \sqrt{a'(0)}\right) < 0,$$

where $w_0(\beta_0) = \int_0^{\sigma_0(\beta_0)} \sqrt{a'(\lambda)}\, d\lambda$, $\sigma_0(\beta_0) > 0$;

b) *one can find a point $\beta_0 > 0$ such that $\sigma_0(\beta_0) = 0$,*

$$-1/2 < \frac{ka'(0)}{\sigma_0'(\beta_0)a''(0)} < 0,\ \sigma_0'(\beta_0) \neq 0, a''(0) \neq 0.$$

The life span X of the corresponding solution in case b) can be estimated as

$$X \leq \bar{x} = -\frac{1}{\sqrt{a'(0)}} \log\left(1 + \frac{2ka'(0)}{\sigma_0'(\beta_0)a''(0)}\right) > 0.$$

We omit the proof of Theorem 9.1 as it can be reduced by differentiation of (9.6) in t to a mixed problem for first-order quasi-linear partial differential equation containing a dissipative term.

To simplify the notation we shall write down the corresponding nonlinear integrodifferential equation which describes the wave of tension propagating "to the right-hand side" in the following form:

$$b(\sigma)\frac{\partial \sigma}{\partial t} + \frac{\partial \sigma}{\partial x} - \int_{-\infty}^{t} K(t-\tau)\frac{\partial \sigma}{\partial x}(\tau, x) = 0, \sigma \in C^1, \tag{9.9}$$

where the function K is called a "kernel of heredity" and σ is the tension of a 1-dimensional solid body in a medium with memory.

The functions $b, K \in C^k(\mathbf{R}^1)$, $k \geq 5$, $b(0) > 0$, $K(-z) = K(z)$, $\forall z$, $K(z) > 0$ everywhere, $K'(z) \leq 0$ for $z \geq 0$, $\lim_{z\to\infty} K(z) = 0$, $K \in L^1(\mathbf{R}^+)$, $\mathbf{R}^+ = \{x \in \mathbf{R}^1 : x > 0\}$.

The equation (9.9) can be rewritten as follows under some suitable assumptions imposed on σ:

$$b(\sigma)\frac{\partial \sigma}{\partial t} + \frac{\partial \sigma}{\partial x} - \frac{\partial}{\partial x}\int_{-\infty}^{t} K(t-\tau)\sigma(\tau, x)d\tau = 0. \tag{9.10}$$

This is possible if σ is bounded and if in a small neighborhood U_{x_0} of each point x_0 one can find a function $p \in L^1$ such that

$$|\sigma_x(t,x)| \leq p(t), \forall x \in U_{x_0},$$

or if σ_x is bounded too.

We are looking for a solution of (9.10) in the form of a travelling wave: $\sigma(t,x) = f(t-x)$, where $f \in C^1(z > 0) \cap C(z \geq 0)$, $f = 0$ for $z \leq 0$, $f \not\equiv 0$.

This way we are simplifying our problem. Usually $\sigma = f(ct - x)$, where c is the velocity of the wave. We do not lose generality by taking $c = 1$, as further on several cases will be studied and they cover the different possible values of c.

We write (9.10) for $t > x$ as follows:

$$\frac{\partial}{\partial t}\int_0^{\sigma(t,x)} b(\lambda)d\lambda + \frac{\partial}{\partial x}\sigma(t,x) - \frac{\partial}{\partial x}\int_x^t K(t-\tau)\sigma(\tau,x)d\tau = 0. \tag{9.11}$$

Define now the function

$$F(z) = \int_0^z K(z-\lambda)f(\lambda)d\lambda, \tag{9.12}$$

where $F \in C^1(\mathbf{R}^1)$, $F(0) = 0$ if f is a bounded function.

The equation (9.11) implies that

$$\frac{\partial}{\partial t}\int_0^{f(t-x)} b(\lambda)d\lambda + \frac{\partial}{\partial x}f(t-x) - \frac{\partial}{\partial x}F(t-x) = 0.$$

Put $\bar{G}(z) = \int_0^z b(\lambda)d\lambda \in C^{k+1}(\mathbf{R}^1)$. Then

$$\frac{\partial}{\partial t}\bar{G}(f(t-x)) + \frac{\partial}{\partial x}f(t-x) - \frac{\partial}{\partial x}F(t-x) = 0 \tag{9.13}$$

for $t > x$.

Having in mind the facts that $\frac{\partial}{\partial x}f(t-x) = -f'(t-x)$ and $f(+0) = \bar{G}(0) = 0$ we obtain

$$\int_0^{f(t-x)} b(\lambda)d\lambda - f(t-x) + \int_0^{t-x} K(t-x-\lambda)f(\lambda)d\lambda = 0$$

for $t \geq x$.

Putting $z = t - x \geq 0$ we rewrite the previous equation as

$$\int_0^{f(z)} b(\lambda)d\lambda - f(z) + \int_0^z K(z-\lambda)f(\lambda)d\lambda = 0, \tag{9.14}$$

$f \in C(z \geq 0) \cap C^1(z > 0)$, $f(0) = 0$, $f \in L^\infty(\mathbf{R}^+)$.

Let $G(z) = \bar{G}(z) - z = \int_0^z b(\lambda)d\lambda - z \in C^{k+1}(\mathbf{R}^1)$. We obtain then the following equation for the travelling wave we are looking for:

$$G(f(z)) + \int_0^z K(z-\lambda)f(\lambda)d\lambda = 0,$$

i.e.,

$$G(f(z)) + K * f(z) = 0. \tag{9.15}$$

Then two different cases appear:

a) $G'(0) = 0$, i.e., $b(0) = 1$.

b) $G'(0) \neq 0$, i.e., $b(0) \neq 1$.

As the case b) is simpler to deal with we shall consider here the case a) only.

These are our main assumptions:

$$(*) \qquad G(z) \leq 0, \forall z \in \mathbf{R}^+, G'(z) < 0, \forall z > 0, G'(0) = 0$$

and there exists $G''(+\infty)$ such that $0 < -G''(\infty) \leq \infty$.

Condition $(*)$ implies that $-G'(\infty) = -G(\infty) = \infty$ and therefore $-G$ is a homeomorphism from $[0,\infty)$ into itself and is a diffeomorphism from $(0,\infty)$ into itself.

Consider the function $H(z) = -\frac{G(z)}{z} \in C^k(\mathbf{R}^1)$. Evidently $H(0) = 0$, $H(z) > 0$ for $z > 0$. We suppose moreover, that

$$(**) \qquad H'(z) > 0 \text{ for } z > 0.$$

According to $(*)$ the function $H(z)$ is a diffeomorphism from $(0,\infty)$ into itself and is a homeomorphism from $[0,\infty)$ into itself.

We shall formulate now our last condition:

$$(***) \qquad G'(0) = \cdots = G^{(r-1)}(0) = 0, (-G)^{(r)}(0) > 0, 3 \le r \le k-1.$$

As the case $G'(0) = 0$, $-G''(0) > 0$ is studied in a similar way we shall not give the details.

This is our main result.

Theorem 9.2. *The equation* (9.9) *possesses a travelling wave solution* $\sigma = f(t-x)$, $f \in C^k(z > 0) \cap C(z \ge 0) \cap L^\infty$, $f(0) = 0$, $f(z) > 0$ *for* $z > 0$, f*-monotonically increasing if the assumptions* $(*)$, $(**)$, $(***)$ *are fulfilled. Moreover,* $f(z) \approx cz^{\frac{1}{r-1}}(1 + o(1))$, $z \to +0$, $c = \text{const} > 0$.

So we conclude that the travelling wave starts with a vertical tangent at $z = 0$ and has a horizontal asymptote at ∞.

Remark 9.3. In the case $G'(0) = 0$, $-G''(0) > 0$ there also exists a solution of (9.9) in the form of a travelling wave and belonging to the same functional class. The only difference is that $f(z)$ has a linear growth near $+0$.

10. Proof of the main result and concluding remarks

At the beginning of this section we shall propose several notes to be useful further on.

1. Let $f \in L^\infty(z \ge 0)$. Then the convolution operator $K * f$ has the following properties:

$$K * f \in C(z \ge 0) \cap L^\infty, (K * f)(0) = 0.$$

2. Let $f \in L^\infty(z \ge 0)$ and $f(z) \ge 0$, $\forall z \ge 0$. Then $(K * f)(z) \ge 0$, $\forall z \ge 0$ and $(K * f)(z) \le |f|_{L^\infty}.I$, where $I = \int_0^\infty K(\lambda)d\lambda > 0$.

So we conclude that the nonlinear operator $A = (-G)^{-1}.K*$ is well defined by the formula $\bar{f}(z) = Af(z) = (-G)^{-1}((K * f))(z)$ for each function $f \in L^\infty(\mathbf{R}^+)$, $f(z) \ge 0$, $z \ge 0$ and $\bar{f} \in C(z \ge 0) \cap L^\infty$, $\bar{f}(z) \ge 0$, $\forall z \ge 0$, $\bar{f}(0) = 0$. Moreover, $0 \le f_1 \le f_2 \in L^\infty(\mathbf{R}^+) \Rightarrow 0 \le Af_1 \le Af_2$.

Consider now two kernels $K_{1,2}$ having the same properties as the kernel K from (9.9). Then $0 \le K_1 \le K_2$ and $0 \le f_1 \le f_2 \in L^\infty(\mathbf{R}^+)$ implies that $0 \le K_1 * f_1 \le K_2 * f_2$ and therefore, $0 \le A_1 f_1 \le A_2 f_2$, where $A_1 = (-G)^{-1}.K_1*$, $A_2 = (-G)^{-1}K_2*$.

The solvability of the equation (9.15) is equivalent to the solvability of the equation $f(z) = Af(z)$, $z \geq 0$ in the cone of the nonnegative bounded functions.

3. Assume that the function $0 \leq f(z) \in L^\infty(\mathbf{R}^+)$ is monotonically increasing. Then $(K * f)(z) \geq 0$, $\forall z \geq 0$ is a monotonically increasing function too and $(K * f)(\infty) = f(\infty).I$, where $f(\infty) = \lim_{z\to\infty} f(z)$.

To prove 3 we use the fact that if $f \in L^\infty(\mathbf{R}^+)$, then $K * f = f * K$, i.e., the convolution is a commutative operator on L^∞. The Lebesgue dominated convergence theorem implies $(K * f)(\infty) = f(\infty)I$.

As a corollary from 3 we see that if $0 \leq f(z) \in L^\infty(\mathbf{R}^+)$ is a monotonically increasing function, then $0 \leq \bar{f} = Af \in C(\mathbf{R}^+) \cap L^\infty$ is a monotonically increasing function and $\bar{f}(\infty) = (-G)^{-1}(If(\infty))$, $\bar{f}(z) \leq (-G)^{-1}(If(z))$, $\forall z \geq 0$. It is evident that $f \in C(z \geq 0) \cap L^\infty \Rightarrow K * f \in C^1(z \geq 0)$.

4. Let $\bar{f}(z) = Af(z)$ and $0 \leq f \in L^\infty \cap C(z \geq 0)$. Suppose that one can find a sequence $z_\nu \to +0$ such that $f(z_\nu) > 0$. Then $\bar{f}(0) = 0$ and $\bar{f} \in C^1(z > 0)$. In fact, the main assumption in 4 implies that $(K * f)(z) > 0$, $\forall z > 0$.

5. Convergence property of the operator A. Let $0 \leq f_n(z) \leq C = \text{const.}$, $f_n \in L^\infty(\mathbf{R}^+)$ and $f_n(z) \to f(z)$ a.e. in $z \geq 0$. Then $Af_n(z) \to Af(z)$, $\forall z \geq 0$. The convergence here is pointwise and not uniform.

6. Let $0 \leq f(z) \in C(z \geq 0) \cap L^\infty$, $f(0) = 0$ be a monotonically increasing solution of the equation $f(z) = Af(z)$, $z \geq 0$. Under the assumption $(***)$ we have that $0 \leq f(z) \leq \text{const.} z^{\frac{1}{r-1}}(K(0))^{\frac{1}{r-1}}$, $z \to +0$, const > 0. The result is interesting if $f(z)$ does not vanish identically on each subinterval having the origin as end point.

7. We shall solve now the equation (9.15) with a constant kernel $K \equiv C > 0$. Obviously, C does not belong to the kernel class introduced after the equation (9.9). Thus, $(-G)(f)(z) = C\int_0^z f(\lambda)d\lambda$. We are looking for a nontrivial solution in each one-side neighborhood of 0 : $f \in C(z \geq 0) \cap C^1(z > 0)$, $f(0) = 0$. By differentiation and after standard calculations we see that the previous integral equation is equivalent to the ODE

$$(-G)'(f(z)).f'(z) = Cf(z). \tag{10.1}$$

The nontrivial solution of (10.1) for $z \geq 0$ is given by the formula $f(z) = p^{-1}(Cz) \in C^1(z > 0) \cap C(z \geq 0)$, where $p(z) = \int_0^z -\frac{G'(\lambda)}{\lambda}d\lambda \in C^k(\mathbf{R}^1)$, $p'(z) > 0$ for $z > 0$ and, according to $(*)$, p is a diffeomorphism from $(0, \infty)$ into itself and is a homeomorphism from $[0, \infty) \to [0, \infty)$, $p(0) = 0$. So $f(z) > 0$ for $z > 0$, $f \in C(z \geq 0) \cap C^1(z > 0)$, f is monotonically increasing for $z \geq 0$. The condition $(***)$ leads us to the conclusion that $f(z) = (Cz)^{\frac{1}{r-1}}(A + o(1))$, $z \to +0$, $A = \text{const.}$ In fact, $w = p(z) = z^{r-1}(M + z^{k-r}B(z))$ for $|z| \ll 1$, $B \in C^2$, $M = \text{const} > 0$.

The inverse function $z = p^{-1}(w) \in C(0 \leq z \ll 1) \cap C^1(z > 0)$ and $z = w^{\frac{1}{r-1}}\left(\frac{1}{M^{\frac{1}{r-1}}} + O(w^{\frac{1}{r-1}})\right)$, $w \to +0$. In a similar way we obtain that if $G'(0) = 0$, $-G''(0) > 0$, then $f(z) = Cz(A + o(1))$, $z \to +0$.

8. Our next step is to solve the equation (9.15) with the special kernel K_1, where

$$K_1(z) = \begin{cases} K(z_0), & 0 \le z \le z_0, \quad z_0 > 0, \\ K(z), & z \ge z_0. \end{cases}$$

Obviously, $K_1 \in C^k(\mathbf{R}^1 \setminus \{z_0\})$. Put $I_1 = \int_0^\infty K_1(z)dz < I$. Then $z_0 K(z_0) < I_1$. Consider now the equation $f(z) = A_1 f(z)$, $A_1 = (-G)^{-1}.K_1*$. The operator A_1 roughly speaking is not a contracting mapping. Because of this reason we shall construct a sequence of successive approximations in the cone of nonnegative, bounded on $\mathbf{R}^+$ and monotonically increasing functions. This approach is influenced by a method proposed by Krasnoselskij in [43], [44], see also [50]. If we start with a first approximation $f_0 \equiv 0$, then all the approximations will be 0 too. So we take the following initial iteration,

$$0 \le f_0(z) = \begin{cases} \tilde{f}_0(z), & 0 \le z \le z_0, \\ H^{-1}(I_1), & z > z_0, \end{cases}$$

and $\tilde{f}_0(z) = p^{-1}(K(z_0).z)$, where H^{-1} is the inverse function of $H = -\frac{G(z)}{z}$. The function $f_0(z)$ is monotonically increasing and with jump at $z = z_0$. In fact, $0 \le z \le z_0 \Rightarrow \tilde{f}_0(z) \le \tilde{f}_0(z_0) \le p^{-1}(K(z_0).z_0)$. On the other hand, $(-G)\tilde{f}_0(z) = K(z_0)\int_0^z \tilde{f}(\lambda)d\lambda$ for $0 \le z \le z_0$. Thus $(-G)\tilde{f}_0(z_0) \le K(z_0).z_0\tilde{f}_0(z_0)$ $\Rightarrow H(\tilde{f}_0(z_0)) \le z_0 K(z_0) I \Rightarrow \tilde{f}_0(z_0) \le H^{-1}(z_0 K(z_0)) < H^{-1}(I_1)$.

Define the second approximation $f_1 = A_1 f_0 \in C(z \ge 0) \cap L^\infty$, which is a monotonically increasing function and moreover, $f_1(z) = \tilde{f}_0(z)$, $0 \le z \le z_0$. Thus, $f_1(z) > 0$, $z > 0$, $f_1(0) = 0$ and according to property 3: $f_1(\infty) = (-G)^{-1}(H^{-1}(I_1).I_1) = f_0(\infty)$ as $I_1 = H(f_0(\infty)) \iff -G(f_0(\infty)) = f_0(\infty).I_1$. Evidently, $0 \le f_1(z) \le f_0(z)$, $\forall z \ge 0$. Put $f_2 = A_1 f_1, \ldots, f_k = A_1 f_{k-1}, \ldots$. The monotonicity of A_1 (see property 2) shows that:

$$f_0(z) \ge f_1(z) \ge f_2(z) \ge \cdots \ge 0, f_n \in C(z \ge 0) \cap L^\infty, n \ge 1,$$

$$f_n(0) = 0, f_0(z) = f_1(z) = \cdots = f_n(z) = \cdots$$

for $0 \le z \le z_0$, $f_n(\infty) = f_0(\infty)$. Certainly, the functions $f_n(z)$ are monotonically increasing.

Define now $f(z) = \lim_{n\to\infty} f_n(z) \Rightarrow f \in L^\infty(\mathbf{R}^+)$, $f(0) = 0$, f is monotonically increasing and $f(z) = f_0(z)$, $0 \le z \le z_0 \Rightarrow f(z) > 0$ for $z > 0$. As the convergence is pointwise it is not obvious that $f(z)$ is a continuous function. Fortunately, the identity $f_n(z) = A_1 f_{n-1}(z)$, $\forall z \ge 0$, $\forall n$ enables us to conclude that $f(z) = A_1 f(z)$, $\forall z \ge 0$. Therefore $f(z) \in C(z \ge 0)$ and the convergence in $\lim_n f_n(z) = f(z)$ is uniform on each compact interval in $\mathbf{R}^+$. Evidently, $f(\infty) = H^{-1}(I_1)$. As the kernel K_1 is not smooth at $z = z_0$ we do not claim that f is smooth for $z = z_0$.

And now we are ready to prove Theorem 9.2.

Put

$$K_n(z) = \begin{cases} K(\frac{1}{n}), & 0 \le z \le \frac{1}{n} \\ K(z), & z \ge \frac{1}{n} \end{cases} \in C(z \ge 0).$$

Thus, $0 \leq K_1(z) \leq K_2(z) \leq \cdots \leq K_n(z) \leq \cdots \to K(z)$, $K(1) \leq \cdots \leq K(\frac{1}{n}) \leq \cdots$, $0 \leq A_n \leq A_{n+1}$, $\forall n$. Fix the integer n. According to property 8 the equation $f^{(n)} = A_n f^{(n)}$, $A_n = (-G)^{-1}.K_n*$ possesses a solution $f^{(n)}$ with the following properties: $0 \leq f^{(n)} \in C(z \geq 0) \cap L^\infty$, $f^{(n)}(0) = 0$, $f^{(n)}(z) > 0$ for $z > 0$, $f^{(n)}$ is monotonically increasing, $f^{(n)}(\infty) = H^{-1}(I_n)$, $I_n = \int_0^\infty K_n(\lambda)d\lambda$. According to the assumption $(***)$ $f^{(n)}(z) = c_{0n} z^{\frac{1}{r-1}}(1 + o(1))$, $z \to +0$. The "zero approximation" for $f^{(n)}(z)$ was given by

$$f_0^{(n)}(z) = \begin{cases} p^{-1}(K(\frac{1}{n})z), & 0 \leq z \leq \frac{1}{n}, \\ H^{-1}(I_n), & z > \frac{1}{n}. \end{cases}$$

Certainly,

$$p^{-1}(K(\frac{1}{n})z) \leq p^{-1}(K(\frac{1}{n+1})z), H^{-1}(I_n) < H^{-1}(I_{n+1}). \tag{10.2}$$

We remind the reader that also $\forall k \geq 1$ $f_0^{(n)} = f_1^{(n)} = \cdots = f_k^{(n)} = \cdots$ $f_k^{(n)}(z) \geq 0$, $f_1^{(n)} = A_n f_0^{(n)}, \ldots, f_k^{(n)} = A_n f_{k-1}^{(n)}$ and $f_0^{(n)} \geq f_1^{(n)} \geq \cdots \geq f_k^{(n)} \geq \cdots \to f^{(n)}$.

Having in mind (10.2) we see that

$$0 \leq f_0^{(1)} \leq f_0^{(2)} \leq \cdots \leq f_0^{(n)} \leq \cdots$$

and property 2 implies that

$$0 \leq f_1^{(1)} \leq f_1^{(2)} \leq \cdots \leq f_1^{(n)} \leq \cdots.$$

Conclusion: $\{f_k^{(m)}\}_{m=1}^\infty$, $k \in \mathbf{N}$ is a monotonically increasing sequence of nonnegative continuous functions for $z \geq 0$. Evidently, $0 \leq f_k^{(n)} \leq f_{k+1}^{(n)}$, n - fixed leads for $k \to \infty$ to the inequality: $0 \leq f^{(n)} \leq f^{(n+1)}$. This way we have constructed a monotonically increasing sequence of nonnegative and continuous monotonically increasing functions:

$$0 \leq f^{(1)} \leq \cdots \leq f^{(n)} \leq \cdots. \tag{10.3}$$

These functions are uniformly bounded as $f^{(n)} \leq f^{(n)}(\infty) = H^{-1}(I_n) < H^{-1}(I)$.

Define now the monotonically increasing bounded function

$$0 \leq f(z) = \lim_n f^{(n)}(z), \forall z \geq 0.$$

Letting $n \to \infty$ in $f^{(n)}(z) = A_n f^{(n)}$, $z \geq 0$ we apply the Lebesgue dominated convergence theorem and conclude that $f(z) = Af(z)$, $\forall z \geq 0$. Thus, $f \in C(z \geq 0) \cap L^\infty$, $f(0) = 0$, $f(+\infty) = H^{-1}(I)$, $f(z) \geq f^{(1)}(z)$, $\forall z \geq 0$, i.e., $f(z) > 0$, $\forall z > 0$. A simple application of the results given by properties 4 and 6 shows that $f \subset C^k(z > 0)$, $c_{01} z^{\frac{1}{r-1}}(1 + o(1)) \leq f(z) \leq \text{const.} z^{\frac{1}{r-1}}(K(0))^{\frac{1}{r-1}}$, $z \to +0$. Theorem 9.2 is proved.

Example. Put $b(\lambda) = 1 - h(\lambda)$, where $h(\lambda) = \sum_{n=1}^{\infty} c_n \lambda^n$ is an entire function, $c_n \geq 0$, $h \not\equiv 0$.

1. Assume that $c_1 = \cdots = c_{r-2} = 0$, $c_{r-1} > 0$, $3 \leq r$. Then there exists a monotonically increasing, bounded, smooth solution of (9.9), $f(z) > 0$, $z > 0$ and $f(z) \approx cz^{\frac{1}{r-1}}$, $z \to +0$, $c > 0$.
2. Suppose that $c_1 > 0$. Then there exists a monotonically increasing smooth solution of (9.9), such that $d_2 z \leq f(z) \leq d_1 z$, $0 \leq z \ll 1$, $d_{1,2} = \text{const} > 0$.

Concluding remarks. Consider now the equation (9.9) with $b = 1 - \sigma^\alpha$, $0 < \alpha < 1$. The solution σ we are interested in will be nonnegative. Repeating the procedure described above we reduce the problem for finding a nontrivial solution of (9.9) in the form of a travelling wave to the solvability of the equation

$$(-G)f = \frac{1}{\alpha+1} f^{\alpha+1}(z) = K * f(z), z \geq 0. \tag{10.4}$$

The solution of (10.4) should be $0 \leq f \in C(z \geq 0) \cap L^\infty$, monotonically increasing, $f(0) = 0$, and $f(z) > 0$ for $z > 0$. Obviously, (10.4) is equivalent to $f(z) = [(\alpha+1)K * f(z)]^{\frac{1}{\alpha+1}}$. The solution of (10.4) (if it exists) can be estimated from above as follows:

$$0 \leq f(z) \leq (\alpha+1)^{\frac{1}{\alpha}} (K(0))^{\frac{1}{\alpha}} z^{\frac{1}{\alpha}}, z \geq 0$$

and therefore there exists $f'(0) = 0$ ($\Rightarrow f \in C^1(z \geq 0)$).

In the next step we shall estimate from below f, $f(z) > 0$ for $z > 0$. By using the fact that K is a monotonically increasing function we obtain

$$f(z) \geq (\alpha+1)^{\frac{1}{\alpha+1}} (K(z))^{\frac{1}{\alpha+1}} \left[\int_0^z f(\lambda) d\lambda\right]^{\frac{1}{\alpha+1}}, z \geq 0. \tag{10.5}$$

Putting $w(z) = \int_0^z f(\lambda) d\lambda \in C^2(z \geq 0)$, $w(z) > 0$ for $z > 0$ we have $w'(z) \geq Aw^{\frac{1}{\alpha+1}} (K(z))^{\frac{1}{\alpha+1}}$, $A = (\alpha+1)^{\frac{1}{\alpha+1}}$. Integrating the last inequality from ε to z, $0 < \varepsilon$ and letting then $\varepsilon \to +0$ we have

$$w(z) \geq B \left(\int_0^z K(\lambda)^{\frac{1}{\alpha+1}} d\lambda\right)^{\frac{\alpha+1}{\alpha}}, B = \left(\frac{\alpha}{\alpha+1} A\right)^{\frac{\alpha+1}{\alpha}}, \forall z \geq 0.$$

Thus, $w(z) \geq \frac{B}{2} z^{\frac{\alpha+1}{\alpha}} (K(0))^{\frac{1}{\alpha}}$, $0 \leq z \ll 1$.

According to (10.5): $f(z) \geq \text{const} z^{\frac{1}{\alpha}}$, $0 \leq z \ll 1$, const > 0. We construct the solution $f(z) \in C^1(z \geq 0) \cap L^\infty$, $f(z) > 0$, $z > 0$, $f'(0) = f(0) = 0$, f monotonically increasing as in Theorem 9.2. The only difference is that in the last case $-G''(\infty) = 0$. The observation that $p(z) = \int_0^z -\frac{G'(\lambda)}{\lambda} d\lambda$ in this case has the property $p(\infty) = \infty$ enables us to complete the proof.

In contrast with the travelling wave studied in Theorem 9.2 the travelling wave in (10.4) starts with a horizontal tangent and $f(\infty) = ((\alpha+1)I)^{\frac{1}{\alpha}}$.

Remark 10.1. Theorem 9.2 remains true for $G \in C^{1+\alpha}$, $0 \le \alpha \le 1$ with condition $(*)'$ instead of $(*)$:

$$(*)' \qquad G(z) \le 0, z \ge 0, G'(z) < 0, \forall z > 0, G'(0) = 0,$$
$$-G'(\infty) = \infty, \int_0^\infty -\frac{G'(\lambda)}{\lambda} d\lambda = \infty$$

and under condition $(**)$.

Consider now the well-known R. Camassa–D. Holm equation

$$u_t - u_{xxt} + 3uu_x = 2u_x u_{xx} + uu_{xxx}, t > 0, x \in \mathbf{R}. \tag{10.6}$$

The function u represents the fluid velocity at time t in the x-direction in appropriate nondimensional units. Put the initial data $u_0(x) = u(0,x) \in H^3(\mathbf{R}^1)$ and suppose that the Cauchy problem possesses a local-in-time solution $u \in C([0,T); H^3) \cap C^1([0,T); H^2)$, $T > 0$. Introducc now the pseudodifferential operator $Q(D_x)$ with symbol $(1+\xi^2)$. Certainly, Q acts in the space variable. As it is shown in [21,22]

$$Q(u_t + uu_x) = -\partial_x \left(u^2 + \frac{1}{2} u_x^2 \right),$$

i.e.,

$$u_t + uu_x = -\partial_x \left(p * \left(u^2 + \frac{1}{2} u_x^2 \right) \right) \quad \text{in } C([0,T); H^1)$$

and $p * f(x) = \int_{-\infty}^{\infty} p(y) f(x-y)$, $p(y) = \frac{1}{2} e^{-|y|}$, $f \in L_2(\mathbf{R}^1)$ as $Q^{-1} f = p * f$.

We define the nonlinear operator

$$F(v) = \frac{1}{2} v^2 + p * \left(v + \frac{1}{2} v_x^2 \right)$$

and rewrite (formally) the equation (10.6) in the form of a conservation law:

$$u_t + (F(u))_x = 0, u(0,x) - u_0(x) \in H^3(\mathbf{R}^1). \tag{10.7}$$

Following [21] we give a definition of the weak solution of the Cauchy problem for (10.6).

Definition 10.2. A function $u \in L_2([0,T); H^1)$ is called a weak solution of the equation (10.6) with Cauchy data $u_0 \in H^1$ if it satisfies the integral identity

$$\int_0^T \int_{\mathbf{R}} (u\Phi_t + F(u)\Phi_x) dx dt + \int_{\mathbf{R}} u_0(x) \Phi(x,0) dx = 0 \tag{10.8}$$

for all $\Phi \in C^1([0,T) \times \mathbf{R})$ being restrictions on $[0,T) \times \mathbf{R}$ of $C^1(\mathbf{R}^2)$ functions compactly supported in the strip $(-T,T) \times \mathbf{R}$.

One can easily see that the travelling wave $u_c = ce^{-|x-ct|}$ is for any $c > 0$ a weak solution of the Cauchy problem (10.6) with initial data $ce^{-|x|}$. To do this we put u_c in (10.8) and compute the corresponding integrals. This example is rather interesting because u_c has a corner point at its peak, i.e., this wave is not C^∞ smooth everywhere.

Consider the Cauchy problem for the following nonlinear nonlocal equation

$$u_t + uu_x + \int_{-\infty}^{\infty} K(x-\xi)u_\xi(t,\xi)d\xi = 0, t \geq 0 \tag{10.9}$$

$$u(0,x) = u_0(x),$$

where the kernel $K \in C(\mathbf{R}) \cap L^1(\mathbf{R})$ is symmetric and monotonically decreasing on $\mathbf{R}_+$, $K \not\equiv 0$ and $u_0 \in H^\infty(\mathbf{R})$. It is proved in [61] that for some $T > 0$ there exists a unique solution of (10.9) in the class $C^\infty([0,T]; H^\infty(\mathbf{R}))$. A proof of the blow up result for (10.9) is given in [22]. More precisely, assume that $u_0 \in H^\infty$ satisfies

$$\inf_{\mathbf{R}} u_0' + \sup_{\mathbf{R}} u_0' \leq -2K(0) < 0.$$

Then u_x blows up for a finite time $t_0 > 0$.

References

[1] Alinhac S., Evolution d'une onde simple pour des équations non-linéaires générales, *Current topics in P.D.E.*, Kinokuniya, Tokyo, (1986), 63–90.

[2] Alinhac S., Interaction d'ondes simples pour des équations complètement non-linéaires, *Ann. Sci. Ec. Norm. Sup. (4 série)*, **21**, (1988), 91–132.

[3] Alinhac S., *Blow up for Nonlinear Hyperbolic Equations,* Progress in nonlinear differential equations and their applications, **v. 17**, Birkhäuser, (1995).

[4] Alinhac S. and Gerard P., *Opérateurs Pseudodifférentiels et Théorème de Nash-Moser*, Inter Editions, Paris, (1991).

[5] Beals M., Self-spreading and strength of singularities for solutions to semi-linear wave equations, *Ann. of Math.*, **118**, (1983), 187–214.

[6] Beals M., Interaction of radially smooth nonlinear waves, in: *Lecture notes in Math.*, **1256**, (1987), 1–27.

[7] Bers L., *Mathematical Aspects of Subsonic and Transonic Gas Dynamics*, Wiley, New York, (1958).

[8] Bony J.-M., Calculus symbolique et propagation des singularités pour les équations aux dérivées partielles non-linéaires, *Ann. Sci. Ec. Norm. Sup. (4 série)*, **14**, (1981), 209–246.

[9] Bony J.-M., *Propagation des Singularités pour les Equations aux Dérivées Partielles Non-linéaires*, Séminaire d'E.D.P., Ecole Polytechnique, Paris, (1979–1980), **No 22**.

[10] Bony J.-M., Second microlocalization and interaction of singularities for non-linear P.D.E., *Hyperbolic Equations and Related Topics*, Mizohata ed., Kinokuniya, (1986), 11–49.

[11] Bony J.-M., *Interaction des singularités pour les équations de Klein-Gordon non-linéaires*, Sem. E.D.P., Ecole Polytechnique, (1983–84), **No 10**.

[12] Bony J.-M., Singularités des solutions de problèmes hyperboliques non-linéaires, *Advances in Microlocal Analysis*, Garnir ed., NATO ASI Series **168**, Reidel, (1985), 15–39.

[13] Bony J.-M., Analyse microlocale des équations aux dérivées partielles non-linéaires, in: *Lecture notes in Math.*, **1495**, (1991), Springer Verlag, 1–45.

[14] Brocker T. and Lander L., *Differentiable Germs and Catastrophes*, London Math. Soc. Lecture Note Series, **17**, (1975), Cambridge Univ. Press.

[15] Camassa R. and Holm D., An integrable shallow water equation with peaked solutions, *Phys. Rev. Letters*, **71**, (1993), 1661–1664.

[16] Chemin J.-Y., Calculus paradifférentiel précisé et applications aux équations aux dérivées partielles non-semilinéaires, *Duke Math. J.*, **56**, (1988), 431–469.

[17] Chemin J.-Y., Interaction de trois ondes dans les équations sémilinéaires strictement hyperbolique d'ordre 2. *Comm. in P.D.E.*, **12**, (1987), 1203–1225.

[18] Chen Shuxing, Propagation of anomalous singularities of solutions to a semilinear hyperbolic equation of higher order, *Northwestern Math. J.*, **1**, (1985), 127–137.

[19] Coifman R. and Meyer Y., *Au delà des Opérateurs Pseudo-différentiels*, Astérisque, Soc. Math. Fr., **57**, (1978).

[20] Colombini F. and Del Santo D., Development of singularities for nonlinear hyperbolic systems with periodic data, *Ann. Univ. Ferrara*, Sez. VII – Sc. Mat., Suppl. vol. **XLI**, (1995), 15–28.

[21] Constantin A. and Escher J., Global existence and blow up for a shallow water equation, *Ann. Sc. Norm. Sup. Pisa, IV Ser.*, **26**, (1998), 303–328.

[22] Constantin A. and Escher J., Wave breaking for nonlinear nonlocal shallow water equation, *Acta Math.*, **181**, (1998), 229–243.

[23] Courant R. and Friedrichs K., *Supersonic Flow and Shock Waves*, N. Y. Interscience Publishers inc. and Interscience Publ. Ltd., (1948).

[24] Courant R. and Hilbert D., *Methods of Mathematical Physics II, Partial Differential Equations by R. Courant*, N. Y. – London, (1962).

[25] Delort J.-M., *Conormalité des Ondes Semi-linéaires le Long des Caustiques*, Séminaire E.D.P., Ecole Polytechnique, (1988–1989), **No 15**.

[26] Dencker N., On the propagation of polarization sets for systems of real principal type, *J. of Funct. Analysis*, **46**, (1982), 351–372.

[27] Egorov Yu.V. and Schulze B.-W., *Pseudodifferential Operators, Singularities, Applications*, Operator Theory: Advances and Applications, **93**, Basel, Birkhäuser, (1997).

[28] Fihtengholtz G., *A Course on Differential and Integral Calculus II,* Physmatgis, Moskow, (1962) (in Russian).

[29] Guillemin V. and Schaeffer D., On a certain class of Fuchsian PDE, *Duke Math. J.*, **44**, (1977), 157–199.

[30] Hong J. and Zuily C., L_p and Hölder estimates for a class of degenerate elliptic boundary value problems, Applications to the Monge-Ampère equation, *Comm. in P.D.E.*, **16**, (1991), 997–1032.

[31] Hörmander L., *Lectures on Nonlinear Hyperbolic Differential Equations*, Ser. Math. & Applications, **26**, Springer Verlag, (1997).

[32] Hörmander L., *The Analysis of Linear Partial Differential Operators*, vol. I–IV, Springer Verlag, New York, (1990).

[33] John F., Formation of singularities in one-dimensional nonlinear wave propagation, *Comm. Pure Appl. Math.*, **27**, (1974), 377–405.

[34] John F., *Nonlinear Wave Equations. Formation of Singularities*, Lehigh Univ. Lecture Series, AMS, Providence, (1990).

[35] Joshi M. and Sa Barreto A., The generation of semilinear singularities by a swallowtail caustic, *American J. of Math.*, **120**, (1998), 529–550.

[36] Iordanov I., Anomalous singularities of semilinear hyperbolic systems, *C. R. Acad. Bulg. Sci.*, **41**, (1988), 13–15.

[37] Iordanov I., Anomalous singularities of semilinear nonstrictly hyperbolic system, *J. Math. Pures et Appl.*, **70**, (1991), 393–426.

[38] Ivrii V., Wave fronts of solutions of symmetric pseudodifferential systems, *Sibirsk. Math. J.*, **20**, (1979), 557–578 (in Russian).

[39] Kamke E., *Differentialgleichungen. Lösungsmethoden und Lösungen I*, Leipzig, (1959).

[40] Keller J., On solutions of nonlinear wave equations, *Comm. Pure Appl. Math.*, **10**, (1957), 523–530.

[41] Keller J. and Lu Ting, Singularities of semilinear waves, *Comm. Pure Appl. Math.*, **46**, (1993), 341–352.

[42] Klainerman S. and Majda A., Formation of singularities for wave equations including the nonlinear vibrating string, *Comm. Pure Appl. Math.*, **33**, (1980), 241–263.

[43] Krasnoselskij M., *Positive Solutions of Operator Equations*, Groningen, Noordhoff, (1964).

[44] Krasnoselskij M. and Zabreiko P., *Geometrical Methods of Nonlinear Analysis*, Springer Verlag, (1984).

[45] Lax P.D., Development of singularities of solutions of nonlinear hyperbolic partial differential equations, *J. Math. Ph.*, **5 (5)**, (1964), 611–613.

[46] Lax P.D., Hyperbolic systems of conservation laws II, *Comm. Pure Appl. Math.*, **10**, (1957), 537–566.

[47] Lebeau G., *Front d'Ondes des Fonctions Non-linéaires et Polynômes*, Séminaire E.D.P., Ecole Polytechnique, (1988–1989), **No 10**.

[48] Li Ta-tsien, *Global Classical Solutions for Quasilinear Hyperbolic Systems*, Research in Applied Mathematics, John Wiley & Sons, Masson, (1994).

[49] Liu T.P. and Xin Z., Overcompressive shock waves, Nonlinear evolution equations that change type, *The IMA volumes in Math. and its Appl.*, **27**, (1991), Springer Verlag, 139–145.

[50] Lokshin A. and Sagomonian E., *Nonlinear Waves in the Mechanics of Solid Bodies*, Edition of Moscow State Univ., (1989) (in Russian).

[51] Majda A., *Compressible Fluid Flow and Systems of Conservation Laws in Several Space Variables*, Springer Verlag, (1984).

[52] Manfrin R., *A Note on the Formation of Singularities for Quasilinear Hyperbolic Systems*, Preprint D.C.A. Ist. Univ. Arch. di Venezia, Febbraio 1998 and a paper in SIAM J. Math. Analysis, **32**, (2000), 261–290.

[53] Melrose R., Conormal rings and semilinear wave equations, *Advances in Microlocal Analysis*, D. Reidel Publishing Company, (1986), 225–251.

[54] Melrose R. and Ritter N., Interaction of nonlinear progressing waves for semilinear equations I, *Ann. of Math.*, **121**, (1985), 187–213.

[55] Melrose R. and Ritter N., Interaction of nonlinear progressing waves for semilinear equations II, *Arkiv for Math.*, **25**, (1987), 91–114.

[56] Messer Th., Propagation of singularities of hyperbolic systems, *Indiana Univ. Math. J.*, **36**, (1987), 45–77.

[57] Micheli L., Propagation of singularities for nonstrictly hyperbolic semilinear systems in one space dimension, *Transactions Amer. Math. Soc.*, **291**, (1985), 451–485.

[58] Mihlin S., *Linear Partial Differential Equations*, Vishaia Shkola, Moscow, (1977) (in Russian).

[59] Moser J., A rapidly convergent iteration method and non-linear differential equations I, *Anali Scuola Norm. Sup. Pisa I*, **20**, (1966), 265–315.

[60] Moser J., A rapidly convergent iteration method and non-linear differential equations II, *Anali Scuola Norm. Sup. Pisa I*, **20**, (1966), 499–535.

[61] Naumkin P. and Shishmarev I., *Nonlinear nonlocal equations in the theory of waves*, Translations of Mathematical Monographs, **133**, American Math. Soc., (1994).

[62] Olver F., *Introduction to Asymptotics and Special Functions*, Academic Press, New York & London, (1974).

[63] Popivanov P., Removable singularities of a class of fully nonlinear partial differential equations, *Nonlinear Analysis*, TMA, **18**, (1992), 851–859.

[64] Popivanov P., Wave fronts of solutions of some classes of nonlinear partial differential equations, *Banach Center Publ.*, **27**, (1992), 361–366.

[65] Popivanov P., Wave fronts of the solutions of fully non-linear symmetric positive systems of partial differential equations, *Bolletino U.M.I.*, **7-B**, (1993), 643–652.

[66] Popivanov P., Nonexistence of isolated singularities for non-linear systems of partial differential equations and some applications, *Journal of Math. Kyoto Univ.*, **37**, (1997), 477–492.

[67] Popivanov P., Travelling waves for a class of non-linear integrodifferential operators arising in solid body mechanics, *Quaderni del Dip. di Mat.*, Univ. di Torino, **52**, (1999), 1–8.

[68] Popivanov P. and Slavova A., Blow up of the solutions for quasilinear hyperbolic homogeneous systems in plane, *C. R. Acad. Bulg. Sci.*, **53**, (2000), 11–14.

[69] Rauch J. and Reed M., Propagation of singularities for semilinear hyperbolic equations in one space variable, *Ann. of Math.*, **111**, (1980), 531–552.

[70] Rauch J. and Reed M., Jump discontinuities of semilinear strictly hyperbolic systems in two variables: creation and propagation, *Comm. Math. Ph.*, **87**, (1981), 203–227.

[71] Rauch J. and Reed M., Propagation of singularities to nonstrictly hyperbolic semilinear systems; examples, *Comm. Pure Appl. Math.*, **35**, (1982), 555–565.

[72] Rauch J. and Reed M., Nonlinear microlocal analysis of semilinear hyperbolic systems in one space dimension, *Duke Math. J.*, **49**, (1982), 397–475.

[73] Reed M., Propagation of singularities for non-linear waves in one space dimension, *Comm. in P.D.E.*, **3**, (1978), 153–199.

[74] Rozhdestvenskij B. and Yanenko N., *Systems of Quasilinear Equations and Their Applications to Gas Dynamics*, Transl. Math. Monographs, **55**, A.M.S., Providence, RI, (1983).

[75] Sa Barreto A., Evolution of semilinear waves with swallow tail singularities, *Duke Math. J.*, **75**, (1994), 645–710.

[76] Shubin M., *Pseudodifferential Operators and Spectral Theory*, "Nauka", Moscow, (1978) (in Russian).

[77] Smirnov M., *Differential Equations of Mixed Type*, Vishaia Shkola, (1985) (in Russian).

[78] Stein E., *Singular Integrals and Differentiability Properties of Functions*, Princeton Univ. Press, (1970).

[79] Vullova V., Semilinear weakly hyperbolic system; anomalous singularities of the solution, *C. R. Acad. Bulg. Sci.*, **46**, (1993), 13–16.

[80] Whittaker E. and Watson G., *A Course of Modern Analysis*, Cambridge Univ. Press, (1927).

[81] Williams M., Spreading of singularities at the boundary in semilinear hyperbolic mixed problems II: crossing and self spreading, *Trans. of the A.M.S.*, **311**, (1989), 291–324.

[82] Xu C.-J., in: *Nonlinear microlocal analysis*, Pitman Res. Notes in Math. Series, **349**, Longman, (1996), 155–182.

P.R. Popivanov
Institute of Mathematics and Informatics
Bulgarian Academy of Sciences
1113 Sofia, Bulgaria

Operator Theory:
Advances and Applications, Vol. 145, 95–155

From Wave to Klein–Gordon Type Decay Rates

Fumihiko Hirosawa and Michael Reissig

Abstract. The goal of the paper is to derive $L_p - L_q$ decay estimates for Klein–Gordon equations with time-dependent coefficients. We explain the influence of the relation between the mass term and the wave propagation speed on $L_p - L_q$ decay estimates. Contrary to the classical Klein–Gordon case we cannot expect in each case a Klein–Gordon type decay rate. One has wave type decay rates, too. Moreover, under certain assumptions no $L_p - L_q$ decay estimates can be proved. In these cases the solution has a Floquet behavior. More precisely, one can show that the energy cannot be estimated from above by time-dependent functions with a suitable growth order if t tends to infinity.

1. Introduction

1.1. Preliminaries on decay properties for hyperbolic equations

There exists a well-known approach to prove global existence of small data solutions for the Cauchy problem (see e.g., [11], [19]). This approach was applied to quite different nonlinear problems, we will give some comments on how to apply it to the nonlinear model problem

$$\begin{cases} u_{tt} - \Delta u = f(\nabla u, u_t, \nabla u_t, \nabla^2 u), & (t,x) \in (0,\infty) \times \mathbb{R}^n , \\ u(0,x) = \varphi(x), \; u_t(0,x) = \psi(x) , & x \in \mathbb{R}^n. \end{cases} \tag{1.1}$$

The basic tool is the so-called Strichartz decay estimate [27] for the solutions of

$$\begin{cases} u_{tt} - \Delta u = 0, & (t,x) \in (0,\infty) \times \mathbb{R}^n , \\ u(0,x) = \varphi(x), \; u_t(0,x) = \psi(x) , & x \in \mathbb{R}^n, \end{cases}$$

which has the following form:

$$\|u_t(t,\cdot)\|_{L_q} + \|\nabla_x u(t,\cdot)\|_{L_q} \le C(1+t)^{-\frac{n-1}{2}(\frac{1}{p}-\frac{1}{q})}\Big(\|\varphi\|_{W_p^{N_p+1}} + \|\psi\|_{W_p^{N_p}}\Big),$$

where $\frac{1}{p}+\frac{1}{q} = 1$, $1 \le p \le 2$, $N_p > n(\frac{1}{p}-\frac{1}{q})$. This estimate expresses two important facts:

2000 *Mathematics Subject Classification.* 35L05, 35L15, 35Q40 .
Key words and phrases. $L_p - L_q$ decay estimates, Cauchy problem, Klein–Gordon equations, Fourier multipliers, Floquet theory, stationary phase method, oscillating integrals .
This work was completed with the support of our TEX-pert.

- If $p=q$, then this estimate follows from conservation of energy. One can even use N_2=0.
- But if we suppose some regularity of the data, then it is possible to estimate the energy based on the L_q-norm by the energy based on the $W_p^{N_p}$-norm and one gains additionally a factor which decays if $n \geq 2$ and $p \neq 2$.

But how can such an estimate be used? It is more or less standard to prove the local existence in time of the solution for problem (1.1) in some evolution space of low regularity $C^0\Big([0,T], W_2^{r+1}(\mathbb{R}^n)\Big) \cap C^1\Big([0,T], W_2^r(\mathbb{R}^n)\Big)$. But if we suppose higher regularity for the data, then this regularity should be transferred to the solution. This can be obtained in principle from the energy estimate of higher order

$$\begin{aligned}&\|u_t(t,\cdot)\|_{W_2^s} + \|\nabla_x u(t,\cdot)\|_{W_2^s}\\ &\leq C_0 \exp\Big(C_1 \int_0^t \Big\|\Big((\nabla,\partial_\tau)^2 u(\tau,\cdot)\Big)\Big\|_{L_\infty} d\tau\Big)\Big(\|\varphi\|_{W_2^{s+1}} + \|\psi\|_{W_2^s}\Big),\end{aligned}$$

where we supposed quadratic growth of the right-hand side from (1.1). Applying Sobolev's embedding theorem and the differential equation one can conclude

$$\|(\nabla,\partial_t)^2 u(t,\cdot)\|_{L_\infty} \leq C_1 \|(\nabla,\partial_t) u(t,\cdot)\|_{W_4^{s_1}}$$

with a suitable $s_1 = s_1(n)$. Now due to the Strichartz decay estimate we have

$$\sup_{0\leq t\leq T} (1+t)^{\frac{n-1}{4}} \|(\nabla,\partial_t) u(t,\cdot)\|_{W_4^{s_1}} \leq C.$$

This means that we can produce a term like $(1+t)^{-\frac{n-1}{4}}$ in the integral appearing in higher order energy estimates. If $n \geq 6$, then this term is integrable over $(0,\infty)$. This allows us to continue the solution step by step for small data. Here the autonomous character of (1.1) is very helpful because the local existence time interval of the solution with respect to t is independent of the hyperplane $t = t_0$ where we prescribe the data.

One of the main observations of this approach is that one can use $L_p - L_q$ decay estimates for suitable linear partial differential equations or systems to prove global-in-time existence results for nonlinear problems. In the case of the wave equation, the Strichartz decay estimate can be obtained from Kirchhoff's explicit representation of solution [19]. Let us demonstrate it in the three-dimensional case for the solution

$$u = u(t,x) = \tfrac{t}{4\pi} \int_{S^2} \psi(x+ty)dy$$

of

$$\begin{cases} u_{tt} - \Delta u = 0, & (t,x) \in (0,\infty)\times\mathbb{R}^3\,, \\ u(0,x) = 0,\ u_t(0,x) = \psi(x)\,, & x \in \mathbb{R}^3. \end{cases} \tag{1.2}$$

To estimate the energy based on the L_∞-norm we have to study

$$\Big|\int_{S^2}\psi(x+ty)dy\Big|,\quad \Big|t\int_{S^2}\nabla\psi(x+ty)dy\Big|.$$

It is useful to split the considerations for $t\in[1,\infty)$ and for $t\in[0,1]$. If $t\in[1,\infty)$, then we use

$$\int_{S^2}\nabla^k\psi(x+ty)dy=-\int_{|y|>t}|y|^{-3}\nabla^k(\nabla\psi(x+y)y)dy\quad (k=0,1).$$

Hence we get in this case

$$\|u_t(t,\cdot)\|_{L_\infty}+\|\nabla_x u(t,\cdot)\|_{L_\infty}\le t^{-1}\|\psi\|_{W_1^2}.$$

If $t\in[0,1]$, then partial integration and transformation of variables yield

$$\begin{aligned} &t^k\int_{S^2}\nabla^k\psi(x+ty)dy\\ &=(-1)^{k+1}\int_{|y|>t}\frac{(|y|-t)^{2-k}}{2^{k-1}|y|^{5-k}}\sum_{|\alpha|=3-k}y_\alpha\left(\partial^\alpha\nabla^k\right)\psi(x+y)dy\quad (k=0,1).\end{aligned}$$

Hence we get in this case

$$\|u_t(t,\cdot)\|_{L_\infty}+\|\nabla_x u(t,\cdot)\|_{L_\infty}\le C\|\psi\|_{W_1^3}.$$

Both estimates together give the L_1-L_∞ estimate

$$\|u_t(t,\cdot)\|_{L_\infty}+\|\nabla_x u(t,\cdot)\|_{L_\infty}\le (1+t)^{-1}\|\psi\|_{W_1^3}.$$

Together with the energy conservation (L_2-L_2 estimate), a general interpolation result from [1] and the information, that the interpolation space $[L_\infty,L_2]_\theta=L_{q_\theta}$, $q_\theta=2/\theta$, $\theta\in[0,1]$, we conclude the Strichartz decay estimate for the solution of (1.2).

We understand that considerations for small t give the necessary regularity of the data and for large t bring the decay rate.

Another way to derive this estimate is to use the Fourier multiplier representation of the solution [1], [18]. It is sufficient to study as a model Fourier multiplier

$$F^{-1}\left(e^{it|\xi|}|\xi|^{-2l}F(\varphi(y))(\xi)\right),\ l\ge 0. \tag{1.3}$$

If we can obtain an L_p-L_q estimate for $t=1$, then the transformations $t\xi=\eta$ and $tz=y$ give immediately L_p-L_q estimates for each $t>0$. First we localize by a suitable cut-off function χ our considerations to small $|\xi|$, that is, we study

$$F^{-1}\left(e^{i|\xi|}|\xi|^{-2l}\chi(|\xi|)F(\varphi(y))(\xi)\right),\ l\ge 0.$$

If $l>0$, then the kernel becomes singular at $|\xi|=0$. The theorem of Hardy–Littlewood [8] tells us, that there exists a relation between the order of singularity

$2l$, the dimension n, and the goal to derive $L_p - L_q$ estimates, namely $\frac{1}{p} - \frac{1}{q} \geq \frac{2l}{n}$, $1 < p \leq 2 \leq q < \infty$. Under these assumptions we have

$$\left\| F^{-1}\left(e^{i|\xi|}|\xi|^{-2l}\chi(|\xi|)F(\varphi(y))(\xi)\right)\right\|_{L_q} \leq C\|\varphi\|_{L_p}.$$

To study

$$F^{-1}\left(e^{i|\xi|}|\xi|^{-2l}(1-\chi(|\xi|))F(\varphi(y))(\xi)\right),\ l \geq 0,$$

we derive at first an estimate for

$$\left\| F^{-1}\left(e^{i|\xi|}|\xi|^{-2l}(1-\chi(|\xi|))\phi_k(\xi)\right)\right\|_{L_\infty},$$

where $\{\phi_k\}_{k\geq 0}$ is a dyadic decomposition of the phase space $\mathbb{R}^n$. Some transformation depending on k transfers, for all k, these integrals to an integral over a compact set in the phase space away from $\xi = 0$. This allows us to apply Littman's lemma [14], where we use that the rank of the Hessian in stationary points of the phase function $x \cdot \xi + |\xi|$ is equal to $n-1$. The obtained L_∞ estimates can be used to derive $L_2 - L_2$ and $L_1 - L_\infty$ estimates and finally after interpolation $L_p - L_q$ estimates of the following form, $\frac{1}{p} + \frac{1}{q} = 1$, $1 \leq p \leq 2$:

$$\left\| F^{-1}\left(e^{i|\xi|}|\xi|^{-2l}(1-\chi(|\xi|))\phi_k(\xi)F(\varphi(y))(\xi)\right)\right\|_{L_q} \leq C 2^{k(\frac{n+1}{2}(\frac{1}{p}-\frac{1}{q})-2l)}\|\varphi\|_{L_p}.$$

From [1] it follows that

$$\left\| F^{-1}\left(e^{i|\xi|}|\xi|^{-2l}(1-\chi(|\xi|))F(\varphi(y))(\xi)\right)\right\|_{L_q} \leq C\|\varphi\|_{L_p}$$

if $2l \geq \frac{n+1}{2}(\frac{1}{p}-\frac{1}{q})$, but now for $\frac{1}{p}+\frac{1}{q} = 1$, $1 < p \leq 2$. Summarizing these estimates gives

$$\left\| F^{-1}\left(e^{i|\xi|}|\xi|^{-2l}F(\varphi(y))(\xi)\right)\right\|_{L_q} \leq C\|\varphi\|_{L_p}$$

if $\frac{n+1}{2}(\frac{1}{p}-\frac{1}{q}) \leq 2l \leq n(\frac{1}{p}-\frac{1}{q})$, $\frac{1}{p}+\frac{1}{q} = 1$, $1 < p \leq 2$. Finally, the above-mentioned transformation to include the t-variable into the phase leads to the $L_p - L_q$ estimate

$$\left\| F^{-1}\left(e^{it|\xi|}|\xi|^{-2l}F(\varphi(y))(\xi)\right)\right\|_{L_q} \leq C|t|^{2l-n(\frac{1}{p}-\frac{1}{q})}\|\varphi\|_{L_p}.$$

If $t \in [0,1]$, then we choose $2l = n(\frac{1}{p}-\frac{1}{q})$, which gives the regularity $N_p \geq n(\frac{1}{p}-\frac{1}{q})$ of the data. If $t \in [1,\infty)$, then we choose $2l = \frac{n+1}{2}(\frac{1}{p}-\frac{1}{q})$, which gives the decay rate $\frac{n-1}{2}(\frac{1}{p}-\frac{1}{q})$.

We understand that considerations for small $|\xi|$ give the necessary regularity of the data and for large $|\xi|$ bring the decay rate.

Naturally, this approach allows more precise considerations and results. Moreover, for more general model equations one cannot expect the existence of explicit representations of solutions, but for a wide class of model equations it is reasonable to ask for representations of its solutions by Fourier multipliers.

If we are interested in $L_p - L_q$ decay estimates for the solutions of the Cauchy problem for the classical Klein–Gordon equation, $u_{tt} - \Delta_n u + m^2 u = 0$, m is a positive constant, then there exist at least two ways to derive those. On the one hand one can transform the Klein–Gordon equation to a wave equation with non-standard data [29]. This can be realized via the transformation

$$v(t, x_1, \dots, x_n, x_{n+1}) := e^{-imx_{n+1}} u(t, x_1, \dots, x_n),$$

where v solves $v_{tt} - \Delta_{n+1} v = 0$. But a new problem arises from the structure of the data

$$\begin{aligned} v(0, x_1, \dots, x_n, x_{n+1}) &= e^{-imx_{n+1}} u(0, x_1, \dots, x_n), \\ v_t(0, x_1, \dots, x_n, x_{n+1}) &= e^{-imx_{n+1}} u_t(0, x_1, \dots, x_n). \end{aligned}$$

Thus one should repeat considerations for wave equations with these non-standard data.

On the other hand one can use Fourier multiplier representations with mass included into the phase [9]. Therefore one has to study the model Fourier multiplier

$$F^{-1}\left(e^{it\langle\xi\rangle_m} \langle\xi\rangle_m^{-2l} F(\varphi(y))(\xi)\right), \; l \geq 0,$$

where $\langle\xi\rangle_m = \sqrt{|\xi|^2 + m^2}$. We will not explain in detail how to handle such a Fourier multiplier. In Sections 6.1 and 6.2 we study more complicated Fourier multipliers appearing from Klein–Gordon type equations. We want only to emphasize, that such a representation has disadvantages; the phase function is not homogeneous, the influence of the Hardy–Littlewood lemma remains even if we have no singularities, but t cannot considered as a parameter. The advantage is that the application of Littman's lemma gives instead of $\frac{n-1}{2}$ the factor $\frac{n}{2}$ in the decay rate, because now the rank of the Hessian is n in a stationary point. The same improvement can be considered by the above described reduction from Klein–Gordon to wave equations.

We want to emphasize once more that the inclusion of the mass term into the phase gives us the possibility to derive the Klein–Gordon decay rate by applying the stationary phase method. But the influence of the Hardy–Littlewood lemma remains.

The second way seems to be more transparent and available and gives e.g., $L_p - L_q$ decay estimates for solutions of elasticity [26], crystal optics [13], thermoelasticity [32].

Summarizing one gets the impression that most of the authors derived $L_p - L_q$ decay estimates for model equations or systems with constant coefficients. If one is interested in studying linear equations or systems with variable coefficients, then new difficulties appear. A first step of investigation toward the solutions is to derive energy estimates ($L_2 - L_2$ estimates) taking into account the influences of mass and dissipation.

The weighted energy estimates which are adopted in [7], [16] and [17] are useful to determine some asymptotic behavior for the L_2 energy to the solution. Let us consider the equation

$$u_{tt} - \Delta u + 2b(t, x)u_t + c(t, x)u = 0 \tag{1.4}$$

in $[0,\infty)\times\mathbb{R}^n$, where non-negative functions $b=b(t,x)$ and $c=c(t,x)$ are dissipation and mass, respectively. Let $g=g(t,x)$ and $h=h(t,x)$ be positive weight functions. Multiplying both sides of (1.4) by gu or hu_t and integrating over $(0,t)\times\mathbb{R}^n$ we arrive at the following identity:

$$\mathcal{E}(t;g,h)+\int_0^t \mathcal{G}(s;g,h)ds=\mathcal{E}(0;g,h)$$

under some suitable conditions to u, g and h as $|x|\to\infty$, where $\mathcal{E}(t;g,h)$ and $\mathcal{G}(t;g,h)$ are weighted energies defined by

$$\mathcal{E}(t;g,h):=\int_{\mathbb{R}^n} h\left(|\nabla_x u|+\left(u_t+\frac{g}{h}u\right)^2\left(c+\frac{2bgh-g_th-g^2}{h^2}\right)u^2\right)dx$$

and

$$\begin{aligned}\mathcal{G}(t;g,h) \quad := \quad & \int_{\mathbb{R}^n}\left((2g-h_t)\left(|\nabla_x u|^2+cu^2\right)+(-2g-h_t+4bh)\,u_t^2\right)dx\\ & -\int_{\mathbb{R}^n}\Big(2b_tg+2bg_t-g_{tt}+c_th\Big)u^2dx,\end{aligned}$$

respectively. Thus the decay order of the energy should be determined by the weight function h. But we also know that the relation between the weight functions g, h and the coefficients b, c are essential for decay, in other words, the weighted energies $\mathcal{E}$ and $\mathcal{G}$ should be positive. Indeed, if b and c are positive constants, then one can choose the weight function h with exponential order growth. On the other hand, if b decreases like t^{-1} and $c=0$, then one cannot choose any faster increasing weight function than those of polynomial order (see also [28], the constant dissipation case is minutely studied in [10]). Constant mass gives no effect for energy decay without dissipation, but the mass effect is essential for the decay order of the energy if the equation has a dissipative term.

Time-dependent propagation speed makes the problem more difficult, especially an oscillating coefficient. Increasing propagation speed has an effect for energy increasing. Therefore, if one proves some decay estimates, then we should consider a balance between the coefficients. Let us consider the equation

$$u_{tt}-a(t)\Delta u+2b(t,x)u_t+c(t,x)u=0 \tag{1.5}$$

which is a generalization of (1.4) to the variable propagation speed case, where $a=a(t)>0$. The Liouville transformation makes the equation (1.4) simpler. Setting $\tau(t):=\int_0^t\sqrt{a(s)}ds$ and $v(\tau(t),x)=u(t,x)$ the equation (1.5) is rewritten as

$$v_{\tau\tau}-\Delta v+2\left(\frac{b}{\sqrt{a}}+\frac{a_t}{4a\sqrt{a}}\right)v_\tau+\frac{c}{a}v=0, \tag{1.6}$$

where $t(\tau) = \tau^{-1}(\tau)$. If a is not oscillating, then (1.6) can be reduced to (1.4). But if a is oscillating, then such a reduction is not always applicable. Moreover, if a vanishes, such a transformation is ineffective (in [5] even some degenerate hyperbolic problems are considered).

One of the merits of the energy method is that one can prove sharp estimates, that is, estimates from above and below, whereas L_2–L_2 estimates are less powerful for dealing with nonlinear problems.

1.2. $L_p - L_q$ decay for hyperbolic equations with variable coefficients

To have the possibility to enlarge the class of problems to which one can apply the above-mentioned approach for proving global small data solutions, the second author began together with *K. Yagdjian (Armenia) from Kansas State University* to study the question for $L_p - L_q$ decay estimates (not only for $L_2 - L_2$ estimates as it was done by several authors before) for wave equations with time-dependent coefficients on the conjugate line ($p + q = pq$). In a series of papers it was shown that this question is quite delicate. Let us explain some of the results for the wave equation with mass term

$$\begin{cases} u_{tt} - a(t) \triangle u + c(t)u = 0, & (t,x) \in (0,\infty) \times \mathbb{R}^n , \\ u(0,x) = \varphi(x), \; u_t(0,x) = \psi(x) , & x \in \mathbb{R}^n, \end{cases}$$

where $a = a(t)$, $c = c(t) \in C^\infty[0,\infty)$, $a(t) > 0$ and $c(t) \geq 0$. It turns out that it is useful to write $a(t) = \lambda(t)^2 b(t)^2$, where $\lambda = \lambda(t)$ describes the increasing behavior (*improving influence on $L_p - L_q$ decay estimates*) and $b = b(t)$ describes the oscillating behavior (*deteriorating influence on $L_p - L_q$ decay estimates*) of the coefficient a.

Then one has to explain the interplay between both parts. This can be realized by the condition

$$|D_t b(t)| \leq \frac{C\lambda(t)}{\Lambda(t)^\beta \Big(\log \Lambda(t) \Big)^\omega} \qquad \text{for large } t , \tag{1.7}$$

where $D_t = -i\frac{\partial}{\partial t}$, $\Lambda(t) = \int_0^t \lambda(s)ds$, $\omega = 0$ if $\beta \in [0,1)$ and $\omega \in (-\infty, 0]$ if $\beta = 1$. If $\beta = 1$ and $\omega \in (-1,0]$, then in [21] it was shown, that under some conditions the dominating influence of the increasing part guarantees an $L_p - L_q$ *decay estimate of wave-type*:

$$\begin{aligned} &\|u_t(t,\cdot)\|_{L_q} + \|\lambda(t)\nabla_x u(t,\cdot)\|_{L_q} \\ &\leq C_\varepsilon \Big(1+\Lambda(t)\Big)^{1+\varepsilon-\frac{n-1}{2}(\frac{1}{p}-\frac{1}{q})} \Big(\|\varphi\|_{W_p^{L+1}} + \|\psi\|_{W_p^L} \Big), \end{aligned} \tag{1.8}$$

where ε is an arbitrary positive real number and L is a positive constant depending on n, p and q. Thus the classical wave decay rate is reproduced (see [27]) up to an unessential error ε. But if $\beta < 1$, or $\beta = 1$ and $\omega < -1$, then one cannot expect an $L_p - L_q$ decay estimate for the wave equation with vanishing mass (see [22]).

From [23] one understands that a mass term can have a stabilizing effect. If $c(t) = \lambda(t)^2 b(t)^2$, then the $L_p - L_q$ *decay estimate of Klein–Gordon type*:

$$\|u_t(t,\cdot)\|_{L_q} + \|\lambda(t)\nabla_x u(t,\cdot)\|_{L_q} \leq C\sqrt{\lambda(t)}\Big(1+\Lambda(t)\Big)^{-\frac{n}{2}(\frac{1}{p}-\frac{1}{q})}\Big(\|\varphi\|_{W_p^{L+1}} + \|\psi\|_{W_p^L}\Big) \tag{1.9}$$

holds if $\beta \in (\frac{1}{2}, 1]$ and $\omega \in (-\infty, 0)$. Thus the classical Klein–Gordon decay rate is reproduced (see [9], [29]). By studying a special example it was shown that the Klein–Gordon decay rate is sharp (see [4]).

The case of wave equations with a dissipation, that is,

$$\begin{cases} u_{tt} - a(t)\triangle u + d(t)u_t = 0, & (t,x) \in (0,\infty)\times\mathbb{R}^n\,, \\ u(0,x) = \varphi(x),\ u_t(0,x) = \psi(x)\,, & x \in \mathbb{R}^n, \end{cases}$$

is studied in [20]. It is shown that there are similar results as in the Klein–Gordon case, only the methods to prove these are different.

The goal of the present paper is to study the Cauchy problem for

$$u_{tt} - \lambda(t)^2 b(t)^2 \triangle u + \frac{\lambda(t)^2 b(t)^2}{(e+\Lambda(t))^{2\gamma}\Big(\log(e+\Lambda(t))\Big)^{2\delta}}\, u = 0 \tag{1.10}$$

under the condition (1.7) for $\beta \in [0,1]$, $\gamma \geq 0$ and $\delta \neq 0$ only for $\gamma = 1$. For $\gamma = 0$ we have (1.9); for $\gamma = +\infty$, due to our arrangement this means no mass, we have (1.8) if $\omega \in (-1, 0]$. It seems to be reasonable and we will follow this strategy that the oscillating behavior in the mass term coincides with that from the main part.

1.3. Some preparations to the approach

What kind of results do the readers expect?

Cases $\beta \in (\frac{1}{2}, 1)$; $\beta = 1$, $\omega < -1$:

- If γ is small, this means we have a bit smaller mass than in the case $\gamma = 0$, then one should expect a Klein–Gordon type decay rate.
- If γ is large, this means we have a small mass, then one cannot expect an $L_p - L_q$ decay estimate; maybe one has a non-decay result of Floquet-type which for a special case was studied in [22].

It should be interesting to describe the change-over from Klein–Gordon to Floquet, to find the critical mass $\gamma_0 = \gamma_0(\beta)$ and to study the influence of the mass on decay estimates (see Table 3 in Section 5 for $\Omega_{1,i}$).

Case $\beta = 1$, $\omega \in (-1, 0]$:

- If γ is small, then one should expect a Klein–Gordon type decay rate.
- If γ is large, then one should expect a wave decay rate.

It should be interesting to describe the change-over from Klein–Gordon to wave decay rates, to find the critical mass $\gamma_0 = \gamma_0(\beta)$ and to study the influence of the mass on decay estimates (see Table 3 in Section 5 for $\Omega_{2,i}$ and $\Omega_{3,i}$).

Case $\beta \in [0, \frac{1}{2})$**:**

- If we have no mass ($\gamma = \infty$), then one cannot expect an $L_p - L_q$ decay estimate.

It should be interesting to study what happens during the change-over from $\gamma = \infty$ *to* $\gamma = 0$. *Does a Floquet-type effect appear? What is the influence of the mass term (see Table* 3 *in Section* 5 *for* $\Omega_{1,i}$*)?*

We try to answer these questions and to describe the change-over to the critical cases. One of the results is the following theorem.

Theorem 1.1: *Let us consider the next Cauchy problem*

$$\begin{cases} u_{tt}(t,x) - (1+t)^{2l} b(t)^2 \Big(\triangle - (1+t)^{-2k} \Big) u(t,x) = 0, & (t,x) \in (0,\infty) \times \mathbb{R}^n, \\ u(0,x) = \varphi(x),\ u_t(0,x) = \psi(x)\,, & x \in \mathbb{R}^n, \end{cases}$$

where $l \geq 0$ *and* $k \geq 0$ *are real numbers and* $b = b(t)$ *is a* C^∞, *positive and periodic function on* $[0,\infty)$. *Then the following Klein–Gordon type decay estimate holds:*

$$\begin{aligned} &\|u_t(t,\cdot)\|_{L_q} + \|(1+t)^l \nabla_x u(t,\cdot)\|_{L_q} \\ &\leq C\sqrt{(1+t)^l}\Big((1+t)^{l+1}\Big)^{r-\frac{n}{2}(\frac{1}{p}-\frac{1}{q})}\Big(\|\varphi\|_{W_p^{N_p+1}} + \|\psi\|_{W_p^{N_p}}\Big), \end{aligned}$$

where $\frac{1}{p} + \frac{1}{q} = 1$, $1 < p \leq 2$, $N_p = \Big[n\Big(\frac{1}{p} - \frac{1}{q}\Big)\Big] + 1$ *and*

- $r = 0$ *if* $l > 1$ *and* $0 \leq k < l - 1$,
- r *is large if* $l \geq 1$ *and* $k = l - 1$.

If $l \geq 0$ *and* $k \geq l$, *then there exist no* $L_p - L_q$ *decay estimates. In this case one can show that the energy has to increase in a special way and that the energy estimate which follows from Gronwall's inequality is near to an optimal one.*

Remark 1.1: If $l \geq 0$ and $k \geq l$ the increasing behavior of the energy is described by Theorem 8.1. We will call this behavior *classical Floquet behavior* (see Definition 8.1). In the exceptional case: $l - 1 < k < l$, we have no result of $L_p - L_q$ decay. However, we know that the L_q energy of the solution is estimated to a better order from above than the classical Floquet estimate from below (see Remark 8.2). We expect an increasing behavior of the energy in the case $l - 1 < k < l$, too. For this reason we shall call such a behavior a *modified Floquet behavior.*

Now let us focus on the following Cauchy problem for second-order hyperbolic equations:

$$\begin{cases} u_{tt}(t,x) - \lambda(t)^2 b(t)^2 \left(\triangle - m(t)^2\right) u(t,x) = 0, & (t,x) \in (0,\infty) \times \mathbb{R}^n\,, \\ u(0,x) = \varphi(x),\ u_t(0,x) = \psi(x)\,, & x \in \mathbb{R}^n\,, \end{cases} \tag{1.11}$$

where $b = b(t)$, $\lambda = \lambda(t)$ and $m = m(t)$ satisfy the following conditions:

b • 1-periodic, non-constant, smooth and strictly positive.

λ • smooth and strictly positive,

• $\lambda'(t) \geq 0$ and there exist positive constants λ_k such that

$$|D_t^k \lambda(t)| \leq \lambda_k \left(\frac{\lambda(t)}{\Lambda(t)}\right)^k \lambda(t)$$

for large t and positive integer k.

m • smooth and strictly positive,

• $m'(t) \leq 0$ and there exist positive constants m_k such that

$$|D_t^k m(t)| \leq m_k \left(\frac{\lambda(t)}{\Lambda(t)}\right)^k m(t)$$

for large t and positive integer k.

Moreover, we introduce the following assumptions for relations between the coefficients:

(A.1) There exist positive constants b_k such that

(i) $|D_t^k b(t)| \leq b_k \left(\frac{\lambda(t)}{\Lambda(t)^\beta}\right)^k, \qquad \beta \in [0,1)$,

or

(ii) $|D_t^k b(t)| \leq b_k \left(\frac{\lambda(t)}{(e_0+\Lambda(t))\left(\log(e_0+\Lambda(t))\right)^\omega}\right)^k, \qquad \omega \in [-1,0]$, for large t and every positive integer k, where e_0 is a real number satisfying $e_0 \geq e$.

(A.2) The function $m = m(t)$ is defined by $\Lambda = \Lambda(t)$ as follows:

(i) $m(t) = \left(e_0 + \Lambda(t)\right)^{-\gamma}, \qquad \gamma \geq 0,\ \gamma \neq 1$,

or

(ii) $m(t) = \left(e_0 + \Lambda(t)\right)^{-1}\left(\log(e_0+\Lambda(t))\right)^{-\delta}, \qquad \delta \in \mathbb{R}$.

Remark 1.2: These functions $m = m(t)$ satisfy the above assumptions. But it is not necessary that $m = m(t)$ is explicitly described as in (A.2). However, we shall consider only this case for simplicity.

Remark 1.3: The restriction in (A.1) to the parameters β and ω is natural from the point of view of $L_p - L_q$ non-decay ([22]). But the periodicity of b is necessary only for the application of Floquet theory. To derive $L_p - L_q$ decay estimates it is sufficient to suppose (A.1).

We shall introduce two typical examples for λ and m which satisfy the preceding conditions.

Example 1.1: Let $\lambda(t) = (1+t)^l$ ($l \geq 0$). Then

$$m(t) \sim C(1+t)^{-\gamma(l+1)} \qquad \text{for (A.2)-(i)},$$
$$m(t) \sim C(1+t)^{-(l+1)}\left(\log(e_0+t)\right)^{-\delta} \qquad \text{for (A.2)-(ii)}.$$

Noting that $\frac{\lambda(t)}{\Lambda(t)^\beta} = (l+1)^\beta(1+t)^{l-\beta(l+1)}$ the assumption (A.1)-(i) holds for $\beta \le \frac{l}{l+1}$, that is, $l \ge \frac{\beta}{1-\beta}$.

Example 1.2: Let $\lambda(t) = e^{t^\alpha}$, $\alpha \in (0,1]$. Then

$$m(t) \sim C(1+t)^{-\gamma(1-\alpha)}e^{-\gamma t^\alpha} \quad \text{for (A.2)-(i)},$$
$$m(t) \sim C(1+t)^{-(1-\alpha)-\alpha\delta}e^{-t^\alpha} \quad \text{for (A.2)-(ii)}.$$

Noting that

$$\frac{\lambda(t)}{(e_0+\Lambda(t))\Big(\log(e_0+\Lambda(t))\Big)^\omega} \sim C(e_0+t)^{\alpha-1-\alpha\omega},$$

the assumption (A.1)-(ii) holds for $\omega \le \frac{\alpha-1}{\alpha}$, that is, $\alpha \ge \frac{1}{1-\omega}$.

Let us separate the following areas of

$$\Omega := \big\{(\beta,\gamma,\omega,\delta) \in [0,1]\times[0,\infty]\times(-\infty,0]\times(-\infty,\infty) \subset \mathbb{R}^4\big\} :$$

(I) $\beta < 1,\ \omega = \delta = 0$:
$\Omega_{1.1} = \{(\beta,\gamma,0,0) \in \Omega\ ;\ \gamma < 2\beta - 1\}$,
$\Omega_{1.2} = \{(\beta,\gamma,0,0) \in \Omega\ ;\ \gamma = 2\beta - 1\}$,
$\Omega_{1.3} = \{(\beta,\gamma,0,0) \in \Omega\ ;\ 2\beta - 1 < \gamma < \beta\}$,
$\Omega_{1.4} = \{(\beta,\gamma,0,0) \in \Omega\ ;\ \beta = \gamma\}$,
$\Omega_{1.5} = \{(\beta,\gamma,0,0) \in \Omega\ ;\ \beta < \gamma\}$.

(II) $\beta = 1,\ \gamma \ne 1, \delta = 0$:
$\Omega_{2.1} = \{(1,\gamma,\omega,0) \in \Omega\ ;\ \gamma > 1,\ \omega = 0\}$,
$\Omega_{2.2} = \{(1,\gamma,\omega,0) \in \Omega\ ;\ \gamma > 1,\ -1 < \omega < 0\}$,
$\Omega_{2.3} = \{(1,\gamma,\omega,0) \in \Omega\ ;\ \gamma > 1,\ \omega = -1\}$,
$\Omega_{2.4} = \{(1,\gamma,\omega,0) \in \Omega\ ;\ \gamma < 1,\ \omega \le 0\}$,
$\Omega_{2.5} = \{(1,\gamma,\omega,0) \in \Omega\ ;\ \gamma > 1,\ \omega < -1\}$.

(III) $\beta = 1,\ \gamma = 1$:
$\Omega_{3.1} = \{(1,1,\omega,\delta) \in \Omega\ ;\ \delta < 2\omega - 1\}$,
$\Omega_{3.2} = \{(1,1,\omega,\delta) \in \Omega\ ;\ 2\omega - 1 \le \delta < 2\omega\}$,
$\Omega_{3.3} = \{(1,1,\omega,\delta) \in \Omega\ ;\ \delta = 2\omega,\ \delta < \omega\}$,
$\Omega_{3.4} = \{(1,1,\omega,\delta) \in \Omega\ ;\ 2\omega < \delta < \omega\}$,
$\Omega_{3.5} = \{(1,1,\omega,\delta) \in \Omega\ ;\ \omega = 0,\ \delta > 0\}$,
$\Omega_{3.6} = \{(1,1,\omega,\delta) \in \Omega\ ;\ -1 < \omega < 0,\ \delta > 0\}$,
$\Omega_{3.7} = \{(1,1,\omega,\delta) \in \Omega\ ;\ \omega = -1,\ \delta \ge 0 \text{ or } \omega > -1,\ \delta = 0\}$,
$\Omega_{3.8} = \{(1,1,\omega,\delta) \in \Omega\ ;\ \omega = \delta < 0\}$,
$\Omega_{3.9} = \{(1,1,\omega,\delta) \in \Omega\ ;\ \omega < -1,\ \omega < \delta\}$,
$\Omega_{3.10} = \{(1,1,\omega,\delta) \in \Omega\ ;\ -1 \le \omega < 0,\ \omega < \delta < 0\}$.

Let us explain, as the end of our Introduction, the character of the problem deriving $L_p - L_q$ decay estimates for solutions of (1.10). From the point of view of the authors this is a degenerate hyperbolic problem although the Cauchy problem itself is of strictly hyperbolic type. But there exists an effect similar to the *loss of*

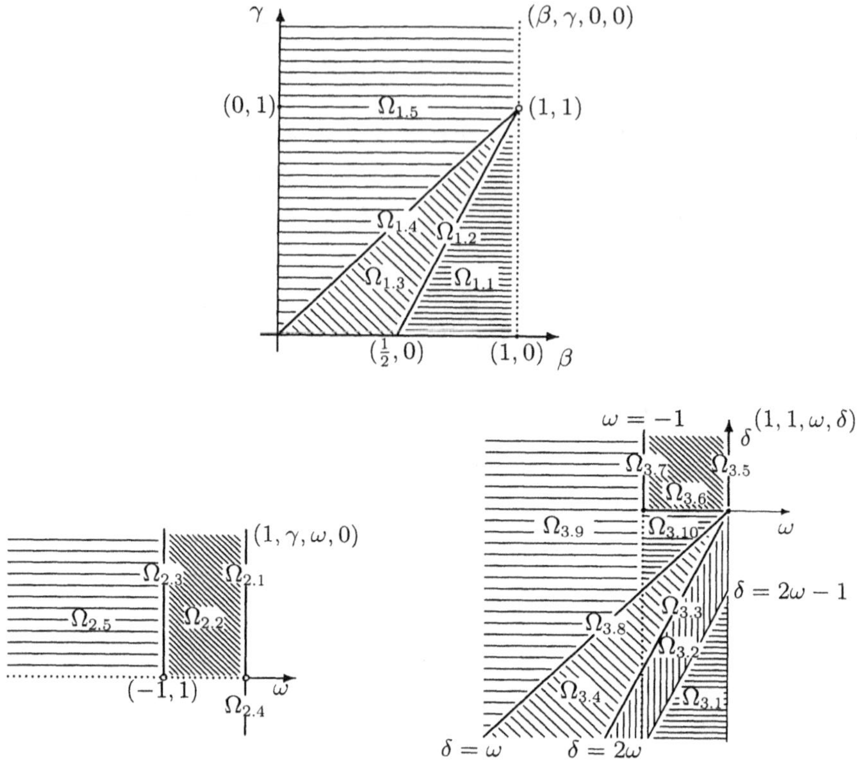

derivatives which is characteristic for such degenerate hyperbolic problems. The time-dependent coefficients don't allow us to consider t as a parameter as in the case of $L_p - L_q$ decay estimates for solutions of Cauchy problems with constant coefficients. Thus t should be included into the phase space, definition of zones, symbol classes etc. The characterization of oscillating behavior (1.7) and main tools of approach are similar to those used in [25] or [30] for weakly hyperbolic Cauchy problems or in [2], [6] or [12] for strictly hyperbolic problems with non-Lipschitz coefficients.

One observation from Table 3 *(Section* 5*) is that in the case of fast oscillations the decay rate differs from the classical decay rate by a constant. This corresponds to the finite loss of derivatives for weakly hyperbolic Cauchy problems and strictly hyperbolic Cauchy problems with non-Lipschitz coefficients. But if the oscillations are slow, then the decay rate differs from the classical one only by an arbitrary small constant* ε.

Especially the last explanations emphasize that all these degenerate hyperbolic problems have a common feature, but the tools should be chosen according to the special problem under consideration.

We will represent the solution and its first derivatives by Fourier multipliers of the form

$$F^{-1}\Big(e^{i\varphi(t,\xi)}a(t,\xi)F(\psi)\Big),$$

where F denotes the partial Fourier transformation with respect to space variables. Sections 2 to 5 are devoted to deriving WKB-solutions of the form $v = v(t,\xi) = e^{i\varphi(t,\xi)}a(t,\xi)v_0(\xi) + \cdots$. In Section 2 we introduce several parts of the phase space, so-called pseudodifferential zone $Z_{pd}(N)$ and hyperbolic zone $Z_{hyp}(N)$. Appropriate classes of amplitudes having symbol character only in $Z_{hyp}(N)$ are introduced. A diagonalization procedure in $Z_{hyp}(N)$ will be presented in Section 3. It gives the structure of φ and of the amplitude a. For Klein–Gordon type equations we include the mass into the phase. A hierarchy of symbol classes allows us to carry out an arbitrary number of steps of perfect diagonalization. This implies relations of the amplitude a and a finite number of its derivatives with respect to ξ to classical symbols. Careful estimates for the amplitude yield decay functions on the one hand and differences to the classical wave- or Klein–Gordon decay rate on the other hand (see Section 4 and Table 3 in Section 5). In Section 6 a special Littman type lemma (Theorem 6.1) allows us to deriving $L_p - L_q$ decay estimates of Klein–Gordon type. The main difficulties appear from the non-standard phase function

$$\varphi(t,\xi) = \int_0^t \lambda(s)b(s)\langle\xi\rangle_{m(s)}\,ds.$$

We have fewer problems with the phase function than we have in the wave-type case. Fourier multipliers with amplitudes supported in $Z_{pd}(N)$ or in $Z_{hyp}(N)$ are studied in Sections 6.3.1 and 6.3.2. The main tools are the Hardy–Littlewood lemma, dyadic decompositions of $Z_{hyp}(N)$, a Littman-type lemma and interpolation arguments. Combining all the results from Sections 2 to 6 gives the desired $L_p - L_q$ decay estimates in Section 7. Finally, counterexamples based on Floquet theory are considered in Section 8.

2. Tools of the approach

The approach to prove $L_p - L_q$ decay (or non-decay) estimates is related to the asymptotic behavior of the product between the increasing part λ and the mass m. In the case of constant mass $\lim\limits_{t\to\infty}\lambda(t)m(t) = 0$ when $m \equiv 0$ and $\lim\limits_{t\to\infty}\lambda(t)m(t) = \infty$ when $m = m_0 > 0$.

In the first case we have a wave-type equation and in the second case a Klein–Gordon type equation. The assumptions (A2)-(i) or (ii) guarantee that the limit $\lim\limits_{t\to\infty}\lambda(t)m(t)$ exists. Thus it seems to be natural to call the equation from (1.11) a *wave-type equation* if $\lim\limits_{t\to\infty}\lambda(t)m(t) = 0$ and a *Klein–Gordon type equation* if $\lim\limits_{t\to\infty}\lambda(t)m(t) = \infty$. In particular, we shall call the equation *of critical type* if $\lim\limits_{t\to\infty}\lambda(t)m(t) = C > 0$. Now let us illustrate this classification for our Examples

1.1 and 1.2 if the assumptions for b from Section 1 are satisfied. In the case of Example 1.1 we see that if

$$(\beta,\gamma,\omega,\delta)\in\Omega_{1.1}\cup\Omega_{1.2}\cup\Omega_{1.3},$$

then the equation is of Klein–Gordon type, if

$$(\beta,\gamma,\omega,\delta)\in\Omega_{1.5},$$

then the equation is of wave type, and if

$$(\beta,\gamma,\omega,\delta)\in\Omega_{1.4},$$

then the equation is of critical type.

In the case of Example 1.2 we see that if

$$(\beta,\gamma,\omega,\delta)\in\Omega_{2.4}\cup\Omega_{3.1}\cup\Omega_{3.2}\cup\Omega_{3.3}\cup\Omega_{3.4},$$

then the equation is of Klein–Gordon type, if

$$(\beta,\gamma,\omega,\delta)\in\Omega_{2.1}\cup\Omega_{2.2}\cup\Omega_{2.3}\cup\Omega_{2.5}\cup\Omega_{3.5}\cup\Omega_{3.6}\cup\Omega_{3.7}\cup\Omega_{3.9}\cup\Omega_{3.10},$$

then the equation is of wave type, and if

$$(\beta,\gamma,\omega,\delta)\in\Omega_{3.8},$$

then the equation is of critical type.

From now on we suppose that $(\beta,\gamma,\omega,\delta)\in\Omega\setminus(\Omega_{1.4}\cup\Omega_{1.5}\cup\Omega_{2.5}\cup\{\Omega_{3.8}:\delta<-1\}\cup\Omega_{3.9})$, because we expect a classical Floquet behavior (see Section 8) in these excluded parts of Ω.

For a given real number $N>1$ and a positive constant $e_0>e$ we define $t_\xi=t(|\xi|;N)$ by the following implicit formula:

$$\Lambda(t_\xi)\Big(\log(e_0+\Lambda(t_\xi))\Big)^{\omega}|\xi|=N$$

for wave-type equations and

$$\Lambda(t_\xi)^{\beta}\Big(\log(e_0+\Lambda(t_\xi))\Big)^{\omega}\langle\xi\rangle_{m(t_\xi)}=N$$

for Klein–Gordon type equations, respectively, where $\langle\xi\rangle_{m(t_\xi)}=\sqrt{|\xi|^2+m(t_\xi)^2}$. Here we note that the definition of t_ξ is equivalent to

$$\Lambda(t_\xi)^{\beta}\langle\xi\rangle_{m(t_\xi)}=N$$

in the case (A.1)-(i) and

$$\Lambda(t_\xi)\Big(\log(e_0+\Lambda(t_\xi))\Big)^{\omega}\langle\xi\rangle_{m(t_\xi)}=N$$

in the case (A.1)-(ii), respectively. Then we easily see that $\lim_{|\xi|\to 0}t(|\xi|;N)=\infty$ for wave-type equations and $\lim_{|\xi|\to 0}t(|\xi|;N)<\infty$ for Klein–Gordon type equations.

Remark 2.1: We want to emphasize that $\lim_{|\xi|\to 0}t(|\xi|;N)=\infty$ for wave-type equations if we adopt the same definition for t_ξ as for Klein–Gordon type equations.

Remark 2.2: The asymptotic behavior of $t_\xi = t(|\xi|;N)$ for $|\xi| \to 0$ gives information about possible results for concrete Cauchy problems of type (1.11). The relation $\lim_{|\xi|\to 0} t(|\xi|;N) < \infty$ leads in a lot of cases to better decay rates than wave-decay rates if $\beta \in [\frac{1}{2}, 1]$ or to a modified Floquet behavior if $\beta \in [0, \frac{1}{2})$. The relation $\lim_{|\xi|\to 0} t(|\xi|;N) = \infty$ leads in a lot of cases to a wave-decay rate if $\beta = 1$, $\omega \in [-1, 0]$ or to a classical Floquet behavior if $\beta \in [0,1)$ or $\beta = 1$, $\omega \in (-\infty, -1)$.

Moreover, we can prove the following property for t_ξ:

Lemma 2.1: *There exist positive constants N_0, e_0 and C_k such that*

$$\left|\frac{d^k t_\xi}{d|\xi|^k}\right| \le \begin{cases} C_k \langle\xi\rangle_{m(t_\xi)}^{-k} \frac{\Lambda(t_\xi)}{\lambda(t_\xi)}, & \text{for Klein–Gordon type equations,} \\ C_k |\xi|^{-k} \frac{\Lambda(t_\xi)}{\lambda(t_\xi)}, & \text{for wave-type equations,} \end{cases}$$

for all $k \ge 1$, $\xi \in \mathbb{R}^n \setminus \{0\}$ and $N \ge N_0$. Moreover, we have $\frac{dt_\xi}{d|\xi|} < 0$ for any $\xi \in \mathbb{R}^n \setminus \{0\}$, that is, t_ξ is monotone decreasing with respect to $|\xi|$.

Proof. We shall prove the statement of the lemma only for Klein–Gordon type equations. Let us consider the following identity:

$$N^2 = \Lambda(t_\xi)^{2\beta}\Big(\log(e_0 + \Lambda(t_\xi))\Big)^{2\omega} \langle\xi\rangle_{m(t_\xi)}^2 , \tag{2.1}$$

where we note that $\beta \in [0,1)$, $\omega = 0$ or $\beta = 1$, $\omega \in (-\infty, 0]$. By differentiating both sides of (2.1) with respect to $|\xi|$, noting $\beta > \gamma$ and $\omega = \delta = 0$ or $\omega > \delta$ we have

$$0 = 2\lambda(t_\xi)\Lambda(t_\xi)^{2\beta-1}\Big(\log(e_0 + \Lambda(t_\xi))\Big)^{2\omega} \frac{dt_\xi}{d|\xi|}\,\Theta_0(\xi) + 2N^2 \frac{|\xi|}{\langle\xi\rangle_{m(t_\xi)}^2} ,$$

where

$$\begin{aligned} \Theta_0(\xi) &= |\xi|^2 \left(\beta + \frac{\omega\Lambda(t_\xi)}{(e_0 + \Lambda(t_\xi))\log(e_0 + \Lambda(t_\xi))}\right) \\ &\quad + m(t_\xi)^2 \left(\beta - \frac{\gamma\Lambda(t_\xi)}{e_0 + \Lambda(t_\xi)} + \frac{(\omega - \delta)\Lambda(t_\xi)}{(e_0 + \Lambda(t_\xi))\log(e_0 + \Lambda(t_\xi))}\right) . \end{aligned}$$

Choosing $e_0 > e^{-\omega}$ yields $\Theta_0(\xi) \ge C_1^{-1}\langle\xi\rangle_{m(t_\xi)}^2$ with a positive constant $C_1 = C_1(\beta,\gamma,\omega,\delta)$. This implies the conclusion of the lemma for $k = 1$. By induction we can prove the statement for all $k > 1$. □

We split the set $[0,\infty) \times (\mathbb{R}^n \setminus \{0\})$ by t_ξ in subdomains which will be called *zones*. For a real number $N > 1$ we define the *pseudodifferential zone* $Z_{pd}(N)$ and the *hyperbolic zone* $Z_{hyp}(N)$ by

$$Z_{pd}(N) := \{(t,\xi) \in [0,\infty) \times (\mathbb{R}^n \setminus \{0\}) \ ; \ 0 \le t \le t_\xi\}$$

and

$$Z_{hyp}(N) := \{(t,\xi) \in [0,\infty) \times (\mathbb{R}^n \setminus \{0\}) \ ; \ t \ge t_\xi\} ,$$

respectively.

For further considerations we need suitable classes of amplitudes which have a *symbol character only in the hyperbolic zone* $Z_{hyp}(N)$.

Definition 2.1: i) For real numbers $r_1, r_2, r_3, \beta \in (0,1]$, $\omega \in [-1,0]$, we denote for Klein–Gordon type equations by $S_{\beta,\omega}\{r_1,r_2,r_3\}$ the set of all amplitudes $a = a(t,\xi) \in C^\infty((0,\infty)\times\mathbb{R}^n\setminus\{0\})$ satisfying

$$|D_t^l D_\xi^\alpha a(t,\xi)| \le C_{\alpha,l}\langle\xi\rangle_{m(t)}^{r_1-|\alpha|}\lambda(t)^{r_2}\left(\frac{\lambda(t)}{\Lambda(t)^\beta\Big(\log(e_0+\Lambda(t))\Big)^\omega}\right)^{r_3+l}$$

for all $(t,\xi)\in Z_{hyp}(N)$, all multi-indices α and all l with constants $C_{\alpha,l}$ independent of N.
ii) By $\tilde{S}_{1,\omega}\{r_1,r_2,r_3\}$ we denote for wave-type equations the set of all amplitudes $a = a(t,\xi)\in C^\infty((0,\infty)\times\mathbb{R}^n\setminus\{0\})$ satisfying

$$|D_t^l D_\xi^\alpha a(t,\xi)| \le C_{\alpha,l}|\xi|^{r_1-|\alpha|}\lambda(t)^{r_2}\left(\frac{\lambda(t)}{\Lambda(t)\Big(\log(e_0+\Lambda(t))\Big)^\omega}\right)^{r_3+l}$$

for all $(t,\xi)\in Z_{hyp}(N)$, all multi-indices α and all l with constants $C_{\alpha,l}$ independent of N.

Lemma 2.2: *The following rules of the symbolic calculus hold:*

(i) $S_{\beta,\omega}\{r_1,r_2,r_3\}\subset S_{\beta,\omega}\{r_1+k,r_2+k,r_3-k\}$ *for* $k\ge 0$;
(ii) *if* $a\in S_{\beta,\omega}\{r_1,r_2,r_3\}$ *and* $b\in S_{\beta,\omega}\{s_1,s_2,s_3\}$, *then* $ab\in S_{\beta,\omega}\{r_1+s_1,r_2+s_2,r_3+s_3\}$;
(iii) *if* $a\in S_{\beta,\omega}\{r_1,r_2,r_3\}$, *then* $D_t a\in S_{\beta,\omega}\{r_1,r_2,r_3+1\}$;
(iv) *if* $a\in S_{\beta,\omega}\{r_1,r_2,r_3\}$, *then* $D_\xi^\alpha a\in S_{\beta,\omega}\{r_1-|\alpha|,r_2,r_3\}$;
(v) *the same rules hold for* $\tilde{S}_{1,\omega}\{r_1,r_2,r_3\}$.

Proof. We restrict ourselves to the case of Klein–Gordon type equations. The rules (ii) $\sim$ (iv) are evident by Definition 2.1 and the Leibniz rule. To show (i) we use the following estimates in $Z_{hyp}(N)$:

$$\begin{aligned}&\langle\xi\rangle_{m(t)}^{r_1}\lambda(t)^{r_2}\left(\frac{\lambda(t)}{\Lambda(t)^\beta\Big(\log(e_0+\Lambda(t))\Big)^\omega}\right)^{r_3}\\&=\langle\xi\rangle_{m(t)}^{r_1+k}\lambda(t)^{r_2+k}\\&\quad\times\left(\frac{\lambda(t)}{\Lambda(t)^\beta\Big(\log(e_0+\Lambda(t))\Big)^\omega}\right)^{r_3-k}\left(\frac{1}{\Lambda(t)^\beta\Big(\log(e_0+\Lambda(t))\Big)^\omega\langle\xi\rangle_{m(t)}}\right)^{k}\\&\le N^{-k}\langle\xi\rangle_{m(t)}^{r_1+k}\lambda(t)^{r_2+k}\left(\frac{\lambda(t)}{\Lambda(t)^\beta\Big(\log(e_0+\Lambda(t))\Big)^\omega}\right)^{r_3-k}.\end{aligned}$$

Consequently, $S_{\beta,\omega}\{r_1, r_2, r_3\} \subset S_{\beta,\omega}\{r_1 + k, r_2 + k, r_3 - k\}$, where we take into consideration, that the higher order derivatives can be estimated in a similar way. $\square$

It is clear that

$$\lambda \in S_{\beta,\omega}\{0,1,0\},\ \tilde{S}_{1,\omega}\{0,1,0\} \quad \text{and} \quad b, m \in S_{\beta,\omega}\{0,0,0\},\ \tilde{S}_{1,\omega}\{0,0,0\}$$

from the conditions to b, m and (A.1).

3. Consideration in the pseudodifferential zone $\mathbf{Z_{pd}(N)}$

We shall derive an estimate for the fundamental solution. After partial Fourier transformation with respect to x, the problem (1.11) is rewritten as the following Cauchy problem for an ordinary differential equation with the parameter ξ:

$$\begin{cases} D_t^2 v(t,\xi) - \lambda(t)^2 b(t)^2 \langle\xi\rangle_{m(t)}^2 v(t,\xi) = 0\ ,\ (t,\xi) \in [0,\infty) \times \mathbb{R}^n\ , \\ v(0,\xi) = v_0(\xi)\ ,\ D_t v(0,\xi) = -iv_1(\xi)\ ,\ \xi \in \mathbb{R}^n. \end{cases} \tag{3.1}$$

3.1. Estimates for Klein–Gordon type equations in $Z_{pd}(N)$

Setting $\mathcal{U} := (\lambda(t)\langle\xi\rangle_{m(t)} v, D_t v)^T$, equation (3.1) can be transformed to the system of first order

$$D_t\,\mathcal{U}(t,\xi) - A(t,\xi)\mathcal{U}(t,\xi) = 0, \tag{3.2}$$

where

$$A(t,\xi) = \begin{pmatrix} \frac{D_t\lambda(t)}{\lambda(t)} + \frac{m(t)D_t m(t)}{\langle\xi\rangle_{m(t)}^2} & \lambda(t)\langle\xi\rangle_{m(t)} \\ \lambda(t)b(t)^2\langle\xi\rangle_{m(t)} & 0 \end{pmatrix}.$$

We are interested in the fundamental solution to the Cauchy problem for that system; this is the matrix-valued solution $U = U(t,s,\xi)$ to the Cauchy problem:

$$\begin{cases} D_t U(t,s,\xi) - A(t,\xi)U(t,s,\xi) = 0, \\ U(s,s,\xi) = I, \end{cases} \tag{3.3}$$

where I is the identity matrix. Using the matrizant we obtain for $U = U(t,s,\xi)$ the explicit representation

$$U(t,s,\xi) = I + \sum_{k=1}^{\infty} \int_s^t A(\tau_1,\xi) \int_s^{\tau_1} A(\tau_2,\xi) \cdots \int_s^{\tau_{k-1}} A(\tau_k,\xi)d\tau_k \cdots d\tau_1\ .$$

Hence we have

$$
\begin{aligned}
\|U(t,0,\xi)\| &\le 1+\sum_{k=1}^{\infty}\frac{1}{k!}\left(\int_0^t \|A(s,\xi)\|ds\right)^k \\
&\le \exp\left(C\int_0^t \lambda(s)\langle\xi\rangle_{m(s)}ds+\int_0^t \frac{\lambda'(s)}{\lambda(s)}ds+\int_0^t \frac{-m'(s)m(s)}{\langle\xi\rangle^2_{m(s)}}ds\right) \\
&\le \exp\left(C\Lambda(t_0)\langle\xi\rangle_{m(t_0)}\tfrac{m(0)}{m(t_0)}+\log\lambda(t_0)+\log\langle\xi\rangle_{m(t_0)}\right) \\
&\le N\lambda(t_0)\Lambda(t_0)^{-\beta}\Big(\log(e_0+\Lambda(t_0))\Big)^{-\omega} \\
&\quad\times\exp\left(CNm(0)\Lambda(t_0)^{1-\beta-\gamma}\Big(\log(e_0+\Lambda(t_0))\Big)^{-\omega-\delta}\right) \\
&\le C_{0,0,N}
\end{aligned}
$$

for any $t \le t_\xi \le t_0$ with a positive constant $C_{0,0,N}$, where λ' and m' are the first-order derivatives of λ and m with respect to t and t_0 denotes t_ξ for $\xi = 0$. The higher-order derivatives of $U = U(t,0,\xi)$ can be estimated in a similar way. Thus we obtain the following proposition:

Proposition 3.1: *The fundamental solution $U = U(t,0,\xi)$ of* (3.3) *satisfies in the case of Klein–Gordon type equations, that is, in $\Omega_{1.1}\cup\Omega_{1.2}\cup\Omega_{1.3}\cup\Omega_{2.4}\cup\Omega_{3.1}\cup\Omega_{3.2}\cup\Omega_{3.3}\cup\Omega_{3.4}$, the following estimates:*

$$
\|D_t^k D_\xi^\alpha U(t,0,\xi)\| \le C_{\alpha,k,N}\langle\xi\rangle_{m(t)}^{-|\alpha|}\Big(\lambda(t)\langle\xi\rangle_{m(t)}\Big)^k
$$

for every k, α and $(t,\xi)\in Z_{pd}(N)$.

3.2. Estimates for wave-type equations in $Z_{pd}(N)$

Setting $\tilde{\mathcal{U}}(t,\xi) := \Big(\lambda(t)|\xi|v, D_t v, \lambda(t)m(t)v\Big)^T$ equation (3.1) can be transformed to the system of first order

$$
D_t\tilde{\mathcal{U}} - \tilde{A}(t,\xi)\,\tilde{\mathcal{U}} = 0,
$$

where

$$
\tilde{A}(t,\xi) = \begin{pmatrix} \frac{D_t\lambda(t)}{\lambda(t)} & \lambda(t)|\xi| & 0 \\ \lambda(t)b(t)^2|\xi| & 0 & -\lambda(t)b(t)^2m(t) \\ 0 & \lambda(t)m(t) & \frac{D_t\lambda(t)}{\lambda(t)}+\frac{D_tm(t)}{m(t)} \end{pmatrix}.
$$

Let $\tilde{U} = \tilde{U}(t,s,\xi)$ be the matrix-valued fundamental solution to the Cauchy problem

$$
\begin{cases} D_t\tilde{U}(t,s,\xi) - \tilde{A}(t,\xi)\tilde{U}(t,s,\xi) = 0, \\ \tilde{U}(s,s,\xi) = I\,. \end{cases} \tag{3.4}
$$

Similar to the considerations for Klein–Gordon type equations we have the following estimates for $\tilde{U}$:

$$\begin{aligned}
&\|\tilde{U}(t,0,\xi)\| \\
&\leq \exp\left(C\left(\int_0^t \lambda(s)|\xi|ds + \int_0^t \lambda(s)m(s)ds\right) + \int_0^t \frac{\lambda'(s)}{\lambda(s)}ds - \int_0^t \frac{m'(s)}{m(s)}ds\right) \\
&\leq \frac{\lambda(t)m(0)}{\lambda(0)m(t)} \exp\left(C\left(\Lambda(t)|\xi| + \int_0^t \frac{\lambda(s)ds}{(e_0+\Lambda(s))^\gamma \left(\log(e_0+\Lambda(s))\right)^\delta}\right)\right) \\
&\leq \begin{cases} C_{\varepsilon,N}\lambda(t)m(t)^{-1}(e_0+\Lambda(t))^\varepsilon & \text{in } \Omega_{2.1}\cup\Omega_{2.2}\cup\Omega_{3.5}\cup\Omega_{3.6}; \\ C_N\lambda(t)m(t)^{-1}(e_0+\Lambda(t))^{\varepsilon_0} & \text{in } \Omega_{2.3}\cup\Omega_{3.7}; \\ C_N\lambda(t)m(t)^{-1}\exp\left(\left(\log(e_0+\Lambda(t))\right)^{-\delta+1}\right) & \text{in } \{\Omega_{3.8}:\delta\in[-1,0)\}\cup\Omega_{3.10}; \end{cases}
\end{aligned}$$

for any given $\varepsilon > 0$, a positive constant ε_0, large t and any $N = N_\varepsilon$, where we use the next estimates in the case $\omega \geq -1$:

$$\Lambda(t)|\xi| \leq N\left(\log(e_0+\Lambda(t))\right)^{-\omega} = \frac{N}{\left(\log(e_0+\Lambda(t))\right)^{1+\omega}}\log(e_0+\Lambda(t))$$

and

$$\int_0^t \frac{\lambda(s)ds}{(e_0+\Lambda(s))^\gamma\left(\log(e_0+\Lambda(s))\right)^\delta} \leq \frac{1}{-\delta+1}\left(\log(e_0+\Lambda(t))\right)^{-\delta+1}.$$

The higher-order derivatives of $\tilde{U} = \tilde{U}(t,0,\xi)$ can be estimated in a similar way. Thus we obtain the following proposition:

Proposition 3.2: *The fundamental solution $\tilde{U} = \tilde{U}(t,0,\xi)$ of (3.4) satisfies in the case of wave-type equations the following estimates for every k, α and $(t,\xi) \in Z_{pd}(N)$:*

$$\|D_t^k D_\xi^\alpha \tilde{U}(t,0,\xi)\| \leq C_{\varepsilon,\alpha,k,N}\lambda(t)m(t)^{-1}(e_0+\Lambda(t))^\varepsilon|\xi|^{-|\alpha|}\left(\lambda(t)|\xi|\right)^k$$

for any $\varepsilon > 0$ in $\Omega_{2.1}\cup\Omega_{2.2}\cup\Omega_{3.5}\cup\Omega_{3.6}$,

$$\|D_t^k D_\xi^\alpha \tilde{U}(t,0,\xi)\| \leq C_{\alpha,k,N}\lambda(t)m(t)^{-1}(e_0+\Lambda(t))^{\varepsilon_0}|\xi|^{-|\alpha|}\left(\lambda(t)|\xi|\right)^k$$

with a positive number ε_0 in $\Omega_{2.3}\cup\Omega_{3.7}$,

$$\|D_t^k D_\xi^\alpha \tilde{U}(t,0,\xi)\| \leq C_{\alpha,k,N}\lambda(t)m(t)^{-1}e^{\left(\log(e_0+\Lambda(t))\right)^{-\delta+1}}|\xi|^{-|\alpha|}\left(\lambda(t)|\xi|\right)^k$$

in $\{\Omega_{3.8}:\delta\in[-1,0)\}\cup\Omega_{3.10}$.

4. Consideration in the hyperbolic zone $Z_{hyp}(N)$

4.1. Estimates for Klein–Gordon type equations in $Z_{hyp}(N)$

We carry out a diagonalization process to get estimates for the solutions to (3.2).

Let us define a diagonalizer for the principal part by

$$M(t)=\frac{1}{2}\sqrt{\frac{\lambda(t)}{b(t)}}\begin{pmatrix} b(t) & -1 \\ b(t) & 1 \end{pmatrix},$$

and set $V_0 = V_0(t,s,\xi) = M(t)U(t,s,\xi)$ for the solution of (3.3). Noting the identities:

$$M(t)\begin{pmatrix} 0 & \lambda(t)\langle\xi\rangle_{m(t)} \\ \lambda(t)b(t)^2\langle\xi\rangle_{m(t)} & 0 \end{pmatrix}M(t)^{-1}=\lambda(t)b(t)\langle\xi\rangle_{m(t)}\begin{pmatrix} -1 & 0 \\ 0 & 1 \end{pmatrix},$$

$$M(t)\begin{pmatrix} 1 & 0 \\ 0 & 0 \end{pmatrix}M(t)^{-1}=\frac{1}{2}\begin{pmatrix} 1 & 0 \\ 0 & 1 \end{pmatrix}+\frac{1}{2}\begin{pmatrix} 0 & 1 \\ 1 & 0 \end{pmatrix}$$

and

$$[M(t),D_t]M(t)^{-1}=-\frac{D_t(\lambda(t)b(t))}{2\lambda(t)b(t)}\begin{pmatrix} 0 & 1 \\ 1 & 0 \end{pmatrix}+\frac{D_t\lambda(t)}{2\lambda(t)}\begin{pmatrix} -1 & 1 \\ 1 & -1 \end{pmatrix},$$

we get the following first-order system for V_0:

$$(D_t-\mathcal{D}(t,\xi)+B(t,\xi))\,V_0(t,s,\xi)=0\,, \tag{4.1}$$

where

$$\mathcal{D}=\mathcal{D}(t,\xi):=\begin{pmatrix} \tau_-(t,\xi) & 0 \\ 0 & \tau_+(t,\xi) \end{pmatrix}$$

with

$$\tau_\pm(t,\xi):=\pm\lambda(t)b(t)\langle\xi\rangle_{m(t)}+\frac{D_t\lambda(t)}{\lambda(t)}+\frac{m(t)D_tm(t)}{2\langle\xi\rangle^2_{m(t)}}$$

and

$$B=B(t,\xi):=-\frac{1}{2}\left(\frac{D_t(\lambda(t)b(t))}{\lambda(t)b(t)}+\frac{m(t)D_tm(t)}{\langle\xi\rangle^2_{m(t)}}\right)\begin{pmatrix} 0 & 1 \\ 1 & 0 \end{pmatrix}.$$

Here we note that $|\tau_+(t,\xi)-\tau_-(t,\xi)|\ge\tau_0\lambda(t)\langle\xi\rangle_{m(t)}$ with a positive constant τ_0, $\mathcal{D}\in S_{\beta,\omega}\{1,1,0\}$ and $B\in S_{\beta,\omega}\{0,0,1\}$ by (A.1) and (A.2).

We will carry out further steps of perfect diagonalization, namely, diagonalization modulo $S_{\beta,\omega}\{-M,-M,M+1\}$ for some given non-negative integer M.

Let us consider the first step (the further steps can be considered inductively).

We define $M_1=M_1(t,\xi)$, $M^{(1)}=M^{(1)}(t,\xi)$ and $R_1=R_1(t,\xi)$ by

$$M_1=I+M^{(1)}:=I+\begin{pmatrix} 0 & \frac{B_{12}}{\tau_--\tau_+} \\ \frac{B_{21}}{\tau_+-\tau_-} & 0 \end{pmatrix}$$

and

$$R_1 = M_1^{-1}(D_t - \mathcal{D} + B)M_1 - (D_t - \mathcal{D}) = M_1^{-1}\Big(D_t M_1 - [\mathcal{D}, M_1] + BM_1\Big)$$

respectively, where $B_{ij} = B(t,\xi)_{ij}$ $(1 \le i,j \le 2)$ denotes the (i,j)-th element of the matrix B. Then we have

$$\begin{aligned} -[\mathcal{D}, M_1] + BM_1 &= \begin{pmatrix} 0 & -B_{12} \\ -B_{21} & 0 \end{pmatrix} + \begin{pmatrix} \frac{B_{12}B_{21}}{\tau_+ - \tau_-} & B_{12} \\ B_{21} & \frac{B_{12}B_{21}}{\tau_- - \tau_+} \end{pmatrix} \\ &= \frac{B_{12}B_{21}}{\tau_+ - \tau_-} \begin{pmatrix} 1 & 0 \\ 0 & -1 \end{pmatrix} \in S_{\beta,\omega}\{-1,-1,2\}\,. \end{aligned}$$

Taking account of $M_1^{-1} \in S_{\beta,\omega}\{0,0,0\}$ (a sufficiently large N will imply this), $D_t M_1 \in S_{\beta,\omega}\{-1,-1,2\}$, we have $R_1 \in S_{\beta,\omega}\{-1,-1,2\}$. In general, we can prove the following proposition:

Proposition 4.1: *For any fixed non-negative integer k there exist matrix-valued functions $M_k = M_k(t,\xi) \in S_{\beta,\omega}\{0,0,0\}$, $F_k = F_k(t,\xi) \in S_{\beta,\omega}\{-1,-1,2\}$, $R_k = R_k(t,\xi) \in S_{\beta,\omega}\{-k,-k,k+1\}$ such that the following operator-valued equality holds in $Z_{hyp}(N)$:*

$$M_k^{-1}(D_t - \mathcal{D} + B)M_k = D_t - \mathcal{D} + F_k + R_k,$$

where F_k is diagonal while the matrices $M_k, M_k^{-1} \in S_{\beta,\omega}\{0,0,0\}$, provided that the parameter N in (2.1) is sufficiently large.

Proof. We already discussed the case $k = 1$. Suppose that M_k and F_k have the following representation:

$$M_k = \sum_{j=0}^{k} M^{(j)}\,, \quad M^{(0)} = I\,, \quad M^{(j+1)} = \begin{pmatrix} 0 & \frac{(M_j R_j)_{12}}{\tau_- - \tau_+} \\ \frac{(M_j R_j)_{21}}{\tau_+ - \tau_-} & 0 \end{pmatrix} \quad (j \ge 0)\,,$$

$$F_k = \sum_{j=0}^{k-1} F^{(j)}\,, \quad F^{(0)} = 0\,, \quad F^{(j)} = \operatorname{diag}(M_j R_j) \quad (j \ge 1)\,.$$

Suppose that $k \ge 1$, $F_k \in S_{\beta,\omega}\{-1,-1,2\}$ and $R_k \in S_{\beta,\omega}\{-k,-k,k+1\}$. Noting the properties:

$$\begin{aligned} M_k R_k - M_k F^{(k)} - [\mathcal{D}, M^{(k+1)}] &= -\Big(M^{(1)} + \cdots + M^{(k)}\Big)F^{(k)} \\ &\in S_{\beta,\omega}\{-k-1,-k-1,k+2\} \end{aligned}$$

and

$$BM^{(k+1)} = \begin{pmatrix} \frac{B_{12}(M_k R_k)_{21}}{\tau_+ - \tau_-} & 0 \\ 0 & \frac{B_{21}(M_k R_k)_{12}}{\tau_- - \tau_+} \end{pmatrix} \in S_{\beta,\omega}\{-k-1,-k-1,k+2\}\,,$$

we get

$$\begin{aligned} M_{k+1}R_{k+1} &= (D_t - \mathcal{D} + B)M_{k+1} - M_{k+1}(D_t - \mathcal{D} + F_{k+1}) \\ &= M_k R_k - M_k F^{(k)} - [\mathcal{D}, M^{(k+1)}] + B\, M^{(k+1)} + D_t M^{(k+1)} \\ &\quad -M^{(k+1)} F_{k+1} \\ &\in S_{\beta,\omega}\{-k-1, -k-1, k+2\}. \end{aligned}$$

Finally, we have $M^{(j)} \in S_{\beta,\omega}\{-j,-j,j\}$, that is, there exists a positive constant C_j independent of N such that

$$\begin{aligned} \|M^{(j)}(t,\xi)\| &\le C_j \left(\frac{1}{\Lambda(t)^\beta \Big(\log(e_0 + \Lambda(t))\Big)^\omega \langle\xi\rangle_{m(t)}} \right)^j \\ &\le C_j \left(\frac{1}{N}\right)^j \end{aligned}$$

for any $(t,\xi) \in Z_{hyp}(N)$ and $j = 1, \ldots, k+1$. This implies

$$\|\sum_{j=1}^{k+1} M^{(j)}(t,\xi)\| \le \sum_{j=1}^{k+1} C_j \left(\frac{1}{N}\right)^j .$$

A sufficiently large N provides $\|M_{k+1} - 1\| \le \frac{1}{2}$ in $Z_{hyp}(N)$ and, in consequence, the statements concerning $M_{k+1}, M_{k+1}^{-1}, R_{k+1}$ and F_{k+1}. □

Let us consider the system $\Big(D_t - \mathcal{D} + F_k + R_k\Big)V_k = 0$. Let us define the matrix-valued functions $E_2 = E_2(t,r,\xi)$ and $\mathcal{R}_k = \mathcal{R}_k(t,r,\xi)$ by

$$\begin{aligned} E_2(t,r,\xi) &= \begin{pmatrix} \exp\left(i \int\limits_r^t \tau_-(s,\xi)ds\right) & 0 \\ 0 & \exp\left(i\int\limits_r^t \tau_+(s,\xi)ds\right) \end{pmatrix} \\ &= \frac{\lambda(t)}{\lambda(r)} \sqrt{\frac{\langle\xi\rangle_{m(t)}}{\langle\xi\rangle_{m(r)}}} \begin{pmatrix} \exp\left(-i\int\limits_r^t \lambda(s)b(s)\langle\xi\rangle_{m(s)}ds\right) & 0 \\ 0 & \exp\left(i\int\limits_r^t \lambda(s)b(s)\langle\xi\rangle_{m(s)}ds\right) \end{pmatrix} \end{aligned}$$

and

$$\mathcal{R}_k(t,r,\xi) = -F_k(t,\xi) - E_2(r,t,\xi)R_k(t,\xi)E_2(t,r,\xi)\,,$$

respectively. Then, we have the following lemma:

Lemma 4.1: *The function $\mathcal{R}_k = \mathcal{R}_k(t,r,\xi)$ satisfies the following estimates:*

$$\begin{aligned}\|\partial_t^l\partial_r^p\partial_\xi^\alpha \mathcal{R}_k(t,r,\xi)\| &\le C_{k,l,p,\alpha}\Big(\lambda(t)\langle\xi\rangle_{m(t)}\Big)^l\Big(\lambda(r)\langle\xi\rangle_{m(r)}\Big)^p\langle\xi\rangle_{m(t)}^{-|\alpha|}\\ &\times\frac{\lambda(t)}{N^l\Lambda(t)^{2\beta}\Big(\log(e_0+\Lambda(t))\Big)^{2\omega}\langle\xi\rangle_{m(t)}}\\ &\times\left[\frac{\Lambda(t)^{|\alpha|}\langle\xi\rangle_{m(t)}^{|\alpha|}}{\Big(\Lambda(t)^\beta\Big(\log(e_0+\Lambda(t))\Big)^\omega\langle\xi\rangle_{m(t)}\Big)^{k-1}}+1\right]\end{aligned}$$

in $\Omega_{1.1}\cup\Omega_{1.2}\cup\Omega_{1.3}\cup\Omega_{2.4}\cup\Omega_{3.1}\cup\Omega_{3.2}\cup\Omega_{3.3}\cup\Omega_{3.4}$ with constants $C_{k,l,p,\alpha}$ independent of N.

Proof. Setting $E_2(t,r,\xi)=\frac{\lambda(t)}{\lambda(r)}\sqrt{\frac{\langle\xi\rangle_{m(t)}}{\langle\xi\rangle_{m(r)}}}\tilde{E}_2(t,r,\xi)$ gives

$$\mathcal{R}_k(t,r,\xi)+F_k(t,\xi)=-\tilde{E}_2(r,t,\xi)R_k(t,\xi)\tilde{E}_2(t,r,\xi).$$

Taking note of

$$\langle\xi\rangle_{m(t)}^{-1}\le N^{-1}\Lambda(t)^\beta\Big(\log(e_0+\Lambda(t))\Big)^\omega\le\Lambda(t),$$

$$\|\partial_t^l\partial_r^p\partial_\xi^\alpha\tilde{E}_2(t,r,\xi)\|\le C_{l,p,\alpha}(\lambda(t)\langle\xi\rangle_{m(t)})^l(\lambda(r)\langle\xi\rangle_{m(r)})^p\Lambda(t)^{|\alpha|}$$

and Proposition 4.1 we obtain

$$\begin{aligned}&\|\partial_t^l\partial_r^p\partial_\xi^\alpha(\mathcal{R}_k(t,r,\xi)+F_k(t,\xi))\|\le C_{k,l,p,\alpha}\lambda(t)\Lambda(t)^{|\alpha|}\\ &\times\Big(\lambda(t)\langle\xi\rangle_{m(t)}\Big)^l\Big(\lambda(r)\langle\xi\rangle_{m(r)}\Big)^p\langle\xi\rangle_{m(t)}^{-k}\Big(\Lambda(t)^\beta\Big(\log(e_0+\Lambda(t))\Big)^\omega\Big)^{-k-1}\end{aligned}$$

for any l, p and α, where the constants $C_{k,l,p,\alpha}$ are independent of N. By Proposition 4.1 we have

$$\begin{aligned}\|\partial_t^l\partial_\xi^\alpha F_k\| &\le C_{k,l,\alpha}(\lambda(t)\langle\xi\rangle_{m(t)})^l\langle\xi\rangle_{m(t)}^{-|\alpha|}\frac{\lambda(t)}{N^l\Lambda(t)^{2\beta}\Big(\log(e_0+\Lambda(t))\Big)^{2\omega}\langle\xi\rangle_{m(t)}}\\ &\le C_{k,l,\alpha}(\lambda(t)\langle\xi\rangle_{m(t)})^l\langle\xi\rangle_{m(t)}^{-|\alpha|}\frac{\lambda(t)}{\Lambda(t)^{2\beta}\Big(\log(e_0+\Lambda(t))\Big)^{2\omega}\langle\xi\rangle_{m(t)}}\end{aligned}$$

for any $N\ge1$. Noting $\|\partial_t^l\partial_r^p\partial_\xi^\alpha\mathcal{R}_k\|\le\|\partial_t^l\partial_r^p\partial_\xi^\alpha(\mathcal{R}_k+F_k)\|+\|\partial_t^l\partial_\xi^\alpha F_k\|$ the lemma is proved. □

By the aid of $\mathcal{R}_k$ we define for $t\ge t_\xi$ the matrix-valued function

$$Q_k(t,t_\xi,\xi):=\sum_{j=1}^\infty i^j\int_{t_\xi}^t\mathcal{R}_k(t_1,t_\xi,\xi)dt_1\int_{t_\xi}^{t_1}\mathcal{R}_k(t_2,t_\xi,\xi)dt_2\cdots\int_{t_\xi}^{t_{j-1}}\mathcal{R}_k(t_j,t_\xi,\xi)dt_j\,.$$

Then the matrix-valued function $V_k = V_k(t,t_\xi,\xi) = E_2(t,t_\xi,\xi)(I+Q_k(t,t_\xi,\xi))$ solves the Cauchy problem

$$\begin{cases} \Big(D_t - \mathcal{D} + F_k + R_k\Big)V_k(t,t_\xi,\xi) = 0, \\ V_k(t_\xi,t_\xi,\xi) = I \quad (t \geq t_\xi). \end{cases}$$

Therefore, we can conclude that the fundamental solution to (3.3) with $s=0$ has the following representation:

$$\begin{aligned} U(t,0,\xi) &= M(t)^{-1}M_k(t,\xi)E_2(t,t_\xi,\xi)(I+Q_k(t,t_\xi,\xi)) \\ &\quad \times M_k(t_\xi,\xi)^{-1}M(t_\xi)U(t_\xi,0,\xi). \end{aligned} \tag{4.2}$$

To estimate the fundamental solution $U(t,0,\xi)$ we note that

$$\|E_2(t,t_\xi,\xi)\| \leq \|E_2(0,t_\xi,\xi)\|\|E_2(t,0,\xi)\|,$$

where besides a careful calculation shows that

$$\|\partial_\xi^\alpha E_2(0,t_\xi,\xi)\| \leq C_\alpha \langle\xi\rangle_{m(t_\xi)}^{-|\alpha|}.$$

It remains to estimate $Q_k = Q_k(t,t_\xi,\xi)$.

Lemma 4.2: *The function $\Theta_{k,\alpha} = \Theta_{k,\alpha}(t,\xi) := \|\partial_\xi^\alpha Q_k(t,t_\xi,\xi)\|$ satisfies the following estimates for any $(t,\xi) \in Z_{hyp}(N)$ and α satisfying $|\alpha| < \frac{k(\beta-\gamma)+\beta-1}{1-\gamma}$ in $\Omega_{1.1}\cup\Omega_{1.2}\cup\Omega_{1.3}\cup\Omega_{2.4}$ and $|\alpha| < \frac{k(\omega-\delta)+\omega-1}{-\delta}$ in $\Omega_{3.1}\cup\Omega_{3.2}\cup\Omega_{3.3}\cup\Omega_{3.4}$, respectively:*

(i) *if $(\beta,\gamma,\omega,\delta) \in \Omega_{1.1}\cup\Omega_{2.4}\cup\Omega_{3.1}$, then there exists a positive constant $C_{k,\alpha}$ such that $\Theta_{k,\alpha}(t,\xi) \leq C_{k,\alpha}\langle\xi\rangle_{m(t)}^{-|\alpha|}$;*

(ii) *if $(\beta,\gamma,\omega,\delta) \in \Omega_{3.2}$, then for any positive constant ε there exists a positive constant $C_{k,\alpha,\varepsilon}$ such that $\Theta_{k,\alpha}(t,\xi) \leq C_{k,\alpha,\varepsilon}\langle\xi\rangle_{m(t)}^{-|\alpha|}(e_0+\Lambda(t))^\varepsilon$;*

(iii) *if $(\beta,\gamma,\omega,\delta) \in \Omega_{1.2}\cup\Omega_{3.3}$, then there exist positive constants $\varepsilon_0 \geq 1$ and $C_{k,\alpha}$ such that $\Theta_{k,\alpha}(t,\xi) \leq C_{k,\alpha}\langle\xi\rangle_{m(t)}^{-|\alpha|}(e_0+\Lambda(t))^{\varepsilon_0}$;*

(iv) *if $(\beta,\gamma,\omega,\delta) \in \Omega_{3.4}$, then there exists a positive constant $C_{k,\alpha}$ such that*

$$\Theta_{k,\alpha}(t,\xi) \leq C_{k,\alpha}\langle\xi\rangle_{m(t)}^{-|\alpha|}\exp\left(\Big(\log(e_0+\Lambda(t))\Big)^{\delta-2\omega+1}\right);$$

(v) *if $(\beta,\gamma,\omega,\delta) \in \Omega_{1.3}$, then there exists a positive constant $C_{k,\alpha}$ such that*

$$\Theta_{k,\alpha}(t,\xi) \leq C_{k,\alpha}\langle\xi\rangle_{m(t)}^{-|\alpha|}\exp\left((e_0+\Lambda(t))^{\gamma-2\beta+1}\right).$$

Proof. Let us prove the statements in the case $|\alpha| = 0$ with $k=1$. In this case it is sufficient to estimate the integral (see Lemma 4.1)

$$\int_{t_1}^{t_2} \frac{\lambda(s)}{\Lambda(s)^{2\beta}\Big(\log(e_0+\Lambda(s))\Big)^{2\omega}\langle\xi\rangle_{m(s)}}\,ds =: \theta_1(t_2,t_1,\xi).$$

Then $\Theta_{k,0}(t,\xi) \le C_0 \exp(\theta_1(t,t_\xi,\xi))\langle\xi\rangle_{m(t)}^{-|\alpha|}$. Noting $\lim_{|\xi|\to 0} t_\xi = t_0\ (<\infty)$ we have

$$\begin{aligned}
\theta_1(t_0,t_\xi,\xi) &\le \frac{1}{\langle\xi\rangle_{m(t_0)}}\int_{t_\xi}^{t_0}\frac{\lambda(s)}{\Lambda(s)^{2\beta}\Big(\log(e_0+\Lambda(s))\Big)^{2\omega}}\,ds \le \frac{\Lambda(t_\xi)^{-2\beta+1}}{(2\beta-1)\langle\xi\rangle_{m(t_0)}}\\
&\le \frac{\langle\xi\rangle_{m(t_\xi)}}{N(2\beta-1)\langle\xi\rangle_{m(t_0)}}\Lambda(t_\xi)^{-\beta+1} \le \frac{\langle\xi\rangle_{m(0)}}{N(2\beta-1)\langle\xi\rangle_{m(t_0)}}\Lambda(t_0)^{-\beta+1}\\
&\le \frac{m(0)\Lambda(t_0)^{-\beta+1}}{N(2\beta-1)\,m(t_0)}
\end{aligned}$$

for $\beta>1/2$, $\omega=\delta=0$ and m satisfies (A.2) (i) or (ii) with $\delta=0$,

$$\begin{aligned}
\theta_1(t_0,t_\xi,\xi) &\le \frac{1}{\Big(\log(e_0+\Lambda(t_0))\Big)^{2\omega}\langle\xi\rangle_{m(t_0)}}\int_{t_\xi}^{t_0}\frac{\lambda(s)}{\Lambda(s)^2}\,ds\\
&\le \frac{1}{\Big(\log(e_0+\Lambda(t_0))\Big)^{2\omega}\langle\xi\rangle_{m(t_0)}}\frac{1}{\Lambda(t_\xi)}\\
&\le \frac{m(0)}{N\Big(\log(e+\Lambda(t_0))\Big)^{2\omega}m(t_0)}
\end{aligned}$$

for $e_0>e$, $\beta=1$, $\omega\le 0$ and m satisfies (A.2) (ii). Furthermore, we obtain

$$\begin{aligned}
\theta_1(t,t_0,\xi) &< \int_{t_0}^{t}\frac{1}{\Lambda(s)^{\gamma}\Big(\log(e_0+\Lambda(s))\Big)^{\delta}m(s)}\frac{\lambda(s)\,ds}{\Lambda(s)^{2\beta-\gamma}\Big(\log(e_0+\Lambda(s))\Big)^{2\omega-\delta}}\\
&\le \left(\frac{e_0+\Lambda(t_0)}{\Lambda(t_0)}\right)^{\gamma}\int_{t_0}^{t}\frac{\lambda(s)\,ds}{\Lambda(s)^{2\beta-\gamma}\Big(\log(e_0+\Lambda(s))\Big)^{2\omega-\delta}},
\end{aligned}$$

here we note that $\frac{e_0+\Lambda(t_0)}{\Lambda(t_0)}$ is estimated by a positive constant independent of $N\ge 1$.

If $(\beta,\gamma,\omega,\delta)\in\Omega_{1.1}\cup\Omega_{2.4}\cup\Omega_{3.1}$, then $\theta_1(t,t_0,\xi)\le\theta_1(\infty,t_0,\xi)<\infty$. Thus, we obtain (i).

If $(\beta,\gamma,\omega,\delta)\in\Omega_{3.2}$, then for any positive constant ε there exists a positive constant C_ε such that

$$\theta_1(t,t_0,\xi)\le\frac{e_0+\Lambda(t_0)}{\Lambda(t_0)}\log\log\left(\frac{e_0+\Lambda(t)}{e_0+\Lambda(t_0)}\right)\le C_\varepsilon+\log(e_0+\Lambda(t))^{\varepsilon}\,.$$

Thus we obtain (ii).

If $(\beta,\gamma,\omega,\delta)\in\Omega_{1.2}\cup\Omega_{3.3}$, then there are positive constants C and ε_0 such that

$$\theta_1(t,t_0,\xi)\le C+\log(e_0+\Lambda(t))^{\varepsilon_0}\,.$$

Thus we obtain (iii).

If $(\beta,\gamma,\omega,\delta)\in\Omega_{3.4}$, then we have

$$\begin{aligned}\theta_1(t,t_0,\xi) &\le \frac{1}{\Lambda(t_0)\Big(\log(e_0+\Lambda(t_0))\Big)^{\delta}m(t_0)}\int\limits_{t_0}^{t}\frac{\lambda(s)\,ds}{\Lambda(s)\Big(\log(e_0+\Lambda(s))\Big)^{2\omega-\delta}}\\ &\le C\Big(\log(e_0+\Lambda(t))\Big)^{-2\omega+\delta+1}.\end{aligned}$$

Thus we obtain (iv).

If $(\beta,\gamma,\omega,\delta)\in\Omega_{1.3}$, then we have

$$\theta_1(t,t_0,\xi)\le C\Lambda(t)^{-2\beta+\gamma+1}\,.$$

Thus we obtain (v).

In the case that $|\alpha|\ge 1$ and $k\ge 1$ we have to estimate among other things the following integral (see Lemma 4.1):

$$\int\limits_{t_\xi}^{t}\frac{\lambda(s)}{\Lambda(s)^{\beta(k+1)-|\alpha|}\Big(\log(e_0+\Lambda(s))\Big)^{\omega(k+1)}\langle\xi\rangle_{m(s)}^{k-|\alpha|}}\,ds=:\theta_2(t,t_\xi,\xi).$$

If $\gamma=0$, that is, m is a positive constant, then $\theta_2(t,t_\xi,\xi)$ is estimated by $\theta_1(t,t_\xi,\xi)$ since $|\alpha|\le\beta(k-1)$ as in Corollary 5.1 from [23].

In the case that $(\beta,\gamma,\omega,\delta)$ belongs to $\Omega_{1.1}\cup\Omega_{1.2}\cup\Omega_{1.3}\cup\Omega_{2.4}$ we have

$$\begin{aligned}\theta_2(t,t_\xi,\xi) &= \int\limits_{t_\xi}^{t}\frac{\lambda(s)}{\Lambda(s)^{\beta(k+1)-|\alpha|}\langle\xi\rangle_{m(s)}^{k-|\alpha|}}ds\\ &\le \frac{1}{\Lambda(t_\xi)^{\gamma(k-|\alpha|)}\langle\xi\rangle_{m(t_\xi)}^{k-|\alpha|}}\int\limits_{t_\xi}^{t}\frac{\lambda(s)}{\Lambda(s)^{\beta(k+1)-|\alpha|-\gamma(k-|\alpha|)}}ds\\ &\le C_{\beta,\gamma,k,\alpha}N^{-k+|\alpha|}\Lambda(t_\xi)^{(1-\beta)(1+|\alpha|)}\le C_{\beta,\gamma,k,\alpha}\Lambda(t_0)^{(1-\beta)(1+|\alpha|)}\end{aligned}$$

for $k\ge|\alpha|$, where we used the inequality $\beta(k+1)-|\alpha|-\gamma(k-|\alpha|)>1$.

In the case that $(\beta,\gamma,\omega,\delta)$ belongs to $\Omega_{3.1}\cup\Omega_{3.2}\cup\Omega_{3.3}\cup\Omega_{3.4}$ we consider the integrals $\theta_2(t_0,t_\xi,\xi)$ and $\theta_2(t,t_0,\xi)$.

For the first integral we have

$$\begin{aligned}
&\theta_2(t_0,t_\xi,\xi)\\
&=\int_{t_\xi}^{t_0}\frac{\Big(\log(e_0+\Lambda(s))\Big)^{-\omega|\alpha|}}{\{\Lambda(s)\Big(\log(e_0+\Lambda(s))\Big)^{\omega}\langle\xi\rangle_{m(s)}\}^{k-|\alpha|-1}}\frac{\lambda(s)ds}{\Lambda(s)^2\ln^{2\omega}(e_0+\Lambda(s))\langle\xi\rangle_{m(s)}}\\
&\le\frac{\Big(\log(e_0+\Lambda(t_0))\Big)^{-\omega|\alpha|}}{N^{k-|\alpha|-1}}\theta_1(t_0,t_\xi,\xi)
\end{aligned}$$

for $k\ge|\alpha|+1$. Moreover, taking account of the inequality $\omega(k+1)-\delta(k-|\alpha|)>1$, we have

$$\begin{aligned}
\theta_2(t,t_0,\xi)\quad\le\quad&\frac{1}{\Big(\Lambda(t_0)\Big(\log(e_0+\Lambda(t_0))\Big)^{\delta}\langle\xi\rangle_{m(t_0)}\Big)^{k-|\alpha|}}\\
&\times\int_{t_0}^{t}\frac{\lambda(s)}{\Lambda(s)\Big(\log(e_0+\Lambda(s))\Big)^{\omega(k+1)-\delta(k-|\alpha|)}}ds\le C_{\omega,\delta,k,\alpha}.
\end{aligned}$$

By the previous estimates for θ_1 and θ_2, Lemma 2.1, Lemma 4.1, the Leibniz formula and monotonicity for $\langle\xi\rangle_{m(t)}^{-1}$, $e_0+\Lambda(t)$ with respect to t, we can also prove the other integrals to get the desired estimates for the higher-order derivatives of $Q_k=Q_k(t,t_\xi,\xi)$ with respect to ξ. □

Remark 4.1: The statements of Lemma 4.2 yield estimates for the amplitudes (appearing in the representations of solutions for (1.10)) in the hyperbolic zone. All these estimates allow L_p-L_q decay estimates besides those for $\Omega_{1.3}\cup\Omega_{3.4}$ which are better than the classical Floquet behavior from Section 8, but worse than the L_p-L_q decay estimates derived in Section 6. Thus it seems to be reasonable to expect a modified Floquet behavior in $\Omega_{1.3}\cup\Omega_{3.4}$.

4.2. Estimates for wave type equations in $Z_{hyp}(N)$

Let $v=v(t,\xi)$ be a solution of (3.1) in the case of wave type equations. Setting $\tilde{\mathcal{U}}:=(\lambda(t))|\xi|v,D_tv)^T$, then (3.1) can be transformed to the following first-order system:

$$D_t\tilde{\mathcal{U}}(t,\xi)-\begin{pmatrix}\frac{D_t\lambda(t)}{\lambda(t)}+\frac{\lambda(t)b(t)^2m(t)^2}{|\xi|} & \lambda(t)|\xi|\\ \lambda(t)b(t)^2|\xi| & 0\end{pmatrix}\tilde{\mathcal{U}}(t,\xi)=0.$$

Let $\tilde{U}=\tilde{U}(t,s,\xi)$ be the matrix-valued solution to the Cauchy problem

$$\begin{cases}D_t\tilde{U}(t,s,\xi)-\begin{pmatrix}\frac{D_t\lambda(t)}{\lambda(t)}+\frac{\lambda(t)b(t)^2m(t)^2}{|\xi|} & \lambda(t)|\xi|\\ \lambda(t)b(t)^2|\xi| & 0\end{pmatrix}\tilde{U}(t,s,\xi)=0\,,\\ \tilde{U}(s,s,\xi)=I\,.\end{cases}\tag{4.3}$$

Using the same diagonalizer M as in the Klein–Gordon case, we have

$$\left(D_t - \tilde{\mathcal{D}}(t,\xi) + \tilde{B}(t,\xi)\right)\tilde{V}_0(t,s,\xi) = 0,$$

where $\tilde{V}_0 := M\tilde{U}$,

$$\tilde{\mathcal{D}} = \tilde{\mathcal{D}}(t,\xi) := \begin{pmatrix} \tilde{\tau}_- & 0 \\ 0 & \tilde{\tau}_+ \end{pmatrix},$$

$$\tilde{B} = \tilde{B}(t,\xi) := -\frac{1}{2}\left(\frac{\lambda(t)b(t)^2m(t)^2}{|\xi|} + \frac{D_t(\lambda(t)b(t))}{\lambda(t)b(t)}\right)\begin{pmatrix} 0 & 1 \\ 1 & 0 \end{pmatrix}$$

and

$$\tilde{\tau}_\pm := \pm\lambda(t)b(t)|\xi| + \frac{D_t\lambda(t)}{\lambda(t)} + \frac{\lambda(t)b(t)^2m(t)^2}{2|\xi|}.$$

Here we note that $\tilde{\tau}_- - \tilde{\tau}_+ = -2\lambda b|\xi| \in \tilde{S}_{1,\omega}\{1,1,0\}$, $(\tilde{\tau}_- - \tilde{\tau}_+)^{-1} \in \tilde{S}_{1,\omega}\{-1,-1,0\}$ and $\lambda b^2m^2/|\xi| \le b_1^2\lambda/\left(N\Lambda(\log(e_0+\Lambda))^\omega\right)$, thus $\lambda b^2m^2/|\xi| \in \tilde{S}_{1,\omega}\{0,0,1\}$.

In analogy to the diagonalization procedure for Klein–Gordon type equations, we obtain the following proposition:

Proposition 4.2: *For any fixed non-negative integer k there exist matrix-valued functions $\tilde{M}_k = \tilde{M}_k(t,\xi) \in \tilde{S}_{1,\omega}\{0,0,0\}$, $\tilde{F}_k = \tilde{F}_k(t,\xi) \in \tilde{S}_{1,\omega}\{-1,-1,2\}$, $\tilde{R}_k = \tilde{R}_k(t,\xi) \in \tilde{S}_{1,\omega}\{-k,-k,k+1\}$ such that the following operator-valued equality holds in $Z_{hyp}(N)$:*

$$\tilde{M}_k^{-1}(D_t - \tilde{\mathcal{D}} + \tilde{B})\tilde{M}_k = D_t - \tilde{\mathcal{D}} + \tilde{F}_k + \tilde{R}_k,$$

where $\tilde{F}_k$ is diagonal while the matrices $\tilde{M}_k, \tilde{M}_k^{-1} \in \tilde{S}_{1,\omega}\{0,0,0\}$ provided that the parameter N in (2.1) is sufficiently large.

Let us define the matrix-valued function

$$\tilde{\mathcal{R}}_k(t,r,\xi) = -\tilde{F}_k(t,\xi) - \tilde{E}_2(r,t,\xi)\tilde{R}_k(t,\xi)\tilde{E}_2(t,r,\xi),$$

where

$$\tilde{E}_2(t,r,\xi) = \frac{\lambda(t)}{\lambda(r)}\begin{pmatrix} \exp\left(i\int_r^t\left(-\lambda(s)b(s)|\xi| + \frac{\lambda(s)b(s)^2m(s)^2}{2|\xi|}\right)ds\right) & 0 \\ 0 & \exp\left(i\int_r^t\left(\lambda(s)b(s)|\xi| + \frac{\lambda(s)b(s)^2m(s)^2}{2|\xi|}\right)ds\right) \end{pmatrix}.$$

Then, we have the next lemma.

Lemma 4.3: *The function $\tilde{\mathcal{R}}_k = \tilde{\mathcal{R}}_k(t,r,\xi)$ satisfies the following estimates:*

$$\|\partial_t^l\partial_r^p\partial_\xi^\alpha\tilde{\mathcal{R}}_k(t,r,\xi)\| \le C_{k,l,p,\alpha}(\lambda(t)|\xi|)^l(\lambda(r)|\xi|)^p|\xi|^{-|\alpha|}$$
$$\times\frac{\lambda(t)}{\Lambda(t)^2(\log(e_0+\Lambda(t)))^{2\omega}|\xi|}\left(\frac{\Lambda(t)^{|\alpha|}|\xi|^{|\alpha|}}{\left(\Lambda(t)\left(\log(e_0+\Lambda(t))\right)^\omega|\xi|\right)^{k-1}}+1\right)$$

with constants $C_{k,l,p,\alpha}$ independent of N.

Proof. For a proof we refer to the proof of Lemma 4.1. □

By the aid of $\tilde{\mathcal{R}}_k$ we define for $t \geq t_\xi$ the matrix-valued function

$$\tilde{Q}_k(t,t_\xi,\xi) := \sum_{j=1}^{\infty} i^j \int_{t_\xi}^{t} \tilde{\mathcal{R}}_k(\tau_1,t_\xi,\xi) \int_{t_\xi}^{\tau_1} \tilde{\mathcal{R}}_k(\tau_2,t_\xi,\xi) \cdots \int_{t_\xi}^{\tau_{j-1}} \tilde{\mathcal{R}}_k(\tau_j,t_\xi,\xi) d\tau_1 \cdots d\tau_j \,.$$

Lemma 4.4: *The function $\tilde{\Theta}_{k,\alpha} = \tilde{\Theta}_{k,\alpha}(t,\xi) := \|\partial_\xi^\alpha \tilde{Q}_k(t,t_\xi,\xi)\|$ satisfies the following estimates for any $(t,\xi) \in Z_{hyp}(N)$ and α satisfying $|\alpha| \leq k-1$:*

(i) *if $(\beta,\gamma,\omega,\delta) \in \Omega_{2.1} \cup \Omega_{2.2} \cup \Omega_{3.5} \cup \Omega_{3.6}$, then for any positive constant ε there exists a positive constant $C_{k,\alpha,\varepsilon}$ such that $\tilde{\Theta}_{k,\alpha}(t,\xi) \leq C_{k,\alpha,\varepsilon}|\xi|^{-|\alpha|}(e_0 + \Lambda(t))^\varepsilon$;*

(ii) *if $(\beta,\gamma,\omega,\delta) \in \Omega_{2.3} \cup \Omega_{3.7} \cup \{\Omega_{3.8} : \delta \in [-1,0)\} \cup \Omega_{3.10}$, then there exist positive constants $\varepsilon_0 \geq 1$ and $C_{k,\alpha}$ such that $\tilde{\Theta}_{k,\alpha}(t,\xi) \leq C_{k,\alpha}|\xi|^{-|\alpha|}(e_0 + \Lambda(t))^{\varepsilon_0}$.*

Proof. Let us consider the case $|\alpha| = 0$ and $k = 1$. We define $\tilde{\theta}_1 = \tilde{\theta}_1(t_2,t_1,\xi)$ by

$$\tilde{\theta}_1(t_2,t_1,\xi) := \int_{t_1}^{t_2} \frac{\lambda(s)}{\Lambda(s)^2\Big(\log(e_0+\Lambda(s))\Big)^{2\omega}|\xi|}\, ds.$$

Then $\tilde{\Theta}_{k,0}(t,\xi) \leq C_0 \exp(\tilde{\theta}_1(t,t_\xi,\xi))|\xi|^{-|\alpha|}$. Let T_0 be a positive constant satisfying $t_\xi < T_0 < t$. We shall separate our considerations to the intervals (t_ξ, T_0) and (T_0, t).

The proof in the cases $\Omega_{2.1}$, $\Omega_{2.2}$, $\Omega_{2.3}$, $\Omega_{3.5}$, $\Omega_{3.7}$, $\{\Omega_{3.8} : \delta \in [-1,0)\}$, $\Omega_{3.10}$ is similar as that for $\Omega_{3.6}$. The function $\tilde{\theta}_1(T_0,t_\xi,\xi)$ is estimated by

$$\begin{aligned}\tilde{\theta}_1(T_0,t_\xi,\xi) &\leq \frac{1}{\Big(\log(e_0+\Lambda(T_0))\Big)^{2\omega}|\xi|}\int_{t_\xi}^{T_0}\frac{\lambda(s)}{\Lambda(s)^2}ds \\ &\leq \frac{1}{\Lambda(t_\xi)\Big(\log(e_0+\Lambda(T_0))\Big)^{2\omega}|\xi|} < \infty.\end{aligned}$$

Integration by parts and monotonicity of Λ gives

$$\begin{aligned}\tilde{\theta}_1(t,T_0,\xi) \leq\ & \frac{1}{N}\Big(\log(e_0+\Lambda(t))\Big)^{-\omega} \\ & + \int_{T_0}^{t} \frac{-2\omega\lambda(s)}{\Lambda(s)(e_0+\Lambda(s))\Big(\log(e_0+\Lambda(s))\Big)^{2\omega+1}|\xi|}\, ds\,.\end{aligned}$$

Using

$$\int_{T_0}^{t} \frac{-2\omega\lambda(s)}{\Lambda(s)(e_0+\Lambda(s))\Big(\log(e_0+\Lambda(s))\Big)^{2\omega+1}|\xi|}\, ds \leq \frac{-2\omega}{N}\log(e_0+\Lambda(t))$$

we obtain to any given positive constant ε with a suitable large $N = N(\varepsilon)$ the estimate

$$\tilde{\theta}_1(t, t_\xi, \xi) \le C_N + \log(e_0 + \Lambda(t))^\varepsilon .$$

Thus, the case $\Omega_{3.6}$ is proved.

In the case $|\alpha| \ge 1$ and $k \ge 1$ it remains to estimate the integral (see Lemma 4.3)

$$\int_{t_\xi}^{t} \frac{\lambda(s)}{\Lambda(s)^{k+1-|\alpha|}\Big(\log(e_0 + \Lambda(s))\Big)^{\omega(k+1)}|\xi|^{k-|\alpha|}} ds =: \tilde{\theta}_2(t, t_\xi, \xi).$$

Using the same steps as in estimating $\tilde{\theta}_2(t_0, t_\xi, \xi)$ in the Klein–Gordon case we get for $\tilde{\theta}_2(t, T_0, \xi)$ the estimate

$$\begin{aligned}\tilde{\theta}_2(t, T_0, \xi) &= \int_{T_0}^{t} \frac{1}{\Big\{\Lambda(s)\Big(\log(e_0 + \Lambda(s))\Big)^{\omega}|\xi|\Big\}^{k-|\alpha|-1}} \\ &\quad \times \frac{\lambda(s)}{\Lambda(s)^2|\xi|}\Big(\log(e_0 + \Lambda(s))\Big)^{-\omega(|\alpha|+2)} ds \\ &\le C_{N,T_0}(e_0 + \Lambda(t))^\varepsilon\end{aligned}$$

for any positive real number $\varepsilon > 0$ with a sufficiently large $N = N(\varepsilon)$.

By the previous estimates for $\tilde{\theta}_1$ and $\tilde{\theta}_2$, Lemma 2.1, Corollary 4.2, the Leibniz formula and monotonicity for $e_0 + \Lambda(t)$ with respect to t, we can also estimate the other integrals which are necessary to derive the desired statements for higher-order derivatives of $Q_k = Q_k(t, t_\xi, \xi)$ with respect to ξ. □

5. Estimates for the amplitudes

By using the approach which was described after the proof of Lemma 4.1 we get WKB-representations for the solutions to (3.1) containing Fourier multipliers of the form $F^{-1}\Big(e^{i\varphi(t,\xi)}a(t,\xi)F(\psi)\Big)$, where $\varphi(t,\xi) = \int_0^t \lambda(s)b(s)\langle\xi\rangle_{m(s)}ds$ for Klein–Gordon type equations and $\varphi(t,\xi) = \int_0^t \lambda(s)b(s)ds|\xi|$ for wave-type equations. Here we applied Lemma 2.1 in the wave case to derive $\|\partial_\xi^\alpha \exp(i\int_0^{t_\xi} \lambda(s)b(s)ds|\xi|)\| \le C_{\alpha,\varepsilon}|\xi|^{\varepsilon-|\alpha|}$.

The following tables express the obtained estimates for the amplitudes. The number "0" in these tables means that the amplitude is uniformly bounded with respect to t, the letter "ε" means that the amplitude is estimated by $C_\varepsilon(e_0+\Lambda(t))^\varepsilon$ for any given positive real number ε, the letter "ε_0" means that the amplitude is estimated by $C(e_0 + \Lambda(t))^{\varepsilon_0}$ for a positive constant ε_0 (it may depend on n), the term "$e^{\log^s}$" means that the amplitude is estimated by $C\exp\Big(\big(\log(\Lambda(t))\big)^s\Big)$, the term "$e^s$" means that the amplitude is estimated by $C\exp(\Lambda(t)^s)$ and finally KG,

TABLE 1. Estimates of $U(t,0,\xi)$ in Z_{pd}

$\backslash i$	1	2	3	4	5	6	7	$8_{(\delta<-1)}$	$8_{(-1\le\delta<0)}$	9	10
$\Omega_{1.i}$	0	0	0	-	-						
$\Omega_{2.i}$	ε	ε	ε_0	0	-						
$\Omega_{3.i}$	0	0	0	0	ε	ε	ε_0	-	$e^{\log^{-\delta+1}}$	-	$e^{\log^{-\delta+1}}$

TABLE 2. Estimates of $\Theta_{k,0}(t,\xi)$ (or $\tilde{\Theta}_{k,0}(t,\xi)$) in Z_{hyp}

$\backslash i$	1	2	3	4	5	6	7	$8_{(\delta<-1)}$	$8_{(-1\le\delta<0)}$	9	10
$\Omega_{1.i}$	0	ε_0	$e^{\gamma-2\beta+1}$	-	-						
$\Omega_{2.i}$	ε	ε	ε_0	0	-						
$\Omega_{3.i}$	0	ε	ε_0	$e^{\log^{\delta-2\omega+1}}$	ε	ε	ε_0	—	ε_0	-	ε_0

TABLE 3. Estimates in $Z_{pd} \cup Z_{hyp}$

$\backslash i$	1	2	3	4	5	6	7	$8_{(\delta<-1)}$	$8_{(-1\le\delta<0)}$	9	10
$\Omega_{1.i}$	0/KG	ε_0/KG	$e^{\gamma-2\beta+1}$	F	F						
$\Omega_{2.i}$	ε/W	ε/W	ε_0/W	0/KG	F						
$\Omega_{3.i}$	0/KG	ε/KG	ε_0/KG	$e^{\log^{\delta-2\omega+1}}$	ε/W	ε/W	ε_0/W	F	$e^{\log^{-\delta+1}}$	F	$e^{\log^{-\delta+1}}$

W and F mean Klein–Gordon type decay, wave-type decay and classical Floquet behavior, respectively.

What can we expect from these tables, especially from Table 3? In $\Omega_{1.1}\cup\Omega_{1.2}\cup\Omega_{2.4}\cup\Omega_{3.1}\cup\Omega_{3.2}\cup\Omega_{3.3}$ we expect an L_p-L_q decay estimate of Klein–Gordon type. But therefore we have to prove a Littman-type lemma for oscillating integrals with phase functions depending on t and ξ but in a more complicated way than for those studied in [23]. This will be done in Section 6.1. In $\Omega_{2.1}\cup\Omega_{2.2}\cup\Omega_{2.3}\cup\Omega_{3.5}\cup\Omega_{3.6}\cup\Omega_{3.7}$ we expect an L_p-L_q decay estimate of wave type. Here the question for a Littman-type lemma is easy to answer, but L_p-L_q decay estimates for Fourier multipliers should be proved. Some model Fourier multipliers will be studied in Section 6.2. In $\Omega_{1.4}\cup\Omega_{1.5}\cup\Omega_{2.5}\cup\{\Omega_{3.8}: \delta<-1\}\cup\Omega_{3.9}$ the solutions show a classical Floquet behavior. Such a behavior was observed and studied in [22] for a special case. The proof, which is based on the considerations of [22], will be given in Section 8.

Between these qualitative quite different behaviors for the solutions we can find a modified Floquet behavior in some parts of $\Omega_{1.3} \cup \Omega_{3.4} \cup \{\Omega_{3.8} : \delta \in [-1,0)\} \cup \Omega_{3.10}$. Up to now we are not able to show that the behavior suggested by Table 3 is optimal in these four parts (see Remark 8.2).

6. Littman-type lemma and $L_p - L_q$ decay estimates for Fourier multipliers

6.1. Littman-type lemma for Klein–Gordon type equations

Here we prove a Littman-type lemma (see [14]) for oscillating integrals having as one part of the phase function $\int_0^t \lambda(s)b(s)\langle\xi\rangle_{m(s)}ds$. The time dependence of m raises new difficulties. In [23] the case in which m is a positive constant is considered.

Theorem 6.1: *Let us suppose that $|\partial_\xi^\alpha a(t,\xi)| \le C\langle\xi\rangle_{m(t)}^{-|\alpha|}$ for $|\alpha| \le n+2$ and $(t,\xi) \in [t_0,\infty)\times\mathbb{R}^n\setminus\{0\}$. Let r be a positive constant and let the function $\psi = \psi(\tau)$ be given from $C_0^\infty(\mathbb{R})$ with $\psi \equiv 1$ for $|\tau| \le r$, $\psi(\tau) \equiv 0$ for $|\tau| \ge 2r$ and $0 \le \psi \le 1$. The mass term is chosen as $m = m(t) = (e_0 + \Lambda(t))^{-\gamma}$ with $\gamma \in [0,\beta)$. Then there exists a constant C such that*

$$\left\| F^{-1}\left(e^{i\int_0^t \lambda(s)b(s)\langle\xi\rangle_{m(s)}ds} a(t,\xi)\psi\Big(\frac{|\xi|}{m(t)}\Big)\right)\right\|_{L_\infty(\mathbb{R}^n)} \le C\Big(1+\Lambda(t)\Big)^{-\frac{n}{2}(1+\gamma)}.$$

Proof. For the stationary points of the phase function we get the relation

$$x + \int_0^t \frac{\lambda(s)b(s)}{\langle\xi\rangle_{m(s)}}\xi ds = 0.$$

Case 1: Let us suppose that to given x and t we have no $\xi \in \mathbb{R}^n \setminus \{0\}$ such that this equality is satisfied. Using spherical polar coordinates it is clear that there exists no ξ: $|\xi| \in \mathbb{R}^+$ fulfilling

$$|x| = \int_0^t \frac{\lambda(s)b(s)|\xi|}{(|\xi|^2 + m(s)^2)^{1/2}}ds\ .$$

The function $f = f(y) = \int_0^t \frac{\lambda(s)b(s)y}{(y^2+m(s)^2)^{1/2}}ds$ is monotone increasing in $y \in (0,\infty)$. Consequently, $|x| \ge \sup_{y\in(0,\infty)} f(y)$. We estimate the function

$$\Phi = \Phi(x,t,\xi) = x + \int_0^t \frac{\lambda(s)b(s)}{\langle\xi\rangle_{m(s)}}\xi ds$$

in the following way:

$$|\Phi| \geq |x| - f(|\xi|) \geq f(|\xi_0|) - f(|\xi|),$$

where $\xi_0 \in \mathbb{R}^n \setminus \{0\}$ will be chosen later. To estimate $f(|\xi_0|) - f(|\xi|)$ we proceed as follows:

$$\begin{aligned} f(|\xi_0|) - f(|\xi|) &= \int_0^t \lambda(s)b(s)\left(\frac{|\xi_0|}{\langle\xi_0\rangle_{m(s)}} - \frac{|\xi|}{\langle\xi\rangle_{m(s)}}\right)ds \\ &\geq C_b \int_0^t \lambda(s)\left(\frac{|\xi_0|}{\langle\xi_0\rangle_{m(s)}} - \frac{|\xi|}{\langle\xi\rangle_{m(s)}}\right)ds. \end{aligned}$$

Taking account of Lemma A.1 from the Appendix we get

$$\int_0^t \lambda(s)\left(\frac{|\xi_0|}{\langle\xi_0\rangle_{m(s)}} - \frac{|\xi|}{\langle\xi\rangle_{m(s)}}\right) \geq \frac{1}{1+\gamma}\Lambda(t)\left(\frac{|\xi_0|}{\langle\xi_0\rangle_{m(t)}} - \frac{|\xi|}{\langle\xi\rangle_{m(t)}}\right).$$

Thus,

$$\begin{aligned} |\Phi| &\geq C_{b,\gamma}\Lambda(t)\left(\frac{|\xi_0|}{\langle\xi_0\rangle_{m(t)}} - \frac{|\xi|}{\langle\xi\rangle_{m(t)}}\right) \\ &= C_{b,\gamma}\Lambda(t)\left(\frac{\frac{|\xi_0|}{m(t)}}{\left(\left(\frac{|\xi_0|}{m(t)}\right)^2+1\right)^{1/2}} - \frac{\frac{|\xi|}{m(t)}}{\left(\left(\frac{|\xi|}{m(t)}\right)^2+1\right)^{1/2}}\right). \end{aligned}$$

For a given t we choose ξ_0 such that $\frac{|\xi_0|}{m(t)} = 3r$. The conditions with respect to ψ guarantee $\frac{|\xi|}{m(t)} \leq 2r$. Hence, $|\Phi| \geq C_{b,\gamma,r}\Lambda(t)$. This allows us to construct the operator $L = |\Phi|^{-2}\sum_{r=1}^{n}\Phi_r D_{\xi_r}$ satisfying

$$L\, e^{ix\cdot\xi + i\int_0^t \lambda(s)b(s)\langle\xi\rangle_{m(s)}ds} = e^{ix\cdot\xi + i\int_0^t \lambda(s)b(s)\langle\xi\rangle_{m(s)}ds}.$$

Integration by parts yields

$$\int_{\frac{|\xi|}{m(t)}\leq 2r} L^N e^{ix\cdot\xi + i\int_0^t \lambda(s)b(s)\langle\xi\rangle_{m(s)}ds} a(t,\xi)\psi\Big(\frac{|\xi|}{m(t)}\Big)d\xi \leq C_{N,b,\gamma,r}\Lambda(t)^{-n\gamma - N(1-\gamma)}$$

for all $t \geq t_0$. Here we need N derivatives of $a = a(t,\xi)$ with respect to ξ. Moreover we use that $|\partial^l_{\xi_r}\frac{\Phi_k}{|\Phi|^2}| \leq C_l\,\Lambda(t)^{-l(1-\gamma)}$; $|\Phi| \geq C_{b,\gamma,r}\,\Lambda(t)$ and supp ψ' belongs to $\Big\{\xi \in \mathbb{R}^n : \frac{|\xi|}{m(t)} \in (r, 2r)\Big\}$.

Using the assumptions of the theorem we have shown in the first case

$$\Big| \int_{\mathbb{R}^n} e^{ix\cdot\xi + i\int_0^t \lambda(s)b(s)\langle\xi\rangle_{m(s)}ds} a(t,\xi)\psi\Big(\frac{|\xi|}{m(t)}\Big)d\xi\Big| \tag{6.1}$$
$$\leq C_{N,b,\gamma,r}\Big(1+\Lambda(t)\Big)^{-n-2(1-\gamma)}.$$

Case 2: Let us suppose that for given x and t we have a stationary point $\xi_0 \in \mathbb{R}^n \setminus \{0\}$ satisfying $\frac{|\xi_0|}{m(t)} \geq 3r$. Then we can follow the ideas from the Case 1 and get an estimate of type (6.1).

Case 3: For given x and t there exists a stationary point $\xi_0 \in \mathbb{R}^n \setminus \{0\}$ fulfilling $\frac{|\xi_0|}{m(t)} \leq 3r$. The stationary point is uniquely determined. We develop the phase function

$$\varphi = \varphi(x,t,\xi) = \varphi(x,t,\xi_0) + (\xi-\xi_0)^T H_\varphi(\rho)(\xi-\xi_0)\,,$$

where $H_\varphi(\rho)$ denotes the Hessian of φ in a point $\rho = \rho(\xi_0,\xi)$ lying "between ξ_0, ξ". The Hessian $H_\varphi(\rho)$ is equal to

$$H_\varphi(\rho) = (H_{jk}(\rho))_{j,k=1}^n = \int_0^t \lambda(s)b(s)\left(\frac{\delta_{jk}}{\langle\rho\rangle_{m(s)}} - \frac{\rho_j\rho_k}{\langle\rho\rangle^3_{m(s)}}\right)ds.$$

Thus it remains to study

$$\int_{\frac{|\xi|}{m(t)}\leq 2r} e^{i\sum_{j,k=1}^n \int_0^t \frac{\lambda(s)b(s)}{\langle\rho\rangle_{m(s)}}\left(\delta_{jk} - \frac{\rho_j\rho_k}{\langle\rho\rangle^2_{m(s)}}\right)ds(\xi-\xi_0)_j(\xi-\xi_0)_k} a(t,\xi)\psi\Big(\frac{|\xi|}{m(t)}\Big)d\xi\,.$$

The exponent is a positive quadratic form, thus

$$\sum_{j,k=1}^n \int_0^t \frac{\lambda(s)b(s)}{\langle\rho\rangle_{m(s)}}\left(\delta_{jk} - \frac{\rho_j\rho_k}{\langle\rho\rangle^2_{m(s)}}\right)ds(\xi-\xi_0)_j(\xi-\xi_0)_k$$
$$\geq C_b \sum_{j,k=1}^n \int_0^t \frac{\lambda(s)}{\langle\rho\rangle_{m(s)}}\left(\delta_{jk} - \frac{\rho_j\rho_k}{\langle\rho\rangle^2_{m(s)}}\right)ds(\xi-\xi_0)_j(\xi-\xi_0)_k$$
$$= C_b \int_0^t \frac{\lambda(s)}{\langle\rho\rangle_{m(s)}} \sum_{j,k=1}^n \left(\delta_{jk} - \frac{\frac{\rho_j}{m(s)}\frac{\rho_k}{m(s)}}{\left(\left|\frac{\rho}{m(s)}\right|^2+1\right)}\right)(\xi-\xi_0)_j(\xi-\xi_0)_k\, ds\,.$$

We know that $|\frac{\rho}{m(s)}| \leq |\frac{\rho}{m(t)}| \leq 3r$. Thus we have a family of positive definite quadratic forms on a compact set with respect to the matrix of coefficients

$$(A_{jk})_{j,k=1}^n = \left(\delta_{jk} - \frac{\frac{\rho_j}{m(s)} \frac{\rho_k}{m(s)}}{\left(\left| \frac{\rho}{m(s)} \right|^2 + 1 \right)} \right)_{j,k=1}^n .$$

With positive constants C_1 and C_2 we have

$$C_1|\xi - \xi_0|^2 \leq \sum_{j,k=1}^n A_{jk}(\xi - \xi_0)_j(\xi - \xi_0)_k \leq C_2|\xi - \xi_0|^2 .$$

Similar to Lemma A.1 one can prove the second inequality in

$$\frac{\Lambda(t)}{\langle\rho\rangle_{m(t)}} \geq \int_0^t \frac{\lambda(s)}{\langle\rho\rangle_{m(s)}} ds \geq \frac{1}{1+\gamma} \frac{\Lambda(t)}{\langle\rho\rangle_{m(t)}} .$$

The first inequality is trivial. Summarizing brings

$$\left| \int_{\frac{|\xi|}{m(t)} \leq 2r} e^{ix\cdot\xi + i\int_0^t \lambda(s)b(s)\langle\xi\rangle_{m(s)}ds} a(t,\xi)\psi\Big(\frac{|\xi|}{m(t)}\Big)d\xi \right|$$

$$= \left| \int_{\frac{|\xi|}{m(t)} \leq 2r} e^{i\sum_{j,k=1}^n m(t)^2 \int_0^t \frac{\lambda(s)b(s)}{\langle\rho\rangle_{m(s)}} \left(\delta_{jk} - \frac{\rho_j\rho_k}{\langle\rho\rangle^2_{m(s)}}\right) ds \left(\frac{\xi-\xi_0}{m(t)}\right)_j \left(\frac{\xi-\xi_0}{m(t)}\right)_k} a(t,\xi)\psi\Big(\frac{|\xi|}{m(t)}\Big)d\xi \right|.$$

The above calculations show that there exists for each $t \geq t_0$ a diffeomorphism with a Jacobian $m(t)^n$ for all t such that

$$\sum_{j,k=1}^n m(t)^2 \int_0^t \frac{\lambda(s)b(s)}{\langle\rho\rangle_{m(s)}} \left(\delta_{jk} - \frac{\rho_j\rho_k}{\langle\rho\rangle^2_{m(s)}}\right) ds \left(\frac{\xi-\xi_0}{m(t)}\right)_j \left(\frac{\xi-\xi_0}{m(t)}\right)_k$$

will be mapped to $m(t)\Lambda(t)|\eta|^2$. It is enough to mention that the integration is over a uniform neighborhood U of 0 with respect to η. By using the assumptions concerning the amplitude $a = a(t,\xi)$ and by [14], [18] (here we need $|\alpha| \leq n+2$) the decay function of

$$\int_{U_\eta(0)} e^{im(t)\Lambda(t)|\eta|^2} a(t,\xi(\eta))\psi(\eta)d\eta$$

is

$$\Big(1 + \Lambda(t)\Big)^{-\frac{n}{2}(1+\gamma)}, \tag{6.2}$$

where $\partial_\eta^\alpha a(t,\xi(\eta))$ brings no difficulties due to the assumptions for $|\partial_\xi^\alpha a(t,\xi)|$, see for example

$$|\partial_\eta a(t,\xi(\eta))| \le C_1 \langle\xi\rangle_{m(t)}^{-1} |\partial_\eta \xi(\eta)| \le C_1 m(t)\Big(|\xi|^2 + m(t)^2\Big)^{-\frac{1}{2}} \le C_1 \,.$$

We are integrating over a compact set with respect to $\frac{\xi}{m(t)}$. If we fix a point $\frac{\tilde{\xi}}{m(t)}$ belonging to this set, then we can find a uniform neighborhood such that the Morse lemma is applicable and leads to the decay function from (6.2). Consequently, a finite number of applications allows us to study our starting integral in the Case 3. Thus we have studied all possible cases. From the estimate (6.1) together with the decay function (6.2) we conclude the statement of the theorem. □

Remark 6.1: In the case of $m(t) = (e_0+\Lambda(t))^{-1}\Big(\log(e_0+\Lambda(t))\Big)^{-\delta}$ with $\delta < \omega < 0$ we can prove by following the same approach under the same assumptions

$$\left\| F^{-1}\left(e^{i\int_0^t \lambda(s)b(s)\langle\xi\rangle_{m(s)}ds} a(t,\xi)\psi\Big(\tfrac{|\xi|}{m(t)}\Big)\right)\right\|_{L_\infty(\mathbb{R}^n)} \le C\frac{\big(\log(e_0+\Lambda(t))\big)^{\frac{\delta n}{2}}}{\big(1+\Lambda(t)\big)^n}.$$

6.2. $L_p - L_q$ decay estimates for Fourier multipliers in the Klein–Gordon case

In this subsection we will study $L_p - L_q$ decay estimates for model Fourier multipliers. From these results we immediately obtain the main results of this paper about $L_p - L_q$ decay estimates for solutions of (1.10) in the Klein–Gordon case which are formulated in the next section (see Theorems 7.1 and 7.2). As we explained in the Introduction the inclusion of the mass into the phase allows us to derive Klein–Gordon decay rates. But the general time-dependent mass term gives the phase function a more complicated structure. Thus new difficulties appear for application of the stationary phase method.

Proposition 6.1: *Let us suppose that $|\partial_\xi^\alpha a(t,\xi)| \le C\langle\xi\rangle_{m(t)}^{-|\alpha|}$ for $|\alpha| \le n+2$ and $(t,\xi) \in [t_0,\infty)\times\mathbb{R}^n\setminus\{0\}$. Then under the assumptions of Theorem 6.1 we have*

$$\left\| F^{-1}\left(e^{i\int_0^t \lambda(s)b(s)\langle\xi\rangle_{m(s)}ds} a(t,\xi)\langle\xi\rangle_{m(t)}^{-2l}\chi\Big(\tfrac{|\xi|}{m(t)}\Big)\right)\right\|_{L_\infty(\mathbb{R}^n)} \le C\Big(1+\Lambda(t)\Big)^{-\frac{n}{2}}$$

if $2l = \frac{n}{2}$, where $\chi = 1-\psi$ and ψ is taken from the previous theorem.

Proof. We choose a non-negative C^∞-function $\phi = \phi(\tau)$ having compact support in $\{\tau \in \mathbb{R} : 2^{m_0} \le \tau \le 2^{m_0+2}\}$. We set $\phi_k(\tau) := \phi(2^{-k}\tau)$ while $\phi_0(\tau) := 1 - \sum_{k=1}^\infty \phi_k(\tau)$. One can find such a function that $\sum_{k=-\infty}^{+\infty}\phi(2^{-k}\tau) = 1,\ \tau \ne 0$. Hence, supp $\phi_0 \subset \{\tau \in \mathbb{R} : |\tau| \le 2^{m_0+2}\}$. We consider for $k \ge 0$ all Fourier multipliers

$$F^{-1}\left(e^{i\int_0^t \lambda(s)b(s)\langle\xi\rangle_{m(s)}ds} a(t,\xi)\langle\xi\rangle_{m(t)}^{-2l}\chi\Big(\tfrac{|\xi|}{m(t)}\Big)\phi_k\Big(\Lambda(t)^\beta\langle\xi\rangle_{m(t)}\Big)F(\varphi)\right).$$

But it is enough to study these multipliers for $k \geq k_0$, k_0 large. The reason is that from $2^{m_0+k} \leq \Lambda(t)^\beta \langle\xi\rangle_{m(t)} \leq 2^{m_0+k+2}$, $t \geq t_0$, it follows that

$$\Lambda(t)^\beta \langle\xi\rangle_{m(t)} = \Lambda(t)^\beta m(t)\left(\left(\frac{|\xi|}{m(t)}\right)^2+1\right)^{\frac{1}{2}} \leq 2^{m_0+k+2}, \quad \frac{|\xi|}{m(t)} \leq C_k,$$

respectively. Thus we can apply the previous theorem. Let us consider the Fourier multipliers for $k \geq k_0$. We use the transformation

$$\Lambda(t)^\beta \langle\xi\rangle_{m(t)} = 2^k \langle\eta\rangle_1 .$$

The function χ vanishes for small $\frac{|\xi|}{m(t)}$. Theorem 6.1 allows us to include this function into the amplitude. The function χ helps us to understand that the transformation $\xi = \xi(\eta)\eta$ in radial direction has a Jacobian which can be estimated by $C(\frac{2^k}{\Lambda(t)^\beta})^n$. Thus, setting

$$I_k := \int_{\mathbb{R}^n} e^{i\,x\cdot\xi + i\int_0^t \lambda(s)b(s)\langle\xi\rangle_{m(s)}ds}\,\frac{a(t,\xi)}{\langle\xi\rangle_{m(t)}^{2l}}\chi\Big(\frac{|\xi|}{m(t)}\Big)\phi_k\Big(\Lambda(t)^\beta\langle\xi\rangle_{m(t)}\Big)d\xi$$

we have

$$\begin{aligned}
|I_k| &= \sup_{x\in\mathbb{R}^n}\left|\int_{\mathbb{R}^n} e^{i\,x\cdot\xi + i\int_0^t \lambda(s)b(s)\langle\xi\rangle_{m(s)}ds}\,\frac{a(t,\xi)}{\langle\xi\rangle_{m(t)}^{2l}}\chi\Big(\frac{|\xi|}{m(t)}\Big)\phi_k\Big(\Lambda(t)^\beta\langle\xi\rangle_{m(t)}\Big)d\xi\right| \\
&= \sup_{y\in\mathbb{R}^n}\left|\int_{\mathbb{R}^n} e^{i\,y\cdot\eta + i\int_0^t \lambda(s)b(s)\left(\frac{2^{2k}\langle\eta\rangle_1^2}{\Lambda(t)^{2\beta}}+m(s)^2-m(t)^2\right)^{\frac{1}{2}}ds}\,\frac{a(t,\xi(\eta))\Lambda(t)^{2\beta l}}{2^{2kl}\langle\eta\rangle_1^{2l}}\right. \\
&\qquad\left.\times\chi\Big(\frac{|\xi(\eta)|}{m(t)}\Big)\phi(\langle\eta\rangle_1)|\det J_\eta(\xi)|d\eta\right| \\
&\leq C\,2^{k(n-2l)}\Lambda(t)^{\beta(2l-n)}\sup_{y\in\mathbb{R}^n}\left|\int_{\mathbb{R}^n} e^{i\,y\cdot\eta + i\int_0^t \lambda(s)b(s)\left(\frac{2^{2k}\langle\eta\rangle_1^2}{\Lambda(t)^{2\beta}}+m(s)^2-m(t)^2\right)^{\frac{1}{2}}ds}\right. \\
&\qquad\left.\times\frac{a(t,\xi(\eta))}{\langle\eta\rangle_1^{2l}}\chi\Big(\frac{|\xi(\eta)|}{m(t)}\Big)\phi(\langle\eta\rangle_1)d\eta\right|.
\end{aligned}$$

Let us study the function $\chi(\frac{|\xi(\eta)|}{m(t)})$. The support of the derivatives lies in the set $\{(t,\xi) : C_1 \leq |\xi|/m(t) \leq C_2\}$ with suitable positive constants C_1 and C_2. Hence,

$$C_1^2 \leq \frac{2^{2k}\langle\eta\rangle_1^2}{\Lambda(t)^{2\beta}m(t)^2} - 1 \leq C_2^2\,, \tag{6.3}$$

that is, all derivatives $\partial_\eta^\alpha \chi$ are bounded. As in the proof of the previous theorem we see that the supposed estimates for $\partial_\xi^\alpha a$ are transferred to $\partial_\eta^\alpha a(t,\xi(\eta))$. Setting

$$v(t,\eta) := \frac{a(t,\xi(\eta))}{\langle\eta\rangle_1^{2l}}\chi\Big(\tfrac{|\xi(\eta)|}{m(t)}\Big)\phi(\langle\eta\rangle_1)$$

it remains to derive an estimate for

$$\sup_{y\in\mathbb{R}^n}\left|\int_K e^{\,i\,y\cdot\eta+i\int_0^t \lambda(s)b(s)\left(\frac{2^{2k}\langle\eta\rangle_1^2}{\Lambda(t)^{2\beta}}+m(s)^2-m(t)^2\right)^{\frac{1}{2}}ds} v(t,\eta)d\eta\right|,$$

where K denotes a compact set with respect to η, which is away from zero and the same for all $t\ge t_0$.

Case 1: Let us suppose that the phase function has a stationary point η_0, this means, for given y and t there exists η_0 with

$$y+\int_0^t \frac{2^{2k}}{\Lambda(t)^{2\beta}}\lambda(s)b(s)g(s,t,\langle\eta_0\rangle_1)^{-1}ds\ \eta_0 = 0\,,$$

where

$$g(s,t,\langle\eta\rangle_1) := \sqrt{\frac{2^{2k}\langle\eta\rangle_1^2}{\Lambda(t)^{2\beta}}+m(s)^2-m(t)^2}.$$

Moreover we suppose that η_0 belongs to a large ball $B_R(0)$.

For the Hessian of the phase function we obtain

$$H_\varphi(\rho) = \left(\int_0^t \frac{2^{2k}}{\Lambda(t)^{2\beta}}\frac{\lambda(s)b(s)}{g(s,t,\langle\rho\rangle_1)}\left(\delta_{jk}-\frac{\frac{2^k}{\Lambda(t)^\beta}\rho_j\frac{2^k}{\Lambda(t)^\beta}\rho_k}{\frac{2^{2k}\langle\rho\rangle_1^2}{\Lambda(t)^{2\beta}}+m(s)^2-m(t)^2}\right)ds\right)_{j,k=1}^n.$$

The boundedness for possible η_0 implies the boundedness for possible ρ. The monotonicity of $m=m(t)$ brings the uniformly positive definiteness of the following quadratic form in $\eta-\eta_0$:

$$\begin{aligned} C_1|\eta-\eta_0|^2 &\le \sum_{j,k=1}^n\left(\delta_{jk}-\frac{\frac{2^k}{\Lambda(t)^\beta}\rho_j\frac{2^k}{\Lambda(t)^\beta}\rho_k}{\frac{2^{2k}\langle\rho\rangle_1^2}{\Lambda(t)^{2\beta}}+m(s)^2-m(t)^2}\right)(\eta-\eta_0)_j(\eta-\eta_0)_k \\ &\le C_2|\eta-\eta_0|^2 \end{aligned}$$

with suitable positive constants C_1 and C_2. The integral

$$\int_0^t \frac{\lambda(s)}{g(s,t,\langle\rho\rangle_1)}ds \quad \text{is equivalent to} \quad \frac{\Lambda(t)}{g(t,t,\langle\rho\rangle_1)} = \frac{\Lambda(t)^{\beta+1}}{2^k\langle\rho\rangle_1}.$$

Thus we get as a large parameter $\frac{2^k}{\Lambda(t)^\beta}$. As in the proof of Theorem 6.1 we apply the Morse Lemma, [14], [18] (here we need $\alpha \leq n+2$) and obtain

$$\sup_{y\in\mathbb{R}^n}\left|\int_K e^{i\,y\cdot\eta+i\int_0^t \lambda(s)b(s)\left(2^{2k}\Lambda(t)^{-2\beta}\langle\eta\rangle_1^2+m(s)^2-m(t)^2\right)^{\frac12}ds} v(t,\eta)d\eta\right|$$
$$\leq C\left(2^k\Lambda(t)^{1-\beta}\right)^{-\frac{n}{2}}.$$

But this gives for I_k the estimate

$$|I_k| \leq C\,\Lambda(t)^{\beta(2l-n)}2^{k(n-2l)}\left(2^k\Lambda(t)^{1-\beta}\right)^{-\frac{n}{2}} \leq C\left(1+\Lambda(t)\right)^{-\frac{n}{2}}$$

if $2l=\frac{n}{2}$.

Case 2: Let us suppose that the phase function has a stationary point η_0 outside the large ball $B_R(0)$. Then we have to estimate

$$\frac{2^{2k}}{\Lambda(t)^{2\beta}}\left(\int_0^t\frac{\lambda(s)ds}{g(s,t,\langle\eta_0\rangle_1)}|\eta_0|-\int_0^t\frac{\lambda(s)ds}{g(s,t,\langle\eta\rangle_1)}|\eta|\right).$$

For this difference we get by using Lemma A.2 the estimate from below

$$\frac{2^{2k}}{\Lambda(t)^{2\beta}}\left(\int_0^t\frac{\lambda(s)ds}{g(s,t,\langle\eta_0\rangle_1)}|\eta_0|-\int_0^t\frac{\lambda(s)ds}{g(s,t,\langle\eta\rangle_1)}|\eta|\right) \geq 2^k\Lambda(t)^{1-\beta}\left(\frac{|\eta_0|}{\langle\eta_0\rangle_1}-\frac{|\eta|}{\langle\eta\rangle_1}\right),$$

where $\eta_0 \notin B_R(0)$ and $\eta \in K$. After choosing the operator L as in the proof of Theorem 6.1 and partial integration leads to

$$\sup_{y\in\mathbb{R}^n}\left|\int_K e^{i\,y\cdot\eta+i\int_0^t \lambda(s)b(s)\left(\frac{2^{2k}\langle\eta\rangle_1^2}{\Lambda(t)^{2\beta}}+m(s)^2-m(t)^2\right)^{\frac12}ds} v(t,\eta)d\eta\right|$$
$$\leq C\,2^{k(n-N)}\Lambda(t)^{-N-\beta(n-N)} \leq C\,\Lambda(t)^{-n}$$

for $N=n$.

Case 3: If the phase function has for given y and t no stationary point, then we use the monotonicity of

$$\int_0^t\frac{\lambda(s)ds|\eta_0|}{g(s,t,\langle\eta_0\rangle_1)}$$

with respect to $|\eta_0|$ and proceed as in the previous case. From the estimates for I_k and with [1] we conclude the desired L_1-L_∞ estimate. □

Remark 6.2: In the case of $m(t) = (e_0+\Lambda(t))^{-1}\Big(\log(e_0+\Lambda(t))\Big)^{-\delta}$ with $\delta < \omega < 0$ we can prove under the same assumptions by following the same approach

$$\Big\|F^{-1}\Big(e^{i\int_0^t \lambda(s)b(s)\langle\xi\rangle_{m(s)}ds}a(t,\xi)\langle\xi\rangle_{m(t)}^{-2l}\chi\Big(\frac{|\xi|}{m(t)}\Big)\Big)\Big\|_{L_\infty(\mathbb{R}^n)} \le C\,\Big(1+\Lambda(t)\Big)^{-\frac{n}{2}}$$

if $2l = \frac{n}{2}$, where $\chi = 1-\psi$ and ψ is taken from the previous theorem.

Straightforwardly, one can derive corresponding L_2-L_2 estimates. These estimates and the L_1-L_∞ estimates from Theorem 6.1, Proposition 6.1 and Remarks 6.1, 6.2 give, after applying an interpolation argument [1], the next results.

Corollary 6.1: *We assume for the mass term $m(t) = (e_0+\Lambda(t))^{-\gamma}$ with $\gamma \in [0,\beta)$, $\beta \in (0,1]$, where β is taken from (1.7). Let us suppose that $|\partial_\xi^\alpha a(t,\xi)| \le C\langle\xi\rangle_{m(t)}^{-|\alpha|}$ for $|\alpha| \le n+2$ and $(t,\xi) \in [t_0,\infty)\times\mathbb{R}^n\setminus\{0\}$. Then we have the L_p-L_q decay estimate*

$$\Big\|F^{-1}\Big(e^{i\int_0^t \lambda(s)b(s)\langle\xi\rangle_{m(s)}ds}a(t,\xi)F(\varphi)(\xi)\Big)\Big\|_{L_q(\mathbb{R}^n)}$$
$$\le C\,\Big(1+\Lambda(t)\Big)^{-\frac{n}{2}(\frac{1}{p}-\frac{1}{q})}\|\varphi\|_{W_p^{N_p}(\mathbb{R}^n)}$$

with $N_p = \Big[\frac{n}{2}\Big(\frac{1}{p}-\frac{1}{q}\Big)\Big]+1$, $\frac{1}{p}+\frac{1}{q}=1$ and $1<p\le 2$.

Corollary 6.2: *We assume for the mass term $m(t) = (e_0+\Lambda(t))^{-1}\Big(\log(e_0+\Lambda(t))\Big)^{-\delta}$ with $\delta < \omega < 0$, where ω is taken from (1.7). Let us suppose that $|\partial_\xi^\alpha a(t,\xi)| \le C\langle\xi\rangle_{m(t)}^{-|\alpha|}$ for $|\alpha| \le n+2$ and $(t,\xi) \in [t_0,\infty)\times\mathbb{R}^n\setminus\{0\}$. Then we have the L_p-L_q decay estimate*

$$\Big\|F^{-1}\Big(e^{i\int_0^t \lambda(s)b(s)\langle\xi\rangle_{m(s)}ds}a(t,\xi)F(\varphi)(\xi)\Big)\Big\|_{L_q(\mathbb{R}^n)}$$
$$\le C\Big(1+\Lambda(t)\Big)^{-\frac{n}{2}(\frac{1}{p}-\frac{1}{q})}\|\varphi\|_{W_p^{N_p}(\mathbb{R}^n)}$$

with $N_p = \Big[\frac{n}{2}\Big(\frac{1}{p}-\frac{1}{q}\Big)\Big]+1$, $\frac{1}{p}+\frac{1}{q}=1$ and $1<p\le 2$.

6.3. $L_p - L_q$ decay estimates for Fourier multipliers in the wave case

One strategy is to generalize the proof of [18] to the Fourier multiplier

$$F^{-1}\Big(e^{i\int_0^t \lambda(s)b(s)ds|\xi|}a(t,\xi)|\xi|^{-2l}F(\varphi)(\xi)\Big).$$

Here we understand under a generalization that it is sufficient to divide the phase space into two zones, the pseudodifferential and the hyperbolic one, contrary to the considerations for Fourier multipliers appearing in the case of Klein–Gordon type equations (cf. with proof of Theorem 7.1). The considerations in the pseudo-differential zone, hyperbolic zone, correspond to the study of (1.3) for small $|\xi|$, large $|\xi|$, respectively. The content of this section is based on the considerations of Section 3.3 of [24]. For the considerations of this section see also [21].

6.3.1. $L_p - L_q$ DECAY ESTIMATES FOR FOURIER MULTIPLIERS WITH AMPLITUDES SUPPORTED IN THE PSEUDODIFFERENTIAL ZONE

Proposition 6.2: *Let the function $\psi = \psi(\eta) \in C_0^\infty(\mathbb{R}^n)$ be the same as in Theorem 6.1 with $r = 1$. Suppose that $|a(t,\xi)| \le C$ for all $(t,\xi) \in Z_{pd}(N)$. Setting $K(t) := N\Lambda(t)^{-1}\Big(\log(e_0+\Lambda(t))\Big)^{-\omega}$, then we have the following $L_p - L_q$ estimate for $t > 0$:*

$$\Big\|F^{-1}\Big(e^{i\int_0^t \lambda(s)b(s)ds|\xi|}|\xi|^{-2l}\psi\Big(\frac{|\xi|}{K(t)}\Big)a(t,\xi)F(\varphi)(\xi)\Big)\Big\|_{L_q(\mathbb{R}^n)}$$
$$\le C_N\Lambda(t)^{2l-n(\frac{1}{p}-\frac{1}{q})}\|\varphi\|_{L_p(\mathbb{R}^n)}$$

provided that $l \ge 0$, $1 < p \le 2$, $\frac{1}{p}+\frac{1}{q} = 1$, $2l \le n\Big(\frac{1}{p}-\frac{1}{q}\Big)$.

Proof. We consider

$$I_0 := \Big\|F^{-1}\Big(|\xi|^{-2l}\psi\Big(\frac{|\xi|}{K(t)}\Big)a(t,\xi)F(\varphi)(\xi)\Big)\Big\|^q_{L_q(\mathbb{R}^n)}.$$

Using the transformations $\eta = \frac{\xi}{K(t)}$, $z = K(t)x$ and setting

$$\tilde a(t,\eta) = |\eta|^{-2l}a(t,K(t)\eta)$$

we see that

$$\begin{aligned} I_0 &= K(t)^{nq-2lq-n}\left\|F^{-1}\left(\psi(|\eta|)|\eta|^{-2l}a(t,K(t)\eta)F(\varphi)(K(t)\eta)\right)\right\|^q_{L_q(\mathbb{R}^n)} \\ &= K(t)^{nq-2lq-n}\Big\|F^{-1}\Big(\psi(|\eta|)\tilde a(t,\eta)F(\varphi)(K(t)\eta)\Big)\Big\|^q_{L_q(\mathbb{R}^n)}. \end{aligned}$$

The point $(t,K(t)\eta) = (t,\xi)$ with $|\eta| = N^{-1}\Lambda(t)\Big(\log(e_0+\Lambda(t))\Big)^{\omega}|\xi| \le 1$, that is $\eta \in \operatorname{supp}\psi$, belongs to $Z_{pd}(N)$. Therefore

$$\tilde a(t,\eta) \le C|\eta|^{-2l} \quad \text{for all} \quad (t,\eta) \in [0,\infty)\times(\mathbb{R}^n\setminus\{0\}).$$

Thus

$$I_0^{\frac{1}{q}} = K(t)^{n-2l-\frac{n}{q}}\Big\|F^{-1}\Big(\psi(|\eta|)\tilde a(t,\eta)F(\varphi)(K(t)\eta)\Big)\Big\|_{L_q(\mathbb{R}^n)}.$$

Now let us denote

$$G_t := F^{-1}\Big(\psi(\eta)\tilde a(t,\eta)\Big).$$

We have

$$\operatorname{meas}\{\eta : |F(G_t) \ge \tau|\} \le \operatorname{meas}\{\eta : C|\eta|^{-2l} \ge \tau\} \le C\tau^{-\frac{n}{2l}}.$$

Due to Theorem 1.11 [8] we have $F(G_t) \in M_p^q$ for all $n\Big(\frac{1}{p}-\frac{1}{q}\Big) \geq 2l$. Here M_p^q denotes the set of Fourier transforms $F(f)$ of distributions $f \in L_p^q$, and L_p^q denotes the space of distributions $f \in \mathcal{S}'$ satisfying

$$\|f * u\|_{L_q(\mathbb{R}^n)} \leq C\|u\|_{L_p(\mathbb{R}^n)}, \quad u \in \mathcal{S},$$

where the constant C is independent of u. Hence, $G_t \in L_p^q$ and

$$\begin{aligned} \Big\|F^{-1}\Big(F(\varphi)(K(t)\eta)\Big)\Big\|_{L_q(\mathbb{R}^n)} &\leq C_p K(t)^{-n}\Big\|\varphi\left(\frac{x}{K(t)}\right)\Big\|_{L_p(\mathbb{R}^n)} \\ &\leq C_p K(t)^{-n+\frac{n}{p}}\|\varphi\|_{L_p(\mathbb{R}^n)} \end{aligned}$$

for all $t \in (0,\infty)$. Thus we have proved

$$I_0^{\frac{1}{q}} \leq CK(t)^{-2l+n(\frac{1}{p}-\frac{1}{q})}\|\varphi\|_{L_p(\mathbb{R}^n)} \leq C_N\Lambda(t)^{2l-n(\frac{1}{p}-\frac{1}{q})}\|\varphi\|_{L_p(\mathbb{R}^n)}$$

by using the assumption $2l \leq n\Big(\frac{1}{p}-\frac{1}{q}\Big)$. Summarizing we get the $L_p - L_q$ decay estimate

$$\begin{aligned} &\Big\|F^{-1}\Big(e^{i\int_0^t \lambda(s)b(s)ds|\xi|}|\xi|^{-2l}\psi\Big(\frac{|\xi|}{K(t)}\Big)a(t,\xi)F(\varphi)(\xi)\Big)\Big\|_{L_q(\mathbb{R}^n)} \\ &\leq C_N\Lambda(t)^{2l-n(\frac{1}{p}-\frac{1}{q})}\|\varphi\|_{L_p(\mathbb{R}^n)}. \end{aligned}$$

Thus the proposition is proved. □

6.3.2. $L_p - L_q$ DECAY ESTIMATES FOR FOURIER MULTIPLIERS WITH AMPLITUDES SUPPORTED IN THE HYPERBOLIC ZONE

Proposition 6.3: *Let us choose $\psi \in C_0^\infty(\mathbb{R}^n)$ as in Proposition 6.2. Suppose that $|\partial_\xi^\alpha a(t,\xi)| \leq C_\alpha|\xi|^{-|\alpha|}$ for all $|\alpha| \leq n+1$, and $(t,\xi) \in Z_{hyp}(N)$. Setting $K(t) := N\Lambda(t)^{-1}\Big(\log(e_0+\Lambda(t))\Big)^{-\omega}$, then to each $\varepsilon > 0$ there exists a positive constant $C_{\varepsilon,l,N}$ such that the following $L_p - L_q$ decay estimate holds for $t \geq 1$:*

$$\begin{aligned} &\Big\|F^{-1}\Big(e^{i\int_0^t \lambda(s)b(s)ds|\xi|}|\xi|^{-2l}\Big(1-\psi\left(\frac{|\xi|}{K(t)}\right)\Big)a(t,\xi)F(\varphi)(\xi)\Big)\Big\|_{L_q(\mathbb{R}^n)} \\ &\leq C_{\varepsilon,l,N}\Lambda(t)^{2l+\varepsilon-n(\frac{1}{p}-\frac{1}{q})}\|\varphi\|_{L_p(\mathbb{R}^n)} \end{aligned}$$

provided that $l \geq 0$, $1 < p \leq 2$, $\frac{1}{p}+\frac{1}{q} = 1$ and $\frac{(n+1)}{2}\Big(\frac{1}{p}-\frac{1}{q}\Big) \leq 2l \leq n\Big(\frac{1}{p}-\frac{1}{q}\Big)$. If $t \in (0,1]$, then the same estimate holds with $\varepsilon = 0$.

Proof. The point (t,ξ) with $\frac{|\xi|}{K(t)} \geq 1$, that is, $(t,\xi) \in \operatorname{supp}\Big(1-\psi(\frac{\xi}{K(t)})\Big)$, belongs to $Z_{hyp}(N)$. Furthermore, we choose a non-negative function $\phi = \phi(\tau)$ having compact support in $\{\tau \in \mathbb{R} : \frac{1}{2} \leq |\tau| \leq 2\}$. We set $\phi_k(\tau) := \phi(2^{-k}\tau)$ while $\phi_0(\tau) := 1-\sum_{k=1}^\infty \phi_k(\tau)$. One can find (see [1]) the function ϕ in such a way that supp $\phi_0 \subset \{\tau \in \mathbb{R} : |\tau| \leq 2\}$ and that $1 = \sum_{k=-\infty}^{\infty} \phi(2^{-k}\tau),\ \tau \neq 0$. This partition of unity allows us to estimate

$$\Big\| F^{-1}\Big(e^{i\int_0^t \lambda(s)b(s)ds|\xi|} |\xi|^{-2l} \Big(1-\psi\Big(\frac{|\xi|}{K(t)}\Big)\Big) a(t,\xi)F(\varphi)(\xi)\Big)\Big\|_{L_q(\mathbb{R}^n)}$$

$$\leq \sum_{k=0}^{\infty} \Big\| F^{-1}\Big(e^{i\int_0^t \lambda(s)b(s)ds|\xi|} \Big(1-\psi\Big(\frac{|\xi|}{K(t)}\Big)\Big) \phi_k\Big(\frac{|\xi|}{K(t)}\Big) \frac{a(t,\xi)}{|\xi|^{-2l}} F(\varphi)(\xi)\Big)\Big\|_{L_q(\mathbb{R}^n)}.$$

Similar to the consideration for Fourier multipliers with amplitudes supported in $Z_{pd}(N)$ we get

$$\Big\| F^{-1}\Big(e^{i\int_0^t \lambda(s)b(s)ds|\xi|} |\xi|^{-2l} \Big(1-\psi\Big(\frac{|\xi|}{K(t)}\Big)\Big) \phi_0\Big(\frac{|\xi|}{K(t)}\Big) a(t,\xi)F(\varphi)(\xi)\Big)\Big\|_{L_q(\mathbb{R}^n)}$$
$$\leq C_N \Lambda(t)^{2l-n(\frac{1}{p}-\frac{1}{q})} \|\varphi\|_{L_p(\mathbb{R}^n)}$$

for $2l \leq n\Big(\frac{1}{p}-\frac{1}{q}\Big)$. Here we use the inequality $|\xi|^{-1} \leq N^{-1}\Lambda(t)\Big(\log(e_0+\Lambda(t))\Big)^{-\omega}$. Let us set $\eta := 2^{-k}\frac{\xi}{K(t)}$. According to Lemma 3 [1] and assumptions to a we have

$$\Big\| F^{-1}\Big(e^{i\int_0^t \lambda(s)b(s)ds|\xi|} |\xi|^{-2l} \Big(1-\psi\Big(\frac{|\xi|}{K(t)}\Big)\Big) \phi_k\Big(\frac{|\xi|}{K(t)}\Big) a(t,\xi)\Big)\Big\|_{L_\infty(\mathbb{R}^n)}$$
$$\leq C 2^{k(n-2l)} K(t)^{n-2l}$$
$$\times \Big\| F^{-1}\Big(e^{i2^k K(t)\int_0^t \lambda(s)b(s)ds|\eta|} |\eta|^{-2l} (1-\psi(2^k|\eta|))\, \phi(|\eta|) a(t, 2^k K(t)\eta)\Big)\Big\|_{L_\infty(\mathbb{R}^n)}$$
$$\leq C 2^{k(n-2l)} K(t)^{n-2l} \left(1 + 2^k K(t) \int_0^t \lambda(s)b(s)ds\right)^{-\frac{n-1}{2}}$$
$$\times \sum_{|\alpha|<n+1} \big\| D_\eta^\alpha \left(|\eta|^{-2l} (1-\psi(2^k|\eta|))\, \phi(|\eta|) a(t, 2^k K(t)\eta)\right)\big\|_{L_\infty(\mathbb{R}^n)}$$

by using [14], [18] (here we need $|\alpha| \leq n+1$). Taking into consideration

$$\big| D_\eta^\alpha \left(|\eta|^{-2l} (1-\psi(2^k|\eta|))\, \phi(|\eta|) a(t, 2^k K(t)\eta)\right)\big|$$
$$\leq \sum_{|\tilde\alpha|=|\alpha|} C_{\tilde\alpha} \big| D_\eta^{\tilde\alpha_1} |\eta|^{-2l} D_\eta^{\tilde\alpha_2}(1-\psi(2^k|\eta|)) D_\eta^{\tilde\alpha_3}\phi(|\eta|) D_\eta^{\tilde\alpha_4} a(t, 2^k K(t)\eta)\big|$$
$$\leq \sum_{|\tilde\alpha|=|\alpha|} C_{\tilde\alpha} |\eta|^{-2l-|\tilde\alpha_1|} \big| D_\eta^{\tilde\alpha_2}(1-\psi(2^k|\eta|))\big|$$
$$\times \big|(\partial_\eta^{\tilde\alpha_3}\phi)(|\eta|)\big| (2^k K(t))^{|\tilde\alpha_4|} \Big| D_\xi^{\tilde\alpha_4} a(t, 2^k K(t)\eta)\Big|,$$

$$\Big| D_\xi^{\tilde\alpha_4} a(t, 2^k K(t)\eta)\Big| \leq C_{\alpha_4} \left(2^k K(t)|\eta|\right)^{-|\tilde\alpha_4|}$$

and

$$\operatorname{supp}\left((1-\psi(2^k|\eta|))\phi(|\eta|)\right) \subset \left\{|\eta| \in \mathbb{R} \, : \tfrac{1}{2} \le |\eta| \le 2\right\},$$

we have

$$\left|D_\eta^\alpha \left(|\eta|^{-2l}\left(1-\psi(2^k|\eta|)\right)\phi(2^k|\eta|)a(t,2^kK(t)\eta)\right)\right| \le C_\alpha.$$

Therefore, it follows that

$$\begin{aligned}
&\left\| F^{-1}\left(e^{i\int_0^t \lambda(s)b(s)ds|\xi|}|\xi|^{-2l}\left(1-\psi\left(\tfrac{|\xi|}{K(t)}\right)\right)\phi_k\left(\tfrac{|\xi|}{K(t)}\right)a(t,\xi)\right)\right\|_{L_\infty(\mathbb{R}^n)} \\
&\le C2^{k(n-2l-\frac{n-1}{2})}K(t)^{n-2l}\left(1+K(t)\int_0^t \lambda(s)b(s)ds\right)^{-\frac{n-1}{2}} \\
&\le C2^{k(-2l+\frac{n+1}{2})}\Lambda(t)^{-n+2l}.
\end{aligned}$$

Finally, this allows us to conclude the L_1-L_∞ estimate

$$\begin{aligned}
&\left\| F^{-1}\left(e^{i\int_0^t \lambda(s)b(s)ds|\xi|}|\xi|^{-2l}\left(1-\psi\left(\tfrac{|\xi|}{K(t)}\right)\right)\phi_k\left(\tfrac{|\xi|}{K(t)}\right)a(t,\xi)F(\varphi)(\xi)\right)\right\|_{L_\infty(\mathbb{R}^n)} \\
&\le C2^{k(-2l+\frac{n+1}{2})}\Lambda(t)^{-n+2l}\|\varphi\|_{L_1(\mathbb{R}^n)}.
\end{aligned}$$

Taking account of

$$\begin{aligned}
&\left\| F^{-1}\left(e^{i\int_0^t \lambda(s)b(s)ds|\xi|}|\xi|^{-2l}\left(1-\psi\left(\tfrac{|\xi|}{K(t)}\right)\right)\phi_k\left(\tfrac{|\xi|}{K(t)}\right)a(t,\xi)\right)\right\|_{L_2(\mathbb{R}^n)} \\
&\le \sup_{2^{k-1}\le|\xi|/K(t)\le 2^{k+1}} \left||\xi|^{-2l}a(t,\xi)\right| \le C2^{2l-2kl}K(t)^{-2l} \le C_{\varepsilon,N}2^{2l-2kl}\Lambda(t)^{2l+\varepsilon}.
\end{aligned}$$

gives the L_2-L_2 estimates

$$\begin{aligned}
&\left\| F^{-1}\left(e^{i\int_0^t \lambda(s)b(s)ds|\xi|}|\xi|^{-2l}\left(1-\psi\left(\tfrac{|\xi|}{K(t)}\right)\right)\phi_k\left(\tfrac{|\xi|}{K(t)}\right)a(t,\xi)F(\varphi)(\xi)\right)\right\|_{L_2(\mathbb{R}^n)} \\
&\le C_{\varepsilon,N}2^{2l-2kl}\Lambda(t)^{2l+\varepsilon}\|\varphi\|_{L_2(\mathbb{R}^n)}.
\end{aligned}$$

Using the previous estimates an interpolation argument [1] yields the L_p-L_q decay estimates

$$\begin{aligned}
&\left\| F^{-1}\left(e^{i\int_0^t \lambda(s)b(s)ds|\xi|}|\xi|^{-2l}\left(1-\psi\left(\tfrac{\xi}{K(t)}\right)\right)\phi_k\left(\tfrac{\xi}{K(t)}\right)a(t,\xi)F(\varphi)(\xi)\right)\right\|_{L_q(\mathbb{R}^n)} \\
&\le C_{\varepsilon,N}2^{2l+k\left(\frac{n+1}{2}(\frac{1}{p}-\frac{1}{q})-2l\right)}\Lambda(t)^{2l+\varepsilon-n(\frac{1}{p}-\frac{1}{q})}\|\varphi\|_{L_p(\mathbb{R}^n)}
\end{aligned}$$

provided that $1<p\le 2$, $\frac{1}{p}+\frac{1}{q}=1$, $2l\le n\left(\frac{1}{p}-\frac{1}{q}\right)$. Using $\frac{(n+1)}{2}\left(\frac{1}{p}-\frac{1}{q}\right)\le 2l\le n\left(\frac{1}{p}-\frac{1}{q}\right)$ the application of Lemma 2 [1] proves the proposition. □

6.3.3. $L_p - L_q$ DECAY ESTIMATES FOR FOURIER MULTIPLIERS IN THE WAVE CASE
Propositions 6.2 and 6.3 bring together the following result for Fourier multipliers with prescribed behavior of the amplitude functions.

Theorem 6.2: *Suppose that for all $t \geq 0$ the amplitude function satisfies*

$$|a(t,\xi)| \leq C, \quad (t,\xi) \in Z_{pd}(N),$$

and

$$\left|\partial_\xi^\alpha a(t,\xi)\right| \leq C_\alpha |\xi|^{-|\alpha|}, \; |\alpha| \leq n+1, \quad (t,\xi) \in Z_{hyp}(N).$$

Then to each $\varepsilon > 0$ there exists a positive constant $C_{\varepsilon,N}$ such that the following $L_p - L_q$ decay estimate holds for $t \geq 0$:

$$\left\|F^{-1}\Big(e^{i\int_0^t \lambda(s)b(s)ds|\xi|} a(t,\xi)F(\varphi)(\xi)\Big)\right\|_{L_q(\mathbb{R}^n)}$$
$$\leq C_{\varepsilon,N}\Big(1+\Lambda(t)\Big)^{\varepsilon - \frac{n-1}{2}\left(\frac{1}{p}-\frac{1}{q}\right)} \|\varphi\|_{W_p^{N_p}(\mathbb{R}^n)}$$

with $N_p = \left[n\left(\frac{1}{p}-\frac{1}{q}\right)\right]+1$, $\frac{1}{p}+\frac{1}{q}=1$ and $1<p\leq 2$.

Proof. Instead of $F^{-1}\Big(e^{i\int_0^t \lambda(s)b(s)ds|\xi|} a(t,\xi)F(\varphi)(\xi)\Big)$ we study

$$F^{-1}\Big(e^{i\int_0^t \lambda(s)b(s)ds|\xi|} |\xi|^{-2l}a(t,\xi)|\xi|^{2l}F(\varphi)(\xi)\Big).$$

If $t \in (0,1]$, then we apply Propositions 6.2 and 6.3 with $2l = n\left(\frac{1}{p}-\frac{1}{q}\right)$. This gives the regularity of the datum φ. If $t > 1$, then we apply Propositions 6.2 and 6.3 with $2l = \frac{(n+1)}{2}\left(\frac{1}{p}-\frac{1}{q}\right)$. This gives the decay rate.

7. $L_p - L_q$ decay estimates for solutions of Klein–Gordon equations

Summarizing all the results from the previous sections we have proved the following $L_p - L_q$ decay estimates for the solutions of our model Cauchy problem

$$u_{tt} - \lambda(t)^2 b(t)^2(\Delta - m(t)^2)u = 0, \quad u(0,x) = \varphi(x), \; u_t(0,x) = \psi(x), \tag{7.1}$$

under the next assumptions:

- The function $\lambda = \lambda(t)$ is smooth, strictly positive and strictly monotone increasing. There exist positive constants λ_k such that

$$|D_t^k \lambda(t)| \leq \lambda_k \Big(\frac{\lambda(t)}{\Lambda(t)}\Big)^k \lambda(t)$$

for large t and every positive integer k. Finally there exist positive constants $\beta \in [1/2, 1)$ and C such that

$$0 < C \le \frac{\lambda(t)}{\Lambda(t)^{\beta}}$$

for large t.

- The function $b = b(t)$ is 1-periodic, non-constant, smooth and strictly positive.
- The function $m = m(t)$ is defined by $m(t) = \frac{1}{(e_0+\Lambda(t))^{\gamma}}$, where $\gamma \ge 0$, $e_0 > e$.

First we focus on the case $\Omega_{1.1} \cup \Omega_{1.2} \cup \Omega_{2.4}$.

Theorem 7.1: *The solution of* (7.1) *satisfies the* $L_p - L_q$ *decay estimate of Klein–Gordon type*

$$\|u_t(t,\cdot)\|_{L_q} + \|\lambda(t)\nabla_x u(t,\cdot)\|_{L_q}$$
$$\le C\sqrt{\lambda(t)}\Big(1+\Lambda(t)\Big)^{r-\frac{n}{2}(\frac{1}{p}-\frac{1}{q})}\Big(\|\varphi\|_{W_p^{N_p+1}} + \|\psi\|_{W_p^{N_p}}\Big),$$

where $\frac{1}{p}+\frac{1}{q}=1$, $1<p\le 2$, $N_p = \Big[n\Big(\frac{1}{p}-\frac{1}{q}\Big)\Big]+1$ *and*

- $r = 0$ *if* $0 \le \gamma < 2\beta - 1$, *that is, in* $\Omega_{1.1} \cup \Omega_{2.4}$;
- r *is a positive constant if* $\gamma = 2\beta - 1$, *that is, in* $\Omega_{1.2}$.

Proof. The representation (4.2) gives us representations for $u_t, \lambda(t)\nabla_x u$ by the aid of Fourier multipliers. Using Lemma 2.1, Proposition 3.1, Proposition 4.1 and Lemma 4.2 we get for each multiplier the corresponding estimates of the amplitudes in $Z_{pd}(N)$ as well as in $Z_{hyp}(N)$. Now we divide the phase space $[0,\infty)\times(\mathbb{R}^n\setminus\{0\})$ into three parts. From Corollary 6.1 we obtain a decay estimate of Klein–Gordon type for all Fourier multipliers and for all $t \ge t_0$, where the constant r in the above estimates appears from the estimates of the amplitudes (cf. with Table 2 from Section 5). It remains to estimate the Fourier multipliers for $t \le t_0$. Here we take into consideration that the influence of the Hardy–Littlewood lemma remains. Therefore one can follow the proofs of Propositions 6.2 and 6.3 with $K(t) = \frac{N}{\Lambda(t)^{\beta}}$. If we choose $2l = n(\frac{1}{p}-\frac{1}{q})$ in the derived estimates and $t \le t_0$, then the statements of this theorem follow immediately. Thus, Corollary 6.1 guarantees the Klein–Gordon decay rate and the application of Propositions 6.2 and 6.3 for $t \le t_0$ yields the necessary regularity of the data. □

The statement of this theorem was proved in [23] for the case $\beta \in [1/2, 1]$, $\gamma = 0$. Now let us suppose that the above conditions are satisfied with

$$0 < C \le \frac{\lambda(t)}{\Lambda(t)}(\log\Lambda(t))^{-\omega}, \quad m(t) = \frac{\Big(\log(e_0+\Lambda(t))\Big)^{-\delta}}{e_0+\Lambda(t)},$$

where $\omega \le 0$, $e_0 > e$ and $\delta \in \mathbb{R}$. Then we can prove in the case $\Omega_{3.1} \cup \Omega_{3.2} \cup \Omega_{3.3}$ the next result.

Theorem 7.2: *The solution of* (7.1) *satisfies the $L_p - L_q$ decay estimate of Klein–Gordon type*

$$\|u_t(t,\cdot)\|_{L_q} + \|\lambda(t)\nabla_x u(t,\cdot)\|_{L_q} \leq C\sqrt{\lambda(t)}\Big(1+\Lambda(t)\Big)^{r-\frac{n}{2}(\frac{1}{p}-\frac{1}{q})}\Big(\|\varphi\|_{W_p^{N_p+1}} + \|\psi\|_{W_p^{N_p}}\Big),$$

where $\frac{1}{p}+\frac{1}{q}=1$, $1<p\leq 2$, $N_p=\Big[n\Big(\frac{1}{p}-\frac{1}{q}\Big)\Big]+1$ *and*

- $r=0$ *if* $\gamma=1$ *and* $\delta<2\omega-1$, *that is, in* $\Omega_{3.1}$;
- $r=\varepsilon$, $C=C(\varepsilon)$, $\varepsilon>0$ *arbitrary, if* $\gamma=1$ *and* $2\omega-1\leq\delta<2\omega$, *that is, in* $\Omega_{3.2}$;
- r *is a positive constant if* $\gamma=1$ *and* $\delta=2\omega$, $\delta<\omega$, *that is, in* $\Omega_{3.3}$.

Proof. The proof coincides with that of Theorem 7.1. □

Finally we are able to derive $L_p - L_q$ decay estimates of wave type in the case $\Omega_{2.1}\cup\Omega_{2.2}\cup\Omega_{2.3}\cup\Omega_{3.5}\cup\Omega_{3.6}\cup\Omega_{3.7}$, where in this case $\beta=1$.

Theorem 7.3: *The solution of* (7.1) *satisfies the $L_p - L_q$ decay estimate of wave type*

$$\|u_t(t,\cdot)\|_{L_q} + \|\lambda(t)\nabla_x u(t,\cdot)\|_{L_q} \leq C\lambda(t)m(t)^{-1}\Big(1+\Lambda(t)\Big)^{r-\frac{n-1}{2}(\frac{1}{p}-\frac{1}{q})}\Big(\|\varphi\|_{W_p^{N_p+1}} + \|\psi\|_{W_p^{N_p}}\Big),$$

where $\frac{1}{p}+\frac{1}{q}=1$, $1<p\leq 2$, $N_p=[n(\frac{1}{p}-\frac{1}{q})]+1$ *and*

- $r=\varepsilon$, $C=C(\varepsilon)$, $\varepsilon>0$ *arbitrary, if* $\{\gamma>1\,,\omega\in(-1,0]\}$ *or* $\{\gamma=1\,,\ \omega=0\,,\ \delta>0\}$ *or* $\{\gamma=1\,,\ \omega\in(-1,0)\,,\ \delta>0\}$, *that is, in* $\Omega_{2.1}\cup\Omega_{2.2}\cup\Omega_{3.5}\cup\Omega_{3.6}$;
- r *is a positive constant if* $\{\gamma>1\,,\ \omega=-1\}$ *or* $\{\gamma=1\,,\ \omega=-1\,,\ \delta\geq 0\}$ *or* $\{\gamma=1\,,\omega>-1\,,\ \delta=0\}$, *that is, in* $\Omega_{2.3}\cup\Omega_{3.7}$.

Proof. The representation (4.2) gives us representations for $u_t, \lambda(t)\nabla_x u$ by the aid of Fourier multipliers. Using Lemma 2.1, Proposition 3.2, Proposition 4.2 and Lemma 4.4 we derive for every Fourier multiplier the corresponding estimates of the amplitudes in $Z_{pd}(N)$ as well as in $Z_{hyp}(N)$. From Theorem 6.2 we obtain a decay estimate of wave type for all Fourier multipliers, where the constant r in the above estimates appears from the estimates of the amplitudes (cf. with Tables 1,2 from Section 5). □

Remark 7.1: For the study of $L_p - L_q$ decay estimates for (7.1) with more general oscillating coefficients b satisfying assumption (A.1) from the Introduction, classifications of oscillations as in special cases described in [20], [21], [22], [23] are useful. That this classification depends heavily on the stabilizing effect of the mass term follows from the Theorems 7.1 to 7.3. In $\Omega_{1.1}\cup\Omega_{2.1}\cup\Omega_{2.2}\cup\Omega_{2.4}\cup\Omega_{3.1}\cup\Omega_{3.2}\cup\Omega_{3.5}\cup\Omega_{3.6}$ we have *slow oscillations*, this means, we have almost the classical Strichartz decay estimate of wave- or Klein–Gordon type. In $\Omega_{1.2}\cup\Omega_{2.3}\cup\Omega_{3.3}\cup\Omega_{3.7}$ we have *fast oscillations*, this means, the decay rates differ by an (eventually large) constant from the classical ones. *This observation for fast oscillations coincides with the well-known loss of derivatives appearing for solutions of weakly hyperbolic Cauchy problems or for solutions of strictly hyperbolic Cauchy problems with non-Lipschitz coefficients.* In all other parts from Ω we expect *very fast oscillations*, this means we cannot expect $L_p - L_q$ decay estimates. That in most of these parts we have really very fast oscillations will be shown in the next section.

8. Classical Floquet behavior

In the previous section we formulated results about $L_p - L_q$ decay estimates for solutions of (7.1). We posed the Cauchy conditions on $t = 0$, but without new difficulties we can derive $L_p - L_q$ decay estimates if we pose the Cauchy conditions on $t = t_0$, $t_0 \neq 0$. In this section we shall introduce counter-examples which show that under certain assumptions one cannot expect $L_p - L_q$ decay estimates for (7.1) with Cauchy conditions on an arbitrary $t = t_0$. The proof of the idea for non-decay is based on the paper [22]. The authors recommend this paper to readers to find some basic ideas for the proof and for applications to more general equations.

Let us consider the following functions $\lambda = \lambda(t)$ and areas $\Omega_{1.4} \cup \Omega_{1.5} \cup \Omega_{2.5} \cup \{\Omega_{3.8} : \delta < -1\} \cup \Omega_{3.9}$:

$$\lambda(t) = \begin{cases} (e_0 + t)^l & (\beta \in [0,1), l = \frac{\beta}{1-\beta}), \\ \exp(t^\alpha) & (\beta = 1,\ \alpha = \frac{1}{1-\omega},\ \omega \le 0), \end{cases}$$

$$\begin{aligned}
\Omega_{1.4} &= \{(\beta,\gamma,0,0) \in \Omega\ ;\ \beta = \gamma\}, \\
\Omega_{1.5} &= \{(\beta,\gamma,0,0) \in \Omega\ ;\ \beta < \gamma\}, \\
\Omega_{2.5} &= \{(1,\gamma,\omega,0) \in \Omega\ ;\ \gamma > 1,\ \omega < -1\}, \\
\Omega_{3.8} &= \{(1,1,\omega,\delta) \in \Omega\ ;\ \omega = \delta < 0\}, \\
\Omega_{3.9} &= \{(1,1,\omega,\delta) \in \Omega\ ;\ \omega < -1,\ \omega < \delta\}.
\end{aligned}$$

Let us consider the Cauchy problem

$$u_{tt} - \lambda(t)^2 b(t)^2 (\Delta - m(t)^2) u = 0, \quad u(t_0,x) = \varphi(x),\ u_t(t_0,x) = \psi(x), \tag{8.1}$$

where $\Lambda(t) = \int_0^t \lambda(s)ds$, $m(t) = (e_0 + \Lambda(t))^{-\gamma}\Big(\log(e_0 + \Lambda(t))\Big)^{-\delta}$ and $b = b(t)$ is a 1-periodic, non-constant, smooth and strictly positive function. For the solutions of (8.1) we can prove the next theorem.

Theorem 8.1: *There are no constants p, q, C_0, C_1, L and $\varepsilon_0 < 2$ such that for every initial time $t_0 \in [T,\infty)$ with a positive constant T and for every initial data $\varphi, \psi \in C_0^\infty(\mathbb{R}^n)$ the estimate*

$$\|u_t(t,\cdot)\|_{L_q} + \|\nabla_x u(t,\cdot)\|_{L_q} \le C_0 \lambda(t)^{C_1 a(t)^{-\varepsilon_0}} \Big(\|\varphi\|_{W_p^{L+1}} + \|\psi\|_{W_p^L}\Big) \tag{8.2}$$

is fulfilled for all $t \in [t_0 + 1, \infty)$ and $(\beta,\gamma,\omega,\delta) \in \Omega_{1.4} \cup \Omega_{1.5} \cup \Omega_{2.5} \cup \{\Omega_{3.8} : \delta < -1\} \cup \Omega_{3.9}$, where $a(t) = \frac{\log \lambda(t)}{\sqrt{t}}$.

Definition 8.1: The behavior of the solution as described in the previous theorem will be called *classical Floquet behavior.*

The notion of *classical Floquet behavior* will be motivated by the fact that one can understand this behavior of solutions for (8.1) after application of the Floquet theorem (see Lemma 8.1 for a special explanation). Contrary to this behavior we are not able to understand the *modified Floquet behavior* (see Remark 1.1) of solutions for (8.1) by Lemma 8.1.

Corollary 8.1: *There are no constants p, q, C_0, C_1, L and $\varepsilon_0 < 2$ such that for every initial time $t_0 \in [T,\infty)$ with a positive constant T and for every initial data $\varphi, \psi \in C_0^\infty(\mathbb{R}^n)$ the following estimate holds:*

$$\|u_t(t,\cdot)\|_{L_q} + \|\nabla_x u(t,\cdot)\|_{L_q} \le C_0 \Lambda(t)^{C_1(\sqrt{t}/\log t)^{\varepsilon_0}} \left(\|\varphi\|_{W_p^{L+1}} + \|\psi\|_{W_p^L}\right)$$

and

$$\|u_t(t,\cdot)\|_{L_q} + \|\nabla_x u(t,\cdot)\|_{L_q} \le C_0 \Lambda(t)^{C_1 t^{\varepsilon_0(1/2-1/(1-\omega))}} \left(\|\varphi\|_{W_p^{L+1}} + \|\psi\|_{W_p^L}\right)$$

in the cases $\lambda(t) = (e_0+t)^{\frac{\beta}{1-\beta}}$ and $\lambda(t) = \exp(t^{\frac{1}{1-\omega}})$, respectively, for $t \in [t_0+1,\infty)$ and $(\beta,\gamma,\omega,\delta) \in \Omega_{1.4} \cup \Omega_{1.5} \cup \Omega_{2.5} \cup \{\Omega_{3.8} : \delta < -1\} \cup \Omega_{3.9}$.

Proof of Theorem 8.1. We shall prove here only the non-critical case, that is, $(\beta,\gamma,\omega,\delta) \in \Omega_{1.5} \cup \Omega_{2.5} \cup \Omega_{3.9}$. For the critical case $(\beta,\gamma,\omega,\delta) \in \Omega_{1.4} \cup \{\Omega_{3.8} : \delta < -1\}$ we refer to Remark 8.1.

1. One lemma for equations with a periodic coefficient – Floquet's theory

Consider the auxiliary ordinary differential equation with a periodic coefficient

$$v_{tt} + \lambda^2 b(t)^2 v = 0\,. \tag{8.3}$$

Let the matrix-valued function $X = X(t,t_0)$ be the solution of the Cauchy problem

$$\frac{d}{dt}X = \begin{pmatrix} 0 & -\lambda^2 b(t)^2 \\ 1 & 0 \end{pmatrix} X, \quad X(t_0,t_0) = \begin{pmatrix} 1 & 0 \\ 0 & 1 \end{pmatrix}. \tag{8.4}$$

Thus, X gives the fundamental solution to the equation (8.3). The following lemma, which concerns the instability of the solutions for Hill's equation with periodic coefficients [3], [15], is very important for our considerations.

Lemma 8.1: *If $b = b(t)$ is a non-constant, positive, smooth function on $\mathbb{R}$ which is 1-periodic, then there exists a $\lambda = \lambda_0 > 0$ such that $X(t_0+1,t_0)$ has the eigenvalues μ_0 and μ_0^{-1} satisfying $|\mu_0| > 1$.*

For the proof of the lemma we refer to Chapter 1 [3], for instance.

It is no loss of generality to regard $\lambda(t) = t^l$ ($l = \frac{\beta}{1-\beta}$) for $\beta \in [0,1)$, the general mass term $m(t) = \Lambda(t)^{-\gamma}(\log \Lambda(t))^{-\delta}$ and $a(t) = \frac{\log \lambda(t)}{\sqrt{t}}$ because we are interested only in the asymptotic behavior for solutions of (8.1) for $t \to \infty$.

Lemma 8.2: *There exist positive constants C and T such that*

$$\lambda(t)^2 - \lambda(t-d)^2 \le Cd\frac{a(t)}{\sqrt{t}}\lambda(t)^2$$

for any $t \ge T$, $0 \le d \le \frac{1}{2}\sqrt{t}a(t)^{\varepsilon-1}$ and $\varepsilon > 0$.

Proof. By Taylor's formula, there exists a constant $\tilde{d} \in (0, d)$ such that

$$\lambda(t)^2 - \lambda(t-d)^2 = 2d\lambda'(t-\tilde{d})\lambda(t-\tilde{d}) = 2d\frac{a(t)}{\sqrt{t}}\lambda(t)^2\frac{t\lambda'(t-\tilde{d})\lambda(t-\tilde{d})}{\lambda(t)^2\log\lambda(t)}.$$

Noting that for $t \ge T$ it holds that $\sqrt{t}a(t)^{\varepsilon-1} \le t$, we have

$$\frac{2t\lambda'(t-\tilde{d})\lambda(t-\tilde{d})}{\lambda(t)^2\log\lambda(t)} = \begin{cases} \frac{2(t-\tilde{d})^{2l-1}}{t^{2l-1}\log t} \le C & (\beta < 1), \\ \frac{2(t-\tilde{d})^{\alpha-1}\lambda(t-\tilde{d})^2}{t^{\alpha-1}\lambda(t)^2} \le C & (\beta = 1, \omega < 0), \end{cases}$$

with a positive constant C. □

2. Some properties of the fundamental matrix and its eigenvalues

Let t_ξ be the solution of the equation

$$\lambda(t_\xi)\langle\xi\rangle_{m(t_\xi)} = \lambda_0, \tag{8.5}$$

where $\lambda_0 > 0$ is taken from Lemma 8.1. Then, t_ξ is decreasing in $|\xi|$, such that

$$t_\xi \to \infty \qquad \text{as} \quad |\xi| \to 0$$

since $\beta < \gamma$ or $\beta = \gamma$ and $\omega < \delta$. This behavior of t_ξ is very important for our further approach. Using Lemma 8.2 we can prove the next result.

Lemma 8.3: *There exists a positive constant C such that*

$$\left|\lambda(t_\xi)^2\langle\xi\rangle^2_{m(t_\xi)} - \lambda(t_\xi - d)^2\langle\xi\rangle^2_{m(t_\xi-d)}\right| \le Ca(t_\xi)^\varepsilon\lambda(t_\xi)^2\langle\xi\rangle^2_{m(t_\xi)}$$

for any small $|\xi|$, $0 \le d \le \frac{1}{2}\sqrt{t_\xi}a(t_\xi)^{\varepsilon-1}$ and $\varepsilon > 0$.

Proof. We shall prove only the case $\beta = 1$ because the other case is evident. By using Lemma 8.2 we have

$$\begin{aligned} &\left|\lambda(t_\xi)^2\langle\xi\rangle^2_{m(t_\xi)} - \lambda(t_\xi - d)^2\langle\xi\rangle^2_{m(t_\xi-d)}\right| \\ \le\ &\left(\lambda(t_\xi)^2 - \lambda(t_\xi-d)^2\right)\langle\xi\rangle^2_{m(t_\xi)} + \lambda(t_\xi-d)^2\left|\langle\xi\rangle^2_{m(t_\xi)} - \langle\xi\rangle^2_{m(t_\xi-d)}\right| \\ \le\ &Cd\frac{a(t_\xi)}{\sqrt{t_\xi}}\lambda(t_\xi)^2\langle\xi\rangle^2_{m(t_\xi)} + \left|\lambda(t_\xi-d)^2\left(m(t_\xi)^2 - m(t_\xi-d)^2\right)\right|. \end{aligned}$$

For the second term we have

$$\begin{aligned}
&\Big|\lambda(t_\xi-d)^2\Big(m(t_\xi)^2-m(t_\xi-d)^2\Big)\Big| \\
=\;& -2d\lambda(t_\xi-d)^2 m'(t_\xi-\tilde d)m(t_\xi-\tilde d) \\
=\;& 2d\lambda(t_\xi-d)^2\frac{\lambda(t_\xi-\tilde d)}{\Lambda(t_\xi-\tilde d)}m(t_\xi-\tilde d)^2\left(1+\frac{\omega}{\log\Lambda(t_\xi-\tilde d)}\right) \\
=\;& 2d\frac{a(t_\xi)}{\sqrt{t_\xi}}\lambda(t_\xi)^2 m(t_\xi)^2 K(t_\xi,\tilde d),
\end{aligned}$$

where

$$K(t_\xi,d):=\frac{\sqrt{t_\xi}}{a(t_\xi)}\left(\frac{\lambda(t_\xi-d)}{\lambda(t_\xi)}\right)^2\left(\frac{m(t_\xi-d)}{m(t_\xi)}\right)^2\frac{\lambda(t_\xi-d)}{\Lambda(t_\xi-d)}.$$

Taking into consideration the definition of λ and m gives

$$\begin{aligned}
K(t_\xi,d)&=t_\xi^{1-\alpha}\left(\frac{\lambda(t_\xi-d)\Lambda(t_\xi)}{\lambda(t_\xi)\Lambda(t_\xi-d)}\right)^2\left(\frac{\log\Lambda(t_\xi)}{\log\Lambda(t_\xi-d)}\right)^{2\delta}\frac{\lambda(t_\xi-d)}{\Lambda(t_\xi-d)} \\
&\le Ct_\xi^{1-\alpha}\left(\frac{t_\xi}{t_\xi-d}\right)^{2(1-\alpha)}\left(\frac{t_\xi^\alpha+(1-\alpha)\log t_\xi}{(t_\xi-d)^\alpha+(1-\alpha)\log(t_\xi-d)}\right)^{2\delta}\frac{1}{(t_\xi-d)^{1-\alpha}} \\
&\le C.
\end{aligned}$$

Summarizing, we obtain

$$\begin{aligned}
&\Big|\lambda(t_\xi)^2\langle\xi\rangle_{m(t_\xi)}^2-\lambda(t_\xi-d)^2\langle\xi\rangle_{m(t_\xi-d)}^2\Big|\le Cd\frac{a(t_\xi)}{\sqrt{t_\xi}}\lambda(t_\xi)^2\langle\xi\rangle_{m(t_\xi)}^2 \\
&\le \tfrac12 Ca(t_\xi)^\varepsilon\lambda(t_\xi)^2\langle\xi\rangle_{m(t_\xi)}^2.
\end{aligned}$$

The statement of the lemma is proved. □

According to the definition of t_ζ and λ_0, we get

$$\Big|\lambda_0^2-\lambda(t_\xi-d)^2\langle\xi\rangle_{m(t_\xi-d)}^2\Big|\le C\lambda_0^2 a(t_\xi)^\varepsilon \tag{8.6}$$

for $0\le d\le\frac12\sqrt{t_\xi}a(t_\xi)^{\varepsilon-1}$ and small $|\xi|$. Let us set

$$X(t_\xi+1,t_\xi):=\begin{pmatrix} x_{11} & x_{12}\\ x_{21} & x_{22}\end{pmatrix},$$

then according to Lemma 8.1 the eigenvalues of this matrix are μ_0 and μ_0^{-1}. Therefore

$$x_{11}+x_{22}=\mu_0+\mu_0^{-1}$$

leads to

$$|x_{11}-\mu_0|+|x_{22}-\mu_0|\ge|\mu_0-\mu_0^{-1}|$$

and implies

$$\max\{|x_{11}-\mu_0|,|x_{22}-\mu_0|\}\ge\tfrac12|\mu_0-\mu_0^{-1}|.$$

We suppose

$$|x_{11} - \mu_0| \geq \tfrac{1}{2}|\mu_0 - \mu_0^{-1}| \tag{8.7}$$

so that

$$|x_{22} - \mu_0^{-1}| \geq \tfrac{1}{2}|\mu_0 - \mu_0^{-1}|.$$

Let us consider for every given integer $n \geq 0$ the equation

$$\frac{d^2}{dt^2}w + \lambda(t_\xi - n + t)^2 \langle\xi\rangle^2_{m(t_\xi - n+t)} b(t_\xi + t)^2 w = 0. \tag{8.8}$$

It can be written as a first-order system which has the fundamental matrix $X_n = X_n(t, t_1)$ solving the Cauchy problems

$$\frac{d}{dt}X = \begin{pmatrix} 0 & -\lambda(t_\xi - n + t)^2 \langle\xi\rangle^2_{m(t_\xi - n+t)} b(t_\xi + t)^2 \\ 1 & 0 \end{pmatrix} X, \quad X(t_1, t_1) = I. \tag{8.9}$$

Due to (8.6) we get the estimate

$$\max_{t\in[0,1]} \|X_n(t,0) - X(t_\xi + t, t_\xi)\| \leq C_1 a(t_\xi)^\varepsilon \tag{8.10}$$

for

$$0 \leq n \leq \tfrac{1}{2}\sqrt{t_\xi}\, a(t_\xi)^{\varepsilon - 1}, \tag{8.11}$$

where the constant C_1 is independent of n and ξ. Then the right-hand side of (8.10) tends to 0 if $|\xi|$ tends to 0. Moreover, it holds that

$$\max_{t,s\in[0,1]} \|X_n(t,s)\| \leq C. \tag{8.12}$$

Indeed, to prove estimate (8.12) we note that for $t \in [0,1]$, $\xi \in \mathbb{R}^n$ with sufficiently small $|\xi|$, and for n satisfying $0 \leq n \leq \frac{1}{2}\sqrt{t_\xi}\, a(t_\xi)^{\varepsilon-1}$ we have $0 \leq t_\xi - n + t$ and, due to Lemma 8.3,

$$\lambda(t_\xi - n + t)^2 \langle\xi\rangle^2_{m(t_\xi - n + t)} \leq C\lambda(t_\xi)^2 \langle\xi\rangle^2_{m(t_\xi)} = C\lambda_0^2.$$

Thus the coefficient matrix of system (8.9) is uniformly bounded for all $t \in [0,1]$, $\xi \in \mathbb{R}^n$ with sufficiently small $|\xi|$, and $1 \leq n \leq \frac{1}{2}\sqrt{t_\xi}\, a(t_\xi)^{\varepsilon-1}$. Therefore, due to Lemma 8.3, we have for these t, ξ, and n the inequality

$$\begin{aligned} &\left|\lambda(t_\xi - n + t)^2 \langle\xi\rangle^2_{m(t_\xi - n+t)} - \lambda(t_\xi - n - 1 + t)^2 \langle\xi\rangle^2_{m(t_\xi - n - 1 + t)}\right| \\ &\leq C\frac{a(t_\xi)}{\sqrt{t_\xi}} \lambda(t_\xi - n + t)^2 \langle\xi\rangle^2_{m(t_\xi - n + t)} \end{aligned}$$

if $0 < n \leq \frac{1}{2}\sqrt{t_\xi}\, a(t_\xi)^{\varepsilon-1}$. Consequently,

$$\|X_{n+1}(1,0) - X_n(1,0)\| \leq C_1 \frac{a(t_\xi)}{\sqrt{t_\xi}} \tag{8.13}$$

for $0 < n \leq \frac{1}{2}\sqrt{t_\xi}\, a(t_\xi)^{\varepsilon-1}$. Using Lemma 8.1, (8.10) and noting the fact that

$$a(t_\xi)^\varepsilon \to 0 \quad \text{for} \quad |\xi| \to 0 \tag{8.14}$$

in the cases $\Omega_{1.5}\cup\Omega_{2.5}\cup\Omega_{3.9}$ the matrix $X_n(1,0)$, with the property $\det X_n(1,0)=1$, has eigenvalues ν_n and ν_n^{-1} which satisfy

$$|\nu_n-\mu_0|\le\kappa \tag{8.15}$$

for any given positive number κ with sufficiently small $|\xi|=|\xi|(\kappa)$. Hence we have

$$|\nu_n|\ge\tfrac{1}{2}(|\mu_0|+1)>1. \tag{8.16}$$

Denoting

$$X_n(1,0):=\begin{pmatrix} x_{11}(n) & x_{12}(n)\\ x_{21}(n) & x_{22}(n)\end{pmatrix}$$

we get

$$|x_{11}(n)-\nu_n|\ge|x_{11}-\mu_0|-\Big(|x_{11}(n)-x_{11}|+|\mu_0-\nu_n|\Big)\ge\tfrac{1}{4}|\mu_0-\mu_0^{-1}|. \tag{8.17}$$

Analogously we prove

$$|x_{22}(n)-\nu_n^{-1}|\ge\tfrac{1}{4}|\mu_0-\mu_0^{-1}|.$$

The estimate (8.13) implies

$$|x_{ij}(n+1)-x_{ij}(n)|\le C\frac{a(t_\xi)}{\sqrt{t_\xi}}. \tag{8.18}$$

Due to estimates (8.18) the eigenvalues ν_n and ν_n^{-1} are distinct for every n. Then, according to (8.18), we get

$$|\nu_{n+1}-\nu_n|\le C\frac{a(t_\xi)}{\sqrt{t_\xi}}.$$

3. Lower bound for the energy of an unstable solution of an auxiliary ordinary differential equation

The calculations of the previous steps serve as a preparation to prove the next lemma.

Lemma 8.4: *Let* n_0 *be the largest integer satisfying* (8.11). *Then there exist positive numbers* C_0 *and* C_1 *such that the solution* $w=w(t,\xi)$ *of*

$$w_{tt}+\lambda(t)^2b(t)^2\langle\xi\rangle^2_{m(t)}w=0$$

with initial data

$$w(t_\xi-n_0-1,\xi)=1,\quad w_t(t_\xi-n_0-1,\xi)=\frac{x_{12}(n_0)}{\nu_{n_0}-x_{11}(n_0)},$$

satisfies

$$|w(t_\xi,\xi)|+|w_t(t_\xi,\xi)|\ge C_0\exp\left(C_1\sqrt{t_\xi}\,a(t_\xi)^{\varepsilon-1}\right) \tag{8.19}$$

for all small $|\xi|$.

Proof. The function $w=w(t_\xi-n_0+t,\xi)$ satisfies equation (8.8). Hence

$$\begin{pmatrix}\frac{d}{dt}w(t_\xi,\xi)\\ w(t_\xi,\xi)\end{pmatrix}=X_0(1,0)X_1(1,0)\cdots X_{n_0-1}(1,0)X_{n_0}(1,0)\begin{pmatrix}\frac{d}{dt}w(t_\xi-n_0-1,\xi)\\ w(t_\xi-n_0-1,\xi)\end{pmatrix}.$$

The matrix

$$B_n = \begin{pmatrix} \frac{x_{12}(n)}{\nu_n - x_{11}(n)} & 1 \\ 1 & \frac{x_{21}(n)}{\nu_n^{-1} - x_{22}(n)} \end{pmatrix}$$

is a diagonalizer for $X_n(1,0)$, that is,

$$X_n(1,0)B_n = B_n \begin{pmatrix} \nu_n & 0 \\ 0 & \nu_n^{-1} \end{pmatrix}.$$

Since $\det X_n(1,0) = 1$ and since the trace of $X_n(1,0)$ is $\nu_n + \nu_n^{-1}$ we get

$$\det B_n = \frac{\nu_n - \nu_n^{-1}}{\nu_n^{-1} - x_{22}(n)}.$$

According to (8.12), (8.16) and (8.17) this leads to

$$|\nu_n^{-1} - x_{22}(n)| \le C, \quad |\det B_n| \ge C > 0$$

for all n with $0 < n \le \frac{1}{2}\sqrt{t_\xi}a(t_\xi)^{\varepsilon-1}$, respectively. Further, taking into account (8.12) we have $|x_{ij}(n)| \le C$ and, consequently,

$$\|B_n\|, \|B_n^{-1}\| \le C, \tag{8.20}$$

for all n with $0 < n \le \frac{1}{2}\sqrt{t_\xi}a(t_\xi)^{\varepsilon-1}$, where C is independent of n. The estimates (8.18) and (8.20) lead to

$$\|B_{n-1}^{-1}B_n - I\| = \|B_{n-1}^{-1}(B_n - B_{n-1})\| \le C\frac{a(t_\xi)}{\sqrt{t_\xi}}. \tag{8.21}$$

Denoting $G_n := B_{n-1}^{-1}B_n - I$, we have

$$\begin{aligned} & X_0(1,0)X_1(1,0)\cdots X_{n_0-1}(1,0)X_{n_0}(1,0) \\ = \; & B_0 \begin{pmatrix} \nu_0 & 0 \\ 0 & \nu_0^{-1} \end{pmatrix} B_0^{-1}B_1 \begin{pmatrix} \nu_1 & 0 \\ 0 & \nu_1^{-1} \end{pmatrix} B_1^{-1}B_2 \cdots B_{n_0-1}^{-1}B_{n_0} \\ & \times \begin{pmatrix} \nu_{n_0} & 0 \\ 0 & \nu_{n_0}^{-1} \end{pmatrix} B_{n_0}^{-1} \\ = \; & B_0 \begin{pmatrix} \nu_0 & 0 \\ 0 & \nu_0^{-1} \end{pmatrix} (I+G_1) \begin{pmatrix} \nu_1 & 0 \\ 0 & \nu_1^{-1} \end{pmatrix} (I+G_2)\cdots(I+G_{n_0}) \\ & \times \begin{pmatrix} \nu_{n_0} & 0 \\ 0 & \nu_{n_0}^{-1} \end{pmatrix} B_{n_0}^{-1}. \end{aligned}$$

From (8.20) we see that in order to prove Lemma 8.4 it is sufficient to show that the $(1,1)$-th element y_{11} of the matrix

$$\begin{pmatrix} \nu_0 & 0 \\ 0 & \nu_0^{-1} \end{pmatrix} (I+G_1) \begin{pmatrix} \nu_1 & 0 \\ 0 & \nu_1^{-1} \end{pmatrix} (I+G_2)\cdots(I+G_{n_0}) \begin{pmatrix} \nu_{n_0} & 0 \\ 0 & \nu_{n_0}^{-1} \end{pmatrix}$$

can be estimated to below with some positive numbers C_0 and C_1 by

$$C_0 \exp\left(C_1\sqrt{t_\xi}a(t_\xi)^{\varepsilon-1}\right).$$

It is evident by (8.21) and Lemma 3.3 of [31] that

$$\Big|y_{11}-\prod_{n=0}^{n_0}\nu_n\Big|\le Ca(t_\xi)^\varepsilon\prod_{n=0}^{n_0}|\nu_n|.$$

Thus, it follows from (8.15) that

$$\begin{aligned}|y_{11}| &\ge (1-Ca(t_\xi)^\varepsilon)\prod_{n=0}^{n_0}|\nu_n|\ge(1-Ca(t_\xi)^\varepsilon)\,|\mu_0-\kappa|^{n_0}\\ &\ge (1-Ca(t_\xi)^\varepsilon)\,|\mu_0-\kappa|^{\frac12\sqrt{t_\xi}a(t_\xi)^{\varepsilon-1}-1}\end{aligned}$$

with $n_0=\max\{n\in\mathbb{N}_+; n\le\frac12\sqrt{t_\xi}a(t_\xi)^{\varepsilon-1}\}$. Recalling (8.14) the lemma is proved. □

4. Construction of unstable solutions

The considerations of this step are based on a family of solutions $u=u(t,x,\xi)$ of (8.1),

$$u_{tt}-\lambda(t)^2b(t)^2\left(\Delta-m(t)^2\right)u=0\,,$$

where $\xi\in\mathbb{R}^n$ is regarded as a parameter. The functions of this family will violate the estimate (8.2),

$$\|u_t(t,\cdot)\|_{L_q}+\|\nabla_x u(t,\cdot)\|_{L_q}\le C_0\lambda(t)^{C_1a(t)^{-\varepsilon_0}}\left(\|\varphi\|_{W_p^{L+1}}+\|\psi\|_{W_p^L}\right)$$

for sufficiently small $|\xi|$. To construct this family we note that there exists a positive number N such that

$$\lim_{t\to\infty}\frac{a(t)^{1-\varepsilon}}{\sqrt t}\exp\left(Na(t)\sqrt t\right)=\begin{cases}\lim\limits_{t\to\infty}t^{Nl-1+\frac{\varepsilon}{2}}(l\log t)^{1-\varepsilon}=\infty & (\beta<1),\\[2ex] \lim\limits_{t\to\infty}t^{(\alpha-\frac12)(1-\varepsilon)-\frac12}e^{Nt^\alpha}=\infty & (\beta=1).\end{cases}\tag{8.22}$$

Further, we fix a cut-off function $\chi\in C_0^\infty(\mathbb{R}^n)$ such that $\chi(x)=1$ when $|x|\le1$, while $\chi(x)=0$ if $|x|\ge2$. Then with the number n_0 from Lemma 8.4 and with N of (8.22) we choose at $t=t_0$, $t_0:=t_\xi-n_0-1$, the initial data φ and ψ as

$$\varphi(x)=e^{ix\cdot\xi}\chi\Big(x\langle\xi\rangle_{m(t_\xi)}^{N+2}\Big),\qquad \psi(x)=e^{ix\cdot\xi}\chi\Big(x\langle\xi\rangle_{m(t_\xi)}^{N+2}\Big)\frac{x_{12}(n_0)}{\nu_{n_0}-x_{11}(n_0)}\tag{8.23}$$

belonging to $C_0^\infty(\mathbb{R}^n)$. Using the theory of the Cauchy problem for strictly hyperbolic equations, especially the existence of a cone of dependence, there exists a unique solution $u=u(t,x,\xi)$ to (8.1), (8.23), where $u=u(t,\cdot,\xi)$ has compact support for every given t and ξ, $t\in[t_0,\infty)$, $\xi\in\mathbb{R}^n$. By (8.22) there exists a positive number T such that

$$4\sqrt t a(t)^{\varepsilon-1}\le C\lambda(t)^N\tag{8.24}$$

for any given C and all $t\ge T$.

5. Cone of dependence and lower bound for the energy

If we take into account the cone of dependence property of the solution, then $u = u(t,x,\xi)$ is representable in $B_1(0) \subset \mathbb{R}^n_x$, in the ball of radius 1 centered at the origin, at time $t = t_\xi$ as

$$u(t_\xi, x, \xi) = e^{ix\cdot\xi} w(t_\xi, \xi), \quad u_t(t_\xi, x, \xi) = e^{ix\cdot\xi} w_t(t_\xi, \xi) \tag{8.25}$$

with the function w from Lemma 8.4. To understand (8.25) let us calculate the lower base at $t = t_0$ of the truncated cone starting from $B_1(0)$ at $t = t_\xi$ with the slope $\lambda(t_\xi)\max_{t\in[0,1]} b(t) = b_1\lambda(t_\xi)$ and the height $t_\xi - t_0 = n_0 + 1$. This lower base is contained in the ball

$$B_{dep} := \{x \in \mathbb{R}^n_x : |x| \le 2b_1\lambda(t_\xi)(n_0+1)\}.$$

On the one hand we have due to (8.24) the inequalities

$$2b_1\lambda(t_\xi)(n_0+1) \le 4b_1\lambda(t_\xi)\sqrt{t_\xi}a(t_\xi)^{\varepsilon-1} \le C\lambda(t_\xi)^{N+1} = C\lambda_0^{N+1}\langle\xi\rangle_{m(t_\xi)}^{-(N+1)}.$$

On the other hand, if $x \in B_{dep}$, then the module of the argument $x\langle\xi\rangle^{N+2}_{m(t_\xi)}$ of the function χ is less than or equal to 1 for sufficiently small $|\xi|$. Hence, we get the identity $\chi\left(x\langle\xi\rangle^{N+2}_{m(t_\xi)}\right) = 1$ on B_{dep}.

The function $e^{ix\cdot\xi}w(t,\xi)$ solves the equation and takes at $t = t_0$ the data (8.23) if $x \in \mathbb{R}^n$, $|x| \le \langle\xi\rangle^{-(N+2)}_{m(t_\xi)}$. The uniqueness of solutions to (8.1), (8.23) implies the representations (8.25). In particular we obtain from this representation

$$\|u_t(t_\xi,\cdot,\xi)\|_{L_q(B_1(0))} + \|\nabla_x u(t_\xi,\cdot,\xi)\|_{L_q(B_1(0))} = \Big(|\xi||w(t_\xi,\xi)| + |w_t(t_\xi,\xi)|\Big)\int_{B_1(0)} dx.$$

Now let us suppose that the inequality (8.2) holds. Then we have

$$\begin{aligned} & \|u_t(t,\cdot,\xi)\|_{L_q(B_1(0))} + \|\nabla_x u(t,\cdot,\xi)\|_{L_q(B_1(0))} \\ \le\ & C\|u_t(t,\cdot,\xi)\|_{L_q} + \|\nabla_x u(t,\cdot,\xi)\|_{L_q} \\ \le\ & C_0\Big(1 + \Big|\frac{x_{12}(n_0)}{\nu_{n_0} - x_{11}(n_0)}\Big|\Big)\lambda(t_\xi)^{C_1 a(t_\xi)^{-\varepsilon_0}}\|e^{ix\cdot\xi}\chi(x\langle\xi\rangle^{N+2}_{m(t_\xi)})\|_{W_p^L}. \end{aligned}$$

Using this inequality and recalling (8.12) and (8.17) we have

$$|w(t_\xi,\xi)| + |w_t(t_\xi,\xi)| \le C_0\lambda(t_\xi)^{C_2 + C_1 a(t_\xi)^{-\varepsilon_0}}$$

for all $\xi \in \mathbb{R}^n$ with sufficiently small $|\xi|$, where the exponent C_2 arises from the corresponding Sobolev norm of the data and the definition (8.5) for t_ξ.

But from Lemma 8.4 we already know that the inequality

$$|w(t_\xi,\xi)| + |w_t(t_\xi,\xi)| \ge C_0\exp\left(C_1\sqrt{t_\xi}a(t_\xi)^{\varepsilon-1}\right) = C_0\lambda(t_\xi)^{C_1 a(t_\xi)^{-2+\varepsilon}}$$

holds with some positive numbers C_0, C_1. According to the conditions of the theorem we have $0 \le \varepsilon < 2$, that is, $0 \le \varepsilon_0 < 2 - \varepsilon < 2$ for any $\varepsilon > 0$. Comparing the last two inequalities leads to a contradiction for sufficiently small $|\xi|$. The theorem is proved. □

Remark 8.1: Let us make some comments on the critical cases $\Omega_{1.4} \cup \{\Omega_{3.8} : \delta < -1\}$. In these cases we have $\lim_{t\to\infty} \lambda(t)m(t) = C > 0$. If we suppose that $b = b(t)$ from (8.1) is 1-periodic, non-constant, smooth and strictly positive, then due to [3] there exists an infinite number of intervals of instability, that is, we have a sequence $\{\lambda_n\}$, $n \in \mathbb{N}$, tending to ∞ such that the eigenvalues μ_n of the corresponding fundamental matrices satisfy $|\mu_n| > 1$. Defining as in (8.5) the function t_ξ it is sufficient to choose a natural number n_0 and the positive constant $\lambda_{n_0} > C$ from the sequence as the right-hand side. Then the function $t_\xi \to \infty$ when $|\xi| \to 0$. But this allows us to repeat the approach to proving Theorem 8.1 and to conclude the same statement.

Remark 8.2: We can expect some intermediate behavior appearing in some parts of the areas $\Omega_{1.3}$, $\Omega_{3.4}$, $\{\Omega_{3.8} : \delta < -1\}$ and $\Omega_{3.10}$. Indeed, by Lemma 4.2 (iv), (v) (see also Table 3), the growth order of the solution is not more rapid than that for classical Floquet cases. For instance, when $\lambda(t) = (1+t)^{\frac{\beta}{\beta-1}}$ and $m(t) = (1+t)^{-\frac{\gamma}{1-\beta}}$, that is the case of Example 1.1 with $l = \frac{\beta}{(1-\beta)}$, the upper bound of the L_q-norm of the solution is $O(\exp(\Lambda(t)^{\gamma-2\beta+1})) \sim O(\exp(t^{\frac{\gamma-2\beta+1}{1-\beta}})) \sim O(\exp(t^{\kappa_0}))$ in $\Omega_{1.3}$ with a fixed $\kappa_0 = \kappa_0(\beta,\gamma) < 1$. On the other hand, by Corollary 8.1, the L_q-norm of the solution cannot be controlled by $O(\exp(t^{\kappa}))$ for any $\kappa < 1$ from below in $\Omega_{1.4} \cup \Omega_{1.5}$. Consequently, the behavior in $\Omega_{1.3}$, which we called *modified Floquet behavior* is better than the classical Floquet behavior in $\Omega_{1.4} \cup \Omega_{1.5}$.

Concluding remark: Now let us give answers to the questions formulated in the Introduction. For this reason we recall the Cauchy problem for our starting equation

$$u_{tt} - \lambda(t)^2 b(t)^2 \triangle u + \frac{\lambda(t)^2 b(t)^2}{(e+\Lambda(t))^{2\gamma}\Big(\log(e+\Lambda(t))\Big)^{2\delta}} u = 0 \tag{8.26}$$

under the assumptions for b, λ and m from Section 1.3.

Case 1: Let us suppose $\beta \in (\frac{1}{2}, 1)$ in (1.7) and $\delta = 0$, $\gamma \geq 0$ in (8.26). Then the change-over from Klein–Gordon behavior to Floquet behavior can be described in the following way:

- $0 \leq \gamma < 2\beta - 1$, in $\Omega_{1.1}$, Klein–Gordon behavior;
- $\gamma = 2\beta - 1$, in $\Omega_{1.2}$, critical Klein–Gordon behavior;
- $2\beta - 1 < \gamma < \beta$, in $\Omega_{1.3}$, modified Floquet behavior;
- $\beta \leq \gamma$, in $\Omega_{1.4} \cup \Omega_{1.5}$, classical Floquet behavior.

Klein–Gordon behavior $\to$ critical Klein–Gordon behavior $\to$ modified Floquet behavior $\to$ classical Floquet behavior

Case 2: Let us suppose $\beta = 1$ in (1.7) and $\delta = 0$, $\gamma > 1$ in (8.26). Then the change-over from wave behavior to Floquet behavior can be described in the following way:

- $\omega \in (-1, 0]$ in (1.7), in $\Omega_{2.1} \cup \Omega_{2.2}$, wave behavior;
- $\omega = -1$ in (1.7), in $\Omega_{2.3}$, critical wave behavior;
- $\omega < -1$ in (1.7), in $\Omega_{2.5}$, classical Floquet behavior.

wave behavior $\rightarrow$ critical wave behavior $\rightarrow$ classical Floquet behavior

Case 3: Let us suppose $\beta = 1$, $\omega = -\frac{1}{2}$ in (1.7) and $\gamma = 1$ in (8.26). Then the change-over from Klein–Gordon behavior to wave behavior can be described in the following way:

- $\delta < -1$, in $\Omega_{3.1} \cup \Omega_{3.2}$, Klein–Gordon behavior;
- $\delta = -1$, in $\Omega_{3.3}$, critical Klein–Gordon behavior;
- $-1 < \delta < 0$, in $\Omega_{3.4} \cup \Omega_{3.8} \cup \Omega_{3.10}$, modified Floquet behavior;
- $0 = \delta$, in $\Omega_{3.7}$, critical wave behavior;
- $0 < \delta$, in $\Omega_{3.6}$, wave behavior.

Klein–Gordon behavior $\rightarrow$ critical Klein–Gordon behavior $\rightarrow$ modified Floquet behavior $\rightarrow$ critical wave behavior $\rightarrow$ wave behavior

9. Appendix

Lemma A.1: *Let $\xi_0, \xi_1 \in \mathbb{R}$. If $|\xi_0| \geq |\xi_1|$, then the following inequality holds:*

$$\int_0^t \lambda(s) \left(\frac{|\xi_0|}{\langle \xi_0 \rangle_{m(s)}} - \frac{|\xi_1|}{\langle \xi_1 \rangle_{m(s)}} \right) ds \geq \frac{1}{1+\gamma} \Lambda(t) \left(\frac{|\xi_0|}{\langle \xi_0 \rangle_{m(t)}} - \frac{|\xi_1|}{\langle \xi_1 \rangle_{m(t)}} \right)$$

for any $t > 0$.

Proof. We note that $\frac{|\xi|}{\langle \xi \rangle_{m(s)}}$ is monotone increasing with respect to $|\xi|$. Let $\xi_0, \xi_1 \in \mathbb{R} \setminus \{0\}$ satisfy $|\xi_0| \geq |\xi_1|$. By partial integration we have

$$\int_0^t \lambda(s) \left(\frac{|\xi_0|}{\langle \xi_0 \rangle_{m(s)}} - \frac{|\xi_1|}{\langle \xi_1 \rangle_{m(s)}} \right) ds = \Lambda(t) \left(\frac{|\xi_0|}{\langle \xi_0 \rangle_{m(t)}} - \frac{|\xi_1|}{\langle \xi_1 \rangle_{m(t)}} \right)$$

$$- \int_0^t \left(\gamma + \frac{\delta}{\log(e_0 + \Lambda(s))} \right) \frac{\Lambda(s)}{e_0 + \Lambda(s)} \lambda(s) \left(\frac{m(s)^2 |\xi_0|}{\langle \xi_0 \rangle^3_{m(s)}} - \frac{m(s)^2 |\xi_1|}{\langle \xi_1 \rangle^3_{m(s)}} \right) ds.$$

Here we remark that the inequality

$$\frac{|\xi_0|}{\langle \xi_0 \rangle_{m(s)}} - \frac{m(s)^2 |\xi_0|}{\langle \xi_0 \rangle^3_{m(s)}} \geq \frac{|\xi_1|}{\langle \xi_1 \rangle_{m(s)}} - \frac{m(s)^2 |\xi_1|}{\langle \xi_1 \rangle^3_{m(s)}}$$

holds for any $s \leq t$, that is, the function $f : |\xi| \rightarrow f(|\xi|)$, where

$$f(|\xi|) := \frac{|\xi|}{\langle \xi \rangle_{m(s)}} - \frac{m(s)^2 |\xi|}{\langle \xi \rangle^3_{m(s)}},$$

is monotone increasing with respect to $|\xi|$. Indeed, we have

$$\begin{aligned}\frac{\partial f(|\xi|)}{\partial|\xi|} &= \langle\xi\rangle_{m(s)}^{-5}\left(\langle\xi\rangle_{m(s)}^{4} - |\xi|^2\langle\xi\rangle_{m(s)}^{2} - m(s)^2\langle\xi\rangle_{m(s)}^{2} + 3m(s)^2|\xi|^2\right)\\ &= 3m(s)^4|\xi|^2\langle\xi\rangle_{m(s)}^{-5} \geq 0.\end{aligned}$$

Therefore, we obtain

$$\begin{aligned}&\int_0^t \lambda(s)\left(\frac{|\xi_0|}{\langle\xi_0\rangle_{m(s)}} - \frac{|\xi_1|}{\langle\xi_1\rangle_{m(s)}}\right)ds\\ &\geq \Lambda(t)\left(\frac{|\xi_0|}{\langle\xi_0\rangle_{m(t)}} - \frac{|\xi_1|}{\langle\xi_1\rangle_{m(t)}}\right) - \gamma\int_0^t \lambda(s)\left(\frac{|\xi_0|}{\langle\xi_0\rangle_{m(s)}} - \frac{|\xi_1|}{\langle\xi_1\rangle_{m(s)}}\right)ds.\end{aligned}$$

Thus the lemma is proved. □

Lemma A.2: *Let t be a positive constant and*

$$E_{t,C_1,C_2} = \{\eta \in \mathbb{R}\setminus\{0\}\ ;\ \eta \text{ satisfies } (6.3)\}$$

with positive constants C_1 and C_2. If $\eta_0, \eta_1 \in E_{t,C_1,C_2}$ satisfy $|\eta_0| \geq |\eta_1|$, then the following inequality holds:

$$\frac{2^k}{\Lambda(t)^\beta}\int_0^t \lambda(s)\left(\frac{|\eta_0|}{g(s,t,\langle\eta_0\rangle_1)} - \frac{|\eta_1|}{g(s,t,\langle\eta_1\rangle_1)}\right)ds \geq \frac{1}{1+\gamma}\Lambda(t)\left(\frac{|\eta_0|}{\langle\eta_0\rangle_1} - \frac{|\eta_1|}{\langle\eta_1\rangle_1}\right).$$

Proof. The proof is similar to that for Lemma A.1. We have only to show that the function $\tilde{f} : |\eta| \to \tilde{f}(|\eta|)$, where

$$\tilde{f}(|\eta|) := \frac{|\eta|}{g(s,t,\langle\eta\rangle_1)} - \frac{m(s)^2|\eta|}{g(s,t,\langle\eta\rangle_1)^3} = \frac{|\eta|}{g(s,t,\langle\eta\rangle_1)}\left(1 - \frac{m(s)^2}{g(s,t,\langle\eta\rangle_1)^2}\right),$$

is monotone increasing with respect to $|\eta|$. Indeed, it follows from (6.3) that $1 - \frac{m(s)^2}{g(s,t,\langle\eta\rangle_1)^2} \geq 0$, hence $\tilde{f}(|\eta|)$ is described by a product of two non-negative monotone increasing functions with respect to $|\eta|$. Thus, the lemma is proved. □

Acknowledgements

The authors began to write this paper during the visit of the first author from November 2000 to February 2001 at the Technische Universität Bergakademie Freiberg. The first author is grateful to the Fakultät für Mathematik und Informatik for the hospitality and to DAAD for financial support. The paper was finished during the work of the second author as a Foreign Professor at the University of Tsukuba in February, March 2002. He is grateful to the staff of the Institute of Mathematics, especially the research groups of Prof. Kajitani, Prof. Taira and Prof. Wakabayashi for their hospitality. Finally, the authors thank Karen Yagdjian for his valuable remarks.

References

[1] P. Brenner, *On $L_p - L_q$ estimates for the wave-equation*, Math. Zeit. **145** (1975), no. 3, 251–254.

[2] F. Colombini, D. DelSanto and M. Reissig, *On the optimal regularity of coefficients in hyperbolic Cauchy problems*, Preprint del Dipartimento di Scienze matematiche dell'Università di Trieste, **n. 517** (2002).

[3] M.S.P. Eastham, *The Spectral Theory of Periodic Differential Equations*, Scottish Academic Press, Edinburgh and London, 1973.

[4] A. Galstian and M. Reissig, *$L_p - L_q$ decay estimates for a Klein-Gordon type model equation*, Eds. H. Begehr et al.,Proceedings of the second ISAAC congress, **vol. 2**, 1355–1369, Kluwer (2000).

[5] F. Hirosawa, *Energy decay for degenerate hyperbolic equations of Klein-Gordon type with dissipative term*, Funkcial. Ekvac. **43** (2000), no. 1, 163–191.

[6] F. Hirosawa, *On the Cauchy problem for second order strictly hyperbolic equations with non-regular coefficients*, to appear.

[7] F. Hirosawa, *Precise energy decay estimates for the dissipative hyperbolic equations in an exterior domain*, Mathematical Research Note 2000-008, Institute of Mathematics, University of Tsukuba.

[8] L. Hörmander, *Translation invariant operators in L^p space*, Acta Math. **104** (1960), 93–140.

[9] L. Hörmander, *Remarks on the Klein-Gordon equation*, Journées Equations aux Dérivées Partielles, Saint Jean de Monts, 1987.

[10] S. Kawashima, M. Nakao and K. Ono, *On the decay property of solutions to the Cauchy problem of the semilinear wave equation with a dissipative term*, J. Math. Soc. Japan **47** (1995), no. 4, 617–653.

[11] S. Klainerman, *Global existence for nonlinear wave equations*, Comm. Pure Appl. Math. **33** (1980), no. 1, 43–101.

[12] A. Kubo and M. Reissig, *Construction of parametrix for hyperbolic equations with fast oscillations in non-Lipschitz coefficients*, Mathematical Research Note 2002-003, Institute of Mathematics, University of Tsukuba.

[13] O. Liess, *Decay estimates for the solutions of the system of crystal optics*, Asymptotic Analysis **4** (1991), no. 1, 61–95.

[14] W. Littman, *Fourier transformations of surface carried measures and differentiability of surface averages*, Bull. Amer. Math. Soc. **69** (1963), 766–770.

[15] W. Magnus and St. Winkler, *Hill's Equation*, Interscience Publishers, New York-London-Sydney, 1966.

[16] A. Matsumura, *Energy decay of solutions of dissipative wave equations*, Proc. Japan Acad. Ser. A Math. Sci. **53** (1977), no. 7, 232–236.

[17] K. Mochizuki and T. Motai, *On energy decay-nondecay problems for wave equations with nonlinear dissipative term in $\mathbb{R}^N$*, J. Math. Soc. Japan **47** (1995), no. 3, 405–421.

[18] H. Pecher, *L_p-Abschätzungen und klassische Lösungen für nichtlineare Wellengleichungen. I*, Math. Zeitschrift **150** (1976), no. 2, 159–183.

[19] R. Racke, *Lectures on nonlinear evolution equations*, Aspects of Mathematics, Vieweg, Braunschweig/Wiesbaden, 1992.

[20] M. Reissig, *Klein-Gordon type decay rates for wave equations with a time-dependent dissipation*, Adv. Math. Sci. Appl. **11** (2001), no. 2, 859–891.

[21] M. Reissig and K. Yagdjian, *About the influence of oscillations on Strichartz-type decay estimates*, Rend. Sem. Mat. Univ. Polit. Torino **58** (2000), no. 3, 117–130.

[22] M. Reissig and K. Yagdjian, *One application of Floquet's theory to $L_p - L_q$ estimates for hyperbolic equations with very fast oscillations*, Math. Meth. Appl. Sci. **22** (1999), no. 11, 937–951.

[23] M. Reissig and K. Yagdjian, *Klein-Gordon type decay rates for wave equations with time-dependent coefficients*, Banach Center Publications, **52** (2000), 189–212.

[24] M. Reissig and K. Yagdjian, *$L_p - L_q$ estimates for the solutions of hyperbolic equations of second order with time-dependent coefficients – Oscillations via growth –*, Preprint 98-5, Fakultät für Mathematik und Informatik, TU Bergakademie Freiberg 1998, ISSN 1433-9307.

[25] M. Reissig and K. Yagdjian, *Weakly hyperbolic equations with fast oscillating coefficients*, Osaka J. Math. **36** (1999), no. 2, 437–464.

[26] M. Stoth, *Globale klassische Lösungen der quasilinearen Elastizitätsgleichungen für kubisch elastische Medien im $\mathbb{R}^2$*, SFB 256 Preprint **157**, Universität Bonn, 1991.

[27] R. Strichartz, *A priori estimates for the wave-equation and some applications*, J. Funct. Anal. **5** (1970), 218–235.

[28] H. Uesaka, *The total energy decay of solutions for the wave equation with a dissipative term*, J. Math. Kyoto Univ. **20** (1980), no. 1, 57–65.

[29] W. v. Wahl, *L^p-decay rates for homogeneous wave-equations*, Math. Zeit. **120** (1971), 93–106.

[30] K. Yagdjian, *The Cauchy problem for hyperbolic operators. Multiple characteristics, micro-local approach*, Math. Topics, Akademie-Verlag, Berlin, 1997.

[31] K. Yagdjian, *Parametric resonance and nonexistence of global solution to nonlinear wave equations*, J. Math. Anal. Appl. **260** (2001), no. 1, 251–268.

[32] S. Zheng and W. Shen, *Global solutions to the Cauchy problem of quasilinear hyperbolic parabolic coupled systems*, Sci. Sinica, Ser. A **30** (1987), no. 11, 1133–1149.

Fumihiko Hirosawa
Institute of Mathematics
University of Tsukuba
Ibaraki 305-8571, Japan
e-mail: `Hirosawa@math.tsukuba.ac.jp`

Michael Reissig
Fakultät für Mathematik und Informatik
TU Bergakademie Freiberg
D-09596 Freiberg, Germany
e-mail: `reissig@math.tu-freiberg.de`

Operator Theory:
Advances and Applications, Vol. 145, 157–205

Local Solutions to Quasi-linear Weakly Hyperbolic Differential Equations

Michael Dreher

Abstract. The purpose of this paper is to investigate weakly hyperbolic equations with degeneracies in the space and time variables. These degeneracies as well as the sharp Levi conditions of C^∞ type are formulated by means of certain weight functions. For Cauchy problems to such quasi-linear weakly hyperbolic equations, the following subjects are studied: local existence of solutions in Sobolev spaces and C^∞, a blow-up criterion, domains of dependence, and C^∞ regularity. The main tools are the transformation of the higher-order equation to a first-order system, a calculus for pseudodifferential operators with non-smooth symbols, and a generalization of Gronwall's lemma to differential inequalities with a singular coefficient.

1. Introduction

Let us consider the differential operator of order m

$$P(x,t,D_x,D_t) = D_t^m + \sum_{j+|\alpha|\le m, j<m} a_{j,\alpha}(x,t) D_x^\alpha D_t^j,$$

where we adopted the usual notation $D_t = -i\partial_t$, $D_x = -i\nabla_x$. This operator P is called *hyperbolic in the direction* t if the roots $\tau_j = \tau_j(x,t,\xi)$ of the equation

$$\tau^m + \sum_{j+|\alpha|=m, j<m} a_{j,\alpha}(x,t)\xi^\alpha \tau^j = 0$$

are real for all real x, t, ξ. The operator P is said to be *strictly hyperbolic in the direction* t, if the roots τ_j are real and *distinct* for $\xi \in \mathbb{R}^n \setminus \{0\}$. If P is hyperbolic, but not (necessarily) strictly hyperbolic, it is called *weakly hyperbolic.*

Hyperbolicity is a necessary condition for C^∞ well-posedness of the Cauchy problem (see [19], [22]). Well-posedness (with respect to chosen topological spaces for the data, right-hand side and the solution) of a Cauchy problem means, as usual, the existence, uniqueness and continuous dependence (in the topologies of the given spaces) of the solution. However, hyperbolicity does not guarantee the

2000 *Mathematics Subject Classification.* 35L70, 35L80.
Key words and phrases. Sobolev solutions, loss of regularity, blow-up criterion, domains of dependence.

well-posedness in, e.g., C^∞ or Sobolev spaces. A sufficient condition for the well-posedness in C^∞ and in Sobolev spaces is the strict hyperbolicity, see [26], [20] and [14].

Therefore it is a natural goal to find classes of weakly hyperbolic Cauchy problems which are C^∞ well posed.

In the weakly hyperbolic case, new phenomena occur which may prevent the Cauchy problem from being well posed. These phenomena are the following:

Oscillations in the coefficients with respect to time

- Colombini, Jannelli, and Spagnolo [6], [7], constructed a smooth function $a(t) \geq 0$ and smooth data $u_0(x)$, $u_1(x)$ with the property that the Cauchy problem

$$u_{tt} - a(t)u_{xx} = 0, \quad u(x,0) = u_0(x), \quad u_t(x,0) = u_1(x)$$

has no solution u in the distribution space $\mathcal{D}'(\mathbb{R} \times [0,1])$. This coefficient $a(t)$ is positive for $t > 0$, oscillating for $t \to 0+0$ and vanishing for $t \leq 0$.
- Let $b(t)$ be a positive, periodic, smooth and non-constant function. Tarama [32] proved that the Cauchy problem

$$\begin{aligned} &u_{tt} - \exp(-2t^{-\alpha})b(t^{-1})^2 u_{xx} = 0, \\ &u(x,0) = u_0(x), \quad u_t(x,0) = u_1(x) \end{aligned}$$

is C^∞ well posed if and only if $\alpha \geq 1/2$.

The influence of lower-order terms

- Ivrii and Petkov [16] showed that necessary conditions for the C^∞ well-posedness of

$$v_{tt} - t^{2l}v_{xx} + t^k v_x = 0, \quad l,k \in \mathbb{N}_0, \tag{1.1}$$

$$u_{tt} - x^{2n}u_{xx} + x^m u_x = 0, \quad n,m \in \mathbb{N}_0, \tag{1.2}$$

are $k \geq l-1$ and $m \geq n$. The sufficiency of these conditions was proved by Oleinik [25].
- If one wants to study well-posedness in Sobolev spaces, one has to pay attention to another phenomenon, which occurs in the border case $k = l-1$ of the C^∞ well-posedness: *the loss of Sobolev regularity.* Qi [27] showed by an explicit representation of the solution to the Cauchy problem

$$\begin{aligned} &u_{tt} - t^2 u_{xx} = b u_x, \\ &u(x,0) = \varphi(x), \quad u_t(x,0) = 0, \quad b = 4m+1, \quad m \in \mathbb{N} \end{aligned} \tag{1.3}$$

that $u(\cdot,t) \in H^{s-m}$ if $\varphi \in H^s$. Or, let us look at it from another point: choose an arbitrary data function $\varphi(x)$ with high Sobolev smoothness $s \gg 1$. Then a number b exists such that there is no classical solution of (1.3). The solution only exists in distribution spaces.

The loss of regularity also occurs for equations of the form

$$u_{tt} - t^{2l}u_{xx} - bt^{l-1}u_x = 0, \tag{1.4}$$
$$u(x,0) = \varphi(x), \quad u_t(x,0) = \psi(x),$$

as shown by Taniguchi and Tozaki [31]. Equations of the type (1.3) and (1.4) are interesting because singularities of their solutions may propagate in a non-standard way, see [12], [13], and [31].

There are different ways to exclude the phenomenon of oscillations and to restrict the influence of the lower order terms. We pick the equation

$$u_{tt} - a(x,t)u_{xx} + b(x,t)u_x + d(x,t)u_t + e(x,t)u = f(x,t)$$

as a model problem. First we consider the oscillations.

- If the degeneracy occurs for $t = 0$ only, we may assume ([25])
$$0 \leq Ca(x,t) \pm \partial_t a(x,t), \quad t \geq 0, \quad C > 0.$$
- We can suppose that the coefficient $a(x,t)$ has the structure
$$a(x,t) = a_0(x,t)\sigma(x)^2\lambda(t)^2$$
with some smooth $a_0(x,t) \geq \alpha > 0$, and $\lambda(0) = 0$, $\lambda'(t) > 0$ $(t > 0)$. The degeneration happens at the zeroes of the product $\sigma\lambda$. The functions σ and λ characterize the *spatial degeneracy* and *time degeneracy*, respectively. Assumptions of this type were made, e.g., in Nersesyan [24], Yagdjian [35], and [10], [11], [13]. *We will follow this idea and generalize it to quasi-linear higher order equations in higher dimensions.*

Second, we consider the lower order terms. Conditions which restrict the influence of these terms are called *Levi conditions*. Our aim is to find conditions which do not exclude the interesting equations (1.3) and (1.4). The following Levi conditions have been used widely in the past:

- If the degeneracy occurs for $t = 0$ only, then we may take the condition
$$Btb(x,t)^2 \leq Aa(x,t) + \partial_t a(x,t), \quad t \geq 0 \tag{1.5}$$
from [25]; A and B are some positive constants. This Levi condition is sharp in the case of finite degeneracy: if one fixes $a(x,t) = x^{2n}t^{2l}$ and $b(x,t) = x^m t^k$, (1.5) implies $m \geq n$, $k \geq l-1$. These are exactly the necessary and sufficient conditions from Ivrii, Petkov and Oleinik. However, this condition is not sharp in the case of time degeneracy of infinite order. There exists an explicit representation of the solutions to
$$u_{tt} - e^{-\frac{2}{t}}\frac{1}{t^4}u_{xx} + be^{-\frac{1}{t}}\frac{1}{t^4}u_x = 0, \quad t \geq 0, \quad b = \text{const}, \tag{1.6}$$
see Aleksandrian [1], which implies that the Cauchy problem for this equation is C^∞ well posed. Yet, the coefficients from (1.6) do not satisfy (1.5). Similarly to (1.3) and (1.4), the solutions to (1.6) lose regularity, too; and their singularities can propagate in an astonishing way, see [1].

- If one wants to include more general degenerations, one may assume the rather general and crude conditions

$$b(x,t)^2 \leq Ca(x,t),$$
$$a_t(x,t) \leq Ca(x,t) \text{ or } a_t(x,t) \geq -Ca(x,t),$$

 or, similarly,

$$Bb(x,t)^2 \leq Aa(x,t) + a_t(x,t), \quad A, B > 0,$$

 compare D'Ancona [8], Manfrin [21]. However, these conditions are not sharp; they exclude (1.3), (1.4) and (1.6).
- It can be presumed $a(x,t) = a_0(x,t)\sigma(x)^2\lambda(t)^2$ with $a_0(x,t) \geq \alpha > 0$ and $|b(x,t)| \leq C|\sigma(x)|\lambda'(t)$. Coefficients $a(x,t)$ and $b(x,t)$ satisfying such a Levi condition include the interesting cases (1.3), (1.4) and (1.6). *We will follow this way and generalize these conditions to higher-order equations.*

Let us list the main results of this paper. We are concerned with the hyperbolic Cauchy problem

$$\begin{aligned} D_t^m u + \sum_{j+|\alpha|=m, j<m} & a_{j,\alpha}(x,t,\{D_x^\beta(c_{k,\beta}(x,t)D_t^k u)\})\lambda(t)^{|\alpha|}D_x^\alpha D_t^j(\sigma(x)^{|\alpha|}u) \quad (1.7) \\ &= f(x,t,\{D_x^\beta(c_{k,\beta}(x,t)D_t^k u)\}|_{k+|\beta|\leq m-1}), \quad m \geq 2, \\ u(x,0) &= \varphi_0(x), \dots, D_t^{m-1}u(x,0) = \varphi_{m-1}(x) \end{aligned}$$

for $(x,t) \in M \times [0,T]$; where M is either $\mathbb{R}^n$ or a smooth closed n-dimensional manifold. The functions $\lambda = \lambda(t)$ and $\sigma = \sigma(x)$ describe the degeneration of the principal part of the differential operator, and the functions $c_{k,\beta} = c_{k,\beta}(x,t)$ are weight functions for the lower order terms and characterize the Levi conditions.

Example. The weight function $\lambda = \lambda(t)$ has to satisfy a certain condition (see Condition 4.1). Examples of such λ are

$$\begin{aligned} &\lambda(t) = t^l, \quad l \in \mathbb{N}, \quad l \geq m-1, \\ &\lambda(t) = \Lambda'(t) \text{ with} \\ &\quad \Lambda(t) = \exp(-|t|^{-r}), \quad r > 0, \\ &\quad \Lambda(t) = \exp(-\exp(\exp(\exp(|t|^{-r})))), \quad r > 0. \end{aligned}$$

There are no restrictions on the choice of σ, any smooth real-valued function $\sigma = \sigma(x)$ is admissible.

The weight functions $c_{k,\beta}$ are connected with λ and σ via the relations (4.11) and (3.22)–(3.25), and special examples are

$$c_{k,\beta}(x,t) = \begin{cases} \lambda(t)^{m-k}\Lambda(t)^{k+|\beta|-m}\sigma(x)^{|\beta|} & : |\beta| > 0, \\ 1 & : |\beta| = 0, \end{cases}$$

where $\Lambda(t) = \int_0^t \lambda(\tau)\,d\tau$.

The following results are proved for such Cauchy problems in this paper.

Local existence in Sobolev spaces: For given data in Sobolev spaces, a solution is found which suffers from the loss of Sobolev regularity, as motivated by Qi's example.

Blow-up criterion: We will prove that a blow-up of the solution in high order Sobolev spaces is only possible if the C^1_* Zygmund norm of certain weighted derivatives (up to the order $m-1$) of the solution blows up. This is a generalization of a similar criterion from the strictly hyperbolic case, see Taylor [33].

Local existence in C^∞: That blow-up criterion leads to the local existence of solutions in C^∞ immediately.

Domains of dependence: A special feature of strictly hyperbolic equations is the finite propagation speed. In other words, the value of the solution at a given point at a given time only depends on the values of the initial data and right-hand side from a certain bounded domain, the so-called domain of dependence.

We will define and study domains of dependence for quasi-linear weakly hyperbolic operators, and use them to prove the local in space and time existence and uniqueness of Sobolev solutions, and their C^∞ regularity provided that the data are from C^∞. Our concept of domains of dependence extends the concept of Alinhac and Metivier [2] from the strictly hyperbolic to the weakly hyperbolic case. Geometrically speaking, these domains can be described by the condition that the principal part of the operator be hyperbolic at each point of the boundary of the domain in the normal direction of the boundary. Since the coefficients of the principal part depend on the solution, the domain of dependence for the solution will be dependent on the solution itself.

Next we give some remarks concerning the used methods and tools.

A crucial step of the investigation of hyperbolic Cauchy problems is an *a priori* estimate of the solution in Sobolev spaces, which is usually proved by means of pseudodifferential operators. However, since the coefficients of the hyperbolic operator depend on the solution and its derivatives itself, and because the solution will be from some Sobolev space, the coefficients of this hyperbolic operator will not have C^∞ smoothness. Thus, the theory of pseudodifferential operators with symbols of *infinite* smoothness seems not to be applicable; hence we present a theory of pseudodifferential operators with symbols of *finite* smoothness (H^s or C^1 or merely C^0) in Section 2. We cite results of Taylor [33] concerning mapping properties, commutator estimates, adjoints and compositions.

Our methods for proving the local existence of a solution to (1.7) are a unification of ideas taken from Taylor [33] who studied quasi-linear strictly hyperbolic equations; and Kajitani and Yagdjian [17] who studied quasi-linear weakly hyperbolic equations with time degeneracy.

In Section 3, we study weakly hyperbolic Cauchy problems with pure spatial degeneracy, i.e., (1.7) without weight function $\lambda = \lambda(t)$. Our approach in this case is as follows. We construct some vector-valued function U^* which contains weighted derivatives of u up to the order $m-1$ and solves a pseudodifferential hyperbolic system of first order

$$\partial_t U^* = K^*(x,t,U^*,D)(\sigma U^*) + F^*(x,t,U^*) \tag{1.8}$$

where K^* is a *strictly* hyperbolic matrix pseudodifferential operator of order 1, and F^* contains the right-hand side and some other terms. We insert some smoothing operators J_ε into (1.8) such that its right-hand side becomes an operator of order 0, and the existence of an approximate solution U_ε^* follows immediately from functional analytic arguments. Next we have to prove independent of ε estimates of U_ε^* and its life-span. These estimates will allow us to show an interesting blow-up criterion:

A blow-up of U^* in the H^s norm is impossible as long as the Zygmund space norm $\|U^*\|_{C^1_*}$ remains bounded.

We are able to extend the results won by Dionne [9] and Taylor [33] from the strictly hyperbolic case to the weakly hyperbolic case.

The general weakly hyperbolic Cauchy problem with spatial and time degeneracy is treated in Section 4. Here we face new difficulties which are typical for the time degeneracy:

One such obstacle is a *singular coefficient in the energy inequality.* Consider, as an example, the weakly hyperbolic equation

$$u_{tt} - \lambda(t)^2 u_{xx} = f(x,t), \quad \lambda(0) = 0, \quad \lambda'(t) > 0 \quad (t > 0).$$

If we choose the energy in the usual way, $E(t) = \|u_t\|^2_{L^2} + \|\lambda(t)u_x\|^2_{L^2}$, then we obtain, after some calculations,

$$E'(t) \le \|f(\cdot,t)\|^2_{L^2} + \frac{\lambda'(t)}{\lambda(t)} E(t).$$

The lemma of Gronwall is not applicable, since the coefficient $\lambda'(t)/\lambda(t)$ becomes unbounded for $t \to 0$. But one can use Nersesyan's lemma (see Lemma 6.2) if the initial data vanish and $\|f(\cdot,t)\|_{L^2}$ has a zero of sufficiently high order at $t = 0$.

Another obstacle is the *loss of regularity.* The example of Qi [27] shows that the solution can lose Sobolev smoothness in comparison with the initial data. The number of lost derivatives depends (in the linear case) on the L^∞-norm of the coefficients of some lower order terms. This makes the investigation of nonlinear Cauchy problems delicate, since the usual fixed point arguments cannot be applied directly. The crucial tool for solving this difficulty is the reduction (Section 4.3) of the Cauchy problem (1.7) to another Cauchy problem which enjoys the so-called *strictly hyperbolic type property*: let L be a weakly hyperbolic operator of order 2 (for simplicity). We say that a Cauchy problem

$$Lu(x,t) = f(x,t), \quad u(x,0) = u_t(x,0) = 0$$

has the *strictly hyperbolic type property* if there is a topological space B and a weight function $\omega(x,t)$ such that $f \in B$ implies $\omega(x,t)\nabla_x u \in B$ and $u_t \in B$. Then the local existence in B of a solution to a *quasi-linear* version of the above Cauchy problem can be proved by standard arguments. In the strictly hyperbolic case, we choose $\omega \equiv 1$ and $B = C([0,T], H^s)$. In the weakly hyperbolic case, ω is chosen according to the degeneracy, and B consists of functions which decay sufficiently fast for $t \to 0$, see Section 4.1. For other applications of such adapted Banach spaces to weakly hyperbolic equations, see [12], [13], and Reissig, Yagdjian [30].

Concerning the investigation of domains of dependence in Section 5, our technique is as follows: we exhaust the domain of dependence with hypersurfaces, and change the variables such that these hypersurfaces become planes of constant time. Outside some small domain, we then change the operator slightly, and transform the equation into a Cauchy problem on a torus which can be treated with the methods of Section 3.

Finally, we introduce some notation.

By M we denote either $\mathbb{R}^n$ or a closed smooth n-dimensional manifold.

The Banach space of functions whose derivatives up to the order k are bounded and continuous is denoted by $C_b^k(M)$, $k \in \mathbb{N}_0$. Similarly, we introduce the Hölder spaces $C_b^s(M)$, $s \in \mathbb{R}^+$, and write $Lip^1(M)$ for the space of Lipschitz continuous functions on M.

Let $C_*^s(M)$ denote the Hölder spaces for $s \notin \mathbb{N}$, and the *Zygmund spaces* for $s \in \mathbb{N}^+$. The Zygmund spaces C_*^s, $s \in \mathbb{N}^+$, consist of all functions u with the property that $u \in C_b^{s-1}$ and (in local coordinates)

$$\sup_{x \neq y} \sum_{|\alpha|=s-1} \frac{|D^\alpha u(x) - 2(D^\alpha u)((x+y)/2) + D^\alpha u(y)|}{|x-y|} < \infty.$$

The spaces C_b^k are continuously embedded in C_*^k, for $k \in \mathbb{N}^+$.

Let $\triangle$ be the Laplace–Beltrami operator on M and set $\langle D \rangle = (1-\triangle)^{1/2}$. In case of $M = \mathbb{R}^n$, $\langle D \rangle$ can be written as a pseudodifferential operator with symbol $\langle \xi \rangle = (1+|\xi|^2)^{1/2}$ (a thorough representation of the theory of pseudodifferential operators can be found in Hörmander [15]). Then we define the Sobolev spaces $H^s(M) = \langle D \rangle^{-s} L^2(M)$ for $s \in \mathbb{R}$, where $L^2(M)$ is the usual Lebesgue space of square integrable functions on M.

Assuming local coordinates $x = (x_1, \ldots, x_n)$ on M, we will employ the multi-index notation:

$$D_x^\alpha = D_{x_1}^{\alpha_1} \ldots D_{x_n}^{\alpha_n}, \quad \alpha = (\alpha_1, \ldots, \alpha_n) \in \mathbb{N}^n, \quad D_{x_j} = \frac{1}{i}\frac{\partial}{\partial x_j}, \quad i^2 = -1.$$

Acknowledgment. I would like to thank Prof. Reissig for many useful discussions and the referee for his careful reading and helpful criticism.

2. Pseudodifferential Operators with Finite Smoothness

2.1. Definition and Mapping Properties

In order to describe the smoothness of functions and pseudodifferential symbols, we introduce some scales $(X^s)_s$ of function spaces:

$$\begin{aligned} X^s &= H^s(M), \quad \frac{n}{2} < s < \infty, \\ X^s &= C_*^s(M), \quad 0 < s < \infty, \\ X^s &= C_b^s(M), \quad 0 \le s < \infty. \end{aligned}$$

Definition 2.1.1 (Space of symbols of finite smoothness). The space $X^sS^m_{1,0}$ consists of all functions $p(x,\xi) : M \times \mathbb{R}^n \to \mathbb{C}$ with

$$\left\|D_\xi^\alpha p(\cdot,\xi)\right\|_{X^s} \le C_\alpha \langle\xi\rangle^{m-|\alpha|} \quad \alpha \ge 0.$$

In other words, for all $N \in \mathbb{N}_0$ it holds that

$$\pi^m_{N,X^s}(p) = \sup\left\{\left\|D_\xi^\alpha p(\cdot,\xi)\right\|_{X^s} \langle\xi\rangle^{-m+|\alpha|} : \xi \in \mathbb{R}^n, |\alpha| \le N\right\} < \infty.$$

Definition 2.1.2 (Classical symbols of finite smoothness). We say that $p(x,\xi) \in X^sS^m_{cl}$ if there is an asymptotic expansion $p(x,\xi) \sim \sum_{j\ge 0}\chi(\xi)p_j(x,\xi)$, where $p_j(x,\xi)$ are positive homogeneous of degree $m-j$ in ξ and $p - \sum_{j=0}^{N-1}\chi p_j \in X^sS^{m-N}_{1,0}$. The function $\chi \in C^\infty(\mathbb{R}^n_\xi)$ vanishes in a neighborhood of 0 and equals 1 for $|\xi| \ge C > 0$.

Definition 2.1.3 (Operators of finite smoothness). Let $M = \mathbb{R}^n$. The operator spaces $OPX^sS^m_{1,0}$, $OPX^sS^m_{cl}$,respectively, consist of all operators $p(x,D)$ mapping $C_0^\infty(M)$ into the space of distributions $\mathcal{D}'(M)$ whose symbols $p(x,\xi)$ belong to $X^sS^m_{1,0}$, $X^sS^m_{cl}$, respectively, and satisfy

$$(p(x,D)u)(x) = (2\pi)^{-n}\int_{\mathbb{R}^n} e^{ix\xi}p(x,\xi)\hat{u}(\xi)d\xi \quad u \in C_0^\infty(\mathbb{R}^n).$$

If M is a C^∞ manifold, then the operator $p(x,D)$ is defined as follows. Let (Ω,κ) be a local chart of M, $\kappa\colon M \supset \Omega \to U \subset \mathbb{R}^n$. Define the pull-back $\kappa^*\colon C_0^\infty(U) \to C_0^\infty(\Omega)$ by $(\kappa^*u)(x) = u(\kappa(x))$, and the push-forward $\kappa_*\colon \mathcal{D}'(\Omega) \to \mathcal{D}'(U)$ by $\langle\kappa_*F,u\rangle = \langle F,\kappa^*u\rangle$, $(F \in \mathcal{D}'(\Omega), u \in C_0^\infty(U))$. Then an operator $P\colon C_0^\infty(M) \to \mathcal{D}'(M)$ belongs to $OPX^sS^m_{1,0}$, $OPX^sS^m_{cl}$ if, for every local chart (Ω,κ), $\kappa_* \circ P \circ \kappa^*\colon C_0^\infty(U) \to \mathcal{D}'(U)$ belongs to $OPX^sS^m_{1,0}$, $OPX^sS^m_{cl}$, respectively.

The following two mapping properties are cited from [33] and [34], Chapter 13.9.

Proposition 2.1.4. *Let $p(x,D) \in OPC^s_*S^m_{1,0}$. Then $p(x,D)$ can be extended to an operator continuously mapping C^{r+m}_* into C^r_* $(-s < r \le s)$ and H^{r+m} into H^r $(-s < r < s)$, respectively.*

In the case of operators with coefficients from Sobolev spaces, we have fewer problems with the borderline case $r = s$:

Proposition 2.1.5. *If $p(x, D) \in OPH^s S^m_{1,0}$, then $p(x, D)$ can be extended to an operator which maps H^{r+m} continuously into H^r for $-s < r \le s$.*

2.2. Special Smoothing Operators

Definition 2.2.1. Denote the Laplace–Beltrami operator of M by $\triangle$. For $0 < \varepsilon \le 1$, we define the smoothing operator $J_\varepsilon = (1 - \varepsilon\triangle)^{-1/2}$.

The proofs of the following lemmas are straightforward.

Lemma 2.2.2. *The operator J_ε is invertible and commutes with $\langle D \rangle$.*

Lemma 2.2.3. *Let X^s be either $H^s(M)$ with $s \in \mathbb{R}$, or $C^s_*(M)$ with $s > 0$. Then there is a constant $C > 0$ such that for every $0 < \varepsilon \le 1$,*

$$\begin{aligned} &\|J_\varepsilon f\|_{X^{s+1}} \le C\varepsilon^{-1} \|f\|_{X^s}, \\ &\|f - J_\varepsilon f\|_{X^{s-t}} \le C\varepsilon^t \|f\|_{X^s}, \quad 0 \le t \le 1. \end{aligned}$$

2.3. Commutator Estimates

We quote some estimates from Coifman, Meyer [5], Kato, Ponce [18] and Taylor [33]:

Proposition 2.3.1. *The following inequalities hold:*

$$\begin{aligned} &\|[P, f]\|_{L^2 \to L^2} \le C \|f\|_{Lip^1}, \quad P \in OPS^1_{1,0}, \quad f \in Lip^1, \\ &\|[P, f]\|_{L^2 \to H^1} \le C \|f\|_{Lip^1}, \quad P \in OPS^0_{1,0}, \quad f \in Lip^1, \\ &\|[P, f]\|_{H^{-1} \to L^2} \le C \|f\|_{Lip^1} \quad P \in OPS^0_{1,0}, \quad f \in Lip^1, \\ &\|P(fu) - fPu\|_{L^2} \le C \|f\|_{Lip^1} \|u\|_{H^{s-1}} + C \|f\|_{H^s} \|u\|_{L^\infty}, \\ &\quad P \in OPS^s_{1,0}, \quad s > 0, \quad f \in Lip^1 \cap H^s, \quad u \in L^\infty \cap H^{s-1}. \end{aligned} \tag{2.1}$$

Lemma 2.3.2. *Let J_ε be the smoothing operator from Definition 2.2.1. Then the assertions of the previous proposition hold for $P = J_\varepsilon$ with a constant C independent of ε, $0 < \varepsilon \le 1$.*

Exploiting the above estimates, we come to the central result of this section: commutator estimates for operators with *non-smooth, classical* symbols.

Proposition 2.3.3. *Let $a(x, D) \in OPC^1_b S^\alpha_{cl}$, $b(x, D) \in OPC^1_b S^\beta_{cl}$ with $\alpha, \beta \in \{0, 1\}$. Then it holds (with some N)*

$$\begin{aligned} &\|[a(x, D), b(x, D)]\|_{H^{\alpha+\beta-1} \to L^2} \\ &\le C \left(\sum_{j=0}^{N-1} \pi^{\alpha-j}_{N,C^1_b}(a_j) + \pi^{\alpha-N}_{N,C^1_b}(r_{a,N}) \right) \left(\sum_{j=0}^{N-1} \pi^{\beta-j}_{N,C^1_b}(b_j) + \pi^{\beta-N}_{N,C^1_b}(r_{b,N}) \right), \end{aligned}$$

where a_j, b_j are the homogeneous components of the expansions of a, b with remainders $r_{a,N}$, $r_{b,N}$, respectively. If $\alpha = \beta = 0$, then we additionally have

$$\begin{aligned}
&\|[a(x,D), b(x,D)]\|_{L^2\to H^1} \\
&\leq C\left(\sum_{j=0}^{N-1} \pi^{\alpha-j}_{N,C^1_b}(a_j) + \pi^{\alpha-N}_{N,C^1_b}(r_{a,N})\right)\left(\sum_{j=0}^{N-1} \pi^{\beta-j}_{N,C^1_b}(b_j) + \pi^{\beta-N}_{N,C^1_b}(r_{b,N})\right).
\end{aligned}$$

The key idea of the proof is the following. Since the symbol $a(x,\xi)$ is *classical*, it allows the expansion (in local coordinates)

$$a(x,\xi) = \sum_{j=0}^{N-1}\sum_{l=0}^{\infty}\sum_{m=1}^{h(l,n)} a_{jlm}(x) Y_{lm}(\xi)\langle\xi\rangle^{\alpha-j} + r_{a,N}(x,\xi),$$

where $Y = Y_{lm}(\xi)$, $0 \leq l < \infty$, $1 \leq m \leq h(l,n)$, are the spherical harmonics, i.e., the eigenfunctions of the Laplace–Beltrami operator on the unit sphere S^{n-1}, which form an orthogonal basis of $L^2(S^{n-1})$. The sequence $\{a_{jlm}\}_{l,m}$ is rapidly decreasing in the sense that

$$\sum_{m=1}^{h(l,n)} \|a_{jlm}\|_{C^1_b} \leq C_k \pi^{\alpha-j}_{N,C^1_b}(a_j)(1+l)^{-k}, \quad k \geq 0,$$

for all $l \geq 0$, and N sufficiently large. Moreover, $r_{a,N} \in C^1_b S^{\alpha-N}_{1,0}$. Plugging this expansion and a similar one for $b(x,\xi)$ into the commutator $[a,b]$, and employing the above commutator estimates from Proposition 2.3.1, one can derive the desired estimates. For details, see [33].

Now we want to generalize (2.1), replacing $f \in Lip^1 \cap H^s$ by $A(x,D) \in OPC^1_b S^\alpha_{cl} \cap OPH^{s_0} S^\alpha_{cl}$ with $s_0 > n/2$, $\alpha \in \mathbb{N}_0$. Here we run into a problem, since an operator P from $OPS^s_{1,0}$ does generally not map $C^s_b(M)$ into $C^0_b(M)$. For this reason we introduce the space $C^\alpha_{\sharp,K_0}$ of all functions u satisfying $\langle D\rangle^\alpha Y_{lm}(D)u \in C^0_b$ for all l, m such that $\sup_{l,m}(1+l)^{-K_0}\|\langle D\rangle^\alpha Y_{lm}(D)u\|_{C^0_b} < \infty$. The constant K_0 is fixed in such a manner that $\|Y_{lm}(D)u\|_{C^0_b} \leq C(1+l)^{K_0}\|u\|_{C^0_b}$ for all $u \in C^\infty_0(M)$. The profit of this definition is that the mapping $B : C^\alpha_{\sharp,K_0} \to C^0_b$ is continuous for all $B \in OPS^\alpha_{cl}$. The embedding $C^{\alpha+\delta}_b \subset C^\alpha_{\sharp,K_0}$ is continuous for any positive δ, see [33], p.126. We have the (set-theoretical) inclusions $C^\alpha_{\sharp,K_0} \subset C^\alpha_b \subset C^\alpha_*$.

Proposition 2.3.4. *Let $P \in OPS^s_{1,0}$, $A(x,D) \in OPC^1_b S^\alpha_{cl} \cap OPH^{s_0} S^\alpha_{cl}$ with $s_0 > n/2$, $0 < s \leq s_0$, $\alpha \in \mathbb{N}_0$ and $K \geq K_0$. Then it holds that*

$$\begin{aligned}
\|[P, A(x,D)]\,u\|_{L^2} &\leq C\left(\sum_{j=0}^{N-1} \pi^{\alpha-j}_{N,C^1_b}(a_j) + \pi^{\alpha-N}_{N,C^1_b}(r_N)\right)\|u\|_{H^{s+\alpha-1}} \\
&\quad + C_K\left(\sum_{j=0}^{N-1} \pi^{-j}_{N,H^s}(a_j) + \pi^{\alpha-N}_{N,H^s}(r_N)\right)\|u\|_{C^\alpha_{\sharp,K}}
\end{aligned}$$

with some constant N and the terms a_j, r_N from the asymptotic expansion of the classical operator A.

A proof can be found in [33]. Now we list properties of the spaces $C^{\alpha}_{\sharp,K}$.

Lemma 2.3.5. *For every $\alpha \in \mathbb{N}_0$, a positive constant C exists such that*

$$\|u\|_{C^{\alpha}_b} + \|\langle D\rangle^{\alpha} u\|_{C^0_b} \leq C\,\|u\|_{C^{\alpha}_{\sharp,K_0}}, \quad u \in C^{\alpha}_{\sharp,K_0}. \tag{2.2}$$

Let $\sigma \in C^{\infty}_b$, $u \in C^{\alpha}_{\sharp,K_0}$ and $K_1 > K_0$ be sufficiently large. Then $\sigma u \in C^{\alpha}_{\sharp,K_1}$ and a constant $C = C(\sigma,\alpha)$ (independently of u) exists with

$$\|\sigma u\|_{C^{\alpha}_{\sharp,K_1}} \leq C\,\|u\|_{C^{\alpha}_{\sharp,K_0}} \tag{2.3}$$

2.4. Adjoint Operators and Compositions

Pseudodifferential operators with symbols of finite smoothness form an algebra in the sense of the following propositions, whose proofs can be found in [33].

Proposition 2.4.1. *Let $K(x,D) \in OPC^1_b S^1_{cl}$ be a matrix pseudodifferential operator. Then the adjoint operator $K^*(x,D)$ satisfies*

$$\operatorname{symb}(K^*(x,D) - R) = \overline{K(x,\xi)^T}, \quad \|RU\|_{L^2} \leq C\pi^1_{N,C^1_b}(K)\,\|U\|_{L^2}.$$

with some operator R and some $N > 0$.

Proposition 2.4.2. *Let $A(x,D) \in OPC^0_b S^j_{cl}$, $B(x,D) \in OPC^1_b S^{1-j}_{cl}$ ($j = 0$ or $j = 1$) be pseudodifferential matrix operators. Then*

$$\begin{aligned} &A(x,D)B(x,D) = C(x,D) + R, \\ &C(x,\xi) = A(x,\xi)B(x,\xi) \in C^0_b S^1_{cl}, \\ &\|RU\|_{L^2} \leq C\pi^j_{N,C^0_b}(A)\pi^{1-j}_{N,C^1_b}(B)\,\|U\|_{L^2}. \end{aligned}$$

3. Weakly Hyperbolic Cauchy Problems with Spatial Degeneracy

3.1. The Linear Case

We are concerned with the linear Cauchy problem

$$\begin{aligned} &D^m_t u + \sum_{j+|\alpha|=m,j<m} a_{j,\alpha}(x,t) D^{\alpha}_x D^j_t\,(\sigma(x)^{\alpha} u) = f(x,t), \\ &u(x,t_0) = \varphi_0(x), \ldots, D^{m-1}_t u(x,t_0) = \varphi_{m-1}(x), \end{aligned} \tag{3.1}$$

under the following condition of hyperbolicity:

Condition 3.1. *We assume that the roots $\tau_j(x,t,\xi)$ of the equation*

$$\tau^m + \sum_{j+|\alpha|=m,j<m} a_{j,\alpha}(x,t)\xi^{\alpha}\tau^j = 0$$

are real and distinct, $|\tau_j(x,t,\xi) - \tau_i(x,t,\xi)| \geq c|\xi|$, $c > 0$, for $i \neq j$, and all (x,t,ξ).

Remark 3.1.1. The space variable x lives on some manifold M, where either $M = \mathbb{R}^n$ or M is a smooth closed n-dimensional manifold, and (3.1) is to be understood in local coordinates. Of special importance is the case of M being a torus, $M = (\mathbb{R}/2\pi)^n$: during the investigation of domains of dependence (in Section 5), we will transfer a hyperbolic equation which is defined in some bounded domain of $\mathbb{R}^n \times [0,T]$ into a hyperbolic equation defined on $(\mathbb{R}/2\pi)^n \times [0,T]$, and bring into play the results to be proved now.

Marking the regularity of the data with subscript "d", we suppose that

$$\varphi_j \in H^{s_d+m-1-j}(M), \quad s_d \geq 0, \tag{3.2}$$

$$f \in C([t_0,T], H^{s_d}(M)). \tag{3.3}$$

The weight function σ is presumed to be real-valued and smooth,

$$\sigma \in C_b^\infty(M, \mathbb{R}). \tag{3.4}$$

In case of M being a closed manifold, we assume

$$a_{j,\alpha} \in C([t_0,T], H^{s_c}(M)) \cap C^1([t_0,T], H^{s_c-1}(M)), \quad s_c > \frac{n}{2}+1, \tag{3.5}$$

where the subscript "c" means "coefficient". And if $M = \mathbb{R}^n$, we assume that there are constants $C_{j,\alpha}$ such that

$$a_{j,\alpha} - C_{j,\alpha} \in C([t_0,T], H^{s_c}(M)) \cap C^1([t_0,T], H^{s_c-1}(M)), \quad s_c > \frac{n}{2}+1. \tag{3.6}$$

The reason for this distinction is that functions of $H^{s_c}(\mathbb{R}^n)$ have to decay at infinity, making Condition 3.1 impossible to hold. For unity of notation, we may define $C_{j,\alpha} = 0$ in the first case, and $A_{j,\alpha} = a_{j,\alpha} - C_{j,\alpha}$ for general M.

Our approach is as follows. We insert a smoothing operator (see Subsection 2.2) into (3.1), such that we obtain an ordinary differential equation for a function with values in a Banach space. Then this equation will be transformed into a first-order pseudodifferential system. An *a priori* estimate and an existence result will be proved for this regularized system, see Proposition 3.1.3 (a), and an *a priori* estimate for the corresponding non-regularized system will be shown in Proposition 3.1.3 (b).

The question of *existence* of a solution to (3.1) will be answered in Section 3.2, *after* we have investigated a quasi-linear version of (3.1).

3.1.1. CONSTRUCTION OF A FIRST-ORDER SYSTEM For $0 < \varepsilon \leq 1$, we consider a regularized version of (3.1):

$$D_t^m u_\varepsilon = f(x,t) - J_\varepsilon \sum_{j+|\alpha|=m, j<m} a_{j,\alpha}(x,t) D_x^\alpha J_\varepsilon^{|\alpha|} D_t^j \left(\sigma(x)^\alpha u_\varepsilon\right), \tag{3.7}$$

$$u_\varepsilon(x,t_0) = \varphi_0(x), \ldots, D_t^{m-1} u_\varepsilon(x,t_0) = \varphi_{m-1}(x).$$

The operator J_ε maps H^r into H^{r+1} for any $r \in \mathbb{R}$ with norm $\mathcal{O}(\varepsilon^{-1})$. This allows us to regard (3.7) as a linear Banach space ODE which is globally solvable, and we acquire a unique solution

$$u_\varepsilon \in C^m([t_0,T], H^{\min(s_c,s_d)}(M)).$$

If we succeed in finding estimates of u_ε which do not depend on ε, then there is hope that the limit $\lim_{\varepsilon\to 0} u_\varepsilon$ in the corresponding topologies exists and is a solution to (3.1). We recall that the $S^0_{1,0}$ seminorms of the pseudodifferential symbol of J_ε can be estimated uniformly in ε, $0 \le \varepsilon \le 1$.

We define the vector of unknowns

$$U_\varepsilon = (U_{\varepsilon,1},\dots,U_{\varepsilon,m})^T, \quad U_{\varepsilon,j} = (\langle D\rangle J_\varepsilon)^{m-j}\left(\sigma^{m-j} D_t^{j-1} u_\varepsilon\right), \tag{3.8}$$

and get the system

$$\begin{aligned} D_t U_{\varepsilon,j} =& \langle D\rangle J_\varepsilon(\sigma U_{\varepsilon,j+1}) + \langle D\rangle J_\varepsilon\left[((\langle D\rangle J_\varepsilon)^{m-j-1},\sigma\right]((\langle D\rangle J_\varepsilon)^{j+1-m} U_{\varepsilon,j+1},\\ D_t U_{\varepsilon,m} =& f - J_\varepsilon \sum_{j+|\alpha|=m, j<m} a_{j,\alpha} P^\alpha \langle D\rangle(\sigma U_{\varepsilon,j+1})\\ & - J_\varepsilon \sum_{j+|\alpha|=m, j<m} a_{j,\alpha} P^\alpha\langle D\rangle\left[((\langle D\rangle J_\varepsilon)^{|\alpha|-1},\sigma\right]((\langle D\rangle J_\varepsilon)^{1-|\alpha|} U_{\varepsilon,j+1} \end{aligned}$$

with $P_j = D_{x_j}\langle D\rangle^{-1}$, $P^\alpha = \prod_{j=1}^n P_j^{\alpha_j}$. This gives

$$\partial_t U_\varepsilon = K_\varepsilon(\sigma U_\varepsilon) + B_\varepsilon U_\varepsilon + F, \quad U_\varepsilon(t_0) = \Phi_{\varepsilon,0}, \tag{3.9}$$

$$K_\varepsilon = J_\varepsilon K_0\langle D\rangle = J_\varepsilon i \begin{pmatrix} 0 & 1 & \dots & 0\\ 0 & 0 & \dots & 0\\ \vdots & \vdots & \ddots & \vdots\\ 0 & 0 & \dots & 1\\ k_0 & k_1 & \dots & k_{m-1}\end{pmatrix}\langle D\rangle \in OPC_b^1 S^1_{cl}, \tag{3.10}$$

$$k_j = -\sum_{|\alpha|=m-j} a_{j,\alpha}P^\alpha,$$

$$B_\varepsilon = i\begin{pmatrix} 0 & b_\varepsilon^{(2)} & 0 & \dots & 0\\ 0 & 0 & b_\varepsilon^{(3)} & \dots & 0\\ \vdots & \vdots & \vdots & \ddots & \vdots\\ 0 & 0 & 0 & \dots & b_\varepsilon^{(m)}\\ b_{\varepsilon,1} & b_{\varepsilon,2} & b_{\varepsilon,3} & \dots & 0\end{pmatrix} \in OPC_b^1 S^0_{cl}, \tag{3.11}$$

$$b_{\varepsilon,k} = -J_\varepsilon \sum_{|\alpha|=m+1-k} a_{k-1,\alpha}P^\alpha\langle D\rangle\left[((\langle D\rangle J_\varepsilon)^{m-k},\sigma\right]((\langle D\rangle J_\varepsilon)^{k-m},$$

$$b_\varepsilon^{(j)} = \langle D\rangle J_\varepsilon\left[((\langle D\rangle J_\varepsilon)^{m-j},\sigma\right]((\langle D\rangle J_\varepsilon)^{j-m},$$

$$F = (0,0,\dots,0,if)^T \in C\left([t_0,T], H^{s_d}\right), \tag{3.12}$$

$$\Phi_{\varepsilon,0} = (((\langle D\rangle J_\varepsilon)^{m-1}(\sigma^{m-1}\varphi_0),\dots,\varphi_{m-1})^T \in H^{s_d}. \tag{3.13}$$

We remark that $K_0\langle D\rangle$ is a *strictly* hyperbolic operator, and does not depend on ε. Next we construct a symmetrizer for K_0, using ideas from Leray [20]. We introduce the notation $p_j = \xi_j \langle\xi\rangle^{-1}$, $p^\alpha = \prod_{j=1}^n p_j^{\alpha_j}$, and denote the eigenvalues of $K_0(x,t,p)$ by $i\tau_j(x,t,p)$. Obviously,

$$K_0\left(1,\tau_j,\dots,\tau_j^{m-1}\right)^T = i\tau_j\left(1,\tau_j,\dots,\tau_j^{m-1}\right)^T.$$

Let $S_0 = V(\tau_1(x,t,p),\dots,\tau_m(x,t,p))$ be the Vandermonde matrix of the numbers $(\tau_1,\dots,\tau_m)$, which satisfies

$$K_0S_0 = iS_0\,\mathrm{diag}(\tau_1,\dots,\tau_m) = iS_0D.$$

We put $s_k(x,t,p) = \sum_{j=1}^m \tau_j(x,t,p)^k$, and see that the matrix

$$S = S_0S_0^T = \begin{pmatrix} s_0 & s_1 & s_2 & \dots & s_{m-1} \\ s_1 & s_2 & s_3 & \dots & s_m \\ s_2 & s_3 & s_4 & \dots & s_{m+1} \\ \vdots & \vdots & \vdots & \ddots & \vdots \\ s_{m-1} & s_m & s_{m+1} & \dots & s_{2m-2} \end{pmatrix}$$

is symmetric and positive definite. Vieta's theorem reveals that the s_k are some polynomials in $a_{j,\alpha}p^\alpha$. The symmetrizer is defined as $R = \det(S)S^{-1}$, and is obviously a symmetric positive definite matrix. It remains to check that RK_0 is symmetric: The matrix K_0S is symmetric since

$$K_0S = K_0S_0S_0^T = iS_0DS_0^T = (iS_0DS_0^T)^T = (K_0S)^T.$$

Setting $c = \det(S)$, we see that R is a symmetrizer for K_0:

$$\begin{aligned} RK_0 &= cS^{-1}K_0 = cS^{-1}(K_0S)S^{-1} = cS^{-1}(K_0S)(S^{-1})^T \\ &= (cS^{-1}(K_0S)(S^{-1})^T)^T = (RK_0)^T. \end{aligned}$$

The components r_{ij} of R are some polynomials of the $a_{j,\alpha}p^\alpha$, that is,

$$r_{ij}(x,t,p) = \sum_{l\in B_{ij}} c_{ijl}\Bigg(\prod_{(j,\alpha)\in D_{ijl}} a_{j,\alpha}(x,t)\Bigg)\Bigg(\prod_{(j,\alpha)\in D_{ijl}} p^\alpha\Bigg), \tag{3.14}$$

with $c_{ijl}\in\mathbb{C}$ and some finite index sets B_{ij} and D_{ijl}. Since the $\tau_k(x,t,p)$ depend on $p_j = \xi_j\langle\xi\rangle^{-1}$, we have $R(x,t,\xi)\in C_b^1S_{cl}^0$. The property of R being a symmetrizer implies (see [20])

$$C^{-1}\left\|V\right\|_{L^2}^2 \le (RV,V)_{L^2(M)} \le C\left\|V\right\|_{L^2}^2, \quad V\in L^2. \tag{3.15}$$

The product structure of the k_{ij} and r_{ij} gives us the estimates

$$\max\{\|K_0\|_{L^2\to L^2}, \|K_0\langle D\rangle\|_{C_\sharp^1\to C_b^0}\} \le C(\max_{j,\alpha}\|a_{j,\alpha}\|_{C_b^0}), \tag{3.16}$$

$$\|R\|_{L^2\to L^2} \le C(\max_{j,\alpha}\|a_{j,\alpha}\|_{C_b^0}), \tag{3.17}$$

where the term $C(\max_{j,\alpha}\|a_{j,\alpha}\|_{C_b^0})$ denotes a universal constant which depends on $\max_{j,\alpha}\|a_{j,\alpha}\|_{C_b^0}$ in a nonlinear way.

Next we consider $(K_0\langle D\rangle)^*R + RK_0\langle D\rangle$. Proposition 2.4.1 gives us an expression of $(K_0\langle D\rangle)^*$, and Proposition 2.4.2 tells us how to compose $(K_0\langle D\rangle)^*$

and R, as well as R and $K_0\langle D\rangle$. This way, we see that the principal symbol of $(K_0\langle D\rangle)^*R + RK_0\langle D\rangle$ falls out, and we obtain

$$\|(K_0\langle D\rangle)^*R + RK_0\langle D\rangle\|_{L^2\to L^2} \le C(\max_{j,\alpha}\|a_{j,\alpha}\|_{C_b^0})\max_{j,\alpha}(\|a_{j,\alpha}\|_{C_b^1}+1). \tag{3.18}$$

Finally, mapping properties of the matrix operator B_ε are studied. The assumption $\sigma \in C_b^\infty$ implies $b_\varepsilon^{(j)} \in OPS_{cl}^0$; hence $\left\|b_\varepsilon^{(j)}v\right\|_{H^s} \le C\|v\|_{H^s}$, uniformly in ε. Similarly, $\|b_{\varepsilon,k}v\|_{L^2} \le C\max_{j,\alpha}\|a_{j,\alpha}\|_{L^\infty}\|v\|_{L^2}$, which yields

$$\|B_\varepsilon U\|_{L^2} \le C(\max_{j,\alpha}\|a_{j,\alpha}\|_{C_b^0})\|U\|_{L^2}\,. \tag{3.19}$$

If $s > 0$, then we can make use of formula (3.1.59) from [33],

$$\|uv\|_{H^s} < C\left(\|u\|_{L^\infty}\|v\|_{H^s} + \|v\|_{L^\infty}\|u\|_{H^s}\right), \quad s>0,$$

and the estimates

$$\left\|P^\alpha\langle D\rangle\left[(\langle D\rangle J_\varepsilon)^{m-k},\sigma\right](\langle D\rangle J_\varepsilon)^{k-m}v\right\|_{H^s} \le C\|v\|_{H^s}\,,$$
$$\left\|P^\alpha\langle D\rangle\left[(\langle D\rangle J_\varepsilon)^{m-k},\sigma\right](\langle D\rangle J_\varepsilon)^{k-m}v\right\|_{L^\infty} \le C\|v\|_{C^0_{\sharp,K_0}}\,,$$

which give us the uniform in ε estimates

$$\|B_\varepsilon U\|_{H^s} \le C(\max_{j,\alpha}\|a_{j,\alpha}\|_{C_b^0})\|U\|_{H^s} + C\max_{j,\alpha}(\|A_{j,\alpha}\|_{H^s}+1)\|U\|_{C_\sharp^0}\,. \tag{3.20}$$

Let us summarize the results:

Proposition 3.1.2. *The regularized linear Cauchy problem* (3.1) *can be transformed into the equivalent system* (3.9) *with* U_ε, K_ε, B_ε, F, $\Phi_{\varepsilon,0}$ *from* (3.8), (3.10), (3.11), (3.12) *and* (3.13), *respectively.*

The matrix operator $K_0\langle D\rangle$ *is a strictly hyperbolic pseudodifferential operator with finite smoothness,* $K_0\langle D\rangle \in OPC_b^1S_{cl}^1$. *In case* $M = \mathbb{R}^n$, *a matrix pseudodifferential operator* $\tilde K$ *exists, whose symbol does not depend on* x, *such that* $K_0\langle D\rangle - \tilde K \in OPH^{s_c}S_{cl}^1$. *In case of a compact manifold* M, *we have* $K_0\langle D\rangle \in OPH^{s_c}S_{cl}^1$.

Furthermore, a symmetrizer R *assigned to* $K_0\langle D\rangle$ *exists, which is a zero-order pseudodifferential operator with symbol of finite smoothness,* $R \in OPC_b^1S_{cl}^0$, *and induces a norm in* L^2 *which is equivalent to the usual norm, see* (3.15).

The operators $K_0\langle D\rangle$, R, $(K_0\langle D\rangle)^*R + RK_0\langle D\rangle$ *and* B_ε *have the mapping properties given in* (3.16)–(3.20), *respectively.*

3.1.2. A PRIORI ESTIMATES Now we have all tools to show an *a priori* estimate of strictly hyperbolic type:

Proposition 3.1.3. **(a)** *The linear system* (3.9) *has a unique global solution* $U_\varepsilon \in C^1([t_0,T], H^{\min(s_c,s_d)}(M))$ *which satisfies the following estimates for* $0 \le s \le$

$\min(s_c, s_d)$:

$$\begin{aligned}\partial_t\left(R\langle D\rangle^s U_\varepsilon, \langle D\rangle^s U_\varepsilon\right) &\le C(\max_{j,\alpha}\|\partial_t a_{j,\alpha}\|_{C_b^0})\,\|U_\varepsilon\|_{H^s}^2 + 2\sqrt{(R\langle D\rangle^s U_\varepsilon, \langle D\rangle^s U_\varepsilon)}\sqrt{(R\langle D\rangle^s F, \langle D\rangle^s F)}\\ &\quad + C(\max_{j,\alpha}\|a_{j,\alpha}\|_{C_b^0})\max_{j,\alpha}(\|a_{j,\alpha}\|_{C_b^1}+1)\,\|U_\varepsilon\|_{H^s}^2\\ &\quad + C(\max_{j,\alpha}\|a_{j,\alpha}\|_{C_b^0})\max_{j,\alpha}(\|A_{j,\alpha}\|_{H^s}+1)\,\|U_\varepsilon\|_{H^s}\,\|U_\varepsilon\|_{C^1_{\sharp,K_0}}.\end{aligned}$$

(b) *Consider the system of type (3.9) which we obtain from (3.7) in the way of Subsection 3.1.1, replacing everywhere J_ε by the identity operator. Let $U \in C([t_0,T], H^{\min(s_c,s_d)}(M)) \cap C^1([t_0,T], H^{\min(s_c,s_d)-1}(M))$ be a solution of such a system and $0 \le s \le \min(s_c, s_d) - 1$. Then*

$$\begin{aligned}\partial_t\left(R\langle D\rangle^s U, \langle D\rangle^s U\right) &\le C(\max_{j,\alpha}\|\partial_t a_{j,\alpha}\|_{C_b^0})\,\|U\|_{H^s}^2 + 2\sqrt{(R\langle D\rangle^s U, \langle D\rangle^s U)}\sqrt{(R\langle D\rangle^s F, \langle D\rangle^s F)}\\ &\quad + C(\max_{j,\alpha}\|a_{j,\alpha}\|_{C_b^0})\max_{j,\alpha}(\|a_{j,\alpha}\|_{C_b^1}+1)\,\|U\|_{H^s}^2\\ &\quad + C(\max_{j,\alpha}\|a_{j,\alpha}\|_{C_b^0})\max_{j,\alpha}(\|A_{j,\alpha}\|_{H^s}+1)\,\|U\|_{H^s}\,\|U\|_{C^1_{\sharp,K_0}}.\end{aligned}$$

Proof of (a) We can write

$$\begin{aligned}&\partial_t\left(R\langle D\rangle^s U_\varepsilon, \langle D\rangle^s U_\varepsilon\right)\\ &= (R_t\langle D\rangle^s U_\varepsilon, \langle D\rangle^s U_\varepsilon) + (R\langle D\rangle^s(K_\varepsilon\sigma U_\varepsilon + B_\varepsilon U_\varepsilon + F), \langle D\rangle^s U_\varepsilon)\\ &\quad + (R\langle D\rangle^s U_\varepsilon, \langle D\rangle^s(K_\varepsilon\sigma U_\varepsilon + B_\varepsilon U_\varepsilon + F)).\end{aligned}$$

It is easy to estimate the first term on the right:

$$\|R_t\langle D\rangle^s U_\varepsilon\|_{L^2} \le C(\max_{j,\alpha}\|\partial_t a_{j,\alpha}\|_{C_b^0})\,\|U_\varepsilon\|_{H^s}.$$

Since $(R\cdot,\cdot)$ is a scalar product of L^2, the Cauchy–Schwarz inequality yields

$$\begin{aligned}&|\,(R\langle D\rangle^s F, \langle D\rangle^s U_\varepsilon) + (R\langle D\rangle^s U_\varepsilon, \langle D\rangle^s F)\,|\\ &\qquad \le 2\sqrt{(R\langle D\rangle^s U_\varepsilon, \langle D\rangle^s U_\varepsilon)}\sqrt{(R\langle D\rangle^s F, \langle D\rangle^s F)}.\end{aligned}$$

From the formulas (3.20) and (3.17) we see that

$$\begin{aligned}&|\,(R\langle D\rangle^s B_\varepsilon U_\varepsilon, \langle D\rangle^s U_\varepsilon) + (R\langle D\rangle^s U_\varepsilon, \langle D\rangle^s B_\varepsilon U_\varepsilon)\,|\\ &\quad \le C(\max_{j,\alpha}\|a_{j,\alpha}\|_{C_b^0})\,\|U_\varepsilon\|_{H^s}^2 + C\max_{j,\alpha}(\|A_{j,\alpha}\|_{H^s}+1)\,\|U_\varepsilon\|_{C^0_{\sharp,K_0}}\,\|U_\varepsilon\|_{H^s}.\end{aligned}$$

It remains to consider the terms

$$I_1 = (R\langle D\rangle^s K_\varepsilon\sigma U_\varepsilon, \langle D\rangle^s U_\varepsilon), \quad I_2 = (R\langle D\rangle^s U_\varepsilon, \langle D\rangle^s K_\varepsilon\sigma U_\varepsilon).$$

The scalar product I_1 can be written in the form

$$\begin{aligned}
I_1 =& I_{11} + I_{12} + I_{13} + I_{14} + I_{15} + I_{16} + I_{17} + I_{18} + I_{19} \\
=& \left(RJ_\varepsilon \left[\langle D\rangle^s, K_0\right] \langle D\rangle \sigma U_\varepsilon, \langle D\rangle^s U_\varepsilon\right) + \left(RJ_\varepsilon K_0 \left[\langle D\rangle^{s+1}, \sigma\right] U_\varepsilon, \langle D\rangle^s U_\varepsilon\right) \\
&+ \left(RJ_\varepsilon \left[K_0, \sigma\right] \langle D\rangle^{s+1} U_\varepsilon, \langle D\rangle^s U_\varepsilon\right) + \left(RJ_\varepsilon \sigma \left[K_0, \langle D\rangle\right] \langle D\rangle^s U_\varepsilon, \langle D\rangle^s U_\varepsilon\right) \\
&+ \left(R\left[J_\varepsilon, \sigma\right] \langle D\rangle K_0 \langle D\rangle^s U_\varepsilon, \langle D\rangle^s U_\varepsilon\right) + \left(\left[R, \sigma\right] \langle D\rangle J_\varepsilon K_0 \langle D\rangle^s U_\varepsilon, \langle D\rangle^s U_\varepsilon\right) \\
&+ \left(\sigma \left[R, J_\varepsilon\right] \langle D\rangle K_0 \langle D\rangle^s U_\varepsilon, \langle D\rangle^s U_\varepsilon\right) + \left(\sigma J_\varepsilon R \left[\langle D\rangle, K_0\right] \langle D\rangle^s U_\varepsilon, \langle D\rangle^s U_\varepsilon\right) \\
&+ \left(\sigma J_\varepsilon R K_0 \langle D\rangle \langle D\rangle^s U_\varepsilon, \langle D\rangle^s U_\varepsilon\right).
\end{aligned}$$

We estimate now $I_{11}, \ldots, I_{18}$. From (3.17), Proposition 2.3.4 and Lemma 2.3.5 it can be deduced that

$$\begin{aligned}
|I_{11}| \leq& C(\max_{j,\alpha} \|a_{j,\alpha}\|_{C_b^0}) \max_{j,\alpha}(\|a_{j,\alpha}\|_{C_b^1} + 1) \|U_\varepsilon\|_{H^s}^2 \\
&+ C(\max_{j,\alpha} \|a_{j,\alpha}\|_{C_b^0}) \max_{j,\alpha}(\|A_{j,\alpha}\|_{H^s} + 1) \|\sigma U_\varepsilon\|_{C^1_{\sharp, K_1}} \|U_\varepsilon\|_{H^s} \\
\leq& C(\max_{j,\alpha} \|a_{j,\alpha}\|_{C_b^0}) \max_{j,\alpha}(\|a_{j,\alpha}\|_{C_b^1} + 1) \|U_\varepsilon\|_{H^s}^2 \\
&+ C(\max_{j,\alpha} \|a_{j,\alpha}\|_{C_b^0}) \max_{j,\alpha}(\|A_{j,\alpha}\|_{H^s} + 1) \|U_\varepsilon\|_{C^1_{\sharp, K_0}} \|U_\varepsilon\|_{H^s}.
\end{aligned}$$

From (3.16), (3.17), and $\left[\langle D\rangle^{s+1}, \sigma\right] \in OPS^s_{1,0}$ we can conclude that

$$|I_{12}| \leq C(\max_{j,\alpha} \|a_{j,\alpha}\|_{C_b^0}) \|U_\varepsilon\|_{H^s}^2.$$

From (3.17) and Proposition 2.3.3 ($\alpha = \beta = 0$) it follows that

$$|I_{13}| \leq C(\max_{j,\alpha} \|a_{j,\alpha}\|_{C_b^0}) \max_{j,\alpha}(\|a_{j,\alpha}\|_{C_b^1} + 1) \|\sigma\|_{C_b^1} \|U_\varepsilon\|_{H^s}^2.$$

By (3.17) and Proposition 2.3.3 ($\alpha = 0$, $\beta = 1$) we have

$$|I_{14}| + |I_{18}| \leq C(\max_{j,\alpha} \|a_{j,\alpha}\|_{C_b^0}) \max_{j,\alpha}(\|a_{j,\alpha}\|_{C_b^1} + 1) \|U_\varepsilon\|_{H^s}^2.$$

We use (3.17), Lemma 2.3.2, Proposition 2.3.3 ($\alpha = \beta = 0$), (3.16), and get

$$\begin{aligned}
&|I_{15}| + |I_{16}| + |I_{17}| \\
&\quad \leq C(\max_{j,\alpha} \|a_{j,\alpha}\|_{C_b^0}) \max_{j,\alpha}(\|a_{j,\alpha}\|_{C_b^1} + 1) \|\langle D\rangle K_0 \langle D\rangle^s U_\varepsilon\|_{H^{-1}} \|U_\varepsilon\|_{H^s} \\
&\quad \leq C(\max_{j,\alpha} \|a_{j,\alpha}\|_{C_b^0}) \max_{j,\alpha}(\|a_{j,\alpha}\|_{C_b^1} + 1) \|U_\varepsilon\|_{H^s}^2.
\end{aligned}$$

Summing up shows

$$\begin{aligned}
|I_1 - I_{19}| \leq& C(\max_{j,\alpha} \|a_{j,\alpha}\|_{C_b^0}) \max_{j,\alpha}(\|a_{j,\alpha}\|_{C_b^1} + 1) \|U_\varepsilon\|_{H^s}^2 \\
&+ C(\max_{j,\alpha} \|a_{j,\alpha}\|_{C_b^0}) \max_{j,\alpha}(\|A_{j,\alpha}\|_{H^s} + 1) \|U_\varepsilon\|_{H^s} \|U_\varepsilon\|_{C^1_{\sharp, K_0}}.
\end{aligned}$$

The scalar product I_2 can be decomposed into

$$\begin{aligned}(R\langle D\rangle^s U_\varepsilon, \langle D\rangle^s J_\varepsilon K_0\langle D\rangle\sigma U_\varepsilon) &= I_{21} + I_{22} + I_{23} + I_{24} \\ &= (R\langle D\rangle^s U_\varepsilon, J_\varepsilon\left[\langle D\rangle^s, K_0\right]\langle D\rangle\sigma U_\varepsilon) + (R\langle D\rangle^s U_\varepsilon, J_\varepsilon K_0\langle D\rangle\left[\langle D\rangle^s, \sigma\right] U_\varepsilon) \\ &\quad + (R\langle D\rangle^s U_\varepsilon, [J_\varepsilon, K_0\langle D\rangle]\,\sigma\langle D\rangle^s U_\varepsilon) + (\sigma J_\varepsilon(K_0\langle D\rangle)^* R\langle D\rangle^s U_\varepsilon, \langle D\rangle^s U_\varepsilon),\end{aligned}$$

where we used the self-adjointness of J_ε and the fact that σ is real-valued. Similarly as above we obtain

$$\begin{aligned}|I_2 - I_{24}| &\le C(\max_{j,\alpha}\|a_{j,\alpha}\|_{C_b^0})\max_{j,\alpha}(\|a_{j,\alpha}\|_{C_b^1}+1)\,\|U_\varepsilon\|_{H^s}^2 \\ &\quad + C(\max_{j,\alpha}\|a_{j,\alpha}\|_{C_b^0})\max_{j,\alpha}(\|A_{j,\alpha}\|_{H^s}+1)\,\|U_\varepsilon\|_{H^s}\,\|U_\varepsilon\|_{C^1_{\sharp,K_0}}.\end{aligned}$$

Finally, (3.18) yields

$$\begin{aligned}|I_{19} + I_{24}| &= |\,(\sigma J_\varepsilon(RK_0\langle D\rangle + (K_0\langle D\rangle)^* R)\langle D\rangle^s U_\varepsilon, \langle D\rangle^s U_\varepsilon)\,| \\ &\le C(\max_{j,\alpha}\|a_{j,\alpha}\|_{C_b^0})\max_{j,\alpha}(\|a_{j,\alpha}\|_{C_b^1}+1)\,\|U_\varepsilon\|_{H^s}^2.\end{aligned}$$

Summing up we obtain the estimate of **(a)**.

Proof of (b). We verify this estimate in a similar way as the previous one replacing the operators J_ε by the identity operator.

The Proposition 3.1.3 is proved. □

Remark 3.1.4. The restriction $s \le \min(s_c, s_d) - 1$ in the part (b) (instead of $s \le \min(s_c, s_d)$ in the part (a)) can be explained as follows: The attempt to estimate $\partial_t\left(R\langle D\rangle^{s_0} U, \langle D\rangle^{s_0} U\right)$ ($s_0 = \min(s_c, s_d)$) leads to a term $(\langle D\rangle^{s_0} K\sigma U, \langle D\rangle^{s_0} U)$ which does in general not exist, if $U(\cdot, t) \in H^{s_0}$.

Or, seen from a different perspective: it is well known [9] that the assumptions (3.2), (3.3), (3.5), (3.6) lead to a solution $U \in C\left([0,T], H^{s_0}\right)$ in the strictly hyperbolic case $\sigma \equiv 1$. Then the energy $E_{s_0}(t) = (R(t)\langle D\rangle^{s_0} U(t), \langle D\rangle^{s_0} U(t))$ is a continuous function of t, but in general not C^1. Hence one cannot expect the estimate from the part (b) to hold for $s = s_0$.

Remark 3.1.5. In case of $s = 0$, we can slightly improve the estimates of Proposition 3.1.3: we may replace (3.20) by (3.19); and in the estimates of I_{11} and I_{21}, we substitute Proposition 2.3.4 with Proposition 2.3.3, leading to

$$\begin{aligned}\partial_t\left(RU, U\right) &\le C(\max_{j,\alpha}\|\partial_t a_{j,\alpha}\|_{C_b^0})\,\|U\|_{L^2}^2 + 2\sqrt{(RU,U)}\sqrt{(RF,F)} \\ &\quad + C(\max_{j,\alpha}\|a_{j,\alpha}\|_{C_b^0})\max_{j,\alpha}(\|a_{j,\alpha}\|_{C_b^1}+1)\,\|U\|_{L^2}^2\end{aligned}$$

for $s_d \ge 1$. The advantage is that no Sobolev norm of the coefficients $a_{j,\alpha}$ appears, and we can weaken the assumptions (3.5), (3.6) to $a_{j,\alpha} \in C_b^1([t_0,T]\times M)$. A similar estimate holds for operators with lower order terms $D_x^\alpha D_t^j(\sigma^{|\alpha|}u)$, $j+|\alpha| \le m-1$, and will be used to study domains of dependence.

3.2. The Quasi-linear Case

Now we consider the quasi-linear Cauchy problem with spatial degeneracy

$$\begin{aligned} D_t^m u + \sum_{j+|\alpha|=m, j<m} & a_{j,\alpha}(x,t,\{D_x^\beta(c_{k,\beta}^0(x,t)D_t^k u)\})D_x^\alpha D_t^j\left(\sigma(x)^{|\alpha|}u\right) \\ &= f(x,t,\{D_x^\beta(c_{k,\beta}^0(x,t)D_t^k u)\}|_{k+|\beta|\le m-1}), \quad m\ge 2, \\ u(x,t_0) &= \varphi_0(x),\ldots,D_t^{m-1}u(x,t_0)=\varphi_{m-1}(x), \end{aligned} \tag{3.21}$$

where the real-valued function $\sigma \in C_b^\infty(M)$ describes the degeneracy, which occurs at the zeroes of σ, and the weight functions $c_{k,\beta}^0$ characterize the Levi conditions as follows:

$$c_{k,\beta}^0 \in C^1([t_0,T],H^{s_c+|\beta|}(M)), \tag{3.22}$$

$$\left\|(\partial_t c_{k,\beta}^0(\cdot,t))v(\cdot)\right\|_{H^{s+|\beta|}} \le C\left\|c_{k,\beta}^0(\cdot,t)v(\cdot)\right\|_{H^{s+|\beta|}}, \tag{3.23}$$

$$\left\|c_{k,\beta}^0(\cdot,t)v(\cdot)\right\|_{H^{s+|\beta|}} \le C\left\|c_{k+1,\beta}^0(\cdot,t)v(\cdot)\right\|_{H^{s+|\beta|}}, \tag{3.24}$$

$$c_{k,\beta}^0(x,t) = \sigma(x)^{|\beta|}, \quad k+|\beta| = m-1, \tag{3.25}$$

for $s_c > n/2+1$, $s_c \ge s \ge 0$.

Example. We give some examples of $c_{k,\beta}^0$, $0 \le k+|\beta| \le m-1$:

- $c_{k,\beta}^0(x,t) = \sigma(x)^{|\beta|}$,
- $c_{k,\beta}^0 \in C^1([t_0,T],H^{s_c+|\beta|}(M))$ with the property that $(\partial_t c_{k,\beta}^0)/c_{k,\beta}^0$ and $c_{k,\beta}^0/c_{k+1,\beta}^0$ belong to $C^1([t_0,T],H^{s_c+|\beta|}(M)) + C^1([t_0,T],C_b^{s_c+|\beta|}(M))$.

However, coefficients $c_{k,\beta}^0$ with non-continuous quotients $(\partial_t c_{k,\beta}^0)/c_{k,\beta}^0$ and $c_{k,\beta}^0/c_{k+1,\beta}^0$ are also possible: let $M = \mathbb{R}/2\pi$ be the unit circle, and set $\sigma(x) = \exp(-(\sin(x))^{-2})$, which has zeroes of infinite order for $x \in \{0,\pi\}$. Then the following weight functions $c_{k,\beta}^0 \in C_b^\infty([0,T]\times M)$ are admissible:

$$c_{k,\beta}^0(x,t) = \begin{cases} \sigma(x)^{|\beta|} & : k+|\beta| = m-1, \\ (t+k+1)\sigma(x)^{|\beta|} & : x\in[0,\pi), \quad k+|\beta|\le m-2, \\ (2t+k+2)\sigma(x)^{|\beta|} & : x\in[\pi,2\pi), \quad k+|\beta|\le m-2. \end{cases}$$

We suppose that the coefficients and the right-hand side are defined in a suitable neighborhood K_G of the initial data,

$$K_G = \{(x,\{v_{k,\beta}\}) \in M\times\mathbb{R}^{n_0} : |v_{k,\beta} - D_x^\beta(c_{k,\beta}^0(x,t_0)\varphi_k(x))| \le G\}, \tag{3.26}$$

for some $G > 0$. Next, we introduce some set $X_G \subset H^{m-1}(M\times[t_0,T])$ containing those functions v which can be reasonably inserted into $a_{j,\alpha}$ and f:

$$v \in X_G \Leftrightarrow (x,\{D_x^\beta(c_{k,\beta}^0(x,t)D_t^k v(x,t))\}) \in K_G \quad \forall (x,t)\in M\times[t_0,T].$$

Our regularity assumptions on the right-hand side f and coefficients $a_{j,\alpha}$ are:

- The mapping

$$\cap_{k=0}^{m-1} C^k([t_0,T], H^{s+m-1-k}(M)) \cap X_G \to C([t_0,T], H^s(M)), \tag{3.27}$$
$$v(x,t) \mapsto f(x,t,\{D_x^\beta(c_{k,\beta}^0(t,x)D_t^k v(t,x))\}),$$

is bounded and continuous for every s with $n/2+1 < s \le s_c$.
- There are constants $C_{j,\alpha}$ such that the mappings

$$\cap_{k=0}^{m-1} C^k([t_0,T], H^{s+m-1-k}) \cap X_G \to \cap_{k=0}^{1} C^k([t_0,T], H^{s-k}), \tag{3.28}$$
$$v(x,t) \mapsto a_{j,\alpha}(x,t,\{D_x^\beta(c_{k,\beta}^0(t,x)D_t^k v(t,x))\}) - C_{j,\alpha},$$

are bounded and continuous for every s with $n/2+1 < s \le s_c$.

Remark 3.2.1. In case of a bounded manifold M, these conditions are satisfied if $f \in C([t_0,T], C^{s_c}(K_G))$ and $a_{j,\alpha} \in C^1([t_0,T], C^{s_c}(K_G))$. If $M = \mathbb{R}^n$, appropriate decays of $f(x,t,\{v_{k,\beta}\})$ and $a_{j,\alpha}(x,t,\{v_{k,\beta}\}) - C_{j,\alpha}$ for $|x| \to \infty$ are required.

The initial data are supposed to satisfy

$$\varphi_j \in H^{s_c+m-1-j}(M), \quad j \le m-1. \tag{3.29}$$

Finally, we assume the hyperbolicity of the Cauchy problem (3.21):

Condition 3.2. *The roots $\tau_j(x,t,v,\xi)$ of*

$$\tau^m + \sum_{j+|\alpha|=m, j<m} a_{j,\alpha}(x,t,v)\xi^\alpha \tau^j = 0$$

are real and distinct, $|\tau_j(x,t,v,\xi) - \tau_i(x,t,v,\xi)| \ge c|\xi|$, $c > 0$, $i \ne j$, for all $(t,x,v,\xi) \in [t_0,T] \times K_G \times \mathbb{R}^n$.

The main result of this subsection is the following theorem:

Theorem 3.2.2. *Under the above assumptions, there is a T_0, $t_0 < T_0 \le T$, such that the Cauchy problem* (3.21) *has a uniquely determined solution u with*

$$\langle D\rangle^k(\sigma^k D_t^{m-1-k}u) \in C([t_0,T_0], H^{s_c}(M)) \cap C^1([t_0,T_0], H^{s_c-1}(M))$$

for $0 \le k \le m-1$. This solution persists as long as the vector of weighted derivatives $(x, \{D_x^\beta(c_{k,\beta}^0(x,t)D_t^k u(x,t))\})$ stays in K_G for all x and the norms $\|\langle D\rangle^k(\sigma^k D_t^{m-k-1}u)\|_{C_*^1}$ remain finite, $0 \le k \le m-1$.

The following well-posedness result in C^∞ is an immediate consequence:

Theorem 3.2.3. *Consider the Cauchy problem* (3.21) *assuming* (3.4) *and $\varphi_j \in H^\infty(M)$. Furthermore, we assume that* (3.22)–(3.25), (3.27), (3.28) *hold for all $s \ge 0$, $s_c > n/2+1$ and that Condition* 3.2 *is satisfied. Then the Cauchy problem* (3.21) *has a unique solution*

$$u \in C^m([t_0,T_0], C^\infty(M)).$$

If the equation is linear, then we have *global* existence:

Remark 3.2.4. Proposition 3.1.3 and Theorem 3.2.2 give us a unique solution u to the linear Cauchy problem (3.1) which cannot blow up, i.e.,

$$\langle D\rangle^k(\sigma^k D_t^{m-1-k}u) \in C([t_0,T], H^{s_0}(M)) \cap C^1([t_0,T], H^{s_0-1}(M))$$

for $0 \leq k \leq m-1$ and $s_0 = \min(s_c, s_d)$.

Of course, we can include lower-order terms $D_x^\beta(c_{k,\beta}^0(x,t)D_t^k u(x,t))$ as long as the equation remains linear.

The proof of Theorem 3.2.2 comprises the Propositions 3.2.5–3.2.8. At first, we construct a regularized hyperbolic first-order system for a vector of weighted derivatives up to the order $m-1$. Employing the ideas from Subsection 3.1 we prove the existence and estimates of the solution U_ε^* to this perturbed system. Then we show that the life-span of the U_ε^* does not tend to zero as ε approaches zero. For $\varepsilon \to 0$, the U_ε converge to a solution U of the asserted regularity. Last, the blow-up criterion will be proved.

3.2.1. Construction of a First-Order System We start with a regularized version of (3.21),

$$\begin{aligned} D_t^m u_\varepsilon + J_\varepsilon & \sum_{j+|\alpha|=m, j<m} a_{j,\alpha}(x,t,\{D_x^\beta(c_{k,\beta}^0 J_\varepsilon^{|\beta|}D_t^k u_\varepsilon)\}) D_x^\alpha J_\varepsilon^\alpha D_t^j\left(\sigma^{|\alpha|}u_\varepsilon\right) \\ &= f(x,t,\{D_x^\beta(c_{k,\beta}^0 J_\varepsilon^{|\beta|}D_t^k u_\varepsilon)\}), \qquad (3.30) \\ u_\varepsilon(x,t_0) &= \varphi_0(x), \ldots, D_t^{m-1}u_\varepsilon(x,t_0) = \varphi_{m-1}(x). \end{aligned}$$

This is a quasi-linear ODE for a function u_ε with values in the Banach space $H^{s_c}(M)$; hence it has a solution

$$u_\varepsilon \in C^m([t_0,T_\varepsilon], H^{s_c}(M)), \quad T_\varepsilon > t_0.$$

We define $U_{\varepsilon,k} = (\langle D\rangle J_\varepsilon)^{m-k}(\sigma^{m-k}D_t^{k-1}u_\varepsilon)$ as in (3.8), and

$$\begin{aligned} U_{\varepsilon,l,k}(x,t) &= (\langle D\rangle J_\varepsilon)^l(\sigma(x)^l D_t^k u_\varepsilon(x,t)), \quad k+l \leq m-2, \\ U_{\varepsilon,k,\beta}(x,t) &= \langle D\rangle^{|\beta|}(c_{k,\beta}^0(x,t)J_\varepsilon^{|\beta|}D_t^k u_\varepsilon(x,t)), \quad k+|\beta| \leq m-2. \end{aligned}$$

Obviously,

$$\partial_t U_{\varepsilon,k,\beta} = \langle D\rangle^{|\beta|}((\partial_t c_{k,\beta}^0)J_\varepsilon^{|\beta|}D_t^k u_\varepsilon) + i\langle D\rangle^{|\beta|}(c_{k,\beta}^0 J_\varepsilon^{|\beta|}D_t^{k+1}u_\varepsilon).$$

Assuming $k+|\beta| \leq m-3$, the H^s norm of the right-hand side is bounded by

$$\begin{aligned} &\left\|(\partial_t c_{k,\beta}^0)J_\varepsilon^{|\beta|}D_t^k u_\varepsilon\right\|_{H^{s+|\beta|}} + \left\|c_{k,\beta}^0 J_\varepsilon^{|\beta|}D_t^{k+1}u_\varepsilon\right\|_{H^{s+|\beta|}} \qquad (3.31) \\ &\leq C\left\|c_{k,\beta}^0 J_\varepsilon^{|\beta|}D_t^k u_\varepsilon\right\|_{H^{s+|\beta|}} + C\left\|c_{k+1,\beta}^0 J_\varepsilon^{|\beta|}D_t^{k+1}u_\varepsilon\right\|_{H^{s+|\beta|}} \\ &= C\left\|U_{\varepsilon,k,\beta}\right\|_{H^s} + C\left\|U_{\varepsilon,k+1,\beta}\right\|_{H^s}. \end{aligned}$$

And for $k + |\beta| = m - 2$, we obtain

$$\begin{aligned}
&\left\|(\partial_t c^0_{k,\beta}) J_\varepsilon^{|\beta|} D_t^k u_\varepsilon\right\|_{H^{s+|\beta|}} + \left\|c^0_{k,\beta} J_\varepsilon^{|\beta|} D_t^{k+1} u_\varepsilon\right\|_{H^{s+|\beta|}} \\
&\leq C \left\|c^0_{k,\beta} J_\varepsilon^{|\beta|} D_t^k u_\varepsilon\right\|_{H^{s+|\beta|}} + C \left\|\sigma^{|\beta|} J_\varepsilon^{|\beta|} D_t^{k+1} u_\varepsilon\right\|_{H^{s+|\beta|}} \\
&\leq C \left\|U_{\varepsilon,k,\beta}\right\|_{H^s} + C \sum_{l\leq|\beta|} \left\|(\langle D\rangle J_\varepsilon)^l (\sigma^l D_t^{k+1} u_\varepsilon)\right\|_{H^s},
\end{aligned} \tag{3.32}$$

see (3.22)–(3.25). We introduce the vector

$$U_\varepsilon^* = (\{U_{\varepsilon,k,\beta}\},\ \{U_{\varepsilon,l,k}\},\ U_\varepsilon^T)^T$$

and obtain

$$\partial_t U_\varepsilon^* = \begin{pmatrix} 0 & 0 \\ 0 & K_\varepsilon(x,t,U_\varepsilon^*,D) \end{pmatrix} (\sigma U_\varepsilon^*) + \begin{pmatrix} G_\varepsilon(x,t,U_\varepsilon^*) \\ F_\varepsilon(x,t,U_\varepsilon^*) \end{pmatrix}$$

with $\|G_\varepsilon\|_{H^s} \leq C \|U_\varepsilon^*\|_{H^s}$, cf. (3.31), (3.32). This system can be written as

$$\partial_t U_\varepsilon^* = K_\varepsilon^*(x,t,U_\varepsilon^*,D)(\sigma U_\varepsilon^*) + F_\varepsilon^*(x,t,U_\varepsilon^*), \quad U_\varepsilon^*(t_0) = \Phi_\varepsilon^*. \tag{3.33}$$

The matrix $R^*(x,t,U^*,D) = \operatorname{diag}(E, R(x,t,U^*,D))$ is a symmetrizer for K^*, where E is the identity matrix and R is the (independent of ε) symmetrizer from Subsection 3.1.

3.2.2. Estimates and Common Existence Interval According to Theorem 3.2.2, the solution $U_\varepsilon^* \in C^1([t_0, T_\varepsilon], H^{s_c})$ persists as long as it stays in K_G and as long as $\|U_\varepsilon^*\|_{H^{s_c}} < \infty$. Applying Proposition 3.1.3 and $C^{1,\delta} \subset C^1_{\sharp,K_0}$ we get

$$\begin{aligned}
&\partial_t \left(R^* \langle D\rangle^s U_\varepsilon^*, \langle D\rangle^s U_\varepsilon^*\right) \\
&\leq C(\max_{j,\alpha} \|\partial_t a_{j,\alpha}\|_{C_b^0}) \|U_\varepsilon^*\|^2_{H^s} \\
&\quad + 2\sqrt{(R^*\langle D\rangle^s U_\varepsilon^*, \langle D\rangle^s U_\varepsilon^*)}\sqrt{(R^*\langle D\rangle^s F_\varepsilon^*, \langle D\rangle^s F_\varepsilon^*)} \\
&\quad + C(\max_{j,\alpha} \|a_{j,\alpha}\|_{C_b^0}) \max_{j,\alpha}(\|a_{j,\alpha}\|_{C_b^1} + 1) \|U_\varepsilon^*\|^2_{H^s} \\
&\quad + C(\max_{j,\alpha} \|a_{j,\alpha}\|_{C_b^0}) \max_{j,\alpha}(\|A_{j,\alpha}\|_{H^s} + 1) \|U_\varepsilon^*\|_{H^s} \|U_\varepsilon^*\|_{C_b^{1,\delta}}
\end{aligned}$$

for $s_c \geq s > n/2 + 1 + \delta$, $0 < \delta < 1$. The Moser-type estimates

$$\begin{aligned}
\|a_{j,\alpha}\|_{C_b^1} &\leq C(\|U_\varepsilon^*\|_{C_b^0})(\|U_\varepsilon^*\|_{C_b^1} + 1), \\
\|A_{j,\alpha}\|_{H^s} &\leq C(\|U_\varepsilon^*\|_{L^\infty})(\|U_\varepsilon^*\|_{H^s} + 1),
\end{aligned}$$

and the embedding inequality $\|U_\varepsilon^*\|_{C_b^{1,\delta}} \leq C \|U_\varepsilon^*\|_{H^s}$ can be applied on the right. Let us consider the term $C(\max_{j,\alpha} \|\partial_t a_{j,\alpha}\|_{C_b^0})$ which denotes some constant that depends in a nonlinear way on $\max_{j,\alpha} \|\partial_t a_{j,\alpha}\|_{C_b^0}$. The computations which lead to this term show that it has the form

$$C(\max_{j,\alpha} \|a_{j,\alpha}\|_{C_b^0}) \cdot \max_{j,\alpha}(\|\partial_t a_{j,\alpha}\|_{C_b^0} + 1).$$

Appealing to Lemma 2.3.5 and (3.16), we have

$$\begin{aligned}\|\partial_t a_{j,\alpha}\|_{C_b^0} &\le C(1+\|\partial_t U_\varepsilon^*\|_{C_b^0}) \le C(1+\|K_\varepsilon^*\sigma U_\varepsilon^*\|_{C_b^0}+\|F_\varepsilon^*\|_{C_b^0})\\ &\le C(1+\|U_\varepsilon^*\|_{C_b^{1,\delta}}).\end{aligned}$$

Taking into account all these inequalities we obtain

$$\begin{aligned}\partial_t\left(R^*\langle D\rangle^s U_\varepsilon^*, \langle D\rangle^s U_\varepsilon^*\right) &\le C(\|U_\varepsilon^*\|_{L^\infty})(\|U_\varepsilon^*\|_{C_b^{1,\alpha}}+1)\,\|U_\varepsilon^*\|_{H^s}^2 \\ &+2\sqrt{(R^*\langle D\rangle^s U_\varepsilon^*, \langle D\rangle^s U_\varepsilon^*)}\sqrt{(R^*\langle D\rangle^s F_\varepsilon^*, \langle D\rangle^s F_\varepsilon^*)}.\end{aligned} \tag{3.34}$$

In the next step we show indirectly that there is a common existence interval of the solutions U_ε^*. We define $t_\varepsilon \in (t_0, T_\varepsilon]$ by the inequality $\|U_\varepsilon^*(t)\|_{H^s} \le 2\sup_{0\le\varepsilon'\le\varepsilon_0}\|\Phi_{\varepsilon'}^*\|_{H^s}+1$, $t_0\le t\le t_\varepsilon$; and by the condition that the components of the vector $U_\varepsilon^*(t)$ be in the interior of the domain of definition of the coefficients $a_{j,\alpha}$ and the right-hand side f, for $t_0 \lesssim t \lesssim t_\varepsilon$.

To obtain a contradiction, let us assume that for every small $\gamma>0$ an $\varepsilon=\varepsilon(\gamma)$ exists with $t_0<t_\varepsilon\le t_0+\gamma$. Now we study estimates of $\|U_\varepsilon^*\|_{H^s}$ and $\|U_\varepsilon^*-\Phi_\varepsilon^*\|_{L^\infty}$.

The norms $\|V\|_{L^2}$ and $\sqrt{(R^*V,V)}$ are equivalent as long as R^* is defined (i.e., for $t\le t_\varepsilon$). From (3.34) and $\|F_\varepsilon^*\|_{H^s}\le C(\|U_\varepsilon^*\|_{L^\infty})(\|U_\varepsilon^*\|_{H^s}+1)$ we see that

$$\partial_t\left(R^*\langle D\rangle^s U_\varepsilon^*, \langle D\rangle^s U_\varepsilon^*\right) \le Q((R^*\langle D\rangle^s U_\varepsilon^*, \langle D\rangle^s U_\varepsilon^*)),$$

where Q is a smooth nonlinear increasing function, independent of ε. Let T_0 be a number with the property that the nonnegative solutions of

$$\begin{aligned}&\partial_t y(t)\le Q(y(t)),\\ &y(t_0)=(R^*(x,t_0,\Phi^*,D)\langle D\rangle^s\Phi_\varepsilon^*, \langle D\rangle^s\Phi_\varepsilon^*)\,,\quad 0\le\varepsilon\le\varepsilon_0,\end{aligned} \tag{3.35}$$

satisfy $y(t)\le 2\sup_{0\le\varepsilon'\le\varepsilon_0}\|\Phi^*\|_{H^s}+1$ for $t_0\le t\le T_0$.

To estimate $\|U_\varepsilon^*-\Phi_\varepsilon^*\|_{L^\infty}$, we write $U_\varepsilon^*=\Phi_\varepsilon^*+V_\varepsilon^*$ and get, by Proposition 3.1.3,

$$\begin{aligned}&\partial_t\left(R^*V_\varepsilon^*, V_\varepsilon^*\right)\le C(\|U_\varepsilon^*\|_{L^\infty})((R^*V_\varepsilon^*,V_\varepsilon^*)+(R^*F_\varepsilon^*,F_\varepsilon^*)),\\ &\left(R^*V_\varepsilon^*,V_\varepsilon^*\right)(t_0)=0.\end{aligned}$$

The norms $\|U_\varepsilon^*\|_{L^\infty}$ are uniformly bounded for $t\le t_\varepsilon$, due to the definition of t_ε. From Gronwall's lemma it can be concluded that

$$\left(R^*V_\varepsilon^*,V_\varepsilon^*\right)(t)\le g(t)^2,\quad t_0\le t\le \min(T_0,t_\varepsilon),$$

$g(t_0)=0$, g continuous and increasing. We obtain $\|V_\varepsilon^*(t)\|_{L^2}^2\le Cg(t)^2$; and an interpolation argument gives us a continuous function g_1, such that

$$\|V_\varepsilon^*(t)\|_{L^\infty}\le g_1(t),\quad g_1(t_0)=0. \tag{3.36}$$

This demonstrates that t_ε cannot come arbitrarily close to t_0, which is a contradiction. Hence there is a common existence interval.

We have proved:

Proposition 3.2.5. *There is a constant $T_0 > t_0$ with the property that the systems* (3.33) *have unique solutions $U_\varepsilon^* \in C^1([t_0, T_0], H^s)$ for $0 < \varepsilon \le \varepsilon_0$ and $s_c \ge s > n/2 + 1$. It holds for all $0 < \varepsilon \le \varepsilon_0$ and all $t_0 \le t \le T_0$ that*

$$\|U_\varepsilon^*(t)\|_{H^s} \le C, \quad \|U_\varepsilon^*(t) - \Phi^*\|_{L^\infty} \le g_1(t)$$

with some continuous function $g_1(t)$, $g_1(t_0) = 0$.

3.2.3. Convergence and Regularity of the Limit Our previous *a priori* estimates allow us to show

$$\|U_\varepsilon^*(\cdot, t) - U_{\varepsilon'}^*(\cdot, t)\|_{L^2}^2 \le C(T_0 - t_0)(\varepsilon + \varepsilon'), \quad t_0 \le t \le T_0.$$

By the uniform bound $\|U_\varepsilon^*\|_{H^{s_c}} \le C$ and interpolation, the following result is easily established:

Proposition 3.2.6. *The above constructed sequence $(U_\varepsilon^*) \subset C^1([t_0, T_0], H^{s_c})$ converges in $C([t_0, T_0], H^s)$ and $C([t_0, T_0], C_b^{1,\delta})$ for any s and δ with $n/2 + 1 + \delta < s < s_c$. The limit U^* belongs to $C^1([t_0, T_0], H^{s-1})$ and is a solution of* (3.33) *with $\varepsilon = 0$.*

It remains to study the regularity of U^*. Here we make use of standard techniques, which can be found e.g., in [28]. The uniform estimate of U_ε^* in H^{s_c} gives $U^* \in L^\infty([t_0, T_0], H^{s_c})$ and $U^* \in Lip^1([t_0, T_0], H^{s_c - 1})$. We want to show that the H^{s_c} norm of U^* is not only bounded, but also continuous:

Proposition 3.2.7. *The above constructed solution U^* to* (3.33) *with $\varepsilon = 0$ belongs to $C([t_0, T_0], H^{s_c}) \cap C^1([t_0, T_0], H^{s_c - 1})$.*

Proof. We fix $t_0 \le t_1 < T_0$ and will show the continuity of $\|U^*(\cdot, t)\|_{H^{s_c}}$ at the point t_1. To this end, we consider the forward Cauchy problem (recycling the variable U_ε^* which we do not need anymore)

$$\partial_t U_\varepsilon^* = K_\varepsilon^*(x, t, U_\varepsilon^*, D)(\sigma U_\varepsilon^*) + F_\varepsilon^*(x, t, U_\varepsilon^*), \quad U_\varepsilon^*(t_1) = U^*(t_1),$$

for some small $\varepsilon > 0$. The only difference to (3.33) is the initial condition. Defining the equivalent norm for the Hilbert space H^{s_c}

$$\|V\|_{H^{s_c}, t_1}^2 = (R^*(t_1)\langle D\rangle^{s_c} V, \langle D\rangle^{s_c} V) \tag{3.37}$$

we deduce from (3.34) that

$$\partial_t \|U_\varepsilon^*(t)\|_{H^{s_c}, t}^2 \le C'(\|U_\varepsilon^*\|_{C_b^{1,\delta}}) \|U_\varepsilon^*(t)\|_{H^{s_c}, t}^2 + C \|F_\varepsilon^*(t)\|_{H^{s_c}}^2.$$

Gronwall's lemma gives, with some $C' = C'(\|U_\varepsilon^*\|_{C_b^{1,\delta}})$,

$$\|U_\varepsilon^*(t)\|_{H^{s_c}, t}^2 \le \|U^*(t_1)\|_{H^{s_c}, t_1}^2 e^{C'(t - t_1)} + C \int_{t_1}^t e^{C'(t - \tau)} \|F_\varepsilon^*(\tau)\|_{H^{s_c}}^2 \, d\tau.$$

The weak compactness of bounded subsets in Hilbert spaces implies ($\varepsilon \to 0$)

$$\|U^*(t)\|_{H^{s_c}, t}^2 \le \|U^*(t_1)\|_{H^{s_c}, t_1}^2 e^{C'(t - t_1)} + C|t - t_1|.$$

The function $\|V\|_{H^{s_c},t}$ defines an equivalent norm in the Hilbert space H^{s_c} if t is fixed. For proving continuity in t we need a norm which does not depend on t. Using the Lipschitz continuity of R^*, we find

$$\|U^*(t)\|^2_{H^{s_c},t_1} \leq \|U^*(t_1)\|^2_{H^{s_c},t_1} e^{C'(t-t_1)} + C''|t-t_1|,$$

resulting in

$$\limsup_{t\to t_1+0} \|U^*(t)\|^2_{H^{s_c},t_1} \leq \|U^*(t_1)\|^2_{H^{s_c},t_1} \leq \liminf_{t\to t_1+0} \|U^*(t)\|^2_{H^{s_c},t_1},$$

which gives the H^{s_c}-continuity of U^* at t_1 from the right. Inverting the time direction the reader can show the continuity from the left. Since t_1 can be chosen arbitrarily, the proof is complete. □

In the last step a criterion for the blow-up is given. The idea of the proof is taken from [33], Proposition 5.1.F.

Proposition 3.2.8. *Let $U^* \in C([t_0,T), H^{s_c}) \cap C^1([t_0,T), H^{s_c-1})$ be a solution of (3.33) ($\varepsilon = 0$) with*

$$\sup_{t\in[t_0,T)} \|U^*(t)\|_{C^1_*} < \infty,$$
$$\inf_{t\in[t_0,T)} \operatorname{dist}((x, \{U_{k,\beta}(x,t)\}), \partial K_G) \geq \delta > 0.$$

Then a constant $T_1 > T$ exists with $U^* \in C([t_0,T_1], H^{s_c}) \cap C^1([t_0,T_1], H^{s_c-1})$.

Proof. We consider the non-regularized version of (3.33), apply J_ε to it, and estimate the terms on the right as in the proof of Proposition 3.1.3. This gives

$$\partial_t \left(R^*\langle D\rangle^{s_c} J_\varepsilon U^*, \langle D\rangle^{s_c} J_\varepsilon U^*\right) \tag{3.38}$$
$$\leq C(\|U^*\|_{L^\infty})(1 + \|U^*\|_{C^1_b} + \|U^*\|_{C^1_{\sharp,K_0}}) \|U^*\|^2_{H^{s_c}} + \|F^*\|^2_{H^{s_c}}.$$

We suppose that the H^{s_c} norm is arranged in such a way that $\|V^*\|_{C^1_*} \leq \|V^*\|_{H^{s_c}}$ holds for every function $V^* \in H^{s_c}$. Then the inequality

$$\|V^*\|_{C^1_{\sharp,K_0}} \leq C\|V^*\|_{C^1_*}\left(1 + \ln\left(\frac{\|V^*\|_{H^{s_c}}}{\|V^*\|_{C^1_*}}\right)\right)$$

can be shown, see [33], (B.2.12). Consequently,

$$\|U^*\|_{C^1_{\sharp,K_0}} \leq C\|U^*\|_{C^1_*}(1 + \ln^+\|U^*\|_{H^{s_c}}) + C.$$

According to Lemma 2.3.5, we have $\|U^*\|_{C^1_b} \leq C\|U^*\|_{C^1_{\sharp,K_0}}$, and from $\|F^*\|^2_{H^{s_c}} \leq C(\|U^*\|_{L^\infty})(e + \|U^*\|^2_{H^{s_c}})$ it follows that

$$\partial_t \left(R^*\langle D\rangle^{s_c} J_\varepsilon U^*, \langle D\rangle^{s_c} J_\varepsilon U^*\right)$$
$$\leq C(\|U^*\|_{L^\infty})(1 + \|U^*\|_{C^1_*})(1 + \ln^+\|U^*\|^2_{H^{s_c}})(e + \|U^*\|^2_{H^{s_c}}).$$

Using the equivalent norm $\|\cdot\|_{H^{s_c},t}$ from (3.37) and $\|U^*\|_{C^1_*} \leq C$, we get

$$\partial_t \|J_\varepsilon U^*(t)\|_{H^{s_c},t} \leq C_0(1 + \ln^+\|U^*(t)\|^2_{H^{s_c},t})(e + \|U^*(t)\|_{H^{s_c},t}).$$

We integrate, let ε tend to 0 and see that

$$\|U^*(t)\|^2_{H^{s_c},t} \leq \|U^*(t_0)\|^2_{H^{s_c},t_0} + C_0 \int_{t_0}^{t} (1 + \ln^+ \|U^*(\tau)\|^2_{H^{s_c},\tau})(e + \|U^*(\tau)\|^2_{H^{s_c},\tau})\, d\tau.$$

Introducing $N(t) = e + \|U^*(t)\|^2_{H^{s_c},t}$ we deduce that

$$N(t) \leq N(t_0) + 2C_0 \int_{t_0}^{t} \ln(N(\tau))N(\tau)\, d\tau = Q(t).$$

The continuity of $N(t)$ yields $Q(t) \in C^1[t_0, T)$. Obviously,

$$Q'(t) = 2C_0 \ln(N(t))N(t) \leq 2C_0 \ln(Q(t))Q(t);$$

hence $Q(t) \leq C$ for all $t_0 \leq t < T$. Taking into account all these inequalities we find $\|U^*(t)\|_{H^{s_c}} \leq C'$ for $t_0 \leq t < T$.

Next we have to extend U^* continuously to some longer time interval. Therefore, we review the regularized Cauchy problems (3.33) with data $U^*_\varepsilon(T - \gamma) = U^*(T - \gamma)$ for some small $\gamma > 0$. The functions $U^*_\varepsilon(t)$ persist as long as their H^{s_c} norms remain bounded and as long as each component of these vectors stays in the domain of the functions $a_{j,\alpha}$ and f. An estimate of the life-span of U^*_ε is provided by (3.35) and (3.36), which are autonomous (differential) inequalities (independent of ε), hence the length of the existence interval only depends on δ and C'. It follows that for small γ the point T is contained in the common existence interval of the U^*_ε, and consequently, of the limit U. □

This completes the proof of Theorem 3.2.2.

Remark 3.2.9. Crucial for the proof of Proposition 3.2.8 was the fact that the right-hand side of (3.38) grows at most linearly in $\|U^*\|_{C^1_*}$ and $\|U^*\|_{C^1_{\sharp,K_0}}$.

Remark 3.2.10. Using similar arguments as above, one can easily show that the solution of a quasi-linear weakly hyperbolic Cauchy problem continuously depends on the data, weight functions, coefficients and right-hand side.

4. Weakly Hyperbolic Cauchy Problems with Spatial and Time Degeneracy

Now we are ready to study the general equation (1.7) which incorporates both types of degeneracy: spatial degeneracy and time degeneracy.

Our approach is divided into three steps:

- First, we study a *linear* Cauchy problem with vanishing initial data, whose right-hand side has a zero of sufficiently high order at $t = 0$. We establish an estimate of *strictly hyperbolic type* in Theorem 4.1.1.

- Secondly, we consider a *quasi-linear* Cauchy problem. Its initial data also vanish, and the right-hand side has a zero of high order at $t = 0$. The estimates of strictly hyperbolic type for the linear problem and the usual iteration procedure imply local existence, see Theorem 4.2.6.
- Thirdly, we transform the general equation (1.7) into a special equation which can be treated with the methods of the second step, see Theorem 4.3.1.

The idea of transforming a weakly hyperbolic problem with *general* right-hand side into another weakly hyperbolic problem with *special* right-hand side has been widely used, for example in Kajitani, Yagdjian [17], Oleinik [25] and Reissig [29].

4.1. A Special Linear Case

We analyze the Cauchy problem

$$\begin{aligned} &D_t^m u + \sum_{j+|\alpha|=m, j<m} a_{j,\alpha}(x,t)\lambda(t)^{|\alpha|} D_x^\alpha D_t^j \left(\sigma(x)^{|\alpha|} u\right) = f(x,t), \qquad (4.1) \\ &u(x,0) = \cdots = D_t^{m-1} u(x,0) = 0 \end{aligned}$$

under the decay assumption

$$\|f(\cdot,t)\|_{H^{s_0}} \le C_f \lambda(t)^p \lambda'(t). \qquad (4.2)$$

Later, the number p will be presumed sufficiently large. Additionally, we suppose Condition 3.1, (3.3), (3.4), and (3.5) or (3.6). For the function $\lambda = \lambda(t)$ we assume:

Condition 4.1. *Fix $\Lambda(t) = \int_0^t \lambda(\tau)\, d\tau$. The function $\lambda = \lambda(t)$ satisfies one of the following conditions:*

- $\lambda(t) = t^l$, $l \in \mathbb{N}$, $l \ge m-1$,
- $\lambda \in C^2([0,T])$, $\lambda(0) = 0$, $\lambda'(t) > 0$ $(t > 0)$, $\lambda''(t) \ge 0$ $(t \ge 0)$,

$$\frac{\lambda(t)}{\Lambda(t)} \le C_\lambda \frac{\lambda'(t)}{\lambda(t)}, \quad \frac{\lambda'(t)}{\lambda(t)} \le C'_\lambda \frac{\lambda(t)}{\Lambda(t)}, \quad C_\lambda < \frac{m}{m-1}.$$

Examples for the second case are $\Lambda(t) = \exp(-|t|^{-r})$, $r > 0$.

The central result is the following theorem:

Theorem 4.1.1. *If the constant p is sufficiently large, then the Cauchy problem (4.1) has a solution u with the property that*

$$\begin{aligned} &U \in C\left([0,T], H^{s_c}\right) \cap C^1\left([0,T], H^{s_c-1}\right), \\ &U = \{\langle D\rangle^{m-i}((\lambda\sigma)^{m-i} D_t^{i-1} u) : i = 1, \ldots, m\}. \end{aligned}$$

There are constants C_1, C_2, C_3 (independent of u) such that the estimate

$$\|U(t)\|_{H^{s_c}} \le C_3 \int_0^t e^{C_1(t-\tau)} \left(\frac{\lambda(t)}{\lambda(\tau)}\right)^{C_2} \|f(\tau)\|_{H^{s_c}}\, d\tau$$

holds for $p > C_2 + 1$.

Proof. We obtain the system

$$\begin{aligned}
\partial_t U_j &= (m-j)\frac{\lambda'}{\lambda} U_j + \lambda i \langle D\rangle(\sigma U_{j+1}) \\
&\qquad + \lambda i \langle D\rangle \left[\langle D\rangle^{m-j-1}, \sigma\right] \langle D\rangle^{j+1-m} U_{j+1}, \quad 1 \le j < m, \\
\partial_t U_m &= -i \sum_{j+|\alpha|=m, j<m} a_{j,\alpha} \lambda^{|\alpha|} D_x^\alpha D_t^j \left(\sigma^{|\alpha|} u\right) + if \\
&= -i\lambda \sum_{j+|\alpha|=m, j<m} a_{j,\alpha} P^\alpha \langle D\rangle (\sigma U_{j+1}) \\
&\quad - i\lambda \sum_{j+|\alpha|=m, j<m} a_{j,\alpha} P^\alpha \langle D\rangle \left[\langle D\rangle^{|\alpha|-1}, \sigma\right] \langle D\rangle^{1-|\alpha|} U_{j+1} + if.
\end{aligned}$$

This leads to

$$\partial_t U = \lambda K_0 \langle D\rangle(\sigma U) + \lambda B U + H \frac{\lambda'}{\lambda} U + F, \quad U(0) = 0 \tag{4.3}$$

with $K_0\langle D\rangle$, B, F as in (3.10), (3.11), (3.12) and $H = \operatorname{diag}(m-1, m-2, \ldots, 1, 0)$. We regularize this system and eliminate the time degeneracy:

$$\begin{aligned}
\partial_t U_{\delta\varepsilon} &= J_\varepsilon(\lambda + \delta) K_0 \langle D\rangle(\sigma U_{\delta\varepsilon}) + \lambda B U_{\delta\varepsilon} + H \frac{\lambda'}{\lambda + \delta} U_{\delta\varepsilon} + F, \\
U_{\delta\varepsilon}(0) &= 0,
\end{aligned} \tag{4.4}$$

where $\delta > 0$ is small. This is a weakly hyperbolic system with pure spatial degeneracy; therefore we can take advantage of the methods of Subsection 3.1. We may choose the same symmetrizer R, since the function $\lambda + \delta$ has no influence on the operator $K_0\langle D\rangle$. This operator does not feel the time degeneracy.

We choose $s = 0$ or $n/2 + 1 < s \le s_c$, and define the norm

$$H_s(U_{\delta\varepsilon}(t)) = \sqrt{(R(t)\langle D\rangle^s U_{\delta\varepsilon}(t), \langle D\rangle^s U_{\delta\varepsilon}(t))}.$$

Then Proposition 3.1.3 gives us the estimate

$$\partial_t H_s(U_{\delta\varepsilon}) \le C_1 (H_s(U_{\delta\varepsilon}) + \|F\|_{H^s}) + C_2 \frac{\lambda'}{\lambda + \delta} H_s(U_{\delta\varepsilon}). \tag{4.5}$$

By Gronwall's lemma we see that

$$\begin{aligned}
H_s(U_{\delta\varepsilon}(t)) &\le \int_0^t e^{C_{2,\delta}(t-\tau)} C_1 \|F(\tau)\|_{H^s} \, d\tau \\
&\le \exp\left(\frac{CT}{\delta}\right) C \int_0^t \lambda'(\tau)\lambda(\tau)^p \, d\tau \le C_\delta \lambda(t)^{p+1}, \\
C_{2,\delta} &= \sup_{t\in[0,T]} \left\{ C_1 + C_2 \tfrac{\lambda'(t)}{\lambda(t)+\delta} \right\} = O(\delta^{-1}).
\end{aligned}$$

This allows us to apply Nersesyan's lemma (see Lemma 6.2) to (4.5) if we assume $p > C_2 + 1$. The result is

$$\begin{aligned} H_s(U_{\delta\varepsilon}(t)) &\leq \int_0^t e^{C_1(t-\tau)} \left(\frac{\lambda(t)}{\lambda(\tau)}\right)^{C_2} C_1 \|F(\tau)\|_{H^s}\, d\tau \\ &\leq C\lambda(t)^{C_2} \int_0^t e^{C_1(t-\tau)} \lambda(\tau)^{p-C_2} \lambda'(\tau) d\tau \\ &\leq \frac{Ce^{C_1 t}}{p - C_2 + 1} \lambda(t)^{p+1}. \end{aligned}$$

We emphasize that this estimate is independent of δ and ε. It is known that $U_{\delta\varepsilon}$ belongs to the spaces $C([0,T], H^{s_c})$ and $C^1([0,T], H^{s_c-1})$. Employing the methods from Subsection 3.2 one can show that there is a limit $U_\delta = \lim_{\varepsilon\to 0} U_{\delta\varepsilon}$ which belongs to the same spaces and solves

$$\partial_t U_\delta - (\lambda+\delta) K_0\langle D\rangle(\sigma U_\delta) + \lambda B U_\delta + H\frac{\lambda'}{\lambda+\delta} U_\delta + F, \quad U_\delta(0) = 0, \tag{4.6}$$

and that the following *a priori* estimate holds for $n/2 + 1 < s \leq s_c$ and $s = 0$:

$$H_s(U_\delta(t)) \leq C_1 \int_0^t e^{C_1(t-\tau)} \left(\frac{\lambda(t)}{\lambda(\tau)}\right)^{C_2} \|F(\tau)\|_{H^s}\, d\tau. \tag{4.7}$$

In the next step we send δ to 0 and study the convergence properties of the sequence (U_δ). The difference $U_\delta - U_{\delta'}$ solves the equation

$$\begin{aligned} &\partial_t(U_\delta - U_{\delta'}) \\ &\quad = (\lambda+\delta') K_0\langle D\rangle(\sigma(U_\delta - U_{\delta'})) + \lambda B(U_\delta - U_{\delta'}) + H\frac{\lambda'}{\lambda+\delta'}(U_\delta - U_{\delta'}) \\ &\qquad + (\delta-\delta')\left(K_0\langle D\rangle(\sigma U_\delta) - H\frac{\lambda'}{(\lambda+\delta)(\lambda+\delta')} U_\delta\right). \end{aligned}$$

From (4.7) with $s = 0$ it can be concluded that

$$H_0((U_\delta - U_{\delta'})(t)) \leq C|\delta - \delta'|\lambda(t)^p.$$

It is standard to verify that the sequence (U_δ) converges to a limit U which is a solution of (4.3) and satisfies

$$U \in C([\gamma, T], H^{s_c}) \cap C^1([\gamma, T], H^{s_c-1}) \quad \forall \gamma > 0.$$

On the other hand, we have $U(0) = 0$ and $\|U(t)\|_{H^{s_c}} \leq C\lambda(t)^{p+1}$. This gives the continuity for $t = 0$ and the theorem is proved. □

4.2. A Special Quasi-linear Case

Now we assume that the initial data vanish and that the right-hand side decays sufficiently fast for $t \to 0$. More precisely, we consider the Cauchy problem

$$\begin{aligned} D_t^m u + &\sum_{j+|\alpha|=m, j<m} a_{j,\alpha}(x,t,\{D_x^\beta c_{k,\beta}(x,t) D_t^k u\}) \lambda(t)^{|\alpha|} D_x^\alpha D_t^j (\sigma(x)^{|\alpha|} u) \\ &= f(x,t,\{D_x^\beta c_{k,\beta}(x,t) D_t^k u\}|_{k+|\beta|\le m-1}), \\ u(x,0) &= \cdots = D_t^{m-1} u(x,0) = 0 \end{aligned} \tag{4.8}$$

with the following asymptotic behavior for f:

$$\|f(\cdot,t,0)\|_{H^{s_c}} \le C_{f0} \lambda(t)^p \lambda'(t), \tag{4.9}$$

$$\sum_{k+|\beta|\le m-1} \left\| \frac{\partial f}{\partial g_{k,\beta}} (\cdot, t, \{g_{k,\beta}\}) \right\|_{L^\infty} \le C_f \tag{4.10}$$

for all $t \in [0,T]$ and all $\{g_{k,\beta}\} \in \mathbb{R}^{n_0}$ from a suitable chosen compact set near zero. Furthermore, we suppose (3.4), (3.27), (3.28), (replace $c^0_{k,\beta}$ by $c_{k,\beta}$, t_0 by 0 and K_G by some compact set near zero) and Condition 3.2. Concerning the weight functions $c_{k,\beta}$, we assume the Levi conditions

$$c_{k,\beta}(x,t) = \begin{cases} \lambda(t)^{m-k} \Lambda(t)^{k+|\beta|-m} c^0_{k,\beta}(x,t) & : |\beta| > 0, \\ c^0_{k,\beta}(x,t) & : |\beta| = 0, \end{cases} \tag{4.11}$$

where $\Lambda(t) = \int_0^t \lambda(\tau)\, d\tau$; and (3.22)–(3.25).

Our intention is to show the local existence of a solution to (4.8), see Theorem 4.2.6. For this purpose we transform the equation into an equivalent system of first order and study a linearized version of this system. In other words, we introduce $U = (U_1, \dots, U_m)^T$, $V = (V_1, \dots, V_m)^T$, $V^* = (V_{k,\beta}^T, V^T)^T$,

$$\begin{aligned} U_j &= \langle D\rangle^{m-j}((\lambda\sigma)^{m-j} D_t^{j-1} u), \qquad V_j = \langle D\rangle^{m-j}((\lambda\sigma)^{m-j} D_t^{j-1} v), \\ V_{k,\beta}(x,t) &= D_x^\beta (c_{k,\beta}(x,t) D_t^k v(x,t)) \end{aligned} \tag{4.12}$$

and get the system

$$\begin{aligned} \partial_t U &= \lambda K_0(x,t,V^*,D)\langle D\rangle \sigma U + \lambda B(x,t,V^*,D) U + F(x,t,V^*) + H \frac{\lambda'}{\lambda} U, \\ U(0) &= 0. \end{aligned}$$

We obtain a mapping $U = \Phi(V)$, defined via $V \mapsto V^* \mapsto U$, and can construct a sequence $\{V^k\}$ by $V^k = \Phi(V^{k-1})$, $V^0 = 0$. This sequence will be shown to converge for large p and small times, after some preparatory estimates.

4.2.1. AUXILIARY ESTIMATES We proceed with estimating V^* in terms of V, and K_0, B, F in terms of V^*.

Lemma 4.2.1. *Let $T < 1$ and $\|V(t)\|_{H^s} \leq C_{Vp}\lambda(t)^{p+1}$. Then there is a constant C_4, independent of V and p, such that*

$$\left\| \langle D\rangle^{|\beta|} c_{k,\beta}(\cdot,t) D_t^k v(\cdot,t) \right\|_{H^s} \leq C_4 C_{Vp}\lambda(t)^p \lambda'(t). \tag{4.13}$$

Proof. The assertion is obvious for $|\beta| = 0$, $k = m-1$. Now let $|\beta| = 0$, $k \leq m-2$. Then (3.23) and (3.24) imply

$$\begin{aligned}\partial_t \left\| c_{k,\beta} D_t^k v \right\|_{H^s} &\leq \left\| (\partial_t c_{k,\beta}) D_t^k v \right\|_{H^s} + \left\| c_{k,\beta} D_t^{k+1} v \right\|_{H^s} \\ &\leq C_c \left\| c_{k,\beta} D_t^k v \right\|_{H^s} + C_c \left\| c_{k+1,\beta} D_t^{k+1} v \right\|_{H^s}.\end{aligned}$$

By Gronwall's lemma and induction, we get (4.13) for $|\beta| = 0$ and $k \leq m-2$. Now $|\beta| > 0$. We choose $k = m-1-|\beta|$ as base of induction, and deduce that

$$\begin{aligned}\left\| \langle D\rangle^{|\beta|} c_{k,\beta}(\cdot,t) D_t^k v \right\|_{H^s} &= \frac{\lambda(t)}{\Lambda(t)} \left\| \lambda(t)^{|\beta|} \langle D\rangle^{|\beta|} (\sigma^{|\beta|} D_t^k v) \right\|_{H^s} \\ &\leq C_\lambda \frac{\lambda'(t)}{\lambda(t)} \|V(t)\|_{H^s} \leq C_\lambda C_{Vp} \lambda(t)^p \lambda'(t).\end{aligned}$$

Let $k + |\beta| \leq m-2$. Making use of $\lambda' \sim \lambda^2 \Lambda$ (see Condition 4.1) and the induction hypothesis, we obtain

$$\begin{aligned}\left\| c^0_{k+1,\beta} D_t^{k+1} v \right\|_{H^{s+|\beta|}} &\leq C C_{Vp} \lambda(t)^{p+k+1-m} \lambda'(t) \Lambda(t)^{m-k-|\beta|-1} \\ &\leq C C_{Vp} \lambda(t)^{p+k-1-m} (\lambda'(t))^2 \Lambda(t)^{m-k-|\beta|}.\end{aligned}$$

By (3.23) and (3.24) we then conclude that

$$\begin{aligned}&\partial_t \left\| c^0_{k,\beta} D_t^k v \right\|_{H^{s+|\beta|}} \\ &\quad \leq C C_{Vp} \lambda(t)^{p+k-1-m} (\lambda'(t))^2 \Lambda(t)^{m-k-|\beta|} + C \left\| c^0_{k,\beta} D_t^k v \right\|_{H^{s+|\beta|}}.\end{aligned}$$

Bringing Gronwall's lemma into play, we then find

$$\begin{aligned}\left\| c^0_{k,\beta} D_t^k v \right\|_{H^{s+|\beta|}} &\leq C C_{Vp} \int_0^t \lambda(\tau)^{p+k-1-m} (\lambda'(\tau))^2 \Lambda(\tau)^{m-k-|\beta|} \, d\tau \\ &\leq C C_{Vp} \lambda(t)^{p+k-m} \Lambda(t)^{m-k-|\beta|} \lambda'(t),\end{aligned}$$

which concludes the proof. □

Remark 4.2.2. The conclusion of this lemma can be sharpened in the following way. If $0 \leq s \leq s_c$, then the estimates

$$\|V_{k,\beta}\|_{H^s} \leq \left\| \langle D\rangle^{|\beta|} (c_{k,\beta} D_t^k v) \right\|_{H^s} \leq \lambda(t)^p \lambda'(t) C_4 \sup_{\tau\in[0,t]} \frac{\|V(\tau)\|_{H^s}}{\lambda(\tau)^{p+1}}$$

hold for $k + |\beta| = m-1$. And for $k + |\beta| \leq m-2$ we have

$$\|V_{k,\beta}\|_{H^s} \leq \lambda(t)^p \lambda'(t) C_4 \sup_{\tau\in[0,t]} \frac{\left\| \langle D\rangle^{|\beta|} (c_{k+1,\beta} D_t^{k+1} v) \right\|_{H^s}}{\lambda(\tau)^p \lambda'(\tau)}.$$

This lemma gives us estimates for U^* and V^* if bounds of U and V are known. An estimate of U in the terms of the right-hand side is given by Theorem 4.1.1. The next lemma will be practical to find an estimate of F in terms of V^*.

Lemma 4.2.3. *Let $K \subset \mathbb{R}^{n_0}$ be compact and M be an n-dimensional smooth closed manifold. Let $f \in C^N(M \times K)$ and $v_i \in H^N(M)$ with $(x, v_1(x), \dots, v_{n_0}(x)) \in M \times K$ for $x \in M$. If N is sufficiently large and $0 \le m < n_0$, then a constant $N_1 < N$ (independent of N) exists with*

$$\begin{aligned}
&\|f(\cdot, v_1(\cdot), \dots, v_{n_0}(\cdot))\|'_{H^N} \\
&\quad \le \varphi_N(\|v_1\|_{C_b^{N_1}}, \dots, \|v_{n_0}\|_{C_b^{N_1}}) \\
&\qquad \times (\|v_1\|_{H^N} + \dots + \|v_m\|_{H^N} + \|v_{m+1}\|_{H^{N-1}} + \dots + \|v_{n_0}\|_{H^{N-1}}) \\
&\qquad + \sum_{j=m+1}^{n_0} \left\| \frac{\partial f}{\partial v_j}(\cdot, v_1(\cdot), \dots, v_{n_0}(\cdot)) \right\|_{L^\infty} \sum_{|\alpha|=N} \|\partial_x^\alpha v_j\|_{L^2} \\
&\qquad + \sum_{|\alpha| \le N} \left\| f^{(\alpha,0,\dots,0)}(\cdot, v_1(\cdot), \dots, v_{n_0}(\cdot)) \right\|_{L^2}.
\end{aligned}$$

Here we used the equivalent norm $\|w\|'_{H^N} = \sum_{|\alpha|\le N} \|\partial_x^\alpha w\|_{L^2}$,

$$C_N^{-1} \|w\|'_{H^N} \le \|w\|_{H^N} \le C_N \|w\|'_{H^N}. \tag{4.14}$$

This lemma generalizes Remark A.1 in [11] and describes precisely the dependence of $\|f(\cdot, v_1, \dots, v_n)\|'_{H^N}$ on the highest orders of some v_j (see the terms $\|\partial_x^\alpha v_j\|_{L^2}$). The proof is omitted. We will use this lemma to determine the loss of Sobolev regularity (it depends on $\|f_{v_j}\|_{L^\infty}$), or, in other words, to determine the space in which the solution exists.

From now on we assume $s_c = N \in \mathbb{N}$ and set $s_1 = N_1$.

Lemma 4.2.4. *Assuming $\|\{V_{k,\beta}\}\|_{C_b^{s_1}} \le 1$, there are constants C_5 and T^*, such that for $0 \le t \le T^*$,*

$$\|f(x, t, \{V_{k,\beta}\})\|_{H^{s_c}} \le C_5 \lambda(t)^p \lambda'(t) \left(C_f \sup_{\tau \in [0,t]} \frac{\|V(\tau)\|_{H^{s_c}}}{\lambda(\tau)^{p+1}} + C_{f0} \right). \tag{4.15}$$

Proof. Lemma 4.2.3 and (4.14) allow us to estimate

$$\begin{aligned}
&\|f(x, t, \{V_{k,\beta}\})\|_{H^{s_c}} \le C_{s_c} \|f(x, t, \{V_{k,\beta}\})\|'_{H^{s_c}} \\
&\quad \le C_{s_c} \varphi_{s_c}(\|\{V_{k,\beta}\}\|_{C_b^{s_1}}) \|\{V_{k,\beta}\}\|_{H^{s_c-1}} + C_{s_c} \sum_{|\alpha| \le s_c} \left\| \frac{\partial^\alpha f}{\partial x^\alpha}(x, t, \{V_{k,\beta}\}) \right\|_{L^2} \\
&\quad + C_{s_c} \sum_{k+|\beta| \le m-1} \left\| \frac{\partial f}{\partial V_{k,\beta}} \right\|_{L^\infty} \sum_{|\alpha|=s_c} \|\partial_x^\alpha V_{k,\beta}\|_{L^2}.
\end{aligned}$$

Repeated application of Remark 4.2.2 yields (for $|\beta| > 0$)

$$\begin{aligned}\|V_{k,\beta}\|_{H^{s_c-1}} &\leq C\lambda(t)^p\lambda'(t) \sup_{\tau\in[0,t]} \frac{\left\|\langle D\rangle^{|\beta|} c_{m-1-|\beta|,\beta} D_t^{m-1-|\beta|} v\right\|_{H^{s_c-1}}}{\lambda(\tau)^p\lambda'(\tau)} \\ &\leq C\lambda(t)^p\lambda'(t) \sup_{\tau\in[0,t]} \lambda(\tau)^{-p} \left\|\langle D\rangle^{|\beta|-1}(\lambda\sigma)^{|\beta|-1} D_t^{m-1-|\beta|} v\right\|_{H^{s_c}}.\end{aligned}$$

Considering the time derivative of the last norm and exploiting Nersesyan's lemma, we find (restricting the time interval)

$$\|\{V_{k,\beta}\}\|_{H^{s_c-1}} \leq \lambda(t)^{p+1} \sup_{\tau\in[0,t]} \frac{\|V(\tau)\|_{H^{s_c}}}{\lambda(\tau)^{p+1}}.$$

Again by Remark 4.2.2,

$$\sum_{|\alpha|=s_c} \|\partial_x^\alpha V_{k,\beta}\|_{L^2} \leq C\lambda(t)^p\lambda'(t) \sup_{\tau\in[0,t]} \frac{\|V(\tau)\|_{H^{s_c}}}{\lambda(\tau)^{p+1}}.$$

Lastly, from Hadamard's formula and (4.9) it can be deduced that

$$\begin{aligned}&\sum_{|\alpha|\leq s_c} \left\|\frac{\partial^\alpha f}{\partial x^\alpha}(x,t,\{V_{k,\beta}\})\right\|_{L^2} \\ &\leq \sum_{|\alpha|\leq s_c} \left\|\frac{\partial^\alpha f}{\partial x^\alpha}(x,t,\{V_{k,\beta}\}) - \frac{\partial^\alpha f}{\partial x^\alpha}(x,t,0)\right\|_{L^2} + \sum_{|\alpha|\leq s_c} \left\|\frac{\partial^\alpha f}{\partial x^\alpha}(x,t,0)\right\|_{L^2} \\ &\leq C\|V^*(t)\|_{H^{s_c-1}} + CC_{f0}\lambda(t)^p\lambda'(t).\end{aligned}$$

Summing up, we can estimate for small t:

$$\begin{aligned}\|f(x,t,\{V_{k,\beta}\})\|_{H^{s_c}} &\leq C_{s_c}(2+\varphi_{s_c}(1))\lambda(t)^{p+1} \sup_{\tau\in[0,t]} \frac{\|V(\tau)\|_{H^{s_c}}}{\lambda(\tau)^{p+1}} \\ &\quad + CC_f\lambda(t)^p\lambda'(t) \sup_{\tau\in[0,t]} \frac{\|V(\tau)\|_{H^{s_c}}}{\lambda(\tau)^{p+1}} + CC_{f0}\lambda(t)^p\lambda'(t) \\ &\leq C_5\lambda(t)^p\lambda'(t)\left(C_f \sup_{\tau\in[0,t]} \frac{\|V(\tau)\|_{H^{s_c}}}{\lambda(\tau)^{p+1}} + C_{f0}\right).\end{aligned}$$

□

4.2.2. ITERATION AND CONVERGENCE Now we have all tools to find a bound for the mapping $V \mapsto V^* \mapsto U$.

Lemma 4.2.5. *We assume that p is sufficiently large and that T^* is sufficiently small. If $\|V(t)\|_{H^{s_c}} \leq \lambda(t)^{p+1}$ for $0 \leq t \leq T^*$, then $\|U(t)\|_{H^{s_c}} \leq \lambda(t)^{p+1}$ for $0 \leq t \leq T^*$.*

Proof. Due to Theorem 4.1.1 and (4.15) we have

$$\begin{aligned}\|U(t)\|_{H^{s_c}} &\le C_3 \int_0^t e^{C_1(t-s)} \left(\frac{\lambda(t)}{\lambda(s)}\right)^{C_2} \|f(s)\|_{H^{s_c}}\,ds \\ &\le C_3 e^{C_1 t} \int_0^t \left(\frac{\lambda(t)}{\lambda(s)}\right)^{C_2} C_5 \lambda(s)^p \lambda'(s) \left(C_f \sup_{\tau\in[0,s]} \frac{\|V(\tau)\|_{H^{s_c}}}{\lambda(\tau)^{p+1}} + C_{f0}\right) ds \\ &\le C_3 C_5 (C_f + C_{f0}) e^{C_1 t} \frac{\lambda(t)^{p+1}}{p - C_2 + 1} \\ &\le \lambda(t)^{p+1}\end{aligned}$$

if $e^{C_1 t} \le 2$ and p is large. □

Now we restrict the constant T^* in such a manner, that the assumption $\|V(t)\|_{H^{s_c}} \le \lambda(t)^{p+1}$ for all $t \in [0, T^*]$ implies $(x, (V_{k,\beta}(x,t))) \in K_G$ for all $(x,t) \in M \times [0, T^*]$. All these results enable us to define a sequence $(V^l) \subset C([0,T^*], H^{s_c}) \cap C^1([0,T^*], H^{s_c-1})$ by $V^0(t) \equiv 0$ and

$$\begin{aligned} &V^l(0) = 0, \quad l \ge 1, \\ &\partial_t V^l = \lambda K_0(x,t,V^{*,l-1}, D)\langle D\rangle \sigma V^l + \lambda B(x,t,V^{*,l-1},D)V^l \\ &\qquad + F(x,t,V^{*,l-1}) + H\frac{\lambda'}{\lambda} V^l, \quad l \ge 1. \end{aligned}$$

Due to Lemma 4.2.5 and Remark 4.2.2 the functions V^l fulfill

$$\left\|V^l(t)\right\|_{H^{s_c}} \le \lambda(t)^{p+1}, \quad \left\|V^l_{k,\beta}(t)\right\|_{H^{s_c}} \le C\lambda(t)^p \lambda'(t).$$

Using the above technique once more and choosing p larger if necessary, we are able to show the estimate

$$\sup_{t\in[0,T^*]} \frac{\left\|V^{l+1}(t) - V^l(t)\right\|_{L^2}}{\lambda(t)^{p+1}} \le \frac{1}{2} \sup_{t\in[0,T^*]} \frac{\left\|V^l(t) - V^{l-1}(t)\right\|_{L^2}}{\lambda(t)^{p+1}}.$$

This confirms that the sequence (V^l) converges in $C([0,T^*], H^0)$. By interpolation we see that (V^l) converges in $C([0,T^*], H^{s_c-1})$, too; and we can prove in a standard way that the limit U is a solution of (4.8). Exploiting the arguments which gave Proposition 3.2.7, we get $U \in C([0,T^*], H^{s_c})$. Thus, we have proved:

Theorem 4.2.6 (Existence). *Let the conditions mentioned at the beginning of this section be fulfilled. Let s_c, $p \in \mathbb{N}$ be sufficiently large and $T^* > 0$ be sufficiently small. Then the Cauchy problem* (4.8) *has a solution u with $U \in C([0,T^*], H^{s_c}) \cap C^1([0,T^*], H^{s_c-1})$.*

4.3. Reduction of a General Quasi-linear Equation to a Quasi-linear Equation with Special Right-Hand Side

In this subsection we reflect upon a general quasi-linear weakly hyperbolic Cauchy problem and find a solution using the technique of the previous subsection. Namely, we will transform the Cauchy problem

$$\begin{aligned}
&D_t^m u + \sum_{j+|\alpha|=m, j<m} a_{j,\alpha}(x,t,\{D_x^\beta c_{k,\beta}(x,t)D_t^k u\})\lambda(t)^{|\alpha|}D_x^\alpha D_t^j(\sigma(x)^{|\alpha|}u) \\
&\quad = f(x,t,\{D_x^\beta c_{k,\beta}(x,t)D_t^k u\}), \qquad (4.16)\\
&u(x,0) = \varphi_0(x), \ldots, D_t^{m-1}u(x,0) = \varphi_{m-1}(x)
\end{aligned}$$

into another Cauchy problem

$$\begin{aligned}
&D_t^m v + \sum_{j+|\alpha|=m, j<m} a_{j,\alpha,p'}(x,t,\{D_x^\beta c_{k,\beta}(x,t)D_t^k v\})\lambda(t)^{|\alpha|}D_x^\alpha D_t^j(\sigma(x)^{|\alpha|}v) \\
&\quad = f_{p'}(x,t,\{D_x^\beta c_{k,\beta}(x,t)D_t^k v\}), \qquad (4.17)\\
&v(x,0) = \cdots = D_t^{m-1}v(x,0) = 0,
\end{aligned}$$

whose right-hand side $f_{p'}$ fulfills the requirements (4.9) and (4.10) with $p = p(p')$. It will be shown that these two Cauchy problems are equivalent in the sense that functions u_1, u_2, ..., $u_{p'}$ exist with

$$u = u_1 + u_2 + \cdots + u_{p'} + v.$$

The functions u_1, u_2, ..., $u_{p'}$ are solutions of ODEs in t with parameter x. If p is large enough, then Theorem 4.2.6 guarantees the existence of a solution to (4.17). This idea has been used in [17] and [29].

The example of Qi [27] shows that a loss of Sobolev regularity (in comparison to the data) must be expected. This corresponds to the fact that the smoothness of the u_j decreases by m, as j increases by 1.

We list the assumptions.

The functions $c_{k,\beta}$ are assumed to satisfy (4.11) together with (3.22)–(3.25). The functions $a_{j,\alpha}$ and the right-hand side f are defined in the set K_G; see (3.26). Finally, we suppose (3.4), (3.27), (3.28), (3.29), Condition 3.2 and Condition 4.1.

Theorem 4.3.1 (Existence). *If $s_c \in \mathbb{N}$ is large enough, then some number $T^* \in (0,T]$ and some $\gamma > 0$ (independent of s_c) exist with the property that there is a solution u of (4.16) with*

$$\begin{aligned}
&U \in C\left([0,T^*], H^{s_c-\gamma}\right) \cap C^1\left([0,T]^*, H^{s_c-\gamma-1}\right),\\
&U = \{\langle D\rangle^{m-i}((\lambda\sigma)^{m-i}D_t^{i-1}u) : i = 1,\ldots,m\}.
\end{aligned}$$

Proof. We define

$$\varepsilon_{l,i,\beta} = \begin{cases} 1 & : |\beta| = 0 \text{ or } l > i, \\ 0 & : |\beta| > 0 \text{ and } l = i \end{cases}$$

and consider the system of ODEs in t with parameter x,

$$\begin{aligned}
&D_t^m u_1(x,t) = f(x,t,\{D_x^\beta c_{k,\beta}(x,t) D_t^k u_1(x,t)\varepsilon_{1,1,\beta}\}),\\
&u_1(x,0) = \varphi_0(x), \dots, D_t^{m-1} u_1(x,0) = \varphi_{m-1}(x),\\
&D_t^m u_l(x,t) = g_l(x,t,u_l(x,t),\dots,D_t^{m-1}u_l(x,t))\\
&\quad = f\Big(x,t,\Big\{D_x^\beta c_{k,\beta}(x,t) D_t^k \sum_{i=1}^{l} \varepsilon_{l,i,\beta} u_i(x,t)\Big\}\Big)\\
&\quad - f\Big(x,t,\Big\{D_x^\beta c_{k,\beta}(x,t) D_t^k \sum_{i=1}^{l-1} \varepsilon_{l-1,i,\beta} u_i(x,t)\Big\}\Big)\\
&\quad - \sum_{j+|\alpha|=m, j<m} \tilde{a}_{j,\alpha,l}(x,t) \sum_{i=1}^{l-1} \lambda(t)^{|\alpha|} D_x^\alpha D_t^j (\sigma(x)^{|\alpha|} u_i(x,t))\\
&\quad + \sum_{j+|\alpha|=m, j<m} \tilde{a}_{j,\alpha,l-1}(x,t) \sum_{i=1}^{l-2} \lambda(t)^{|\alpha|} D_x^\alpha D_t^j (\sigma(x)^{|\alpha|} u_i(x,t)),\\
&u_l(x,0) = \dots = D_t^{m-1} u_l(x,0) = 0, \quad l = 2,\dots,p.
\end{aligned}$$

The following abbreviations have been used here:

$$\tilde{a}_{j,\alpha,l}(x,t) = a_{j,\alpha}\Big(x,t,\Big\{D_x^\beta c_{k,\beta}(x,t) D_t^k \sum_{i=1}^{l} u_i(x,t)\varepsilon_{l,i,\beta}\Big\}\Big), \quad \sum_{i=1}^{0} = 0.$$

These equations can be solved step by step, and they possess solutions $u_l \in C^m\left([0,T_l], H^{s_c - m(l-1)}\right)$, $s_c - ml > n/2 + 1$. For a proof, see Theorem 3.2.2 with $\sigma \equiv 0$.

The functions u_l have a special asymptotical behavior for $t \to 0$. If λ satisfies the second set of assumptions of Condition 4.1, we fix some positive number ε with $2\varepsilon < m - (m-1)C_\lambda$. Otherwise, $\varepsilon = 0$. Inductively, we prove

$$\sum_{j=0}^{m-1} \left\| D_t^j u_l(\cdot,t) \right\|_{H^{s_c - m(l-1)}} \le C_6 \left(t^{1/2} \lambda(t)^\varepsilon\right)^{l-1}$$

with $s_c - ml > n/2 + 1$. This is obvious for $l = 1$. To prove a corresponding estimate for u_{l+1}, we consider $\|g_{l+1}(t)\|_{H^{s_c - ml}}$. Hadamard's formula yields

$$\begin{aligned}
&f\Big(x,t,\Big\{D_x^\beta c_{k,\beta} D_t^k \sum_{i=1}^{l+1} \varepsilon_{l+1,i,\beta} u_i\Big\}\Big) - f\Big(x,t,\Big\{D_x^\beta c_{k,\beta} D_t^k \sum_{i=1}^{l} \varepsilon_{l,i,\beta} u_i\Big\}\Big)\\
&\quad = \sum_k d_{1lk}(x,t) c_{k,0} D_t^k u_{l+1} + \sum_{|\beta|>0,k} d_{2lk\beta}(x,t) D_x^\beta c_{k,\beta} D_t^k u_l.
\end{aligned}$$

The $H^{s_c - ml}$-norms of the functions d_{1lk}, $d_{2lk\beta}$ are bounded, by the induction assumption and $u_{l+1} \in C^m\left([0,T_{l+1}], H^{s_c - ml}\right)$. The function $\lambda(t)^{m-2\varepsilon} \Lambda(t)^{1-m}$ is monotonically increasing due to the choice of ε, see Condition 4.1. This implies

$$\|c_{k,\beta}(\cdot,t)\|_{H^s} \le C_s \lambda(t)^{2\varepsilon}, \quad s \in \mathbb{R}^+, \quad |\beta| > 0.$$

The other contributions to g_{l+1} can be discussed similarly. We conclude that u_{l+1} is a solution of

$$D_t^m u_{l+1} + \sum_{k<m} d_{3lk}(x,t) D_t^k u_{l+1} = h_{l+1}(x,t)$$

with

$$\|h_{l+1}(\cdot,t)\|_{H^{s_c-ml}} \le C_l \left(t^{1/2}\lambda(t)^{\varepsilon}\right)^{l-1} \lambda(t)^{2\varepsilon}, \quad \|d_{3lk}(\cdot,t)\|_{H^{s_c-ml}} \le C_l. \tag{4.18}$$

Utilizing a standard technique one shows (restricting the time interval)

$$\sum_{j=0}^{m-1} \left\| D_t^j u_{l+1}(\cdot,t) \right\|_{H^{s_0-ml}} \le C_l \int_0^t \|h_{l+1}(\cdot,\tau)\|_{H^{s_c-ml}}\, d\tau \le \left(t^{1/2}\lambda(t)^{\varepsilon}\right)^l .$$

This is the desired estimate. Summing up the differential equations for $u_1, \ldots, u_l$ we deduce that

$$\begin{aligned} D_t^m(u_1 + \cdots + u_l) &= f\Big(x,t,\Big\{D_x^\beta c_{k,\beta} D_t^k \sum_{i=1}^l \varepsilon_{l,i,\beta} u_i\Big\}\Big) \\ &\quad - \sum_{j+|\alpha|=m, j<m} \tilde{a}_{j,\alpha,l} \sum_{i=1}^{l-1} \lambda^{|\alpha|} D_x^\alpha D_t^j(\sigma^{|\alpha|} u_i), \\ D_t^j(u_1 + \cdots + u_l)(x,0) &= \varphi_j(x), \quad j = 0, \ldots, m-1. \end{aligned}$$

We define

$$a_{j,\alpha,l}(x,t,\{D_x^\beta c_{k,\beta} D_t^k v\}) = a_{j,\alpha}\Big(x,t,\Big\{D_x^\beta c_{k,\beta} D_t^k \Big(\sum_{i=1}^l u_i + v\Big)\Big\}\Big),$$

$$\begin{aligned} & f_l(x,t,\{D_x^\beta c_{k,\beta} D_t^k v\}) \\ &= f\Big(x,t,\Big\{D_x^\beta c_{k,\beta} D_t^k \Big(\sum_{i=1}^l u_i + v\Big)\Big\}\Big) - f\Big(x,t,\Big\{D_x^\beta c_{k,\beta} D_t^k \sum_{i=1}^l \varepsilon_{l,i,\beta} u_i\Big\}\Big) \\ &- \sum_{j+|\alpha|=m, j<m} a_{j,\alpha}\Big(x,t,\Big\{D_x^\beta c_{k,\beta} D_t^k \Big(\sum_{i=1}^l u_i + v\Big)\Big\}\Big) \lambda^{|\alpha|} D_x^\alpha D_t^j \Big(\sigma^{|\alpha|} \sum_{i=1}^l u_i\Big) \\ &+ \sum_{j+|\alpha|=m, j<m} a_{j,\alpha}\Big(x,t,\Big\{D_x^\beta c_{k,\beta} D_t^k \sum_{i=1}^l \varepsilon_{l,i,\beta} u_i\Big\}\Big) \lambda^{|\alpha|} D_x^\alpha D_t^j \Big(\sigma^{|\alpha|} \sum_{i=1}^{l-1} u_i\Big). \end{aligned}$$

If a function v has homogeneous initial data and satisfies

$$D_t^m v + \sum_{j+|\alpha|=m, j<m} a_{j,\alpha,l}(x,t,\{D_x^\beta c_{k,\beta} D_t^k v\}) \lambda^{|\alpha|} D_x^\alpha D_t^j(\sigma^{|\alpha|} v) = f_l(x,t,\{D_x^\beta c_{k,\beta} D_t^k v\})$$

(which is (4.17) with $p' = l$), then the function $u = \sum_{i=1}^l u_i + v$ solves (4.16).

It remains to verify the conditions (4.9) and (4.10) for this function f_l: We fix $\delta > 0$ and restrict all the intervals $[0, T_l]$ in such a way that

$$\left\| D_t^j \sum_{i=1}^{l} u_i(\cdot, t) - \varphi_j(\cdot) \right\|_{H^{s_c - ml}} \leq \delta, \quad 0 \leq t \leq T_l, \quad 0 \leq j \leq m-1.$$

The condition (4.10) is obviously satisfied for $t \leq T_l$ and the constant C_f only depends on δ, but not on l. In a similar way as in the proof of (4.18) one shows

$$\|f_l(\cdot, t, 0)\|_{H^{s_c - ml}} \leq C \left(t^{1/2} \lambda(t)^{\varepsilon} \right)^l,$$

which proves (4.9).

Theorem 4.2.6 gives the existence of a solution v to (4.17) which satisfies $\langle D \rangle^k ((\lambda\sigma)^k v) \in C\left([0, T^*], H^{s_c - ml}\right)$, if p and $s_c - ml$ are large enough. □

Remark 4.3.2. Since the time degeneracy occurs only for $t = 0$, the blow-up criterion of Proposition 3.2.8 is still valid; we only have to take into account the number of lost derivatives.

5. Domains of Dependence

We will construct so-called *domains of dependence*. It turns out that our definition generalizes the definition of [2] from the strictly hyperbolic case to the weakly hyperbolic case. These domains can be exhausted with hypersurfaces, and the Cauchy problem is weakly hyperbolic in the normal direction at each point of each hypersurface, see Definition 5.1.2. One example of such domains is a cone, whose slope does not exceed some critical value. This concept will be applied to prove some results of uniqueness, finite propagation speed and regularity:

Global uniqueness for linear equations: A solution to a linear Cauchy problem is unique in any domain of dependence, see Theorem 5.2.1.

Local uniqueness for quasi-linear equations: For every ball in the initial plane one can find a cone (with suitably small slope) over this ball with the property that a solution is unique in this cone, see Theorem 5.3.2.

Local existence for quasi-linear equations: For every rectangle in the initial plane one can find a rectangular parallelepipedon over the rectangle with the property that a Sobolev solution of the quasi-linear equation exists in this parallelepipedon, cf. Theorem 5.3.1. This solution exists in the whole domain of dependence if the equation is linear, cf. Corollary 5.3.4.

C^∞ **regularity:** We consider a quasi-linear Cauchy problem, whose coefficients, right-hand side, weight functions and initial data are C^∞. Let us be given a Sobolev solution in some domain of dependence. Then this solution is C^∞ in this domain, cf. Theorem 5.4.2.

5.1. Definition of Domains of Dependence

We come to the definition of a domain of dependence, see also [2].

We consider the Cauchy problem

$$D_t^m u + \sum_{j+|\alpha|\le m, j<m} a_{j,\alpha}(x,t) D_x^\alpha D_t^j \left(\sigma(x)^{|\alpha|} u\right) = f(x,t), \tag{5.1}$$

$$u(x,0) = \varphi_0(x), \ldots, D_t^{m-1} u(x,0) = \varphi_{m-1}(x)$$

for $(x,t) \in \Omega_0 \times [0,T] \subset \mathbb{R}_x^n \times \mathbb{R}_t$; Ω_0 is an open and bounded domain with smooth boundary.

We suppose Condition 3.1 and

$$\sigma \in C_b^\infty(\Omega_0), \tag{5.2}$$

$$a_{j,\alpha} \in \begin{cases} C_b^1(\overline{\Omega_0} \times [0,T]) & : j + |\alpha| = m, \\ C_b^0(\overline{\Omega_0} \times [0,T]) & : j + |\alpha| < m. \end{cases} \tag{5.3}$$

The principal part of the operator on the left is

$$P_{m,\sigma}(x,t,D_x,D_t) = D_t^m + \sum_{j+|\alpha|=m, j<m} a_{j,\alpha}(x,t)\sigma(x)^{|\alpha|} D_x^\alpha D_t^j.$$

To this operator we assign the strictly hyperbolic operator

$$P_{m,1}(x,t,D_x,D_t) = D_t^m + \sum_{j+|\alpha|=m, j<m} a_{j,\alpha}(x,t) D_x^\alpha D_t^j.$$

The domain of dependence Ω over a bounded domain $\Omega_0 \subset \mathbb{R}^n$ has to satisfy the following conditions. At first,

$$\Omega = \Omega' \cap \{(x,t) : t \ge 0\}, \quad \Omega' \Subset \mathbb{R}^{n+1}, \quad \Omega' \text{ open}, \tag{5.4}$$

$$\Omega_0 = \Omega' \cap \{(x,t) : t = 0\}. \tag{5.5}$$

Next, the projections $\pi : (x,t) \mapsto x$ of the level sets $\Omega_{t_0} = \Omega' \cap \{(x,t) : t = t_0\}$ shall become "smaller" for increasing t_0,

$$\pi\Omega_{t_1} \Subset \pi\Omega_{t_0} \quad 0 \le t_0 < t_1 \le T. \tag{5.6}$$

The set Ω can be exhausted with hypersurfaces S_r,

$$\Omega = \bigcup_{0\le r<r^*} S_r = \bigcup_{0\le r<r^*} \{(x,t) : g(x,t) = r\}. \tag{5.7}$$

We suppose $g \in C_b^\infty(\Omega_0 \times [0,T])$ and

$$\frac{\partial g}{\partial t} > 0 \quad \text{in } \Omega. \tag{5.8}$$

Furthermore, we assume that each hypersurface S_r intersects with the initial domain Ω_0 and

$$\left(\Omega_0 \cap \bigcup_{0\le r\le r_0} S_r\right) \Subset \left(\Omega_0 \cap \bigcup_{0\le r\le r_1} S_r\right) \quad 0 \le r_0 < r_1 < r^*. \tag{5.9}$$

Finally, we need a connection between the slope of the normal vector to S_r at the point (x,t) and the largest characteristic root of $P_{m,\sigma}$ at that point. Let Q_m be the principal part of a hyperbolic in direction τ operator. Then the largest characteristic root $\lambda_{max}(x,t;Q_m)$ of Q_m at the point (x,t) is defined by

$$\lambda_{max}(x,t;Q_m) = \sup\{|\tau| : Q_m(x,t,\tau,\xi) = 0, |\xi| = 1\}.$$

The slope of the normal vector and the largest characteristic roots satisfy

$$\lambda_{max}(x,t;P_{m,\sigma}) \left| \frac{\nabla_x g(x,t)}{g_t(x,t)} \right| < 1. \tag{5.10}$$

This condition can be interpreted in the way that the polynomial $P_{m,\sigma}$ is weakly hyperbolic at the point (x,t) in the normal direction of S_r.

For technical reasons we assume the following condition:

$$\text{The domain } \Omega_0 \text{ has the } J\text{-extension property defined below.} \tag{5.11}$$

Definition 5.1.1. We say that a domain Ω_0 has the J-extension property if for every small $\varepsilon > 0$ there is an operator $\mathcal{E}_\varepsilon : C_b(\Omega_0) \longrightarrow C_b(\mathbb{R}^n)$ such that

- $(\mathcal{E}_\varepsilon u)(x) = u(x)$ for $x \in \Omega_0$, $\operatorname{dist}(x, \partial\Omega_0) > \varepsilon$,
- there is a C^∞ mapping $\Psi : \mathbb{R}^n \to \Omega_0 \cap \{x\colon \operatorname{dist}(x,\partial\Omega_0) > \varepsilon\}$ such that $|(\mathcal{E}_\varepsilon u)(x) - u(\Psi(x))| < \varepsilon$ for all $x \in \mathbb{R}^n$.

Example. Every star-shaped domain has the J-extension property.

Definition 5.1.2. A set Ω is called a *domain of dependence over* Ω_0 *for the operator* $P_{m,\sigma}$ if the conditions (5.4)–(5.11) are satisfied.

Example (Characteristic cone). The characteristic cone $K(B)$ for the ball $B = B(x^*, d)$ in the initial plane is defined by

$$K(B) = \left\{ (x,t) : |x - x^*| < d - \lambda'_{max,\sigma} t, \quad 0 \le t < \frac{d}{\lambda'_{max,\sigma}} \right\}$$

with

$$\lambda'_{max,\sigma} = \|\sigma\|_{L^\infty(B)} \sup \left\{ \lambda_{max}(x,t;P_{m,1}) : \quad (x,t) \in \overline{B} \times [0,T] \right\}.$$

In this section we will also examine the quasi-linear Cauchy problem

$$\begin{aligned} D_t^m u + \sum_{j+|\alpha|\le m, j<m} & a_{j,\alpha}(x,t,\{D_x^\beta \sigma(x)^{|\beta|} D_t^k u\}) D_x^\alpha D_t^j \left(\sigma(x)^{|\alpha|} u\right) \\ &= f(x,t,\{D_x^\beta \sigma(x)^{|\beta|} D_t^k u\}|_{k+|\beta|\le m-1}), \\ u(x,0) &= \varphi_0(x), \dots, D_t^{m-1} u(x,0) = \varphi_{m-1}(x). \end{aligned} \tag{5.12}$$

This equation will be written as $P_{m,\sigma}^{(u)} u = f^{(u)}$.

If u solves this Cauchy problem, then one can define a domain of dependence $\Omega^{(u)}$, which itself depends on u, since the coefficients of the principal part depend on u. For this Cauchy problem we will assume almost the same conditions as in Section 3, see (3.27), (3.28) and Remark 3.2.1:

$$a_{j,\alpha} \in C^1([0,T], C^{s_c}(\Omega_0 \times \mathbb{R}^{n_0})), \quad s_c > \frac{n}{2} + 1, \tag{5.13}$$

$$\varphi_j \in H^{s_c+m-1-j}(\Omega_0), \tag{5.14}$$

$$f \in C([0,T], C^{s_c}(\Omega_0 \times \mathbb{R}^{n_0})). \tag{5.15}$$

And we modify Condition 3.2 in an obvious way:

Condition 5.1. *The roots $\tau_j(x,t,v,\xi)$ of*

$$\tau^m + \sum_{j+|\alpha|=m, j<m} a_{j,\alpha}(x,t,v)\xi^\alpha \tau^j = 0$$

are real and distinct, $|\tau_j(x,t,v,\xi) - \tau_i(x,t,v,\xi)| \geq c|\xi|$, $c > 0$, $i \neq j$, for all $(x,t,v,\xi) \in \Omega_0 \times [0,T] \times \mathbb{R}^{n_0} \times \mathbb{R}^n$.

5.2. Uniqueness for Linear Equations

There is another way to define domains of dependence. A set $\Omega \subset \mathbb{R}_+ \times \mathbb{R}^n$ is called *domain of dependence over* $\Omega_0 \subset \mathbb{R}^n$ for some hyperbolic operator if the vanishing of data in Ω_0 and the vanishing of the right-hand side in Ω imply the vanishing of the solution to the Cauchy problem in Ω. The following theorem states that the two definitions are in concordance.

Theorem 5.2.1 (Uniqueness). *We suppose* (5.2), (5.3) *and Condition* 3.1. *Let $\Omega \subset \Omega_0 \times [0,T]$ be a domain of dependence over Ω_0 for the operator $P_{m,\sigma}$. Assume $\varphi_0 \equiv \cdots \equiv \varphi_{m-1} \equiv 0$ in Ω_0 and $f \equiv 0$ in Ω. Then $u \equiv 0$ in Ω for every solution u of* (5.1) *with*

$$D_t^{m-|\alpha|} D_x^\alpha \sigma^{|\alpha|} u \in C(\Omega), \quad |\alpha| \leq m.$$

Proof. Let us sketch the proof. The solution is defined in Ω, the coefficients $a_{j,\alpha}$ are defined in $\overline{\Omega_0} \times [0,T]$. In a first step we extend u and $a_{j,\alpha}$ to the domain $\Omega_0 \times (-\infty, 0]$; the function u vanishes there. This gives a new Cauchy problem with solution u and vanishing initial values. In a second step we transform the variables. The domain Ω is mapped to some domain $\tilde{\Omega} = \{(y,r) : y \in \Omega_0, g(y,0) \leq r < r^*\}$, and v, the image of u, has compact support with respect to the spatial variable y if the time variable is frozen. We extend the transformed coefficients $\tilde{a}_{j,\alpha}$ and the weight σ to $\mathbb{R}^n \times [0,T]$, apply an energy estimate, and obtain $v \equiv 0$. This will prove the theorem.

The coefficients $a_{j,\alpha}$ satisfy Condition 3.1, (5.3) and are defined in $\Omega_0 \times [0,T]$. For $0 < t \leq \varepsilon$ we set $a_{j,\alpha}(x,-t) = 2a_{j,\alpha}(x,0) - a_{j,\alpha}(x,t)$. Then we change $a_{j,\alpha}(x,t)$ appropriately for $-\varepsilon \leq t \leq -\varepsilon/2$, set $a_{j,\alpha}(x,t) = a_{j,\alpha}(x,-\varepsilon)$ for $-\infty < t < -\varepsilon$, and conclude that $a_{j,\alpha} \in C_b^1(\Omega_0 \times (-\infty, T])$ for $j + |\alpha| = m$ and $a_{j,\alpha} \in C_b^0(\Omega_0 \times (-\infty, T])$ for $j + |\alpha| < m$. Condition 3.1 is true in this domain if ε is small.

The same method can be used to extend the derivative $g_t(x,t)$ of the function $g(x,t)$ to $\Omega_0 \times (-\infty, T]$. The result is $g \in C^\infty(\Omega_0 \times (-\infty, T])$, $g_t \in C_b^\infty(\Omega_0 \times$

$(-\infty,T])$ $g_t(x,t)=g_t(x,-\varepsilon)$, $t\le -\varepsilon$. We see that the function g takes arbitrarily small values and conclude that

$$\Omega\cup Z=\Omega\cup(\Omega_0\times(-\infty,0])=\bigcup_{-\infty<r<r^*} S_r.$$

We extend the solution u by $u(x,t)=0$, $(x,t)\in Z$, the letter Z stands for "zero". Then u solves (5.1) in Z with homogeneous data for $t=-1$, $f\equiv 0$ in Z, and

$$D_t^{m-|\alpha|}D_x^\alpha\sigma^{|\alpha|}u\in C(\Omega\cup Z),\quad |\alpha|\le m.$$

We apply a Holmgren type transform to change the variables,

$$y=x,\quad r=g(x,t),\quad \nabla_x=\nabla_y+(\nabla_x g)\partial_r,\quad \partial_t=(\partial_t g)\partial_r,$$
$$v(y,r)=u(x,t).$$

The dual variables fulfill

$$\xi=\eta+(\nabla_x g)\varrho=\eta+c(y,r)\varrho,\quad \tau=g_t\varrho=c_0(y,r)\varrho.$$

The domain Ω is mapped to the set

$$\tilde\Omega=\{(y,r):y\in\Omega_0,g(y,0)\le r<r^*\},$$

and $\tilde Z$ is the image of Z:

$$\tilde Z=\{(y,r):y\in\Omega_0,-\infty<r\le g(y,0)\}.$$

It is easy to verify that $D_r^{m-|\alpha|}D_y^\alpha\sigma(y)^{|\alpha|}v(y,r)\in C(\tilde\Omega\cup\tilde Z)$ for $|\alpha|\le m$. The function v is a solution of

$$D_r^m v+\sum_{j+|\alpha|\le m,j<m}\tilde a_{j,\alpha}(y,r)D_y^\alpha D_r^j\left(\sigma(y)^{|\alpha|}v\right)=0,$$
$$v(y,0)=\dots=D_r^{m-1}v(y,0)=0,\quad (y,r)\in\Omega_0\times[0,r^*),$$

$\tilde a_{j,\alpha}(y,r)$ are the transformed coefficients. We have to check whether this Cauchy problem is weakly hyperbolic and satisfies the Levi conditions. We remember the definition of (strict) hyperbolicity, see [23]:

Definition 5.2.2. A differential operator $Q(z,D_z)=\sum_{|\alpha|=m}a_\alpha(z)D_z^\alpha$ is called *hyperbolic* at the point z_0 in the direction $N\ne 0$ if

- $Q(z_0,N)\ne 0$,
- $Q(z_0,\tau N+\zeta)=0$ has only real roots τ for every $\zeta\ne 0$.

A differential operator $Q(z,D_z)$ is called *strictly hyperbolic* at (z_0,N) if it is hyperbolic at (z_0,N) and if $Q(z_0,\tau N+\zeta)=0$ has m real and distinct roots τ for every $\zeta\perp N$, $\zeta\ne 0$.

By definition, the operator $P_{m,\sigma}(x,t,D_x,D_t)$ is hyperbolic in the direction $N=(1,0,\dots,0)\in\mathbb{R}^{1+n}$ and the operator $P_{m,1}(x,t,D_x,D_t)$ is strictly hyperbolic in this direction N. The symbol of the principal part $\tilde P_{m,\sigma}$ of the transformed operator is

$$\tilde P_{m,\sigma}(y,r,\eta,\varrho)=P_{m,\sigma}(x,t,\eta+c\varrho,c_0\varrho).$$

To be able to apply the results of Section 3, we have to verify that the operator $\tilde{P}_{m,\sigma}$ is hyperbolic in the direction $\tilde{N}$ and that the operator $\tilde{P}_{m,1}$ is strictly hyperbolic in the direction $\tilde{N}$. Here $\tilde{N} = (1, 0, \dots, 0) \in \mathbb{R}^1_\varrho \times \mathbb{R}^n_y$ is the normal direction of the hypersurfaces $r = \text{const}$.

It can be seen that

$$\tilde{P}_{m,\sigma}(y, r, 0, 1) = P_{m,\sigma}(x, t, c, c_0) = |c|^m P_{m,\sigma}\left(x, t, \frac{c}{|c|}, \frac{c_0}{|c|}\right).$$

From (5.10) and the definition of $\lambda_{max}(x, t; P_{m,\sigma})$ we then conclude that

$$\tilde{P}_{m,\sigma}(y, r, \tilde{N}) = \tilde{P}_{m,\sigma}(y, r, 0, 1) \neq 0.$$

In the next step we show that the equation $\tilde{P}_{m,\sigma}(y, r, \eta, \varrho) = 0$ has m real roots $\varrho_1, \dots, \varrho_m$ for every $\eta \neq 0$. It holds that

$$\tilde{P}_{m,\sigma}(y, r, \eta, \varrho) = P_{m,\sigma}(x, t, \eta + c\varrho, c_0\varrho) = c_0^m P_{m,1}\left(x, t, \frac{\sigma}{c_0}(\eta + c\varrho), \varrho\right).$$

If $\sigma(x) = 0$, then the only roots are $\varrho_1 = \dots = \varrho_m = 0$. If $\sigma(x) \neq 0$, then

$$\tilde{P}_{m,\sigma}(y, r, \eta, \varrho) = \sigma^m P_{m,1}\left(x, t, \eta + c\varrho, \frac{c_0}{\sigma}\varrho\right).$$

The polynomial $P_{m,1}(x, t, \eta, c_0\varrho/\sigma)$ is strictly hyperbolic in the direction N. From (5.10) we gather

$$\left|\frac{\sigma(x)c(x,t)}{c_0(x,t)}\right| < \frac{1}{\lambda_{max}(x, t; P_{m,1})},$$

and Proposition 6.1 reveals the strict hyperbolicity of the polynomial $P_{m,1}(x, t, \eta + c\varrho, (c_0/\sigma)\varrho)$ in the direction $N + \sigma c/c_0$. We get that $\tilde{P}_{m,\sigma}(y, r, \eta, \varrho) = 0$ has m real roots, which are distinct if $\sigma(y) \neq 0$.

Exploiting Proposition 6.1 once more, and making use of

$$\tilde{P}_{m,1}(y, r, \eta, \varrho) = P_{m,1}(x, t, \eta + \sigma c\varrho, c_0\varrho),$$

it is easy to verify that $\tilde{P}_{m,1}$ is strictly hyperbolic.

We come back to the function v. Our aim is to show that $v \equiv 0$ in $\tilde{\Omega}$, by the aid of Remark 3.1.5 and Gronwall's lemma. But first the coefficients $\tilde{a}_{j,\alpha}$ and the weight σ have to be extended to the whole $\mathbb{R}^n_x \times [0, T]$, $\mathbb{R}^n_x$, respectively, since they are only defined on $\Omega_0 \times (-\infty, T]$ and Ω_0.

We fix some arbitrary $0 < r_0 < r^*$ and will show $v(y, r) = 0$ in $\Omega_0 \times [0, r_0]$. The function $v(y, r)$ vanishes for y close to $\partial\Omega_0$; hence we are allowed to change the coefficients $\tilde{a}_{j,\alpha}$ and the weight σ there. We extend these functions by

$$\tilde{a}_{j,\alpha}(y, t) = (\mathcal{E}_\varepsilon \tilde{a}_{j,\alpha})(y, t), \quad \sigma(y) = (\mathcal{E}_\varepsilon \sigma)(y), \quad y \notin \Omega_0,$$

where $\mathcal{E}_\varepsilon$ is the extension operator of Definition 5.1.1. Condition 3.1 is satisfied everywhere in $\mathbb{R}^n_y \times [0, r_0]$ if ε is small enough.

We get coefficients $a^*_{j,\alpha} \in C^1_b(\mathbb{R}^n \times [0, r_0])$ for $j + |\alpha| = m$ and $a^*_{j,\alpha} \in C^0_b(\mathbb{R}^n \times [0, r_0])$ for $j + |\alpha| < m$. The function v can be extended by zero, and solves

$$D^m_t v + \sum_{j+\alpha=m, j<m} a^*_{j,\alpha}(y,r) D^\alpha_y D^j_r(\sigma(y)^{|\alpha|} v) = 0, \quad (y,r) \in \mathbb{R}^n \times [0, r_0],$$

$$v(y, 0) = \cdots = D^{m-1}_r v(y, 0) = 0, \quad y \in \mathbb{R}^n.$$

Remark 3.1.5 and Gronwall's lemma yield $v(y, r) = 0$ for $y \in \mathbb{R}^n$, $0 \le r \le r_0$. Since $r_0 < r^*$ can be chosen arbitrarily, we have $v = 0$ in $\tilde{\Omega}$, hence $u = 0$ in Ω. The theorem is proved. □

5.3. Existence and Uniqueness for Solutions of Quasi-linear Equations

Theorem 5.3.1 (Local existence). *We suppose* (5.13)–(5.15) *and Condition* 5.1. *Let* $Q_0 = \Pi^n_{i=1}[a_i, b_i]$ *be a rectangular parallelepiped (RP for short),* $Q_0 \Subset \Omega_0$. *Then a constant* $0 < T_0 \le T$ *and a solution* u *of* (5.12) *exist with*

$$D^j_t \sigma^{m-1-j} u \in C([0, T_0], H^{s_c+m-1-j}(Q_0)), \quad j = 0, \ldots, m-1.$$

Proof. We take a cut-off function $\varphi(x)$ which is supported in a neighborhood of Q_0 and is identical to 1 on Q_0. Then we replace the functions $\sigma(x)$, $f(x, t, \{V_{k,\beta}\})$, $\varphi_j(x)$ by $\varphi(x)\sigma(x)$, $\varphi(x) f(x, t, \{V_{k,\beta}\})$ and $\varphi(x)\varphi_j(x)$.

Let Q be an RP with $\operatorname{supp} \varphi \Subset Q$. Using suitable "reflections" (see the proof of Theorem 5.2.1), we can extend the coefficients $a_{j,\alpha}$ from $Q \times [0,T] \times \mathbb{R}^{n_0}$ to the larger set $Q' \times [0,T] \times \mathbb{R}^{n_0}$, Q' being an *RP* with twice the edge lengths of Q which can be regarded as a torus. We get a Cauchy problem on Q'. Theorem 3.2.2 shows that a solution u exists with the desired smoothness on the torus Q'. This function is a solution on $Q_0 \times [0, T_0]$, since $\varphi \equiv 1$ on Q_0. □

So far, nothing has been said about the *uniqueness* of this solution u. This question is taken up in the next theorem.

Theorem 5.3.2 (Local uniqueness). *Let the conditions of Theorem* 5.3.1 *be satisfied. Let* $B_0 \Subset \Omega_0$ *be a sufficiently small ball. Then a number* $T_0 > 0$ *and a cone* Ω *over* B_0 *exist such that a uniquely determined solution* u *with*

$$D^j_t \sigma^{m-1-j} u \in \mathcal{C}([0, T_0], H^{s_c+m-1-j}(\Omega_t)), \quad j = 0, \ldots, m-1,$$

exists in Ω. *The notation* $v \in \mathcal{C}([0,T], H^s(\Omega_t))$ *means that*

- $\sup_{t\in[0,T]} \|v(\cdot,t)\|_{H^s(\Omega_t)} < \infty$,
- *for all* T', Ω'_0 *with* $\Omega'_0 \times [0, T'] \Subset \Omega$ *it holds that* $v \in C([0,T'], H^s(\Omega'))$.

Here $H^s(\Omega_t)$ *denotes the usual Sobolev space on the bounded domain* Ω_t.

Proof. Theorem 5.3.1 shows that a small T_0 and a solution u exist with

$$D^j_t \sigma^{m-1-j} u \in C([0, T_0], H^{s_c+m-1-j}(B_0)), \quad j = 0, \ldots, m-1,$$

$$P^{(u)}_{m,\sigma} u = f^{(u)} \text{ in } B_0 \times [0, T_0], \quad D^j_t u(\cdot, t) = \varphi_j(\cdot) \text{ in } B_0.$$

We define the cone $\Omega = K(B_0)$ over B_0 as given in the Example following Definition 5.1.2 with

$$P_{m,1}(x,t,\tau,\xi) = \tau^m + \sum_{j+|\alpha|=m, j<m} a_{j,\alpha}(x,t,\{D_x^\beta \sigma^{|\beta|} D_t^k u\})\xi^\alpha \tau^j.$$

Then it follows that

$$D_t^j \sigma^{m-1-j} u \in \mathcal{C}([0,T_0], H^{s_c+m-1-j}(K(B_0)_t)), \quad j = 0, \ldots, m-1.$$

Let U^* be another solution in Ω, i.e.,

$$P_{m,\sigma}^{(u^*)} u^* = f^{(u^*)} \text{ in } B_0 \times [0,T_0], \quad D_t^j u^* = \varphi_j \text{ in } B_0.$$

The difference $u - u^*$ then solves

$$P_{m,\sigma}^{(u)}(u-u^*) = \sum_{k+|\beta|\le m-1} g_{k,\beta}(x,t) D_x^\beta (\sigma^{|\beta|} D_t^k (u-u^*))$$

with $g_{k,\beta} \in C(\Omega)$, and Theorem 5.2.1 applied to the homogeneous Cauchy problem reveals $u - u^* \equiv 0$ in Ω. □

Remark 5.3.3. The techniques of Section 5.4 enable us to show that solutions to quasi-linear Cauchy problems are unique not only locally, but also globally in domains of dependence. Since the proof is quite similar to that of Theorem 5.4.2, it is dropped.

If the equation is *linear*, we even have *global* existence:

Corollary 5.3.4 (Global existence). *Let us consider the Cauchy problem* (5.1). *We suppose Condition* 3.1 *and*

$$\sigma \in C^\infty(\Omega_0^*), \quad \Omega_0^* \Supset \Omega_0,$$

$$a_{j,\alpha} \in \begin{cases} C^1([0,T], H^{s_c}(\Omega_0^*)) & : j+|\alpha| = m, \\ C([0,T], H^{s_c}(\Omega_0^*)) & : j+|\alpha| < m, \end{cases} \qquad s_c > \frac{n}{2}+1,$$

$$\varphi_j \in H^{s_c+m-1-j}(\Omega_0^*), \quad f \in C([0,T], H^{s_c}(\Omega_0^*)).$$

Let Ω be a domain of dependence for the operator $P_{m,\sigma}$ over the domain Ω_0. Then a unique solution u exists with

$$D_t^j \sigma^{m-1-j} u \in \mathcal{C}([0,T], H^{s_c+m-1-j}(\Omega_t)).$$

Proof. We pick a cut-off function $\varphi(x)$ that is identical to 1 in a neighborhood of Ω_0 and supported in Ω_0^*. Then a cube Q with $\operatorname{supp}\varphi \Subset Q$ is chosen. We replace the functions $\sigma(x)$, $f(x,t)$, $\varphi_j(x)$ by $\varphi(x)\sigma(x)$, $\varphi(x)f(x,t)$ and $\varphi(x)\varphi_j(x)$. The coefficients $a_{j,\alpha}(\cdot,t)$ are extended from Ω_0^* to $Q \setminus \Omega_0^*$ by the aid of the procedure from the end of the proof of Theorem 5.2.1. We acquire a linear Cauchy problem on a torus. Remark 3.2.4 gives us a solution defined in $Q \times [0,T]$ which has the desired smoothness. Theorem 5.2.1 shows that this solution is the only solution in Ω. □

5.4. C^∞ regularity

First, we show a *local* regularity result in C^∞.

Lemma 5.4.1 (Local C^∞-regularity). *Let u be a function defined in $\Omega^{(u)}$ which is a domain of dependence over Ω_0 for the operator $P^{(u)}_{m,\sigma}$. Assume that u with*

$$D_t^j \sigma^{m-1-j} u \in \mathcal{C}([0,T], H^{s_c+m-1-j}(\Omega_t)), \quad j = 0, \dots, m-1,$$

is a solution of (5.12). The coefficients $a_{j,\alpha}$ and the right-hand side are supposed to be C^∞ with respect to all their arguments. Finally, we assume that

$$u \in C^\infty(\Omega^{(u)} \cap \{(x,t) : t \le t_0\})$$

for some $t_0 \ge 0$. Let $B(x_0, d) \Subset \Omega_{t_0}$ be a ball. Then a number $t_1 = t_1(x_0, t_0, d) > t_0$ exists with

$$u \in C^\infty(B(x_0,d) \times [t_0, t_1]), \quad B(x_0,d) \times [t_0,t_1] \Subset \Omega^{(u)}.$$

Proof. We apply the procedure given in the proof of Theorem 5.3.1 to extend the Cauchy problem from $B(x_0,d) \times [t_0,T]$ to the set $Q' \times [t_0,T]$, with Q' being a torus. We get a quasi-linear weakly hyperbolic Cauchy problem on a torus with C^∞ coefficients and C^∞ data. Theorem 3.2.3 gives us a local C^∞ solution, and Theorem 5.3.2 shows that this solution is unique in some domain of dependence which contains some cylindrical set $B(x_0,d)\times[t_0,t_1]$ with small $t_1 - t_0$. This implies $u \in C^\infty(B(x_0,d) \times [t_0,t_1])$. □

This lemma is the key tool to prove the following theorem:

Theorem 5.4.2 (Global C^∞ regularity). *Let u be a function defined in $\Omega^{(u)}$ which is a domain of dependence over Ω_0 for the operator $P^{(u)}_{m,\sigma}$. We suppose that u with*

$$D_t^j \sigma^{m-1-j} u \in \mathcal{C}([0,T], H^{s_c+m-1-j}(\Omega_t^{(u)})), \quad j = 0, \dots, m-1,$$

is a solution of (5.12). Moreover, we assume $a_{j,\alpha}, f \in C_b^\infty(\Omega_0 \times [0,T] \times \mathbb{R}^{n_0})$, $\varphi_j \in C_b^\infty(\Omega_0)$. Then $u \in C^\infty(\Omega^{(u)})$.

Proof. If $B(x_0,d) \Subset \Omega_0$ is a ball, Lemma 5.4.1 gives us a number $t_1(x_0,d)$ with

$$u \in C^\infty(B(x_0,d) \times [0, t_1(x_0,d)]).$$

This implies that a smooth function $h = h(x) > 0$ and a set M_0 exist with

$$M_0 = \{(x,t) : x \in \Omega_0, 0 \le t < h(x)\}, \quad u \in C^\infty(M).$$

The domain $\Omega^{(u)}$ can be exhausted with hypersurfaces S_r. We transform the variables in the same way as in the proof of Theorem 5.2.1. This results in a quasi-linear weakly hyperbolic initial value problem for the function v,

$$\tilde{P}^{(v)}_{m,\sigma} v = \tilde{f}(y, r, \{D_y^\beta \sigma(y)^{|\beta|} D_r^k v\}), \quad y \in \Omega_0, \quad g(y,0) \le r < r^*,$$
$$v(y, g(y,0)) = \psi_0(y), \dots, D_r^{m-1} v(y, g(y,0)) = \psi_{m-1}(y).$$

Denoting the images of M_0 and $\Omega^{(u)}$ under the transformation of variables by $\tilde{M}_0$ and $\tilde{\Omega}^{(u)}$, we know that v is C^∞ on the sets

$$\tilde{M}_0, \quad \tilde{M}(r_0) = \tilde{\Omega}^{(u)} \cap \{(y,r)\colon y \in \Omega_0, r \leq r_0\},$$

for some $r_0 > 0$. The proof will be complete if we verify two properties:

A: If $v \in C^\infty(\tilde{M}(r))$, then $v \in C^\infty(\tilde{M}(r'))$ for some $r' > r$.

B: If $v \in C^\infty(\tilde{M}(r))$ for all r with $r_0 \leq r < r_1$, then $v \in C^\infty(\tilde{M}(r_1))$.

For **A**: The set $\partial\tilde{M}(r_0) = \tilde{\Omega}^{(u)} \cap \{(y,r)\colon y \in \Omega_0\}$ can be covered by $\tilde{M}_0$ and finitely many balls. For each ball, we apply Lemma 5.4.1.

For **B**: We cover the set $\partial\tilde{M}(r_1)$ by $\tilde{M}_0$ and finitely many balls $B(y_0,d)$. For each such ball, we consider a Cauchy problem with initial data prescribed at $B(y_0,d) \times \{r = r_1 - \varepsilon\}$, $\varepsilon > 0$ small. This Cauchy problem has the solution v, which is locally C^∞. Proposition 3.2.8 and Remark 3.2.10 tell us that the life-span of this solution cannot approach zero for $\varepsilon \to 0$. This shows that v is C^∞ on $B(y_0,d) \times \{r \leq r_1\}$. □

6. Appendix

Here we provide some auxiliary results.

Proposition 6.1 ([3], [4]). *Let the homogeneous differential operator $P(x,D)$ of order m be strictly hyperbolic at the point x_0 in the direction N, $|N| = 1$. By $\lambda_{\max}$ we denote the largest absolute value of the characteristic roots, i.e.,*

$$\lambda_{\max} = \sup\{|\tau| : P(x_0, \tau N + \xi) = 0, |\xi| = 1, \xi \perp N\}.$$

Then P is strictly hyperbolic at x_0 in any direction $N+e$ with $N \perp e$, $|e|^{-1} > \lambda_{\max}$.

The following lemma is a generalization of Gronwall's lemma to differential inequalities with a singular coefficient.

Lemma 6.2 ([24]). *Let $y(t) \in C([0,T]) \cap C^1((0,T))$ be a solution of the differential inequality*

$$y'(t) \leq K(t)y(t) + f(t), \quad 0 < t < T,$$

where the functions $K(t)$ and $f(t)$ belong to $C(0,T)$. We assume for every $t \in (0,T)$ and every $\delta \in (0,t)$ that

$$\int_0^\delta K(\tau)\,d\tau = \infty, \quad \int_\delta^T K(\tau)\,d\tau < \infty,$$

$$\lim_{\delta \to +0} \int_\delta^t \exp\left(\int_s^t K(\tau)\,d\tau\right) f(s)\,ds \quad \textit{exists},$$

$$\lim_{\delta \to +0} y(\delta) \exp\left(\int_s^t K(\tau)\,d\tau\right) = 0.$$

Then it holds that

$$y(t) \leq \int_0^t \exp\left(\int_s^t K(\tau)\,d\tau\right) f(s)\,ds.$$

References

[1] G. Aleksandrian. Parametrix and propagation of the wave front of a solution to a Cauchy problem for a model hyperbolic equation (in Russian). *Izv. Akad. Nauk Arm. SSR*, 19(3):219–232, 1984.

[2] S. Alinhac and G. Metivier. Propagation de l'analyticité des solutions de systèmes hyperboliques non-linéaires. *Inv. math.*, 75:189–204, 1984.

[3] M.F. Atiyah, R. Bott, and L. Gårding. Lacunas for hyperbolic differential operators with constant coefficients. I. *Acta Math.*, 124:109–189, 1970.

[4] M.F. Atiyah, R. Bott, and L. Gårding. Lacunas for hyperbolic differential operators with constant coefficients. II. *Acta Math.*, 131:145–206, 1973.

[5] R. Coifman and Y. Meyer. Commutateurs d'intégrales singulières et opérateurs multilinéaires. *Ann. Inst. Fourier (Grenoble)*, 28(3):177–202, 1978.

[6] F. Colombini, E. Jannelli, and S. Spagnolo. Well-posedness in the Gevrey classes of the Cauchy problem for a non-strictly hyperbolic equation with coefficients depending on time. *Ann. Scuola Norm. Sup. Pisa IV*, 10:291–312, 1983.

[7] F. Colombini and S. Spagnolo. An example of a weakly hyperbolic Cauchy problem not well posed in C^∞. *Acta Math.*, 148:243–253, 1982.

[8] P. D'Ancona. Well-posedness in C^∞ for a weakly hyperbolic second order equation. *Rend. Sem. Mat. Univ. Padova*, 91:65–83, 1994.

[9] P.A. Dionne. Sur les problèmes hyperboliques bien posés. *J. Analyse Math.*, 10:1–90, 1962.

[10] M. Dreher and M. Reissig. About the C^∞-well-posedness of fully nonlinear weakly hyperbolic equations of second order with spatial degeneracy. *Adv. Diff. Eq.*, 2(6):1029–1058, 1997.

[11] M. Dreher and M. Reissig. Local solutions of fully nonlinear weakly hyperbolic differential equations in Sobolev spaces. *Hokk. Math. J.*, 27(2):337–381, 1998.

[12] M. Dreher and M. Reissig. Propagation of mild singularities for semilinear weakly hyperbolic equations. *J. Analyse Math.*, 82:233–266, 2000.

[13] M. Dreher and I. Witt. Edge Sobolev spaces and weakly hyperbolic equations. *Ann. mat. pura et appl.*, 180:451–482, 2002.

[14] L. Gårding. *Cauchy's Problem for Hyperbolic Equations.* University Chicago, 1957.

[15] L. Hörmander. *The Analysis of Linear Partial Differential Operators.* Springer, 1985.

[16] V. Ivrii and V. Petkov. Necessary conditions for the Cauchy problem for non-strictly hyperbolic equations to be well posed. *Russian Math. Surveys*, 29(5):1–70, 1974.

[17] K. Kajitani and K. Yagdjian. Quasi-linear hyperbolic operators with the characteristics of variable multiplicity. *Tsukuba J. Math.*, 22(1):49–85, 1998.

[18] T. Kato and G. Ponce. Commutator estimates and the Euler and Navier–Stokes equations. *Comm. Pure Appl. M.*, 41:891–907, 1988.

[19] P.D. Lax. Asymptotic solutions of oscillatory initial value problems. *Duke Math. J.*, 24(4):627–646, 1957.

[20] J. Leray. *Hyperbolic Differential Equations.* Inst. Adv. Study, Princeton, 1954.

[21] R. Manfrin. Analytic regularity for a class of semi-linear weakly hyperbolic equations of second order. *NoDEA*, 3(3):371–394, 1996.

[22] S. Mizohata. Some remarks on the Cauchy problem. *J. Math. Kyoto Univ.*, 1(1):109–127, 1961.

[23] S. Mizohata. *The Theory of Partial Differential Equations.* Cambridge University Press, 1973.

[24] A. Nersesyan. On a Cauchy problem for degenerate hyperbolic equations of second order (in Russian). *Dokl. Akad. Nauk SSSR*, 166(6):1288–1291, 1966.

[25] O. Oleinik. On the Cauchy problem for weakly hyperbolic equations. *Comm. Pure Appl. M.*, 23:569–586, 1970.

[26] I.G. Petrovskij. On the Cauchy problem for systems of linear partial differential equations. *Bull. Univ. Mosk. Ser. Int. Mat. Mekh.*, 1(7):1–74, 1938.

[27] M.-Y. Qi. On the Cauchy problem for a class of hyperbolic equations with initial data on the parabolic degenerating line. *Acta Math. Sinica*, 8:521–529, 1958.

[28] R. Racke. *Lectures on Nonlinear Evolution Equations. Initial Value Problems.* Vieweg Verlag, Braunschweig et al., 1992.

[29] M. Reissig. Weakly hyperbolic equations with time degeneracy in Sobolev spaces. *Abstract Appl. Anal.*, 2(3,4):239–256, 1997.

[30] M. Reissig and K. Yagdjian. Weakly hyperbolic equations with fast oscillating coefficients. *Osaka J. Math.*, 36(2):437–464, 1999.

[31] K. Taniguchi and Y. Tozaki. A hyperbolic equation with double characteristics which has a solution with branching singularities. *Math. Japonica*, 25(3):279–300, 1980.

[32] S. Tarama. On the second order hyperbolic equations degenerating in the infinite order. – example –. *Math. Japonica*, 42(3):523–533, 1995.

[33] M.E. Taylor. *Pseudodifferential Operators and Nonlinear PDE.* Birkhäuser, Boston, 1991.

[34] M.E. Taylor. *Partial Differential Equations III. Nonlinear Equations.* Springer, 1996.

[35] K. Yagdjian. *The Cauchy Problem for Hyperbolic Operators. Multiple Characteristics, Micro-Local Approach*, volume 12 of *Math. Topics.* Akademie Verlag, Berlin, 1997.

Michael Dreher
Faculty of Mathematics and Computer Science
Freiberg University of Mining and Technology
Agricola-Straße 1
D-09596 Freiberg, Germany
e-mail: dreher@math.tu-freiberg.de

Operator Theory:
Advances and Applications, Vol. 145, 207–248

An Approach to a Version of the $S(M,g)$-pseudo-differential Calculus on Manifolds

F. Baldus

Abstract. For appropriate triples $(\mathcal{M}, g, M)$, where $\mathcal{M}$ is an (in general non-compact) manifold, g is a metric on $T^*\mathcal{M}$, and M is a weight function on $T^*\mathcal{M}$, we develop a pseudo-differential calculus on $\mathcal{M}$ which is based on the $S(M,g)$-calculus of L. Hörmander [27] in local models. In order to do so, we generalize the concept of E. Schrohe [41] of so-called SG-compatible manifolds. In the final section we give an outlook onto topological properties of the algebras of pseudo-differential operators. We state the existence of "order reducing operators" and that the algebra of operators of order zero is a submultiplicative Ψ^*-algebra in the sense of B. Gramsch [18] in $\mathcal{L}\left(L^2(\mathcal{M})\right)$.

1. Introduction

When pseudo-differential operators were introduced in 1965 (cf. J.J. Kohn and L. Nirenberg [29], A. Unterberger and J. Bokobza [51], [52], L. Hörmander [24], [25]), there arose naturally a desire to prove invariance under changes of variables and thus, to define pseudo-differential operators on manifolds. Having developed a pseudo-differential calculus on $\mathbb{R}^n$ or on a manifold, properties of the algebra Ψ^0 of pseudo-differential operators of order zero as a topological algebra came to be of interest. With the help of "order reducing operators", many questions of the full filtered algebra of pseudo-differential operators reduce to this algebra Ψ^0. In a series of articles (cf. R. Beals [4], [5], B. Gramsch [19], H.O. Cordes [12], J. Ueberberg [48], E. Schrohe [43], [44]. B. Gramsch, J. Ueberberg, and K. Wagner [22], J.-M. Bony and J.-Y. Chemin [9], and [1], [2]), spectral invariance of Ψ^0 in the Banach algebra $\mathcal{B}$ of bounded operators on L^2 or, more generally, on L^p- and Orlicz-spaces and on the corresponding (possibly weighted) Sobolev spaces was proved in many situations, i.e.,

$$\Psi^0 \cap \mathcal{B}^{-1} = (\Psi^0)^{-1}$$

2000 *Mathematics Subject Classification.* AMS 1991 Subject Classification: Primary 35S05; 58G15 Secondary 47A60; 47A10; 47D30; 46F05.
Key words and phrases. pseudo-differential operators on manifolds; Weyl-Hörmander calculus; SG-calculus; spectral invariance; submultiplicativity.

holds for the groups of invertible elements. As a matter of fact, in many cases Ψ^0 is a Ψ- or Ψ^*-algebra in $\mathcal{B}$ in the sense of B. Gramsch [18], [19], i.e., a continuously embedded Fréchet subalgebra of $\mathcal{B}$ with the same unit element, such that for the groups of invertible elements the above equation holds, and which is symmetric in the "$*$-case". In addition to implying an open group of invertible elements and continuous inversion, the properties of a Ψ-algebra guarantee stability with respect to the holomorphic functional calculus of L. Waelbroeck [53]. In addition to [19], a rather detailed discussion of the concept of Ψ-algebras can be found in [31]. In connection with non-commutative cohomology and Oka's principle, submultiplicativity of the algebra of operators of order zero is of importance (cf. B. Gramsch [20] and B. Gramsch, W. Kaballo [21]).

Originally, the calculus of pseudo-differential operators on manifolds was based on symbols which satisfy local estimates, i.e., estimates which are only uniform for x in a compact set (cf. [51], [52], [24], [25]). On $\mathbb{R}^n$, symbols satisfying global estimates were considered (cf. L. Hörmander [25], M.A. Shubin [45], C. Parenti [36], H. Kumano-go and K. Taniguchi [30], R. Beals and C. Fefferman [6], R. Beals [3], H.O. Cordes [11], [13], B. Helffer [23]). In connection with the Weyl calculus, L. Hörmander [27] (cf. also [28, Section 18.5]) presented a vast extension of these classes: To a metric g on $T^*\mathbb{R}^n$ and a weight function M on $T^*\mathbb{R}^n$ he associated the symbol class $S(M,g)$ and developed the corresponding pseudo-differential calculus. For essential contributions to the $S(M,g)$-calculus we wish also to refer to A. Unterberger [49], [50], R. Beals [5], N. Dencker [15], J.-M. Bony and N. Lerner [10], J.-M. Bony and J.-Y. Chemin [9], J.-M. Bony [8].These data g and M also make sense on the cotangent bundle of a manifold but, for a generalization of the $S(M,g)$-calculus, we have to make assumptions concerning both the manifold, expressed by the existence of a suitable atlas, and the data M and g. In [27] L. Hörmander investigated already the behavior of operators associated to metrics not too far from the standard ρ,δ-metric under certain canonical transformations, introduced Fourier integral operators, and proved a generalization of Egorov's theorem.

In the present paper we generalize the concept of E. Schrohe (cf. [40], [42], [44]) of so-called SG-compatible manifolds which are appropriate in order to transfer the SG-calculus of M.A. Shubin [45], C. Parenti [36], and H.O. Cordes [11]. In so doing, we develop an $S(M,g)$-based pseudo-differential calculus on manifolds $\mathcal{M}$, at least for appropriate combinations of $(\mathcal{M},g,M)$. We introduce conditions expressed by the existence of appropriate local models and compatible changes of coordinates (cf. Section 2). Associated to M and g, we define – up to negligible terms – pseudo-differential operators on $\mathcal{M}$ as operators $\mathcal{C}_0^\infty(\mathcal{M}) \to \mathcal{D}'(\mathcal{M})$ possessing representations as pseudo-differential operators in local coordinates (in the corresponding classes). For the study of transformations under diffeomorphisms we apply the proof of Kuranishi (cf. [17, Appendix of Chapter I], [35, p. 155], [26, Proposition 2.1.3]). Concerning the negligible terms we assume a behavior based on characterizations with the help of mapping properties of iterated commutators with distinguished operators in a scale of Sobolev spaces corresponding to g as in

[1, Theorem 3.6.6]. More precisely, we define the negligible class as the set of all operators such that these iterated commutators have order $-\infty$ in this scale of Sobolev spaces. Moreover, for an appropriate choice of cut-off functions, we have representations of the negligible terms as integral operators with rapidly decreasing kernels.

H. Widom [54] constructed a complete symbolic calculus associated to pseudo-differential operators on compact manifolds, defined with the help of a connection. Yu. Safarov [39] proceeded in this direction to define ρ, δ-type classes of pseudo-differential operators without the restriction $1-\rho \leq \delta$ which is known to be necessary in order to define pseudo-differential operators through their representations in local coordinates. For different approaches to pseudo-differential operators on non-compact manifolds we wish also to refer to H.O. Cordes and S.H. Doong [14], R. Beals and P. Greiner [7] and M.A. Shubin [47]. In [37], [38], B. A. Plamenevskij and V.N. Senichkin develop an approach to a pseudo-differential calculus on stratified manifolds and study irreducible representations of the C*-algebra generated by pseudo-differential operators of order zero. The stratified manifolds are locally represented by "models" located in $\mathbb{R}^n$ and "admissible diffeomorphisms" between the local representations. The SG-calculus is closely related to the scattering calculus of R.B. Melrose (cf. [33, Section 4] and [34, Section 6.3]). Moreover, Y.V. Egorov and B.-W. Schulze apply SG-calculus to define a pseudo-differential calculus on the infinite stretched cone $\mathbb{R}_+ \times X$ over a closed compact $\mathcal{C}^\infty$-manifold X (cf. [16, Section 8.2]). For a groupoid approach to spectral questions on non-compact Riemannian manifolds we refer to R. Lauter and V. Nistor [32].

Under more restrictive assumptions, spectral invariance and submultiplicativity can be proved. Common tools for these two topics are commutator methods for the characterization of pseudo-differential operators. In order to derive spectral invariance and submultiplicativity, the existence of "order reducing operators" is of essential importance. We state these results without proof in Section 8.

Instead of explaining the organization of this paper at this point, we refer to brief introductions at the beginning of each section.

This paper is part of the author's Ph. D. thesis [2] written under the supervision of Professor B. Gramsch at the University of Mainz, Germany.

2. Admissible coordinates

There is a natural way to generalize g and M and thus, $S(M,g)$ to cotangent bundles over manifolds. But, on the other hand, it is well known that the $S^m_{\rho,\delta}$-classes are invariant under (local) diffeomorphisms – and thus can be considered on the cotangent bundles of manifolds – if and only if $0 \leq 1-\rho \leq \delta \leq \rho \leq 1$, $\delta < 1$. Therefore, for an extension of the $S(M,g)$-calculus, we have to make assumptions concerning both the manifold, expressed by the existence of a suitable atlas, and the data M and g. In [41] E. Schrohe transfers a calculus of M.A. Shubin [45], C. Parenti [36], and H.O. Cordes [11], which is based on global estimates, to so-called SG-compatible manifolds (cf. also [40], [42], [44]). The local charts are assumed to satisfy conditions which allow us to prove invariance of the pseudo-differential calculus with respect to the changes of coordinates. Here we generalize these conditions in order to treat classes of operators associated to more general calculi on $\mathbb{R}^n$. For instance, we can transfer the general Cordes' classes corresponding to metrics

$$g_{(x,\xi)} := \frac{<\xi>^{2\delta}}{<x>^{2\rho_1}} \|dx\|_2^2 + \frac{\|d\xi\|_2^2}{<\xi>^{2\rho_2}}, \quad 0 \leq \rho_1, \rho_2 \leq 1, 1-\rho_2 \leq \delta \leq \rho_2, \delta < 1$$

and classes corresponding to metrics

$$g_{(x,\xi)} := <x,\xi>^{2\delta} \|dx\|_2^2 + \frac{\|d\xi\|_2^2}{<x,\xi>^{2\rho}}, \quad 0 \leq \rho \leq 1, 1-\rho \leq \delta \leq \rho, \delta < 1$$

to SG-compatible manifolds (cf. 7). In contrast to most of the already existing calculi, our approach allows us also to patch different local models together, provided that we have compatible changes of coordinates (cf. 7.1). Moreover, we can admit anisotropic versions of the above examples. In order to treat them, we make the following definition:

Definition 2.1. Let $\Gamma := (\gamma_\alpha)_{\alpha \in A}$ be a family of positive definite quadratic forms on $\mathbb{R}^m$.

1. We call a decomposition $\mathbb{R}^m = E_1 \oplus^\perp \cdots \oplus^\perp E_l$ Γ-*admissible* if
   - $\forall j \in \{1, \ldots, l\}$, $\exists C_j > 0$, $\forall t_1, t_2 \in E_j \setminus \{0\}$, $\forall \alpha \in A$, $\frac{\gamma_\alpha(t_1)}{\|t_1\|_2^2} \cdot \frac{\|t_2\|_2^2}{\gamma_\alpha(t_2)} \leq C_j$.
   - $\exists C > 0$, $\forall t_1 \in E_1, \ldots, \forall t_l \in E_l$, $\forall \alpha \in A$, $\sum_{j=1}^{l} \gamma_\alpha(t_j) \leq C\gamma_\alpha(t_1 + \cdots + t_l)$.
2. We call Γ *quasi-decomposable* if there exists a Γ-admissible decomposition.
3. We call an orthonormal (with respect to the Euclidean metric) basis $\mathcal{B} = \{t_1, \ldots, t_m\}$ Γ-*admissible* if there exists a Γ-admissible decomposition $\mathbb{R}^m = E_1 \oplus^\perp \cdots \oplus^\perp E_l$ and for each $j \in \{1, \ldots, m\}$ a constant $\nu_j \in \{1, \ldots, l\}$ such that $t_j \in E_{\nu_j}$.

Remark 2.2.

1. Let $\Gamma := (\gamma_\alpha)_{\alpha \in A}$, $\tilde{\Gamma} := (\tilde{\gamma}_\alpha)_{\alpha \in A}$ be two families of positive definite quadratic forms on $\mathbb{R}^m$ such that with a suitable constant $C > 0$ for all $\alpha \in A$ the estimate $C^{-1}\gamma_\alpha(\cdot) \leq \tilde{\gamma}_\alpha(\cdot) \leq C\gamma_\alpha(\cdot)$ holds. Then Γ is quasi-decomposable if and only if $\tilde{\Gamma}$ is quasi-decomposable, and $\mathbb{R}^m = E_1 \oplus^\perp \cdots \oplus^\perp E_l$ is a Γ-admissible decomposition if and only if it is a $\tilde{\Gamma}$-admissible decomposition.
2. Let $\Gamma := (\gamma_\alpha)_{\alpha \in A}$ be a family of positive definite quadratic forms on $\mathbb{R}^m$ and let $\mathbb{R}^m = E_1 \oplus^\perp \cdots \oplus^\perp E_l$ be a decomposition such that for all $j \in \{1, \ldots, l\}$ and all $\alpha \in A$, E_j is contained in an eigenspace of γ_α. Then this decomposition is Γ-admissible.

3. Let $\Gamma := (\gamma_\alpha)_{\alpha\in A}$ be a quasi-decomposable family of positive definite quadratic forms on $\mathbb{R}^m$ and let $\mathbb{R}^m = E_1 \oplus^\perp \cdots \oplus^\perp E_l$ be a Γ-admissible decomposition. For $j \in \{1,\dots,l\}$ let $\widehat{t_j} \in E_j$ with $\|\widehat{t_j}\|_2 = 1$. For $\alpha \in A$, we define $\widehat{\gamma}_\alpha(t_1 + \cdots + t_l) := \sum_{j=1}^{l} \gamma_\alpha(\widehat{t_j}) \|t_j\|_2^2$ ($t_j \in E_j, j = 1,\dots,l$). Then, with a suitable constant $C > 0$, for all $\alpha \in A$ we have $C^{-1}\,\widehat{\gamma}_\alpha \le \gamma_\alpha \le C\,\widehat{\gamma}_\alpha$.

Lemma 2.3. *Let $\Gamma := (\gamma_\alpha)_{\alpha\in A}$ be a quasi-decomposable family of positive definite quadratic forms on $\mathbb{R}^m$. Then there exists a unique coarsest Γ-admissible decomposition.*

Next we introduce canonical weight functions which play an important role in the following:

Definition 2.4. Let $\mathcal{U}$ be an open subset of $\mathbb{R}^n$ and g a metric on $\mathcal{U} \times \mathbb{R}^n$.

1. For $x \in \mathcal{U}$ and $\xi \in \mathbb{R}^n$ let $h_g(x,\xi) := \left(\sup_{T\in\mathbb{R}^{2n}\setminus\{0\}} \frac{g_{(x,\xi)}(T)}{g^\sigma_{(x,\xi)}(T)}\right)^{1/2}$ where $g^\sigma_{(x,\xi)}(T) := \sup_{W\in\mathbb{R}^{2n}\setminus\{0\}} \frac{\sigma(W,T)^2}{g_{(x,\xi)}(W)}$.
2. Let $\mathcal{B}$ be a basis of $\mathbb{R}^n$. For $x \in \mathcal{U}$ and $\xi \in \mathbb{R}^n$ we define
$$H_{g,\mathcal{B}}(x,\xi) := \max_{t\in\mathcal{B}} \sqrt{g_{(x,0)}(t,0) g_{(x,\xi)}(0,t)}\ .$$

Now we are prepared to define admissible local models and metrics on these models which are suitable in order to generalize the corresponding calculus on $\mathbb{R}^n$ to manifolds. Apart from some technical conditions concerning the metric, e.g., a generalization of the assumption $1-\rho \le \delta$ (cf. 2.6), we require that the diffeomorphisms have differentials which are matrix-valued symbols in the calculus and that they extend to larger domains which are chosen in such a way that the existence of admissible cut-off functions and partitions of unity can be proved (cf. 4.15). The conditions concerning the manifold $\mathcal{M}$ with atlas $\mathcal{A}$ depend on two metrics g and $\mathcal{G}$ on $T^*\mathcal{M}$ and $\mathcal{M}$, respectively, which determine the pseudo-differential calculus to be defined in Section 5. In the SG-calculus we have $\mathcal{G} = \mathcal{R}_1(g)$, but, as the examples in Section 7 show, in some cases it is reasonable to consider $\mathcal{G} \ne \mathcal{R}_1(g)$. Consequently, we describe admissible (in order to develop a pseudo-differential calculus) structures by the data $(\mathcal{M},\mathcal{A},\mathcal{G},g)$ (cf. 4.8 and 4.10). Moreover, in 2.6 we introduce conditions on weight functions and diffeomorphisms.

Definition 2.5.

1. Let $\mathcal{U}, U$ be open subsets of $\mathbb{R}^n$ with $\overline{U} \subseteq \mathcal{U}$, and let $\mathcal{G}$ be a smooth metric on $\mathcal{U}$. We call $(\mathcal{U}, U, \mathcal{G})$ an *admissible model* if the following conditions are satisfied:
   - $\exists r_\mathcal{G} > 0,\ \forall x \in \overline{U},\ \forall y \in \mathbb{R}^n,\ \mathcal{G}_x(x-y) \le r_\mathcal{G}^2 \implies y \in \mathcal{U}$.
   - $\exists C_\mathcal{G} > 0,\ \forall x,y \in \mathcal{U},\ \mathcal{G}_x(x-y) \le C_\mathcal{G}^{-1} \implies C_\mathcal{G}^{-1}\mathcal{G}_x(\cdot) \le \mathcal{G}_y(\cdot) \le C_\mathcal{G}\mathcal{G}_x(\cdot)$.
   - The family of forms $(\mathcal{G}_x)_{x\in\mathcal{U}}$ is quasi-decomposable.
2. Let $\mathcal{U}_1$ be an open subset of $\mathbb{R}^n$, let $(\mathcal{U}_2, U_2, \mathcal{G}_2)$ be an admissible model, and let $\kappa : \mathcal{U}_1 \to \mathcal{U}_2$ be a C^∞-diffeomorphism. κ is called $(\mathcal{U}_2, U_2, \mathcal{G}_2)$*-compatible* if the following statements hold:

- $(\mathcal{U}_1, U_1, \mathcal{G}_1)$ is an admissible model, where $U_1 := \kappa^{-1}(U_2)$ and $\mathcal{G}_1 := \kappa^* \mathcal{G}_2$, i.e., $\mathcal{G}_{1,x}(t) = \mathcal{G}_{2,\kappa(x)}(\kappa'(x)(t))$.
- $[\mathcal{U}_1 \ni x \mapsto d\kappa(x)] \in S(1, \mathcal{G}_1; \mathcal{L}(\mathbb{R}^n))$ and[1] $[\mathcal{U}_2 \ni y \mapsto d(\kappa^{-1})(y)] \in S(1, \mathcal{G}_2; \mathcal{L}(\mathbb{R}^n))$.
- $\exists C > 0,\ \forall x \in \mathcal{U}_1,\ \forall y \in \mathcal{U}_2,\quad \|\kappa(x)\|_2 \leq C <x>,\ \|\kappa^{-1}(y)\|_2 \leq C<y>$.
- For $i = 1, 2$ there exists a $(\mathcal{G}_{i,x})_{x\in\mathcal{U}_i}$-admissible decomposition $\mathbb{R}^n = E_1^{(i)} \oplus^\perp \cdots \oplus^\perp E_l^{(i)}$ such that we have $E_j^{(1)} = d\kappa(x)^{-1}\left(E_j^{(2)}\right)$ for all $x \in \mathcal{U}_1$ and $j \in \{1, \ldots, l\}$.

Definition 2.6.

1. We call a split Hörmander metric on $\mathbb{R}^{2n}$ *of type* (s) if there exists a constant $C > 0$ such that $\forall x, \xi, t, \tau \in \mathbb{R}^n$,
$$g_{(x,0)}(t, 0) \leq C g_{(x,\xi)}(t, 0)\ ,\ g_{(x,0)}(0, \tau) \geq C^{-1} g_{(x,\xi)}(0, \tau)$$
and there exists a basis of $\mathbb{R}^n$ which is orthogonal with respect to $g^{(1)}_{(x,\xi)}$ and $g^{(2)}_{(x,\xi)}$ for all $(x, \xi) \in \mathbb{R}^{2n}$, where $g^{(1)}_{(x,\xi)}(t) := g_{(x,\xi)}(t, 0)$ and $g^{(2)}_{(x,\xi)}(\tau) := g_{(x,\xi)}(0, \tau)$.
2. Let $(\mathcal{U}, U, \mathcal{G})$ be an admissible model and let g be a smooth metric on $\mathcal{U} \times \mathbb{R}^n$. We call $(\mathcal{U}, U, \mathcal{G}, g)$ an admissible quadruple of *type* (p), if the following conditions are satisfied: Concerning both $\mathcal{G}$ and g, we suppose that the following conditions hold:
   - For all $x \in \mathcal{U}$ we have $\mathcal{G}_x(\cdot) \leq \mathcal{R}_1(g)_x(\cdot)$, where $\mathcal{R}_1(g)_x(t) := g_{(x,0)}(t, 0)$. This definition generalizes naturally to metrics on the cotangent bundle over a manifold.
   - $\exists C > 0,\ \forall x \in \mathcal{U},\ \forall t, \xi \in \mathbb{R}^n,\quad \mathcal{G}_x(t)\ g_{(x,\xi)}(0, \xi) \leq C g_{(x,\xi)}(t, 0)$.
   - The family of forms $\left(\mathcal{G}_x, g^{(1)}_{(x,\xi)}, g^{(2)}_{(x,\xi)}\right)_{x\in\mathcal{U},\xi\in\mathbb{R}^n}$ is quasi-decomposable.

   We continue with conditions concerning g:
   - There exists a Hörmander metric g_s of type (s) on $\mathbb{R}^{2n}$ such that with a suitable constant $C_s > 0$ for all $x \in \mathcal{U}$ and all $\xi \in \mathbb{R}^n$ the estimate $C_s^{-1} g_{s,(x,\xi)}(\cdot) \leq g_{(x,\xi)}(\cdot) \leq C_s g_{s,(x,\xi)}(\cdot)$ holds.
   - $\exists C_g > 0,\ \forall x, y \in \mathcal{U}, \forall \xi \in \mathbb{R}^n,\ g_{(x,0)}(x - y, 0) \leq C_g^{-1} \implies C_g^{-1} g_{(x,\xi)}(\cdot) \leq g_{(y,\xi)}(\cdot) \leq C_g g_{(x,\xi)}(\cdot)$.
   - $\forall K > 0,\ \exists C_K > 0, \forall x \in \mathcal{U}, \forall \xi \in \mathbb{R}^n, \forall A \in \mathcal{L}(\mathbb{R}^n)^{-1}$ such that $\|A\|_{\mathcal{L}(\mathbb{R}^n)} \leq K$ and $\|A^{-1}\|_{\mathcal{L}(\mathbb{R}^n)} \leq K$, $C_K^{-1} g_{(x,\xi)}(\cdot) \leq g_{(x,A\xi)}(\cdot) \leq C_K g_{(x,\xi)}(\cdot)$.
   - $\exists C > 0,\ \forall x \in \mathcal{U},\ \forall t \in \mathbb{R}^n,\quad \frac{\|t\|_2^2}{<x>^2} \leq C g_{(x,0)}(t, 0)$.

[1]We recall that, for a metric g on an open set $\mathcal{U}$, a weight function M on $\mathcal{U}$ and a Banach space E, $S(M, g; E)$ is the space of all $\mathcal{C}^\infty$-functions such that for all $l \in \mathbb{N}_0$, $\sup \frac{\left\|\partial_{T_1} \ldots \partial_{T_l} a(X)\right\|_E}{M(X) \prod_{j=1}^l \sqrt{g_X(T_j)}} < \infty$, where the supremum is taken over all $X \in \mathcal{U}$ and all $T_1, \ldots, T_l \in \mathbb{R}^{2n} \setminus \{0\}$.

Moreover, with a g-admissible basis $\mathcal{B}$ the following conditions are satisfied:

- $\exists C>0,\ \exists N \in \mathbb{N},\ \forall x \in \mathcal{U},\ \forall \xi \in \mathbb{R}^n, \quad <\xi> \le C H_{g,\mathcal{B}}(x,\xi)^{-N}$.
- $\forall T \in \mathbb{R}^{2n},\ \exists C>0,\ \exists N \in \mathbb{N},\ \forall x \in \mathcal{U},\ \forall \xi \in \mathbb{R}^n, \quad C^{-1} H_{g,\mathcal{B}}(x,\xi)^N \le g_{(x,\xi)}(T) \le C H_{g,\mathcal{B}}(x,\xi)^{-N}$.

Finally, with a $g,\mathcal{G}$-admissible basis the following condition holds:

$$\exists C>0,\ \exists N \in \mathbb{N},\ \forall x \in \mathcal{U} \setminus U,\ \forall \xi \in \mathbb{R}^n,$$
$$<x> \le C \left(\max_{t \in \mathcal{B}} \sqrt{\mathcal{G}_x(t)}\ H_{g,\mathcal{B}}(x,\xi)^N \right)^{-1} .$$

3. Let $(\mathcal{U},U,\mathcal{G},g)$ be an admissible quadruple of type (p) and let $M : \mathcal{U} \times \mathbb{R}^n \to (0,\infty)$ be a continuous function. We call M a *$(\mathcal{U},U,g)$-admissible weight function of type* (p), if
   - for each split Hörmander metric g_s as in 2.6 there exists a g_s-admissible weight function $\tilde{M} : \mathbb{R}^{2n} \to (0,\infty)$ and a constant $\tilde{C}>0$ such that for all $(x,\xi) \in \mathcal{U} \times \mathbb{R}^n$ we have $\tilde{C}^{-1}\tilde{M}(x,\xi) \le M(x,\xi) \le \tilde{C}\tilde{M}(x,\xi)$.
   - $\forall K>0,\ \exists C_K>0, \forall x \in \mathcal{U}, \forall \xi \in \mathbb{R}^n, \forall A \in \mathcal{L}(\mathbb{R}^n)^{-1}$ such that $\|A\|_{\mathcal{L}(\mathbb{R}^n)} \le K$ and $\|A^{-1}\|_{\mathcal{L}(\mathbb{R}^n)} \le K, \quad C_K^{-1} M(x,\xi) \le M(x,A\xi) \le C_K M(x,\xi)$.
   - $\forall x,y \in \mathcal{U},\ \forall \xi \in \mathbb{R}^n,\ g_{(x,0)}(x-y,0) \le C_g^{-1} \implies C_M^{-1} M(x,\xi) \le M(y,\xi) \le C_M M(x,\xi)$.
4. Let $\mathcal{U}_1$ be an open subset of $\mathbb{R}^n$, let $(\mathcal{U}_2,U_2,\mathcal{G}_2,g_2)$ be an admissible quadruple of type (p), and let $\kappa : \mathcal{U}_1 \to \mathcal{U}_2$ be a $(\mathcal{U}_2,U_2,\mathcal{G}_2)$-compatible diffeomorphism. κ is called *$(\mathcal{U}_2,U_2,\mathcal{G}_2,g_2)$-compatible of type* (p) if the following statements hold:
   - $(\mathcal{U}_1,U_1,\mathcal{G}_1,g_1)$ is an admissible quadruple of type (p), where $U_1 := \kappa^{-1}(U_2)$, $\mathcal{G}_1 := \kappa^*\mathcal{G}_2$, $g_1 := \chi^* g_2$, $\chi : \mathcal{U}_1 \times \mathbb{R}^n \to \mathcal{U}_2 \times \mathbb{R}^n$, $(x,\xi) \mapsto \left(\kappa(x), {}^t d\kappa(x)^{-1}\xi\right)$.
   - For $i=1,2$ there exists a $(\mathcal{G}_{i,x}, g^{(1)}_{i,(x,\xi)}, g^{(2)}_{i,(x,\xi)})_{(x,\xi)\in \mathcal{U}_i \times \mathbb{R}^n}$-admissible decomposition $\mathbb{R}^n = E_1^{(i)} \oplus^\perp \cdots \oplus^\perp E_l^{(i)}$ such that we have $E_j^{(1)} = d\kappa(x)^{-1}\left(E_j^{(2)}\right)$ for all $x \in \mathcal{U}_1$ and $j \in \{1,\dots,l\}$.

   If moreover M_2 is a $(\mathcal{U}_2,U_2,g_2)$-admissible weight function of type (p) and $M_1 := \chi^* M_2$ is a $(\mathcal{U}_1,U_1,g_1)$-admissible weight function of type (p), then we call κ $(\mathcal{U}_2,U_2,\mathcal{G}_2,g_2,M_2)$-compatible *of type* (p).

Remark 2.7.

1. Let $(\mathcal{U},U,\mathcal{G})$ be an admissible model. In the following we will suppose that $C_\mathcal{G} \ge 1$ and $r_\mathcal{G} \le 1/\sqrt{C_\mathcal{G}}$. If $(\mathcal{U},U,\mathcal{G},g)$ is an admissible quadruple of type (p), we suppose that $C_g \ge C_\mathcal{G}$, C_g is chosen such that for $x,y \in \mathcal{U}$ and $\xi,\eta \in \mathbb{R}^n$,

$$g_{(x,\xi)}(x-y,\xi-\eta) \le C_g^{-1} \implies C_g^{-1} g_{(x,\xi)}(\cdot) \le g_{(y,\eta)}(\cdot) \le C_g g_{(x,\xi)}(\cdot)\ ,$$

   and $r_g \le \min\{C_g^{-1/2}, r_\mathcal{G}\}$.

2. Let $(\mathcal{U}, U, \mathcal{G}, g)$ be an admissible quadruple of type (p). We call an orthonormal (with respect to the Euclidean metric) basis $\mathcal{B}$ of $\mathbb{R}^n$ *g-admissible*, respectively, *$g, \mathcal{G}$-admissible*, if $\mathcal{B}$ is $\left(g^{(1)}_{(x,\xi)}, g^{(2)}_{(x,\xi)}\right)_{(x,\xi)\in\mathcal{U}\times\mathbb{R}^n}$-admissible, respectively, $\left(g^{(1)}_{(x,\xi)}, g^{(2)}_{(x,\xi)}, \mathcal{G}_x\right)_{(x,\xi)\in\mathcal{U}\times\mathbb{R}^n}$-admissible.
3. Let $\mathcal{U}_1 \subseteq \mathbb{R}^n$ be an open set, let $(\mathcal{U}_2, U_2, \mathcal{G}_2, g_2)$ be an admissible quadruple of type (p), and let $\kappa : \mathcal{U}_1 \to \mathcal{U}_2$ be a $(\mathcal{U}_2, U_2, \mathcal{G}_2)$-compatible diffeomorphism. For $i = 1, 2$ let $\mathbb{R}^n = E^{(i)}_1 \oplus^\perp \cdots \oplus^\perp E^{(i)}_l$ be a $(\mathcal{G}_{i,x})_{x\in\mathcal{U}_i}$-admissible decomposition. Then there exists an orthogonal map $\mathcal{T} : \mathbb{R}^n \to \mathbb{R}^n$ such that $\mathcal{T}\left(E^{(1)}_i\right) = E^{(2)}_j$ $(j = 1, \ldots, l)$. If κ is $(\mathcal{U}_2, U_2, \mathcal{G}_2, g_2)$-compatible of type (p), then we choose the decompositions $\mathbb{R}^n = E^{(i)}_1 \oplus^\perp \cdots \oplus^\perp E^{(i)}_l$ $\left(\mathcal{G}_{i,x}, g^{(1)}_{i,(x,\xi)}, g^{(2)}_{i,(x,\xi)}\right)_{(x,\xi)\in\mathcal{U}_i\times\mathbb{R}^n}$-admissible $(i = 1, 2)$.
4. $\mathcal{B}_2$ can be chosen as a $\mathcal{G}_2$-admissible, respectively, $g_2, \mathcal{G}_2$-admissible basis such that $\mathcal{B}_1 := \mathcal{T}^{-1}(\mathcal{B}_2)$ is $\mathcal{G}_1$-admissible, respectively, $g_1, \mathcal{G}_1$-admissible, too.

3. Transformation of pseudo-differential operators

Now we investigate the behavior of pseudo-differential operators under consideration when changing coordinates. In so doing, we make use of Kuranishi's method (cf. [17, Appendix of Chapter I], [35, p. 155], [26, Proposition 2.1.3]). More precisely, as usual, we split a general pseudo-differential operator into an operator with Schwartz kernel being supported in a neighborhood of the diagonal and a remainder term. As far as the examination of the first term is concerned, applying approximation arguments (cf. 3.12) we only have to apply Kuranishi's method to pseudo-differential operators associated to amplitudes in $C^\infty_0(\mathbb{R}^{3n}_{x,\xi,y})$ with Schwartz kernels being supported near the diagonal (cf. 3.11). Afterwards we must be able to "reduce amplitudes into symbols" within the classes under consideration (cf. 3.8). For this reduction of an amplitude $a(x, \xi, y)$ to a symbol $b(x, \xi)$ we freeze the x-variable in a and consider $a(x, \cdot, \cdot\cdot)$ as a "right symbol" which can be transformed into a "left symbol" (cf. [28, Theorem 18.5.10]; cf. also 3.10).

Definition 3.1. Let $(\mathcal{U}, U, \mathcal{G}, g)$ be an admissible quadruple of type (p). For $z \in \mathcal{U}$ and $\rho > 0$ let

$$U_{z,\rho,\mathcal{G}} := \left\{x \in \mathbb{R}^n \mid \mathcal{G}_z(x - z) \leq \rho^2\right\} .$$

For $0 < \rho \leq r_\mathcal{G}$ let $U_{\rho,\mathcal{G}} := \bigcup_{z\in\overline{U}} U_{z,\rho,\mathcal{G}}$ $(\subseteq \mathcal{U})$. For abbreviation, we will write $U_{z,\rho}$ and U_ρ in case no confusion can arise. Moreover, if $\mathcal{G} = \mathcal{R}_1(g)$, then we let $U_{z,\rho,g} := U_{z,\rho,\mathcal{G}}$, $U_{\rho,g} := U_{\rho,\mathcal{G}}$.

Theorem 3.2. *Let $(\mathcal{U}, U, \mathcal{G}, g)$ be an admissible quadruple of type (p). Let $0 < \rho_1 < \rho_2 < \rho_3 < \rho_4 \leq r_\mathcal{G}$ and $\rho > 0$ such that $\rho_i + 3C_\mathcal{G}\ \rho \leq \rho_{i+1}$, $i = 1, 2, 3$. Then, for each $z \in U_{\rho_2}$, there exist functions $\varphi_z \in C^\infty_0(U_{z,\rho})$ and $\psi_z \in C^\infty_0(U_{z,2\rho})$ with the following properties:*

1. *The families* $(\varphi_z)_{z\in U_{\rho_2}}$ *and* $(\psi_z)_{z\in U_{\rho_2}}$ *are bounded in* $S(1,\mathcal{G})$.
2. *For* $z\in U_{\rho_2}$ *we have supp* $(\varphi_z)\subseteq U_{\rho_3}$ $(\subseteq\mathcal{U})$ *and supp* $(\psi_z)\subseteq U_{\rho_3}$ $(\subseteq\mathcal{U})$.
3. *For* $x\in\mathbb{R}^n$ *we define*[2] $\varphi(x):=\int_{U_{\rho_2}}\varphi_z(x)\,|\mathcal{G}_z|^{1/2}dz$. *Then we have* $0\le\varphi\le1$, *supp* $(\varphi)\subseteq U_{\rho_3}$, *and for* $x\in U_{\rho_1}$ *we have* $\varphi(x)=1$.
4. *For* $z\in U_{\rho_2}$ *and* $x\in$ *supp* (φ_z) *we have* $\psi_z(x)=1$, *i.e.,* $\psi_z\cdot\varphi_z=\varphi_z$.
5. *The functions* $U_{\rho_2}\ni z\mapsto\varphi_z\in S(1,\mathcal{G})$ *and* $U_{\rho_2}\ni z\mapsto\psi_z\in S(1,\mathcal{G})$ *are continuous.*

Proof. Let $\varphi\in C_0^\infty([-1,1])$ such that $0\le\varphi_0\le1$ and $\varphi_0(s)=1$ for $s\in[0,1/4]$. For $z\in\mathcal{U}$ we define

$$\varphi_{z,\rho}(x):=\varphi_0\left(\frac{\mathcal{G}_z(x-z)}{\rho^2}\right)\ ,\ \psi_{z,\rho}(x):=\varphi_0\left(\frac{\mathcal{G}_z(x-z)}{4\rho^2}\right)\ .$$

Moreover, let $\Gamma(x):=\int_{\mathcal{U}}\varphi_{z,\rho}(x)\ |\mathcal{G}_z|^{1/2}dz$ and finally, for $x\in\mathcal{U}$ and $z\in U_{\rho_2}$,

$$\varphi_z(x):=\frac{\varphi_{z,\rho}(x)}{\Gamma(x)}\ ,\ \psi_z(x):=\psi_{z,\rho}(x)\ .$$

□

Definition 3.3. Let $\mathcal{U}$ be an open subset of $\mathbb{R}^n$ and let g be a metric on $\mathcal{U}\times\mathbb{R}^n$. Moreover, let $M:\mathcal{U}\times\mathbb{R}^n\to(0,\infty)$. We define the *space of amplitudes* by

$$A(M,g):=S\left(\mathcal{D}(M),\mathcal{D}(g)\right)\quad\text{and}\quad\|a\|_{k,A(M,g)}:=\|a\|_{k,S(\mathcal{D}(g),\mathcal{D}(M))}\ (k\in\mathbb{N}_0)$$

where

$$\begin{aligned}\mathcal{D}(g)_{(x_1,\xi,x_2)}(t_1,\tau,t_2)\ &:=\ g_{(x_1,\xi)}(t_1,0)+g_{(x_2,\xi)}(0,\tau)+g_{(x_2,\xi)}(t_2,0)\ ,\\ \mathcal{D}(M)(x_1,\xi,x_2)\ &:=\ M(x_2,\xi)\end{aligned}$$

on $\mathcal{U}\times\mathbb{R}^n\times\mathcal{U}$.

Lemma 3.4. *Let* $\mathcal{U}_1\subseteq\mathbb{R}^n$ *be an open set, let* $(\mathcal{U}_2,U_2,\mathcal{G}_2,g_2)$ *be an admissible quadruple of type (p), and let* $\kappa:\mathcal{U}_1\to\mathcal{U}_2$ *be a* $(\mathcal{U}_2,U_2,\mathcal{G}_2,g_2)$*-compatible diffeomorphism of type (p). Let* $\chi:\mathcal{U}_1\times\mathbb{R}^n\to\mathcal{U}_2\times\mathbb{R}^n$, $(x,\xi)\mapsto\left(\kappa(x),{}^t d\kappa(x)^{-1}\xi\right)$, $U_1:=\kappa^{-1}(U_2)$, *and* $g_1:=\chi^*g_2$. *Then the following statements hold:*

1. *For* $x\in U_1$ *and* $y\in\mathbb{R}^n$ *with* $g_{1,(x,0)}(x-y,0)\le r_{g_1}^2$ *we define*

$$k(x,y)\ :=\ \int_0^1(d\kappa)\left(y+\theta(x-y)\right)d\theta\ \left(\in\mathcal{L}\left(\mathbb{R}^n\right)\right)\ .$$

Then: $\exists C>0,\ \exists 0<r\le r_{g_1},\ \forall x\in U_1,\ \forall y\in\mathbb{R}^n$ *with* $g_{1,(x,0)}(x-y,0)\le r^2$, $k(x,y)\in\mathcal{L}\left(\mathbb{R}^n\right)^{-1}$ *and* $\|k(x,y)\|_{\mathcal{L}(\mathbb{R}^n)}\le C$, $\left\|k(x,y)^{-1}\right\|_{\mathcal{L}(\mathbb{R}^n)}\le C$ *hold.*

[2]We use the notation $|\mathcal{G}_z|:=\det(G_z)$ where $\mathcal{G}_z(t)=\langle G_zt,t\rangle$ for $t\in\mathbb{R}^n$.

2. *There exists an open subset* $W_{1,r}$ *of* $\mathbb{R}^{3n}$ *with*
   (a) $W_{1,r} \subseteq U_1 \times \mathbb{R}^n \times \mathcal{U}_1$.
   (b) $(x,\xi,y) \in W_{1,r} \implies g_{1,(x,0)}(x-y,0) \leq r^2$.
   (c) $\exists 0 < r_0 \leq r,\ \forall x \in U_1,\ \forall y \in \mathbb{R}^n,\ g_{1,(x,0)}(x-y,0) \leq r_0^2,\ \forall \xi \in \mathbb{R}^n,$ $(x,\xi,y) \in W_{1,r}$.
3. *Now we restrict* $\mathcal{D}(g_1)$ *to* $W_{1,r}$. *Then we have*
   (a) $\left[W_{1,r} \ni (x,\xi,y) \mapsto k(x,y), {}^t k(x,y)^{-1}\right] \in S\left(1, \mathcal{D}(g_1); \mathcal{L}\left(\mathbb{R}^n\right)\right)$.
   (b) $\left[W_{1,r} \ni (x,\xi,y) \mapsto |\det(k(x,y))|^{\pm 1}\right] \in S\left(1, \mathcal{D}(g_1)\right)$.

Proof. First we fix $x \in U_1$ and let $y \in \mathbb{R}^n$ with $g_{1,(x,0)}(x-y,0) \leq r_{g_1}^2$. With suitable constants and with the definition $x_\theta := y + \theta(x-y)$, for $t \in \mathbb{R}^n$ the estimate

$$\begin{aligned}\|\langle t, \nabla_y\rangle\, k(x,y)\|_{\mathcal{L}(\mathbb{R}^n)} &\leq \int_0^1 (1-\theta)\, \|(\partial_t d\kappa)\,(x_\theta)\|_{\mathcal{L}(\mathbb{R}^n)}\, d\theta \\ &\leq C_1 \int_0^1 \sqrt{g_{1,(x_\theta,0)}(t,0)}\, d\theta \ \leq\ C_2 \sqrt{g_{1,(x,0)}(t,0)}\end{aligned}$$

holds. By virtue of the mean value theorem we obtain

$$\begin{aligned}\|k(x,y) - k(x,x)\|_{\mathcal{L}(\mathbb{R}^n)} &= \left\| \int_0^1 \langle x-y, \nabla_y\rangle\, k(x,y)_{|y=x_s}\, ds \right\|_{\mathcal{L}(\mathbb{R}^n)} \\ &\leq C_2 \sqrt{g_{1,(x,0)}(x-y,0)}\ .\end{aligned}$$

We set $D := \sup_{x\in\mathcal{U}_1} \left\|d\kappa(x)^{-1}\right\|_{\mathcal{L}(\mathbb{R}^n)}$ and $r := \min\{(2DC_2)^{-1}, r_{g_1}\}$. Then, for $x \in U_1$ and $y \in \mathbb{R}^n$ with $g_{1,(x,0)}(x-y,0) \leq r^2$, we have

$$\|k(x,y) - k(x,x)\|_{\mathcal{L}(\mathbb{R}^n)} \leq \frac{1}{2\,\|k(x,x)^{-1}\|_{\mathcal{L}(\mathbb{R}^n)}}\ ,$$

where $k(x,x) = d\kappa(x) \in \mathcal{L}\left(\mathbb{R}^n\right)^{-1}$. This yields $k(x,y) \in \mathcal{L}\left(\mathbb{R}^n\right)^{-1}$ and

$$\left\|k(x,y)^{-1} - k(x,x)^{-1}\right\|_{\mathcal{L}(\mathbb{R}^n)} \ \leq\ D\ ,$$

hence $\|k(x,y)\|_{\mathcal{L}(\mathbb{R}^n)} \leq C_3$ and $\left\|k(x,y)^{-1}\right\|_{\mathcal{L}(\mathbb{R}^n)} \leq C_3$.

For $x \in U_1$ there exists $0 < \varepsilon_x \leq r/\left(2\sqrt{C_{g_1}}\right)$ $(\leq r_{g_1})$ such that $x' \in \mathbb{R}^n$ with $g_{1,(x,0)}(x-x',0) \leq \varepsilon_x^2$ implies $x' \in U_1$. Let $\Delta_r(x)$ be the set of all $(x',y) \in \mathbb{R}^{2n}$ such that $g_{1,(x,0)}(x-x',0) < \varepsilon_x^2$ and $g_{1,(x,0)}(x-y,0) < \left(r/\left(2\sqrt{C_{g_1}}\right)\right)^2$. Then $\Delta_r(x)$ is open and for $(x',y) \in \Delta_r(x)$ we have

$$\sqrt{g_{1,(x',0)}(x'-y,0)} \leq r \leq r_{g_1}\ .$$

In particular, we have $\Delta_r(x) \subseteq U_1 \times \mathcal{U}_1$. Moreover, we define $\Delta_r := \bigcup_{x\in U_1} \Delta_r(x)$ $(\subseteq U_1 \times \mathcal{U}_1)$ and $W_{1,r} := \{(x,\xi,y) \in U_1 \times \mathbb{R}^n \times \mathbb{R}^n \mid (x,y) \in \Delta_r\}$. Then $W_{1,r}$ is an open subset of $\mathbb{R}^{3n}$ with $W_{1,r} \subseteq U_1 \times \mathbb{R}^n \times \mathcal{U}_1$.

The last assertion is obvious. □

Lemma 3.5. *We apply the notation from 3.4.*

1. $\forall k \in \mathbb{N}_0,\ \exists C_k,\ \forall (x,\xi) \in \mathcal{U}_1 \times \mathbb{R}^n,\ \forall T_1,\dots,T_k \in \mathbb{R}^{2n},$

$$g_{2,\chi(x,\xi)}\left(\chi^{(k)}(x,\xi)(T_1,...,T_k)\right) \quad \le \quad C_k \prod_{j=1}^{k} g_{1,(x,\xi)}(T_j)\ .$$

2. *For $x \in U_1$, $y \in \mathbb{R}^n$ with $g_{1,(x,0)}(x-y,0) \le r^2$ (cf. 3.4), and $\xi \in \mathbb{R}^n$ let*

$$K(x,\xi,y) \quad := \quad \left(\kappa(x), {}^t k(x,y)^{-1}\xi, \kappa(y)\right)\ .$$

Then:

$$\exists 0 < r_0 \le r,\ \forall k \in \mathbb{N}_0,\ \exists C_k > 0,\ \forall x \in U_1,\ \forall y \in \mathbb{R}^n$$

with

$$g_{1,(x,0)}(x-y,0) \le r_0^2,\ \forall \xi \in \mathbb{R}^n,\ \forall T_1,\dots,T_k \in \mathbb{R}^{3n},$$

$$\mathcal{D}(g_2)_{K(x,\xi,y)}\left(K^{(k)}(x,\xi,y)\,(T_1,\dots,T_k)\right) \le C_k \prod_{j=1}^{k} \mathcal{D}\,(g_1)_{(x,\xi,y)}\,(T_j)\ .$$

Proof. The proof is elementary. For details we refer to [2, Lemma 2.3.6]. □

Corollary 3.6. *We apply the notation from* 3.4 *and* 3.5*. Moreover, let M_2 be $(\mathcal{U}_2, U_2, g_2)$-admissible of type (p). We define $M_1 := \chi^* M_2$.*

1. *Then we have $\kappa^* S(1,\mathcal{G}_2) \hookrightarrow S(1,\mathcal{G}_1)$ and $\chi^* S(M_2,g_2) \hookrightarrow S(M_1,g_1)$.*
2. *We restrict $\mathcal{D}(g_1)$, $\mathcal{D}(M_1)$ to $W_{1,r}$ and $\mathcal{D}(g_2)$, $\mathcal{D}(M_2)$ to the set $\{(\kappa(x),\xi,\kappa(y)) \mid (x,\xi,y) \in W_{1,r}\}$. Then we have $K^* S\left(\mathcal{D}(M_2),\mathcal{D}(g_2)\right) \hookrightarrow S\left(\mathcal{D}(M_1),\mathcal{D}(g_1)\right)$.*

Definition 3.7. If $(\mathcal{U},U,\mathcal{G},g)$ is an admissible quadruple of type (p), then we define $(0 < \rho_1, \rho_2 \le r_g)$

$$\Delta_{g,\rho_1,\rho_2} := \left\{(x,\xi,x') \in U_{\rho_1} \times \mathbb{R}^n \times U_{\rho_1} \mid g_{(x,0)}(x-x',0) \le \rho_2^2\right\}$$

and $\Delta_{g,r_g} := \Delta_{g,r_g,r_g}$ (cf. 3.1 for the notation).

Theorem 3.8. *Let $(\mathcal{U},U,\mathcal{G},g)$ be an admissible quadruple of type (p) and let M be a $(\mathcal{U},U,g)$-admissible weight function of type (p). For $a \in A(M,g)$ with support in $U_{r_g} \times \mathbb{R}^n \times U_{r_g}$ and $x,x' \in \mathcal{U}$, $\xi \in \mathbb{R}^n$ we define $a_x(x',\xi) := a(x,\xi,x')$. Then, for each $x \in \mathcal{U}$, we have $a_x \in S(\tilde{M},g_s)$ (cf. 2.6) and thus we can define $((x',\xi) \in \mathcal{U} \times \mathbb{R}^n)$*

$$\begin{aligned}\mathcal{R}(a)(x,\xi) \quad &:= \quad \exp\left(i\left\langle D_{x'},D_\xi\right\rangle\right) a_x(x',\xi)_{|x'=x},\\ \mathcal{R}_1(a)(x,\xi) \quad &:= \quad a(x,\xi,x) - \mathcal{R}(a)(x,\xi)\ .\end{aligned}$$

Then the following statements hold:

1. *For each* $a \in A(M,g)$ *with* $supp(a) \subseteq \Delta_{g,r_g}$ *we have* $\mathcal{R}(a) \in S(M,g)$, $\mathcal{R}_1(a) \in S(Mh_g, g)$, *and* $supp(\mathcal{R}(a)), supp(\mathcal{R}_1(a)) \subseteq U_{r_g} \times \mathbb{R}^n$. *More precisely, for each* $k \in \mathbb{N}_0$ *there exist constants* $C_k > 0$ *and* $l_k \in \mathbb{N}$, *depending on* M *only through the constant* C_M *in* 2.6 *and the global constants of a fixed* g_s*-admissible weight function* $\tilde{M}$ *as in* 2.6, *such that*
$$\|\mathcal{R}(a)\|_{k,S(M,g)} \leq C_k \|a\|_{l_k,A(M,g)} \ , \ \|\mathcal{R}_1(a)\|_{k,S(Mh_g,g)} \leq C_k \|a\|_{l_k,A(M,g)} \, .$$
2. *If* $(a_k)_{k\in\mathbb{N}} \subseteq A(M,g)$ *is a bounded sequence such that* $supp(a_k) \subseteq \Delta_{g,r_g}$ *for each* $k \in \mathbb{N}$, *and if furthermore* $(a_k)_{k\in\mathbb{N}}$ *converges in* $\mathcal{C}^\infty(\mathbb{R}^{3n})$ *to* a, *then* $(\mathcal{R}(a_k))_{k\in\mathbb{N}}$ *converges in* $\mathcal{C}^\infty(\mathbb{R}^{2n})$ *to* $\mathcal{R}(a)$ *and* $(\mathcal{R}_1(a_k))_{k\in\mathbb{N}}$ *converges in* $\mathcal{C}^\infty(\mathbb{R}^{2n})$ *to* $\mathcal{R}_1(a)$.

Proof.

1. First we observe that $[x \mapsto a_x] \in \mathcal{C}^\infty\left(\mathbb{R}^n, \mathcal{C}^\infty(\mathbb{R}^{2n})\right)$. Moreover, since $\operatorname{supp}(a) \subseteq \Delta_{g,r_g}$, for all $k \in \mathbb{N}_0$ and all $t_1, \ldots, t_k \in \mathbb{R}^n \setminus \{0\}$, the function $\mathbb{R}^n \to S\left(\tilde{M}(t_1,\ldots,t_k), g_s\right)$,
$$x \mapsto \langle t_1, \nabla_x\rangle \ldots \langle t_k, \nabla_x\rangle \, a_x = \left(\langle t_1, \nabla_x\rangle \ldots \langle t_k, \nabla_x\rangle \, a\right)_x$$
is bounded, where $\tilde{M}(t_1,\ldots,t_k)(x',\xi) := \tilde{M}(x',\xi)\prod_{j=1}^k \sqrt{g_{s,(x',\xi)}(t_j,0)}$. Together with the continuity of $\exp(i\langle D_{x'}, D_\xi\rangle)$ on bounded subsets of $S(M_0, g_s)$ with respect to the $\mathcal{C}^\infty$-topology, where M_0 is a g_s-admissible weight function, by virtue of the mean value theorem we obtain that
$$[(x,\xi,x') \mapsto \exp(i\langle D_{x'}, D_\xi\rangle) a_x(x',\xi)] \in \mathcal{C}^\infty(\mathbb{R}^{3n})$$
and thus in particular $\mathcal{R}(a) \in \mathcal{C}^\infty(\mathbb{R}^{2n})$ with
$$\begin{aligned} \partial_{x_j}\mathcal{R}(a)(x,\xi) &= \exp(i\langle D_{x'}, D_\xi\rangle)\left(\partial_{x_j} a\right)_x (x',\xi)_{|x=x'} \\ &\quad + \partial_{x_j} \exp(i\langle D_{x'}, D_\xi\rangle) a(x',\xi)_{|x=x'} \end{aligned}$$
($j = 1, \ldots, n$). Thus, we obtain $\mathcal{R}(a) \in S(M,g)$ and the asserted estimates for $\mathcal{R}(a)$. For the corresponding statements concerning $\mathcal{R}_1(a)$ we apply [28, Theorem 18.5.10]).
2. By assumption, for each $x \in \mathbb{R}^n$ we have $[(a_k)_x \to a_x] \in \mathcal{C}^\infty(\mathbb{R}^{2n})$. This yields $\exp(i\langle D_{x'}, D_\xi\rangle)(a_k)_x(x',\xi) \to \exp(i\langle D_{x'}, D_\xi\rangle) a_x(x',\xi)$ for all $x, x', \xi \in \mathbb{R}^n$. In particular, this yields $\mathcal{R}(a_k)(x,\xi) \to \mathcal{R}(a)(x,\xi)$ for all $x, \xi \in \mathbb{R}^n$. $(\mathcal{R}(a_k))_{k\in\mathbb{N}}$ is bounded in $S\left(\tilde{M}, g_s\right)$ and thus in particular in $\mathcal{C}^\infty(\mathbb{R}^{2n})$. Since $\mathcal{C}^\infty(\mathbb{R}^{2n})$ is Montel, the standard indirect argument, taking into account the pointwise limits, yields the convergence of $\mathcal{R}(a_k)$ to $\mathcal{R}(a)$ in $\mathcal{C}^\infty(\mathbb{R}^{2n})$. For $\mathcal{R}_1(a)$ the proof is analogous.

□

Definition 3.9. For $a \in \mathcal{C}_0^\infty(\mathbb{R}^{3n})$, $u \in \mathcal{S}(\mathbb{R}^n)$, and $x \in \mathbb{R}^n$ we define
$$a(x,D,x')u(x) := \frac{1}{(2\pi)^n} \iint e^{i\langle x-y,\xi\rangle} a(x,\xi,y) u(y) dy d\xi \, .$$

Remark 3.10. We assume the situation of the preceding definition. By virtue of [28, Theorem 18.5.10] we obtain

$$\begin{aligned} a(x,D,x')u(x) &= \frac{1}{(2\pi)^n}\iint e^{i\langle x-y,\xi\rangle}a_x(y,\xi)u(y)dyd\xi \\ = \left(\exp\left(i\langle D_{x'},D_\xi\rangle\right)a_x(x',\xi)_{|x'=x}\right)(x,D)u(x) &= \mathcal{R}(a)(x,D)u(x)\ . \end{aligned}$$

The following lemma makes use of Kuranishi's method.

Lemma 3.11. *Let $\mathcal{U}_1 \subseteq \mathbb{R}^n$ be open, let $(\mathcal{U}_2, U_2, \mathcal{G}_2, g_2)$ be an admissible quadruple of type (p), let M_2 be a $(\mathcal{U}_2, U_2, g_2)$-admissible weight function of type (p), and let $\kappa : \mathcal{U}_1 \to \mathcal{U}_2$ be $(\mathcal{U}_2, U_2, \mathcal{G}_2, g_2, M_2)$-compatible of type (p). There exist constants $0 < r_K, \rho_K < r_{g_2}$ such that for $a \in C_0^\infty(\mathbb{R}^{3n})$ with supp$(a) \subseteq \Delta_{g_2,r_K,\rho_K}$ and $u \in \mathcal{S}(\mathbb{R}^n)$ with supp$(u) \subseteq \mathcal{U}_1$ we have $(x \in \mathcal{U}_1)$*

$$\begin{aligned} \kappa^*a(x,D,x')\ (\kappa^{-1})^*u(x) &= b(x,D,x')u(x)\ , \quad \textit{where} \\ b(x,\xi,x') &:= a\left(\kappa(x),\ {}^t\!k(x,x')^{-1}\xi,\ \kappa(x')\right)\frac{|\det(d\kappa(x'))|}{|\det(k(x,x'))|}\ , \\ k(x,x') &:= \int_0^1 (d\kappa)(x'+\theta(x-x'))d\theta\ . \end{aligned}$$

In particular, $b \in C_0^\infty(\mathbb{R}^{3n})$.

Proof. For $0 < r$ sufficiently small, $(\mathcal{U}, U_{2,r}, \mathcal{G}_2, g_2)$ is admissible of type (p) and κ is $(\mathcal{U}, U_{2,r}, \mathcal{G}_2, g_2)$-compatible of type (p). There exists a constant $\rho > 0$ such that for $x, x' \in \mathcal{U}_1$ and $\xi \in \mathbb{R}^n$ with $(\kappa(x), \xi, \kappa(x')) \in \Delta_{g_2,r,\rho}$ we have $(x,\xi,x') \in W_{1,r'}$ with a suitable constant $r' > 0$, $W_{1,r'}$ as in 3.4 with $(\mathcal{U}_2, U_{2,r}, \mathcal{G}_2, g_2)$ instead of $(\mathcal{U}_2, U_2, \mathcal{G}_2, g_2)$, and thus there exists $k(x,x')^{-1}$, and $k(x,x')$ and $k(x,x')^{-1}$ are uniformly bounded with respect to those x, x'. Then, applying transformation formula twice, we have for $x \in \mathcal{U}_1$,

$$\begin{aligned} &(2\pi)^n\kappa^*a(x,D,x')\ (\kappa^{-1})^*u(x) \\ =\ &\iint e^{i\langle\kappa(x)-\kappa(x'),\xi\rangle}a\left(\kappa(x),\xi,\kappa(x')\right)u(x')|\det(d\kappa(x'))|dx'd\xi \\ =\ &\iint e^{i\langle x-x',{}^tk(x,x')\xi\rangle}a\left(\kappa(x),\xi,\kappa(x')\right)u(x')|\det(d\kappa(x'))|dx'd\xi \\ =\ &\iint e^{i\langle x-x',\xi\rangle}a\left(\kappa(x),\ {}^t\!k(x,x')^{-1}\xi,\ \kappa(x')\right)u(x')\frac{|\det(d\kappa(x'))|}{|\det({}^tk(x,x'))|}dx'd\xi\ . \end{aligned}$$

□

Theorem 3.12. *We apply the notation of 3.11. Let $\varphi, \psi \in C^\infty(\mathbb{R}^n)$ such that*

- *supp$((x,\xi,y) \mapsto \varphi(x)\cdot\psi(y)) \subseteq \Delta_{g_2,r_K,\rho_K}$.*
- *$[(x,\xi) \mapsto \varphi(x)], [(x,\xi) \mapsto \psi(x)] \in S(1,g_2)$.*

Then, for each $a \in S(M_2,g_2)$ and $u \in \mathcal{S}(\mathbb{R}^n)$ with supp$(u) \subseteq \mathcal{U}_1$, we have $(x \in \mathcal{U}_1)$

$$\kappa^*\left[(\varphi a)(x,D)\ \left(\psi\cdot(\kappa^{-1})^*(u)\right)\right](x) = b(x,D)u(x)\ ,$$

where

$$b(x,\xi) \quad := \quad \mathcal{R}\left(\varphi(\kappa(x))\ a\left(\kappa(x),\ {}^t k(x,x')^{-1}\xi\right)\ \psi(\kappa(x'))\ \frac{|\det(d\kappa(x'))|}{|\det(k(x,x'))|}\right),$$

$$k(x,x') \quad := \quad \int_0^1 (d\kappa)(x'+\theta(x-x'))d\theta\ .$$

In particular, we have $b \in S(M_1, g_1)$. *Moreover, if* $U \subseteq \mathbb{R}^n$ *such that* $supp\,(a) \subseteq U \times \mathbb{R}^n$, *then we have* $supp\,(b) \subseteq \kappa^{-1}(U) \times \mathbb{R}^n$. *Finally, the definition*

$$c(x,\xi) := b(x,\xi) - \varphi(\kappa(x))a\left(\kappa(x),{}^t d\kappa(x)^{-1}\xi\right)\psi(\kappa(x))$$

yields $c \in S(M_1 h_{g_1}, g_1)$.

Proof. By assumption we have $\mathrm{supp}\,(\varphi), \mathrm{supp}\,(\psi) \subseteq U_{2,r_K}$ and $\psi \in \mathcal{C}_0^\infty(\mathbb{R}^n)$. 3.2 yields the existence of functions $\tilde\varphi_1, \tilde\varphi_2 \in \mathcal{C}^\infty(\mathbb{R}^n)$ with $\mathrm{supp}\,(\tilde\varphi_i) \subseteq \mathcal{U}_2$ $(i=1,2)$ such that $[(x,\xi) \mapsto \tilde\varphi_i(x)] \in S(1,g_2)$, $\tilde\varphi_1(x) = 1$ for all $x \in U_{2,r_K}$ and $\tilde\varphi_2(x) = 1$ for all $x \in \mathrm{supp}\,(\tilde\varphi_1)$. Defining $\tilde a(x,\xi) := \tilde\varphi_1(x)\cdot a(x,\xi)$ $(x,\xi \in \mathbb{R}^n)$ we have $\tilde a \in S(\tilde M_2, g_{2,s})$. There exists a sequence $(\tilde a_\nu)_{\nu\in\mathbb{N}} \subseteq \mathcal{C}_0^\infty(\mathbb{R}^{2n})$ which is bounded in $S(\tilde M_2, g_{2,s})$ such that $\tilde a_\nu \to \tilde a$ in $\mathcal{C}^\infty(\mathbb{R}^{2n})$. For $\nu \in \mathbb{N}$ and $x,\xi,x' \in \mathbb{R}^n$ we define $a_\nu(x,\xi,x') := \varphi(x)\tilde a_\nu(x,\xi)\psi(x')$. Noting the continuity of the mappings $[\mathcal{S}(\mathbb{R}^n) \ni u \mapsto \kappa^*(\tilde\varphi_2\cdot u) \in \mathcal{S}(\mathbb{R}^n)]$ and $[\mathcal{S}(\mathbb{R}^n) \ni u \mapsto (\kappa^{-1})^*(\kappa^*(\tilde\varphi_2)\cdot u) \in \mathcal{S}(\mathbb{R}^n)]$ we have

$$\begin{aligned} & \kappa^*\left[(\varphi a)(x,D)\ \ (\psi\cdot(\kappa^{-1})^*(u))\right](x) \\ = \quad & \kappa^*\left[\tilde\varphi_2\cdot\varphi\cdot\tilde a(x,D)\ \ (\psi\cdot(\kappa^{-1})^*(\kappa^*\tilde\varphi_2\cdot u))\right](x) \\ \overset{\mathcal{S}(\mathbb{R}^n)}{\longleftarrow} \quad & \kappa^*\left[\tilde\varphi_2\cdot\varphi\ \tilde a_\nu(x,D)\ \ (\psi\cdot(\kappa^{-1})^*(\kappa^*\tilde\varphi_2\cdot u))\right](x) \\ = \quad & \kappa^*\left[a_\nu(x,D,x')\ \ ((\kappa^{-1})^*u)\right](x)\ . \end{aligned}$$

By virtue of 3.10 and 3.11 we have

$$\kappa^*\left[a_\nu(x,D,x')\ \ ((\kappa^{-1})^*u)\right](x) \quad = \quad b_\nu(x,D)u(x)$$

where

$$b_\nu \quad = \quad \mathcal{R}\left(\varphi(\kappa(x))\ \tilde a_\nu\left(\kappa(x),\ {}^t k(x,x')^{-1}\xi\right)\ \psi(\kappa(x'))\ \frac{|\det(d\kappa(x'))|}{|\det(k(x,x'))|}\right).$$

We denote the argument in the bracket by c_ν and without ν simply with c. With the notation from the proof of 3.11 we have $K^{-1}(\Delta_{g_2,r_K,\rho_K}) \subseteq W_{1,r'}$, where $K : W_{1,r'} \to \mathbb{R}^{3n}$, $(x,\xi,y) \mapsto \left(\kappa(x),{}^t k(x,y)^{-1},\kappa(y)\right)$. Noting that K maps compact subsets of $W_{1,r'}$ onto compact sets, by virtue of the chain rule we have $c_\nu \to c$ in $\mathcal{C}^\infty(\mathbb{R}^{3n})$. Furthermore, by virtue of 3.4 and 3.6, the sequence $(c_\nu)_{\nu\in\mathbb{N}}$ is bounded in $A(M_1,g_1)$. By virtue of 3.8 this yields $b_\nu = \mathcal{R}(c_\nu) \to \mathcal{R}(c) = b$ in $\mathcal{C}^\infty(\mathbb{R}^{2n})$ and $(\mathcal{R}(c_\nu))_{\nu\in\mathbb{N}}$ is bounded in $S(\tilde M_1, g_{1,s})$, thus $b_\nu(x,D)u \to b(x,D)u$ in $\mathcal{S}(\mathbb{R}^n)$. The assertion concerning c is a consequence of 3.8. □

In the following proposition, we study the behavior of a pseudo-differential operator under compatible diffeomorphisms and show that the transformed operator is the sum of an operator with symbol in the pull-back of the original symbol class and a remainder part which is a bounded operator on $L^2(\mathbb{R}^n)$. The first summand comes from the part of the original operator with Schwartz kernel supported near the diagonal to which we can apply Kuranishi's method (cf. 3.12). The second summand is in fact much better and is even of order $-\infty$ in the original global symbolic calculus. But, so far we have no description of this smoothing class invariant under compatible diffeomorphisms which is the case for $\mathcal{L}\left(L^2(\mathbb{R}^n)\right)$. This proposition will be applied in the investigation of the behavior of local Sobolev spaces under changes of coordinates in the following section (cf. 4.7). In 5.12 we will apply similar arguments and hence give a detailed proof of this proposition and will omit the proof of 5.12.

Proposition 3.13. *Let $\mathcal{U}_1 \subseteq \mathbb{R}^n$ be open, let $(\mathcal{U}_2, U_2, \mathcal{G}_2, g_2)$ be an admissible quadruple of type (p), and let $\kappa : \mathcal{U}_1 \to \mathcal{U}_2$ be $(\mathcal{U}_2, U_2, \mathcal{G}_2, g_2)$-compatible of type (p). We define $M_2 := H^{-1}_{g_2,\mathcal{B}_2}$ (cf. 2.4), where $\mathcal{B}_2$ is a g_2-admissible basis of $\mathbb{R}^n$. Let $\rho_K, r_K \leq r_{g_2}$ be the constant from 3.12 and let $C > 0$ be chosen such that $\kappa^{-1}(U_{2,z,\rho}) \subseteq U_{1,\kappa^{-1}(z),C\rho}$ for $z \in U_2$ and $0 < \rho < \min\{r_{g_1}, r_{g_2}\}/C$. Moreover, let $0 < \rho_1 < \rho_2 < \rho_3 < \rho_4 := \min\{r_K, r_{g_1}/C\}$ and let $0 < \rho \leq \rho_K/(2\sqrt{C_{g_2}})$ be sufficiently small such that $\rho_i + 3C_{g_2}\,\rho \leq \rho_{i+1}$ $(i = 1,2,3)$. Let φ be the function from Theorem 3.2 applied with $(\mathcal{U}_2, U_2, \mathcal{R}_1(g_2))$ instead of $(\mathcal{U}, U, \mathcal{G})$. Furthermore, let $\psi \in C^\infty(\mathbb{R}^n)$ be a function with supp$(\psi) \subset \mathcal{U}_1$ such that $[\mathbb{R}^{2n} \ni (x,\xi) \mapsto \psi(x)] \in S(1, g_1)$. Then, for $s \in \mathbb{R}$, $a \in S(M_2^s, g_2)$, and $u \in L^2(\mathbb{R}^n)$ we have*

$$\kappa^*[(\varphi a)(x,D)]\ (\kappa^{-1})^*(\psi \cdot u) = A_1 u + A_2 u$$

where $A_1 \in \Psi(\tilde{M}_1^s, g_{1,s})$, $g_1 := \chi^* g_2$, $M_1 := \chi^* M_2$ (cf. 2.6), and $A_2 \in \mathcal{L}\left(L^2(\mathbb{R}^n)\right)$.

Proof. Let $\tilde{\varphi}_1, \varphi_2 \in C^\infty(\mathbb{R}^n)$ with supp$(\tilde{\varphi}_i) \subset \mathcal{U}_2$ $(i = 1,2)$, $\tilde{\varphi}_1(x) = 1$ for all $x \in U_{2,\rho_3}$, $\tilde{\varphi}_2(x) = 1$ for all $x \in$ supp$(\tilde{\varphi}_1)$, and $[(x,\xi) \mapsto \tilde{\varphi}_i(x)] \in S(1, g_2)$ $(i = 1,2)$ (cf. 3.2). Let $\tilde{a} := \tilde{\varphi}_1 \cdot a$. Then, for $f \in \mathcal{S}(\mathbb{R}^n)$ and $x \in \mathbb{R}^n$ we have

$$\begin{aligned}
&\varphi(x,D)\tilde{a}(x,D)f(x)\\
&= \int_{U_{2,\rho_2}} \varphi_z(x,D)\tilde{a}(x,D)\psi_z(x,D)f(x)\ |\mathcal{R}_1(g_2)_z|^{1/2}dz\\
&\quad + \int_{U_{2,\rho_2}} \varphi_z(x,D)\tilde{a}(x,D)(1-\psi_z)(x,D)f(x)\ |\mathcal{R}_1(g_2)_z|^{1/2}dz\ .
\end{aligned}$$

Here we note that for $u \in \mathcal{S}(\mathbb{R}^n)$

$$\begin{aligned}
&\left(\kappa^* \int_{U_{2,\rho_2}} [\varphi_z(x,D)\tilde{a}(x,D)\psi_z(x,D)]\ (\kappa^{-1})^*(\psi \cdot u)\ |\mathcal{R}_1(g_2)_z|^{1/2}dz\right)(x)\\
&= \int_{U_{2,\rho_2}} \kappa^*\left[\varphi_z(x,D)\tilde{a}(x,D)\psi_z(x,D)\right]\ (\kappa^{-1})^*(\psi \cdot u)(x)\ |\mathcal{R}_1(g_2)_z|^{1/2}dz\ .
\end{aligned}$$

For $z \in U_{2,\rho_2}$ and $x, y \in U_{z,\rho,g_2}$ we have $g_{2,(x,0)}(x-y,0) \leq 4C_{g_2}\rho^2$ and

$$\operatorname{supp}((x,y) \mapsto \varphi_z(x) \cdot \psi_z(y)) \subseteq U_{2,\rho_3} \times U_{2,\rho_3} \subseteq U_{2,r_K} \times U_{2,r_K},$$

hence $\operatorname{supp}((x,\xi,y) \mapsto \varphi_z(x) \cdot \psi_z(y)) \subseteq \Delta_{g_2,r_K,\rho_K}$. Thus we find for $z \in U_{2,\rho_2}$ (cf. 3.12)

$$\kappa^* [\varphi_z(x,D)\tilde{a}(x,D)\psi_z(x,D)] \; (\kappa^{-1})^*(\psi \cdot u) = b_z(x,D)(\psi \cdot u)$$

with $b_z \in S(M_1^s, g_1)$. More precisely, 3.6, 3.2, 3.8, and 3.12 yield that $U_{2,\rho_2} \ni z \mapsto b_z \in S(M_1^s, g_1)$ is continuous, bounded, and $\operatorname{supp}(b_z) \subseteq \kappa^{-1}(U_{2,z,\rho}) \times \mathbb{R}^n$. For $z \in \kappa^{-1}(U_{2,\rho_2})$ let $\tilde{b}_z := b_{\kappa(z)}$. Then we have $\operatorname{supp}(\tilde{b}_z) \subseteq \kappa^{-1}(U_{2,\kappa(z),\rho}) \times \mathbb{R}^n \subseteq U_{1,z,C\rho} \times \mathbb{R}^n$ and $\kappa^{-1}(U_{2,\rho_2}) \subseteq U_{1,C\rho_2}$. The definition

$$b(x,\xi) := \int_{U_{2,\rho_2}} b_z(x,\xi) \; |\mathcal{R}_1(g_2)_z|^{1/2} dz = \int_{\kappa^{-1}(U_{2,\rho_2})} \tilde{b}_z(x,\xi) \; |\mathcal{R}_1(g_1)_z|^{1/2} dz$$

yields a symbol $b \in S(M_1^s, g_1)$ with $\operatorname{supp}(b) \subseteq \kappa^{-1}(U_{2,\rho_3}) \times \mathbb{R}^n \subseteq \mathcal{U}_1 \times \mathbb{R}^n$, and we have

$$\begin{aligned} &\left(\kappa^* \int_{U_{2,\rho_2}} [\varphi_z(x,D)\tilde{a}(x,D)\psi_z(x,D)] \; (\kappa^{-1})^*(\psi \cdot u) \; |\mathcal{R}_1(g_2)_z|^{1/2} dz\right) \\ &= b(x,D)(\psi \cdot u) \; . \end{aligned}$$

Furthermore, $\varphi_z(x,D)\tilde{a}(x,D)(1-\psi_z)(x,D) = c_z(x,D)$ where $U_{2,\rho_2} \ni z \mapsto c_z \in S(1,g_2)$ is continuous. Defining $c := \int_{U_{2,\rho_2}} c_z(x,\xi) \; |\mathcal{R}_1(g_2)_z|^{1/2} dz$, we have $c \in S(1,g)$ and

$$\begin{aligned} &\int_{U_{2,\rho_2}} \varphi_z(x,D)\tilde{a}(x,D)(1-\psi_z)(x,D)f \; |\mathcal{R}_1(g_2)_z|^{1/2} dz \\ = \; &\tilde{\varphi}_2 \int_{U_{2,\rho_2}} \varphi_z(x,D)\tilde{a}(x,D)(1-\psi_z)(x,D)f \; |\mathcal{R}_1(g_2)_z|^{1/2} dz = \tilde{\varphi}_2 \; c(x,D)f \; . \end{aligned}$$

This yields the assertion for $u \in \mathcal{S}(\mathbb{R}^n)$. For $u \in L^2(\mathbb{R}^n)$ we make use of a sequence $(u_\nu)_{\nu \in \mathbb{N}} \subset \mathcal{S}(\mathbb{R}^n)$ with $u_\nu \to u$ in $L^2(\mathbb{R}^n) \hookrightarrow \mathcal{S}'(\mathbb{R}^n)$. □

4. Sobolev spaces

We will describe negligible terms in the calculus of pseudo-differential operators on manifolds (cf. 5.4, 5.9) with the help of mapping properties of iterated commutators with distinguished vector fields and multipliers between suitable global Sobolev spaces on the manifold (cf. 4.20). Here we first define and discuss the local models for these spaces. For an admissible triple $(\mathcal{U}, U, g)$ and a $(\mathcal{U}, U, g)$-admissible weight function M we define the associated Sobolev space with the help of a Hörmander metric g_s and an admissible weight function $\tilde{M}$ as in 2.6. Nevertheless, 4.4 gives descriptions of the local Sobolev spaces which consider g_s and $\tilde{M}$ only on $\mathcal{U}$ and thus, the definition of $H(g, M; U')$ is independent of the choice of g_s and $\tilde{M}$. In 4.7 we apply these descriptions together with 3.13 to show that the local Sobolev spaces associated to the standard weights h_g and $H_{g,B}$ behave well under changes of coordinates. Aside from the just mentioned results, this theorem is based on the fact that the L^2-spaces behave well under compatible diffeomorphism.

Then we define and discuss the Sobolev spaces on a generalized SG-compatible manifold $\mathcal{M}$ which are associated to powers of the weight H_g (cf. 4.16). More precisely, a distribution $u \in \mathcal{D}'(\mathcal{M})$ is said to be in $H\left(H_g^s, g; \mathcal{M}, \mathcal{A}\right)$ (cf. 4.20) if for all admissible charts the localization of u with an admissible cut-off function is in the corresponding local Sobolev space on $\mathbb{R}^n$. In 4.22 we show that these localizations are in $H\left(H_g^s, g; \mathcal{M}, \mathcal{A}\right)$ if and only if they are pull-backs of elements of the local Sobolev space. In the proof we make use of 4.7.

At this stage we avoid endowing the global Sobolev space with a topology. To have nevertheless meaningful extensions of operators $\mathcal{C}_0^\infty(\mathcal{M}) \to \mathcal{D}'(\mathcal{M})$ to operators between suitable Sobolev spaces, we introduce the notion "bounded in charts" (cf. 4.25).

We start with a repetition of the definition of the global Sobolev spaces $H(M,g)$. For an investigation of these weighted L^2-Sobolev spaces adapted to the $S(M,g)$-calculus we refer to R. Beals [5] and J.-M. Bony and J.-Y. Chemin [9].

Definition 4.1. Let g be a Hörmander metric and M a g-admissible weight function. We define

$$H(M,g) := \left\{u \in \mathcal{S}'(\mathbb{R}^n) \mid Au \in L^2(\mathbb{R}^n) \text{ for all } A \in \Psi(M,g)\right\} .$$

Remark 4.2. By virtue of [9, Théorème 6.4], $H(M,g)$ is a Hilbert space and we have $\mathcal{S}(\mathbb{R}^n) \hookrightarrow H(M,g) \hookrightarrow \mathcal{S}'(\mathbb{R}^n)$ with dense embeddings, where $\mathcal{S}'(\mathbb{R}^n)$ is endowed with the w^*-topology. Moreover, for two g-admissible weight functions M_1 and M_2 with $M_1 \leq M_2$ we have $H(M_2,g) \hookrightarrow H(M_1,g)$.

Definition 4.3. Let g be a split Hörmander metric and M a g-admissible weight function. For an open set $U \subseteq \mathbb{R}^n$ we define

$$H(M,g;U) := \{u \in H(M,g) \mid \operatorname{supp}(u) \subseteq U\} .$$

Lemma 4.4. *Let g be a split Hörmander metric of type (s). With an orthonormal (with respect to the Euclidean metric) basis $\mathcal{B}$, we define $M := H_{g,\mathcal{B}}^{-1}$ (cf. 2.4). Let U be an open subset of $\mathbb{R}^n$. Let $\varphi_1, \varphi_2, \varphi_3 \in \mathcal{C}^\infty(\mathbb{R}^n)$ such that $[(x,\xi) \mapsto \varphi_i(x)] \in S(1,g)$ for $i = 1,2,3$ and, moreover, $\varphi_1 = 1$ on U and $\varphi_{i+1} = 1$ on supp(φ_i) for $i = 1,2$. Let $s \geq 0$. Then the following statements hold:*

1. (a) *$u \in H(M^s,g;U) \iff u \in L^2(\mathbb{R}^n)$, supp$(u) \subseteq U$, and $\varphi_2 A(\varphi_1 u) \in L^2(\mathbb{R}^n)$ for all $A \in \Psi(M^s,g)$.*
   (b) *If $(u_\nu)_{\nu\in\mathbb{N}_0} \subset H(M^s,g;U)$, then we have $u_\nu \to u_0$ in $H(M^s,g) \iff u_\nu \to u_0$ in $L^2(\mathbb{R}^n)$ and $\varphi_2 A(\varphi_1 u_\nu) \to \varphi_2 A(\varphi_1 u_0)$ in $L^2(\mathbb{R}^n)$ for all $A \in \Psi(M^s,g)$.*
2. (a) *$u \in H(M^{-s},g;U) \iff u = \varphi_3 A(\varphi_2 u_1) + u_2$ where $A \in \Psi(M^s,g)$ and $u_1, u_2 \in L^2(\mathbb{R}^n)$, and supp$(u) \subseteq U$.*

(b) *For* $u \in H(M^{-s}, g; U)$ *the functions* $u_1, u_2 \in L^2(\mathbb{R}^n)$ *can be chosen depending continuously on* $u \in H(M^{-s}, g)$. *Thus, for* $(u_\nu)_{\nu\in\mathbb{N}_0} \subset H(M^{-s}, g; U)$, *we have* $u_\nu \to u_0$ *in* $H(M^{-s}, g) \iff u_{\nu,i} \to u_{0,i}$ *in* $L^2(\mathbb{R}^n)$ $(i = 1, 2)$.

Proof.

1. Since $M \geq 1$ we have $H(M^s, g; U) \subseteq H(M^s, g) \subseteq L^2(\mathbb{R}^n)$, where the last embedding is continuous. For $A \in \Psi(M^s, g)$ we have $\varphi_2 A \varphi_1 \in \Psi(M^s, g)$, and thus, $\varphi_2 A \varphi_1$ maps $H(M^s, g)$ continuously to $L^2(\mathbb{R}^n)$. On the other hand, let $u \in L^2(\mathbb{R}^n)$ with $\operatorname{supp}(u) \subseteq U$ such that $\varphi_2 A \varphi_1 u \in L^2(\mathbb{R}^n)$ for all $A \in \Psi(M^s, g)$. Let $A \in \Psi(M^s, g)$. Then we have
$$Au = A(\varphi_1 u) = \varphi_2 A(\varphi_1 u) + (1 - \varphi_2) A(\varphi_1 u) ,$$
where $\varphi_2 A(\varphi_1 u) \in L^2(\mathbb{R}^n)$ by assumption and $(1 - \varphi_2) A \varphi_1 \in \Psi(1, g)$, thus $(1 - \varphi_2) A(\varphi_1 u) \in L^2(\mathbb{R}^n)$. This implies $Au \in L^2(\mathbb{R}^n)$ and consequently $u \in H(M^s, g; U)$. Now let $(u_\nu)_{\nu\in\mathbb{N}_0} \subset H(M^s, g; U)$ with $u_\nu \to u_0$ in $L^2(\mathbb{R}^n)$ and $\varphi_2 A(\varphi_1 u_\nu) \to \varphi_2 A(\varphi_1 u_0)$ in $L^2(\mathbb{R}^n)$ for all $A \in \Psi(M^s, g)$. We choose $A \in \Psi(M^s, g)$ as an isomorphism $H(M^s, g) \to L^2(\mathbb{R}^n)$. Then the assumptions imply that $Au_\nu \to Au_0$ in $L^2(\mathbb{R}^n)$ and thus $u_\nu \to u_0$ in $H(M^s, g)$.
2. Let $u_1 \in L^2(\mathbb{R}^n)$ $(\subseteq H(M^{-s}, g))$ and $A \in \Psi(M^s, g)$. Then we have $\varphi_2 A(\varphi_1 u_1) \in H(M^{-s}, g)$. On the other hand, let $u \in H(M^{-s}, g; U)$. Let $J_s \in \Psi(M^s, g)$ such that there exists $J_{-s} := J_s^{-1} \in \Psi(M^{-s}, g)$. Then we have
$$\begin{aligned} u &= J_s(\varphi_2 J_{-s}(\varphi_1 u)) + J_s((1 - \varphi_2) J_{-s}(\varphi_1 u)) \\ &= \varphi_3 J_s(\varphi_2 \underbrace{J_{-s}(\varphi_1 u)}_{:=u_1}) + \underbrace{((1 - \varphi_3) J_s \varphi_2) J_{-s}(\varphi_1 u)}_{=:v} \\ &\quad + \underbrace{J_s((1 - \varphi_2) J_{-s} \varphi_1) u}_{=:w} . \end{aligned}$$
Since $J_{-s}\varphi_1 \in \Psi(M^{-s}, g)$ and $u \in H(M^{-s}, g)$ we have $u_1 \in L^2(\mathbb{R}^n)$. Moreover, we have $(1 - \varphi_3) J_s \varphi_2 \in \Psi(1, g)$ and $(1 - \varphi_2) J_{-s} \varphi_1 \in \Psi(M^{-2s}, g)$ and thus $v = (1 - \varphi_3) J_s(\varphi_2 u_1) \in L^2(\mathbb{R}^n)$ and $w \in L^2(\mathbb{R}^n)$.

□

Definition 4.5. Let $(\mathcal{U}, U, \mathcal{G}, g)$ be an admissible quadruple of type (p). Let $\mathcal{B}$ be a g-admissible basis (cf. 2.7). Let $r < r_g$ and $s \in \mathbb{R}$. Let $U' \subseteq U_r$ be an open subset. Applying the notation from 2.6, we define $H(H^s_{g,\mathcal{B}}, g; U') := H(H^s_{g_s,\mathcal{B}}, g_s; U')$. Moreover, we endow $H(H^s_{g,\mathcal{B}}, g; U')$ with the topology induced by $H(H^s_{g_s,\mathcal{B}}, g_s)$.

Remark 4.6. By virtue of 4.4 and 3.2, the definition of $H(H^s_g, g; U')$ is independent of the choice of g_s and $\mathcal{B}$.

Theorem 4.7. *Let* $\mathcal{U}_1 \subseteq \mathbb{R}^n$ *be open, let* $(\mathcal{U}_2, U_2, \mathcal{G}_2, g_2)$ *be an admissible quadruple of type* (p), *and let* $\kappa : \mathcal{U}_1 \to \mathcal{U}_2$ *be* $(\mathcal{U}_2, U_2, \mathcal{G}_2, g_2)$*- compatible of type* (p). *Let* $M_2 := H^{-1}_{g_2,\mathcal{B}_2}$ *(*$\mathcal{B}_2$ *a* g_2*-admissible basis such that* $\mathcal{T}^{-1}(\mathcal{B}_2)$ *is* g_1*-admissible,* $\mathcal{T}$ *as* *in 2.7. There exist constants $K_S, r_S > 0$ with $r_S < r_{g_2}$ and $K_S r_S < r_{g_1}$ such that the following statements hold:*

1. $\kappa^{-1}(U_{2,r_S}) \subseteq U_{1,K_S r_S}$.
2. *Let $g_1 := \chi^* g_2$, $M_1 := \chi^* M_2$, and $\mathcal{B}_1 := \mathcal{T}^{-1}\mathcal{B}_2$ (for the notation cf. 2.6 and 2.7). Then there exists a constant $C > 0$ such that for all $x \in \mathcal{U}_1$ and all $\xi \in \mathbb{R}^n$ the estimate $C^{-1}M_1(x,\xi) \le H^{-1}_{g_1,\mathcal{B}_1}(x,\xi) \le CM_1(x,\xi)$ holds.*
3. *Let $U_2' \subseteq U_{2,r_S}$ and $U_1' := \kappa^{-1}(U_2')$. Let $s \in \mathbb{R}$. Then we have*

$$(\kappa^{-1})^* \left(H\left(M_1^s, g_1; U_1'\right)\right) = H\left(M_2^s, g_2; U_2'\right) \ .$$

Proof. The proof of the first two assertions is elementary. Part 3 is a consequence of 3.13 and 4.4. □

Now we introduce generalized SG-compatible manifolds and admissible structures.

Definition 4.8. Let $\mathcal{M}$ be a $\mathcal{C}^\infty$-manifold of dimension n and $\mathcal{G}$ a metric on $\mathcal{M}$.

1. Let $(\mathcal{X},\kappa)$ be a chart of $\mathcal{M}$, $\kappa : \mathcal{M} \supseteq \mathcal{X} \to \kappa(\mathcal{X}) \subseteq \mathbb{R}^n$, and let $X \subseteq \mathcal{X}$ be an open subset. We call $(\mathcal{X}, X, \kappa)$ an *admissible chart* if $\left(\kappa(\mathcal{X}), \kappa(X), (\kappa^{-1})^*\mathcal{G}_{|\mathcal{X}}\right)$ is an admissible model[3] (cf. 2.5).
2. We call $\mathcal{A}$ a *generalized SG-compatible atlas* of $\mathcal{M}$ if the following conditions are satisfied:
   - $\mathcal{A}$ is finite and consists of admissible charts $(\mathcal{X}_1, X_1, \kappa_1), \ldots, (\mathcal{X}_m, X_m, \kappa_m)$.
   - $\mathcal{M} = \bigcup_{j=1,\ldots,m} X_j$.
   - For each $i,j \in \{1,\ldots,m\}$, $(\kappa_j(\mathcal{X}_i \cap \mathcal{X}_j), \kappa_j(X_i \cap X_j), (\kappa_j^{-1})^*\mathcal{G})$ is an admissible model and the diffeomorphism $\kappa_j \circ \kappa_i^{-1} : \kappa_i(\mathcal{X}_i \cap \mathcal{X}_j) \to \kappa_j(\mathcal{X}_i \cap \mathcal{X}_j)$ is $(\kappa_j(\mathcal{X}_i \cap \mathcal{X}_j), \kappa_j(X_i \cap X_j), (\kappa_j^{-1})^*\mathcal{G})$-compatible.

   Moreover, we call $(\mathcal{M},\mathcal{A},\mathcal{G})$ a *generalized SG-compatible manifold.*

Remark 4.9. For $i = 1,2$ let $(\mathcal{M}^{(i)}, \mathcal{A}^{(i)}, \mathcal{G}^{(i)})$ be a generalized SG-compatible manifold of dimension n_i, $\mathcal{A}^{(i)} = \left\{(\mathcal{X}_j^{(i)}, X_j^{(i)}, \kappa_j^{(i)}) \mid j \in \{1,\ldots,m_i\}\right\}$. On $\mathcal{M} := \mathcal{M}^{(1)} \times \mathcal{M}^{(2)}$ we define $\left(\mathcal{G}^{(1)} \otimes \mathcal{G}^{(2)}\right)_{(x_1,x_2)}(t_1,t_2) := \mathcal{G}^{(1)}_{x_1}(t_1) + \mathcal{G}^{(2)}_{x_2}(t_2)$ Moreover, let $\mathcal{A}$ consist of all charts $\left(\mathcal{X}_i^{(1)} \times \mathcal{X}_j^{(2)}, X_1^{(1)} \times X_j^{(2)}, (\kappa_i^{(1)}, \kappa_j^{(2)})\right)$, $i \in \{1,\ldots,m_1\}$, $j \in \{1,\ldots,m_2\}$. Then $(\mathcal{M},\mathcal{A},\mathcal{G})$ is a generalized SG-compatible manifold of dimension $n_1 + n_2$.

Definition 4.10. Let $(\mathcal{M},\mathcal{A},\mathcal{G})$ be a generalized SG-compatible manifold with atlas $\mathcal{A} = \{(\mathcal{X}_j, X_j, \kappa_j) \mid j = 1,\ldots,m\}$.

1. We call a metric g on the cotangent bundle $T^*\mathcal{M}$ over $\mathcal{M}$ $(\mathcal{M},\mathcal{A},\mathcal{G})$-*admissible of type* (p) if the following conditions are satisfied:

[3] In the following, we will omit the restriction of metrics and functions to chart regions when applying pull-backs.

- For all $j \in \{1, \ldots, m\}$, $\big(\kappa_j(\mathcal{X}_j), \kappa_j(X_j), (\kappa_j^{-1})^*\mathcal{G}, (\chi_j^{-1})^* g\big)$ is an admissible quadruple of type (p), where we define $\chi_j : T^*\mathcal{X}_j \to T^*\mathbb{R}^n \simeq \mathbb{R}^{2n}$, $T^*_x\mathcal{X}_j \ni \xi \mapsto (\kappa_j(x), {}^t d\kappa_j(x)^{-1}\xi)$.
- For each $i, j \in \{1, \ldots, m\}$, $(\kappa_j(\mathcal{X}_i \cap \mathcal{X}_j), \kappa_j(X_i \cap X_j), (\kappa_j^{-1})^*\mathcal{G}, (\chi_j^{-1})^* g)$ is an admissible quadruple of type (p) and the diffeomorphism $\kappa_j \circ \kappa_i^{-1} : \kappa_i(\mathcal{X}_i \cap \mathcal{X}_j) \to \kappa_j(\mathcal{X}_i \cap \mathcal{X}_j)$ is $(\kappa_j(\mathcal{X}_i \cap \mathcal{X}_j), \kappa_j(X_i \cap X_j), (\kappa_j^{-1})^*\mathcal{G}, (\chi_j^{-1})^* g)$-compatible of type (p).

If $(\mathcal{M}, \mathcal{A}, \mathcal{G})$ is a generalized SG-compatible manifold and g is $(\mathcal{M}, \mathcal{A}, \mathcal{G})$-admissible of type (p), then we call $(\mathcal{M}, \mathcal{A}, \mathcal{G}, g)$ an *admissible structure of type* (p).

2. Now let $(\mathcal{M}, \mathcal{A}, \mathcal{G}, g)$ be an admissible structure of type (p). Let $M : T^*\mathcal{M} \to (0, \infty)$. We call M an $(\mathcal{M}, \mathcal{A}, g)$-*admissible weight function of type* (p), if for each $(\mathcal{X}, X, \kappa) \in \mathcal{A}$, $(\chi^{-1})^* M$ is a $\big(\kappa(\mathcal{X}), \kappa(X), (\chi^{-1})^* g\big)$-admissible weight function of type (p).

For abbreviation we introduce some notation which will be used in the following.

Definition 4.11.

1. Let $(\mathcal{M}, \mathcal{A}, \mathcal{G})$ be a generalized SG-compatible manifold, $\mathcal{A} = \{(\mathcal{X}_j, X_j, \kappa_j) \mid j \in \{1, \ldots, m\}\}$. For $i, j \in \{1, \ldots, m\}$ we define $\mathcal{U}_j := \kappa_j(\mathcal{X}_j)$, $U_j := \kappa_j(X_j)$, $\mathcal{G}_j := (\kappa_j^{-1})^*\mathcal{G}$, $\mathcal{U}_{i,j} := \kappa_j(\mathcal{X}_i \cap \mathcal{X}_j)$, $U_{i,j} := \kappa_j(X_i \cap X_j)$, $\mathcal{G}_{i,j} := (\kappa_j^{-1})^*\mathcal{G}$, and $\kappa_{i,j} := \kappa_j \circ \kappa_i^{-1} : \kappa_i(\mathcal{X}_i \cap \mathcal{X}_j) \to \kappa_j(\mathcal{X}_i \cap \mathcal{X}_j)$. Moreover, let $r_{\mathcal{G}} := \min_{i,j=1}^m \{r_{\mathcal{G}_j}, r_{\mathcal{G}_{i,j}}\}$ and $C_{\mathcal{G}} := \max_{j=1}^m C_{\mathcal{G}_j}$. For each $i, j \in \{1, \ldots, m\}$ there exists a constant $C_{i,j} > 0$ such that for $0 < \rho < r_{\mathcal{G}}/C_{i,j}$ we have $\kappa_{i,j}^{-1}\left(U_{i,j,\rho,\mathcal{G}_{i,j}}\right) \subseteq U_{j,i,C_{i,j}\rho,\mathcal{G}_{j,i}}$. More precisely, we suppose $C_{i,j}$ $(i, j = 1, \ldots, m)$ being chosen sufficiently large such that for $x, y \in \mathcal{U}_{j,i}$ with the property that $x + t(y - x) \in \mathcal{U}_{j,i}$ for all $t \in [0, 1]$ the estimate
$$\mathcal{G}_{i,j,\kappa_{i,j}(x)}(\kappa_{i,j}(x) - \kappa_{i,j}(y)) \le C_{i,j}\mathcal{G}_{j,i,x}(x - y)$$
holds. We define $C_{\mathcal{A}} := \max_{i,j=1}^m C_{i,j}$. Furthermore, let $r_{\mathcal{A}} < r_{\mathcal{G}}$ be fixed. For $(\mathcal{X}, X, \kappa) \in \mathcal{A}$ we define $\widehat{X}_{\mathcal{G}} := \widehat{X} := \kappa^{-1}\left(U_{r_{\mathcal{A}},\mathcal{G}_j}\right)^{\circ}$ (cf. 3.1), where $U := \kappa(X)$.
2. Let $(\mathcal{M}, \mathcal{A}, \mathcal{G}, g)$ be an admissible structure of type (p), $\mathcal{A} = \{(\mathcal{X}_j, X_j, \kappa_j) \mid j \in \{1, \ldots, m\}\}$, and let M be an $(\mathcal{M}, \mathcal{A}, g)$-admissible weight function of type (p). We suppose that the constants introduced for $(\mathcal{U}, U, \mathcal{G})$ are also applicable for $(\mathcal{U}, U, \mathcal{R}_1(g))$. For $i, j \in \{1, \ldots, m\}$ we define $g_j := (\chi_j^{-1})^* g$, $g_{i,j} := (\chi_j^{-1})^* g$, and $M_j := (\chi_j^{-1})^* M$, $M_{i,j} := (\chi_j^{-1})^* M$. Moreover, let $r_g := \min_{i,j=1}^m \{r_{g_j}, r_{g_{i,j}}\}$ and $C_g := \max_{j=1}^m C_{g_j}$. Thus, for an admissible structure $(\mathcal{M}, \mathcal{A}, \mathcal{G}, g)$, we have $r_g \le r_{\mathcal{G}}$ and $C_g \ge C_{\mathcal{G}}$ (cf. 2.7). Finally, for $(\mathcal{X}, X, \kappa) \in \mathcal{A}$, let $\widehat{X}_g := \widehat{X}_{\mathcal{R}_1(g)}$.

Definition 4.12. Let $(\mathcal{M}, \mathcal{A}, \mathcal{G})$ and $(\mathcal{M}, \tilde{\mathcal{A}}, \mathcal{G})$ be generalized SG-compatible manifolds. We write $\mathcal{A} \le \tilde{\mathcal{A}}$ if $\mathcal{A} = \{(\mathcal{X}_j, X_j, \kappa_j) \mid j = 1, \ldots, m\}$ and $\tilde{\mathcal{A}} = \{(\mathcal{X}_j, \tilde{X}_j, \kappa_j)$

$|j = 1,\dots,m\}$ with $X_j \subseteq \tilde{X}_j$. In this case we choose the constants in 4.11 simultaneously for $(\mathcal{M},\mathcal{A},\mathcal{G})$ and $(\mathcal{M},\tilde{\mathcal{A}},\mathcal{G})$.

Lemma 4.13.

1. *Let $(\mathcal{M},\mathcal{A},\mathcal{G})$ be a generalized SG-compatible manifold. We apply the notation from the preceding numbers. Then*

$$\widehat{\mathcal{A}} := \left\{ \left(\mathcal{X}_j, \widehat{X}_{j,\mathcal{G}}, \kappa_j\right) \mid j \in \{1,\dots,m\} \right\}$$

   is a generalized SG-compatible atlas, $\mathcal{A} \le \widehat{\mathcal{A}}$, and $(\mathcal{M},\widehat{\mathcal{A}},\mathcal{G})$ is a generalized SG-compatible manifold. In particular, if $(\mathcal{M},\mathcal{A},\mathcal{G},g)$ is an admissible structure of type (p), then $(\mathcal{M},\widehat{\mathcal{A}},\mathcal{G},g)$ also is an admissible structure of type (p).
2. *Let $\overline{\mathcal{G}}$ be another metric on $\mathcal{M}$ such that $(\mathcal{M},\mathcal{A},\overline{\mathcal{G}})$ is a generalized SG-compatible manifold with $\mathcal{G} \le \overline{\mathcal{G}}$. We suppose that $r_{\mathcal{A}} < \min\{r_{\mathcal{G}}, r_{\overline{\mathcal{G}}}\}$ and let $\underline{\widehat{\mathcal{A}}} := \left\{ \left(\mathcal{X}_j, \widehat{X}_{j,\overline{\mathcal{G}}}, \kappa_j\right) \,\middle|\, j \in \{1,\dots,m\} \right\}$. Then $(\mathcal{M},\underline{\widehat{\mathcal{A}}},\overline{\mathcal{G}})$ also is a generalized SG-compatible manifold and we have $\underline{\widehat{\mathcal{A}}} \le \widehat{\mathcal{A}}$.*

Proof. This proof is a generalization of [41, Lemma 3.3]. For details we refer also to [2, 2.8.7]. □

We proceed with the definition of a canonical scale of Sobolev spaces on generalized SG-compatible manifolds. As a preparation, we first introduce some notations concerning cut-off functions and canonical weight functions which we will apply in the following.

Definition 4.14. Let $(\mathcal{M},\mathcal{A},\mathcal{G})$ be a generalized SG-compatible manifold with atlas $\mathcal{A} = \{(\mathcal{X}_j, X_j, \kappa_j) \mid j = 1,\dots,m\}$.

1. (a) Let $(\mathcal{X}, X, \kappa) \in \mathcal{A}$. Let $0 < r < r_{\mathcal{A}}$ (cf. 4.11). We call a function $\varphi : \mathcal{M} \to [0,1]$ an $(\mathcal{X},X,\kappa,\mathcal{G},r)$-*admissible cut-off function* (or $(\mathcal{X},X,\kappa,\mathcal{G})$-*admissible* or $(\mathcal{X},X,\kappa)$-*admissible* for short) if the following conditions are satisfied:
      - $\operatorname{supp}(\varphi) \subseteq \kappa^{-1}\left(\kappa(X)_r\right)^\circ$ $(\subseteq \mathcal{X})$.
      - $\varphi \circ \kappa^{-1} \in S\left(1, (\kappa^{-1})^*\mathcal{G}\right)$.

   (b) We call a partition of unity $(\varphi_i)_{i=1,\dots,l}$ on $\mathcal{M}$ $(\mathcal{M},\mathcal{A},\mathcal{G})$-*admissible* if for each $i \in \{1,\dots,l\}$ there exists $j_i \in \{1,\dots,m\}$ such that φ_i is an $(\mathcal{X}_{j_i}, X_{j_i}, \kappa_{j_i})$-admissible cut-off function. More precisely, for $0 < r < r_{\mathcal{A}}$ we call $(\varphi_i)_{i=1,\dots,l}$ $(\mathcal{M},\mathcal{A},\mathcal{G},r)$-admissible if for each $i \in \{1,\dots,l\}$, φ_i is an $(\mathcal{X}_{j_i}, X_{j_i}, \kappa_{j_i}, \mathcal{G}, r)$-admissible cut-off function.

   If $(\mathcal{M},\mathcal{A},\mathcal{G},g)$ is an admissible structure, then we call an $(\mathcal{X},X,\kappa,\mathcal{R}_1(g),r)$-admissible cut-off function $(\mathcal{X},X,\kappa,g,r)$-admissible or $(\mathcal{X},X,\kappa,g)$-admissible and an $(\mathcal{M},\mathcal{A},\mathcal{R}_1(g),r)$-admissible partition of unity $(\mathcal{M},\mathcal{A},g,r)$-admissible or $(\mathcal{M},\mathcal{A},g)$-admissible.

2. Let $(\mathcal{X}, X, \kappa) \in \mathcal{A}$ and let φ be an $(\mathcal{X}, X, \kappa)$-admissible cut-off function. We call an $(\mathcal{X}, X, \kappa)$-admissible cut-off function ψ an *associated cut-off function* if $\psi = 1$ on $\operatorname{supp}(\varphi)$.

Lemma 4.15. *We apply the notation from above. Then there exists an $(\mathcal{M}, \mathcal{A}, \mathcal{G})$-admissible partition of unity with associated system of cut-off functions.*

Proof. Let $j \in \{1, \ldots, m\}$. Let $0 < \rho_1 < \rho_2 < \rho_3 < \rho_4 := r/2$ and $0 < \rho$ such that $\rho_i + 3C_{\mathcal{G}} \leq \rho_{i+1}$ $(i = 1, 2, 3)$. Let $\tilde{\phi}_j$ be the function φ from 3.2 with $(\kappa_j(\mathcal{X}_j), \kappa_j(X_j), (\kappa_j^{-1})^*\mathcal{G}_{|\mathcal{X}_j})$ instead of $(\mathcal{U}, U, \mathcal{G})$. We define $\phi_j := \tilde{\phi}_j \circ \kappa_j$. In particular, ϕ_j is an $(\mathcal{X}_j, X_j, \kappa_j)$-admissible cut-off function and $\phi_j = 1$ on X_j. We define $\varphi_j := \frac{\phi_j}{\sum_{i=1}^m \phi_i}$. By virtue of 3.6, φ_j is an $(\mathcal{X}_j, X_j, \kappa_j)$-admissible cut-off function. 3.2 yields the existence of ψ_j for $j = 1, \ldots, m$. □

Definition 4.16.

1. We apply the notation from above. Let $(\varphi_i)_{i=1,\ldots,l}$ be an $(\mathcal{M}, \mathcal{A}, g)$-admissible partition of unity. For $x \in \mathcal{M}$ and $\xi \in T_x^*\mathcal{M}$ we define
$$h_g(x, \xi) := \sum_{i=1}^{l} h_{g_{j_i}}(\chi_{j_i}(x, \xi)) \cdot \varphi_i$$
and
$$H_g(x, \xi) := \sum_{i=1}^{l} H_{g_{j_i}, \mathcal{B}_{j_i}}(\chi_{j_i}(x, \xi)) \cdot \varphi_i(x),$$
where $\mathcal{B}_{j_i}$ is a g_{j_i}-admissible basis (cf. 2.7).
2. Let M be an $(\mathcal{M}, \mathcal{A}, g)$-admissible weight function of type (p). We call M *h_g^{-1}-adapted*, respectively, *H_g^{-1}-adapted*, if with suitable constants $C > 0$ and $N \in \mathbb{N}$ the estimate $C^{-1}h_g^N \leq M \leq Ch_g^{-N}$, respectively, $C^{-1}H_g^N \leq M \leq CH_g^{-N}$, holds.

Remark 4.17. If $(\mathcal{X}, X, \kappa) \in \mathcal{A}$, then there exists a constant $C > 0$ such that for all $x \in \mathcal{X}$ and all $\xi \in T_x^*\mathcal{X}$ the estimate $C^{-1}h_g(x, \xi) \leq h_{(\chi^{-1})^*g}(\chi(x, \xi)) \leq Ch_g(x, \xi)$ holds. If $\tilde{h}_g$ is defined with another $(\mathcal{M}, \mathcal{A}, g)$-admissible partition of unity, then, with a suitable constant $C > 0$, we have $C^{-1}h_g \leq \tilde{h}_g \leq Ch_g$. Moreover, h_g is an $(\mathcal{M}, \mathcal{A}, g)$-admissible weight function of type (p). The same remark holds for H_g. Moreover, up to multiplicative constants, the definition of H_g does not depend on the choice of $\mathcal{B}_{j_i}$ for $i = 1, \ldots, l$.

Definition 4.18. Let $\mathcal{M}$ be a $\mathcal{C}^\infty$-manifold with a complete atlas $\mathcal{A}$. If to every coordinate system $(X_\kappa, \kappa) \in \mathcal{A}$ we are given a distribution $u_\kappa \in \mathcal{D}'(\kappa(X_\kappa))$ such that
$$u_{\tilde{\kappa}} = (\kappa \circ \tilde{\kappa}^{-1})^* u_\kappa \quad \text{in} \quad \tilde{\kappa}(X_\kappa \cap X_{\tilde{\kappa}})$$
$((X_\kappa, \kappa), (X_{\tilde{\kappa}}, \tilde{\kappa}) \in \mathcal{A})$, we call the system a *distribution* u in $\mathcal{M}$ and write $(\kappa^{-1})^*u := u_\kappa$. The set of all distributions in $\mathcal{M}$ is denoted by $\mathcal{D}'(\mathcal{M})$.

Remark 4.19. Let $(X_j,\kappa_j)_{j=1,\dots,m}$ be a system of charts of $\mathcal{M}$ such that $\mathcal{M} = \bigcup_{j=1}^{m} X_j$. Let $(\varphi_i)_{i=1,\dots,l}$ be a partition of unity subordinate to $(X_j)_{j=1,\dots,m}$ (i.e., $\sum_{i=1}^{l} \varphi_j = 1$ and for all $i \in \{1,\dots,l\}$ there exists $j_i \in \{1,\dots,m\}$ such that $\operatorname{supp}(\varphi_i) \subset X_{j_i}$). Then the following statements are equivalent:

$$\begin{aligned} u \in \mathcal{D}'(\mathcal{M}) &\iff \varphi_i u \in \mathcal{D}'(\mathcal{M}) \quad \text{for} \quad i=1,\dots,l \\ &\iff \varphi_i u \in \mathcal{D}'(X_{j_i}) \quad \text{for} \quad i=1,\dots,l \\ &\iff (\kappa_{j_i}^{-1})^*(\varphi_i u) \in \mathcal{D}'(\kappa_{j_i}(X_{j_i})) \quad \text{for} \quad i=1,\dots,l\,. \end{aligned}$$

In the last equivalence we have applied [28, Theorem 6.3.4].

Definition 4.20. Let $(\mathcal{M},\mathcal{A},\mathcal{G},g)$ be an admissible structure of type (p). Let H_g be as in 4.16. For $s \in \mathbb{R}$ we define $H\left(H_g^s, g; \mathcal{M},\mathcal{A}\right)$ as the set of all $u \in \mathcal{D}'(\mathcal{M})$ such that

$$(\kappa^{-1})^*(\varphi u) \in H\left((\chi^{-1})^*(H_g^s), (\chi^{-1})^* g; \kappa(\widehat{X}_g)\right)$$

for each $(\mathcal{X},X,\kappa) \in \mathcal{A}$ and each $(\mathcal{X},X,\kappa)$-admissible cut-off function φ, where we have applied the notation from 4.10, 4.11, and 4.14.

Remark 4.21. Obviously $H\left(H_g^s, g; \mathcal{M},\mathcal{A}\right)$ is a vector space and we have

$$H\left(H_g^{s'}, g; \mathcal{M},\mathcal{A}\right) \subseteq H\left(H_g^s, g; \mathcal{M},\mathcal{A}\right)$$

for $s' \leq s$.

Lemma 4.22. *We apply the notation of the preceding definition. Let $u \in \mathcal{D}'(\mathcal{M})$, $(\mathcal{X},X,\kappa) \in \mathcal{A}$, and let φ be an $(\mathcal{X},X,\kappa)$-admissible cut-off function. If we have*

$$(\kappa^{-1})^*(\varphi u) \in H\left((\chi^{-1})^*(H_g^s), (\chi^{-1})^* g; \kappa(\widehat{X}_g)\right)\,,$$

then $\varphi u \in H\left(H_g^s, g; \mathcal{M},\mathcal{A}\right)$.

Proof. This is a consequence of 4.7. □

Corollary 4.23. *We apply the notation from the preceding definition and from 4.14. Let $(\varphi_i)_{i=1,\dots,l}$ be an $(\mathcal{M},\mathcal{A},g)$-admissible partition of unity. For $u \in \mathcal{D}'(\mathcal{M})$ we have $u \in H\left(H_g^s, g; \mathcal{M},\mathcal{A}\right)$ if and only if for all $i=1,\dots,l$,*

$$(\kappa_{j_i}^{-1})^*(\varphi_i u) \in H\left((\chi_{j_i}^{-1})^*(H_g^s), (\chi_{j_i}^{-1})^* g; \kappa_{j_i}(\widehat{X}_{j_i,g})\right)\,.$$

Corollary 4.24. *Let $(\mathcal{M},\mathcal{A},\mathcal{G},g)$, $(\mathcal{M},\tilde{\mathcal{A}},\mathcal{G},g)$, and $(\mathcal{M},\mathcal{A}',\mathcal{G},g)$ be admissible structures of type (p). If $\mathcal{A} \leq \tilde{\mathcal{A}}$ and $\mathcal{A} \subseteq \mathcal{A}'$, then we have $H\left(H_g^s, g; \mathcal{M},\mathcal{A}\right) = H\left(H_g^s, g; \mathcal{M},\tilde{\mathcal{A}}\right) = H\left(H_g^s, g; \mathcal{M},\mathcal{A}'\right)$ for each $s \in \mathbb{R}$.*

Proof. This is a consequence of the preceding corollary together with 4.4. □

Definition 4.25. Let $(\mathcal{M}, \mathcal{A}, \mathcal{G}, g)$ be an admissible structure of type (p). Let $r, s \in \mathbb{R}$. We call a linear operator $P : H\left(H_g^r, g; \mathcal{M}, \mathcal{A}\right) \to H\left(H_g^s, g; \mathcal{M}, \mathcal{A}\right)$ *bounded in charts* if for all $(\mathcal{X}, X, \kappa), (\tilde{\mathcal{X}}, \tilde{X}, \tilde{\kappa}) \in \mathcal{A}$, all $(\mathcal{X}, X, \kappa)$-admissible cut-off functions φ, and all $(\tilde{\mathcal{X}}, \tilde{X}, \tilde{\kappa})$-admissible cut-off functions $\tilde{\varphi}$ the operator

$$\begin{array}{rcl} H\left((\tilde{\chi}^{-1})^*(H_g^r), (\tilde{\chi}^{-1})^* g; \tilde{\kappa}(\widehat{\tilde{X}}_g)\right) & \to & H\left((\chi^{-1})^*(H_g^s), (\chi^{-1})^* g; \kappa(\widehat{X}_g)\right) \\ u & \mapsto & (\kappa^{-1})^* \left[\varphi P\left(\tilde{\varphi} \cdot \tilde{\kappa}^* u\right)\right] \end{array}$$

(which is well defined by virtue of 4.22) is continuous.

Remark 4.26.

1. Let $(\mathcal{M}, \mathcal{A}, \mathcal{G}, g)$ be an admissible structure of type (p). Let $r, s, t \in \mathbb{R}$. Moreover, let $P : H\left(H_g^r, g; \mathcal{M}, \mathcal{A}\right) \to H\left(H_g^s, g; \mathcal{M}, \mathcal{A}\right)$ and $Q : H\left(H_g^s, g; \mathcal{M}, \mathcal{A}\right) \to H\left(H_g^t, g; \mathcal{M}, \mathcal{A}\right)$ be bounded in charts. Then $QP : H\left(H_g^r, g; \mathcal{M}, \mathcal{A}\right) \to H\left(H_g^t, g; \mathcal{M}, \mathcal{A}\right)$ is bounded in charts, too.
2. If $P : H\left(H_g^r, g; \mathcal{M}, \mathcal{A}\right) \to H\left(H_g^s, g; \mathcal{M}, \mathcal{A}\right)$ is bounded in charts, then P is uniquely determined by its actions on $\mathcal{C}_0^\infty(\mathcal{M})$.

Lemma 4.27. *Let $(\mathcal{M}, \mathcal{A}, \mathcal{G}, g)$ be an admissible structure of type (p), let $(\mathcal{X}, X, \kappa)$, $(\tilde{\mathcal{X}}, \tilde{X}, \tilde{\kappa}) \in \mathcal{A}$, and let φ be an $(\mathcal{X}, X, \kappa)$-admissible, $\tilde{\varphi}$ be a $(\tilde{\mathcal{X}}, \tilde{X}, \tilde{\kappa})$-admissible cut-off function. Let $r, s \in \mathbb{R}$ and let $P : \mathcal{C}_0^\infty(\mathcal{M}) \to \mathcal{D}'(\mathcal{M})$ be an operator such that*

$$\begin{array}{rcl} H\left((\tilde{\chi}^{-1})^*(H_g^r), (\tilde{\chi}^{-1})^* g; \tilde{\kappa}(\widehat{\tilde{X}}_g)\right) & \to & H\left((\chi^{-1})^*(H_g^s), (\chi^{-1})^* g; \kappa(\widehat{X}_g)\right) \\ u & \mapsto & (\kappa^{-1})^* \left[\varphi P\left(\tilde{\varphi} \cdot \tilde{\kappa}^* u\right)\right] \end{array}$$

is continuous and extends to a continuous map $\mathcal{S}'(\mathbb{R}^n) \to \mathcal{S}'(\mathbb{R}^n)$. Then $\varphi P \tilde{\varphi} : H\left(H_g^r, g; \mathcal{M}, \mathcal{A}\right) \to H\left(H_g^s, g; \mathcal{M}, \mathcal{A}\right)$ is bounded in charts.

Proof. This follows from 4.22 and the closed graph theorem. □

Definition 4.28. Let $(\mathcal{M}, \mathcal{A}, \mathcal{G}, g)$ be an admissible structure of type (p). By $\mathcal{L}(g; \mathcal{M}, \mathcal{A})^{-\infty}$ we denote the set of all operators $P : \mathcal{C}_0^\infty(\mathcal{M}) \to \mathcal{D}'(\mathcal{M})$ which have for each $s \in \mathbb{R}$ an extension P_s mapping $H\left(H_g^s, g; \mathcal{M}, \mathcal{A}\right)$ bounded in charts to $H\left(H_g^{s+s'}, g; \mathcal{M}, \mathcal{A}\right)$ for all $s' \in \mathbb{R}$.

Lemma 4.29. *Let $(\mathcal{M}, \mathcal{A}, \mathcal{G}, g)$ be an admissible structure of type (p). Let $(\mathcal{X}, X, \kappa) \in \mathcal{A}$ and let φ be an $(\mathcal{X}, X, \kappa)$-admissible cut-off function with associated cut-off function ψ (cf. 4.14). Let $P : \mathcal{C}_0^\infty(\mathcal{M}) \to \mathcal{D}'(\mathcal{M})$, let $s' \in \mathbb{R}$, and let $p \in S\left((\chi^{-1})^*(H_g^{s'}), (\chi^{-1})^* g\right)$ such that*

$$Pu = \kappa^* \left[\left((\varphi \circ \kappa^{-1}) \cdot p\right)(x, D)(\kappa^{-1})^* (\psi u)\right]$$

for $u \in \mathcal{C}_0^\infty(\mathcal{M})$. Then, for each $s \in \mathbb{R}$, the operator P extends to an operator

$$P_s : H\left(H_g^s, g; \mathcal{M}, \mathcal{A}\right) \to H\left(H_g^{s-s'}, g; \mathcal{M}, \mathcal{A}\right)$$

which is bounded on charts.

Proof. The assertion is a consequence of 4.27. □

As canonical domain of the pseudo-differential operators to be introduced in the following section we define the space $\mathcal{S}(\mathcal{M})$.

Definition 4.30. Let $(\mathcal{M}, \mathcal{A}, \mathcal{G})$ be a generalized SG-compatible manifold. We define $\mathcal{S}(\mathcal{M})$ as the set of all functions $u \in \mathcal{C}^\infty(\mathcal{M})$ such that, for all $(\mathcal{X}, X, \kappa) \in \mathcal{A}$ and all functions $\varphi \in \mathcal{C}^\infty(\mathcal{M})$ with the properties that

- $\operatorname{supp}(\varphi) \subseteq \mathcal{X}$ and
- $\varphi \circ \kappa^{-1}$ is a function on $\mathbb{R}^n$ with all derivatives of order ≥ 0 being at most of polynomial growth,

we have $(\kappa^{-1})^*(\varphi \cdot u) \in \mathcal{S}(\mathbb{R}^n)$.

5. Pseudo-differential operators on manifolds

Following the usual approach, we define $\Psi(M, g; \mathcal{M}, \mathcal{A}, \mathcal{G})$ – up to negligible terms in $\Psi(g; \mathcal{M}, \mathcal{A}, \mathcal{G})^{-\infty}$ – as the set of all operators $\mathcal{C}_0^\infty(\mathcal{M}) \to \mathcal{D}'(\mathcal{M})$ with representations as pseudo-differential operators in local coordinates (in the corresponding classes). One important result and tool in the following is the conclusion from one local representation to those in other coordinates (cf. 5.12). In the proof we mainly apply the cut-off functions from 3.2 and furthermore Theorem 3.12 for the transformation of the part of the operator with Schwartz kernel being supported near the diagonal. In this theorem we also calculate the principal symbol of the operator (cf. also section 6). Moreover, we discuss mapping properties of pseudo-differential operators in the scale $H\left(H_g^s, g; \mathcal{M}, \mathcal{A}\right)$ $(s \in \mathbb{R})$ of Sobolev spaces (cf. 5.14), and prove that the composition of two pseudo-differential operators is such an operator, too (cf. 5.17). We have decided to describe the operators in $\Psi(g; \mathcal{M}, \mathcal{A}, \mathcal{G})^{-\infty}$ with the help of mapping properties of iterated commutators with distinguished vector fields and multipliers as operators of order $-\infty$ in the scale of Sobolev spaces $H\left(H_g^s, g; \mathcal{M}, \mathcal{A}\right)$ $(s \in \mathbb{R})$ (cf. 5.4). This description is motivated by characterizations of pseudo-differential operators as in [1, 3.6.6]. Nevertheless, we can prove that for an appropriate choice of the cut-off functions the remainder terms have integral representations with rapidly decreasing kernels (cf. 5.18, 5.19; cf. also 6.13).

Definition 5.1. Let $(\mathcal{M}, \mathcal{A}, \mathcal{G})$ be a generalized SG-compatible manifold. We define $\mathcal{M}_{c,\mathcal{A}} := \bigcup_{(\mathcal{X},X,\kappa)\in\mathcal{A}} \mathcal{X} \setminus X$.

Definition 5.2. Let $(\mathcal{M}, \mathcal{A}, \mathcal{G}, g)$ be an admissible structure of type (p).

1. By $\mathcal{V}_m(\mathcal{M}, \mathcal{A}, \mathcal{G})$ we denote the set of all operators of the form
$$\kappa^* \left[M_{x_j}(\kappa^{-1})^*(\phi \cdot u)\right]$$
where $u \in \mathcal{C}^\infty(\mathcal{M})$, $j \in \{1, \ldots, n\}$, $(M_{x_j} f)(x) = x_j f(x)$ for $f : \mathbb{R}^n \to \mathbb{C}$, $(\mathcal{X}, X, \kappa) \in \mathcal{A}$, and $\phi \in \mathcal{C}^\infty(\mathcal{M})$ is a function satisfying the following properties:

- There exists an atlas $\tilde{\mathcal{A}}$ such that $(\mathcal{M}, \tilde{\mathcal{A}}, \mathcal{G}, g)$ is an admissible structure of type (p), too, $\mathcal{A} \leq \tilde{\mathcal{A}}$, and ϕ is an $(\mathcal{X}, \tilde{X}, \kappa, \mathcal{G})$-admissible cut-off function, where $(\mathcal{X}, \tilde{X}, \kappa) \in \tilde{\mathcal{A}}$.
- $d\phi = 0$ on $\mathcal{M} \setminus \mathcal{M}_{c,\mathcal{A}}$.

2. By $\mathcal{V}_d(\mathcal{M}, \mathcal{A}, \mathcal{G})$ we denote the set of all operators of the form

$$\kappa^* \left[D_{x_j} (\kappa^{-1})^* (\phi \cdot u) \right]$$

where we have applied the notation from 1.

3. We define $\mathcal{V}(\mathcal{M}, \mathcal{A}, \mathcal{G}) = \mathcal{V}_m(\mathcal{M}, \mathcal{A}, \mathcal{G}) \cup \mathcal{V}_d(\mathcal{M}, \mathcal{A}, \mathcal{G})$.

Lemma 5.3. *Let $(\mathcal{M}, \mathcal{A}, \mathcal{G}, g)$ be an admissible structure of type (p).*

1. *For $u \in \mathcal{D}'(\mathcal{M})$, $\varphi \in \mathcal{S}(\mathcal{M})$ and, $V \in \mathcal{V}(\mathcal{M}, \mathcal{A}, \mathcal{G})$ we have $Vu \in \mathcal{D}'(\mathcal{M})$ and $V\varphi \in \mathcal{S}(\mathcal{M})$.*
2. *Let $u \in \mathcal{D}'(\mathcal{M})$ such that $V_1 \dots V_N u \in H(1, g; \mathcal{M}, \mathcal{A})$ for all $N \in \mathbb{N}_0$ and all $V_1, \dots, V_N \in \mathcal{V}(\mathcal{M}, \mathcal{A}, \mathcal{G})$. Then we have $u \in \mathcal{S}(\mathcal{M})$. More precisely, if $u \in \mathcal{D}'(\mathcal{M})$ such that $V_1 \dots V_k \tilde{V}_1 \dots \tilde{V}_l u \in H(1, g; \mathcal{M}, \mathcal{A})$ for all $k, l \in \mathbb{N}_0$ and all operators $V_1 \dots V_k \in \mathcal{V}_d(\mathcal{M}, \mathcal{A}, \mathcal{G})$ and all operators $\tilde{V}_1 \dots \tilde{V}_l \in \mathcal{V}_m(\mathcal{M}, \mathcal{A}, \mathcal{G})$, then $u \in \mathcal{S}(\mathcal{M})$.*

Proof. For the proof of Part 2 we apply Sobolev's embedding theorem. □

Definition 5.4. Let $(\mathcal{M}, \mathcal{A}, \mathcal{G}, g)$ be an admissible structure of type (p). We define the class of negligible $\Psi(g; \mathcal{M}, \mathcal{A}, \mathcal{G})^{-\infty}$ as the set of all operators $P : C_0^\infty(\mathcal{M}) \to \mathcal{D}'(\mathcal{M})$ with the property that for all $N \in \mathbb{N}_0$ and all $V_1, \dots, V_N \in \mathcal{V}(\mathcal{M}, \mathcal{A}, \mathcal{G})$ the operator[4] $\operatorname{ad}(V_1) \dots \operatorname{ad}(V_N) P : C_0^\infty(\mathcal{M}) \to \mathcal{D}'(\mathcal{M})$ is in $\mathcal{L}(g; \mathcal{M}, \mathcal{A})^{-\infty}$. In the case $\mathcal{G} = \mathcal{R}_1(g)$ we write $\Psi(g; \mathcal{M}, \mathcal{A})^{-\infty}$.

Lemma 5.5. *Let $(\mathcal{M}, \mathcal{A}, \mathcal{G}, g)$ and $(\mathcal{M}, \tilde{\mathcal{A}}, \mathcal{G}, g)$ be admissible structures of type (p) such that $\mathcal{A} \leq \tilde{\mathcal{A}}$. Let $P \in \Psi(g; \mathcal{M}, \mathcal{A}, \mathcal{G})^{-\infty}$. Moreover, let $(\mathcal{X}, \tilde{X}, \kappa) \in \tilde{\mathcal{A}}$ and let φ, ψ be $(\mathcal{X}, \tilde{X}, \kappa)$-admissible cut-off functions. Then there exists $p \in \bigcap_{q \in \mathbb{N}} S\left((\chi^{-1})^*(H_g^q), (\chi^{-1})^* g\right)$ such that*

$$\varphi P(\psi u) = \kappa^* \left[\left((\varphi \circ \kappa^{-1}) \cdot p \right)(x, D) (\kappa^{-1})^* (\psi u) \right]$$

for $u \in C_0^\infty(\mathcal{M})$.

Proof. This is a consequence of the definition together with [1, 3.6.6]. □

Lemma 5.6. *Let $(\mathcal{M}, \mathcal{A}, \mathcal{G}, g)$ and $(\mathcal{M}, \underline{\mathcal{A}}, \mathcal{G}, g)$ be admissible structures of type (p) with $\underline{\mathcal{A}} \leq \mathcal{A}$. Let $(\mathcal{X}, X, \kappa) \in \mathcal{A}$ and let φ be an $(\mathcal{X}, X, \kappa, g)$-admissible cut-off function with associated cut-off function ψ (cf. 4.14). Let $P : C_0^\infty(\mathcal{M}) \to \mathcal{D}'(\mathcal{M})$, let $s \in \mathbb{R}$, and let $p \in S\left((\chi^{-1})^*(H_g^s), (\chi^{-1})^* g\right)$ such that*

$$Pu = \kappa^* \left[\left((\varphi \circ \kappa^{-1}) \cdot p \right)(x, D) (\kappa^{-1})^* (\psi u) \right]$$

[4] As usual, we define $\operatorname{ad}(V)P := VP - PV : C_0^\infty(\mathcal{M}) \to \mathcal{D}'(\mathcal{M})$ and apply this notation also to unique extensions.

for $u \in C_0^\infty(\mathcal{M})$. *Then, for all* $N \in \mathbb{N}$ *and all* $V_1, \dots, V_N \in \mathcal{V}(\mathcal{M}, \underline{\mathcal{A}}, \mathcal{G})$, *there exist* $s_N \in \mathbb{R}$ *and* $p(V_1, \dots, V_N) \in S\left((\chi^{-1})^*(H_g^{s-s_N})_{|T^*\mathcal{X}}, (\chi^{-1})^* g_{|T^*\mathcal{X}}\right)$ *such that*

$$ad(V_1) \dots ad(V_N) Pu = \kappa^* \left[\left((\varphi_1 \circ \kappa^{-1}) \cdot p(V_1, \dots, V_N) \right)(x, D)(\kappa^{-1})^* (\psi_1 u) \right]$$

for $u \in C_0^\infty(\mathcal{M})$ *and for all* $(\mathcal{X}, X, \mathcal{A}, g)$*-admissible cut-off functions* φ_1 *with associated cut-off functions* ψ_1 *such that* $\varphi_1 = 1$ *on* $supp\,(\varphi)$ *and* $\psi_1 = 1$ *on* $supp\,(\psi)$.

Proof. For one commutator this follows from the behavior of the remainder term in the asymptotic expansion of the product symbol. More precisely, in the case of an operator in $\mathcal{V}_m(\mathcal{M}, \mathcal{A}, \mathcal{G})$ one uses an approximation argument. Then the complete assertion follows inductively. For a detailed proof we refer to [2, 2.10.10]. □

Corollary 5.7. *We apply the notation from* 5.6. *Let* $P : C_0^\infty(\mathcal{M}) \to \mathcal{D}'(\mathcal{M})$, *and* $p \in \bigcap_{q \in \mathbb{N}} S\left((\chi^{-1})^*(H_g^q), (\chi^{-1})^* g\right)$ *such that*

$$Pu = \kappa^* \left[\left((\varphi \circ \kappa^{-1}) \cdot p \right)(x, D)(\kappa^{-1})^* (\psi u) \right]$$

for $u \in C_0^\infty(\mathcal{M})$. *Then we have* $P \in \Psi\left(g; \mathcal{M}, \underline{\mathcal{A}}, \mathcal{G}\right)^{-\infty}$.

Proof. This is a consequence of 5.6, 4.29, and the fact that $\mathcal{L}(g; \mathcal{M}, \mathcal{A})^{-\infty} \subseteq \mathcal{L}(g; \mathcal{M}, \underline{\mathcal{A}})^{-\infty}$. □

Lemma 5.8. *Let* $(\mathcal{M}, \mathcal{A}, \mathcal{G}, g)$ *be an admissible structure of type* (p). *Moreover, let* $(\mathcal{X}, X, \kappa), (\tilde{\mathcal{X}}, \tilde{X}, \tilde{\kappa}) \in \mathcal{A}$, *let* φ *be an* $(\mathcal{X}, X, \kappa)$*-admissible cut-off function, let* $\tilde{\varphi}$ *be a* $(\tilde{\mathcal{X}}, \tilde{X}, \tilde{\kappa})$*-admissible cut-off function, and let* $k \in \mathcal{S}(\mathbb{R}^{2n})$. *We define* ($u \in C_0^\infty(\mathcal{M})$)

$$Pu := \kappa^* \left[\varphi \circ \kappa^{-1} \cdot \int_{\mathbb{R}^n} k(\cdot, y) \cdot (\tilde{\varphi} \cdot u)(\tilde{\kappa}^{-1}(y)) dy \right].$$

Then we have $P \in \Psi(g; \mathcal{M}, \mathcal{A}, \mathcal{G})^{-\infty}$.

Proof. The assertion follows with 4.27. □

Now we define pseudo-differential operators, associated to a metric g and a weight function M, on $\mathcal{M}$ or, more precisely, on the generalized SG-compatible manifold $(\mathcal{M}, \mathcal{A}, \mathcal{G})$.

Definition 5.9. Let $(\mathcal{M}, \mathcal{A}, \mathcal{G}, g)$ be an admissible structure of type (p). Let M be an $(\mathcal{M}, \mathcal{A}, g)$-admissible weight function of type (p) which is H_g^{-1}-adapted (cf. 4.16). By $\Psi(M, g; \mathcal{M}, \mathcal{A}, \mathcal{G})$ we denote the set of all operators $P : C_0^\infty(\mathcal{M}) \to \mathcal{D}'(\mathcal{M})$ such that for all $(\mathcal{X}, X, \kappa) \in \mathcal{A}$, for each $(\mathcal{X}, X, \kappa, g)$-admissible cut-off function φ, and for each associated cut-off function ψ there exists a symbol $p \in S\left((\chi^{-1})^* M, (\chi^{-1})^* g\right)$ such that

$$\begin{aligned} \varphi P(\psi u) &= \kappa^* \left[\left((\varphi \circ \kappa^{-1}) \cdot p \right)(x, D)(\kappa^{-1})^* (\psi u) \right] \quad (u \in C_0^\infty(\mathcal{M})), \\ \varphi P(1 - \psi) &\in \Psi(g; \mathcal{M}, \mathcal{A}, \mathcal{G})^{-\infty}. \end{aligned}$$

In the case $\mathcal{G} = \mathcal{R}_1(g)$ we write $\Psi(M, g; \mathcal{M}, \mathcal{A})$.

Remark 5.10. Obviously $\Psi(M, g; \mathcal{M}, \mathcal{A}, \mathcal{G})$ is a vector space and we have $\Psi(g; \mathcal{M}, \mathcal{A}, \mathcal{G})^{-\infty} \subseteq \Psi(M, g; \mathcal{M}, \mathcal{A}, \mathcal{G})$.

Corollary 5.11. *Let $P \in \Psi(M, g; \mathcal{M}, \mathcal{A}, \mathcal{G})$. Moreover, let $\tilde{\varphi}, \tilde{\psi} \in \mathcal{C}^\infty(\mathcal{M})$ with $\tilde{\varphi} \cdot \tilde{\psi} = 0$ such that $[(x,\xi) \mapsto \tilde{\varphi} \circ \kappa^{-1}(x)]$, $[(x,\xi) \mapsto \tilde{\psi} \circ \kappa^{-1}(x)] \in S\left(1, (\chi^{-1})^* g_{|T^*\mathcal{X}}\right)$ for all $(\mathcal{X}, X, \kappa) \in \mathcal{A}$. Then we have $\tilde{\varphi} P \tilde{\psi} \in \Psi(g; \mathcal{M}, \mathcal{A}, \mathcal{G})^{-\infty}$.*

Proof. Fixing an $(\mathcal{M}, \mathcal{A}, g)$-admissible partition of unity with associated system of cut-off functions, 5.7 yields the assertion. □

The following theorem is the "work horse" in our discussion of the operators in $\Psi(M, g; \mathcal{M}, \mathcal{A}, \mathcal{G})$.

Theorem 5.12. *Let $(\mathcal{M}, \mathcal{A}, \mathcal{G}, g)$ be an admissible structure of type (p). Let M be an $(\mathcal{M}, \mathcal{A}, g)$-admissible weight function of type (p) which is H_g^{-1}-adapted. Let $(\mathcal{X}, X, \kappa) \in \mathcal{A}$, let φ be an $(\mathcal{X}, X, \kappa, g)$-admissible cut-off function, and let ψ be an associated cut-off function. Let $P : \mathcal{C}_0^\infty(\mathcal{M}) \to \mathcal{D}'(\mathcal{M})$ be an operator such that there exists a symbol $p \in S\left((\chi^{-1})^* M, (\chi^{-1})^* g\right)$ with*

$$\begin{aligned} \varphi P(\psi u) &= \kappa^* \left[\left((\varphi \circ \kappa^{-1}) \cdot p \right)(x, D) (\kappa^{-1})^* (\psi u) \right] \quad (u \in \mathcal{C}_0^\infty(\mathcal{M})), \\ \varphi P(1 - \psi) &\in \Psi(g; \mathcal{M}, \mathcal{A}, \mathcal{G})^{-\infty}. \end{aligned}$$

Then $\varphi P \in \Psi(M, g; \mathcal{M}, \mathcal{A}, \mathcal{G})$. Moreover, if $(\tilde{\mathcal{X}}, \tilde{X}, \tilde{\kappa}) \in \mathcal{A}$ and $\tilde{\varphi}$ is a $(\tilde{\mathcal{X}}, \tilde{X}, \tilde{\kappa}, g)$-admissible cut-off function with associated cut-off function $\tilde{\psi}$, then there exists $\tilde{p} \in S\left((\tilde{\chi}^{-1})^* M, (\tilde{\chi}^{-1})^* g\right)$ such that for each $u \in \mathcal{C}_0^\infty(\mathcal{M})$ we have

$$\tilde{\varphi} \varphi P(\tilde{\psi} u) = \tilde{\kappa}^* \left[\left((\tilde{\varphi} \circ \tilde{\kappa}^{-1}) \cdot \tilde{p} \right)(x, D) (\tilde{\kappa}^{-1})^* \left(\tilde{\psi} u \right) \right]$$

where

$$\tilde{\varphi}(\tilde{\kappa}^{-1}(x)) \cdot \tilde{p}(x, \xi) = (\varphi \cdot \tilde{\varphi})(\tilde{\kappa}^{-1}(x)) \cdot p\left((\kappa \circ \tilde{\kappa}^{-1})(x), {}^t d\left(\kappa \circ \tilde{\kappa}^{-1}(x)\right)^{-1} \xi \right) + r(x, \xi)$$

$((x, \xi) \in \tilde{\kappa}(\tilde{\mathcal{X}}) \times \mathbb{R}^n)$ with $r \in S\left((\tilde{\chi}^{-1})^* (M h_g), (\tilde{\chi}^{-1})^* g\right)$.

Proof. As in the proof of 3.13, applying 3.6, 3.2, 3.8, and 3.12 we show that $\varphi P \psi \in \Psi(M, g; \mathcal{M}, \mathcal{A}, \mathcal{G})$ and thus, $\varphi P \in \Psi(M, g; \mathcal{M}, \mathcal{A}, \mathcal{G})$. For the last assertion we have to apply 3.12 once more. For details we refer to [2, 2.10.27]. □

The preceding theorem allows us to characterize pseudo-differential operators on an admissible structure $(\mathcal{M}, \mathcal{A}, \mathcal{G}, g)$ of type (p) by the help of a fixed $(\mathcal{M}, \mathcal{A}, g)$-admissible partition of unity and an associated system of cut-off functions.

Lemma 5.13. *Let $(\mathcal{M}, \mathcal{A}, \mathcal{G}, g)$ and $(\mathcal{M}, \tilde{\mathcal{A}}, \mathcal{G}, g)$ be admissible structures of type (p) with $\mathcal{A} \leq \tilde{\mathcal{A}}$. Let M be an $(\mathcal{M}, \tilde{\mathcal{A}}, g)$-admissible weight function of type (p)[5] which is H_g^{-1}-adapted. Let $(\mathcal{X}, X, \kappa) \in \mathcal{A}$, $(\mathcal{X}, \tilde{X}, \kappa) \in \tilde{\mathcal{A}}$, $X \subseteq \tilde{X}$, let φ be an $(\mathcal{X}, X, \kappa)$-admissible cut-off function, let $\tilde{\psi}$ be an $(\mathcal{X}, \tilde{X}, \kappa)$-admissible cut-off function, let $p \in S((\chi^{-1})^* M, (\chi^{-1})^* g)$, and let $P : \mathcal{C}_0^\infty(\mathcal{M}) \to \mathcal{D}'(\mathcal{M})$ be an operator such that*

$$\varphi P(\tilde{\psi} u) = \kappa^* \left[\left((\varphi \circ \kappa^{-1}) \cdot p \right)(x, D) (\kappa^{-1})^* (\tilde{\psi} u) \right] .$$

Then we have $\varphi P \tilde{\psi} \in \Psi(M, g; \mathcal{M}, \mathcal{A}, \mathcal{G})$.

[5] Then M also is an $(\mathcal{M}, \mathcal{A}, g)$-admissible weight function of type (p).

Proof. Let ψ be an $(\mathcal{X}, X, \kappa)$-admissible cut-off function with $\psi = 1$ on $\operatorname{supp}(\varphi)$. Then, by virtue of 5.12, we have $\varphi P \tilde{\psi} \psi \in \Psi(M, g; \mathcal{M}, \mathcal{A}, \mathcal{G})$. Moreover, again by virtue of 5.12, we have $\varphi P \tilde{\psi} \in \Psi(M, g; \mathcal{M}, \tilde{\mathcal{A}}, \mathcal{G})$ and thus, by virtue of 5.11, $\varphi P \tilde{\psi}(1-\psi) \in \Psi(g; \mathcal{M}, \tilde{\mathcal{A}}, \mathcal{G})^{-\infty}$. Then 5.5 and 5.7 yield $\varphi P \tilde{\psi}(1-\psi) \in \Psi(g; \mathcal{M}, \mathcal{A}, \mathcal{G})^{-\infty}$. □

Remark 5.14. Let $(\mathcal{M}, \mathcal{A}, \mathcal{G}, g)$ be an admissible structure of type (p). Let $s_M \in \mathbb{R}$ and let $P \in \Psi(H_g^{s_M}, g; \mathcal{M}, \mathcal{A}, \mathcal{G})$. Then, by virtue of 4.29, for each $s \in \mathbb{R}$, P extends to an operator $H\left(H_g^s, g; \mathcal{M}, \mathcal{A}\right) \to H\left(H_g^{s-s_M}, g; \mathcal{M}, \mathcal{A}\right)$ which is bounded in charts. Moreover, P extends canonically to an operator $\mathcal{S}(\mathcal{M}) \to \mathcal{S}(\mathcal{M})$.

Lemma 5.15. *Let $(\mathcal{M}, \mathcal{A}, \mathcal{G}, g)$ be an admissible structure of type (p). Let $s \in \mathbb{R}$, $P \in \Psi(H_g^s, g; \mathcal{M}, \mathcal{A}, \mathcal{G})$, $N \in \mathbb{N}$, and let $V_1, \ldots, V_N \in \mathcal{V}(\mathcal{M}, \mathcal{A}, \mathcal{G})$. Then there exists s_N such that $ad(V_1) \ldots ad(V_N) P \in \Psi(H_g^{s-s_N}, g; \mathcal{M}, \mathcal{A}, \mathcal{G})$.*

Proof. This is a consequence of 5.6 and 5.12. □

Corollary 5.16. *Let $(\mathcal{M}, \mathcal{A}, \mathcal{G}, g)$ be an admissible structure of type (p). Let $s \in \mathbb{R}$, $P \in \Psi(H_g^s, g; \mathcal{M}, \mathcal{A}, \mathcal{G})$, and $Q \in \Psi(g; \mathcal{M}, \mathcal{A}, \mathcal{G})^{-\infty}$. Then we have $PQ, QP \in \Psi(g; \mathcal{M}, \mathcal{A}, \mathcal{G})^{-\infty}$.*

Theorem 5.17. *Let $(\mathcal{M}, \mathcal{A}, \mathcal{G}, g)$ be an admissible structure of type (p). For $i = 1, 2$ let M_i be an $(\mathcal{M}, \mathcal{A}, g)$-admissible weight function of type (p) which is H_g^{-1}-adapted. Moreover, for $i = 1, 2$ let $P_i \in \Psi(M_i, g; \mathcal{M}, \mathcal{A}, \mathcal{G})$. Then we have $P_1 P_2 \in \Psi(M_1 M_2, g; \mathcal{M}, \mathcal{A}, \mathcal{G})$.*

Proof. Let $(\varphi_i)_{i=1,\ldots,l}$ be an $(\mathcal{M}, \mathcal{A}, g)$-admissible partition of unity with associated system of cut-off functions $(\psi_i)_{i=1,\ldots,l}$. By virtue of 5.11 and 5.16 it suffices to consider $\varphi_i P_1 \psi_i \varphi_k P_2 \psi_k$ for $i, k \in \{1, \ldots, l\}$. With the notation from 4.14, let $\tilde{\varphi}_k$ be an $(\mathcal{X}_{j_k}, X_{j_k}, \kappa_{j_k}, g)$-admissible cut-off function with associated cut-off function $\tilde{\psi}_k$ such that $\psi_k = 1$ on $\operatorname{supp}(\varphi_k)$ and $\tilde{\varphi}_k = 1$ on $\operatorname{supp}(\tilde{\psi}_k)$. Then we have

$$\varphi_i P_1 \left(\psi_i \varphi_k P_2 \psi_k\right) = \left(\tilde{\varphi}_k \varphi_i P_1 \psi_i \tilde{\psi}_k + (1-\tilde{\varphi}_k) \varphi_i P_1 \psi_i \tilde{\psi}_k\right) \varphi_k P_2 \psi_k \ .$$

By virtue of 5.11 and 5.16 we have $\left((1-\tilde{\varphi}_k)\varphi_i P_1 \psi_i \tilde{\psi}_k\right)(\varphi_k P_2 \psi_k) \in \Psi(g; \mathcal{M}, \mathcal{A}, \mathcal{G})^{-\infty}$. The remaining part $\left(\tilde{\varphi}_k \varphi_i P_1 \psi_i \tilde{\psi}_k\right)(\varphi_k P_2 \psi_k)$ is seen to be in $\Psi(M_1 M_2, g; \mathcal{M}, \mathcal{A}, \mathcal{G})$ when being considered in local coordinates (cf. 5.12). □

We finish this section with a refined discussion of the smoothing terms appearing in the calculus. More precisely, we show that we can restrict ourselves to operators having kernel representations in local coordinates with rapidly decreasing kernels when admitting only special cut-off functions.

Theorem 5.18. *Let $(\mathcal{M}, \mathcal{A}, \mathcal{G}, g)$ be an admissible structure of type (p). There exists $0 < r_0 < r_{\mathcal{A}}$ and for all $(\mathcal{X}, X, \kappa) \in \mathcal{A}$ there exists a function $\widehat{\psi} \in \mathcal{C}^\infty(\mathcal{M})$ such that for $0 < r < r_0$ and for all $(\mathcal{X}, X, \kappa, g, r)$-admissible cut-off functions φ the following statements hold:*

- $\widehat{\psi} = 1$ *on* $supp(\varphi)$ *and* $supp(\widehat{\psi}) \subseteq \kappa^{-1}\left(\kappa(\underline{\mathcal{X}})_{\widehat{\rho},(\chi^{-1})^* g}\right)^\circ$ $(\subseteq \mathcal{X})$ *where* $\underline{\mathcal{X}} := \kappa^{-1}\left(\kappa(X)_{\rho,(\kappa^{-1})^*\mathcal{G}}\right)^\circ$ *and* $\rho, \widehat{\rho} > 0$ *are sufficiently small*[6].
- $[(x,\xi) \mapsto \widehat{\psi}(\kappa^{-1}(x))] \in S\left(1, (\chi^{-1})^* g\right)$.
- $\forall P \in \Psi(g; \mathcal{M}, \mathcal{A}, \mathcal{G})^{-\infty}$, $\forall(\tilde{\mathcal{X}}, \tilde{X}, \tilde{\kappa}) \in \mathcal{A}$, $\forall \tilde{\varphi}$ $(\tilde{\mathcal{X}}, \tilde{X}, \tilde{\kappa}, g)$*-admissible cut-off function,* $\exists k_1, k_2 \in \mathcal{S}(\mathbb{R}^{2n})$, $\forall u \in \mathcal{C}_0^\infty(\mathcal{M})$,

$$\begin{aligned}\varphi P\left((1-\widehat{\psi})(\tilde{\varphi}u)\right) &= \kappa^*\left[\varphi(\cdot)\cdot\int_{\mathbb{R}^n} k_1(\cdot,y)\cdot(\tilde{\varphi}\cdot u)(\tilde{\kappa}^{-1}(y))dy\right],\\ \tilde{\varphi}(1-\widehat{\psi})P(\varphi u) &= \tilde{\kappa}^*\left[\tilde{\varphi}(\cdot)\cdot\int_{\mathbb{R}^n} k_2(\cdot,y)\cdot(\varphi\cdot u)(\kappa^{-1}(y))dy\right].\end{aligned}$$

Proof. Let $\mathcal{A} = \{(\mathcal{X}_j, X_j, \kappa_j) \mid j = 1,\ldots,m\}$. We apply the notation from 4.11. Let $0 < \rho < r_{\mathcal{A}}/2$. For $j \in \{1,\ldots,m\}$ we define $\widehat{X}_j := \kappa_j^{-1}\left(\kappa_j(X_j)_{\rho,\mathcal{G}_j}\right)^\circ$ and $\underline{\mathcal{X}}_j := \kappa_j^{-1}\left(\kappa_j(X_j)_{2\rho,\mathcal{G}_j}\right)^\circ$. Moreover, let $\tilde{\mathcal{A}} := \{(\mathcal{X}_j, \underline{\mathcal{X}}_j, \kappa_j) \mid j = 1,\ldots,m\}$. By virtue of 4.13, $(\mathcal{M}, \tilde{\mathcal{A}}, \mathcal{G})$ is a generalized SG-compatible manifold and thus, $(\mathcal{M}, \tilde{\mathcal{A}}, \mathcal{G}, g)$ is an admissible structure of type (p). Let $r_{\tilde{\mathcal{A}}}$ be the corresponding constant (cf. 4.11) and let $0 < \tilde{\rho} < r_{\tilde{\mathcal{A}}}$. For $j \in \{1,\ldots,l\}$ let $\widehat{\underline{\mathcal{X}}}_j := \kappa_j^{-1}\left(\kappa_j(\underline{\mathcal{X}}_j)_{\tilde{\rho},g_j}\right)^\circ$. We fix $j \in \{1,\ldots,l\}$ and let $\kappa := \kappa_j$, $X := X_j$, $\widehat{X} := \widehat{X}_j$, $\underline{\mathcal{X}} := \underline{\mathcal{X}}_j$, and $\widehat{\underline{\mathcal{X}}} := \widehat{\underline{\mathcal{X}}}_j$. Then we have $X \subseteq \kappa^{-1}\left(\kappa(X)_{\rho,(\chi^{-1})^* g}\right)^\circ \subseteq \widehat{X} \subseteq \underline{\mathcal{X}} \subseteq \widehat{\underline{\mathcal{X}}} \subseteq \mathcal{X}$. By virtue of 3.2 there exist an $(\mathcal{X}, X, \kappa, g)$-admissible cut-off function φ with $\operatorname{supp}(\varphi) \subseteq \widehat{X}$, an $(\mathcal{X}, X, \kappa, \mathcal{G})$-admissible cut-off function ϕ with $\operatorname{supp}(\phi) \subseteq \underline{\mathcal{X}}$, $\phi = 1$ on $\widehat{X}$, and an $(\mathcal{X}, \underline{\mathcal{X}}, \kappa, g)$-admissible cut-off function $\widehat{\psi}$ with $\operatorname{supp}(\widehat{\psi}) \subseteq \widehat{\underline{\mathcal{X}}}$ and $\widehat{\psi} = 1$ on $\operatorname{supp}(\phi)$. Moreover, let $(\tilde{\mathcal{X}}, \tilde{X}, \tilde{\kappa}) \in \mathcal{A}$, let $\tilde{\varphi}$ be a $(\tilde{\mathcal{X}}, \tilde{X}, \tilde{\kappa}, g)$-admissible cut-off function, and let $\underline{\phi}$ be a $(\tilde{\mathcal{X}}, \tilde{X}, \tilde{\kappa}, \mathcal{G})$-admissible cut-off function with $\underline{\phi} = 1$ on $\operatorname{supp}(\tilde{\varphi}) \cup \tilde{X}$. We define $\tilde{\phi} := (1-\phi)\cdot\underline{\phi}$. Then we have $1 - \phi = 1$ on $\operatorname{supp}(1-\widehat{\psi})$, in particular on $\operatorname{supp}((1-\widehat{\psi})\cdot\tilde{\varphi})$, and $\underline{\phi} = 1$ on $\operatorname{supp}(\tilde{\varphi})$, in particular on $\operatorname{supp}((1-\widehat{\psi})\cdot\tilde{\varphi})$. Thus $\tilde{\phi} = 1$ on $\operatorname{supp}((1-\widehat{\psi})\cdot\tilde{\varphi})$, $\tilde{\phi} = 0$ on $\operatorname{supp}(\varphi)$, and $\tilde{\phi}$ is a $(\tilde{\mathcal{X}}, \tilde{X}, \tilde{\kappa}, \mathcal{G})$-admissible cut-off function. Moreover, we have $d\tilde{\phi} = 0$ on $\mathcal{M} \setminus \mathcal{M}_{c,\mathcal{A}}$. By definition, for all $N \in \mathbb{N}_0$ and all $V_1,\ldots,V_N \in \mathcal{V}(\mathcal{M}, \mathcal{A}, \mathcal{G})$, the operator

$$L^2(\mathbb{R}^n) \ni u \mapsto (\kappa^{-1})^*\left[\operatorname{ad}(V_1)\ldots\operatorname{ad}(V_N)\left(\varphi P\left((1-\widehat{\psi})\tilde{\varphi}\right)\tilde{\kappa}^* u\right)\right] \in L^2(\mathbb{R}^n)$$

is bounded. For $j \in \{1,\ldots,n\}$ and $u \in \mathcal{C}^\infty(\mathcal{M})$ let

$$\begin{aligned}V_1^{(j)}u &:= \kappa^*\left[M_{x_j}(\kappa^{-1})^*(\phi\cdot u)\right], \; V_2^{(j)}u := \kappa^*\left[D_{x_j}(\kappa^{-1})^*(\phi\cdot u)\right],\\ \tilde{V}_1^{(j)}u &:= \tilde{\kappa}^*\left[M_{x_j}(\tilde{\kappa}^{-1})^*(\tilde{\phi}\cdot u)\right], \; \tilde{V}_2^{(j)}u := \tilde{\kappa}^*\left[D_{x_j}(\tilde{\kappa}^{-1})^*(\tilde{\phi}\cdot u)\right]\end{aligned}$$

[6] In particular, there exists an atlas $\tilde{\mathcal{A}}$ such that $(\mathcal{M}, \tilde{\mathcal{A}}, \mathcal{G}, g)$ is an admissible structure of type (p), $\mathcal{A} \leq \tilde{\mathcal{A}}$, $(\mathcal{X}, \underline{\mathcal{X}}, \kappa) \in \tilde{\mathcal{A}}$, and $\widehat{\psi}$ is an $(\mathcal{X}, \underline{\mathcal{X}}, \kappa, g)$-admissible cut-off function.

and $L_1^{(j)} := M_{x_j}$, $L_2^{(j)} := D_{x_j}$. Let $m_1, \tilde{m}_1, m_2, \tilde{m}_2 \in \mathbb{N}_0$. Then we have

$$\begin{aligned}
&\left(\mathrm{Id} + \sum_{j=1}^{n} \left(\mathrm{ad}\left(V_1^{(j)}\right)\right)^2\right)^{m_1} \left(\mathrm{Id} + \sum_{j=1}^{n} \left(\mathrm{ad}\left(V_2^{(j)}\right)\right)^2\right)^{m_2} \\
&\quad \times \left(\mathrm{Id} + \sum_{j=1}^{n} \left(\mathrm{ad}\left(\tilde{V}_2^{(j)}\right)\right)^2\right)^{\tilde{m}_2} \left(\mathrm{Id} + \sum_{j=1}^{n} \left(\mathrm{ad}\left(\tilde{V}_1^{(j)}\right)\right)^2\right)^{\tilde{m}_1} \left(\varphi P \left(1 - \widehat{\psi}\right) \tilde{\varphi}\right) \\
= &\left(\mathrm{Id} + \sum_{j=1}^{n} \left(V_1^{(j)}\right)^2\right)^{m_1} \left(\mathrm{Id} + \sum_{j=1}^{n} \left(V_2^{(j)}\right)^2\right)^{m_2} \left(\varphi P \left(1 - \widehat{\psi}\right) \tilde{\varphi}\right) \\
&\quad \times \left(\mathrm{Id} + \sum_{j=1}^{n} \left(\tilde{V}_1^{(j)}\right)^2\right)^{\tilde{m}_1} \left(\mathrm{Id} + \sum_{j=1}^{n} \left(\tilde{V}_2^{(j)}\right)^2\right)^{\tilde{m}_2} .
\end{aligned}$$

Thus, the operator $L^2(\mathbb{R}^n) \to L^2(\mathbb{R}^n)$,

$$u \mapsto <M_x>^{2m_1} <D_x>^{2m_2} (\kappa^{-1})^* \left[\varphi P \left(\left(1 - \widehat{\psi}\right) \tilde{\varphi}\tilde{\kappa}^* \left[<M_x>^{2\tilde{m}_1} <D_x>^{2\tilde{m}_2} u\right]\right)\right],$$

is bounded. Hence, by virtue of [1, 3.6.6], the operator

$$\tilde{P}u := (\kappa^{-1})^* \left[\varphi P \left(\left(1 - \widehat{\psi}\right) \tilde{\varphi}\tilde{\kappa}^* u\right)\right]$$

($u \in \mathcal{S}(\mathbb{R}^n)$) is in $\bigcap_{q_1,q_2 \in \mathbb{N}} \Psi\left(M_1^{-q_1} M_2^{-q_2}, g^{1,1}\right)$ where

$$g^{1,1}_{(x,\xi)} := \frac{\|dx\|_2^2}{<x>^2} + \frac{\|d\xi\|_2^2}{<\xi>^2}, \quad M_1(x,\xi) := <x>, \quad \text{and} \quad M_2(x,\xi) := <\xi> .$$

Since $\bigcap_{q_1,q_2 \in \mathbb{N}} S\left(M_1^{-q_1} M_2^{-q_2}, g^{1,1}\right) = \mathcal{S}(\mathbb{R}^{2n})$, this yields the assertion concerning $\varphi P(1 - \widehat{\psi})\tilde{\varphi}$. For $\tilde{\varphi}(1 - \widehat{\psi})P\varphi$ the proof is analogous. □

Corollary 5.19. *Let $(\mathcal{M}, \mathcal{A}, \mathcal{G}, g)$ be an admissible structure of type (p). Let M be an $(\mathcal{M}, \mathcal{A}, g)$-admissible weight function of type (p) which is H_g^{-1}-adapted. Let $(\varphi_i)_{i=1,\dots,l}$ be an $(\mathcal{M}, \mathcal{A}, g, r)$-admissible partition of unity, where $0 < r < r_0$ (cf.* 5.18*).*

1. *For $i \in \{1, \dots, l\}$ let $\widehat{\psi}_i \in \mathcal{C}^\infty(\mathcal{M})$ be the function from* 5.18 *corresponding to φ_i. Then: $\forall P \in \Psi(M, g; \mathcal{M}, \mathcal{A}, \mathcal{G})$, $\exists p_i \in S\left((\chi_{j_i}^{-1})^* M, (\chi_{j_i}^{-1})^* g\right)$, $\forall \nu \in \{1, \dots, l\}$, $\exists k_{i,\nu} \in \mathcal{S}(\mathbb{R}^{2n})$, $\forall u \in \mathcal{C}_0^\infty(\mathcal{M})$*

$$\varphi_i P(\widehat{\psi}_i u) = \varphi_i \cdot \kappa_{j_i}^* \left[p_i(x,D)(\kappa_{j_i}^{-1})^*(\widehat{\psi}_i \cdot u)\right], \tag{5.1}$$

$$\varphi_i P((1 - \widehat{\psi}_i)u) = \sum_{\nu=1}^{l} \varphi_i \cdot \kappa_{j_i}^* \left[\int_{\mathbb{R}^n} k_{i,\nu}(\cdot, y) \cdot (\varphi_\nu \cdot u)\left(\kappa_{j_\nu}^{-1}(y)\right) dy\right]. \tag{5.2}$$

2. *Let $\widehat{\psi}_i$, $i = 1, \dots, l$, be the functions from part (a). If $P : \mathcal{C}_0^\infty(\mathcal{M}) \to \mathcal{D}'(\mathcal{M})$ has a representation* (5.1), (5.2)*, then we have $P \in \Psi(M, g; \mathcal{M}, \mathcal{A}, \mathcal{G})$.*

Proof. The first assertion follows with 5.18 and for the second assertion we apply 5.8 and 5.13. □

Corollary 5.20. *Let* $(\mathcal{M}, \mathcal{A}, \mathcal{G}, g)$, $(\mathcal{M}, \tilde{\mathcal{A}}, \mathcal{G}, g)$, *and* $(\mathcal{M}, \mathcal{A}', \mathcal{G}, g)$ *be admissible structures of type* (p) *with* $\mathcal{A} \leq \tilde{\mathcal{A}}$ *and* $\mathcal{A} \subseteq \mathcal{A}'$. *Moreover, let* M *be an* $(\mathcal{M}, \tilde{\mathcal{A}}, g)$*-admissible weight function of type* (p)[7] *which is* H_g^{-1}*-adapted. Then we have* $\Psi(M, g; \mathcal{M}, \mathcal{A}, \mathcal{G}) = \Psi(M, g; \mathcal{M}, \tilde{\mathcal{A}}, \mathcal{G})$. *If* M *is an* $(\mathcal{M}, \mathcal{A}', g)$*-admissible weight function of type* (p) *which is* H_g^{-1}*-adapted, then we have* $\Psi(M, g; \mathcal{M}, \mathcal{A}, \mathcal{G}) = \Psi(M, g; \mathcal{M}, \mathcal{A}', \mathcal{G})$.

Proof. The fact that $\Psi(g; \mathcal{M}, \mathcal{A}, \mathcal{G})^{-\infty} \subseteq \Psi(g; \mathcal{M}, \tilde{\mathcal{A}}, \mathcal{G})^{-\infty}$ yields $\Psi(M, g; \mathcal{M}, \mathcal{A}, \mathcal{G}) \subseteq \Psi(M, g; \mathcal{M}, \tilde{\mathcal{A}}, \mathcal{G})$. Conversely, 5.19 together with 5.8 and 5.13 yields $\Psi(M, g; \mathcal{M}, \tilde{\mathcal{A}}, \mathcal{G}) \subseteq \Psi(M, g; \mathcal{M}, \mathcal{A}, \mathcal{G})$. The equation $\Psi(M, g; \mathcal{M}, \mathcal{A}, \mathcal{G}) = \Psi(M, g; \mathcal{M}, \mathcal{A}', \mathcal{G})$ is a consequence of 5.19 together with 5.8 and 5.13, too. □

Corollary 5.21. *Let* $(\mathcal{M}, \mathcal{A}, \mathcal{G}, g)$ *and* $(\mathcal{M}, \mathcal{A}, \mathcal{G}, \underline{g})$ *be admissible structures of type* (p) *such that* $\underline{g} \leq g$. *Let* M *be an* $(\mathcal{M}, \mathcal{A}, g)$*- and* $(\mathcal{M}, \mathcal{A}, \underline{g})$*-admissible weight function of type* (p) *which is* H_g^{-1}*- and* $H_{\underline{g}}^{-1}$*-adapted. Then we have* $\Psi(M, \underline{g}; \mathcal{M}, \mathcal{A}, \mathcal{G}) \subseteq \Psi(M, g; \mathcal{M}, \mathcal{A}, \mathcal{G})$.

Proof. By virtue of 4.13 there exists an atlas $\tilde{\mathcal{A}}$ such that $(\mathcal{M}, \tilde{\mathcal{A}}, \mathcal{G}, g)$ is an admissible structure of type (p), $\mathcal{A} \leq \tilde{\mathcal{A}}$, and for each $(\mathcal{X}, X, \kappa) \in \mathcal{A}$, each $(\mathcal{X}, X, \kappa, \underline{g})$-admissible cut-off function is an $(\mathcal{X}, \tilde{X}, \kappa, g)$-admissible cut-off function where $(\mathcal{X}, \tilde{X}, \kappa) \in \tilde{\mathcal{A}}$. Then the assertion is a consequence of 5.19 together with 5.8, 5.13, and 5.20. □

6. Symbol-homomorphism and $*$-structure

In this section, we prove the existence of a symbol-homomorphism mapping an operator P to its principal symbol $\sigma_{p,M}(P)$ with the property that $\sigma_{p,M_1 M_2}(P_1 P_2) = \sigma_{p,M_1}(P_1) \cdot \sigma_{p,M_2}(P_2)$ (cf. 6.5 and 6.6).

Furthermore, we introduce an invariant measure α_G on $\mathcal{M}$ such that $H(1, g; \mathcal{M}, \mathcal{A}) = L^2(\mathcal{M}, \alpha_G)$, chosen in such a way that $P \in \Psi(1, g; \mathcal{M}, \mathcal{A}, \mathcal{G})$ implies $P^* \in \Psi(1, g; \mathcal{M}, \mathcal{A}, \mathcal{G})$ (cf. 6.14). The density α_G is induced by a so-called $(\mathcal{M}, \mathcal{A}, g)$-admissible Riemannian metric G (cf. 6.7, 6.10). With the help of this measure we can write the remainder term of a pseudo-differential operator as an integral operator with rapidly decreasing kernel (cf. 6.13).

Definition 6.1. Let $(\mathcal{M}, \mathcal{A}, \mathcal{G}, g)$ be an admissible structure of type (p) and let M be an $(\mathcal{M}, \mathcal{A}, g)$-admissible weight function of type (p). We define $S(T^*\mathcal{M}, \mathcal{A}; M, g)$ as the set of all functions $a \in \mathcal{C}^\infty(T^*\mathcal{M})$ such that we have

$$(\chi^{-1})^* a \in S\left((\chi^{-1})^* M, (\chi^{-1})^* g\right)$$

for all $(\mathcal{X}, X, \kappa) \in \mathcal{A}$, where

$$\chi : T^*\mathcal{X} \to T^*\mathbb{R}^n \simeq \mathbb{R}^{2n}, \quad T_x^*\mathcal{X} \ni \xi \mapsto (\kappa(x), {}^t d\kappa(x)^{-1} \xi).$$

[7] Then M also is an $(\mathcal{M}, \mathcal{A}, g)$-admissible weight function of type (p).

Theorem 6.2. *Let $(\mathcal{M},\mathcal{A},\mathcal{G},g)$ be an admissible structure of type (p) and let M be an $(\mathcal{M},\mathcal{A},g)$-admissible weight function of type (p) which is H_g^{-1}-adapted.*

1. *To each operator $P\in\Psi(M,g;\mathcal{M},\mathcal{A},\mathcal{G})$ there exists $p\in S\left(T^*\mathcal{M},\mathcal{A};M,g\right)$ with the property that for $(\mathcal{X},X,\kappa)\in\mathcal{A}$ and for an $(\mathcal{X},X,\mathcal{A},g)$-admissible cut-off function φ with associated cut-off function ψ there exist symbols*
$$q\in S\left((\chi^{-1})^*M,(\chi^{-1})^*g\right)\quad\text{and}\quad r\in S\left((\chi^{-1})^*(Mh_g),(\chi^{-1})^*g\right)$$
such that
$$\varphi P(\psi u)=\kappa^*\left[((\varphi\circ\kappa^{-1})\cdot q)(x,D)(\kappa^{-1})^*\left(\psi u\right)\right]$$
for $u\in C_0^\infty(\mathcal{M})$, and for $(x,\xi)\in\kappa(\mathcal{X})\times\mathbb{R}^n$ we have
$$\varphi(\kappa^{-1}(x))\cdot q(x,\xi)=\varphi(\kappa^{-1}(x))\cdot p\left(\kappa^{-1}(x),{}^t d\kappa^{-1}(x)^{-1}\xi\right)+r(x,\xi)\ .$$
2. *If $p_1,p_2\in S\left(T^*\mathcal{M},\mathcal{A};M,g\right)$ have the properties of p, then*
$$p_1-p_2\in S\left(T^*\mathcal{M},\mathcal{A};Mh_g,g\right).$$
3. *If for $i=1,2$, M_i is an $(\mathcal{M},\mathcal{A},g)$-admissible weight function of type (p) which is H_g^{-1}-adapted, $P_i\in\Psi\left(M_i,g;\mathcal{M},\mathcal{A},\mathcal{G}\right)$, $p_i\in S\left(T^*\mathcal{M},\mathcal{A};M_i,g\right)$ is the corresponding symbol of Part 1, and if $c\in\mathbb{C}$, then*
$$cp_1+p_2\in S\left(T^*\mathcal{M},\mathcal{A};M_1+M_2,g\right)$$
is such a symbol for the operator cP_1+P_2.

Proof. Let $(\varphi_i)_{i=1,\dots,l}$ be an $(\mathcal{M},\mathcal{A},g)$-admissible partition of unity with associated system of cut-off functions $(\psi_i)_{i=1,\dots,l}$. We apply the notation from 4.14. For $i\in\{1,\dots,l\}$ there exists $q_i\in S\left((\chi_{j_i}^{-1})^*M,(\chi_{j_i}^{-1})^*g\right)$ such that we have for $u\in C_0^\infty(\mathcal{M})$:
$$\varphi_iP(\psi_iu)=\kappa_{j_i}^*\left[((\varphi_i\circ\kappa_{j_i}^{-1})\cdot q_i)(x,D)(\kappa_{j_i}^{-1})^*\left(\psi_iu\right)\right]\ .$$
We define $p_i:=\chi_{j_i}^*\left((\varphi_i\circ\kappa_{j_i}^{-1})\cdot q_i\right)$ $(\in S\left(T^*\mathcal{M},\mathcal{A};M,g\right))$. By virtue of 5.12, p_i has the asserted properties with φ_iP instead of P. Then $p:=\sum_{i=1}^l p_i$ has the asserted properties. The other assertions are obvious. □

Definition 6.3. The theorem induces a well-defined linear mapping
$$\Psi(M,g;\mathcal{M},\mathcal{A},\mathcal{G})\to S\left(T^*\mathcal{M},\mathcal{A};M,g\right)/S\left(T^*\mathcal{M},\mathcal{A};Mh_g,g\right)\ ,\ P\mapsto\sigma_{p,M}(P)$$
of an operator P to its *principal symbol* $\sigma_{p,M}(P)$.

Lemma 6.4. *Let $(\mathcal{M},\mathcal{A},\mathcal{G},g)$ be an admissible structure of type (p) and let M be an $(\mathcal{M},\mathcal{A},g)$-admissible weight function of type (p) which is H_g^{-1}-adapted.*

1. *For $P\in\Psi(M,g;\mathcal{M},\mathcal{A},\mathcal{G})$ we have*
$$\sigma_{p,M}(P)=0\quad\text{if and only if}\quad P\in\Psi(Mh_g,g;\mathcal{M},\mathcal{A},\mathcal{G}).$$

2. *Let $p \in S(T^*\mathcal{M}, \mathcal{A}; M, g)$. Then there exists $P \in \Psi(M, g; \mathcal{M}, \mathcal{A}, \mathcal{G})$ such that p has the properties of* 6.2 1. *More precisely, P can be chosen such that the following statement holds: For each $(\mathcal{X}, X, \kappa) \in \mathcal{A}$ and each $(\mathcal{X}, X, \kappa, g)$-admissible cut-off function φ, there exists an $(\mathcal{X}, \widehat{X}_g, \kappa, g)$-admissible cut-off function $\widehat{\varphi}$ (cf.* 4.11, 4.13) *such that $\widehat{\varphi} = 1$ on $supp(\varphi)$ and $\varphi P = \varphi P \widehat{\varphi}$, $P\varphi = \widehat{\varphi} P \varphi$.*

Proof. The first assertion is obvious. To prove Part 2, let $p \in S(T^*\mathcal{M}, \mathcal{A}; M, g)$. We apply the notations from 3.2 with $(\mathcal{U}_j, U_j, \mathcal{R}_1(g_j))$ (cf. 4.11) instead of $(\mathcal{U}, U, \mathcal{G})$, the constants $\rho, \rho_1, \ldots, \rho_4$ being chosen simultaneously for all $j \in \{1, \ldots, m\}$. Moreover, we suppose that $\rho_4 < r_{\mathcal{A}}$ and $2C_{\mathcal{A}}\sqrt{C_g}\rho < \widehat{r}_{\mathcal{A}}$ ($\widehat{r}_{\mathcal{A}}$ the constant from 4.11 corresponding to $\widehat{\mathcal{A}}$ in 4.13). Let $2C_{\mathcal{A}}\sqrt{C_g}\rho < r_0 < \widehat{r}_{\mathcal{A}}$. Let $(\varphi_{j,z})_{z \in U_{j,\rho_2,g_j}}$, $(\psi_{j,z})_{z \in U_{j,\rho_2,g_j}}$ $(j = 1, \ldots, m)$ be the corresponding families and let

$$\tilde{\varphi}_j := \int_{U_{j,\rho_2,g_j}} \varphi_{j,z} \, |\mathcal{R}_1(g_j)_z|^{1/2} dz.$$

Let $(\varphi_j)_{j=1,\ldots,m}$ be an $(\mathcal{M}, \mathcal{A}, g, \rho_1)$-admissible partition of unity with associated system of cut-off functions $(\psi_j)_{j=1,\ldots,m}$ (cf. 4.15). Furthermore, for $j = 1, \ldots, m$ and $u \in \mathcal{C}_0^\infty(\mathcal{M})$, we define

$$\begin{aligned} \tilde{p}_j &:= \int_{U_{j,\rho_2,g_j}} \left(\varphi_{j,z} \cdot (\chi_j^{-1})^* p\right) \;\#_{\mathrm{KN}}\; \psi_{j,z} \, |\mathcal{R}_1(g_j)_z|^{1/2} dz, \\ P_j u &:= \kappa_j^* \left[\tilde{p}_j(x, D)(\kappa_j^{-1})^*(\psi_j \cdot u)\right] . \end{aligned}$$

By virtue of 5.12, $\varphi_j \cdot p$ plays the role of p in 6.2 1. with $\varphi_j P_j$ instead of P. Let $P := \sum_{j=1}^m \varphi_j P_j$. Then P has the asserted properties. For details we refer to [2, 2.11.5]. □

Corollary 6.5. *Let $(\mathcal{M}, \mathcal{A}, \mathcal{G}, g)$ be an admissible structure of type (p) and let M be an $(\mathcal{M}, \mathcal{A}, g)$-admissible weight function of type (p) which is H_g^{-1}-adapted. Then the following statements hold:*

1. *The mapping*

$$\begin{aligned} \Psi(M, g; \mathcal{M}, \mathcal{A}, \mathcal{G}) &\to S(T^*\mathcal{M}, \mathcal{A}; M, g) / S(T^*\mathcal{M}, \mathcal{A}; M h_g, g) , \\ P &\mapsto \sigma_{p,M}(P) \end{aligned}$$

is surjective.

2. *$Ker(\sigma_{p,M}) = \Psi(M h_g, g; \mathcal{M}, \mathcal{A}, \mathcal{G})$. This induces a bijective linear mapping*

$$\Psi(M,g;\mathcal{M},\mathcal{A},\mathcal{G})/\Psi(Mh_g,g;\mathcal{M},\mathcal{A},\mathcal{G}) \to S(T^*\mathcal{M},\mathcal{A};M,g)/S(T^*\mathcal{M},\mathcal{A};Mh_g,g).$$

Theorem 6.6. *Let $(\mathcal{M}, \mathcal{A}, \mathcal{G}, g)$ be an admissible structure of type (p) and let M_1, M_2 be $(\mathcal{M}, \mathcal{A}, g)$-admissible weight functions of type (p) which are H_g^{-1}-adapted. For $P_i \in \Psi(M_i, g; \mathcal{M}, \mathcal{A}, \mathcal{G})$ $(i = 1, 2)$ we have*

$$\sigma_{p,M_1M_2}(P_1P_2) = \sigma_{p,M_1}(P_1) \cdot \sigma_{p,M_2}(P_2)$$

which belongs to $S(T^*\mathcal{M}, \mathcal{A}; M_1M_2, g) / S(T^*\mathcal{M}, \mathcal{A}; M_1M_2h_g, g)$.

Proof. Taking into account the behavior of the remainder term in the asymptotic expansion for the product symbol of the calculus on $\mathbb{R}^n$, this assertion is proved directly. For details cf. [2, 2.11.7]. □

Definition 6.7.

1. Let $(\mathcal{M},\mathcal{A},\mathcal{G})$ be a generalized SG-compatible manifold. We call a Riemannian metric G on $\mathcal{M}$ $(\mathcal{M},\mathcal{A},\mathcal{G})$*-admissible* if we have
$$((\kappa^{-1})^*G)_x(\xi_1,\xi_2)=\left\langle G^{(\kappa)}(x)\xi_1,\xi_2\right\rangle \quad (x\in\kappa(\mathcal{X}),\xi_1,\xi_2\in\mathbb{R}^n)$$
for all $(\mathcal{X},X,\kappa)\in\mathcal{A}$ with a positive definite matrix $G^{(\kappa)}(x)$ such that
$$\left[x\mapsto G^{(\kappa)}(x)\right],\left[x\mapsto\left(G^{(\kappa)}(x)\right)^{-1}\right]\in S\left(1,(\kappa^{-1})^*\mathcal{G}_{|\mathcal{X}};\mathcal{L}(\mathbb{R}^n)\right).$$
2. If $(\mathcal{M},\mathcal{A},\mathcal{G},g)$ is an admissible structure of type (p), then we call an $(\mathcal{M},\mathcal{A},\mathcal{R}_1(g))$-admissible Riemannian metric $(\mathcal{M},\mathcal{A},g)$*-admissible.*

Lemma 6.8. *If $(\mathcal{M},\mathcal{A},\mathcal{G})$ is a generalized SG-compatible manifold, then there exists an $(\mathcal{M},\mathcal{A},\mathcal{G})$-admissible Riemannian metric.*

Proof. Let $\mathcal{A}:=\{(\mathcal{X}_j,X_j,\kappa_j)\mid j\in\{1,\ldots,m\}\}$ and let $(\varphi_i)_{i=1,\ldots,l}$ be an $(\mathcal{M},\mathcal{A},\mathcal{G})$-admissible partition of unity. Applying the notation from 4.14, for $x\in\mathcal{M}$ and $\xi_1,\xi_2\in T_x\mathcal{M}$ we define $G_x(\xi_1,\xi_2):=\sum_{i=1}^{l}\varphi_i(x)\left\langle d\kappa_{j_i}(x)\xi_1,d\kappa_{j_i}(x)\xi_2\right\rangle$. This yields an $(\mathcal{M},\mathcal{A},\mathcal{G})$-admissible Riemannian metric. □

Remark 6.9. If G is an $(\mathcal{M},\mathcal{A},\mathcal{G})$-admissible Riemannian metric, then, for all $(\mathcal{X},X,\kappa)\in\mathcal{A}$, we have
$$\left[x\mapsto|G^{(\kappa)}(x)|:=\det(G^{(\kappa)}(x))\right],\left[x\mapsto|G^{(\kappa)}(x)|^{-1}\right]\in S\left(1,(\kappa^{-1})^*\mathcal{G}_{|\mathcal{X}}\right)$$
and thus $\left[x\mapsto|G^{(\kappa)}(x)|^{1/2}\right],\left[x\mapsto|G^{(\kappa)}(x)|^{-1/2}\right]\in S\left(1,(\kappa^{-1})^*\mathcal{G}_{|\mathcal{X}}\right)$.

Remark 6.10. It is well known that each Riemannian C^∞-manifold $(\mathcal{M},G)$ has a canonical non-vanishing density α_G defining an invariant integral on $\mathcal{M}$.

Remark 6.11. Let $(\mathcal{M},\mathcal{A},\mathcal{G},g)$ be an admissible structure of type (p) and let G be an $(\mathcal{M},\mathcal{A},g)$-admissible Riemannian metric. Then we have $H(1,g;\mathcal{M},\mathcal{A})=L^2(\mathcal{M},\alpha_G)$. In particular, $H(1,g;\mathcal{M},\mathcal{A})$ is a Hilbert space.

Lemma 6.12. *Let $(\mathcal{M},\mathcal{A},\mathcal{G},g)$ be an admissible structure of type (p). Let G be an $(\mathcal{M},\mathcal{A},g)$-admissible Riemannian metric. Let $K\in\mathcal{S}(\mathcal{M}\times\mathcal{M})$ (cf. 4.9) and $\mathcal{K}:C_0^\infty(\mathcal{M})\to\mathcal{D}'(\mathcal{M})$ be defined by*
$$\mathcal{K}u(x):=\int_{\mathcal{M}}K(x,y)u(y)\alpha_G(y).$$
Then we have $\mathcal{K}\in\Psi(g;\mathcal{M},\mathcal{A},\mathcal{G})^{-\infty}$.

Proof. Applying an $(\mathcal{M},\mathcal{A},g)$-admissible partition of unity, the assertion follows from 5.8. □

Theorem 6.13. *Let $(\mathcal{M}, \mathcal{A}, \mathcal{G}, g)$ be an admissible structure of type (p). Let M be an $(\mathcal{M}, \mathcal{A}, g)$-admissible weight function of type (p) which is H_g^{-1}-adapted. Let G be an admissible Riemannian metric. Let r_0 as in* 5.18, *let $0 < r < r_0$, and let $(\varphi_i)_{i=1,\ldots,l}$ be an $(\mathcal{M}, \mathcal{A}, g, r)$-admissible partition of unity. Let $(\widehat{\psi}_i)_{i=1,\ldots,l}$ be an associated family of functions as in* 5.19. *Then, applying the notation from* 4.14, *for an operator $P : \mathcal{C}_0^\infty(\mathcal{M}) \to \mathcal{D}'(\mathcal{M})$ the following statements are equivalent:*

1. $P \in \Psi(M, g; \mathcal{M}, \mathcal{A}, \mathcal{G})$.
2. $\exists K \in \mathcal{S}(\mathcal{M} \times \mathcal{M})$, $\forall i \in \{1, \ldots, l\}$, $\exists p_i \in S\left((\chi_{j_i}^{-1})^* M, (\chi_{j_i}^{-1})^* g\right)$, $\forall u \in \mathcal{C}_0^\infty(\mathcal{M})$,

$$\begin{aligned} Pu &= \sum_{i=1}^{l} \varphi_i P\left(\widehat{\psi}_i u\right) + \int_{\mathcal{M}} K(\cdot, y) u(y) \alpha_G(y) \ , \ \textit{where} \\ \varphi_i P\left(\widehat{\psi}_i u\right) &= \kappa_{j_i}^* \left[\left((\varphi_i \circ \kappa_{j_i}^{-1}) \cdot p_i\right)(x, D)(\kappa_{j_i}^{-1})^* (\widehat{\psi}_i u)\right] . \end{aligned}$$

Proof. To prove "1. ⇒ 2.", let $k_{i,\nu} \in \mathcal{S}(\mathbb{R}^n \times \mathbb{R}^n)$ as in 5.19 ($i, \nu \in \{1, \ldots, l\}$). For $x, y \in \mathcal{M}$ we define $K(x, y) := \sum_{i,\nu=1}^{l} K_{i,\nu}(x, y)$ where

$$K_{i,\nu}(x, y) := \varphi_i(x) \cdot k_{i,\nu}\left(\kappa_{j_i}(x), \kappa_{j_\nu}(y)\right) \cdot \varphi_\nu(y) \ |G^{(\kappa_{j_\nu})}(\kappa_{j_\nu}(y))|^{-1/2} .$$

The implication "2. ⇒ 1." is a consequence of 5.13 and 6.12. □

Theorem 6.14. *Let $(\mathcal{M}, \mathcal{A}, \mathcal{G}, g)$ be an admissible structure of type (p). Let M be an $(\mathcal{M}, \mathcal{A}, g)$-admissible weight function of type (p) which is H_g^{-1}-adapted. Let G be an $(\mathcal{M}, \mathcal{A}, g)$-admissible Riemannian metric. For $P \in \Psi(M, g; \mathcal{M}, \mathcal{A}, \mathcal{G})$ there exists an operator $P^* \in \Psi(M, g; \mathcal{M}, \mathcal{A}, \mathcal{G})$ with the property that for $u, v \in \mathcal{S}(\mathcal{M})$ the equation*

$$\langle Pu, v \rangle_{L^2(\mathcal{M}, \alpha_g)} = \langle u, P^* v \rangle_{L^2(\mathcal{M}, \alpha_g)}$$

holds. The operator P^* is called the formal adjoint (with respect to $\langle \cdot, \cdot\cdot \rangle_{L^2(\mathcal{M}, \alpha_g)}$).

Proof. The assertion is a consequence of 6.13 together with the corresponding result from the global calculus on $\mathbb{R}^n$ (cf. [28, 18.5]), where we have to take into account 5.12 and 5.20. □

7. Examples of algebras of pseudo-differential operators on manifolds

Example. Let $\mathcal{M}$ be an SG-compatible manifold in the sense of E. Schrohe [41] with an atlas $\mathcal{A} = \{(\mathcal{X}_j, X_j, \kappa_j) \mid j = 1, \ldots, m\}$. For $j \in \{1, \ldots, m\}$ let $\mathcal{U}_j := \kappa_j(\mathcal{X}_j)$ and $\chi_j : T^*\mathcal{X}_j \to T^*\mathbb{R}^n \simeq \mathbb{R}^{2n}$, $T_x^*\mathcal{X}_j \ni \xi \mapsto (\kappa(x), {}^t d\kappa(x)^{-1}\xi)$. Let

$$g_{(x,\xi)}^{\rho_1,\rho_2,\delta} \quad := \quad \frac{<\xi>^{2\delta}}{<x>^{2\rho_1}} \|dx\|_2^2 + \frac{\|d\xi\|_2^2}{<\xi>^{2\rho_2}}$$

where $0 \leq \rho_1 \leq 1, 0 < \rho_2 \leq 1, 1-\rho_2 \leq \delta \leq \rho_2, \delta < 1$ and let

$$g^{\rho,\delta}_{(x,\xi)} \quad := \quad <x,\xi>^{2\delta} \|dx\|_2^2 + \frac{\|d\xi\|_2^2}{<x,\xi>^{2\rho}}$$

where $0 < \rho \leq 1, 1-\rho \leq \delta < \rho$ (cf. [13], [23], [46]). With the help of a partition of unity we can define metrics $g_1 := g^{[\rho_1,\rho_2,\delta]}$ and $g_2 := g^{[\rho,\delta]}$ on $T(T^*\mathcal{M})$ such that $\forall j \in \{1,\dots,m\},\ \exists C_j > 0,\ \forall (x,\xi) \in \mathcal{U}_j \times \mathbb{R}^n$,

$$\begin{array}{rcccl} C_j^{-1} g^{\rho_1,\rho_2,\delta}_{(x,\xi)}(\cdot) & \leq & (\chi_j^{-1})^* g_{1,(x,\xi)}(\cdot) & \leq & C_j g^{\rho_1,\rho_2,\delta}_{(x,\xi)}(\cdot) \\ C_j^{-1} g^{\rho,\delta}_{(x,\xi)}(\cdot), & \leq & (\chi_j^{-1})^* g_{2,(x,\xi)}(\cdot) & \leq & C_j g^{\rho,\delta}_{(x,\xi)}(\cdot)\ . \end{array}$$

Moreover, let $\mathcal{G} = \mathcal{R}_1(g^{[1,1,0]})$. Then the following statements hold:

1. $(\mathcal{M},\mathcal{A},\mathcal{G})$ is a generalized SG-compatible manifold.
2. $(\mathcal{M},\mathcal{A},\mathcal{G},g_1)$ and $(\mathcal{M},\mathcal{A},\mathcal{G},g_2)$ are admissible structures of type (p).
3. $(\mathcal{M},\mathcal{A},\mathcal{R}_1(g^{[1,\delta]}),g^{[1,\delta]})$ $(0 \leq \delta < 1)$ is an admissible structure of type (p).
4. If $\rho_1 > 0$, then $(\mathcal{M},\mathcal{A},g_1)$ is an admissible structure of type (p), too.

We continue with examples of anisotropic metrics on products of manifolds.

Example. Let $\mathcal{M}^{(1)}$ and $\mathcal{M}^{(2)}$ be two compact manifolds. For $i = 1,2$ let $\mathcal{G}^{(i)}$ be a continuous metric on $\mathcal{M}^{(i)}$ and let $\mathcal{A}^{(i)}$ such that $(\mathcal{M}^{(i)},\mathcal{A}^{(i)},\mathcal{G}^{(i)})$ is an SG-compatible manifold. Let $\mathcal{M}$, $\mathcal{A}$, and $\mathcal{G}$ be as in 4.9. Moreover, let $\kappa_{j_1,j_2} := (\kappa^{(1)}_{j_1},\kappa^{(2)}_{j_2})$, and $\chi_{j_1,j_2} : T^*\left(\mathcal{X}^{(1)}_{j_1} \times \mathcal{X}^{(2)}_{j_2}\right) \to \mathbb{R}^{n_1+n_2}$,

$$(x,\xi) \mapsto \left(\kappa_{j_1,j_2}(x), {}^t d\kappa_{j_1,j_2}(x)^{-1}\xi\right)$$

$(j_i \in \{1,\dots,m_i\}, i = 1,2)$. For[8] $0 < \rho_1,\rho_2 \leq 1$, $1 - \min\{\rho_1,\rho_2\} \leq \delta_1 \leq \rho_1$, $\delta_1 < 1$, $1-\min\{\rho_1,\rho_2\} \leq \delta_2 \leq \rho_2$, $\delta_2 < 1$, and $x,\xi \in \mathbb{R}^n$, $t_1,\tau_1 \in \mathbb{R}^{n_1}$, and $t_2,\tau_2 \in \mathbb{R}^{n_2}$ let

$$g^{\rho_1,\rho_2,\delta_1,\delta_2}_{(x,\xi)}(t_1,t_2,\tau_1,\tau_2) := <\xi>^{2\delta_1} \|t_1\|_2^2 + <\xi>^{2\delta_2} \|t_2\|_2^2 + \frac{\|\tau_1\|_2^2}{<\xi>^{2\rho_1}} + \frac{\|\tau_2\|_2^2}{<\xi>^{2\rho_2}}.$$

With the help of a partition of unity we can construct a metric g on $T(T^*\mathcal{M})$ such that $\forall j_i \in \{1,\dots,m_i\}$ $(i = 1,2)$, $\exists C$, $\forall x \in \kappa_{j_1,j_2}\left(\mathcal{X}^{(1)}_{j_1} \times \mathcal{X}^{(2)}_{j_2}\right)$, $\forall \xi \in \mathbb{R}^{n_1+n_2}$,

$$C^{-1}\ g^{\rho_1,\rho_2,\delta_1,\delta_2}_{(x,\xi)}(\cdot) \leq (\chi^{-1}_{j_1,j_2})^* g_{(x,\xi)}(\cdot) \leq C\ g^{\rho_1,\rho_2,\delta_1,\delta_2}_{(x,\xi)}(\cdot)\ .$$

Moreover, $(\mathcal{M},\mathcal{A},\mathcal{G},g)$ is an admissible structure of type (p).

Example. Let $(\mathcal{M}^{(1)},\mathcal{A}^{(1)},\mathcal{G}^{(1)})$ and $(\mathcal{M}^{(2)},\mathcal{A}^{(2)},\mathcal{G}^{(2)})$ be two SG-compatible manifolds of dimension n_1 and n_2, respectively. Moreover, for $0 < \rho_1,\rho_2 \leq 1$,

[8] Obviously $\delta_1 = \rho_1 = 1/2$, $\delta_2 = \rho_2 = 2/3$ fulfill the assumptions.

$1-\min\{\rho_1,\rho_2\} \le \delta_1 < \rho_1$, $1-\min\{\rho_1,\rho_2\} \le \delta_2 < \rho_2$, and $x,\xi \in \mathbb{R}^n$, $t_1,\tau_1 \in \mathbb{R}^{n_1}$, and $t_2,\tau_2 \in \mathbb{R}^{n_2}$ let

$$\begin{aligned} g_{(x,\xi)}^{\rho_1,\rho_2,\delta_1,\delta_2}(t_1,t_2,\tau_1,\tau_2) \quad &:= \quad <x,\xi>^{2\delta_1} \|t_1\|_2^2 + <x,\xi>^{2\delta_2} \|t_2\|_2^2 \\ &\quad + \frac{\|\tau_1\|_2^2}{<x,\xi>^{2\rho_1}} + \frac{\|\tau_2\|_2^2}{<x,\xi>^{2\rho_2}} \,. \end{aligned}$$

We apply the notation analogous to that in the preceding example. There exists a metric g on $T(T^*\mathcal{M})$ such that $\forall j_i \in \{1,\ldots,m_i\}$ $(i=1,2)$, $\exists C$, $\forall x \in \kappa_{j_1,j_2}\left(\mathcal{X}_{j_1}^{(1)} \times \mathcal{X}_{j_2}^{(2)}\right)$, $\forall \xi \in \mathbb{R}^{n_1+n_2}$,

$$C^{-1}\, g_{(x,\xi)}^{\rho_1,\rho_2,\delta_1,\delta_2}(\cdot) \le (\chi_{j_1,j_2}^{-1})^* g_{(x,\xi)}(\cdot) \le C\, g_{(x,\xi)}^{\rho_1,\rho_2,\delta_1,\delta_2}(\cdot) \,.$$

Moreover, $(\mathcal{M},\mathcal{A},\mathcal{G},g)$ is an admissible structure of type (p).

Remark 7.1. Let $\mathcal{M} = \mathcal{M}_0 \,\dot\cup\, \mathcal{M}_1 \,\dot\cup\, \mathcal{M}_2 \,\dot\cup\, \partial\mathcal{M}_1 \,\dot\cup\, \partial\mathcal{M}_2$, where $\mathcal{M}_0,\mathcal{M}_1,\mathcal{M}_2$ are n-dimensional submanifolds, $\partial\mathcal{M}_1,\partial\mathcal{M}_2$ are connected $(n-1)$-dimensional submanifolds, $\mathcal{M}_0$ is relatively compact, $\partial\mathcal{M}_0 = \partial\mathcal{M}_1 \,\dot\cup\, \partial\mathcal{M}_N$, for $j \in \{1,\ldots,N\}$, $\mathcal{M}_j$ is diffeomorphic to $\partial\mathcal{M}_j \times (1,\infty)$, and let $0 \le \rho_i \le 1$ $(i \in \{0,1,2\})$, $0 < \rho_2 \le 1$, $1-\rho_2 \le \delta < \rho_2$. Applying the notation from the preceding example, with the help of a partition of unity we can construct a metric g on $\mathcal{M}$ such that the following statements hold:

- $(\mathcal{M},\mathcal{A},\mathcal{G},g)$ is an admissible structure of type (p).
- $\exists C>0$, $\forall i \in \{0,1,2\}$, $\forall x \in \mathcal{M}_i$, $\forall \xi \in T_x^*\mathcal{M}$, $C^{-1} g_{(x,\xi)}^{[\rho_i,\rho_2,\delta]} \le g_{(x,\xi)} \le C g_{(x,\xi)}^{[\rho_i,\rho_2,\delta]}$.

In particular, this example shows that situations appear where different Hörmander metrics occur in different local models, i.e., in the admissible quadruples.

8. Topological properties of the algebras of pseudo-differential operators

In this section we give an outlook and present results concerning the topological properties of the algebras of pseudo-differential operators constructed so far. More precisely, in order to prove these results, we have to impose stronger conditions on the admissible structures than we have done so far. But the following results are true for all examples given in Section 7 if we always assume that $\delta < \rho$, respectively, $\delta_1 < \rho_1$ and $\delta_2 < \rho_2$. We omit the proofs and refer instead to [2, Sections 3.8, 5.4, and 5.5]. In fact, the conditions concerning the admissible structures imposed in [2] exclude the anisotropic examples. But in the situations occuring there the results can be proved with minor modifications. The most important modification consists of working out the characterizations with $\mathcal{V}(\mathcal{M}_1,\mathcal{A}_1,\mathcal{G}_1) \cup \mathcal{V}(\mathcal{M}_2,\mathcal{A}_2,\mathcal{G}_2)$ instead of $\mathcal{V}(\mathcal{M},\mathcal{A},\mathcal{G})$.

In the following, let g be one of the metrics in the examples in Section 7 and let M be an admissible weight function which is in local coordinates of the form $(x,\xi) \mapsto \left(\tilde{g}_{(x,\xi)}(T)\right)^s$, where $s \in \mathbb{R}$, $T \in \mathbb{R}^{2n} \setminus \{0\}$, and $\tilde{g}$ is the local model corresponding to g.

The first theorem states the existence of order reducing operators. Due to this result we can restrict ourselves in the following to the algebra of operators of order zero.

Theorem 8.1. *There exist operators $P \in \Psi(M,g;\mathcal{M},\mathcal{A},\mathcal{G})$ and $Q \in \Psi(1/M,g;\mathcal{M},\mathcal{A},\mathcal{G})$ such that $PQ = Id = QP$.*

Remark 8.2. In particular, P is an isomorphism $H(M,g;\mathcal{M},\mathcal{A}) \to H(1,g;\mathcal{M},\mathcal{A})$ where the Sobolev space $H(M,g;\mathcal{M},\mathcal{A})$ can be defined as the space of all $u \in \bigcup_{s\in\mathbb{R}} H(H_g^s,g;\mathcal{M},\mathcal{A})$ such that $Au \in H(1,g;\mathcal{M},\mathcal{A})$ for all $A \in \Psi(M,g;\mathcal{M},\mathcal{A},\mathcal{G})$. Since $H(1,g;\mathcal{M},\mathcal{A}) = L(\mathcal{M},\alpha_G)$ (cf. Section 6), $H(M,g;\mathcal{M},\mathcal{A})$ is a Banach space.

For weight functions M Beal's- and Corde's-type characterizations for $\Psi(M,g;\mathcal{M},\mathcal{A},\mathcal{G})$ via commutators with operators in $\mathcal{V}(\mathcal{M},\mathcal{A},\mathcal{G})$ can be proved. These, together with the existence of order reducing operators, yield the following two theorems:

Theorem 8.3. *$\Psi(1,g;\mathcal{M},\mathcal{A},\mathcal{G})$ is spectrally invariant in $\mathcal{L}\left(L^2(\mathbb{R})\right)$, i.e.,*

$$\Psi(1,g;\mathcal{M},\mathcal{A},\mathcal{G}) \cap \mathcal{L}\left(L^2(\mathbb{R})\right)^{-1} = \Psi(1,g;\mathcal{M},\mathcal{A},\mathcal{G})^{-1}.$$

Theorem 8.4. *For $k \in \mathbb{N}_0$ there exists a Banach algebra $\Psi_k(g;\mathcal{M},\mathcal{A},\mathcal{G})$ such that $\Psi_{k+1}(g;\mathcal{M},\mathcal{A},\mathcal{G}) \hookrightarrow \Psi_k(g;\mathcal{M},\mathcal{A},\mathcal{G})$, $\Psi_0(g;\mathcal{M},\mathcal{A},\mathcal{G}) = \mathcal{L}\left(L^2(\mathbb{R})\right)$, and*

$$\Psi(1,g;\mathcal{M},\mathcal{A},\mathcal{G}) = \bigcap_{k\in\mathbb{N}_0} \Psi_k(g;\mathcal{M},\mathcal{A},\mathcal{G}) .$$

In particular, $\Psi(1,g;\mathcal{M},\mathcal{A},\mathcal{G})$ is a submultiplicative Fréchet algebra which is continuously embedded in $\mathcal{L}\left(L^2(\mathbb{R})\right)$.

Summing up, 6.14, 8.3, and 8.4 yield that $\Psi(1,g;\mathcal{M},\mathcal{A},\mathcal{G})$ is a submultiplicative Ψ^*-algebra in the sense of B. Gramsch [18], [19].

Acknowledgement

I am very grateful to Professor B. Gramsch for proposing to me to generalize the $S(M,g)$-calculus to manifolds and to investigate spectral invariance and submultiplicativity as well as for valuable support and helpful suggestions. I wish to thank Professor E. Schrohe for several hints, explicitly for explaining to me how to simplify an argument in the proof of 4.13 which allowed me to dispense with an additional assumption concerning the admissible manifolds. Moreover, I wish to thank Dr. R. Lauter for many fruitful discussions and some good advice. Finally, I thank Dr. J. Lutgen, Dr. J. Møller, and Dr. O. Caps for useful comments and for correcting my English.

References

[1] F. Baldus. Weyl-Hörmander-Quantisierung auf dem $\mathbb{R}^n$, Spektralinvarianz und Submultiplikativität der durch Kommutatoren definierten Fréchetalgebren. Diplomarbeit, Fachbereich 17-Mathematik, Johannes Gutenberg-Universität Mainz, 1996.

[2] F. Baldus. *$S(M, g)$-pseudo-differential calculus with spectral invariance on $\mathbb{R}^n$ and manifolds for Banach function spaces.* PhD thesis, Johannes Gutenberg-Universität Mainz, 2001, Logos Verlag, Berlin, 2001. 300 pages.

[3] R. Beals. A general calculus of pseudodifferential operators. *Duke Math. J.*, 42:1–42, 1975.

[4] R. Beals. Characterization of pseudodifferential operators and applications. *Duke Math. J.*, 44:45–57, 1977. ibid. 46:215, 1979.

[5] R. Beals. Weighted distribution spaces and pseudodifferential operators. *J. An. Math.*, 39:130–187, 1981.

[6] R. Beals and C. Fefferman. Spatially inhomogeneous pseudodifferential operators I. *Comm. Pure Appl. Math.*, 27:1–24, 1974.

[7] R. Beals and P. Greiner. *Calculus on Heisenberg manifolds*, volume 119 of *Annals of Mathematics Studies.* Princeton University Press, Princeton, NJ, 1988.

[8] J.-M. Bony. Fourier integral operators and Weyl-Hörmander calculus. In J.-M. Bony and M. Morimoto, editors, *New trends in microlocal analysis*, pages 3–22. Springer-Verlag, Tokyo – Berlin – Heidelberg – New York, 1997.

[9] J.-M. Bony and J.-Y. Chemin. Espaces fonctionnels associés au calcul de Weyl-Hörmander. *Bull. Soc. Math. France*, 122:77–118, 1994.

[10] J.-M. Bony and N. Lerner. Quantification asymptotique et microlocalisations d'ordre supérieur I. *Ann. Sci. Ecole Norm. Sup. 4e série*, 22:377–433, 1989.

[11] H.O. Cordes. A global parametrix for pseudodifferential operators over $\mathbb{R}^n$. Preprint No. 90, SFB 72, Bonn, 1976.

[12] H.O. Cordes. *On some C^*-algebras and Fréchet*-algebras of pseudodifferential operators*, volume 43 of *Proc. Symp. in Pure Math. – Pseudodifferential operators*, pages 79–104. Amer. Math. Soc., Providence, Rhode Island, 1985.

[13] H.O. Cordes. *The technique of pseudodifferential operators*, volume 202 of *London Mathematical Society, Lecture Note Series.* Cambridge University Press, Cambridge – London – New York, 1995.

[14] H.O. Cordes and S.H. Doong. The Laplace comparison algebra of spaces with conical and cylindrical ends. In H.O. Cordes, B. Gramsch, and H. Widom, editors, *Pseudo-differential operators*, volume 1256 of *Lect. Notes Math.*, pages 55–90, Berlin – Heidelberg – New York – London – Paris – Tokyo, 1987. Proc. Conf., Oberwolfach/Ger. 1986, Springer-Verlag.

[15] N. Dencker. The Weyl calculus with locally temperate metrics and weights. *Ark. Mat.*, 24:59–79, 1986.

[16] Y.V. Egorov and B.-W. Schulze. *Pseudo-differential Operators, Singularities, Applications*, volume 93 of *Operator Theory, Advances and Applications.* Birkhäuser Verlag, Basel – Boston – Berlin, 1997.

[17] K.O. Friedrichs. *Pseudo-Differential Operators, An Introduction.* Lecture Notes, Courant Inst. Math. Sci. New York University, 1968.

[18] B. Gramsch. Some homogeneous spaces in the operator theory and Ψ-algebras. In *Tagungsbericht Oberwolfach 42/81 – Funktionalanalysis: C^*-Algebren*, 1981.

[19] B. Gramsch. Relative Inversion in der Störungstheorie von Operatoren und Ψ-Algebren. *Math. Annalen*, 269:27–71, 1984.

[20] B. Gramsch. Oka's principle for special Fréchet Lie groups and homogeneous manifolds in topological algebras of the microlocal analysis. In *Banach algebras 97 Proceedings*, pages 189–204, Berlin – New York, 1998. de Gruyter.

[21] B. Gramsch and W. Kaballo. Multiplicative decompositions of holomorphic Fredholm functions and Ψ^*-algebras. *Math. Nachr.*, 204:83–100, 1999.

[22] B. Gramsch, J. Ueberberg, and K. Wagner. Spectral invariance and submultiplicativity for Fréchet algebras with applications to pseudo-differential operators and Ψ^*-quantization. In *Operator theory: Advances and Applications,* vol. 57, pages 71–98, Birkhäuser, Basel, 1992.

[23] B. Helffer. *Théorie spectrale pour des opérateurs globalement elliptiques*, volume 112 of *Astérisque*. Société Mathématique de France, 1984.

[24] L. Hörmander. Pseudo-differential operators. *Comm. Pure Appl. Math.*, 18:501–517, 1965.

[25] L. Hörmander. *Pseudo-differential operators and hypoelliptic equations*, volume X of *Proc. Symp. in Pure Math. – Singular Integrals*, pages 138–183. Amer. Math. Soc., Providence, Rhode Island, 1966.

[26] L. Hörmander. Fourier integral operators I. *Acta Math.*, 127:79–183, 1971.

[27] L. Hörmander. The Weyl calculus of pseudo-differential operators. *Comm. Pure Appl. Math.*, 32:359–443, 1979.

[28] L. Hörmander. *The analysis of linear partial differential operators,* vol. 3. Springer-Verlag, Berlin – Heidelberg – New York, 1985.

[29] J.J. Kohn and L. Nirenberg. An algebra of pseudo-differential operators. *Comm. Pure Appl. Math.*, 18:269–305, 1965.

[30] H. Kumano-go and K. Taniguchi. Oscillatory integrals of symbols of pseudo-differential operators on $\mathbb{R}^n$ and operators of Fredholm type. *Proc. Japan Acad.*, 49:397–402, 1973.

[31] R. Lauter. *Holomorphic functional calculus in several variables and Ψ^*-algebras of totally characteristic operators on manifolds with boundary.* PhD thesis, Johannes Gutenberg-Universität Mainz, 1996, Shaker Verlag, Aachen, 1997, 241 pages.

[32] R. Lauter and V. Nistor. Analysis of geometric operators on open manifolds: a groupoid approach. Heft 108, Universität Münster, SFB 478 *Geometrische Strukturen in der Mathematik*, May 2000.

[33] R.B. Melrose. Spectral and scattering theory for the Laplacian on asymptotically Euclidean space. In M. Ikawa, editor, *Spectral and Scattering Theory*, volume 162 of *Lecture Notes in Pure and Applied Mathematics*, pages 85–130, New York, 1994. Marcel Dekker Inc. Proceedings of the Taniguchi International Workshop held in Sanda, November 1992.

[34] R.B. Melrose. *Geometric scattering theory.* Cambridge University Press, 1995.

[35] L. Nirenberg. Pseudo-differential operators. *Amer. Math. Soc. Symp. Pure Math.*, 16:149–167, 1970.

[36] C. Parenti. Operatori pseudo-differenziali in $\mathbb{R}^n$ e applicazioni. *Ann. Mat. Pura Appl. (4)*, 93:359–389, 1972.

[37] B.A. Plamenevskij and V.N. Senichkin. Pseudodifferential operators on manifolds with singularities. *Funct. Anal. Appl.*, 33(2):154–156, 1999.

[38] B.A. Plamenevskij and V.N. Senichkin. A class of pseudodifferential operators in $\mathbb{R}^m$ and on stratified manifolds. *Mat. Sb.*, 191(5):725–757, 2000.

[39] Yu. Safarov. Pseudodifferential operators and linear connections. Proc. London Math. Soc., 74(3):379–416, 1997.

[40] E. Schrohe. *Komplexe Potenzen elliptischer Pseudodifferentialoperatoren.* PhD thesis, Johannes Gutenberg-Universität Mainz, 1986.

[41] E. Schrohe. Spaces of weighted symbols and weighted Sobolev spaces on manifolds. In H.O. Cordes, B. Gramsch, and H. Widom, editors, *Pseudo-differential operators*, volume 1256 of *Lect. Notes Math.*, pages 360–377, Berlin – Heidelberg – New York – London – Paris – Tokyo, 1987. Proc. Conf., Oberwolfach/Ger. 1986, Springer-Verlag.

[42] E. Schrohe. A Ψ^*-algebra of pseudo-differential operators on noncompact manifolds. *Arch. Math.*, 51:81–86, 1988.

[43] E. Schrohe. Boundedness and spectral invariance for standard pseudo-differential operators on anisotropically weighted L_p-Sobolev-spaces. *Integral equations operator theory*, 13(2):271–284, 1990.

[44] E. Schrohe. Fréchet algebra techniques for boundary value problems on noncompact manifolds: Fredholm criteria and functional calculus via spectral invariance. *Math. Nachrichten*, 199:145–185, 1999.

[45] M.A. Shubin. Pseudodifferential operators in $\mathbb{R}^n$. *Sov. Math. Dokl.*, 12, Ser. N1, pages 147–151, 1971.

[46] M.A. Shubin. *Pseudodifferential Operators and Spectral Theory.* Springer-Verlag, Berlin – Heidelberg – New York – London – Paris – Tokyo, 1987.

[47] M.A. Shubin. Spectral theory of elliptic operators on non-compact manifolds. In *Méthodes semi-classiques, Volume 1, École d'Éte, Nantes 1991*, volume 207 of *Astérisque*, pages 35–108. Société Mathématique de France, 1992.

[48] J. Ueberberg. Zur Spektralinvarianz von Algebren von Pseudo-Differential Operatoren in der L^p-Theorie. *Manuscripta Math.*, 61:459–475, 1988.

[49] A. Unterberger. Encore des classes de symboles. In *Sém. Goulaouic-Schwartz 1977–1978*, Ecole Polytechnique, Paris, 1977, 17 pages.

[50] A. Unterberger. Oscillateur harmonique et opérateurs pseudo-différentiels. *Ann. Inst. Fourier*, 29-3:201–221, 1979.

[51] A. Unterberger and J. Bokobza. Les opérateurs de Calderon-Zygmund précisés. *C. R. Acad. Sci. Paris*, 259:1612–1614, 1964.

[52] A. Unterberger and J. Bokobza. Sur les opérateurs pseudo-différentiels d'ordre variable. *C. R. Acad. Sci. Paris*, 261:2271–2273, 1965.

[53] L. Waelbroeck. Le calcul symbolique dans les algèbres commutatives. *J. Math. Pures Appl.*, 33(9):147–186, 1954.

[54] H. Widom. A complete symbolic calculus for pseudodifferential operators. *Bull. Sc. math., 2e série*, 104:19–63, 1980.

F. Baldus
Fachbereich Mathematik
Johannes Gutenberg-Universität Mainz
Staudingerweg 9
D-55099 Mainz, Germany
e-mail: baldus@mathematik.uni-mainz.de

Operator Theory:
Advances and Applications, Vol. 145, 249–280

Spectral Invariance and Submultiplicativity for the Algebras of $S(M,g)$-pseudo-differential Operators on Manifolds

F. Baldus

Abstract. For appropriate triples $(\mathcal{M}, g, M)$, where $\mathcal{M}$ is an (in general non-compact) manifold, g is a metric on $T^*\mathcal{M}$, and M is a weight function on $T^*\mathcal{M}$, we developed in [5] a pseudo-differential calculus on $\mathcal{M}$ which is based on the $S(M,g)$-calculus of L. Hörmander [30] in local models. Here we prove that the algebra of operators of order zero is a submultiplicative Ψ^*-algebra in the sense of B. Gramsch [21] in $\mathcal{L}\left(L^2(\mathcal{M})\right)$. For the basic calculus we generalized the concept of E. Schrohe [40] of so-called SG-compatible manifolds. In the proof of the existence of "order reducing operators" we apply a method from [4], and the proof of spectral invariance and submultiplicativity uses methods and results from B. Gramsch, J. Ueberberg, and K. Wagner [26] building on Beals'-type characterizations.

1. Introduction

Already in 1967, R.T. Seeley proved in connection with a holomorphic functional calculus for pseudo-differential operators that the algebra of classical pseudo-differential operators of order zero on a compact manifold M is spectrally invariant in $H^0(M)$ (cf. [45, Theorem 5]; cf. also [46]). Since the seventies great effort has been made to prove more and more general results concerning spectral invariance for algebras of pseudo-differential operators in $\mathcal{L}\left(L^2(M)\right)$ and $\mathcal{L}\left(L^2(\mathbb{R}^n)\right)$ (cf. R. Beals [7], [8], J. Dunau [20], F. Bruyant [13], H.O. Cordes [15], [16], B. Gramsch [22], B. Gramsch, J. Ueberberg, and K. Wagner [26], E. Schrohe [41], [43], J.-M. Bony and J.-Y. Chemin [11], J. Sjöstrand [49]) or, more generally, on $L^p(\mathbb{R}^n)$ (cf. J. Ueberberg [50], cf. also [2]) and weighted Sobolev spaces over $L^p(\mathbb{R}^n)$ (cf. E. Schrohe [42]). Spectral invariance was analyzed for boundary value problems by E. Schrohe [44]. On the other hand, as e.g., H. Widom [55] and E.B. Davies, B. Simon, and M. Taylor [18] pointed out, spectral invariance

---

2000 *Mathematics Subject Classification.* Primary 35S05; 58J40 Secondary 47A60; 47A10; 47L15; 46F05.
Key words and phrases. pseudo-differential operators on manifolds; Weyl–Hörmander calculus; SG-calculus; spectral invariance; submultiplicativity.

fails in slightly different situations. B. Gramsch [21], [22] stressed the importance of spectral invariance for Fréchet operator algebras in the perturbation and homotopy theory of Fredholm functions related to pseudo-differential analysis. For further results in this direction we refer to B. Gramsch [23] and B. Gramsch and W. Kaballo [24]. Moreover, spectral invariance plays a decisive role in [4].

Originally, the calculus of pseudo-differential operators on manifolds was based on symbols which satisfy local estimates, i.e., estimates which are only uniform for x in a compact set (cf. [53], [54], [28], [29]). On $\mathbb{R}^n$ symbols satisfying global estimates were considered (cf. L. Hörmander [29], M.A. Shubin [47], C. Parenti [37], H. Kumano-go and K. Taniguchi [32], R. Beals and C. Fefferman [9], R. Beals [6], H.O. Cordes [14], [17], B. Helffer [27]). In connection with the Weyl calculus, L. Hörmander [30] (cf. also [31, Section 18.5]) presented a vast extension of these classes: To a metric g on $T^*\mathbb{R}^n$ and a weight function M on $T^*\mathbb{R}^n$ he associated the symbol class $S(M, g)$ and developed the corresponding pseudo-differential calculus. For essential contributions to the $S(M, g)$-calculus we wish also to refer to A. Unterberger [51], [52], R. Beals [8], N. Dencker [19], J.-M. Bony and N. Lerner [12], J.-M. Bony and J.-Y. Chemin [11], J.-M. Bony [10]. These data g and M also make sense on the cotangent bundle of a manifold, but, for a generalization of the $S(M, g)$-calculus, we have to make assumptions concerning both the manifold, expressed by the existence of a suitable atlas, and the data M and g. In [30] L. Hörmander investigated already the behavior of operators associated to metrics not too far from the standard ρ, δ-metric under certain canonical transformations, introduced Fourier integral operators, and proved a generalization of Egorov's theorem.

In [5] we generalized the concept of E. Schrohe (cf. [39], [41], [44]) of so-called SG-compatible manifolds which are appropriate in order to transfer the SG-calculus of M.A. Shubin [47], C. Parenti [37], and H.O. Cordes [14]. In so doing, we developed an $S(M, g)$-based pseudo-differential calculus on manifolds $\mathcal{M}$, at least for appropriate combinations of $(\mathcal{M}, g, M)$.

Having provided the basic calculus on manifolds – including composition, mapping properties in the canonical scale of Sobolev spaces, and the existence of a symbol homomorphism – in a quite general setting, in this article we impose more restrictive assumptions in order to prove spectral invariance and submultiplicativity for the algebra of operators of order zero. Thus, it is a submultiplicative Ψ^*-algebra in the sense of B. Gramsch [21] in $\mathcal{L}\left(L^2(\mathcal{M})\right)$.

For the proofs of spectral invariance and submultiplicativity, the existence of "order reducing operators" is of essential importance. To show their existence we apply ellipticity-arguments within the $S(M, g)$-calculus. We used this method already in [4]. Moreover, in the proof of spectral invariance we apply Beals'-type characterizations with commutator methods for the operators in our calculus on manifolds. Here methods and results from B. Gramsch, J. Ueberberg, and K. Wagner [26] play a decisive role (cf. 4.13, 4.15). In connection with submultiplicativity we apply a method from [26] starting from commutator-characterizations which was also used by B. Gramsch and E. Schrohe [25] and in [2]. By the way, it is an open problem whether every Ψ^*-algebra is submultiplicative (cf. B. Gramsch, List of open problems, p. 552 in [1]). For the commutative case we refer to W. Żelazko [56, Theorem C], for the non-commutative case cf. also [56, Theorem 3]. For an application of submultiplicativity in connection with non-abelian cohomology and Oka's principle we refer to B. Gramsch [23] and B. Gramsch, W. Kaballo [24].

Our approach to $S(M, g)$-pseudo-differential calculus on manifolds developed so far allows already – at least in simple situations – to piece different local models together (cf. 6). To go further in this direction, e.g., to piece different ρ, δ-type calculi together, in [3] we defined algebras of pseudo-differential operators with the help of commutator methods (cf. [26]) and proved spectral invariance in $\mathcal{L}\left(L^2(\mathcal{M})\right)$. Besides modifications of the characterizations in 3.20, for the proof of spectral invariance we have to admit conditions for the commutators with pseudo-differential operators of sufficiently small order in the definition of the algebras. In the case of the algebras constructed in [5], these conditions are automatically fulfilled. All these conditions make the construction rather technical. Thus, we decided not to present them in this article but to refer the reader to [3, Sections 5.6 and 7.5]. Here we restrict ourselves to a slightly different approach using methods from [26] which we describe in Section 7. Since their methods are similar, at this point we wish to refer to results of R. Lauter [33], [34], [35] and F. Mantlik [36] in connection with manifolds with singularities.

Instead of explaining the organization of this paper at this point, we refer to brief introductions at the beginning of each section.

This paper is part of the author's Ph. D. thesis [3] written under the supervision of Professor B. Gramsch at the University of Mainz, Germany.

2. Admissible coordinates

There is a natural way to generalize g and M and thus, $S(M, g)$ to cotangent bundles over manifolds. But, on the other hand, it is well known that the $S^m_{\rho,\delta}$-classes are invariant under (local) diffeomorphisms – and thus can be considered on the cotangent bundles of manifolds – if and only if $0 \leq 1 - \rho \leq \delta \leq \rho \leq 1$, $\delta < 1$. Therefore, for an extension of the $S(M, g)$-calculus, we have to make assumptions concerning both the manifold, expressed by the existence of a suitable atlas, and the data M and g. In [40] E. Schrohe transfers a calculus of M.A. Shubin [47], C. Parenti [37], and H.O. Cordes [14], which is based on global estimates, to so-called SG-compatible manifolds (cf. also [39], [41], [44]). The local charts are assumed to satisfy conditions which allow us to prove invariance of the pseudo-differential calculus with respect to the changes of coordinates. Here we generalize these conditions in order to treat classes of operators associated to more general calculi on $\mathbb{R}^n$. Apart from some technical conditions concerning the metric, e.g., a generalization of the assumption $1 - \rho \leq \delta$, we require that the diffeomorphisms have differentials which are matrix valued symbols in the calculus and that they extend to larger domains which are chosen in such a way that the existence of admissible cut-off functions and partitions of unity can be proved. The conditions concerning the manifold $\mathcal{M}$ with atlas $\mathcal{A}$ depend on two metrics g and $\mathcal{G}$ on $T^*\mathcal{M}$ and $\mathcal{M}$, respectively, which determine the pseudo-differential calculus. In the SG-calculus we have $\mathcal{G} = \mathcal{R}_1(g)$, but, as the examples in Section 6 show, in some cases it is reasonable to consider $\mathcal{G} \neq \mathcal{R}_1(g)$. Consequently, we describe admissible (in order to develop a pseudo-differential calculus) structures by the data $(\mathcal{M}, \mathcal{A}, \mathcal{G}, g)$. In contrast to most of the already existing calculi, our approach allows us to patch different local models together, provided that we have compatible changes of coordinates (cf. 6).

In this section we introduce the underlying local models, metrics and weight functions on these models, and diffeomorphisms between them. The assumptions imposed on these data are more restrictive than in [5]. For the formulation of the conditions we need the following weight functions:

Definition 2.1. Let $\mathcal{U}$ be an open subset of $\mathbb{R}^n$ and g a metric on $\mathcal{U} \times \mathbb{R}^n$.

1. For $x \in \mathcal{U}$ and $\xi \in \mathbb{R}^n$ let $h_g(x,\xi) := \left(\sup_{T\in\mathbb{R}^{2n}\setminus\{0\}} \frac{g_{(x,\xi)}(T)}{g^\sigma_{(x,\xi)}(T)}\right)^{1/2}$ where $g^\sigma_{(x,\xi)}(T) := \sup_{W\in\mathbb{R}^{2n}\setminus\{0\}} \frac{\sigma(W,T)^2}{g_{(x,\xi)}(W)}$.
2. For $x \in \mathcal{U}$ and $\xi, t \in \mathbb{R}^n$, $t \neq 0$, we define

$$H_{g,t}(x,\xi) := \sqrt{g_{(x,0)}(t,0)g_{(x,\xi)}(0,t)}.$$

Definition 2.2.

1. Let $\mathcal{U}, U$ be open subsets of $\mathbb{R}^n$ with $\overline{U} \subseteq \mathcal{U}$, and let $\mathcal{G}$ be a smooth metric on $\mathcal{U}$. We call $(\mathcal{U}, U, \mathcal{G})$ an *admissible model* if the following conditions are satisfied:
   - $\exists r_\mathcal{G} > 0,\ \forall x \in \overline{U},\ \forall y \in \mathbb{R}^n,\ \mathcal{G}_x(x-y) \le r_\mathcal{G}^2 \implies y \in \mathcal{U}$.
   - $\exists C_\mathcal{G} > 0,\ \forall x, y \in \mathcal{U},\ \mathcal{G}_x(x-y) \le C_\mathcal{G}^{-1} \implies C_\mathcal{G}^{-1}\mathcal{G}_x(\cdot) \le \mathcal{G}_y(\cdot) \le C_\mathcal{G}\mathcal{G}_x(\cdot)$.
   - $\exists C > 0,\ \forall t_1, t_2 \in \mathbb{R} \setminus \{0\},\ \forall x \in \mathcal{U},\quad \frac{\mathcal{G}_x(t_1)}{\|t_1\|_2^2}\frac{\|t_2\|_2^2}{\mathcal{G}_x(t_2)} \le C.$
2. Let $\mathcal{U}_1$ be an open subset of $\mathbb{R}^n$, let $(\mathcal{U}_2, U_2, \mathcal{G}_2)$ be an admissible model, and let $\kappa : \mathcal{U}_1 \to \mathcal{U}_2$ be a $\mathcal{C}^\infty$-diffeomorphism. κ is called $(\mathcal{U}_2, U_2, \mathcal{G}_2)$-*compatible* if the following statements hold:
   - $(\mathcal{U}_1, U_1, \mathcal{G}_1)$ is an admissible model, where $U_1 := \kappa^{-1}(U_2)$ and $\mathcal{G}_1 := \kappa^*\mathcal{G}_2$, i.e., $\mathcal{G}_{1,x}(t) = \mathcal{G}_{2,\kappa(x)}(\kappa'(x)(t))$.
   - $[\mathcal{U}_1 \ni x \mapsto d\kappa(x)] \in S(1, \mathcal{G}_1; \mathcal{L}(\mathbb{R}^n))$ and $[\mathcal{U}_2 \ni y \mapsto d(\kappa^{-1})(y)] \in S(1, \mathcal{G}_2; \mathcal{L}(\mathbb{R}^n))$.
   - $\exists C > 0,\ \forall x \in \mathcal{U}_1,\ \forall y \in \mathcal{U}_2,\quad \|\kappa(x)\|_2 \le C\langle x\rangle,\ \|\kappa^{-1}(y)\|_2 \le C\langle y\rangle$.

Definition 2.3. We call 'split' a Hörmander metric g on $\mathbb{R}^{2n}$ with $g_{(x,\xi)} = \frac{\|dx\|_2^2}{\varphi(x,\xi)^2} + \frac{\|d\xi\|_2^2}{\phi(x,\xi)^2}$ $((x,\xi) \in \mathbb{R}^{2n})$ *of type* (t) if the following conditions are satisfied:

- There exists a constant $C > 0$ such that

$$\forall x, \xi, t, \tau \in \mathbb{R}^n,\quad g_{(x,0)}(t,0) \le C g_{(x,\xi)}(t,0)\ ,\ g_{(x,0)}(0,\tau) \ge C^{-1} g_{(x,\xi)}(0,\tau)\ .$$

- $\exists C_g > 0,\ \forall x, y, \xi \in \mathbb{R}^n,\ g_{(x,0)}(x-y, 0) \leq C_g^{-1} \implies C_g^{-1} g_{(x,\xi)}(\cdot) \leq g_{(y,\xi)}(\cdot) \leq C_g g_{(x,\xi)}(\cdot)$.
- $\forall K > 0,\ \exists C_K > 0,\ \forall x, \xi \in \mathbb{R}^n,\ \forall A \in \mathcal{L}(\mathbb{R}^n)^{-1}$ with $\|A\|_{\mathcal{L}(\mathbb{R}^n)} \leq K$ and $\|A^{-1}\|_{\mathcal{L}(\mathbb{R}^n)} \leq K, \quad C_K^{-1} g_{(x,\xi)}(\cdot) \leq g_{(x,A\xi)}(\cdot) \leq C_K g_{(x,\xi)}(\cdot)$.
- $\exists C > 0,\ \exists 0 \leq \delta < 1,\ \forall x, \xi \in \mathbb{R}^n,\ \forall t, \tau \in \mathbb{R}^n$

$$\begin{aligned} C^{-1} \frac{\|t\|_2^2}{\langle x, \xi \rangle^2} &\leq g_{(x,\xi)}(t, 0) \leq C \langle x, \xi \rangle^{2\delta} \|t\|_2^2 \\ C^{-1} \frac{\|\tau\|_2^2}{\langle x, \xi \rangle^2} &\leq g_{(x,\xi)}(0, \tau) \leq C \|\tau\|_2^2 . \end{aligned}$$

Moreover, with a vector $t \in \mathbb{R}^n \setminus \{0\}$ the following assumptions hold:

- $\exists C > 0,\ \exists N \in \mathbb{N},\ \forall x, \xi \in \mathbb{R}^n, \quad \langle \xi \rangle \leq C H_{g,t}(x, \xi)^{-N}$.
- $\forall T \in \mathbb{R}^{2n},\ \exists C > 0,\ \exists N \in \mathbb{N},\ \forall x, \xi \in \mathbb{R}^n, \quad C^{-1} H_{g,t}(x, \xi)^N \leq g_{(x,\xi)}(T) \leq C H_{g,t}(x, \xi)^{-N}$.

Definition 2.4.

1. Let $(\mathcal{U}, U, \mathcal{G})$ be an admissible model and let g be a smooth metric on $\mathcal{U} \times \mathbb{R}^n$. We call $(\mathcal{U}, U, \mathcal{G}, g)$ an admissible quadruple *of type* (ψ), if the following conditions are satisfied: We start with the condition concerning $\mathcal{G}$:
   - There exists $C > 0$ such that we have $\mathcal{G}_x(t) \leq C \|t\|_2^2$ for all $x \in \mathcal{U}$ and all $t \in \mathbb{R}$, and $\mathcal{G}_x(t) \leq C \langle x \rangle^{-2} \|t\|_2^2$ for all $x \in \mathcal{U} \setminus U$ and all $t \in \mathbb{R}$.

   We continue with conditions concerning g:
   - There exists a Hörmander metric g_s of type (t) on $\mathbb{R}^{2n}$ such that with a suitable constant $C_s > 0$ for all $x \in \mathcal{U}$ and all $\xi \in \mathbb{R}^n$ the estimate $C_s^{-1} g_{s,(x,\xi)}(\cdot) \leq g_{(x,\xi)}(\cdot) \leq C_s g_{s,(x,\xi)}(\cdot)$ holds.
   - $\exists C > 0,\ \exists N \in \mathbb{N},\ \forall x \in \mathcal{U},\ \forall \xi \in \mathbb{R}^n, \quad \langle \xi \rangle \leq C h_g(x, \xi)^{-N}$.
   - $\forall T \in \mathbb{R}^{2n},\ \exists C > 0,\ \exists N \in \mathbb{N},\ \forall x \in \mathcal{U},\ \forall \xi \in \mathbb{R}^n, \quad C^{-1} h_g(x, \xi)^N \leq g_{(x,\xi)}(T) \leq C h_g(x, \xi)^{-N}$.

   Next, concerning both $\mathcal{G}$ and g, we suppose that the following conditions hold:
   - For all $x \in \mathcal{U}$ we have $\mathcal{G}_x(\cdot) \leq \mathcal{R}_1(g)_x(\cdot)$, where $\mathcal{R}_1(g)_x(t) := g_{(x,0)}(t, 0)$.
   - $\exists C > 0,\ \forall x \in \mathcal{U},\ \forall t, \xi \in \mathbb{R}^n, \quad \mathcal{G}_x(t)\ g_{(x,\xi)}(0, \xi) \leq C g_{(x,\xi)}(t, 0)$.
2. We apply the notation from the preceding point. We call $(\mathcal{U}, U, \mathcal{G}, g)$ an admissible quadruple *of type* $(\psi)_s$, if additionally the following condition is satisfied:
   - There exists $C > 0$ such that $g_{(x,\xi)}(t, \tau) \leq C \|(t, \tau)\|_2^2$ for all $(x, \xi) \in \mathcal{U} \times \mathbb{R}^n$ and all $t, \tau \in \mathbb{R}^n$.

Remark 2.5. Let $(\mathcal{U}, U, \mathcal{G})$ be an admissible model. In the following we will suppose that $C_{\mathcal{G}} \geq 1$ and $r_{\mathcal{G}} \leq 1/\sqrt{C_{\mathcal{G}}}$. If $(\mathcal{U}, U, \mathcal{G}, g)$ is an admissible quadruple of type (ψ), we suppose that $C_g \geq C_\mathcal{G}$, C_g is chosen such that for $x, y \in \mathcal{U}$ and $\xi, \eta \in \mathbb{R}^n$,

$$g_{(x,\xi)}(x-y, \xi-\eta) \leq C_g^{-1} \implies C_g^{-1} g_{(x,\xi)}(\cdot) \leq g_{(y,\eta)}(\cdot) \leq C_g g_{(x,\xi)}(\cdot) ,$$

and $r_g \leq \min\{C_g^{-1/2}, r_\mathcal{G}\}$.

Definition 2.6. Let $(\mathcal{U}, U, \mathcal{G}, g)$ be an admissible quadruple of type (ψ) and let $M : \mathcal{U} \times \mathbb{R}^n \to (0, \infty)$ be a continuous function. We call M a $(\mathcal{U}, U, g)$*-admissible weight function of type* (ψ), if

- For each split Hörmander metric g_s as in 2.4 there exists a g_s-admissible weight function $\tilde{M} : \mathbb{R}^{2n} \to (0, \infty)$ and a constant $\tilde{C} > 0$ such that for all $(x, \xi) \in \mathcal{U} \times \mathbb{R}^n$ we have $\tilde{C}^{-1}\tilde{M}(x,\xi) \leq M(x,\xi) \leq \tilde{C}\tilde{M}(x,\xi)$.
- $\forall K > 0$, $\exists C_K > 0$, $\forall x \in \mathcal{U}$, $\forall \xi \in \mathbb{R}^n$, $\forall A \in \mathcal{L}(\mathbb{R}^n)^{-1}$ such that $\|A\|_{\mathcal{L}(\mathbb{R}^n)} \leq K$ and $\|A^{-1}\|_{\mathcal{L}(\mathbb{R}^n)} \leq K$, $\quad C_K^{-1} M(x,\xi) \leq M(x, A\xi) \leq \ C_K M(x,\xi)$.
- $\forall x, y \in \mathcal{U}$, $\forall \xi \in \mathbb{R}^n$, $g_{(x,0)}(x-y, 0) \leq C_g^{-1} \implies C_M^{-1} M(x,\xi) \leq M(y,\xi) \leq C_M M(x,\xi)$.

Definition 2.7. Let $\mathcal{U}_1$ be an open subset of $\mathbb{R}^n$, let $(\mathcal{U}_2, U_2, \mathcal{G}_2, g_2)$ be an admissible quadruple of type (ψ), and let $\kappa : \mathcal{U}_1 \to \mathcal{U}_2$ be a $(\mathcal{U}_2, U_2, \mathcal{G}_2)$-compatible diffeomorphism. κ is called $(\mathcal{U}_2, U_2, \mathcal{G}_2, g_2)$*-compatible of type* (ψ) if $(\mathcal{U}_1, U_1, \mathcal{G}_1, g_1)$ is an admissible quadruple of type (ψ), where $U_1 := \kappa^{-1}(U_2)$, $\mathcal{G}_1 := \kappa^*\mathcal{G}_2$, $g_1 := \chi^* g_2$, $\chi : \mathcal{U}_1 \times \mathbb{R}^n \to \mathcal{U}_2 \times \mathbb{R}^n$, $(x,\xi) \mapsto \left(\kappa(x), {}^t d\kappa(x)^{-1}\xi\right)$.

If we replace the definitions in [5] concerning the local models by the preceding definitions, we obtain more restrictive conditions on generalized SG-compatible manifolds and admissible structures which we now call admissible structures of type (ψ). We omit to repeat the precise definition in this article and apply the notation introduced in [5] in connection with the basic calculus.

3. Characterizations of pseudo-differential operators on manifolds

In this section we prove characterizations of operators in $\Psi(M, g; \mathcal{M}, \mathcal{A}, \mathcal{G})$, expressed by mapping properties of iterated commutators with operators in $\mathcal{V}(\mathcal{M}, \mathcal{A}, \mathcal{G})$ between appropriate Sobolev spaces (cf. 3.20). In the proof of 3.20 we make use of the corresponding result on $\mathbb{R}^n$ (cf. [2, 3.6.6]). In order to formulate and prove these characterizations, we first have to introduce Sobolev spaces $H(M, g; \mathcal{M}, \mathcal{A})$, associated to general weight functions of the calculus, and endow these spaces with a suitable topology such that pseudo-differential operators are continuous maps between appropriate spaces (cf. 3.14). As in the case of the scale of spaces associated to H_g, we define the Sobolev spaces with the help of our local models (cf. 3.5). In Section 4 we will prove the existence of "order reducing operators" with the help of which $H(M, g; \mathcal{M}, \mathcal{A})$ become Banach spaces (cf. 4.21).

Lemma 3.1. *Let g be a Hörmander metric of type (t), let $U \subseteq \mathcal{U} \subseteq \mathbb{R}^n$ be open subsets, and let $\mathcal{G}$ be a metric on $\mathcal{U}$ such that $(\mathcal{U}, U, \mathcal{G}, g_{|\mathcal{U}\times\mathbb{R}^n})$ is an admissible quadruple of type (ψ). Let $t \in \mathbb{R}^n \setminus \{0\}$. Let $0 < \rho < r_{\mathcal{G}}/2$ and let $\widehat{U} := (U_{\rho,\mathcal{G}})^\circ$, $\underline{\mathcal{U}} := (U_{2\rho,\mathcal{G}})^\circ$. Then $(\mathcal{U}, \underline{\mathcal{U}}, \mathcal{G})$ is an admissible model. Let $\tilde{r}_{\mathcal{G}}$ be the corresponding constant from 2.2, let $0 < \tilde{\rho} < \tilde{r}_{\mathcal{G}}$, and let $\widehat{\underline{\mathcal{U}}} := (\underline{\mathcal{U}}_{\tilde{\rho},\mathcal{G}})^\circ$. Moreover, let $\varphi, \widehat{\psi} \in \mathcal{C}^\infty(\mathbb{R}^n)$ be functions satisfying the following properties:*

- $[(x,\xi) \mapsto \varphi(x)], [(x,\xi) \mapsto \widehat{\psi}(x)] \in S(1,g)$.
- *supp* $(\varphi) \subseteq \widehat{U}$.
- *supp* $(\widehat{\psi}) \subseteq \widehat{\underline{\mathcal{U}}}$, $\widehat{\psi} = 1$ *on* $\underline{\mathcal{U}}$.

Then, for all $s \in \mathbb{R}$ and all $p \in S\left(H^s_{g,t}, g\right)$, there exists $q \in \mathcal{S}(\mathbb{R}^{2n})$ such that

$$(1 - \widehat{\psi})p(x,D)\varphi = q(x,D) \ .$$

Proof. There exists a function $\phi \in S(1,\mathcal{G})$ with $\operatorname{supp}(\phi) \subseteq \underline{\mathcal{U}}$ and $\phi = 1$ on $\widehat{U}$. Then we have $(1-\widehat{\psi})p(x,D)\varphi \in \bigcap_{\mu\in\mathbb{N}} \Psi\left(H^\mu_{g,t}, g\right)$. For $j \in \{1,\ldots,n\}$ and $u \in \mathcal{S}(\mathbb{R}^n)$ let

$$\begin{aligned} V_1^{(j)}u &:= M_{x_j}(\phi\cdot u) \ , \ \tilde{V}_1^{(j)}u := M_{x_j}((1-\phi)\cdot u) \ , \\ V_2^{(j)}u &:= D_{x_j}(\phi\cdot u) \ , \ \tilde{V}_2^{(j)}u := D_{x_j}((1-\phi)\cdot u) \ . \end{aligned}$$

For $m_1, \tilde{m}_1, m_2, \tilde{m}_2 \in \mathbb{N}_0$ we obtain

$$\begin{aligned} &\mathcal{L}(L^2(\mathbb{R}^n)) \\ \ni\ & \left(\mathrm{Id} + \sum_{j=1}^n \left(\mathrm{ad}(\tilde{V}_1^{(j)})\right)^2\right)^{\tilde{m}_1} \left(\mathrm{Id} + \sum_{j=1}^n \left(\mathrm{ad}(\tilde{V}_2^{(j)})\right)^2\right)^{\tilde{m}_2} \\ & \times \left(\mathrm{Id} + \sum_{j=1}^n \left(\mathrm{ad}(V_2^{(j)})\right)^2\right)^{m_2} \left(\mathrm{Id} + \sum_{j=1}^n \left(\mathrm{ad}(V_1^{(j)})\right)^2\right)^{m_1} (1-\widehat{\psi})p(x,D)\varphi \\ =\ & \langle M_x\rangle^{2\tilde{m}_1}\langle D_x\rangle^{2\tilde{m}_2}(1-\widehat{\psi})p(x,D)\varphi\langle M_x\rangle^{2m_1}\langle D_x\rangle^{2m_2} \ . \end{aligned}$$

By virtue of [2, 3.6.6] this yields

$$(1-\widehat{\psi})p(x,D)\varphi \in \bigcap_{\mu_1,\mu_2\in\mathbb{N}} \Psi\left(M_1^{-\mu_1}M_2^{-\mu_2}, g^{1,1}\right) \ ,$$

where we define $g^{1,1}_{(x,\xi)} := \frac{\|dx\|_2^2}{\langle x\rangle^2} + \frac{\|d\xi\|_2^2}{\langle \xi\rangle^2}$, $M_1(x,\xi) := \langle x\rangle$, and $M_2(x,\xi) := \langle\xi\rangle$ ($x,\xi \in \mathbb{R}^n$). Since $\bigcap_{\mu_1,\mu_2\in\mathbb{N}} S\left(M_1^{-\mu_1}M_2^{-\mu_2}, g^{1,1}\right) = \mathcal{S}(\mathbb{R}^{2n})$, this yields the assertion. ☐

Lemma 3.2. *Let $(\mathcal{U}, U, \mathcal{G}, g)$ be an admissible quadruple of type (ψ). Let $t \in \mathbb{R}^n\setminus\{0\}$. Let M be a $(\mathcal{U}, U, g)$-admissible weight function of type (ψ) such that with suitable constants $C > 0$ and $N \in \mathbb{N}$ for all $(x,\xi) \in \mathcal{U}\times\mathbb{R}^n$ the estimate*

$$C^{-1}H_{g,t}(x,\xi)^N \le M(x,\xi) \le CH_{g,t}(x,\xi)^{-N}$$

holds. Let $0 < \rho < r_{\mathcal{G}}/2$ *and let* $\varphi, \widehat{\psi}$ *be functions as in* 3.1 *with* $\varphi = 1$ *on* $U_{\rho/2,\mathcal{G}}$. *Finally, let* g_s *be as in* 2.4 *and let* $\tilde{M}$ *be as in* 2.6. *We define* $M_{N,g_s,t} := \min(\tilde{M}, H_{g_s,t}^{-N})$ *(which is a* g_s*-admissible weight function).*

1. *For* $u \in \mathcal{S}'(\mathbb{R}^n)$ *with* $supp\,(u) \subseteq U_{\rho/2,\mathcal{G}}$ *we have*

$$u \in H\left(M_{N,g_s,t}, g\right) \iff \forall A \in \Psi(M_{N,g_s,t}, g_s),\ \widehat{\psi}A(\varphi u) \in L^2(\mathbb{R}^n)\ .$$

2. *If* $(u_\nu)_{\nu \in \mathbb{N}_0} \subseteq H\left(M_{N,g_s,t}, g\right)$ *with* $supp\,(u_\nu) \subseteq U_{\rho/2,\mathcal{G}}$, *then we have*

$$u_\nu \to u_0 \text{ in } H\left(M_{N,g_s,t}, g_s\right) \iff \forall A \in \Psi(M_{N,g_s,t}, g_s),\ \widehat{\psi}A(\varphi u_\nu) \to \widehat{\psi}A(\varphi u_0) \text{ in } L^2(\mathbb{R}^n).$$

Proof. In Part 1 we apply 3.1. To prove Part 2, let $(u_\nu)_{\nu \in \mathbb{N}} \subset H\left(M_{N,g_s,t}, g\right)$ with $\operatorname{supp}(u_\nu) \subseteq U_{\rho/2,\mathcal{G}}$ such that $\widehat{\psi}A(\varphi u_\nu) \to 0$ in $L^2(\mathbb{R}^n)$ for all $A \in \Psi(M_{N,g_s,t}, g_s)$. There exist constants $C > 0$ and $\tilde{N} \in \mathbb{N}$ such that $\langle x,\xi\rangle^{-2\tilde{N}} \leq C M_{N,g_s,t}(x,\xi)$ for all $x, \xi \in \mathbb{R}^n$. Thus, by assumption and by virtue of [27, Proposition 1.7.5, Théorème 1.8.11 and Corollaire 1.9.5], there exists $q \in S\left(M_{N,g_s,t}, g_s\right)$ such that $\langle x, D_x\rangle^{2\tilde{N}} q(x,D) = \mathrm{Id}$. Since $\langle x, D_x\rangle^{2\tilde{N}} := \left(1 + \sum_{j=1}^n (x_j^2 + D_{x_j}^2)\right)^{\tilde{N}}$ is a local operator, we obtain

$$\begin{aligned} u_\nu &= \varphi u_\nu = \varphi\langle x, D_x\rangle^{2\tilde{N}} q(x,D) u_\nu = \varphi\langle x, D_x\rangle^{2\tilde{N}} \big(\underbrace{\widehat{\psi} q(x,D)(\varphi u_\nu)}_{\to 0 \text{ in } L^2(\mathbb{R}^n)}\big) \\ &\to 0 \text{ in } \mathcal{S}'(\mathbb{R}^n)\ . \end{aligned}$$

Thus, applying 3.1 once more, for $A \in \Psi(M_{N,g_s,t}, g_s)$ we have

$$Au_\nu = \underbrace{\widehat{\psi}A(\varphi u_\nu)}_{\to 0 \text{ in } L^2(\mathbb{R}^n)} + \underbrace{(1-\widehat{\psi})A(\varphi u_\nu)}_{\to 0 \text{ in } \mathcal{S}(\mathbb{R}^n)} \to 0 \text{ in } L^2(\mathbb{R}^n)\ .$$

This yields $u_\nu \to 0$ in $H\left(M_{N,g_s,t}, g\right)$. □

Definition 3.3. Let $(\mathcal{U}, U, \mathcal{G}, g)$ be an admissible quadruple of type (ψ) and M as in 3.2. With the notation from 3.2, for an open set $U' \subseteq U_{\rho/2,\mathcal{G}}$ we define $H(M,g;U')$ as the space of all $u \in H\left(M_{N,g_s,t}, g_s\right)$ with $\operatorname{supp}(u) \subseteq U'$ and endow $H(M,g;U')$ with the topology inherited by $H\left(M_{N,g_s,t}, g_s\right)$.

Remark 3.4. By virtue of 3.2, the definition of $H(M,g;U')$ is independent of the choice of g_s, $\tilde{M}$, and t.

Definition 3.5. Let $(\mathcal{M}, \mathcal{A}, \mathcal{G}, g)$ be an admissible structure of type (ψ) and let M be an $(\mathcal{M}, \mathcal{A}, g)$-admissible weight function of type (ψ) which is H_g^{-1}-admissible. We define $H\left(M, g; \mathcal{M}, \mathcal{A}\right)$ as the set of all $u \in \mathcal{D}'(\mathcal{M})$ such that $(\kappa^{-1})^*(\varphi u) \in H\left((\chi^{-1})^* M, (\chi^{-1})^* g; \kappa(\widehat{X}_g)\right)$ for all $(\mathcal{X}, X, \kappa) \in \mathcal{A}$ and all $(\mathcal{X}, X, \kappa, g)$-admissible cut-off functions φ.

Definition 3.6. Let $(\mathcal{M}, \mathcal{A}, \mathcal{G}, g)$ be an admissible structure of type (ψ). Then we define $H(g; \mathcal{M}, \mathcal{A})^{-\infty} := \bigcup_{s \in \mathbb{R}} H\left(H_g^s, g; \mathcal{M}, \mathcal{A}\right)$.

Lemma 3.7. *Let $(\mathcal{M}, \mathcal{A}, \mathcal{G}, g)$ be an admissible structure of type (ψ) and let M be an $(\mathcal{M}, \mathcal{A}, g)$-admissible weight function of type (ψ) which is H_g^{-1}-admissible. Let $u \in H(g; \mathcal{M}, \mathcal{A})^{-\infty}$. Then we have*

$$u \in H(M, g; \mathcal{M}, \mathcal{A}) \iff \forall P \in \Psi(M, g; \mathcal{M}, \mathcal{A}, \mathcal{G}), \ Pu \in H(1, g; \mathcal{M}, \mathcal{A}) .$$

Proof. The direction "$\Longrightarrow$" is proved elementary. For "$\Longleftarrow$" we apply 3.2. □

Remark 3.8. Let $(\mathcal{M}, \mathcal{A}, \mathcal{G}, g)$ be an admissible structure of type (ψ) and let M, M' be $(\mathcal{M}, \mathcal{A}, g)$-admissible weight functions of type (ψ) which are H_g^{-1}-admissible. Then, if $P \in \Psi(M, g; \mathcal{M}, \mathcal{A}, \mathcal{G})$, we have $P(H(M', g; \mathcal{M}, \mathcal{A})) \subseteq (M'/M, g; \mathcal{M}, \mathcal{A})$.

Definition 3.9. Let $(\mathcal{M}, \mathcal{A}, \mathcal{G}, g)$ be an admissible structure of type (ψ) and let $(\varphi_i)_{i=1,\dots,l}$ be an $(\mathcal{M}, \mathcal{A}, g)$-admissible partition of unity. For $u \in H(1, g; \mathcal{M}, \mathcal{A})$ we define

$$\|u\|_{H(1,g;\mathcal{M},\mathcal{A})} := \sum_{i=1}^{l} \left\|(\kappa_{j_i}^{-1})^*(\varphi_i u)\right\|_{L^2(\mathbb{R}^n)} .$$

Remark 3.10.

1. Let $(\mathcal{X}, X, \kappa) \in \mathcal{A}$ and let φ be an $(\mathcal{X}, X, \kappa, g)$-admissible cut-off function. There exists a constant $C > 0$ such that for all $u \in H(1, g; \mathcal{M}, \mathcal{A})$ the estimate
$$C^{-1} \|\varphi u\|_{H(1,g;\mathcal{M},\mathcal{A})} \le \left\|(\kappa^{-1})^*(\varphi u)\right\|_{L^2(\mathbb{R}^n)} \le C \|\varphi u\|_{H(1,g;\mathcal{M},\mathcal{A})} \le C \|u\|_{H(1,g;\mathcal{M},\mathcal{A})}$$
holds.
2. In particular, up to multiplicative constants, the norm $\|\cdot\|_{H(1,g;\mathcal{M},\mathcal{A})}$ is independent of the choice of the $(\mathcal{M}, \mathcal{A}, \mathcal{G})$-admissible partition of unity.
3. Moreover, a sequence $(u_\nu)_{\nu\in\mathbb{N}} \subseteq H(1, g; \mathcal{M}, \mathcal{A})$ is convergent, respectively, a Cauchy sequence, with respect to $\|\cdot\|_{H(1,g;\mathcal{M},\mathcal{A})}$, if and only if for all $(\mathcal{X}, X, \kappa) \in \mathcal{A}$ and all $(\mathcal{X}, X, \kappa, g)$-admissible cut-off functions φ the sequence $((\kappa^{-1})^*(\varphi u_\nu))$ is convergent, respectively, a Cauchy sequence, in $L^2(\mathbb{R}^n)$.
4. Consequently, $(H(1, g; Mm, \mathcal{A}, \|\cdot\|_{H(1,g;\mathcal{M},\mathcal{A})})$ is a Banach space and $\mathcal{C}_0^\infty(\mathcal{M})$ is dense in $H(1, g; \mathcal{M}, \mathcal{A})$.
5. Let G be an $(\mathcal{M}, \mathcal{A}, g)$-admissible Riemannian metric. There exists $C > 0$ such that for $u \in H(1, g; \mathcal{M}, \mathcal{A}) = L^2(\mathcal{M}, \alpha_G)$ the estimate
$$C^{-1} \|u\|_{H(1,g;\mathcal{M},\mathcal{A})} \le \|u\|_{L^2(\mathcal{M},\alpha_G)} \le C \|u\|_{H(1,g;\mathcal{M},\mathcal{A})}$$
holds.

Corollary 3.11. *Let $(\mathcal{M}, \mathcal{A}, \mathcal{G}, g)$ be an admissible structure. Then, for a linear operator $P : H(1, g; \mathcal{M}, \mathcal{A}) \to H(1, g; \mathcal{M}, \mathcal{A})$, the following statements are equivalent:*

1. *P is bounded in charts.*
2. $P \in \mathcal{L}\Big(H(1, g; \mathcal{M}, \mathcal{A}), \|\cdot\|_{H(1,g;\mathcal{M},\mathcal{A})}\Big)$.

Definition 3.12. Let $(\mathcal{M}, \mathcal{A}, \mathcal{G}, g)$ be an admissible structure of type (ψ) and let M be an $(\mathcal{M}, \mathcal{A}, g)$-admissible weight function of type (ψ) which is H_g^{-1}-adapted. We endow $H(M, g; \mathcal{M}, \mathcal{A})$ with the coarsest topology such that for each $P \in \Psi(M, g; \mathcal{M}, \mathcal{A}, \mathcal{G})$ the operator

$$P : H(M, g; \mathcal{M}, \mathcal{A}) \to \left(H(1, g; \mathcal{M}, \mathcal{A}), \|\cdot\|_{H(1,g;\mathcal{M},\mathcal{A})}\right)$$

is continuous.

Remark 3.13.

1. On $H(1, g; \mathcal{M}, \mathcal{A})$ this topology coincides with the topology induced by $\|\cdot\|_{H(1,g;\mathcal{M},\mathcal{A})}$.
2. Let $(\mathcal{M}, \tilde{\mathcal{A}}, \mathcal{G}, g)$ be an admissible structure of type (ψ) such that $\mathcal{A} \leq \tilde{\mathcal{A}}$. Then M is $(\mathcal{M}, \tilde{\mathcal{A}}, g)$-admissible of type (ψ), too. We have $H(g; \mathcal{M}, \mathcal{A})^{-\infty} = H(g; \mathcal{M}, \tilde{\mathcal{A}})^{-\infty}$ and $H(M, g; \mathcal{M}, \mathcal{A}) = H\left(M, g; \mathcal{M}, \tilde{\mathcal{A}}\right)$. Moreover, by definition, the topologies on $H(M, g; \mathcal{M}, \mathcal{A})$ and $H\left(M, g; \mathcal{M}, \tilde{\mathcal{A}}\right)$ coincide.
3. The topology on $H(M, g; \mathcal{M}, \mathcal{A})$ is the projective topology with respect to the mappings in $\Psi(M, g; \mathcal{M}, \mathcal{A}, \mathcal{G})$. Thus, this topology is induced by the semi-norms $p_P(u) := \|Pu\|_{H(1,g;\mathcal{M},\mathcal{A})}$ $(P \in \Psi(M, g; \mathcal{M}, \mathcal{A}, \mathcal{G}))$. Hence, sets of the form $\bigcap_{i \in I} P_i^{-1}(V_i)$, where I is a finite index set, $P_i \in \Psi(M, g; \mathcal{M}, \mathcal{A}, \mathcal{G})$, and $V_i \subseteq H(1, g; \mathcal{M}, \mathcal{A})$ is an open set, form a base of topology (cf. [38, Chapter II 5]).

Lemma 3.14. *Let $(\mathcal{M}, \mathcal{A}, \mathcal{G}, g)$ be an admissible structure of type (ψ) and let M, M' be two $(\mathcal{M}, \mathcal{A}, g)$-admissible weight functions of type (ψ) which are H_g^{-1}-adapted. Let $P \in \Psi(M, g; \mathcal{M}, \mathcal{A}, \mathcal{G})$. Then the mapping $P : H(M', g; \mathcal{M}, \mathcal{A}) \to H(M'/M, g; \mathcal{M}, \mathcal{A})$ (cf. 3.8) is continuous.*

Proof. Let $V \subseteq H(M'/M, g; \mathcal{M}, \mathcal{A})$ be open. Hence, there exists an index set Λ, for each $\lambda \in \Lambda$ a finite index set I_λ, and, for all $\lambda \in \Lambda$, $i \in I_\lambda$, an open set $U_{\lambda,i} \subseteq H(1, g; \mathcal{M}, \mathcal{A})$ and operators $B_{\lambda,i} \in \Psi(M'/M, g; \mathcal{M}, \mathcal{A}, \mathcal{G})$ such that $V = \bigcup_{\lambda \in \Lambda} V_\lambda$ where $V_\lambda := \bigcap_{i \in I_\lambda} B_{\lambda,i}^{-1}(U_{\lambda,i})$. This yields

$$A^{-1}(V) = \bigcup_{\lambda \in \Lambda} A^{-1}(V_\lambda) = \bigcup_{\lambda \in \Lambda} \bigcap_{i \in I_\lambda} A^{-1}(B_{\lambda,i}^{-1}(U_{\lambda,i})) = \bigcup_{\lambda \in \Lambda} \bigcap_{i \in I_\lambda} (B_{\lambda,i}A)^{-1}(U_{\lambda,i}) .$$

□

Corollary 3.15. *If $M \leq M'$, then we have $H(M', g; \mathcal{M}, \mathcal{A}) \hookrightarrow H(M, g; \mathcal{M}, \mathcal{A})$.*

Lemma 3.16. *Let $(\mathcal{M}, \mathcal{A}, \mathcal{G}, g)$ be an admissible structure of type (ψ) and let M be an $(\mathcal{M}, \mathcal{A}, g)$-admissible weight function of type (ψ) which is H_g^{-1}-adapted.*

1. *Let $(u_\nu)_{\nu \in \mathbb{N}} \subseteq H(M, g; \mathcal{M}, \mathcal{A})$. Then the following statements are equivalent:*
   (a) *$u_\nu \to 0$ in $H(M, g; \mathcal{M}, \mathcal{A})$.*
   (b) *For all $(\mathcal{X}, X, \kappa) \in \mathcal{A}$ and for all $(\mathcal{X}, X, \kappa, g)$-admissible cut-off functions φ we have $(\kappa^{-1})^*(\varphi u_\nu) \to 0$ in $H\left((\chi^{-1})^* M, (\chi^{-1})^* g; \kappa(\widehat{X}_g)\right)$.*

2. *Let $(\mathcal{X}, X, \kappa) \in \mathcal{A}$ and let φ be an $(\mathcal{X}, X, \kappa, g)$-admissible cut-off function. Moreover, let $(u_\nu)_{\nu\in\mathbb{N}} \subseteq H\left((\chi^{-1})^*M, (\chi^{-1})^*g; \kappa(\widehat{X}_g)\right)$, $u_\nu \to 0$. Then we have $(\varphi \cdot \kappa^* u_\nu)_{\nu\in\mathbb{N}} \subseteq H(M, g; \mathcal{M}, \mathcal{A})$ and $\varphi \cdot \kappa^* u_\nu \to 0$ in $H(M, g; \mathcal{M}, \mathcal{A})$.*

Proof.

1. For the implication "$\Longrightarrow$" we apply 3.2. For the inverse implication note that for $(u_\nu)_{\nu\in\mathbb{N}} \subseteq H(M,g;\mathcal{M},\mathcal{A})$, in order to show that $u_\nu \to 0$ in $H(M, g; \mathcal{M}, \mathcal{A})$, we have to prove that $Pu_\nu \to 0$ in $H(1,g;\mathcal{M},\mathcal{A})$ for all $P \in \Psi(M,g;\mathcal{M},\mathcal{A},\mathcal{G})$.
2. In addition to the last observation we apply 3.7.

□

Remark 3.17. If $P : H(M, g; \mathcal{M}, \mathcal{A}) \to H(M', g; \mathcal{M}, \mathcal{A})$ is continuous, then P is uniquely determined by its action on $\mathcal{C}_0^\infty(\mathcal{M})$.

Definition 3.18. Let $(\mathcal{M}, \mathcal{A}, \mathcal{G}, g)$ be an admissible structure of type (ψ). Let $j \in \{1, \ldots, m\}$ and let $t_j \in \mathbb{R}^n$ with $\|t_j\|_2 = 1$. Let $\Phi_j(x,\xi) := \left(g_{j,(x,\xi)}(t_j, 0)\right)^{1/2}$ and $\Psi_j(x,\xi) := \left(g_{j,(x,\xi)}(0, t_j)\right)^{1/2}$ $((x,\xi) \in \kappa_j(\mathcal{X}_j) \times \mathbb{R}^n)$. Now let $(\varphi_i)_{i=1,\ldots,l}$ be an $(\mathcal{M}, \mathcal{A}, g)$-admissible partition of unity. For $x \in \mathcal{M}$ and $\xi \in T_x^*\mathcal{M}$ we define

$$\begin{aligned} M_g(1,0)(x,\xi) &:= \sum_{i=1}^{l} \varphi_i(x) \cdot \Phi_{j_i}(\chi_{j_i}(x,\xi)), \\ M_g(0,1)(x,\xi) &:= \sum_{i=1}^{l} \varphi_i(x) \cdot \Psi_{j_i}(\chi_{j_i}(x,\xi)). \end{aligned}$$

For $m_1, m_2 \in \mathbb{N}_0$ let $M_g(m_1, m_2) := M_g(1,0)^{m_1} M_g(0,1)^{m_2}$.

Remark 3.19.

1. For $(\mathcal{X}, X, \kappa) \in \mathcal{A}$ there exists a constant $C > 0$ such that for all $x \in \mathcal{X}$, all $\xi \in T_x^*\mathcal{X}$, and all $t \in \mathbb{R}$ with $\|t\|_2 = 1$ the estimates

$$\begin{aligned} C^{-1}\left(M_g(1,0)(x,\xi)\right)^2 &\le \left((\chi^{-1})^* g_{|T^*\mathcal{X}}\right)_{\chi(x,\xi)}(t,0) \le C\left(M_g(1,0)(x,\xi)\right)^2 \\ C^{-1}\left(M_g(0,1)(x,\xi)\right)^2, &\le \left((\chi^{-1})^* g_{|T^*\mathcal{X}}\right)_{\chi(x,\xi)}(0,t) \le C\left(M_g(0,1)(x,\xi)\right)^2 \end{aligned}$$

hold. If $\tilde{M}_g(1,0)$ and $\tilde{M}_g(0,1)$ are defined with another $(\mathcal{M}, \mathcal{A}, g)$-admissible partition of unity, then, with a suitable constant $C > 0$, we have

$$\begin{aligned} C^{-1}M_g(1,0) &\le \tilde{M}_g(1,0) \le CM_g(1,0)\ , \\ C^{-1}M_g(0,1) &\le \tilde{M}_g(0,1) \le CM_g(0,1)\ . \end{aligned}$$

Finally, $M_g(1,0)$ and $M_g(0,1)$ are $(\mathcal{M}, \mathcal{A}, g)$-admissible weight functions of type (ψ). By assumption, they are H_g^{-1}-adapted and there exists a constant $C > 0$ such that $M_g(0,1) \le C$.

Theorem 3.20. *Let $(\mathcal{M}, \mathcal{A}, \mathcal{G}, g)$ be an admissible structure of type (ψ) and let M be an $(\mathcal{M}, \mathcal{A}, g)$-admissible weight function of type (ψ) which is H_g^{-1}-adapted.*

1. *Let $P : \mathcal{C}_0^\infty(\mathcal{M}) \to \mathcal{D}'(\mathcal{M})$. Then the following statements are equivalent:*
   (a) $P \in \Psi(M, g; \mathcal{M}, \mathcal{A}, \mathcal{G})$.
   (b) *For all $l, k \in \mathbb{N}_0$ with $l \leq k$, all $V_1, \ldots, V_l \in \mathcal{V}_m(\mathcal{M}, \mathcal{A}, \mathcal{G})$, all $V_{l+1}, \ldots, V_k \in \mathcal{V}_d(\mathcal{M}, \mathcal{A}, \mathcal{G})$, all permutations π of $\{1, \ldots, k\}$, and all $(\mathcal{M}, \mathcal{A}, g)$-admissible weight functions M' of type (ψ) which are H_g^{-1}-adapted the operator*
   $$ad\left(V_{\pi(1)}\right) \ldots ad\left(V_{\pi(k)}\right) P : \mathcal{C}_0^\infty(\mathcal{M}) \to \mathcal{D}'(\mathcal{M})$$
   has a continuous extension to an operator
   $$H\left(M', g; \mathcal{M}, \mathcal{A}\right) \to H\left(M' M^{-1} M_g(k-l, l)^{-1}, g; \mathcal{M}, \mathcal{A}\right) .$$
   (c) *For all $l, k, k_1, k_2 \in \mathbb{N}_0$ with $l \leq k$, all $V_1, \ldots, V_l \in \mathcal{V}_m(\mathcal{M}, \mathcal{A}, \mathcal{G})$, all $V_{l+1}, \ldots, V_k \in \mathcal{V}_d(\mathcal{M}, \mathcal{A}, \mathcal{G})$, and all permutations π of $\{1, \ldots, k\}$ the operator*
   $$ad\left(V_{\pi(1)}\right) \ldots ad\left(V_{\pi(k)}\right) P : \mathcal{C}_0^\infty(\mathcal{M}) \to \mathcal{D}'(\mathcal{M})$$
   has a continuous extension to an operator
   $$H\left(M_g(k_1, k_2)^{-1}, g; \mathcal{M}, \mathcal{A}\right) \to H\left(M^{-1} M_g(k_1, k_2)^{-1} M_g(k-l, l)^{-1}, g; \mathcal{M}, \mathcal{A}\right).$$
2. *More precisely, as far as the implication $\Longrightarrow$ "(a) $\Longrightarrow$ (b)" is concerned: If $P \in \Psi(M, g; \mathcal{M}, \mathcal{A}, \mathcal{G})$, $l, k \in \mathbb{N}_0$ with*
   $$l \leq k, \quad V_1, \ldots, V_l \in \mathcal{V}_m(\mathcal{M}, \mathcal{A}, \mathcal{G}), \quad V_{l+1}, \ldots, V_k \in \mathcal{V}_d(\mathcal{M}, \mathcal{A}, \mathcal{G}),$$
   and π is a permutation of $\{1, \ldots, k\}$, then we have
   $$ad\left(V_{\pi(1)}\right) \ldots ad\left(V_{\pi(k)}\right) P \in \Psi\left(M M_g(k-l, l), g; \mathcal{M}, \mathcal{A}, \mathcal{G}\right) .$$
3. *More precisely, as far as the implication "(c) $\Longrightarrow$ (a)" is concerned, there exist finite subsets*
   $$\mathcal{W}_m(\mathcal{M}, \mathcal{A}, \mathcal{G}) \subseteq \mathcal{V}_m(\mathcal{M}, \mathcal{A}, \mathcal{G}) \quad \text{and} \quad \mathcal{W}_d(\mathcal{M}, \mathcal{A}, \mathcal{G}) \subseteq \mathcal{V}_d(\mathcal{M}, \mathcal{A}, \mathcal{G})$$
   such that the property that for all $l, k, k_1, k_2 \in \mathbb{N}_0$ with $l \leq k$, all $V_1, \ldots, V_l \in \mathcal{W}_m(\mathcal{M}, \mathcal{A}, \mathcal{G})$, all $V_{l+1}, \ldots, V_k \in \mathcal{W}_d(\mathcal{M}, \mathcal{A}, \mathcal{G})$, and all permutations π of $\{1, \ldots, k\}$ the operator
   $$ad\left(V_{\pi(1)}\right) \ldots ad\left(V_{\pi(k)}\right) P : \mathcal{C}_0^\infty(\mathcal{M}) \to \mathcal{D}'(\mathcal{M})$$
   has a continuous extension to an operator
   $$H\left(M_g(k_1, k_2)^{-1}, g; \mathcal{M}, \mathcal{A}\right) \to H\left(M^{-1} M_g(k_1, k_2)^{-1} M_g(k-l, l)^{-1}, g; \mathcal{M}, \mathcal{A}\right)$$
   implies $P \in \Psi(M, g; \mathcal{M}, \mathcal{A}, \mathcal{G})$.

Proof. We start with the proof of Part 2. Let $(\mathcal{X}, X, \kappa) \in \mathcal{A}$, let ϕ be as in the definition of $\mathcal{V}(\mathcal{M}, \mathcal{A}, \mathcal{G})$ in [5], let $j \in \{1, \ldots, n\}$, let $L_j := M_{x_j}$ or $L_j := D_{x_j}$, and let $Vu := \kappa^* \left[L_j \left(\kappa^{-1}\right)^* (\phi u)\right]$ $(u \in \mathcal{C}_0^\infty(\mathcal{M}))$. By assumption there exists an atlas $\mathcal{A} = \left\{\left(\tilde{\mathcal{X}}, \widehat{\tilde{X}}, \kappa\right) \mid \left(\tilde{\mathcal{X}}, \tilde{X}, \kappa\right) \in \mathcal{A}\right\}$ such that $(\mathcal{M}, \widehat{\mathcal{A}}, \mathcal{G}, g)$ is an admissible structure of type (ψ), $\mathcal{A} \leq \widehat{\mathcal{A}}$, and $\operatorname{supp}(\phi) \subseteq \widehat{X}$ where $(\mathcal{X}, \widehat{X}, \kappa) \in \widehat{\mathcal{A}}$. Let φ be an $(\mathcal{X}, \widehat{X}, \kappa, g)$-admissible cut-off function which equals 1 on $\widehat{X}$. Moreover, let $\widehat{\psi}$

be a corresponding function as in [5, Theorem 5.18] applied with $\widehat{\mathcal{A}}$ instead of $\mathcal{A}$. For $u \in \mathcal{C}_0^\infty(\mathcal{M})$ we have

$$\begin{aligned}
\operatorname{ad}(V)Pu &= V(\varphi Pu) - P(\widehat{\psi}Vu) \\
&= V(\varphi P(\widehat{\psi}u)) + V(\varphi P((1-\widehat{\psi})u)) - \widehat{\psi}P(\widehat{\psi}Vu) - (1-\widehat{\psi})P(\widehat{\psi}Vu) \\
&= V(\widehat{\psi}P(\widehat{\psi}u)) - \widehat{\psi}P(\widehat{\psi}Vu) + V(\varphi P((1-\widehat{\psi})u)) - (1-\widehat{\psi})P(\varphi Vu),
\end{aligned}$$

where $V\varphi P(1-\widehat{\psi}), (1-\widehat{\psi})P\varphi V \in \Psi(g;\mathcal{M},\mathcal{A},\mathcal{G})^{-\infty}$.

A precise study of the behavior of the remainder term in asymptotic expansions of the product symbol in the calculus on $\mathbb{R}^n$ shows that $\operatorname{ad}(V)\left(\widehat{\psi}P\widehat{\psi}\right) \in \Psi(MM_g(m_V), g;\mathcal{M},\mathcal{A},\mathcal{G})$ where $m_V := (0,1)$ if $L_j = M_{x_j}$ and $m_V := (1,0)$ if $L_j = D_{x_j}$. For details in this proof we refer to [3, 3.8.19]. Together this yields $\operatorname{ad}(V)P \in \Psi(MM_g(m_V), g;\mathcal{M},\mathcal{A},\mathcal{G})$ and thus, by induction, Part 2.

For the proof of Part 1 it remains to show that (c) implies (a). Let P be an operator which satisfies (c). We apply the notation from [5, 5.18 and 5.19]. There exists an atlas $\tilde{\mathcal{A}}$ such that $(\mathcal{M},\tilde{\mathcal{A}},\mathcal{G},g)$ is an admissible structure of type (c), $\mathcal{A} \le \tilde{\mathcal{A}}$, and for $i \in \{1,\ldots,l\}$, φ_i and $\widehat{\psi}_i$ are $(\mathcal{X}_{j_i},\tilde{X}_{j_i},\kappa_{j_i},g)$-admissible cut-off functions where $(\mathcal{X}_{j_i},\tilde{X}_{j_i},\kappa_{j_i}) \in \tilde{\mathcal{A}}$. By virtue of [2, 3.6.6], for each $i \in \{1,\ldots,l\}$ there exists a symbol $p_i \in S\left((\chi_{j_i}^{-1})^*M, (\chi_{j_i}^{-1})^*g\right)$ such that we have

$$\varphi_i P(\widehat{\psi}_i u) = \kappa_{j_i}^* \left[\left((\varphi_i \circ \kappa_{j_i}^{-1}) \cdot p_i\right)(x,D)(\kappa_{j_i}^{-1})^*(\widehat{\psi}_i \cdot u)\right]$$

for $u \in \mathcal{C}_0^\infty(\mathcal{M})$. Moreover, applying the notation from 3.18 and 2.2, and the methods from the proof of 3.1, for $m, m_1, \tilde{m}_1, m_2, \tilde{m}_2 \in \mathbb{N}_0$, $i,\mu \in \{1,\ldots,l\}$, and all $\lambda > 0$ the operator

$$\begin{aligned}
L^2(\mathbb{R}^n) \ni\ & u \mapsto \langle x, D_x\rangle^{2m}\langle M_x\rangle^{2m_1}\langle D_x\rangle^{2m_2} \\
& (\kappa_{j_i}^{-1})^*\left[\varphi_i P\left((1-\widehat{\psi}_i)\cdot\varphi_\mu\cdot(\kappa_{j_\mu})^*\left[\langle M_x\rangle^{2\tilde{m}_1}\langle D_x\rangle^{2\tilde{m}_2}u\right]\right)\right] \\
\in\ & H\left(\left(1+\tilde{\Phi}_{j_i}\right)^{-(m_2+m+\tilde{m}_2)}\left(1+\tilde{\Psi}_{j_i}\right)^{-(m_1+m+\tilde{m}_1)}\tilde{M}_{j_i}^{-1}, g_{j_i,s}\right)
\end{aligned}$$

is bounded. By assumption, there exist constants $C > 0$ and $0 \le \delta < 1$ such that

$$\tilde{M}_{j_i}(x,\xi) \le C\langle x,\xi\rangle^{2N},\quad 1+\tilde{\Phi}_{j_i}(x,\xi) \le C\langle x,\xi\rangle^{2\delta},\quad 1+\tilde{\Psi}_{j_i}(x,\xi) \le C$$

for all $x, \xi\ in\mathbb{R}^n$. Since $\frac{2(m-N)}{2(m_2+m+\tilde{m}_2)\delta} \to \delta^{-1}$ as $m \to \infty$ there exists $m \in \mathbb{N}$ such that $2m \ge 2(m_2+m+\tilde{m}_2)\delta + 2N$ and thus,

$$\langle x,\xi\rangle^{-2m} \le \tilde{C}\left(1+\tilde{\Phi}_{j_i}\right)^{-(m_2+m+\tilde{m}_2)}\tilde{M}_{j_i}^{-1}.$$

By virtue of [27, Proposition 1.7.5, Théorème 1.8.11, and Corollaire 1.9.5] there exists $q \in S\left(\left(1+\tilde{\Phi}_{j_i}\right)^{-(m_2+m+\tilde{m}_2)}\tilde{M}_{j_i}^{-1}, g_{j_i,s}\right)$ such that $q(x,D)\langle x, D_x\rangle^{2m} = \mathrm{Id}$.

Thus, defining

$$\begin{aligned}&P(m_1,\tilde{m}_1,m_2,\tilde{m}_2)(u)\\&\quad:=\langle M_x\rangle^{2m_1}\langle D_x\rangle^{2m_2}(\kappa_{j_i}^{-1})^*\\&\qquad\left[\varphi_i P\left((1-\widehat{\psi}_i)\cdot\varphi_\mu\cdot(\kappa_{j_\mu})^*\left[\langle M_x\rangle^{2\tilde{m}_1}\langle D_x\rangle^{2\tilde{m}_2}u\right]\right)\right],\end{aligned}$$

for all $J\in\Psi\left(\left(1+\tilde{\Phi}_{j_i}\right)^{m_2+m+\tilde{m}_2}\tilde{M}_{j_i},g_{j_i,s}\right)$ the operator

$$\begin{aligned}JP(m_1,\tilde{m}_1,m_2,\tilde{m}_2)&=J\,q(x,D)\langle x,D_x\rangle^{2m}P(m_1,\tilde{m}_1,m_2,\tilde{m}_2)\\&:\ L^2(\mathbb{R}^n)\to H\left(\left(1+\tilde{\Phi}_{j_i}\right)^{-(m_2+m+\tilde{m}_2)}\tilde{M}_{j_i}^{-1},g_{j_i,s}\right)\end{aligned}$$

is continuous and thus, $P(m_1,\tilde{m}_1,m_2,\tilde{m}_2)\in\mathcal{L}\left(L^2(\mathbb{R}^n)\right)$. As in the proof of 3.1, there exists $k_{i,\mu}\in\mathcal{S}(\mathbb{R}^{2n})$ such that we have

$$\varphi_i P\left((1-\widehat{\psi}_i)\cdot\varphi_\mu\cdot u\right)=\kappa_{j_i}^*\left[(\varphi_i\circ\kappa_{j_i}^{-1})\cdot\int_{\mathbb{R}^n}k_{i,\mu}(\cdot,y)\cdot(\varphi_\mu\cdot u)(\kappa_{j_\mu}^{-1}(y))dy\right]$$

for $u\in C_0^\infty(\mathcal{M})$. This yields $P\in\Psi(M,g;\mathcal{M},\mathcal{A},\mathcal{G})$. In all these conclusions, only iterated commutators with a finite number of operators in $\mathcal{V}(\mathcal{M},\mathcal{A},\mathcal{G})$ have to be taken into account. □

4. Spectral invariance for $\Psi(1,g;\mathcal{M},\mathcal{A},\mathcal{G})$ in $\mathcal{L}\left(L^2(\mathcal{M})\right)$

In this section, we prove spectral invariance for the algebra $\Psi(1,g;\mathcal{M},\mathcal{A},\mathcal{G})$ in $\mathcal{L}\left(H\left(1,g;\mathcal{M},\mathcal{A}\right)\right)$ $(=\mathcal{L}\left(L^2(\mathcal{M},\alpha_G)\right))$. More precisely, to do so, we have to assume that $(\mathcal{M},\mathcal{A},\mathcal{G},g)$ is of type (ψ) and there exists an admissible structure $(\mathcal{M},\mathcal{A},\mathcal{G},\underline{g})$ of type $(\psi)_s$ such that $\underline{g}$ is a minorant of g (cf. 4.22). The assumptions arise from the conclusion that, with $P\in\Psi(1,g;\mathcal{M},\mathcal{A},\mathcal{G})\cap\mathcal{L}\left(H\left(1,g;\mathcal{M},\mathcal{A}\right)\right)^{-1}$, also P^{-1} fulfills the characterizations from 3.20, expressed by the mapping properties of iterated commutators with operators in $\mathcal{V}(\mathcal{M},\mathcal{A},\mathcal{G})$ between appropriate Sobolev spaces. For this conclusion we have to justify the formula $\mathrm{ad}(V)\left(P^{-1}\right)=-P^{-1}\left(\mathrm{ad}(V)P\right)P^{-1}$ and therefore prove that P^{-1} maps $\mathcal{S}(\mathcal{M})$ to $\mathcal{S}(\mathcal{M})$ and behaves well between the Sobolev spaces which have to be taken into consideration in the characterizations (cf. 4.15 and 4.20). To do so we need "order reducing operators", the existence of which we can show under the just mentioned conditions (cf. 4.21). We prove their existence in three steps: In the first step, for $M\geq 1$, we prove the existence of an operator $P\in\Psi(M,g;\mathcal{M},\mathcal{A},\mathcal{G})$ such that $P: H\left(M,g;\mathcal{M},\mathcal{A}\right)\to H\left(1,g;\mathcal{M},\mathcal{A}\right)$ is a topological isomorphism (cf. 4.5, 4.6). Then $H\left(M,g;\mathcal{M},\mathcal{A}\right)$ can be endowed with a norm and, for $M,M'\geq 1$, we can apply the same proof as before to show the existence of an operator $P\in\Psi(M,g;\mathcal{M},\mathcal{A},\mathcal{G})$ such that $P: H\left(MM',g;\mathcal{M},\mathcal{A}\right)\to H\left(M',g;\mathcal{M},\mathcal{A}\right)$ is a topological isomorphism (cf. 4.8). This allows us in the third step to prove spectral invariance of $\Psi(1,g;\mathcal{M},\mathcal{A},\mathcal{G})$ in $\mathcal{L}\left(H\left(1,g;\mathcal{M},\mathcal{A}\right)\right)$, where $(\mathcal{M},\mathcal{A},\mathcal{G},g)$

is supposed to be of type $(\psi)_s$ (cf. 4.16). As essential tools, we make use of results from B. Gramsch, J. Ueberberg, and K. Wagner [26] (cf. 4.13, 4.15) and of the characterizations from 3.20. In particular, for $M \geq 1$ there exist operators $P \in \Psi(M, g; \mathcal{M}, \mathcal{A}, \mathcal{G})$ and $Q \in \Psi(1/M, g; \mathcal{M}, \mathcal{A}, \mathcal{G})$ such that $PQ = \mathrm{Id} = QP$. With the help of this result we obtain "order reducing operators" in the general case of weight functions under consideration (cf. 4.21). With these preparations we can – applying 3.20 again – finally prove spectral invariance of $\Psi(1, g; \mathcal{M}, \mathcal{A}, \mathcal{G})$ in $\mathcal{L}(H(1, g; \mathcal{M}, \mathcal{A}))$ under the already stated assumptions.

For the Ψ^*-property in the special case of the SG-calculus we refer to E. Schrohe [41].

The following definitions and the succeeding lemma originate from B. Gramsch [22, Definition 5.1 and Lemma 5.3].

Definition 4.1. Let $\mathcal{B}$ be a Banach algebra with unit element e and let $\mathcal{A}$ be a subalgebra of $\mathcal{B}$ with $e \in \mathcal{A}$.

1. $\mathcal{A}$ is called *locally spectrally invariant* in $\mathcal{B}$ if there exists an $\varepsilon > 0$ such that
$$\{a \in \mathcal{A} \mid \|e - a\|_{\mathcal{B}} < \varepsilon\} \subseteq \mathcal{A}^{-1} .$$
2. $\mathcal{A}$ is called *spectrally invariant* in $\mathcal{B}$ if
$$\mathcal{A} \cap \mathcal{B}^{-1} = \mathcal{A}^{-1} .$$

Definition 4.2.

1. Let $\mathcal{B}$ be a Banach algebra with unit element e. A continuously embedded Fréchet subalgebra $\mathcal{A}$ of $\mathcal{B}$ is called a *Ψ-algebra* if $e \in \mathcal{A}$ and $\mathcal{A}$ is spectrally invariant in $\mathcal{B}$.
2. If $\mathcal{B}$ is a C*-algebra with unit element and $\mathcal{A}$ is a symmetric[1] Ψ-subalgebra, then $\mathcal{A}$ is called a *Ψ^*-algebra.*

Lemma 4.3. *Let $\mathcal{B}$ be a unital Banach algebra and let $\mathcal{A}$ be a locally spectrally invariant subalgebra. Let $\overline{\mathcal{A}}$ be the closure of $\mathcal{A}$ in $\mathcal{B}$. Then $\mathcal{A}$ is spectrally invariant in $\overline{\mathcal{A}}$.*

Corollary 4.4. *Let $\mathcal{B}$ be a C*-algebra with unit element and let $\mathcal{A}$ be a locally spectrally invariant subalgebra. We define $\mathcal{A}^* := \{a \in \mathcal{A} \mid a^* \in \mathcal{A}\}$. Then $\mathcal{A}^*$ is a symmetric spectrally invariant subalgebra of $\mathcal{B}$.*

Theorem 4.5. *Let $(\mathcal{M}, \mathcal{A}, \mathcal{G}, g)$ be an admissible structure of type (ψ) and let M be an $(\mathcal{M}, \mathcal{A}, g)$-admissible weight function of type (ψ) which is h_g^{-1}-adapted and which satisfies $M \geq 1$. Then there exist operators $P \in \Psi(M, g; \mathcal{M}, \mathcal{A}, \mathcal{G})$ and $Q \in \Psi(1/M, g; \mathcal{M}, \mathcal{A}, \mathcal{G})$ such that $PQ, QP \in \mathcal{L}(H(1, g; \mathcal{M}, \mathcal{A}))^{-1}$.*

Proof. With the help of an $(\mathcal{M}, \mathcal{A}, g)$-admissible partition of unity, we can assume that $M \in S(T^*\mathcal{M}, \mathcal{A}; M, g)$. For $j \in \{1, \ldots, m\}$ let M_j and g_j be as in [5] and $\tilde{M}_j$ and $g_{j,s}$ be as in 2.4, 2.6. Since M is h_g^{-1}-adapted there exist $N \in \mathbb{N}$ and $C > 0$

[1] I.e., $a \in \mathcal{A}$ implies $a^* \in \mathcal{A}$.

such that $M_j \le C h_{g_{j,s}}^{-N}$ for all $j \in \{1, \dots, m\}$. We define $\widehat{M_j} := \min\left(h_{g_{j,s}}^{-N}, \tilde{M}_j\right)$ (which is a $g_{j,s}$-admissible weight function). We apply the notation from [5, 3.2] with $(\mathcal{U}_j, U_j, \mathcal{R}_1(g_j))$ instead of $(\mathcal{U}, U, \mathcal{G})$, the constants $\rho, \rho_1, \dots, \rho_4$ being chosen simultaneously for all $j \in \{1, \dots, m\}$. Moreover, we suppose that $\rho_4 \le r_{\mathcal{A}}/2$ and $2\sqrt{C_g}\rho < \widehat{r}_{\mathcal{A}}$. Let $(\varphi_{j,z})_{z \in U_{j,\rho_2,g_j}}$, $(\psi_{j,z})_{z \in U_{j,\rho_2,g_j}}$ $(j = 1, \dots, m)$ be the corresponding families and let $\tilde{\varphi}_j := \int_{U_{j,\rho_2,g_j}} \varphi_{j,z} \, |\mathcal{R}_1(g)_z|^{1/2} dz$. Let $(\varphi_j)_{j=1,\dots,m}$ be an $(\mathcal{M}, \mathcal{A}, g, \rho_1)$-admissible partition of unity with associated system of cut-off functions $(\psi_j)_{j=1,\dots,m}$. For $j \in \{1, \dots, m\}$ let $\widehat{\varphi}_j$, $\widehat{\psi}_j$ be $(\mathcal{X}_j, X_j, \kappa_j, g)$-admissible cut-off functions such that $\widehat{\varphi}_j \circ \kappa_j^{-1} = 1$ on $\operatorname{supp}(\varphi_{j,z})$, $\widehat{\psi}_j \circ \kappa_j^{-1} = 1$ on $\operatorname{supp}(\psi_{j,z})$ for all $z \in U_{j,\rho_2,g_j}$, and, moreover, $\widehat{\psi}_j = 1$ on $\operatorname{supp}(\varphi_j)$. For $\lambda > 0$ and $j \in \{1, \dots, m\}$ let

$$q_{\lambda,j}(x, \xi, y) := \int_{U_{j,\rho_2,g_j}} \frac{\varphi_{j,z}(x)}{M_j(x, \xi) + \lambda} \, \psi_{j,z}(y) \, |\mathcal{R}_1(g_j)_z|^{1/2} dz$$

$((x, \xi, y) \in \mathcal{U}_j \times \mathbb{R}^n \times \mathcal{U}_j)$. This yields an amplitude $q_{\lambda,j} \in A((M_j + \lambda)^{-1}, g_j)$ with estimates independent of $\lambda > 0$. For $u \in \mathcal{S}(\mathbb{R}^n)$ we define

$$\tilde{Q}_{\lambda,j} u := \int_{U_{j,\rho_2,g_j}} \left(\frac{\varphi_{j,z}}{M_j + \lambda}\right)(x, D) \, (\psi_{j,z} \cdot u) \, |\mathcal{R}_1(g_j)_z|^{1/2} dz$$

and, for $v \in C_0^\infty(\mathcal{M})$, $Q_{\lambda,j} v := \kappa_j^* \left[\tilde{Q}_{\lambda,j} (\kappa_j^{-1})^* \left(\widehat{\psi}_j \cdot v\right)\right]$. Let $Q_\lambda := \sum_{j=1}^m \varphi_j Q_{\lambda,j}$. Then we have $\tilde{Q}_{\lambda,j} = \mathcal{R}(q_{\lambda,j})(x, D)$ and $Q_{\lambda,j} \in \Psi\left(\frac{1}{M+\lambda}, g; \mathcal{M}, \mathcal{A}, \mathcal{G}\right) \subseteq \Psi\left(\frac{1}{M}, g; \mathcal{M}, \mathcal{A}, \mathcal{G}\right)$. Moreover, for $(x, \xi, y) \in \mathcal{U}_j \times \mathbb{R}^n \times \mathcal{U}_j$ we have

$$\begin{aligned}
&(M_j(x, \xi) + \lambda) \cdot \varphi_j(\kappa_j^{-1}(x)) \cdot q_{\lambda,j}(x, \xi, y) \\
&\quad = \varphi_j(\kappa_j^{-1}(x)) \cdot \int_{U_{j,\rho_2,g_j}} \varphi_{j,z}(x) \psi_{j,z}(y) \, |\mathcal{R}_1(g_j)_z|^{1/2} dz
\end{aligned}$$

and thus,

$$\begin{aligned}
&\mathcal{R}\left((M_j + \lambda) \cdot (\varphi_j \circ \kappa_j^{-1}) \cdot q_{\lambda,j}\right)(x, \xi) \\
= \; &\varphi_j(\kappa_j^{-1}(x)) \cdot \int_{U_{j,\rho_2,g_j}} \varphi_{j,z}(x) \psi_{j,z}(x) \, |\mathcal{R}_1(g_j)_z|^{1/2} dz \, .
\end{aligned}$$

Now let $P \in \Psi(M, g; \mathcal{M}, \mathcal{A}, \mathcal{G})$ be an operator as in [5, Lemma 6.4] with M instead of p. For $\lambda > 0$ we have

$$(P + \lambda) Q_\lambda - \mathrm{Id} \;=\; \sum_{j=1}^m (\phi_j (P + \lambda) \varphi_j Q_{\lambda,j} - \varphi_j)$$

where ϕ_j is an $(\mathcal{X}_j, \widehat{X}_{j,g}, \kappa, g)$-admissible cut-off function like $\widehat{\varphi}$ in [5, Lemma 6.4]. For $u \in \mathcal{C}_0^\infty(\mathcal{M})$ and $j \in \{1, \dots, m\}$ we have

$$\begin{aligned}
&\phi_j(P+\lambda)\,(\varphi_j Q_{\lambda,j} u)\\
&\quad = \kappa_j^* \left[\left((\phi_j \circ \kappa_j^{-1}) \cdot (M_j + \lambda + r_j)\ \#_{\mathrm{KN}}\ \left((\varphi_j \circ \kappa_j^{-1}) \cdot \mathcal{R}(q_{\lambda,j})\right)\right)\right.\\
&\qquad\qquad \left.(x, D)(\kappa_j^{-1})^* \left(\widehat{\psi}_j \cdot u\right)\right]
\end{aligned}$$

where $r_j \in S\left(M_j h_{g_j}, g_j\right)$. Defining $\Gamma_j := \{(x, \xi, x) \in \mathcal{U}_j \times \mathbb{R}^n \times \mathcal{U}_j\}$ we observe

$$\begin{aligned}
& (\phi_j \circ \kappa_j^{-1}) \cdot (M_j + \lambda)\ \#_{\mathrm{KN}}\ \left((\varphi_j \circ \kappa_j^{-1}) \cdot \mathcal{R}(q_{\lambda,j})\right) - \varphi_j \circ \kappa_j^{-1}\\
= \;& (\phi_j \circ \kappa_j^{-1}) \cdot (M_j + \lambda)\ \#_{\mathrm{KN}}\ \left((\varphi_j \circ \kappa_j^{-1}) \cdot \mathcal{R}(q_{\lambda,j})\right)\\
& - \mathcal{R}\left((\phi_j \circ \kappa_j^{-1}) \cdot (M_j + \lambda) \cdot (\varphi_j \circ \kappa_j^{-1}) \cdot q_{\lambda,j}\right)\\
& - \mathcal{R}\left((\phi_j \circ \kappa_j^{-1}) \cdot M_j \cdot (\varphi_j \circ \kappa_j^{-1}) \cdot q_{\lambda,j}\right)\\
& -(\phi_j \circ \kappa_j^{-1}) \cdot \lambda \cdot (\varphi_j \circ \kappa_j^{-1}) \cdot \mathcal{R}\left(q_{\lambda,j}\right)\\
= \;& (\phi_j \circ \kappa_j^{-1}) \cdot M_j\ \#_{\mathrm{KN}}\ \left((\varphi_j \circ \kappa_j^{-1}) \cdot \mathcal{R}(q_{\lambda,j})\right)\\
& -\mathcal{R}\left((\phi_j \circ \kappa_j^{-1}) \cdot M_j \cdot (\varphi_j \circ \kappa_j^{-1}) \cdot q_{\lambda,j}\right)\\
= \;& (\phi_j \circ \kappa_j^{-1}) \cdot M_j\ \#_{\mathrm{KN}}\ \left((\varphi_j \circ \kappa_j^{-1}) \cdot \left[(q_{\lambda,j})_{|\Gamma_j} - \mathcal{R}_1(q_{\lambda,j})\right]\right)\\
& - \left[(\phi_j \circ \kappa_j^{-1}) \cdot M_j \cdot (\varphi_j \circ \kappa_j^{-1}) \cdot (q_{\lambda,j})_{|\Gamma_j}\right.\\
& \left. -\mathcal{R}_1\left((\phi_j \circ \kappa_j^{-1}) \cdot M_j \cdot (\varphi_j \circ \kappa_j^{-1}) \cdot q_{\lambda,j}\right)\right]\\
= \;& (\phi_j \circ \kappa_j^{-1}) \cdot M_j\ \#_{\mathrm{KN}}\ \left((\varphi_j \circ \kappa_j^{-1}) \cdot (q_{\lambda,j})_{|\Gamma_j}\right)\\
& -(\phi_j \circ \kappa_j^{-1}) \cdot M_j \cdot (\varphi_j \circ \kappa_j^{-1}) \cdot (q_{\lambda,j})_{|\Gamma_j}\\
& - (\phi_j \circ \kappa_j^{-1}) \cdot M_j\ \#_{\mathrm{KN}}\ \left((\varphi_j \circ \kappa_j^{-1}) \cdot \mathcal{R}_1(q_{\lambda,j})\right)\\
& +\mathcal{R}_1\left((\phi_j \circ \kappa_j^{-1}) \cdot M_j \cdot (\varphi_j \circ \kappa_j^{-1}) \cdot q_{\lambda,j}\right) .
\end{aligned}$$

Each of the three terms is in $S\left(\frac{\widehat{M}_j}{\lambda+\widehat{M}_j} \cdot h_{g_{j,s}}, g_{j,s}\right)$ with estimates independent of $\lambda > 0$. Moreover,

$$(\phi_j \circ \kappa_j^{-1}) \cdot r_j\ \#_{\mathrm{KN}}\ \left((\varphi_j \circ \kappa_j^{-1}) \cdot \mathcal{R}(q_{\lambda,j})\right) \in S\left(\frac{\widehat{M}_j}{\lambda + \widehat{M}_j} \cdot h_{g_{j,s}}, g_{j,s}\right)$$

with estimates independent of $\lambda > 0$. Together we have proved that

$$\begin{aligned}
r_{\lambda,j} \;&:=\; (\phi_j \circ \kappa_j^{-1}) \cdot (M_j + \lambda + r_j)\ \#_{\mathrm{KN}}\ \left((\varphi_j \circ \kappa_j^{-1}) \cdot \mathcal{R}(q_{\lambda,j})\right) - \varphi_j \circ \kappa_j^{-1}\\
&\in\; S\left(\frac{\widehat{M}_j}{\lambda + \widehat{M}_j} \cdot h_{g_{j,s}}, g_{j,s}\right)
\end{aligned}$$

with estimates independent of $\lambda > 0$.

Now let $j \in \{1, \ldots, m\}$ and $0 < \varepsilon \leq 1/N$. By the definition of $\widehat{M_j}$ we have $h_{g_{j,s}} \leq \widehat{M_j}^{-1/N}$ and thus, for $X \in \mathbb{R}^{2n}$, the estimate

$$\frac{h_{g_{j,s}}(X) \cdot \widehat{M_j}(X)}{\widehat{M_j}(X) + \lambda} \leq \frac{\widehat{M_j}(X)^{1-\varepsilon}}{\widehat{M_j}(X) + \lambda} \leq \frac{1}{\lambda^\varepsilon}$$

holds. This yields for $X \in \mathbb{R}^{2n}$, $k \in \mathbb{N}_0$, $l \leq k$, $T_1, \ldots, T_l \in \mathbb{R}^{2n} \setminus \{0\}$,

$$\frac{|\partial_{T_1} \ldots \partial_{T_l} r_{\lambda,j}(X)|}{\prod_{\mu=1}^{l} \sqrt{g_{j,s,X}(T_j)}} \leq \frac{1}{\lambda^\varepsilon} \|r_{\lambda,j}\|_{k,S\left(\frac{\widehat{M_j}}{\lambda+\widehat{M_j}} \cdot h_{g_{j,s}}, g_{j,s}\right)} \leq C(k) \left(\frac{1}{\lambda}\right)^\varepsilon$$

where the constant $C(k) > 0$ can be chosen independently of λ. Hence,

$$\|r_{\lambda,j}\|_{k,S(1,g_{j,s})} \leq C(k) \left(\frac{1}{\lambda}\right)^\varepsilon .$$

Defining

$$\tilde{R}_{\lambda,j} := r_{\lambda,j}(x, D) , \ R_{\lambda,j} := \kappa_j^* \left[\tilde{R}_{\lambda,j}(\kappa_j^{-1})^* \left(\widehat{\psi}_j \cdot u\right)\right] \quad (u \in C_0^\infty(\mathcal{M})) ,$$

with suitable constants independent of $\lambda > 0$ we obtain $(u \in C_0^\infty(\mathcal{M}))$

$$\left\|\tilde{R}_{\lambda,j}(\kappa_j^{-1})^* \left(\widehat{\psi}_j \cdot u\right)\right\|_{L^2(\mathbb{R}^n)} \underset{3.10}{\leq} C_2 \left(\frac{1}{\lambda}\right)^\varepsilon \|u\|_{H(1,g;\mathcal{M},\mathcal{A})} .$$

Moreover, noting that $\|\cdot\|_{H(1,g;\mathcal{M},\mathcal{A})} = \|\cdot\|_{H\left(1,g;\mathcal{M},\widehat{\mathcal{A}}\right)}$, we have

$$\|R_{\lambda,j}u\|_{H(1,g;\mathcal{M},\mathcal{A})} \underset{3.10}{\leq} C_2 C_3 \left(\frac{1}{\lambda}\right)^\varepsilon \|u\|_{H(1,g;\mathcal{M},\mathcal{A})} .$$

Since $C_0^\infty(\mathcal{M})$ is dense in $H(1, g; \mathcal{M}, \mathcal{A})$ (cf. 3.10) this implies

$$\begin{aligned} \|(P+\lambda)Q_\lambda - \mathrm{Id}\|_{\mathcal{L}(H(1,g;\mathcal{M},\mathcal{A}))} &\leq \sum_{j=1}^{m} \|\phi_j(P+\lambda)\varphi_j Q_{\lambda,j} - \varphi_j\|_{\mathcal{L}(H(1,g;\mathcal{M},\mathcal{A}))} \\ &= \sum_{j=1}^{m} \|R_{\lambda,j}\|_{\mathcal{L}(H(1,g;\mathcal{M},\mathcal{A}))} \leq C_4 \left(\frac{1}{\lambda}\right)^\varepsilon . \end{aligned}$$

Thus, for $\lambda > 0$ sufficiently large, we have

$$(P+\lambda)Q_\lambda \in \mathcal{L}(H(1, g; \mathcal{M}, \mathcal{A}))^{-1} .$$

Analogously, we show that

$$Q_\lambda(P+\lambda) \in \mathcal{L}(H(1, g; \mathcal{M}, \mathcal{A}))^{-1}$$

for $\lambda > 0$ sufficiently large. □

Corollary 4.6. *We apply the notation from the theorem. Then* $P : H(M, g; \mathcal{M}, \mathcal{A}) \to H(1, g; \mathcal{M}, \mathcal{A})$ *is invertible and* $P^{-1} : H(1, g; \mathcal{M}, \mathcal{A}) \to H(M, g; \mathcal{M}, \mathcal{A})$ *is continuous. In particular,* $H(M, g; \mathcal{M}, \mathcal{A})$ *is a Banach space.*

Remark 4.7. We apply the assumptions and notation from the preceding corollary and define

$$\|u\|_{H(M,g;\mathcal{M},\mathcal{A})} := \|Pu\|_{H(1,g;\mathcal{M},\mathcal{A})} \,.$$

Then, for all $(\mathcal{X}, X, \kappa) \in \mathcal{A}$ and all $(\mathcal{X}, X, \kappa, g)$-admissible cut-off functions φ, there exists a constant $C > 0$ (which depends on P) such that for all $u \in H(M, g; \mathcal{M}, \mathcal{A})$ the estimate

$$\begin{aligned} C^{-1} \|\varphi u\|_{H(M,g;\mathcal{M},\mathcal{A})} &\leq \left\|(\kappa^{-1})^*(\varphi u)\right\|_{H((\chi^{-1})^*M,(\chi^{-1})^*g;\kappa(\widehat{X}_g))} \\ &\leq C\|\varphi u\|_{H(M,g;\mathcal{M},\mathcal{A})} \leq C^2 \|u\|_{H(M,g;\mathcal{M},\mathcal{A})} \end{aligned}$$

holds.

Corollary 4.8. *Let $(\mathcal{M}, \mathcal{A}, \mathcal{G}, g)$ be an admissible structure of type (ψ) and let M, M' be $(\mathcal{M}, \mathcal{A}, g)$-admissible weight functions of type (ψ) which are h_g^{-1}-adapted and which satisfy $M, M' \geq 1$. Then there exist operators $P \in \Psi(M, g; \mathcal{M}, \mathcal{A}, \mathcal{G})$ and $Q \in \Psi(1/M, g; \mathcal{M}, \mathcal{A}, \mathcal{G})$ such that $PQ, QP \in \mathcal{L}(H(M', g; \mathcal{M}, \mathcal{A}))^{-1}$. In particular, $P : H(M'M, g; \mathcal{M}, \mathcal{A}) \to H(M', g; \mathcal{M}, \mathcal{A})$ is invertible and $P^{-1} : H(M', g; \mathcal{M}, \mathcal{A}) \to H(M'M, g; \mathcal{M}, \mathcal{A})$ is continuous.*

Proof. Applying 4.7, this is proved as in 4.5 and 4.6. □

Lemma 4.9. *Let $(\mathcal{M}, \mathcal{A}, \mathcal{G}, g)$ be an admissible structure of type (ψ) and let M be an $(\mathcal{M}, \mathcal{A}, g)$-admissible weight function of type (ψ) which is H_g^{-1}-adapted. Let $Z \in \mathcal{V}(\mathcal{M}, \mathcal{A}, \mathcal{G}) \cup \Psi(M, g; \mathcal{M}, \mathcal{A}, \mathcal{G})$ and $D(Z) := \{u \in H(1, g; \mathcal{M}, \mathcal{A}) \mid Zu \in H(1, g; \mathcal{M}, \mathcal{A})\}$. Then we have $\mathcal{S}(\mathcal{M}) \subseteq D(Z)$ and $Z : D(Z) \to H(1, g; \mathcal{M}, \mathcal{A})$ is closed.*

Proof. Let $(u_\nu)_{\nu\in\mathbb{N}} \subseteq D(Z)$ and $u, v \in H(1, g; \mathcal{M}, \mathcal{A})$ such that $u_\nu \to u$ in $H(1, g; \mathcal{M}, \mathcal{A})$ and $Zu_\nu \to v$ in $H(1, g; \mathcal{M}, \mathcal{A})$. Consideration in local coordinates shows that $Zu = v \in H(1, g; \mathcal{M}, \mathcal{A})$ and thus, in particular, $u \in D(Z)$. □

In the following we will apply results from [26]. As a preparation, we will repeat some notation introduced in [26] in a more general situation.

Definition 4.10. Let $(\mathcal{M}, \mathcal{A}, \mathcal{G}, g)$ be an admissible structure of type (ψ). Moreover, let G be an $(\mathcal{M}, \mathcal{A}, g)$-admissible Riemannian metric. With the *-operation induced by $\langle\cdot,\cdot\cdot\rangle_{L^2(\mathcal{M},\alpha_G)}$, where α_G is the corresponding density, $\mathcal{L}(H(1, g; \mathcal{M}, \mathcal{A}))$ is a C*-algebra.

- For $0 < \tau \leq 1$ let $\mathcal{Z}_\tau := \mathcal{V}_m(\mathcal{M}, \mathcal{A}, \mathcal{G}) \cup \Psi(h_g^{-\tau}, g; \mathcal{M}, \mathcal{A}, \mathcal{G})$ and let $\mathcal{Z} := \mathcal{Z}_1$.
- For $Z \in \mathcal{Z}$ let

$$\begin{aligned} D(Z) &:= \{u \in H(1, g; \mathcal{M}, \mathcal{A}) \mid Zu \in H(1, g; \mathcal{M}, \mathcal{A})\}, \\ \mathcal{T}(Z) &:= \{A \in \mathcal{L}(H(1, g; \mathcal{M}, \mathcal{A})) \mid A(D(Z)) \subseteq D(Z)\} \,. \end{aligned}$$

- For $Z \in \mathcal{Z}$, $A \in \mathcal{T}(Z)$, and $u \in D(Z)$ let $\delta_Z(A)u := ZAu - AZu$.
- For $Z \in \mathcal{Z}$ let $D(\delta_Z) := \mathcal{A}(Z)$ be the set of all operators $A \in \mathcal{T}(Z)$ such that $\delta_Z(A) : D(Z) \to H(1, g; \mathcal{M}, \mathcal{A})$ extends continuously to an element of $\mathcal{L}(H(1, g; \mathcal{M}, \mathcal{A}))$.

- Let $l \in \mathbb{N}$ and $Z_1, \ldots, Z_l \in \mathcal{Z}$. For $k \in \mathbb{N}$ and $j_1, \ldots, j_k \in \{1, \ldots, l\}$ let
$$D\left(\delta_{Z_{j_k}} \ldots \delta_{Z_{j_1}}\right) := \left\{A \in D(\delta_{Z_{j_1}}) \mid \delta_{Z_{j_\mu}} \ldots \delta_{Z_{j_1}}(A) \in D(\delta_{Z_{j_{\mu+1}}}), 1 \leq \mu \leq k-1\right\}.$$
Moreover, for $n \in \mathbb{N}$ let
$$\Psi_0 := \Psi_0(Z_1, \ldots, Z_l) := \mathcal{L}\left(H\left(1, g; \mathcal{M}, \mathcal{A}\right)\right),$$
$$\Psi_n(Z_1, \ldots, Z_l) := \Big\{A \in \Psi_0 \mid A \in D\left(\delta_{Z_{j_k}} \ldots \delta_{Z_{j_1}}\right), \quad k \leq n,\ j_1, \ldots, j_k \in \{1, \ldots, l\}\Big\}$$
By $\Psi_n(\mathcal{Z}_\tau)$ $(0 < \tau \leq 1)$ we denote the set of all operators $P \in \Psi_0$ such that $P \in \Psi_n(Z_1, \ldots, Z_l)$ for all $l \in \mathbb{N}$ and all $Z_1, \ldots, Z_l \in \mathcal{Z}_\tau$ and define $\Psi(\mathcal{Z}_\tau) := \bigcap_{n \in \mathbb{N}} \Psi_n(\mathcal{Z}_\tau)$.
- For $Z_1, \ldots, Z_l \in \mathcal{Z}$, $n \in \mathbb{N}_0$ let $\Psi_n^*(Z_1, \ldots, Z_l) := \{A \in \Psi_n(Z_1, \ldots, Z_l) \mid ,\ A^* \in \Psi_n(Z_1, \ldots, Z_l)\}$. Analogously, for $0 < \tau \leq 1$ let
$$\Psi_n^*(\mathcal{Z}_\tau) := \{A \in \Psi_n(\mathcal{Z}_\tau) \mid A^* \in \Psi_n(\mathcal{Z}_\tau)\} \quad \text{and}$$
$$\Psi^*(\mathcal{Z}_\tau) := \{A \in \Psi(\mathcal{Z}_\tau) \mid A^* \in \Psi(\mathcal{Z}_\tau)\} = \bigcap_{n \in \mathbb{N}} \Psi_n^*(\mathcal{Z}_\tau).$$
- For $n \in \mathbb{N}$ let $\overline{\Psi}_n(Z_1, \ldots, Z_l)$ be the closure of $\Psi_n(Z_1, \ldots, Z_l)$ in $\mathcal{L}\left(H(1, g; \mathcal{M}, \mathcal{A})\right)$.

Remark 4.11.

1. If $A \in \Psi_n(Z_1, \ldots, Z_l)$ and $j \in \{1, \ldots, l\}$, then $\delta_{Z_j}(A) \in \Psi_{n-1}(Z_1, \ldots, Z_l)$.
2. For all $n, l \in \mathbb{N}$ and all $Z_1, \ldots, Z_l \in \mathcal{Z}$ we have
$$\Psi(1, g; \mathcal{M}, \mathcal{A}, \mathcal{G}) \subseteq \Psi_n^*(Z_1, \ldots, Z_l)\ .$$

Lemma 4.12. *Let $(\mathcal{M}, \mathcal{A}, \mathcal{G}, g)$ be an admissible structure of type (ψ). Let $0 < \tau \leq 1$ and $A \in \Psi(\mathcal{Z}_\tau)$.*

1. *Let M be an $(\mathcal{M}, \mathcal{A}, g)$-admissible weight function of type (ψ) which is h_g^{-1}-adapted and which satisfies $M \geq 1$. Then $A_{|H(M,g;\mathcal{M},\mathcal{A})} : H\left(M, g; \mathcal{M}, \mathcal{A}\right) \to H\left(M, g; \mathcal{M}, \mathcal{A}\right)$ is well defined and continuous.*
2. $A\left(\mathcal{S}(\mathcal{M})\right) \subseteq \mathcal{S}(\mathcal{M})$.

Proof.

1. By assumption, there exist $N \in \mathbb{N}$ and $C > 0$ such that $1 \leq M \leq C h_g^{-N\tau}$. By virtue of 4.8, for $\nu \in \{1, \ldots, N\}$ there exists an operator
$$Z_\nu \in \Psi(M^{1/N}, g; \mathcal{M}, \mathcal{A}, \mathcal{G})$$
such that
$$Z_\nu : H\left(M^{\nu/N}, g; \mathcal{M}, \mathcal{A}\right) \to H\left(M^{(\nu-1)/N}, g; \mathcal{M}, \mathcal{A}\right)$$

is an isomorphism. By assumption we have $A \in \Psi_N(Z_1,\dots,Z_N)$. For $u \in H(M,g;\mathcal{M},\mathcal{A})\,(\subseteq D(Z_N))$ we have $Z_N A u = \delta_{Z_N}(A)u + AZ_N u$. Inductively we obtain

$$Z_1 \dots Z_N A u = \sum \delta_{Z_{\nu_1}} \dots \delta_{Z_{\nu_l}}(A)\left(Z_{\mu_1} \dots Z_{\mu_{N-l}} u\right)$$

where the sum runs over all $0 \le l \le N$, $1 \le \nu_1 < \dots < \nu_l \le N$, $1 \le \mu_1 < \dots < \mu_{N-l} \le N$, $\{\nu_1,\dots,\nu_l\} \cap \{\mu_1,\dots,\mu_{N-l}\} = \emptyset$. This yields $Z_1 \dots Z_N Au \in H(1,g;\mathcal{M},\mathcal{A})$ and therefore, since $Z_1 \dots Z_N : H(M,g;\mathcal{M},\mathcal{A}) \to H(1,g;\mathcal{M},\mathcal{A})$ is an isomorphism, we have $Au \in H(M,g;\mathcal{M},\mathcal{A})$. Moreover, since $H(M,g;\mathcal{M},\mathcal{A}) \hookrightarrow H(1,g;\mathcal{M},\mathcal{A})$ (cf. 3.15) and $A \in \mathcal{L}(H(1,g;\mathcal{M},\mathcal{A}))$, by the closed graph theorem the operator $A_{|H(M,g;\mathcal{M},\mathcal{A})} : H(M,g;\mathcal{M},\mathcal{A}) \to H(M,g;\mathcal{M},\mathcal{A})$ is continuous.

2. Let $V_1,\dots,V_k \in \mathcal{V}_d(\mathcal{M},\mathcal{A},\mathcal{G})$, $\tilde{V}_1,\dots,\tilde{V}_l \in \mathcal{V}_m(\mathcal{M},\mathcal{A},\mathcal{G})$. By assumption, there exists $N \in \mathbb{N}$ such that $V_1 \dots V_k \in \Psi\left(h_g^{-N},g;\mathcal{M},\mathcal{A};\mathcal{G}\right)$. Let $\Lambda \in \Psi\left(h_g^{-N},g;\mathcal{M},\mathcal{A};\mathcal{G}\right)$ be an isomorphism $\Lambda : H\left(h_g^{-N},g;\mathcal{M},\mathcal{A}\right) \to H(1,g;\mathcal{M},\mathcal{A})$ as in part 1. As before, for $u \in \mathcal{S}(\mathcal{M})$ we have $\Lambda\tilde{V}_1 \dots \tilde{V}_l A u \in H(1,g;\mathcal{M},\mathcal{A})$. Moreover, $V_1 \dots V_k \Lambda^{-1} \in \mathcal{L}(H(1,g;\mathcal{M},\mathcal{A}))$ and thus, $V_1 \dots V_k \tilde{V}_1 \dots \tilde{V}_l A u = V_1 \dots V_k \Lambda^{-1}\Lambda \tilde{V}_1 \dots \tilde{V}_l A u \in H(1,g;\mathcal{M},\mathcal{A})$. Then Sobolev's embedding theorem yields $Au \in \mathcal{S}(\mathcal{M})$.

□

Theorem 4.13. *Applying the notation from 4.10, for $l,n \in \mathbb{N}$ and $Z_1,\dots,Z_l \in \mathcal{Z}$ we have*

$$\overline{\Psi}_n(Z_1,\dots,Z_l)^{-1} \cap \Psi_n(Z_1,\dots,Z_l) = \Psi_n(Z_1,\dots,Z_l)^{-1}\ .$$

Proof. Cf. [26, Theorem 2.10]. □

Corollary 4.14. *For $l,n \in \mathbb{N}$ and $Z_1,\dots,Z_l \in \mathcal{Z}$ we have*

$$\mathcal{L}(H(1,g;\mathcal{M},\mathcal{A}))^{-1} \cap \Psi_n^*(Z_1,\dots,Z_l) = \Psi_n^*(Z_1,\dots,Z_l)^{-1}\ .$$

Thus, in particular, for $0 < \tau \le 1$ we have

$$\mathcal{L}(H(1,g;\mathcal{M},\mathcal{A}))^{-1} \cap \Psi^*(\mathcal{Z}_\tau) = \Psi^*(\mathcal{Z}_\tau)^{-1}\ .$$

Proof. This is a consequence of the preceding theorem together with 4.4. □

Corollary 4.15. *Let $(\mathcal{M},\mathcal{A},\mathcal{G},g)$ be an admissible structure of type (ψ). Furthermore, let $P \in \Psi(1,g;\mathcal{M},\mathcal{A};\mathcal{G}) \cap \mathcal{L}(H(1,g;\mathcal{M},\mathcal{A}))^{-1}$.*

1. *Let M be an $(\mathcal{M},\mathcal{A},g)$-admissible weight function of type (ψ) which is h_g^{-1}-adapted and which satisfies $M \ge 1$. Then $(P^{-1})_{|H(M,g;\mathcal{M},\mathcal{A})} : H(M,g;\mathcal{M},\mathcal{A}) \to H(M,g;\mathcal{M},\mathcal{A})$ is well defined and continuous.*
2. *$P^{-1}(\mathcal{S}(\mathcal{M})) \subseteq \mathcal{S}(\mathcal{M})$.*
3. *For $u \in \mathcal{S}(\mathcal{M})$ and $Z \in \mathcal{V}(\mathcal{M},\mathcal{A},\mathcal{G})$ we have*

$$[Z,P^{-1}]u = -P^{-1}[Z,P]P^{-1}u.$$

Proof. This is a consequence of 4.11, 4.12, and 4.14. □

Corollary 4.16. *Let $(\mathcal{M},\mathcal{A},\mathcal{G},g)$ be an admissible structure of type $(\psi)_s$.*

1. *We have*

$$\Psi(1,g;\mathcal{M},\mathcal{A},\mathcal{G})\cap\mathcal{L}\left(H\left(1,g;\mathcal{M},\mathcal{A}\right)\right)^{-1}=\Psi(1,g;\mathcal{M},\mathcal{A},\mathcal{G})^{-1}\ .$$

2. *Let M be an $(\mathcal{M},\mathcal{A},g)$-admissible weight function of type (ψ) which is h_g^{-1}-adapted and which satisfies $M\geq 1$. Then there exist operators $P\in\Psi(M,g;\mathcal{M},\mathcal{A},\mathcal{G})$ and $Q\in\Psi(1/M,g;\mathcal{M},\mathcal{A},\mathcal{G})$ such that $PQ=Id=QP$. In particular, $H\left(M,g;\mathcal{M},\mathcal{A}\right)$ and $H\left(1/M,g;\mathcal{M},\mathcal{A}\right)$ are Banach spaces.*

Proof. Part 1 is a consequence of the preceding corollary together with 3.20. Then Part 2 follows together with 4.5. □

The following proof is an adaption of methods from R. Beals [7] (cf. also [50]).

Lemma 4.17. *Let $(\mathcal{M},\mathcal{A},\mathcal{G},g)$ be an admissible structure of type (ψ) and let $(\mathcal{M},\mathcal{A},\mathcal{G},\underline{g})$ be an admissible structure of type $(\psi)_s$ such that $\underline{g}\leq g$. Let M be an $(\mathcal{M},\mathcal{A},g)$- and $(\mathcal{M},\mathcal{A},\underline{g})$-admissible weight function of type (ψ) which is h_g^{-1}- and $h_{\underline{g}}^{-1}$-adapted and which satisfies $M\geq 1$. Then, for each $P\in\Psi(1,g;\mathcal{M},\mathcal{A},\mathcal{G})\cap\mathcal{L}\left(H\left(1,g;\mathcal{M},\mathcal{A}\right)\right)^{-1}$ and all $t\in\mathbb{R}$, the operator $P:H\left(M^t,g;\mathcal{M},\mathcal{A}\right)\to H(M^t,g;\mathcal{M},\mathcal{A})$ is invertible.*

Proof. For $t>0$ let $\Lambda_t\in\Psi\left(M^t,\underline{g};\mathcal{M},\mathcal{A},\mathcal{G}\right)$ $(\subseteq\Psi\left(M^t,g;\mathcal{M},\mathcal{A},\mathcal{G}\right))$ such that there exists $\Lambda_{-t}:=\Lambda_t^{-1}\in\Psi\left(M^{-t},\underline{g};\mathcal{M},\mathcal{A},\mathcal{G}\right)$. For $t\geq 0$ the assertion is already proved in 4.15. By assumption, there exist $s_0>0$ and $C>0$ such that $M^{s_0}\leq Ch_g^{-1}$.

First we show surjectivity of $P:H\left(M^{-t},g;\mathcal{M},\mathcal{A}\right)\to H\left(M^{-t},g;\mathcal{M},\mathcal{A}\right)$ for all $t\geq 0$. Let $t_0\geq 0$ be chosen such that P is surjective on $H\left(M^{-t},g;\mathcal{M},\mathcal{A}\right)$ for all $0\leq t\leq t_0$. Let $s\in[0,s_0]$. Then we have $[\Lambda_s,P]\in\Psi(1,g;\mathcal{M},\mathcal{A},\mathcal{G})$. Let $t\in[0,t_0]$ and $v\in H\left(M^{-t-s},g;\mathcal{M},\mathcal{A}\right)$.

$$\begin{aligned}
&\Longrightarrow && \Lambda_{-s}v\ \in\ H\left(M^{-t},g;\mathcal{M},\mathcal{A}\right)\\
&\Longrightarrow && \exists u\ \in\ H\left(M^{-t},g;\mathcal{M},\mathcal{A}\right)\ ,\ Pu\ =\ \Lambda_{-s}v\\
&\Longrightarrow && v\ =\ \Lambda_sPu\ =\ [\Lambda_s,P]u+P\Lambda_su\\
&\Longrightarrow && v-P\Lambda_su\ =\ [\Lambda_s,P]u\ \in\ H\left(M^{-t},g;\mathcal{M},\mathcal{A}\right)\\
&\Longrightarrow && \exists w\ \in\ H\left(M^{-t},g;\mathcal{M},\mathcal{A}\right)\ ,\ Pw\ =\ v-P\Lambda_su\\
&\Longrightarrow && v\ =\ Pw+P\Lambda_su=P(w+\Lambda_su)
\end{aligned}$$

where $w+\Lambda_su\in H\left(M^{-t-s},g;\mathcal{M},\mathcal{A}\right)$.

To show injectivity of P on $H\left(M^{-t},g;\mathcal{M},\mathcal{A}\right)$ for all $t\geq 0$, let $t_0\geq 0$ be chosen such that $P\in\mathcal{L}\left(H\left(M^{-t},g;\mathcal{M},\mathcal{A}\right)\right)^{-1}$ for all $0\leq t\leq t_0$. For $s\in[0,s_0]$ we have $[P,\Lambda_{-s}]\in\Psi\left(M^{-2s},g;\mathcal{M},\mathcal{A},\mathcal{G}\right)$. Let $t\in[0,t_0]$ and $u\in H\left(M^{-t-s},g;\mathcal{M},\mathcal{A}\right)$

with $Pu = 0$.

$$\begin{aligned}
&\Longrightarrow \quad P\Lambda_{-s}u \;=\; P\Lambda_{-s}u - \Lambda_{-s}Pu \;=\; [P,\Lambda_{-s}]u \;\in\; H\left(M^{-t+s}, g; \mathcal{M}, \mathcal{A}\right)\\
&\Longrightarrow \quad \exists w \;\in\; H\left(M^{-t+s}, g; \mathcal{M}, \mathcal{A}\right) \;\left(\subseteq H\left(M^{-t}, g; \mathcal{M}, \mathcal{A}\right)\right)\;,\; Pw = P\Lambda_{-s}u\\
&\Longrightarrow \quad P(w - \Lambda_{-s}u) = 0 \quad \text{where} \quad w - \Lambda_{-s}u \in H\left(M^{-t}, g; \mathcal{M}, \mathcal{A}\right)\\
&\Longrightarrow \quad \Lambda_{-s}u \;=\; w \;\in\; H\left(M^{-t+s}, g; \mathcal{M}, \mathcal{A}\right)\\
&\Longrightarrow \quad u \;\in\; H\left(M^{-t}, g; \mathcal{M}, \mathcal{A}\right)
\end{aligned}$$

and thus, by the choice of t, $u = 0$. □

Definition 4.18. Let $(\mathcal{M}, \mathcal{A}, \mathcal{G}, g)$ and $(\mathcal{M}, \mathcal{A}, \mathcal{G}, \underline{g})$ be admissible structures of type (ψ). We call $\underline{g}$ a minorant of g if

- $\underline{g}_{(x,\xi)} \leq g_{(x,\xi)}$ for all $x \in \mathcal{M}$ and all $\xi \in T^*_x\mathcal{M}$.
- h_g, $M_g(1,0)$, and $M_g(0,1)$ are $(\mathcal{M}, \mathcal{A}, \underline{g})$-admissible weight functions of type (ψ).

Remark 4.19. If $\underline{g}$ is a minorant of g, then h_g, $M_g(1,0)$, and $M_g(0,1)$ are $h_{\underline{g}}^{-1}$-adapted.

Corollary 4.20. *Let $(\mathcal{M}, \mathcal{A}, \mathcal{G}, g)$ be an admissible structure of type (ψ) and let $(\mathcal{M}, \mathcal{A}, \mathcal{G}, \underline{g})$ be an admissible structure of type $(\psi)_s$ such that $\underline{g}$ is a minorant of g. Let M be an $(\mathcal{M}, \mathcal{A}, g)$- and $(\mathcal{M}, \mathcal{A}, \underline{g})$-admissible weight function of type (ψ) which is h_g^{-1}- and $h_{\underline{g}}^{-1}$-adapted. Then, for each $P \in \Psi(1, g; \mathcal{M}, \mathcal{A}, \mathcal{G}) \cap \mathcal{L}\left(H(1, g; \mathcal{M}, \mathcal{A})\right)^{-1}$, the operator $P : H\left(M, g; \mathcal{M}, \mathcal{A}\right) \to H\left(M, g; \mathcal{M}, \mathcal{A}\right)$ is invertible. Moreover, $H\left(M, g; \mathcal{M}, \mathcal{A}\right)$ is a Banach space and there exists a constant $C > 0$ such that for all $u \in \mathcal{S}(\mathcal{M})$ the estimate $\left\|P^{-1}u\right\|_{H(M,g;\mathcal{M},\mathcal{A})} \leq C\left\|u\right\|_{H(M,g;\mathcal{M},\mathcal{A})}$ holds.*

Proof. By assumption, there exist $C > 0$ and $N \in \mathbb{N}$ such that $Mh_{\underline{g}}^N < C$. 4.16 yields operators

$$\begin{aligned}
&\Lambda\left(Mh_{\underline{g}}^N\right) \in \Psi\left(Mh_{\underline{g}}^N, \underline{g}; \mathcal{M}, \mathcal{A}, \mathcal{G}\right) \;\left(\subseteq \Psi\left(Mh_{\underline{g}}^N, g; \mathcal{M}, \mathcal{A}, \mathcal{G}\right)\right),\\
&\Lambda\left(M^{-1}h_{\underline{g}}^{-N}\right) \in \Psi\left(M^{-1}h_{\underline{g}}^{-N}, g; \mathcal{M}, \mathcal{A}, \mathcal{G}\right),\\
&\Lambda\left(h_{\underline{g}}^{N}\right) \in \Psi\left(h_{\underline{g}}^{N}, g; \mathcal{M}, \mathcal{A}, \mathcal{G}\right), \qquad \text{and}\\
&\Lambda\left(h_{\underline{g}}^{-N}\right) \in \Psi\left(h_{\underline{g}}^{-N}, g; \mathcal{M}, \mathcal{A}, \mathcal{G}\right)
\end{aligned}$$

such that

$$\begin{aligned}
&\Lambda\left(Mh_{\underline{g}}^N\right)\Lambda\left(M^{-1}h_{\underline{g}}^{-N}\right) = \mathrm{Id} = \Lambda\left(M^{-1}h_{\underline{g}}^{-N}\right)\Lambda\left(Mh_{\underline{g}}^N\right),\\
&\Lambda\left(h_{\underline{g}}^{N}\right)\Lambda\left(h_{\underline{g}}^{-N}\right) = \mathrm{Id} = \Lambda\left(h_{\underline{g}}^{-N}\right)\Lambda\left(h_{\underline{g}}^{N}\right).
\end{aligned}$$

By virtue of 4.17, $\Lambda\left(h_{\underline{g}}^{-N}\right)P\Lambda\left(h_{\underline{g}}^{N}\right) \in \Psi(1, g; \mathcal{M}, \mathcal{A}, \mathcal{G}) \cap \mathcal{L}\left(H\left(1, g; \mathcal{M}, \mathcal{A}\right)\right)^{-1}$. Applying 4.17 again, we have

$$\begin{aligned}
&\Lambda\left(Mh_{\underline{g}}^N\right)\Lambda\left(h_{\underline{g}}^{-N}\right)P\Lambda\left(h_{\underline{g}}^{N}\right)\Lambda\left(M^{-1}h_{\underline{g}}^{-N}\right)\\
&\qquad \in \Psi(1, g; \mathcal{M}, \mathcal{A}, \mathcal{G}) \cap \mathcal{L}\left(H\left(1, g; \mathcal{M}, \mathcal{A}\right)\right)^{-1}.
\end{aligned}$$

Since $\Lambda\left(Mh_g^N\right)\Lambda\left(h_g^{-N}\right) : H(M,g;\mathcal{M},\mathcal{A}) \to H(1,g;\mathcal{M},\mathcal{A})$ is an isomorphism, this yields $P(M) \in \mathcal{L}(H(M,g;\mathcal{M},\mathcal{A}))^{-1}$, where $P(M)$ denotes the extension of $P : \mathcal{S}(\mathcal{M}) \to \mathcal{S}(\mathcal{M})$ to $H(M,g;\mathcal{M},\mathcal{A})$. Thus, there exists a constant $C > 0$ such that for all $u \in H(M,g;\mathcal{M},\mathcal{A})$ the estimate

$$\left\|P(M)^{-1}u\right\|_{H(M,g;\mathcal{M},\mathcal{A})} \leq C\left\|u\right\|_{H(M,g;\mathcal{M},\mathcal{A})}$$

holds. □

Remark 4.21. In the proof of the preceding corollary we have shown – under the assumptions of the corollary – that there exist "order reducing operators", i.e., operators $\Lambda(M) \in \Psi(M,g;\mathcal{M},\mathcal{A},\mathcal{G})$ and $\Lambda(M^{-1}) \in \Psi(M^{-1},g;\mathcal{M},\mathcal{A},\mathcal{G})$ such that $\Lambda(M)\Lambda(M^{-1}) = \mathrm{Id} = \Lambda(M^{-1})\Lambda(M)$. In particular, $\Lambda(M) : H(M,g;\mathcal{M},\mathcal{A}) \to H(1,g;\mathcal{M},\mathcal{A})$ is an isomorphism and thus, $H(M,g;\mathcal{M},\mathcal{A})$ is a Banach space and $\mathcal{S}(\mathcal{M})$ is dense in $H(M,g;\mathcal{M},\mathcal{A})$.

Theorem 4.22. *Let $(\mathcal{M},\mathcal{A},\mathcal{G},g)$ be an admissible structure of type (ψ) and let $(\mathcal{M},\mathcal{A},\mathcal{G},\underline{g})$ be an admissible structure of type $(\psi)_s$ such that $\underline{g}$ is a minorant of g. Then we have*

$$\Psi(1,g;\mathcal{M},\mathcal{A},\mathcal{G}) \cap \mathcal{L}(H(1,g;\mathcal{M},\mathcal{A}))^{-1} = \Psi(1,g;\mathcal{M},\mathcal{A},\mathcal{G})^{-1}\ .$$

Proof. This is a consequence of the preceding corollary together with 4.15 and 3.20. □

5. Submultiplicative Fréchet topology for $\Psi(1,g;\mathcal{M},\mathcal{A},\mathcal{G})$

So far, we have not endowed $\Psi(1,g;\mathcal{M},\mathcal{A};\mathcal{G})$ with any topology. Under the assumptions in 4.20 we proved that $H(M,g;\mathcal{M},\mathcal{A})$ is a Banach space (cf. 4.21). Together with 3.20 this allows us to define a submultiplicative Fréchet topology on $\Psi(1,g;\mathcal{M},\mathcal{A};\mathcal{G})$ (cf. 5.5). This result generalizes [2, Satz 3.8.5] in the case of $\mathbb{R}^n$ to manifolds. On the other hand, in [2] we proved submultiplicativity for all $\Psi(1,g)$-classes in the Weyl–Hörmander calculus without technical restrictions in contrast to the current situation. The method which we apply was developed by B. Gramsch, J. Ueberberg, K. Wagner [26] to prove submultiplicativity of $\Psi^0_{\rho,\delta}$ and applied by B. Gramsch, E. Schrohe [25] to prove submultiplicativity of Boutet de Monvel's algebra. In [1, p. 552] B. Gramsch formulates the question whether every Ψ^*-algebra is submultiplicative as an open problem. For the commutative case we refer to W. Żelazko [56, Theorem C], for the non-commutative case cf. also [56, Theorem 3]. B. Gramsch [23] and B. Gramsch, W. Kaballo [24] applied submultiplicativity in connection with non-abelian cohomology and Oka's principle.

Definition 5.1. Let $(\mathcal{M},\mathcal{A},\mathcal{G},g)$ be an admissible structure of type (ψ) and $(\mathcal{M},\mathcal{A},\mathcal{G},\underline{g})$ an admissible structure of type $(\psi)_s$ such that $\underline{g}$ is a minorant of g.

1. For $x \in \mathcal{M}$ and $\xi \in T_x^*\mathcal{M}$ we define $M_g(0,0)(x,\xi) := 1$. Applying notation from 3.18, for $k_1, k_2, l_1, l_2 \in \mathbb{N}_0$ we define
$$\mathcal{L}_g(k_1,k_2,l_1,l_2) :=$$
$$\mathcal{L}\left(H\left(M_g(k_1,k_2)^{-1}, g; \mathcal{M}, \mathcal{A}\right), H\left(M_g(k_1,k_2)^{-1} M_g(l_1,l_2)^{-1}, g; \mathcal{M}, \mathcal{A}\right)\right).$$
2. We fix finite sets $\mathcal{W}_m(\mathcal{M},\mathcal{A},\mathcal{G}) \subseteq \mathcal{V}_m(\mathcal{M},\mathcal{A},\mathcal{G})$ and $\mathcal{W}_d(\mathcal{M},\mathcal{A},\mathcal{G}) \subseteq \mathcal{V}_d(\mathcal{M},\mathcal{A},\mathcal{G})$ as in 3.20. For $k \in \mathbb{N}_0$ let $\mathcal{P}_k(\mathcal{M},\mathcal{A},\mathcal{G},g)$ be the set of all operators such that for all $k_1, k_2, l_1, l_2 \in \mathbb{N}_0$ with $k_1 + k_2 + l_1 + l_2 \le k$, all $V_1, \ldots, V_{l_1} \in \mathcal{W}_d(\mathcal{M},\mathcal{A},\mathcal{G})$, $V_{l_1+1}, \ldots, V_{l_1+l_2} \in \mathcal{W}_m(\mathcal{M},\mathcal{A},\mathcal{G})$, and all permutations π of $\{1, \ldots, l_1 + l_2\}$ the operator
$$\operatorname{ad}\left(V_{\pi(1)}\right) \ldots \operatorname{ad}\left(V_{\pi(l_1+l_2)}\right) P : \mathcal{C}_0^\infty(\mathcal{M}) \to \mathcal{D}'(\mathcal{M})$$
has an extension to an operator in $\mathcal{L}_g(k_1,k_2,l_1,l_2)$. Moreover, for $k \in \mathbb{N}_0$ let
$$\|P\|_{\mathcal{P}_k(\mathcal{M},\mathcal{A},\mathcal{G},g)} := \max \|P\|_{\mathcal{L}_g(k_1,k_2,l_1,l_2)}$$
where the maximum is taken over all $k_1, k_2, l_1, l_2 \in \mathbb{N}_0$ with $k_1 + k_2 + l_1 + l_2 \le k$, all $V_1, \ldots, V_{l_1} \in \mathcal{W}_d(\mathcal{M},\mathcal{A},\mathcal{G})$, $V_{l_1+1}, \ldots, V_{l_1+l_2} \in \mathcal{W}_m(\mathcal{M},\mathcal{A},\mathcal{G})$, and all permutations π of $\{1, \ldots, l_1 + l_2\}$.

Remark 5.2. Theorem 3.20 implies $\Psi(1, g; \mathcal{M}, \mathcal{A}, \mathcal{G}) = \bigcap_{k \in \mathbb{N}_0} \mathcal{P}_k(\mathcal{M},\mathcal{A},\mathcal{G},g)$.

Lemma 5.3. *Let $(\mathcal{M},\mathcal{A},\mathcal{G},g)$ be an admissible structure of type (ψ) and $(\mathcal{M},\mathcal{A},\mathcal{G},\underline{g})$ an admissible structure of type $(\psi)_s$ such that $\underline{g}$ is a minorant of g. For each $k \in \mathbb{N}_0$, $\mathcal{P}_k(\mathcal{M},\mathcal{A},\mathcal{G},g)$ is a Banach space.*

Proof. Let $k \in \mathbb{N}_0$ and let $(A_\nu)_{\nu \in \mathbb{N}}$ be a Cauchy sequence in
$$\left(\mathcal{P}_k(\mathcal{M},\mathcal{A},\mathcal{G},g), \|\cdot\|_{\mathcal{P}_k(\mathcal{M},\mathcal{A},\mathcal{G},g)}\right).$$
Let $k_1, k_2, l_1, l_2 \in \mathbb{N}_0$ with $k_1 + k_2 + l_1 + l_2 \le k$, $V_1, \ldots, V_{l_1} \in \mathcal{W}_d(\mathcal{M},\mathcal{A},\mathcal{G})$, $V_{l_1+1}, \ldots, V_{l_1+l_2} \in \mathcal{W}_m(\mathcal{M},\mathcal{A},\mathcal{G})$, and let π be a permutation of $\{1, \ldots, l_1 + l_2\}$. Then, by definition of $\|\cdot\|_{\mathcal{P}_k(\mathcal{M},\mathcal{A},\mathcal{G},g)}$, $\left(\operatorname{ad}(V_{\pi(1)}) \ldots \operatorname{ad}(V_{\pi(l_1+l_2)}) A_\nu\right)_{\nu \in \mathbb{N}}$ is a Cauchy sequence in $\mathcal{L}_g(k_1,k_2,l_1,l_2)$ and thus, converges to an operator $A(k_1,k_2,\pi,V_1,\ldots,V_{l_1+l_2}) \in \mathcal{L}_g(k_1,k_2,l_1,l_2)$. For all $k_1, k_2 \in \mathbb{N}_0$ with $k_1 + k_2 + l_1 + l_2 \le k$ the operators $A(k_1,k_2,\pi,V_1,\ldots,V_{l_1+l_2})$ coincide on $\mathcal{S}(\mathcal{M})$ and thus, coincide with an operator $A(\pi,V_1,\ldots,V_{l_1+l_2}) : \mathcal{S}(\mathcal{M}) \to \mathcal{D}'(\mathcal{M})$ with continuous extensions in $\mathcal{L}_g(k_1,k_2,l_1,l_2)$ for all $k_1, k_2 \in \mathbb{N}_0$ with $k_1 + k_2 + l_1 + l_2 \le k$. Let $A := A(\emptyset)$. Thus, $A_\nu \to A$ in $\mathcal{L}(H(1, g; \mathcal{M}, \mathcal{A}))$. Consideration in local coordinates shows that $\operatorname{ad}(V_{\pi(1)}) \ldots \operatorname{ad}(V_{\pi(l_1+l_2)}) Au = A(\pi,V_1,\ldots,V_{l_1+l_2})u$. This yields $A \in \mathcal{P}_k(\mathcal{M},\mathcal{A},\mathcal{G},g)$ and $A_\nu \to A$ in $\mathcal{P}_k(\mathcal{M},\mathcal{A},\mathcal{G},g)$. □

Lemma 5.4. *Let $(\mathcal{M},\mathcal{A},\mathcal{G},g)$ be an admissible structure of type (ψ) and $(\mathcal{M},\mathcal{A},\mathcal{G},\underline{g})$ an admissible structure of type $(\psi)_s$ such that $\underline{g}$ is a minorant of g. For $k \in \mathbb{N}_0$ there exists $C_k > 0$ such that we have*
$$\|PQ\|_{\mathcal{P}_k(\mathcal{M},\mathcal{A},\mathcal{G},g)} \le C_k \|P\|_{\mathcal{P}_k(\mathcal{M},\mathcal{A},\mathcal{G},g)} \|Q\|_{\mathcal{P}_k(\mathcal{M},\mathcal{A},\mathcal{G},g)}$$
for all $P, Q \in \Psi(1, g; \mathcal{M}, \mathcal{A}, \mathcal{G})$.

Proof. Let M_1, M_2 be two $(\mathcal{M}, \mathcal{A}, g)$-admissible weight functions of type (ψ) which are H_g^{-1}-adapted. For $P \in \Psi(M_1, g; \mathcal{M}, \mathcal{A}, \mathcal{G})$, $Q \in \Psi(M_2, g; \mathcal{M}, \mathcal{A}, \mathcal{G})$, $l \in \mathbb{N}$, and $V_1, \ldots, V_l \in \mathcal{V}(\mathcal{M}, \mathcal{A}, \mathcal{G})$, we have

$$\operatorname{ad}(V_1)\ldots\operatorname{ad}(V_l)(PQ) = \sum [\operatorname{ad}(V_{i_1})\ldots\operatorname{ad}(V_{i_\nu})P]\left[\operatorname{ad}(V_{j_1})\ldots\operatorname{ad}(V_{j_{l-\nu}})Q\right]$$

as an operator $\mathcal{S}(\mathcal{M}) \to \mathcal{S}(\mathcal{M})$, where the sum runs over all $0 \le \nu \le l$, $1 \le i_1 < \cdots < i_\nu \le l$, $1 \le j_1 < \cdots < j_{l-\nu} \le l$ with $\{i_1, \ldots, i_l\} \cap \{j_1, \ldots, j_{l-\nu}\} = \emptyset$. □

Theorem 5.5. *Let $(\mathcal{M}, \mathcal{A}, \mathcal{G}, g)$ be an admissible structure of type (ψ) and $(\mathcal{M}, \mathcal{A}, \mathcal{G}, \underline{g})$ an admissible structure of type $(\psi)_s$ such that $\underline{g}$ is a minorant of g. For $k \in \mathbb{N}_0$ let $\Psi_k(g; \mathcal{M}, \mathcal{A}, \mathcal{G})$ be the closure of $\Psi(1, g; \mathcal{M}, \mathcal{A}, \mathcal{G})$ in*

$$\left(\mathcal{P}_k(\mathcal{M}, \mathcal{A}, \mathcal{G}, g), \|\cdot\|_{\mathcal{P}_k(\mathcal{M}, \mathcal{A}, \mathcal{G}, g)}\right).$$

Then the following statements hold:

1. *For $k \in \mathbb{N}_0$, $\Psi_k(g; \mathcal{M}, \mathcal{A}, \mathcal{G})$ is a Banach algebra.*
2. $\Psi(1, g; \mathcal{M}, \mathcal{A}, \mathcal{G}) = \bigcap_{k \in \mathbb{N}} \Psi_k(g; \mathcal{M}, \mathcal{A}, \mathcal{G})$.

In particular, endowed with the obvious topology induced by Part 2, $\Psi(1, g; \mathcal{M}, \mathcal{A}, \mathcal{G})$ is a submultiplicative Fréchet algebra.

Proof. Part 1 is a consequence of 5.3 and 5.4 and the continuous embedding of $\mathcal{P}_k(\mathcal{M}, \mathcal{A}, \mathcal{G}, g)$ in $\mathcal{L}(H(1, g; \mathcal{M}, \mathcal{A}))$. Part 2 is a consequence of

$$\Psi(1, g; \mathcal{M}, \mathcal{A}, \mathcal{G}) \subseteq \Psi_k(g; \mathcal{M}, \mathcal{A}, \mathcal{G}) \subseteq \mathcal{P}_k(\mathcal{M}, \mathcal{A}, \mathcal{G}, g)$$

$(k \in \mathbb{N}_0)$ and 5.2. □

6. Examples of algebras of pseudo-differential operators on manifolds

Example. Let $\mathcal{M}$ be an SG-compatible manifold in the sense of E. Schrohe [40] with an atlas $\mathcal{A} = \{(\mathcal{X}_j, X_j, \kappa_j) \mid j = 1, \ldots, m\}$ For $j \in \{1, \ldots, m\}$ let $\mathcal{U}_j := \kappa_j(\mathcal{X}_j)$ and $\chi_j : T^*\mathcal{X}_j \to T^*\mathbb{R}^n \simeq \mathbb{R}^{2n}$, $T_x^*\mathcal{X}_j \ni \xi \mapsto (\kappa(x), {}^t d\kappa(x)^{-1}\xi)$. Let

$$g_{(x,\xi)}^{\rho_1,\rho_2,\delta} := \frac{\langle\xi\rangle^{2\delta}}{\langle x\rangle^{2\rho_1}}\|dx\|_2^2 + \frac{\|d\xi\|_2^2}{\langle\xi\rangle^{2\rho_2}}$$

where $0 \le \rho_1 \le 1, 0 < \rho_2 \le 1, 1 - \rho_2 \le \delta < \rho_2$ and let

$$g_{(x,\xi)}^{\rho,\delta} := \langle x,\xi\rangle^{2\delta}\|dx\|_2^2 + \frac{\|d\xi\|_2^2}{\langle x,\xi\rangle^{2\rho}}$$

where $0 < \rho \le 1, 1 - \rho \le \delta < \rho$ (cf. [17], [27], [48]). With the help of a partition of unity we can define metrics $g_1 := g^{[\rho_1,\rho_2,\delta]}$ and $g_2 := g^{[\rho,\delta]}$ on $T(T^*\mathcal{M})$ such that $\forall j \in \{1, \ldots, m\}$, $\exists C_j > 0$, $\forall (x,\xi) \in \mathcal{U}_j \times \mathbb{R}^n$,

$$\begin{aligned} C_j^{-1} g_{(x,\xi)}^{\rho_1,\rho_2,\delta}(\cdot) &\le (\chi_j^{-1})^* g_{1,(x,\xi)}(\cdot) \le C_j g_{(x,\xi)}^{\rho_1,\rho_2,\delta}(\cdot), \\ C_j^{-1} g_{(x,\xi)}^{\rho,\delta}(\cdot) &\le (\chi_j^{-1})^* g_{2,(x,\xi)}(\cdot) \le C_j g_{(x,\xi)}^{\rho,\delta}(\cdot). \end{aligned}$$

Moreover, let $\mathcal{G} = \mathcal{R}_1(g^{[1,1,0]})$. Then the following statements hold:

1. $(\mathcal{M}, \mathcal{A}, \mathcal{G})$ is a generalized SG-compatible manifold.
2. $(\mathcal{M}, \mathcal{A}, \mathcal{G}, g_1)$ and $(\mathcal{M}, \mathcal{A}, \mathcal{G}, g_2)$ are admissible structures of type (ψ). Moreover, defining $\underline{g}_1 := g^{[1,1,0]}$ and $\underline{g}_2 := g^{[1,0]}$, then $(\mathcal{M}, \mathcal{A}, \mathcal{G}, \underline{g}_i)$ is an admissible structure of type $(\psi)_s$ and $\underline{g}_i$ is a minorant of g_i, $i = 1, 2$.

Example. Let $\mathcal{M} = \mathcal{M}_0 \dot\cup \mathcal{M}_1 \dot\cup \mathcal{M}_2 \dot\cup \partial\mathcal{M}_1 \dot\cup \partial\mathcal{M}_2$, where $\mathcal{M}_0, \mathcal{M}_1, \mathcal{M}_2$ are n-dimensional submanifolds, $\partial\mathcal{M}_1, \partial\mathcal{M}_2$ are connected $(n-1)$-dimensional submanifolds, $\mathcal{M}_0$ is relatively compact, $\partial\mathcal{M}_0 = \partial\mathcal{M}_1 \dot\cup \partial\mathcal{M}_N$, for $j \in \{1, \ldots, N\}$, $\mathcal{M}_j$ is diffeomorphic to $\partial\mathcal{M}_j \times (1, \infty)$, and let $0 \le \rho_i \le 1$ $(i \in \{0, 1, 2\})$, $0 < \rho_2 \le 1$, $1 - \rho_2 \le \delta < \rho_2$. Applying the notation from the preceding example, with the help of a partition of unity we can construct a metric g on $\mathcal{M}$ such that the following statements hold:

- $(\mathcal{M}, \mathcal{A}, \mathcal{G}, g)$ is an admissible structure of type (ψ).
- $\exists C > 0$, $\forall i \in \{0, 1, 2\}$, $\forall x \in \mathcal{M}_i$, $\forall \xi \in T_x^*\mathcal{M}$, $C^{-1} g_{(x,\xi)}^{[\rho_i, \rho_2, \delta]} \le g_{(x,\xi)} \le C g_{(x,\xi)}^{[\rho_i, \rho_2, \delta]}$.
- $(\mathcal{M}, \mathcal{A}, \mathcal{G}, g^{[1,1,0]})$ is an admissible structure of type $(\psi)_s$ and $g^{[1,1,0]}$ is a minorant of g.

In particular, this example shows that situations appear where different Hörmander metrics occur in different local models, i.e., in the admissible quadruples.

The following anisotropic examples are excluded by the assumptions which we impose on admissible structures of type (ψ). But they fulfill the assumptions of quadruples which we call *of type* (p) and for which we introduced the general $S(M, g)$-pseudo-differential calculus in [3]. Moreover, the results stated below can be shown with minor modifications of the proofs which we have presented so far. The most important modification consists of working out the characterizations with $\mathcal{V}(\mathcal{M}_1, \mathcal{A}_1, \mathcal{G}_1) \sqcup \mathcal{V}(\mathcal{M}_2, \mathcal{A}_2, \mathcal{G}_2)$ instead of $\mathcal{V}(\mathcal{M}, \mathcal{A}, \mathcal{G})$.

Example. Let $\mathcal{M}^{(1)}$ and $\mathcal{M}^{(2)}$ be two compact manifolds. For $i = 1, 2$ let $\mathcal{G}^{(i)}$ be a continuous metric on $\mathcal{M}^{(i)}$ and let $\mathcal{A}^{(i)}$ be such that $(\mathcal{M}^{(i)}, \mathcal{A}^{(i)}, \mathcal{G}^{(i)})$ is an SC compatible manifold. Let $\mathcal{M}$, $\mathcal{A}$, and $\mathcal{G}$ be the obviously defined objects on the product. Moreover, let $\kappa_{j_1,j_2} := (\kappa_{j_1}^{(1)}, \kappa_{j_2}^{(2)})$, and $\chi_{j_1,j_2} : T^*\left(\mathcal{X}_{j_1}^{(1)} \times \mathcal{X}_{j_2}^{(2)}\right) \to \mathbb{R}^{n_1+n_2}$, $(x, \xi) \mapsto \left(\kappa_{j_1,j_2}(x), {}^t d\kappa_{j_1,j_2}(x)^{-1}\xi\right)$ $(j_i \in \{1, \ldots, m_i\}, i = 1, 2)$. For $0 < \rho_1, \rho_2 \le 1$, $1 - \min\{\rho_1, \rho_2\} \le \delta_1 < \rho_1$, $1 - \min\{\rho_1, \rho_2\} \le \delta_2 < \rho_2$, and $x, \xi \in \mathbb{R}^n$, $t_1, \tau_1 \in \mathbb{R}^{n_1}$, and $t_2, \tau_2 \in \mathbb{R}^{n_2}$, let

$$g_{(x,\xi)}^{\rho_1,\rho_2,\delta_1,\delta_2}(t_1, t_2, \tau_1, \tau_2) := \langle\xi\rangle^{2\delta_1} \|t_1\|_2^2 + \langle\xi\rangle^{2\delta_2} \|t_2\|_2^2 + \frac{\|\tau_1\|_2^2}{\langle\xi\rangle^{2\rho_1}} + \frac{\|\tau_2\|_2^2}{\langle\xi\rangle^{2\rho_2}} .$$

With the help of a partition of unity we can construct a metric g on $T(T^*\mathcal{M})$ such that $\forall j_i \in \{1, \ldots, m_i\}$ $(i = 1, 2)$, $\exists C$, $\forall x \in \kappa_{j_1,j_2}\left(\mathcal{X}_{j_1}^{(1)} \times \mathcal{X}_{j_2}^{(2)}\right)$, $\forall \xi \in \mathbb{R}^{n_1+n_2}$,

$$C^{-1}\, g_{(x,\xi)}^{\rho_1,\rho_2,\delta_1,\delta_2}(\cdot) \le (\chi_{j_1,j_2}^{-1})^* g_{(x,\xi)}(\cdot) \le C\, g_{(x,\xi)}^{\rho_1,\rho_2,\delta_1,\delta_2}(\cdot) .$$

For weight functions which are in local coordinates of the form $(x,\xi) \mapsto \langle\xi\rangle^s$, $s \in \mathbb{R}$, we have order reducing operators, and $\Psi(1, g; \mathcal{M}, \mathcal{A}, \mathcal{G})$ is a submultiplicative Ψ^*-algebra in $\mathcal{L}\left(L^2(\mathbb{R}^n)\right)$.

Example. Let $\left(\mathcal{M}^{(1)}, \mathcal{A}^{(1)}, \mathcal{G}^{(1)}\right)$ and $\left(\mathcal{M}^{(2)}, \mathcal{A}^{(2)}, \mathcal{G}^{(2)}\right)$ be two SG-compatible manifolds of dimension n_1 and n_2, respectively. Moreover, for $0 < \rho_1, \rho_2 \leq 1$, $1 - \min\{\rho_1, \rho_2\} \leq \delta_1 < \rho_1$, $1 - \min\{\rho_1, \rho_2\} \leq \delta_2 < \rho_2$, and $x, \xi \in \mathbb{R}^n$, $t_1, \tau_1 \in \mathbb{R}^{n_1}$, and $t_2, \tau_2 \in \mathbb{R}^{n_2}$, let

$$\begin{aligned} g_{(x,\xi)}^{\rho_1,\rho_2,\delta_1,\delta_2}(t_1,t_2,\tau_1,\tau_2) \quad &:= \quad \langle x,\xi\rangle^{2\delta_1} \|t_1\|_2^2 + \langle x,\xi\rangle^{2\delta_2} \|t_2\|_2^2 \\ &\quad + \frac{\|\tau_1\|_2^2}{\langle x,\xi\rangle^{2\rho_1}} + \frac{\|\tau_2\|_2^2}{\langle x,\xi\rangle^{2\rho_2}} \; . \end{aligned}$$

We apply notation analogous to that in the preceding example. There exists a metric g on $T(T^*\mathcal{M})$ such that

$$\forall j_i \in \{1,\ldots,m_i\} \ (i=1,2),\ \exists C,\ \forall x \in \kappa_{j_1,j_2}\left(\mathcal{X}_{j_1}^{(1)} \times \mathcal{X}_{j_2}^{(2)}\right),\ \forall \xi \in \mathbb{R}^{n_1+n_2},$$

$$C^{-1}\ g_{(x,\xi)}^{\rho_1,\rho_2,\delta_1,\delta_2}(\cdot) \leq (\chi_{j_1,j_2}^{-1})^* g_{(x,\xi)}(\cdot) \leq C\ g_{(x,\xi)}^{\rho_1,\rho_2,\delta_1,\delta_2}(\cdot) \; .$$

For weight functions which are in local coordinates of the form $(x,\xi) \mapsto \langle x,\xi\rangle^s$, $s \in \mathbb{R}$, we have order reducing operators, and $\Psi(1, g; \mathcal{M}, \mathcal{A}, \mathcal{G})$ is a submultiplicative Ψ^*-algebra in $\mathcal{L}\left(L^2(\mathbb{R}^n)\right)$.

7. Generalized algebras of pseudo-differential operators

Combining the methods from [26], Chapters 2 and 3, we can define *submultiplicative* Ψ^*-algebras which cover different pseudo-differential calculi in different local models when we make choices concerning the operators to commute with. This procedure uses arguments analogous to those presented in Sections 4 and 5. More precisely, we fix a Riemannian metric on $\mathcal{M}$ and let Δ be the Friedrichs' extension of the corresponding Laplace–Beltrami operator starting from $\mathcal{C}_0^\infty(\mathcal{M})$ if $\mathcal{M}$ is non-compact. Let $\varepsilon > 0$ (e.g., $\varepsilon \leq 1/2$) and $\Lambda := (\mathrm{Id} - \Delta)^\varepsilon$. In the case of the classes $\Psi_{\rho,\delta}^0$ on $\mathbb{R}^n$ with $0 \leq \delta \leq \rho \leq 1$, $\delta < 1$, we can take $\varepsilon := 1 - \delta$ (cf. [50, 4.1]) and thus, in the case of manifolds of the form $\mathcal{M} = \mathcal{M}_1 \,\dot\cup\, \tilde{\mathcal{M}}_1 \,\dot\cup\, \mathcal{M}_2 \,\dot\cup\, \partial\tilde{\mathcal{M}}_1$ where

- $\mathcal{M}_j$ ($j \in \{1,\ldots,N\}$) is diffeomorphic to $\mathbb{R}^n \setminus \{x \in \mathbb{R}^n | \ \|x\|_2 \leq 1\}$,
- $\tilde{\mathcal{M}}_j$ ($j \in \{1,\ldots,N-1\}$) is relatively compact,
- $\partial\tilde{\mathcal{M}}_j = \partial\mathcal{M}_j \,\dot\cup\, \partial\mathcal{M}_{j+1}$, $j \in \{1,\ldots,N-1\}$,

we may think of $0 < \varepsilon \leq \min\{\varepsilon_1, \varepsilon_2\}$ with $\varepsilon_1 := 1 - \delta_1$ and $\varepsilon_2 := 1 - \delta_2$ where $0 \leq \delta_1 \leq \rho_1 \leq 1$, $\delta_1 < 1$, and $0 \leq \delta_2 \leq \rho_2 \leq 1$, $\delta_2 < 1$. The invariance with respect to compactly supported diffeomorphisms for $1-\rho \leq \delta \leq \rho$ is not needed. Moreover, let $N \in \mathbb{N}$ and, for $j = 1,\ldots,N$, let $m_j \in \mathcal{C}^\infty(\mathcal{M})$ be a real-valued function such that, (e.g.) with a suitable constant $C > 0$, $|m_j(x)| \leq C\,(\mathrm{dist}(x,x_0) + 1)$, where $x_0 \in \mathcal{M}$ is fixed and possibly using the Levi-Civita connection for decay estimates of "derivatives" of m_j. Starting with the operators Λ and $M_j : \varphi \mapsto m_j \cdot \varphi$, with the methods from [26, Chapter 2] we can define a submultiplicative Ψ^*-algebra $\mathcal{A}_0$ in $\mathcal{L}\left(L^2(\mathcal{M})\right)$ such that there exists a dense subspace $\mathcal{D}$ of $L^2(\mathcal{M})$ with $\mathcal{C}_0^\infty(\mathcal{M}) \subseteq \mathcal{D}$ which is invariant under all operators in $\mathcal{A}_0$ and under Λ and M_j, $j = 1, \ldots, N$ (cf. 4.10–4.16). The commutator conditions occurring in the definition of this algebra $\mathcal{A}_0$ are formulated without order shift. However, $\mathcal{A}_0$ might not even be pseudo-local.

With the methods from [26, Chapter 3] $\mathcal{A}_0$ is refined with the help of commutator conditions with order shift. For this purpose, we consider the scale of Sobolev spaces $H^s_\Lambda(\mathcal{M})$, $s \in \mathbb{R}$, associated to Λ and define commutator conditions within this scale of spaces. On different non-compact parts of the manifold (e.g., $\mathcal{M}_1$ and $\mathcal{M}_2$) we might require commutator conditions with different shift properties for vector fields supported on these parts. This procedure maintains the Ψ-property since all operations can be considered on $\mathcal{D} \subseteq \bigcap_s H^s_\Lambda(\mathcal{M})$ (cf. 4.17). Moreover, we can endow the algebra $\mathcal{A}_\infty$ obtained in this way with a submultiplicative Fréchet topology (cf. [26, (3.5) and Theorem 3.8]).

Acknowledgement. I am very grateful to Professor B. Gramsch for proposing to me to generalize the $S(M,g)$-calculus to manifolds and to investigate spectral invariance and submultiplicativity as well as for valuable support and helpful suggestions. Moreover, I wish to thank Dr. R. Lauter for many fruitful discussions and some good advice. Finally, I thank Dr. J. Lutgen, Dr. J. Møller, and Dr. O. Caps for useful comments and for correcting my English.

References

[1] E. Albrecht and M. Mathieu, editors. *Banach Algebras '97*, Berlin – New York, 1998. Proc. of the 13th International Conference on Banach algebras held in Blaubeuren, July 20 – August 3, 1997, Walter de Gruyter.

[2] F. Baldus. Weyl-Hörmander-Quantisierung auf dem $\mathbb{R}^n$, Spektralinvarianz und Submultiplikativität der durch Kommutatoren definierten Fréchetalgebren. Diplomarbeit, Fachbereich 17-Mathematik, Johannes Gutenberg-Universität Mainz, 1996.

[3] F. Baldus. *$S(M,g)$ pseudo-differential calculus with spectral invariance on $\mathbb{R}^n$ and manifolds for Banach function spaces.* PhD thesis, Johannes Gutenberg-Universität Mainz, 2001, Logos Verlag, Berlin, 2001. 300 pages.

[4] F. Baldus. Application of the Weyl-Hörmander calculus to generators of Feller semigroups. in preparation.

[5] F. Baldus. An approach to a version of the $S(M,g)$-pseudo-differential-calculus on manifolds. This volume, 207–248.

[6] R. Beals. A general calculus of pseudodifferential operators. *Duke Math. J.*, 42:1–42, 1975.

[7] R. Beals. Characterization of pseudodifferential operators and applications. *Duke Math. J.*, 44:45–57, 1977. ibid. 46:215, 1979.

[8] R. Beals. Weighted distribution spaces and pseudodifferential operators. *J. An. Math.*, 39:130–187, 1981.

[9] R. Beals and C. Fefferman. Spatially inhomogeneous pseudodifferential operators I. *Comm. Pure Appl. Math.*, 27:1–24, 1974.

[10] J.-M. Bony. Fourier integral operators and Weyl-Hörmander calculus. In J.-M. Bony and M. Morimoto, editors, *New trends in microlocal analysis*, pages 3–22. Springer-Verlag, Tokyo – Berlin – Heidelberg – New York, 1997.

[11] J.-M. Bony and J.-Y. Chemin. Espaces fonctionnels associés au calcul de Weyl-Hörmander. *Bull. Soc. Math. France*, 122:77–118, 1994.

[12] J.-M. Bony and N. Lerner. Quantification asymptotique et microlocalisations d'ordre supérieur I. *Ann. Sci. Ecole Norm. Sup.* 4e *série*, 22:377–433, 1989.

[13] F. Bruyant. *Estimations pour la composition d'un grand nombre d'opérateurs pseudo-différentiels et applications.* PhD thesis, Univ. Reims, 1979.

[14] H.O. Cordes. A global parametrix for pseudodifferential operators over $\mathbb{R}^n$. Preprint No. 90, SFB 72, Bonn, 1976.

[15] H.O. Cordes. On pseudodifferential operators and smoothness of special Lie-group representations. *Manuscripta Math.*, 28:51–69, 1979.

[16] H.O. Cordes. *On some C^*-algebras and Fréchet*-algebras of pseudodifferential operators*, volume 43 of *Proc. Symp. in Pure Math. – Pseudodifferential operators*, pages 79–104. Amer. Math. Soc., Providence, Rhode Island, 1985.

[17] H.O. Cordes. *The technique of pseudodifferential operators*, volume 202 of *London Mathematical Society, Lecture Note Series.* Cambridge University Press, Cambridge – London – New York, 1995.

[18] E.B. Davies, B. Simon, and M. Taylor. L^p spectral theory of Kleinian groups. *J. Funct. Anal.*, 78:116–136, 1988.

[19] N. Dencker. The Weyl calculus with locally temperate metrics and weights. *Ark. Mat.*, 24:59–79, 1986.

[20] J. Dunau. Fonctions d'un opérator elliptique sur une variété compacte. *J. Math. pures et appl.*, 56:367–391, 1977.

[21] B. Gramsch. Some homogeneous spaces in the operator theory and Ψ-algebras. In *Tagungsbericht Oberwolfach* 42/81 – *Funktionalanalysis: C^*-Algebren*, 1981.

[22] B. Gramsch. Relative Inversion in der Störungstheorie von Operatoren und Ψ-Algebren. *Math. Annalen*, 269:27–71, 1984.

[23] B. Gramsch. Oka's principle for special Fréchet Lie groups and homogeneous manifolds in topological algebras of the microlocal analysis. In *Banach algebras 97 Proceedings*, pages 189–204, Berlin – New York, 1998. de Gruyter.

[24] B. Gramsch and W. Kaballo. Multiplicative decompositions of holomorphic Fredholm functions and Ψ^*-algebras. *Math. Nachr.*, 204:83–100, 1999.

[25] B. Gramsch and E. Schrohe. *Submultiplicativity of Boutet de Monvel's algebra for boundary value problems*, pages 235–258. Mathematical topics – Advances in Partial Differential Equations. Akademie-Verlag, Berlin, 1994.

[26] B. Gramsch, J. Ueberberg, and K. Wagner. Spectral invariance and submultiplicativity for Fréchet algebras with applications to pseudo-differential operators and Ψ^*-quantization. In *Operator theory: Advances and Applications,* vol. 57, pages 71–98. Birkhäuser, Basel, 1992.

[27] B. Helffer. *Théorie spectrale pour des opérateurs globalement elliptiques*, volume 112 of *Astérisque*. Société Mathématique de France, 1984.

[28] L. Hörmander. Pseudo-differential operators. *Comm. Pure Appl. Math.*, 18:501–517, 1965.

[29] L. Hörmander. *Pseudo-differential operators and hypoelliptic equations*, volume X of *Proc. Symp. in Pure Math. – Singular Integrals*, pages 138–183. Amer. Math. Soc., Providence, Rhode Island, 1966.

[30] L. Hörmander. The Weyl calculus of pseudo-differential operators. *Comm. Pure Appl. Math.*, 32:359–443, 1979.

[31] L. Hörmander. *The analysis of linear partial differential operators,* vol. 3. Springer-Verlag, Berlin – Heidelberg – New York, 1985.

[32] H. Kumano-go and K. Taniguchi. Oscillatory integrals of symbols of pseudo-differential operators on $\mathbb{R}^n$ and operators of Fredholm type. *Proc. Japan Acad.*, 49:397–402, 1973.

[33] R. Lauter. *Holomorphic functional calculus in several variables and Ψ^*-algebras of totally characteristic operators on manifolds with boundary.* PhD thesis, Johannes Gutenberg-Universität Mainz, 1996, Shaker Verlag, Aachen, 1997. 241 pages.

[34] R. Lauter. On the existence and structure of Ψ^*-algebras of totally characteristic operators on compact manifolds with boundary. *J. Funct. Anal.*, 169(1):81–120, 1999.

[35] R. Lauter. Pseudodifferential analysis on conformally compact spaces. Fachbereich Mathematik, volume 163, number 777, of *Mem. Amer. Math. Soc.*, Providence, Rhode Island, 2003.

[36] F. Mantlik. Norm closures of operator algebras with symbolic structure. *Math. Nach.*, 201:91–116, 1999.

[37] C. Parenti. Operatori pseudo-differenziali in $\mathbb{R}^n$ e applicazioni. *Ann. Mat. Pura Appl. (4)*, 93:359–389, 1972.

[38] H.H. Schaefer. *Topological vector spaces*. Springer Verlag, Berlin – Heidelberg – New York, 1971.

[39] E. Schrohe. *Komplexe Potenzen elliptischer Pseudodifferentialoperatoren.* PhD thesis, Johannes Gutenberg-Universität Mainz, 1986.

[40] E. Schrohe. Spaces of weighted symbols and weighted Sobolev spaces on manifolds. In H.O. Cordes, B. Gramsch, and H. Widom, editors, *Pseudo-differential operators*, volume 1256 of *Lect. Notes Math.*, pages 360–377, Berlin – Heidelberg – New York – London – Paris – Tokyo, 1987. Proc. Conf., Oberwolfach/Ger. 1986, Springer-Verlag.

[41] E. Schrohe. A Ψ^*-algebra of pseudo-differential operators on noncompact manifolds. *Arch. Math.*, 51:81–86, 1988.

[42] E. Schrohe. Boundedness and spectral invariance for standard pseudo-differential operators on anisotropically weighted L_p-Sobolev-spaces. *Integral equations operator theory*, 13(2):271–284, 1990.

[43] E. Schrohe. Spectral invariance, ellipticity, and the Fredholm property for pseudo-differential operators on weighted Sobolev spaces. *Ann. Global Anal. and Geometry*, 10:237–254, 1992.

[44] E. Schrohe. Fréchet algebra techniques for boundary value problems on noncompact manifolds: Fredholm criteria and functional calculus via spectral invariance. *Math. Nachrichten*, 199:145–185, 1999.

[45] R.T. Seeley. *Complex powers of an elliptic operator*, volume X of *Proc. Symp. in Pure Math. – Singular Integrals*, pages 288–307. Amer. Math. Soc., Providence, Rhode Island, 1967.

[46] R.T. Seeley. Topics in pseudo-differential operators. In L. Nirenberg, editor, *Pseudo-differential operators*. Roma, 1968. CIME, Edizione Cremonese, 1969.

[47] M.A. Shubin. Pseudodifferential operators in $\mathbb{R}^n$. *Sov. Math. Dokl.,* 12, Ser. N1, pages 147–151, 1971.

[48] M.A. Shubin. *Pseudodifferential Operators and Spectral Theory*. Springer-Verlag, Berlin – Heidelberg – New York – London – Paris – Tokyo, 1987.

[49] J. Sjöstrand. Wiener type algebras of pseudo-differential operators. In *Séminaire Equations aux Dérivées Partielles 1994–1995, Exposé n^0 IV*, Centre de Mathématiques, Ecole Polytechnique, Palaiseau, December 1995. 19 pages.

[50] J. Ueberberg. Zur Spektralinvarianz von Algebren von Pseudo-Differential Operatoren in der L^p-Theorie. *Manuscripta Math.*, 61:459–475, 1988.

[51] A. Unterberger. Encore des classes de symboles. In *Sém. Goulaouic-Schwartz 1977–1978*, Ecole Polytechnique, Paris, 1977. 17 pages.

[52] A. Unterberger. Oscillateur harmonique et opérateurs pseudo-différentiels. *Ann. Inst. Fourier*, 29-3:201–221, 1979.

[53] A. Unterberger and J. Bokobza. Les opérateurs de Calderon-Zygmund précisés. *C. R. Acad. Sci. Paris*, 259:1612–1614, 1964.

[54] A. Unterberger and J. Bokobza. Sur les opérateurs pseudo-différentiels d'ordre variable. *C. R. Acad. Sci. Paris*, 261:2271–2273, 1965.

[55] H. Widom. Singular integral equations in L^p. *Transactions Am. Math. Soc.*, 97:131–160, 1960.

[56] W. Żelazko. Concerning entire functions in B_0-algebras. *Studia Math.*, 110:283–290, 1994.

F. Baldus
Fachbereich Mathematik
Johannes Gutenberg-Universität Mainz
Staudingerweg 9
D-55099 Mainz, Germany
e-mail: `baldus@mathematik.uni-mainz.de`

Operator Theory:
Advances and Applications, Vol. 145, 281–362

Domain Perturbations and Capacity in General Hilbert Spaces and Applications to Spectral Theory

André Noll

1. The significance of the bottom eigenvalue in mathematical physics

In this section we explain the significance of the bottom eigenvalue of a self-adjoint operator in various areas of mathematical physics, so this will be a motivation for investigating and estimating the bottom eigenvalue. The main role is played by the Laplacian because this is one of the most important self-adjoint operators in mathematical physics as it plays a fundamental role in quantum mechanics, theory of heat, theory of vibrations and other areas.

On the other hand, higher-order differential operators are important as well. For instance the dynamics of the clamped plate is described by the bi-potential equation

$$\Delta^2 u = 0,$$

where Δ^2 is the biharmonic operator, subject to Dirichlet boundary conditions. Another example of a relevant higher-order differential operator is the analysis of the vertices of incompressible fluids which also leads to the biharmonic equation.

Most of the known estimates for the bottom eigenvalue only apply to second-order differential operators and some of them [Szn98], [Tay79], [Oza81], [MR84], can only handle the Laplacian. This is due to the fact that in the case of second-order differential operators there is an interplay between analysis and stochastics via the theory of Dirichlet forms. One can prove that for certain second-order differential operators, see Appendix C.1 for a precise statement, there is a stochastic process associated to the operator. It turns out that the existence of such a process is closely related to positivity preserving properties of the semigroup (and the resolvent) and to the maximum principle, [BH86, FOT94, MR92, DC]. This allows us to treat analytic problems, e.g., estimating eigenvalues, with stochastic methods or by using the powerful tools of potential analysis and vice versa.

It is known that in contrast to the second-order case there is no such interplay for higher-order differential operators. Therefore the proofs of the eigenvalue estimates which rely on stochastic processes or the maximum principle typically do not carry over to the higher-order case.

In order to include higher-order differential operators one needs a more general method. The results for self-adjoint operators in abstract Hilbert spaces (i.e., not necessarily L^2-spaces) which are described in Section 5 turn out to be an appropriate tool in estimating eigenvalues of a wide class of self-adjoint operators. For instance they are sufficient to handle eigenvalue estimates for higher-order differential operators.

1.1. Preliminaries on the Dirichlet Laplacian on a bounded domain

Let Ω be an open and bounded subset of $\mathbb{R}^d$. The following Poincaré inequality is valid for all $u \in H_0^1(\Omega)$,

$$\int_\Omega |u|^2 \mathrm{d}x \leq c_\Omega \int_\Omega |\nabla u|^2 \mathrm{d}x,$$

where c_Ω is a geometric constant and may be taken four times the diameter of Ω, see [Joh71], page 99. From the functional analytic point of view this inequality simply states that the (negative) Dirichlet Laplacian $-\Delta_\Omega$ in $L^2(\Omega)$ is a strictly positive operator, i.e., there is a constant $c > 0$ such that

$$\langle -\Delta_\Omega u, u\rangle \geq c\|u\|^2$$

for all $u \in \mathrm{dom}(H)$. Here $-\Delta_\Omega$ is defined as the unique self-adjoint operator which is associated to the closure of the quadratic form

$$\mathcal{E}[u] := \int_\Omega |\nabla u|^2 \mathrm{d}x,$$

with u in the initial form domain $C_c^\infty(\Omega)$, the set of smooth functions with compact support in Ω, see Section 2.

The boundedness of Ω implies that the resolvent $(-\Delta_\Omega - z)^{-1}$, z in the resolvent set of $-\Delta_\Omega$, is a compact operator in $L^2(\Omega)$, see [RS78], p. 255. Since compactness of the resolvent is equivalent to discreteness of the spectrum, $\sigma(-\Delta_\Omega)$ consists only of isolated eigenvalues and each of them is of finite multiplicity. If additionally Ω is connected, i.e., Δ_Ω is the Dirichlet Laplacian on a bounded domain, it is known that the bottom eigenvalue λ_1 is simple, see [Goe77]. In fact this non-degeneracy of the ground-state holds in great generality for second-order elliptic differential operators even in the more general case of Riemannian manifolds. This can be proved with the aid of a Perron–Frobenius argument which uses the positivity preserving property of the semigroup $(e^{t\Delta_\Omega})_{t>0}$. A proof can be found in [Dav89], Theorem 5.2.1 and Proposition 1.4.3. In contrast to the second-order case, higher-order operators may have a degenerate bottom eigenvalue. For example the ground-state of the Bi-Laplacian Δ^2 on a punctured disc turns out to be degenerate, see [CD92], or the survey article [Dav97].

1.2. Interpretation of the bottom eigenvalue for the heat equation

With the notion of the previous section, look at Ω as a body in Euclidean space which carries an initial heat distribution $f \in L^2(\Omega)$. As time goes by the heat will distribute itself among Ω. According to the laws of physics the time evolution of the system is given by the *heat equation*

$$\frac{\partial}{\partial t}u(t,\cdot)=\sum_{j=1}^{d}\frac{\partial^2}{\partial x_j^2}u(t,\cdot)=\Delta_\Omega u(t,\cdot),\quad u(0,\cdot)=f(\cdot),\quad t>0. \tag{1.1}$$

The choice of Dirichlet boundary conditions describes the physical situation that the boundary of Ω is held on temperature zero. The unique solution of the heat equation (1.1) is given by the semigroup $(e^{t\Delta_\Omega})_{t>0}$, i.e., we have

$$u(t,\cdot)=e^{t\Delta_\Omega}f,\quad t>0.$$

Let us arrange the eigenvalues of $-\Delta_\Omega$ in increasing order

$$0<\lambda_1<\lambda_2\leq\lambda_3\leq\cdots,$$

where each λ_j has to be repeated according to its multiplicity. Let $(\phi_j)_{j\in\mathbb{N}}$ be a corresponding sequence of normalized eigenfunctions.

The *heat content* stored in the system at time t and with initial distribution f is

$$Q(t,f):=\int_\Omega u(t,x)\mathrm{d}x=\int_\Omega e^{t\Delta_\Omega}f\mathrm{d}x=\int_\Omega\sum_{j=1}^{\infty}e^{-\lambda_j t}\langle\phi_j,f\rangle\phi_j. \tag{1.2}$$

If $f\equiv 1$, i.e., in the beginning the body is at constant temperature 1, then $Q(t):=Q(t,1)$ will be a decreasing function of t because of the Dirichlet boundary conditions. This can also be read off from equation (1.3) below. It is therefore natural to ask for the speed of decay of $Q(t)$. From (1.2) we get

$$Q(t)=\int_\Omega e^{t\Delta_\Omega}1\mathrm{d}x=\int_\Omega\sum_{j=1}^{\infty}e^{-\lambda_j t}\langle\phi_j,1\rangle\phi_j=\sum_{j=1}^{\infty}e^{-\lambda_j t}\left|\int_\Omega\phi_j\,\mathrm{d}x\right|^2. \tag{1.3}$$

From this equality we immediately deduce the estimates

$$e^{-\lambda_1 t}\left|\int_\Omega\phi_1\,\mathrm{d}x\right|^2\leq Q(t)\leq|\Omega|\sum_{j=1}^{\infty}e^{-\lambda_j t}=|\Omega|\mathrm{tr}(e^{t\Delta_\Omega}),$$

where $\mathrm{tr}(e^{t\Delta_\Omega})$ denotes the trace of the semigroup $(e^{t\Delta_\Omega})$. Clearly $e^{-\lambda_1 t}$ in the above sum is the most significant term. Moreover the norm of $e^{t\Delta_\Omega}$ as an operator in $L^2(\Omega)$ equals $e^{-\lambda_1 t}$ by the functional calculus. Putting all together we agree that the bottom eigenvalue λ_1 measures how quickly the system cools out.

1.3. The significance of the bottom eigenvalue in quantum mechanics

The states of a quantum mechanical system are described by the non-zero vectors of a separable Hilbert space $\mathcal{H}$ over the complex field. Two vectors $u,v\in\mathcal{H}$ describe the same state if $u=\lambda v$ for some scalar $\lambda\neq 0$. Hence we may always assume that the vector $u\in\mathcal{H}$ has unit length.

Physical quantities which can be determined experimentally are called *observables* and it is a basic postulate of quantum mechanics that each observable corresponds to a unique self-adjoint operator in $\mathcal{H}$. One very important observable is the energy. The corresponding self-adjoint operator H is called the *Hamiltonian* or the *Schrödinger operator*. Since arbitrarily high energies should be possible, the Hamiltonian turns out to be always an unbounded operator. Hence the domain $\mathrm{dom}(H)$ of H is a dense subspace of $\mathcal{H}$ and we always have $\mathrm{dom}(H) \neq \mathcal{H}$ (otherwise H would be bounded because of the closed graph theorem). The energy of a state $u \in \mathrm{dom}(H)$ is given by the expectation value

$$E(u) = \langle u, Hu \rangle$$

and is a real number since H is self-adjoint. Observe that $E(u) = E(\lambda u)$ for each scalar λ of modulus 1.

Assume that at time $t = 0$ the system is in state $u_0 \in \mathcal{H}$. Then at time t the state of the system is represented by some vector $u(t) \in \mathcal{H}$ and it is again a postulate of quantum mechanics that the $\mathcal{H}$-valued function $t \mapsto u(t)$ satisfies *Schrödinger's equation*

$$i\hbar \frac{d}{dt} u(t) = Hu(t), \quad u(0) = u_0,$$

where $\hbar = 1.055 \cdot 10^{-34}$Js is the Planck constant. Clearly $\hbar$ enters only as a physical constant without mathematical relevance. We shall therefore assume from now on that $\hbar = 1$.

The solution of the Schrödinger equation is given by the unitary group $(e^{-itH})_{t>0}$, i.e., we have

$$u(t) = e^{-itH} u_0, \quad t \geq 0.$$

If u_0 is an eigenvector of H with eigenvalue λ, then clearly

$$u(t) = e^{-itH} u_0 = e^{-it\lambda} u_0. \tag{1.4}$$

Since u_0 and $e^{-it\lambda} u_0$ describe the same state, the eigenvectors of H are often called *bound-states* of the system. The converse can also be proved easily: If the state is time-independent, then it must be an eigenvector of the Hamiltonian. From equation (1.4) we see that

$$E(u(t)) = E(u_0) = \lambda \|u_0\|^2 = \lambda.$$

Thus the energy of a bound-state equals the corresponding eigenvalue of the Hamiltonian. Of particular physical interest is the *ground-state* of the system (if it exists), i.e., the state that minimizes the energy. This minimal energy is called the *ground-state energy* and is exactly the bottom eigenvalue λ_1 of H. One can compute λ_1 explicitly in the case of the hydrogen atom, for the quantum mechanical harmonic oscillator and some other systems, see [AGHKH88] for a comprehensive treatment of solvable models in quantum mechanics. Nevertheless it is far from obvious to compute the ground-state energy of more complicated systems. It is therefore of physical interest to have estimates for λ_1.

1.4. Domain perturbations in mathematical physics

Let Ω be a "pitcher filled with liquid", by which we simply mean an open and bounded subset of $\mathbb{R}^d$. Let $-\Delta_\Omega$ be the Dirichlet Laplacian in $L^2(\Omega)$, defined as the Friedrichs extension of the Laplacian, initially defined on $C_c^\infty(\Omega)$, cf. Section 2.1. As described in Section 1.2, the semigroup $(e^{t\Delta_\Omega})_{t>0}$ determines the solution $u(t,\cdot)$ of the heat equation

$$\frac{\partial}{\partial t}u(t,\cdot) = \sum_{j=1}^{d} \frac{\partial^2}{\partial x_j^2}u(t,\cdot) = \Delta_\Omega u(t,\cdot), \quad u(0,\cdot) = f, \quad t > 0. \tag{1.5}$$

Moreover $u(t,\cdot)$ behaves asymptotically like $ce^{-t\lambda_1(\Omega)}$ as $t \to \infty$, where $\lambda_1(\Omega)$ is the bottom eigenvalue of $-\Delta_\Omega$. Hence the size of $\lambda_1(\Omega)$ is a measure of the efficiency of cooling.

Now let K be a compact subset of Ω and consider the Dirichlet Laplacian $-\Delta_{\Omega\setminus K}$ in $L^2(\Omega \setminus K)$ which may be defined by the same method we used to define $-\Delta_\Omega$, see Figure 1. Physically speaking, $-\Delta_{\Omega\setminus K}$ describes the situation that the

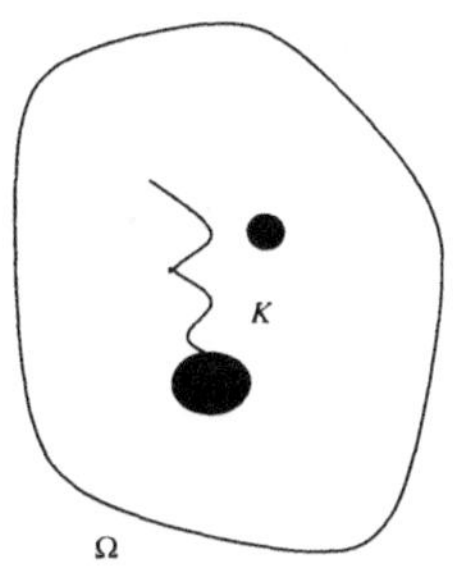

FIGURE 1. Domain perturbations

"pitcher" Ω is held on temperature zero in K. Thus K models a cooler maintained at temperature zero. It is clear for physical reasons as well as from the fact that the form domain of $-\Delta_{\Omega\setminus K}$ may be viewed as a subspace of the form domain of $-\Delta_\Omega$, that the bottom eigenvalue $\lambda_1(\Omega \setminus K)$ of $-\Delta_{\Omega\setminus K}$ is greater than or equal to $\lambda_1(\Omega)$. If K consists of many small pieces one may look at K as crushed ice. From the mathematical point of view it is an interesting problem to determine the improvement in cooling efficiency resulting from the extra cooler K, i.e., the question is how the difference $\lambda_1(\Omega\setminus K) - \lambda(\Omega)$ depends on K. Crude guesses might be that the surface of K measures this improvement of cooling efficiency but this turns out to be wrong as was proved in [Tay76]. See also [RT75a], [Rau75], [RT75b], as well as [Sim79], Section 22 for a general survey on the crushed ice problem. In fact the capacity of the set K determines the shift of the bottom eigenvalue. This is a surprising fact because the capacity of a ball of radius r in $\mathbb{R}^3$ is given by some constant times r, and the above result yields

$$c_0 r \leq \lambda_1(\Omega \setminus K) - \lambda(\Omega) \leq c_1 r$$

in this particular case. This means that the radius of the ball is the decisive quantity rather than its surface area.

Let us finally mention that there are many other physically relevant questions which lead to the same mathematical problem: Give estimates on the shift of the bottom eigenvalue of a self-adjoint operator which is restricted to a smaller domain. A further example is the "fireman's pole problem" which consists in describing the effect of the presence of a perturbation for the propagation of sound. Another example is to explain rigorously why a cloud of small conductors sprayed into the air appear solid on a radar screen.

2. Defining of operators by quadratic forms

In this section we recall the method of defining self-adjoint operators by the quadratic form technique. The first subsection summarizes without proofs the necessary facts on this rather standard topic whereas Section 2.2 emphasizes several advantages of this method. In particular we give an easy example of operators H and V in $L^2(\mathbb{R})$ for which $\mathrm{dom}(H) \cap \mathrm{dom}(V) = \emptyset$, but the intersection of the according form domains happens to be dense in $L^2(\mathbb{R})$.

2.1. The one-to-one correspondence between quadratic forms and self-adjoint operators

Let $\mathcal{H}$ be a real or complex Hilbert space. If B is a bounded operator in $\mathcal{H}$, the sesquilinear form

$$\mathcal{E}(u,v) := \langle u, Bv\rangle, \quad u,v \in \mathcal{H}$$

satisfies, by Cauchy–Schwarz inequality

$$|\mathcal{E}(u,v)| \le C\|u\|\|v\|, \tag{2.1}$$

where $C = \|B\|$. Conversely, let $\mathcal{E}$ be a sesquilinear form on $\mathcal{H} \times \mathcal{H}$ satisfying inequality (2.1). Then for each $u \in \mathcal{H}$ the linear functional

$$F_u(v) := \mathcal{E}(u,v)$$

is bounded. By the Riesz representation theorem there is a unique vector $Bu \in \mathcal{H}$ such that

$$F_u(v) = \langle Bu, v\rangle$$

and it is easy to see that the map $u \mapsto Bu$ defines a bounded linear operator in $\mathcal{H}$ with $\|B\| \le C$. Moreover B is self-adjoint if and only if $\mathcal{E}$ is hermitian, i.e., $\mathcal{E}(u,v) = \overline{\mathcal{E}(v,u)}$ for all $u,v \in \mathcal{H}$.

The situation becomes much more complicated if the operators/forms are no longer bounded, but it turns out that the one-to-one correspondence described above carries over to certain unbounded operators. This allows us to define self-adjoint operators by means of hermitian sesquilinear forms or by quadratic forms since the quadratic form of a hermitian sesquilinear form already determines the sesquilinear form by the polarization identity.

Since all theorems of this section are standard results which can be found in many text books, for instance in [Kat80] or [Wei76], we only state the results without proof.

Definition 2.1. Let $\mathcal{F}$ be a dense subspace of $\mathcal{H}$ and let $\mathcal{E}$ be a hermitian sesquilinear form on $\mathcal{F} \times \mathcal{F}$.

(a) $(\mathcal{E}, \mathcal{F})$ is called *semi-bounded from below with bound* $\gamma \in \mathbb{R}$ if

$$\mathcal{E}[u] := \mathcal{E}(u,u) \geq \gamma \|u\|^2$$

for all $u \in \mathcal{F}$. If, in particular, $\gamma = 0$, then $(\mathcal{E}, \mathcal{F})$ is called *non-negative.*

(b) $(\mathcal{E}, \mathcal{F})$ is called *closed* if $\mathcal{F}$ together with the scalar product

$$\mathcal{E}_{1-\gamma}(u,v) := \mathcal{E}(u,v) + (1-\gamma)\langle u, v\rangle$$

is a Hilbert space.

Now the one-to-one correspondence between forms and operators reads as follows.

Theorem 2.2.

(a) *Let $\mathcal{F}$ be a dense subspace of $\mathcal{H}$ and let $\mathcal{E}$ be a hermitian sesquilinear form on $\mathcal{F} \times \mathcal{F}$ which is semi-bounded from below and closed. Then there is a unique self-adjoint operator H in $\mathcal{H}$ such that*

$$\mathrm{dom}(H) \subset \mathcal{F} \text{ and } \langle u, Hv\rangle = \mathcal{E}(u,v)$$

for all $u \in \mathcal{F}$ and all $v \in \mathrm{dom}(H)$. One has

$$\mathrm{dom}(H) = \{u \in \mathcal{F} : \exists v \in \mathcal{H},\ \ \forall w \in \mathcal{F} : \mathcal{E}(w,u) = \langle w, v\rangle\}.$$

In fact this v is unique and satisfies $Hu = v$.

(b) *Let H be a self-adjoint operator in $\mathcal{H}$ such that $\langle u, Hu\rangle \geq \gamma\|u\|^2$ for all $u \in \mathrm{dom}(H)$. Define $\mathcal{F} := \mathrm{dom}((H-\gamma)^{1/2})$ and*

$$\mathcal{E}(u,v) := \langle (H-\gamma)^{1/2}u, (H-\gamma)^{1/2}v\rangle + \gamma\langle u, v\rangle$$

for $u, v \in \mathcal{F}$. Then $(\mathcal{E}, \mathcal{F})$ is a closed form which is bounded below with bound γ. Moreover the unique self-adjoint operator associated to $(\mathcal{E}, \mathcal{F})$ according to part (a) *coincides with H.*

In practice it is difficult to determine a dense subspace $\mathcal{F}$ that makes $(\mathcal{E}, \mathcal{F})$ closed. This leads to the following definition.

Definition 2.3. Let $(\mathcal{E}, \mathcal{F})$ be a densely defined hermitian sesquilinear form which is semi-bounded from below. $(\mathcal{E}, \mathcal{F})$ is called *closable* if for any Cauchy sequence $(u_n)_{n\in\mathbb{N}}$ in the pre-Hilbert space $(\mathcal{F}, \mathcal{E}_{1-\gamma}(\cdot,\cdot))$ with $u_n \to 0$ in $\mathcal{H}$ we have $u_n \to 0$ in $(\mathcal{F}, \mathcal{E}_{1-\gamma}(\cdot,\cdot))$.

Remark 2.4. Let $(\overline{\mathcal{F}}, \overline{\mathcal{E}_{1-\gamma}}(\cdot,\cdot))$ be the completion of $(\mathcal{F}, \mathcal{E}_{1-\gamma}(\cdot,\cdot))$. Then the embedding $J : (\mathcal{F}, \mathcal{E}_{1-\gamma}(\cdot,\cdot)) \hookrightarrow \mathcal{H}$ extends to a bounded linear operator

$$\overline{J} : (\overline{\mathcal{F}}, \overline{\mathcal{E}_{1-\gamma}}(\cdot,\cdot)) \hookrightarrow \mathcal{H}.$$

The form $(\mathcal{E}, \mathcal{F})$ is closable if and only if $\overline{J}$ is injective.

The following Lemma justifies the name "closable".

Lemma 2.5. *Let $(\mathcal{E}, \mathcal{F})$ be a closable form. Then there is a unique smallest closed extension of $(\mathcal{E}, \mathcal{F})$ which is called the* closure *of $(\mathcal{E}, \mathcal{F})$.*

An important application of the theory of quadratic forms is the *Friedrichs extension* of a symmetric operator.

Theorem 2.6. *Let S be a symmetric operator in $\mathcal{H}$ satisfying*

$$\langle u, Su \rangle \geq \gamma \|u\|^2$$

for all $u \in \operatorname{dom}(S)$. Then the hermitian sesquilinear form

$$\mathcal{E}(u, v) := \langle u, Sv \rangle, \quad u, v \in \mathcal{F} := \operatorname{dom}(S)$$

is closable.

By Theorem 2.2 there is a unique self-adjoint operator associated to the closure of $(\mathcal{E}, \mathcal{F})$ which is called the *Friedrichs extension* of S.

2.2. Advantages of the quadratic form technique

Defining a self-adjoint operator via the quadratic form technique has several advantages. We illustrate some of them here: Firstly the sum of two quadratic forms coming from self-adjoint operators H_1, H_2 again defines a self-adjoint operator $H_1 \dotplus H_2$ if the intersection of the two form domains is dense which often can be checked very easily. $H_1 \dotplus H_2$ is called the *form sum of H_1 and H_2* and may differ from the *operator sum*

$$\operatorname{dom}(H_1 + H_2) := \operatorname{dom}(H_1) \cap \operatorname{dom}(H_2), \quad (H_1 + H_2)u = H_1 u + H_2 u,$$

which may happen to be defined only for the zero vector although the form sum is densely defined. We next give an example for this phenomenon which is taken from [Bra94].

Example. Let $(x_n)_{n \in \mathbb{N}}$ be an enumeration of the rational numbers in $\mathbb{R}$. Define for each $n \in \mathbb{N}$ the function $V_n \in L^1(\mathbb{R})$ by

$$V_n(x) := \sum_{j \leq n} 2^{-j} |x - x_j|^{-1/2} e^{-|x - x_j|}, \quad x \in \mathbb{R} \setminus \{x_1, \ldots, x_n\},$$

see Figure 2. An easy calculation shows that $(V_n)_{n \in \mathbb{N}}$ is a Cauchy sequence in $L^1(\mathbb{R})$. Hence the L^1-limit

$$V := \lim_{n \to \infty} V_n$$

belongs to $L^1(\mathbb{R})$ and is a non-negative function. Therefore the form sum $-\frac{\mathrm{d}^2}{\mathrm{d}x^2} \dotplus V$ defines a self-adjoint operator in $L^2(\mathbb{R}^d)$. On the other hand

$$\int_\Omega V(x)^2 \mathrm{d}x = \infty$$

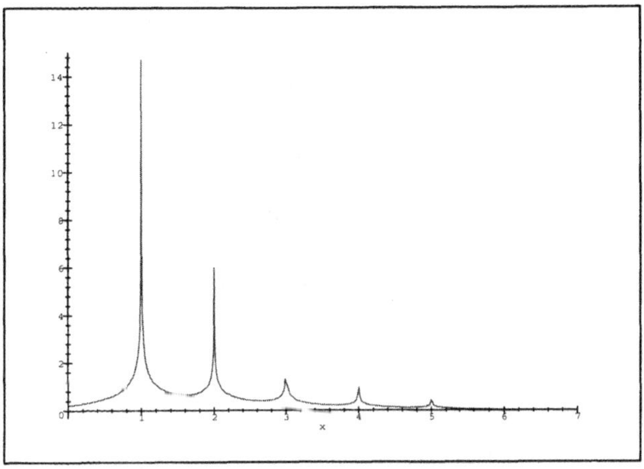

FIGURE 2. The function V_n for $n = 5$ and $x_j = j$. V_n has integrable poles at each x_j. As n increases, the set of poles becomes dense in $\mathbb{R}$ but the limit $\lim_{n\to\infty} V_n$ exists as a function in $L^1(\mathbb{R})$.

for any non-void open set $\Omega \subset \mathbb{R}$. Thus $Vf \notin L^2(\mathbb{R})$ for any continuous function $f \neq 0$, i.e., the only continuous function in the domain of the multiplication operator corresponding to V is the zero function. Since

$$\operatorname{dom}\left(-\frac{\mathrm{d}^2}{\mathrm{d}x^2}\right) = H^2(\mathbb{R}) \subset C(\mathbb{R})$$

by the Sobolev embedding theorem, cf. [Ada78], Theorem 5.4, we have

$$\operatorname{dom}\left(-\frac{\mathrm{d}^2}{\mathrm{d}x^2}\right) \cap \operatorname{dom}(V) = \emptyset,$$

i.e., the operator sum $-\frac{\mathrm{d}^2}{\mathrm{d}x^2} + V$ does not define a self-adjoint operator in $L^2(\mathbb{R})$.

Secondly there is a concept for monotonicity of quadratic forms which allows one to handle limiting processes in a fairly simple way. To be concrete, let $(\mathcal{E}_j, \mathcal{F}_j)_{j\in\mathbb{N}}$ be an increasing sequence of quadratic forms associated to self-adjoint operators H_j. This means that the form domains $\mathcal{F}_j$ are decreasing and for all $u \in \mathcal{F} := \cap_{j\in\mathbb{N}}\mathcal{F}_j$ the sequence $(\mathcal{E}_j[u])_{j\in\mathbb{N}}$ is increasing. Then the limit form

$$\mathcal{E}[u] := \lim_{j\to\infty} \mathcal{E}_j[u],$$

defined for all $u \in \mathcal{F}$ where this limit is finite, is associated to a self-adjoint operator H if it is densely defined, see [RS80] Theorem S.14. Moreover we have $H_j \to H$ in strong resolvent sense. There is no analogous result for (unbounded) operators.

Finally we give another advantage of the quadratic form technique in the case of elliptic differential operators which is the most striking aspect from our point of view. For simplicity we only consider the Dirichlet Laplacian on a bounded set Ω in $\mathbb{R}^d$. One can *always* define $-\Delta_\Omega$ via the closure of the quadratic form

$$\mathcal{E}[u] := \int_\Omega |\nabla u|^2 \mathrm{d}x, \quad u \in C_c^\infty(\Omega). \tag{2.2}$$

In particular, domains with highly singular boundary such as fractal domains can be treated. For such domains it is difficult to define the operator $-\Delta_\Omega$ directly because it is difficult to describe in which sense the functions in $\mathrm{dom}(-\Delta_\Omega)$ should vanish on the boundary.

3. Domain perturbations

Considering a formally self-adjoint differential expression (e.g., the Laplacian) in two different Hilbert spaces leads to the question of comparing the spectra of the two resulting self-adjoint operators in terms of these Hilbert spaces. See Section 1.4 for a motivation of this kind of problem in mathematical physics.

In Section 3.1 we only consider local operators in L^2-spaces and indicate the difficulties in defining self-adjoint realizations of the restricted operator directly by means of the operator. These difficulties can be overcome by using the quadratic form technique to define the domain perturbations.

This general method is described in Section 3.2 and it is illustrated how the perturbed operators appear in the case of differential and integral operators.

Section 3.3 treats another method of defining domain perturbations in the case of regular Dirichlet forms. With the aid of the associated Hunt process one can define a new process by killing the sample paths when leaving a prescribed set. This new process turns out to be again a Hunt process, hence it gives rise to a unique self-adjoint operator. As we shall see, this operator coincides with the domain perturbations in the sense of Section 3.2.

Finally, in Section 3.4 we collect some simple properties of the perturbed operator which immediately follow from the construction.

3.1. The L^2-case

Here we explain the difficulties of directly defining domain perturbations in an L^2-space by means of the operator itself. Let $(X, \mathcal{A}, m_0)$ be a measure space, that is, X is an arbitrary non-empty set, $\mathcal{A}$ is a σ-algebra on X and m_0 is a measure defined on $\mathcal{A}$. If X is a topological space there is always a natural σ-algebra $\mathcal{B}_0$ on X, namely the smallest σ-algebra that contains all open subsets of X. $\mathcal{B}_0$ is called the *Borel σ-algebra of X*. Let $\mathcal{B}$ be the *completion of $\mathcal{B}_0$ with respect to m_0* and let m be the extension of m_0 to $\mathcal{B}$, that is

$$\begin{aligned} \mathcal{B} &:= \{A \subset X : \exists B_1, B_2 \in \mathcal{B}_0 \text{ such that } B_1 \subset A \subset B_2\}, \\ m(A) &:= m_0(B_1). \end{aligned}$$

Passing from $(X,\mathcal{B}_0,m_0)$ to $(X,\mathcal{B},m)$ has the advantage that every subset of a null-set is measurable and has of course measure zero. When dealing with a topological space X we always take $\mathcal{B}$ as the σ-algebra and simply write (X,m) for $(X,\mathcal{B},m)$.

The set $L^2(X,m)$ consisting of all measurable and real-valued (complex-valued) equivalence classes of square-integrable functions on X becomes a real (complex) Hilbert space with respect to the scalar product

$$\langle u,v\rangle_{L^2(X,m)} := \int_X \bar{u}v\,dm, \tag{3.1}$$

where the bar is meaningless in the real case. We shall omit the subscript in (3.1) if it is clear from the context which underlying measure space is meant. If X is some open subset of $\mathbb{R}^d$ and m is the Lebesgue measure, we will further simplify our notation by writing $L^2(X)$ for $L^2(X,m)$.

Suppose that u is a measurable function on the topological space X. If u is continuous, the *support* of u is the set

$$\operatorname{supp}(u) := \overline{\{x\in X : u(x)\neq 0\}}.$$

For functions which are merely measurable this definition of the support can cause trouble as it depends on the representative in the equivalence class of u. For instance the zero function on $\mathbb{R}$ may be written as $u(x)=1$ if x is rational and zero elsewhere. We therefore define the support of u for any measurable function u on X slightly differently, namely

$$\operatorname{supp}(u) := X\setminus\bigcup\Big\{U : U\subset X, U \text{ open},\ \ u(x)=0\ \ m\text{-a.e. on } U\Big\} \tag{3.2}$$

It is easy to see that the two definitions of the support coincide if u is continuous. Note that the right-hand side of equation (3.2) is exactly the definition of the support of the measure $\mu_{|u|}$ on $\mathcal{B}$ which is defined by

$$\mu_{|u|}(B) := \int_B |u|\,dm.$$

A (bounded or unbounded) operator H in $L^2(X,m)$ is called *local* if

$$\operatorname{supp}(Hu)\subset\operatorname{supp}(u)$$

for all $u\in\operatorname{dom}(H)$. Standard examples for local operators are differential operators which shall be treated in detail below. If Y is any measurable subset of X, the restriction of m to the *trace-σ-algebra*

$$\mathcal{B}' := \{Y\cap B : B\in\mathcal{A}\}$$

yields a measure m' on $\mathcal{B}'$. In this way we obtain a Hilbert space $L^2(Y,m')$ which may be viewed as a subspace of $L^2(X,m)$ in the obvious way. In what follows we omit the dashes and use also the letter m for the restriction m'.

Suppose now that H is a self-adjoint and local operator in $L^2(X,m)$ and Y is an open subset of X. One can try to define a self-adjoint operator H^Y in $L^2(Y,m)$ by starting from the restriction of H to the space of all functions in $\operatorname{dom}(H)$ which are contained in $L^2(Y,m)$, i.e., which have support contained in Y. Clearly this yields a hermitian operator in $L^2(Y,m)$. If X and Y are open subsets of $\mathbb{R}^d$, m is the Lebesgue measure and $C_c^\infty(X)$ is contained in the domain of H, it is also clear that H^Y is even symmetric. Hence H^Y is closable. Nevertheless this construction does not yield a self-adjoint operator in $L^2(Y,m)$ as is seen from the following easy example.

Example. Let $X = (-1,1) \subset \mathbb{R}$ and let H be the Dirichlet Laplacian in $L^2(X)$. Further let $Y = (0,1)$ and denote by H^Y the restriction of H to the space of all functions in the domain of H with support in Y. If $u \in \operatorname{dom}(H^Y)$, then all derivatives of u vanish on the boundary points of Y. Consequently the closure of H^Y is not a self-adjoint operator, i.e., H^Y is not essentially self-adjoint.

The crucial point of this example is of course that the restriction of the operator domain already determines the value of all derivatives of u at the boundary of Y if $u \in \operatorname{dom}(H^Y)$. This makes $\operatorname{dom}(H^Y)$ too small for essential self-adjointness. In order to obtain a self-adjoint operator one has to introduce suitable boundary conditions which turns out to be a difficult procedure in the general case of differential operators of higher order, see [RSS94], Section 2 and [Agr97]. However, in the next subsection we shall see that there is a much simpler way to define domain perturbations which is based on quadratic form techniques and avoids the above-mentioned troubles. Moreover this method of defining self-adjoint operators makes even sense in a general Hilbert space setting.

3.2. The general Hilbert space case

Let $(\mathcal{H}, \langle\cdot,\cdot\rangle)$ be an arbitrary real or complex Hilbert space and let H be a self-adjoint operator in $\mathcal{H}$ which is semi-bounded from below with spectral bound $\lambda := \inf \sigma(H)$. Let $(\mathcal{E}, \mathcal{F})$ be the non-negative closed quadratic form which corresponds to $H - \lambda$ in the sense of Theorem 2.2. We then have $\mathcal{F} = \operatorname{dom}((H-\lambda)^{1/2})$ and

$$\mathcal{E}(u,v) := \langle (H-\lambda)^{1/2}u, (H-\lambda)^{1/2}v\rangle \quad u,v \in \mathcal{F}. \tag{3.3}$$

In what follows we will use the abbreviations

$$\mathcal{E}_1(u,v) := \mathcal{E}(u,v) + \langle u,v\rangle, \quad \mathcal{E}[u] := \mathcal{E}(u,u), \quad \mathcal{E}_1[u] := \mathcal{E}_1(u,u). \tag{3.4}$$

Since $(\mathcal{E}, \mathcal{F})$ is closed, the space $(\mathcal{F}, \mathcal{E}_1(\cdot,\cdot))$ is a Hilbert space. Observe also that $\mathcal{E}$ depends on λ, a dependence which is suppressed in our notation.

Consider the following situation: Suppose that $\mathcal{G}$ is a closed subspace of the Hilbert space $(\mathcal{F}, \mathcal{E}_1(\cdot,\cdot))$. Then the form $(\mathcal{E}+\lambda, \mathcal{G})$ is (of course) semi-bounded from below and closed. Moreover it is densely defined in $\mathcal{H}^\mathcal{G} := \overline{\mathcal{G}}$, the closure being taken with respect to the weaker topology of $\mathcal{H}$. Hence there is a unique self-adjoint operator $H^\mathcal{G}$ in $\mathcal{H}^\mathcal{G}$ that corresponds to $(\mathcal{E}+\lambda, \mathcal{G})$ in the sense of Theorem 2.2.

Definition 3.1. We call the operator $H^\mathcal{G}$ the *domain perturbation of H with respect to the subspace* $\mathcal{G}$.

Let us illustrate this construction in the case of the Dirichlet Laplacian.

Example. Let Ω and Λ be open subsets of $\mathbb{R}^d$ with $\Lambda \subset \Omega$ and let $H = -\Delta_\Omega$ be the Dirichlet Laplacian in $\mathcal{H} = L^2(\Omega)$. The form $(\mathcal{E}, \mathcal{F})$ which corresponds to H is given by

$$\mathcal{F} = H_0^1(\Omega), \quad \mathcal{E}[u] = \int_\Omega |\nabla u|^2 \mathrm{d}x \quad \text{for } u \in \mathcal{F},$$

see Section 1.1. Clearly the Sobolev space $\mathcal{G} := H_0^1(\Lambda)$ may be viewed as a closed subspace of $\mathcal{F}$. Hence we can apply the above construction. By definition the Hilbert space $\mathcal{H}^\mathcal{G}$ equals the L^2-closure of $\mathcal{G}$, i.e., we have $\mathcal{H}^\mathcal{G} = L^2(\Lambda)$. Moreover the restriction of $(\mathcal{E}, \mathcal{F})$ to $\mathcal{G}$ is obviously just the form which is associated to the Dirichlet Laplacian in $L^2(\Lambda)$. Hence $H^\mathcal{G} = -\Delta_\Lambda$.

In general the space $\mathrm{dom}(H) \cap \mathcal{H}^\mathcal{G}$ is not an invariant subspace for H but nevertheless $H^\mathcal{G}$ defined via the quadratic form $(\mathcal{E}, \mathcal{G})$ is an operator in $\mathcal{H}^\mathcal{G}$. It is therefore natural to ask for a characterization of $\mathrm{dom}(H^\mathcal{G})$ as well as for the action of $H^\mathcal{G}$ on this domain. This question is partially answered by the following proposition.

Proposition 3.2. *We have* $\mathrm{dom}(H) \cap \mathcal{H}^\mathcal{G} \subset \mathrm{dom}(H^\mathcal{G})$ *and* $H^\mathcal{G} u = \Pi^\mathcal{G} H u$ *for* $u \in \mathrm{dom}(H) \cap \mathcal{H}^\mathcal{G}$, *where* $\Pi^\mathcal{G} \in B(\mathcal{H})$ *denotes the orthogonal projection onto* $\mathcal{H}^\mathcal{G}$.

Proof. By Theorem 2.2 (a) the domain of $H^\mathcal{G}$ is given by

$$\mathrm{dom}(H^\mathcal{G}) = \mathrm{dom}(H^\mathcal{G} - \lambda) = \{u \in \mathcal{G} : \exists v \in \mathcal{H}^\mathcal{G} \ \forall w \in \mathcal{G} : \ \mathcal{E}(u, w) = \langle v, w \rangle\},$$

and $(H^\mathcal{G} - \lambda)u = v$ in this case. Therefore let $u \in \mathrm{dom}(H) \cap \mathcal{H}^\mathcal{G}$, put $v := \Pi^\mathcal{G}(H - \lambda)u$ and let $w \in \mathcal{G}$ be arbitrary. Then

$$\mathcal{E}(u, w) = \langle (H - \lambda)u, w \rangle = \langle (H - \lambda)u, \Pi^\mathcal{G} w \rangle = \langle \Pi^\mathcal{G}(H - \lambda)u, w \rangle = \langle v, w \rangle.$$

Consequently $u \in \mathrm{dom}(H^\mathcal{G})$ and $(H^G - \lambda)u = v = \Pi^\mathcal{G}(H - \lambda)u = \Pi^\mathcal{G} H u - \lambda u$, which completes the proof. □

Remark 3.3. If H is bounded, then $\mathcal{F} = \mathrm{dom}(H) = \mathcal{H}$ and the topology induced by the scalar product $\mathcal{E}_1(\cdot, \cdot)$ coincides with the topology of $\mathcal{H}$. We conclude that $\mathcal{G} = \mathcal{H}^\mathcal{G}$. Therefore Proposition 3.2 gives us

$$\mathcal{G} = \mathrm{dom}(H) \cap \mathcal{H}^\mathcal{G} \subset \mathrm{dom}(H^\mathcal{G}) \subset \mathcal{G}.$$

Hence $\mathrm{dom}(H^\mathcal{G}) = \mathcal{H}^\mathcal{G}$ and $H^\mathcal{G}$ is exactly the restriction of $\Pi^\mathcal{G} H$ to $\mathcal{H}^\mathcal{G}$.

The following example shows that the domain perturbation of a bounded integral operator in $L^2(X, \mathcal{A}, m)$ to a subspace of the form $L^2(Y, \mathcal{A}, m)$ is just the integral operator which arises from the truncated kernel.

Example. Let $(X, \mathcal{A}, m)$ be a measure space and $k : X \times X \to \mathbb{C}$ be an integral kernel which induces an everywhere defined integral operator K, that is:

(i) For all $f \in L^2(X, \mathcal{A}, m)$ and almost all $x \in X$ the function $k(x, \cdot)f(\cdot)$ belongs to $L^1(X, \mathcal{A}, m)$.
(ii) For all $f \in L^2(X, \mathcal{A}, m)$ the function $\int_X k(\cdot, y)f(y)\mathrm{d}y$ belongs to the space $L^2(X, \mathcal{A}, m)$.

The induced integral operator is then given by

$$Kf = \int_X k(\cdot, y) f(y) \mathrm{d}y. \tag{3.5}$$

Because of the first condition the integral in equality (3.5) is well defined for all $f \in L^2(X, \mathcal{A}, m)$ and the second condition guarantees that K maps $L^2(X, \mathcal{A}, m)$ into itself. One can prove, see [HS78] Theorem 3.10., that every integral operator K with $\mathrm{dom}(K) = L^2(X, \mathcal{A}, m)$ is closed. Hence, by the closed graph theorem, K is bounded. In order to have self-adjointness we have to assume additionally that $k(x, y) = \overline{k(y, x)}$ for all $x, y \in X$.

Now let Y be a measurable subset of X. Denote the restriction of k to $Y \times Y$ by k^Y. Clearly k^Y induces a bounded and self-adjoint operator K^Y in $L^2(Y, \mathcal{A}, m)$. We claim that K^Y coincides with the operator we obtain by the procedure described in the beginning of this section. For the proof first note that $\mathcal{F} = L^2(X, \mathcal{A}, m)$ since K is bounded. Hence $\mathcal{G} := L^2(Y, \mathcal{A}, m)$ is a closed subspace of $\mathcal{F}$ and we may define the self-adjoint operator $K^{\mathcal{G}}$ according to Definition 3.1. By Remark 3.3 we have

$$\mathrm{dom}(K^{\mathcal{G}}) = L^2(Y, \mathcal{A}, m)$$

and $K^{\mathcal{G}} u = \Pi^{\mathcal{G}} K u$ for all $u \in L^2(Y, \mathcal{A}, m)$, where $\Pi^{\mathcal{G}} \in B(L^2(X, \mathcal{A}, m))$ is the orthogonal projection onto $L^2(Y, \mathcal{A}, m)$. We write $\Pi^{\mathcal{G}} u = \chi_Y u$ with χ_Y being the characteristic function of Y. Putting all these facts together we obtain for $u \in L^2(Y, \mathcal{A}, m)$,

$$\begin{aligned} K^{\mathcal{G}} u &= \chi_Y K u \\ &= \chi_Y \int_X k(\cdot, y) u(y) \mathrm{d}y \\ &= \int_Y \chi_Y k(\cdot, y) u(y) \mathrm{d}y \\ &= \int_Y k^Y(\cdot, y) u(y) \mathrm{d}y \\ &= K^Y u. \end{aligned}$$

Remark 3.4. The preceding example suggests the following method for estimating the shift of the eigenvalues of differential operators which are subjected to a domain perturbation: Suppose that H is a self-adjoint and semi-bounded differential operator in $L^2(\Omega)$ with Ω an open subset of $\mathbb{R}^d$. By the functional calculus the spectrum of the semigroup $(e^{-tH})_{t>0}$ is given by $\sigma(e^{-tH}) = e^{-t\sigma(H)}$. In particular, if $\sigma(H)$ is discrete, then so is the spectrum of e^{-tH} in each interval (ε, ∞). Therefore the semigroup carries the same spectral information as H and it seems to be easier to analyze the semigroup rather than the differential operator H itself, since then one has to deal only with bounded operators.

If, in addition, the semigroup is ultracontractive, i.e., each e^{-tH} maps $L^2(\Omega)$ boundedly into $L^\infty(\Omega)$, one can easily conclude from the Dunford–Pettis Theorem, cf. [Tre67], that each e^{-tH} is an integral operator whose kernel $k_t(\cdot,\cdot)$ satisfies

$$\|e^{-tH}\|_{B(L^2,L^\infty)} = \sup_{x\in\Omega}\sqrt{\int_X |k_t(x,y)|^2 \mathrm{d}y}.$$

This result is sometimes called Korotkov's Theorem [Kor65] although the Dunford–Pettis Theorem is much older. It is well known, cf. [Sim82], that ultracontractivity holds in great generality for a large class of Schrödinger operators, in particular it holds for the Laplacian.

The idea for estimating the perturbed eigenvalues is the following: If λ is an eigenvalue of H, then $e^{-t\lambda}$ is an eigenvalue of e^{-tH}. Hence it might be easier to treat the domain perturbation of the bounded integral operator e^{-tH} than the perturbation of H itself while also giving estimates for λ. But this does not work since the perturbed semigroup is *not* the semigroup which corresponds to the perturbed operator, i.e., with our notation we have

$$e^{-tH^{\mathcal{G}}} \neq \left(e^{-tH}\right)^{\mathcal{G}}. \tag{3.6}$$

In the case of the Dirichlet Laplacians $H = -\Delta_\Omega$, $H^{\mathcal{G}} = -\Delta_\Lambda$ with open and bounded sets $\Lambda, \Omega \subset \mathbb{R}^d$ satisfying $\overline{\Lambda} \subset \Omega$, cf. Example 3.2, this can be seen by the stochastic interpretation of the semigroup via the Hunt process which corresponds to Brownian motion in $\mathbb{R}^d$, see Appendix C. Given any subset A of $\mathbb{R}^d$, define the first hitting time of A by

$$\tau_A := \inf\{t > 0 : X_t \in A\}.$$

For $x \in \Lambda$ we have

$$\begin{aligned}\left(e^{-tH^{\mathcal{G}}}\chi_\Lambda\right)(x) &= (e^{t\Delta_\Lambda}\chi_\Lambda)(x)\\ &= \mathbb{E}_x\{\chi_\Lambda(X_t) : \tau_{\Lambda^c} > t\}\\ &= \mathbb{P}_x\{\tau_{\Lambda^c} > t, X_t \in \Lambda\}\\ &= \mathbb{P}_x\{\tau_{\Lambda^c} > t, X_t \in \Lambda, \tau_{\Omega^c} > t\}.\end{aligned}$$

Our assumption $\overline{\Lambda} \subset \Omega$ implies that $\mathbb{P}_x\{\tau_{\Lambda^c} \le t, X_t \in \Lambda, \tau_{\Omega^c} > t\} > 0$. Hence the calculation continues:

$$\begin{aligned}\left(e^{-tH^{\mathcal{G}}}\chi_\Lambda\right)(x) &< \mathbb{P}_x\{\tau_{\Lambda^c} > t, X_t \in \Lambda, \tau_{\Omega^c} > t\} + \mathbb{P}_x\{\tau_{\Lambda^c} \le t, X_t \in \Lambda, \tau_{\Omega^c} > t\}\\ &= \mathbb{P}_x\{\tau_{\Omega^c} > t, X_t \in \Lambda\}\\ &= \mathbb{E}_x\{\chi_\Lambda(X_t) : \tau_{\Omega^c} > t\}\\ &= \left(e^{t\Delta_\Omega}\chi_\Lambda\right)(x).\end{aligned}$$

Since $x \in \Lambda$, we end up with

$$\left(e^{-tH^{\mathcal{G}}}\chi_\Lambda\right)(x) < \chi_\Lambda(x)\left(e^{t\Delta_\Omega}\chi_\Lambda\right)(x) = \left(\left(e^{-tH}\right)^{\mathcal{G}}\chi_\Lambda\right)(x).$$

In particular we get (3.6). Therefore we will not discuss this idea further.

Finally we prove a generalization of Dynkin's formula, which has been published in [Nol97b]. To this aim we first take a closer look at Dynkin's formula from a functional analytic point of view. Let $(\mathcal{E}, \mathcal{F})$ be a regular Dirichlet form in $L^2(X, m)$ with corresponding self-adjoint operator H. See Appendix C.1 for a brief introduction to regular Dirichlet forms. Further let Y be an open subset of X such that its complement $K = X \setminus Y$ is compact. By Theorem C.20 (a) the restriction of $\mathcal{E}$ to the space

$$\mathcal{F}_K := \{u \in \mathcal{F} : \tilde{u} = 0 \text{ q.e. on } K\}$$

corresponds to a unique self-adjoint and non-negative operator H^Y in $L^2(Y, m)$. Dynkin's formula (see Theorem C.23) entails that

$$(H+1)^{-1} f = (H^Y + 1)^{-1}(f_{|Y}) + U_1 \mu_K^{(H+1)^{-1} f}, \tag{3.7}$$

valid for all non-negative $f \in L^2(X, m)$. Here $\mu_K^{(H+1)^{-1} f}$ is the unique Radon measure of finite energy integral which is associated to K and $(H+1)^{-1} f$ in the sense of Theorem C.23 (a) and $U_1 \mu_K^{(H+1)^{-1} f}$ denotes its 1-potential. In view of Theorem C.20 (b) we have

$$U_1 \mu_K^{(H+1)^{-1} f} = P_{\mathcal{F}_K^\perp}(H+1)^{-1} f,$$

hence equality (3.7) becomes

$$(H+1)^{-1} f = (H^Y + 1)^{-1}(f_{|Y}) + P_{\mathcal{F}_K^\perp}(H+1)^{-1} f.$$

This equation can be generalized to our Hilbert space setting of domain perturbations.

Theorem 3.5. *Let H be an arbitrary self-adjoint and non-negative operator in $\mathcal{H}$ with corresponding form $(\mathcal{E}, \mathcal{F})$. Further let $\mathcal{G}$ be any closed subspace of $(\mathcal{F}, \mathcal{E}_1(\cdot,\cdot))$ and $H^{\mathcal{G}}$ its associated domain perturbation in the sense of Definition* 3.1. *Then, for any $f \in \mathcal{H}$ the following assertions hold true.*

(a) *Let $\mathcal{H}^{\mathcal{G}} = \overline{\mathcal{G}}$ be the Hilbert space in which $H^{\mathcal{G}}$ acts and denote by $\Pi^{\mathcal{G}} \in B(\mathcal{H})$ the orthogonal projection onto $\mathcal{H}^G$. Then*

$$(H+1)^{-1} f = (H^{\mathcal{G}} + 1)^{-1} \Pi^{\mathcal{G}} f + P_{\mathcal{G}^\perp}(H+1)^{-1} f,$$

(b) $\mathcal{E}_1\left((H^{\mathcal{G}} + 1)^{-1} f, P_{\mathcal{G}^\perp}(H+1)^{-1} f\right) = 0.$

Proof. (a) Put $u := P_{\mathcal{G}}(H+1)^{-1} f$ and let $v \in \mathcal{G}$ be arbitrary. Then

$$\mathcal{E}_1(u, v) = \mathcal{E}_1(P_{\mathcal{G}}(H+1)^{-1} f, v) = \mathcal{E}_1((H+1)^{-1} f, v) = \langle f, v \rangle = \langle \Pi^{\mathcal{G}} f, v \rangle.$$

This implies $u \in \operatorname{dom}(H^{\mathcal{G}})$ and $(H^{\mathcal{G}} + 1)u = \Pi^{\mathcal{G}} f$, i.e.,

$$\begin{aligned} (H+1)^{-1} f &= P_{\mathcal{G}}(H+1)^{-1} f + P_{\mathcal{G}^\perp}(H+1)^{-1} f \\ &= u + P_{\mathcal{G}^\perp}(H+1)^{-1} f \\ &= (H^{\mathcal{G}} + 1)^{-1} \Pi^{\mathcal{G}} f + P_{\mathcal{G}^\perp}(H+1)^{-1} f. \end{aligned}$$

Part (b) is clear since $(H^{\mathcal{G}} + 1)^{-1} f \in \mathcal{G}$ and $P_{\mathcal{G}^\perp}(H+1)^{-1} f \in \mathcal{G}^\perp$. □

3.3. Defining domain perturbations by means of the associated stochastic process

Only for the sake of completeness we mention another method of defining domain perturbations which works in the context of regular Dirichlet forms. This construction uses the fact that there is a one-to-one correspondence between regular Dirichlet forms and Hunt processes, see Theorem C.18. Hence one may use the process to define domain perturbations. The idea is to define a new process by "killing" the sample paths when leaving the complement of the set K on which the domain perturbation takes place. Therefore this process is sometimes called the *process killed on* K.

To be precise, assume that $(\mathcal{E},\mathcal{F})$ is a regular Dirichlet form in $L^2(X,m)$ with associated self-adjoint operator H, see Appendix C.1. Let

$$M=(\Omega,\mathcal{M},X_t,(\mathbb{P}_x)_{x\in X_\triangle})$$

be the corresponding m-symmetric Hunt process on X. Given a closed subset K of X, the operation of killing the sample paths upon leaving $Y := X\setminus K$ is defined as follows. Let $Y_\triangle$ be the one-point compactification of Y. When Y is already compact, $\triangle$ is adjoint as an isolated point. Put

$$X_t^Y(\omega):=\begin{cases} X_t(\omega), & \text{if } 0\le t\le \inf\{t\ge 0: X_t\in K\},\\ \triangle, & \text{otherwise.}\end{cases}$$

The quadruple $M^Y := (\Omega,\mathcal{M},X_t^Y,(\mathbb{P}_x)_{x\in Y_\triangle})$ is called the *part of* M *on the subset* Y. It turns out that this construction again yields a Hunt process whose transition function can be given explicitly.

Theorem 3.6. *M^Y is a Hunt process on Y with transition function*

$$p_t^Y(x,E)=\mathbb{P}_x\{X_t\in E,\sigma_K>t\},\quad E\in\mathcal{B}(Y_\triangle)$$

where

$$\sigma_K=\inf\{t>0: X_t\in K\}$$

is the first hitting time of K.

For a proof we refer to [FOT94], Theorem A.2.10.

Denote by $(\mathcal{E}^Y,\mathcal{F}^Y)$ the regular Dirichlet form of M^Y and let H^Y be the self-adjoint operator in $L^2(Y,m)$ which corresponds to $(\mathcal{E}^Y,\mathcal{F}^Y)$ in the sense of Theorem 2.2, see also the discussion after Theorem C.17 for details. Hence the operation of killing the sample paths upon leaving Y is another method of defining domain perturbations. It is therefore a natural question to ask for an analytic description of H^Y. The following theorem gives the answer.

Theorem 3.7. *In the situation just described let*

$$\mathcal{F}_K := \{u\in\mathcal{F}:\tilde{u}=0 \text{ q.e. on } K\}.$$

Then $H^Y = H^{\mathcal{F}_K}$, i.e., the self-adjoint operator H^Y defined by killing the sample paths upon leaving Y coincides with the domain perturbation $H^{\mathcal{F}_K}$ of H with respect to the subspace $\mathcal{F}_K$, see Definition 3.1.

The proof is carried out in [FOT94], Theorem 4.4.3.

3.4. Preliminary remarks on the spectrum of the perturbed operator

Let $H^{\mathcal{G}}$ be the domain perturbation with respect to the subspace $\mathcal{G}$ which was introduced in Definition 3.1. Even without further specifying the unperturbed operator H one can obtain some information about the spectrum of $H^{\mathcal{G}}$. The minimax principle (see Appendix A) gives the following information on the discreteness of the spectrum (recall that the spectrum is said to be purely discrete if it consists only of isolated eigenvalues of finite multiplicity): $H^{\mathcal{G}}$ has purely discrete spectrum if and only if the sequence

$$\mu_n := \inf\left\{\sup\left\{\frac{\mathcal{E}[u]}{\|u\|^2} : u \in \mathcal{L}\setminus\{0\}\right\} : \mathcal{L} \subset \mathcal{G}, \dim\mathcal{L} = n\right\}, \quad n \geq 1 \tag{3.8}$$

tends to infinity as $n \to \infty$. Clearly this happens if H has purely discrete spectrum.

From equation (3.8) we also conclude that $\mu_n \geq \lambda_n$, where

$$\lambda_n := \inf\left\{\sup\left\{\frac{\mathcal{E}[u]}{\|u\|^2} : u \in \mathcal{L}\setminus\{0\}\right\} : \mathcal{L} \subset \mathcal{F}, \dim\mathcal{L} = n\right\}, \quad n \geq 1$$

is the corresponding sequence for H. This means that the eigenvalues of H (if there are eigenvalues at all) can only be shifted to the right by passing from H to $H^{\mathcal{G}}$. It is therefore a natural question to ask for the magnitude of this shift in terms of the subspace $\mathcal{G}$. If $\mathcal{H} = L^2(X, \mathcal{A}, m)$ and $\mathcal{G}$ is such that $\mathcal{H}^{\mathcal{G}} = L^2(Y, \mathcal{A}, m)$ with some measurable subset Y of X, this leads to the question of giving estimates on the shift of eigenvalues by geometric properties of Y. In the special case of the Dirichlet Laplacians $H = -\Delta_\Omega$ and $H^{\mathcal{G}} = -\Delta_\Lambda$, see Example 3.2, we retain the physical problem of measuring the improvement in cooling efficiency resulting from the cooler $\Omega \setminus \Lambda$ which served in Section 1.4 as a motivation for considering this kind of problems.

As we already noted in Section 1.4 it is the capacity which plays the decisive role in estimating the perturbed eigenvalues of the Dirichlet Laplacian and we are heading towards a generalization of the results stated there to our general type of domain perturbations. To this aim we of course need a notation of capacity appropriate to this level of generality.

4. Several approaches to capacity

In this section we introduce several notions of capacity and compare them with each other. As a motivation the first subsection gives a rather heuristic definition of capacity which is motivated by an electrostatic consideration. The following Section 4.2 treats the capacity of a regular Dirichlet form. It is shown that one may look at this capacity as an orthogonal projection in the associated Dirichlet space. This observation is used in Section 4.3 as the starting point for defining the capacity of a subspace in a general Hilbert space setting. We also show that this capacity may be expressed as an infimum in the case of L^2-spaces. As a consequence we prove that the general capacity coincides with the capacity of a regular Dirichlet form in this special case.

Finally in Section 4.4 we give a brief overview on certain other capacities, but we restrict ourselves to the capacity of a non-regular, non-symmetric Dirichlet form and to the higher-order capacity in the sense of Maz'ja.

The following table gives a short overview of the different capacities and their associated operators which are treated in this thesis.

Section	Hilbert space	Operators
4.1	$L^2(\mathbb{R}^d)$	$-\Delta$
4.2	real $L^2(X, m)$, X a locally compact metric space	H which corresponds to a regular symmetric Dirichlet form
4.3	$(\mathcal{H}, \langle\cdot,\cdot\rangle)$ arbitrary, real or complex	H self-adjoint and semi-bounded
4.4.1	real $L^2(X, m)$, X a Hausdorff space	H which corresponds to a (non-regular, non-symmetric) Dirichlet form
4.4.2	complex $L^2(\Omega)$, $\Omega \subset \mathbb{R}^d$ open	H an elliptic differential operator of arbitrary order

4.1. Definition of the electrostatic capacity

Let $d \geq 3$ and consider a compact and connected subset K of $\mathbb{R}^d$ together with an electric charge distribution μ on K. This means that μ is a Radon measure supported on K with finite total charge $Q := \mu(K)$. Then μ induces an electrostatic potential U^μ and a force $F = -\nabla U^\mu$. According to the laws of physics, see e.g., [Jac62] Section 1, U^μ satisfies Poisson's equation

$$\Delta U^\mu = \mu$$

which has to be understood in the sense of distributions. Therefore $U^\mu = \mu * E_d$, where E_d is any fundamental solution of the Laplacian. One usually takes

$$E_d(x) = c_d|x|^{2-d}.$$

Here c_d is a constant whose exact value is given by

$$c_d = \frac{\omega_d}{2-d} = \frac{\Gamma\left(\frac{d}{2}\right)}{2(2-d)\pi^{d/2}},$$

and ω_d is the surface area of the unit ball in $\mathbb{R}^d$. See [PS78], Section 3.1 or [Joh71], p. 99 for further information. The total energy $E(\mu)$ stored in the system is given by

$$E(\mu) = \int_{\mathbb{R}^d} |\nabla U^\mu|^2 \mathrm{d}x.$$

If $|\nabla U^\mu|^2$ does not belong to $L^1(\mathbb{R}^d)$ the convention $E(\mu) = \infty$ applies. Let s be any real number. By elliptic regularity, see Appendix B or [Rud73], we have $U^\mu \in H^{2+s}_{loc}(\mathbb{R}^d)$, whenever $\mu \in H^s_{loc}(\mathbb{R}^d)$. If, in particular, $\mu = \rho\mathrm{d}x$ with some bounded function ρ, supported in K, we have $U^\mu \in H^2_{loc}(\mathbb{R}^d)$. Hence $|\nabla U^\mu|^2 \in L^1_{loc}(\mathbb{R}^d)$.

Moreover it is easy to see that $|\nabla U^\mu|$ decays like $|x|^{1-d}$ as x tends to infinity which implies $|\nabla U^\mu|^2 \in L^1(\mathbb{R}^d)$ since $d \geq 3$.

If K is a conductor it is clear for physical reasons that the charge μ will distribute itself freely on K to become the *equilibrium measure* $\nu_{K,Q}$ which will only depend on the shape of K and on the total charge Q but is independent of μ. Of course we have $\nu_{K,Q}(K) = Q$. The equilibrium measure $\nu_{K,Q}$ determines the equilibrium potential $e_{K,Q} := U^{\nu_{K,Q}} = \nu_{K,Q} * E_d$. If the system is in its equilibrium state a unit test charge at $x \in K$ does not experience any force, i.e., $\nabla e_{K,Q} = 0$ on K. Therefore $e_{K,Q}$ must be constant on K. Denote this constant value by $\alpha(K, Q)$ and define the (zero-order-)*capacity* of K by

$$\mathrm{Cap}_0(K) := \frac{1}{\alpha(K,1)}.$$

We use the index 0 to distinguish this capacity from the more general capacities we shall define later on. If $Q = \mathrm{Cap}_0(K)$ and $x \in K$, then

$$e_{K,Q}(x) = \alpha(K, \mathrm{Cap}_0(K)) = \mathrm{Cap}_0(K)\alpha(K,1) = 1.$$

Thus $\mathrm{Cap}_0(K)$ is the charge one has to put on K in order to make the equilibrium potential equal to one on K. By convention one normalizes the equilibrium potential in such a way that its total mass equals the capacity, i.e., instead of $\nu_{K,Q}$ one considers $\nu_K := \nu_{K,\mathrm{Cap}(K)}$ and $e_K := \nu_K * E_d$. Then

$$E(\nu_K) = \int_{\mathbb{R}^d} |\nabla e_K|^2 \mathrm{d}x = \int_{\mathbb{R}^d} e_K \mathrm{d}\nu_K = \nu_K(K) = \mathrm{Cap}_0(K).$$

Therefore the capacity of K equals the energy of the system being in its equilibrium state. On the other hand the initial charge distribution μ will distribute itself in such a way that minimizes the energy stored in the system. Thus

$$\mathrm{Cap}_0(K) = \inf\{E(\mu) : U^\mu \geq 1 \text{ on } K\} = \inf\left\{\int_{\mathbb{R}^d} |\nabla U^\mu|^2 \mathrm{d}x : U^\mu \geq 1 \text{ on } K\right\},$$

and one can prove that in the last expression the potentials U^μ may be replaced by smooth functions. Thus

$$\mathrm{Cap}_0(K) = \inf\left\{\int_{\mathbb{R}^d} |\nabla \phi|^2 \mathrm{d}x : \phi \in C_c^\infty(\mathbb{R}^d), \phi \geq 1 \text{ on } K\right\}. \tag{4.1}$$

In fact the unique minimizer of this infimum turns out to be U^{ν_K}.

4.2. The capacity of a regular Dirichlet form

In this section we recall the definition of the capacity associated to a regular Dirichlet form. With the aid of the corresponding Hunt process we show that one may look at the capacity as an orthogonal projection in the form domain. This observation serves as a motivation for defining the capacity in general Hilbert spaces which is done in Section 4.3.

Let X be a locally compact separable metric space and let m be a strictly positive Radon measure on X. Fix a self-adjoint and semi-bounded operator H in $L^2(X, m)$. By Theorem 2.2 there is a unique quadratic form $(\mathcal{E}, \mathcal{F})$ that corresponds to H. Throughout this subsection we assume that $(\mathcal{E}, \mathcal{F})$ is a regular Dirichlet form in $L^2(X, m)$. A comprehensive treatment of the theory of Dirichlet forms can be found in [Fuk80], [FOT94], [BH91] or [MR92]. The most important results from the spectral theoretic point of view are summarized in Appendix C.1.

The (first-order) capacity of an open subset U of X is defined by

$$\mathrm{Cap}(U) := \inf\{\mathcal{E}_1[u] : u \in \mathcal{F},\ u \geq 1 \ m\text{-a.e. on } U\}, \tag{4.2}$$

where $\mathcal{E}_1[u] := \mathcal{E}(u, u) + \|u\|^2_{L^2(X,m)}$. If there is no such u the convention $\mathrm{Cap}(U) = \infty$ applies. This definition is extended to arbitrary subsets of X by

$$\mathrm{Cap}(A) := \inf\{\mathrm{Cap}(U) : U \supset A, U \text{ open}\}.$$

Recall that a statement is said to hold *quasi-everywhere* (q.e.), if it holds outside a set of zero capacity. Similarly an element $u \in \mathcal{F}$ is called *quasi-continuous* if for each $\varepsilon > 0$ there is an open set $U_\varepsilon \subset X$ such that $\mathrm{Cap}(U_\varepsilon) < \varepsilon$ and the restriction of u to U_ε is continuous. It is an important consequence of the regularity of $(\mathcal{E}, \mathcal{F})$ that each $u \in \mathcal{F}$ admits a quasi-continuous m-version $\tilde{u}$, i.e., there is a quasi-continuous function $\tilde{u}$ that coincides with u m-almost everywhere. In view of Theorem C.17 (d) the capacity of an arbitrary subset A of X may be expressed as

$$\mathrm{Cap}(A) = \inf\{\mathcal{E}_1[u] : u \in \mathcal{F}, \tilde{u} \geq 1 \text{ q.e. on } A\},$$

and by Theorem C.17 (e) this infimum is attained for some unique element $e_A \in \mathcal{F}$. The function e_A is called the (first-order) *equilibrium potential* of the set A.

According to Theorem C.18 there is always a unique Hunt process $M = (\Omega, \mathcal{M}, (X_t)_{t>0}, (\mathbb{P}_x)_{x\in X})$ which corresponds to $(\mathcal{E}, \mathcal{F})$ and H in the sense that

$$(e^{-tH} f)(x) = \mathbb{E}_x\{f(X_t)\}$$

for all $f \in L^2(X, m)$, where $\mathbb{E}_x$ denotes the expectation with respect to $\mathbb{P}_x$. See Appendix C for a brief introduction to the theory of Markov processes and Hunt processes. We next show that M can be used to give a stochastic interpretation of the equilibrium potential. The following theorem was published in [DMN97] and its proof is based on ideas taken from [DC].

Theorem 4.1. *Let $B \subset X$ be a Borel set of finite capacity and let*

$$\tau_B := \inf\{t > 0 : X_t \in B\}$$

be the first hitting time of B. Then

$$e_B(x) = \mathbb{E}_x\{e^{-\tau_B}\}. \tag{4.3}$$

Proof. We first show that $v_B(x) := \mathbb{E}_x\{e^{-\tau_B} : \tau_B < \infty\}$ is a 1-excessive function, i.e., $e^{-tH}v_B \leq e^t v_B$ for each $t > 0$. This is seen from

$$\begin{aligned}(e^{-tH}v_B)(x) &= \mathbb{E}_x\{v_B(X_t)\}\\ &= \mathbb{E}_x\{v_B(X_t) : \tau_B \leq t\} + \mathbb{E}_x\{v_B(X_t) : t < \tau_B < \infty\}\\ &\leq \mathbb{P}_x\{\tau_B \leq t\} + \mathbb{E}_x\{v_B(X_t) : t < \tau_B < \infty\}\\ &\leq e^t\,\mathbb{E}_x\{e^{-\tau_B} : \tau_B \leq t\} + \mathbb{E}_x\{\mathbb{E}_{X_t}\{e^{-\tau_B} : \tau_B < \infty\} : t < \tau_B < \infty\}\\ &\leq e^t\,\mathbb{E}_x\{e^{-\tau_B} : \tau_B \leq t\} + \mathbb{E}_x\{e^{-(\tau_B - t)} : t < \tau_B < \infty\}\\ &= e^t v_B(x).\end{aligned}$$

According to Theorem C.17 (e) the equilibrium potential e_B is also 1-excessive. Because of Theorem C.13 (b) it suffices to show

$$v_B \leq e_B \tag{4.4}$$

and

$$v_B = 1 \text{ q.e. on } B. \tag{4.5}$$

Then we obtain $v_B \in \mathcal{F}$ as well as $\mathcal{E}_1[v_B] \leq \mathcal{E}_1[e_B]$, and we can conclude from (4.5) and Theorem C.17 (e) that $e_B = v_B$. That means it remains to prove (4.4) and (4.5).

To get (4.5) suppose that there is some $x_0 \in B$ such that $v_B(x_0) \neq 1$. Then $\mathbb{P}_{x_0}\{\tau_B > 0\} > 0$, i.e., x_0 is an irregular point of B. (4.5) follows since the set of irregular points is a set of zero capacity by Theorem C.25.

For the proof of (4.4) choose an open set $U \subset X$ and consider the random variables

$$Y_t := e^{-t} e_U(X_t), \quad (t \geq 0).$$

We claim that for $x \in X$ fixed, $(\Omega, \mathcal{M}, \mathcal{M}_t, Y_t, \mathbb{P}_x)$ is a supermartingale, see Definition C.6 (a). This follows from the Markov property and the fact that e_U is 1-excessive by

$$\begin{aligned}\mathbb{E}_x\{Y_s|\mathcal{M}_t\} &= e^{-s}\mathbb{E}_x\{e_U(X_s)|\mathcal{M}_t\}\\ &= e^{-s}\mathbb{E}_{X_t}\{e_U(X_{s-t})\}\\ &= e^{-s}\left(e^{-(s-t)H}e_U\right)(X_t)\\ &\leq e^{-s}e^{s-t}e_U(X_t)\\ &= Y_t,\end{aligned}$$

where $s \geq t \geq 0$. Since this supermartingale is clearly of class (DL), see Definition C.6 (b), we get from Doob's optional sampling theorem (Theorem C.7)

$$\mathbb{E}_x\{Y_{s\wedge\tau}|\mathcal{M}_t\} \leq Y_{t\wedge\tau},$$

valid for any $(\mathcal{M}_t)$-stopping time τ. In particular we may choose $\tau = \tau_U$, $t = 0$ and get

$$\mathbb{E}_x\{Y_{s\wedge\tau_U}\} = \mathbb{E}_x\{\mathbb{E}_x\{Y_{s\wedge\tau_U}|\mathcal{M}_0\}\} \leq \mathbb{E}_x\{Y_0\} = e_U(x).$$

By letting s tend to infinity the desired inequality (4.4) follows for open subsets of X. For the general case choose a decreasing sequence $(U_n)_{n\in\mathbb{N}}$ of open subsets of X such that $U_n \supset B$ and $\mathrm{Cap}(U_n) \to \mathrm{Cap}(B)$. Then $e_{U_n} \to e_B$ in $(\mathcal{F}, \mathcal{E}_1(\cdot,\cdot))$ since

$$\begin{aligned}\mathcal{E}_1[e_{U_n} - e_B] &= \mathcal{E}_1[e_{U_n}] + \mathcal{E}_1[e_B] - 2\mathcal{E}_1(e_{U_n}, e_B)\\ &= \mathrm{Cap}(U_n) + \mathrm{Cap}(B) - 2\int_X \tilde{e_{U_n}}\, d\nu_B\\ &= \mathrm{Cap}(U_n) - \mathrm{Cap}(B),\end{aligned}$$

where ν_B is the equilibrium measure of B, see Theorem C.17 (f) and (C.4). Consequently the inequality

$$v_B \le v_{U_n} \le e_{U_n}$$

carries over to the limit and yields (4.4), completing the proof. □

We are now heading towards a description of the capacity as an orthogonal projection in the Hilbert space $(\mathcal{F}, \mathcal{E}_1(\cdot,\cdot))$. Clearly

$$\mathcal{F}_B := \{u \in \mathcal{F} : \tilde{u} = 0 \text{ q.e. on } B\} \tag{4.6}$$

is a closed subspace of $(\mathcal{F}, \mathcal{E}_1(\cdot,\cdot))$. Therefore we have the orthogonal decomposition

$$\mathcal{F} = \mathcal{F}_B \oplus \mathcal{F}_B^{\perp}. \tag{4.7}$$

The orthogonal projection $P_{\mathcal{F}_B^{\perp}}$ onto $\mathcal{F}_B^{\perp}$ admits a description via M: By Theorem C.20 we have for each $u \in \mathcal{F}$,

$$(P_{\mathcal{F}_B^{\perp}} u)(\cdot) = \mathbb{E}_{(\cdot)}\{e^{-\tau_B}\tilde{u}(X_{\tau_B})\}.$$

Using the nice continuity properties of the sample paths of M, see Definition C.3, it is then easy to see that

$$(P_{\mathcal{F}_B^{\perp}} u)(\cdot) = \mathbb{E}_{(\cdot)}\{e^{-\tau_B}\}, \tag{4.8}$$

provided that $u = 1$ on some neighborhood of B. But the right-hand side of identity (4.8) is exactly the description (4.3) of the equilibrium potential e_B. Hence $e_B = P_{\mathcal{F}_B^{\perp}} u$ and

$$\mathrm{Cap}(B) = \mathcal{E}_1[e_B] = \mathcal{E}_1[P_{\mathcal{F}_B^{\perp}} u]. \tag{4.9}$$

Equation (4.9) entails that in order to compute the capacity of the set B one can do the following: Consider the space $\mathcal{F}_B$ as in (4.6), define $\mathcal{F}_B^{\perp} := \mathcal{F} \ominus \mathcal{F}_B$ and choose any function u in the form domain which is equal to 1 in some neighborhood of B. Then evaluate the quadratic form $\mathcal{E}_1$ at the projection of u to that subspace. From this point of view it is no longer necessary to restrict oneself to regular Dirichlet forms. Hence one can define a more general version of capacity. The next subsection treats this problem in detail.

4.3. Capacity in general Hilbert spaces

Before we come to the main part of the article, let us briefly illustrate the scope of regular Dirichlet forms. The well-known Beurling–Deny formulae, see [FOT94] Theorem 3.2.1, [Fuk80], Section 2.2. or [MR92], Theorem II.2.8 state that every regular Dirichlet form $(\mathcal{E}, \mathcal{F})$ can uniquely be expressed as the sum of three forms which are called the *diffusion part, jump part* and *killing part* respectively. These names come from the stochastic interpretation of the Dirichlet form. For example the paths are continuous if and only if the jump part equals zero. If $X = \Omega$ is an open subset of $\mathbb{R}^d$ and $\mathcal{F} \subset C_c^\infty(\Omega)$, one can even prove that

$$\begin{aligned} \mathcal{E}(u,v) &= \sum_{i,j=1}^{d} \int_\Omega \frac{\partial u}{\partial x_i} \frac{\partial v}{\partial x_j} \mathrm{d}\nu_{ij} \\ &\quad + \int_{(\Omega\times\Omega)\setminus\triangle} (u(x) - u(y))(v(x) - v(y))\, \mathrm{d}J(x,y) + \int_\Omega uv\, \mathrm{d}k, \end{aligned} \tag{4.10}$$

where ν_{ij} and k are uniquely determined Radon measures on Ω, $\triangle = \{(x,x) : x \in \Omega\}$ is the diagonal of $\Omega \times \Omega$, and J is a symmetric Radon measure on the product space $(\Omega \times \Omega) \setminus \triangle$ which is also uniquely determined by $(\mathcal{E}, \mathcal{F})$.

This representation shows that the class of operators which can be treated by using regular Dirichlet forms is very restricted. In particular differential operators of order greater than 2 cannot be included.

We now want to generalize the concept of capacity to the general Hilbert space setting. For this we assume as in Section 3.2 that $(\mathcal{H}, \langle\cdot,\cdot\rangle)$ is an arbitrary real or complex Hilbert space and H is a self-adjoint operator in $\mathcal{H}$ which is semi-bounded from below with spectral bound $\lambda := \inf \sigma(H)$. Define $(\mathcal{E}, \mathcal{F})$ as the non-negative closed quadratic form which corresponds to $H - \lambda$. Using the abbreviations (3.4) the definition of the abstract capacity reads as follows.

Definition 4.2. Let $\mathcal{G}$ be a closed subspace of $(\mathcal{F}, \mathcal{E}_1(\cdot,\cdot))$ and let $u \in \mathcal{F}$. The *u-capacity of $\mathcal{G}$* is defined by

$$\mathrm{Cap}_u(\mathcal{G}) := \mathcal{E}_1[P_\mathcal{G} u],$$

where $P_\mathcal{G}$ is the orthogonal projection onto $\mathcal{G}$ in $(\mathcal{F}, \mathcal{E}_1(\cdot,\cdot))$.

In view of equation (4.9) this is a quite natural definition and we have the following result.

Proposition 4.3. *If H is such that $(\mathcal{E}, \mathcal{F})$ is a regular Dirichlet form in $L^2(X, m)$, $B \subset X$ is any Borel set and $u \in \mathcal{F}$ is such that $u = 1$ in some neighborhood of B, then*

$$\mathrm{Cap}(B) = \mathrm{Cap}_u(\mathcal{F}_B^\perp),$$

where $\mathcal{F}_B$ and $\mathcal{F}_B^\perp$ are as in (4.6) and (4.7).

Unlike the zero-order capacity (4.1) the capacity of Definition 4.2 does not have any scaling properties. In Section 7 we shall define a zero-order version in the general Hilbert space setting. In this context is proved a scaling property for this capacity in the case of homogeneous differential operators.

Let us summarize some simple facts about the u-capacity which follow immediately from its definition.

Remark 4.4. Let $\mathcal{G}$ be a closed subspace of $(\mathcal{F}, \mathcal{E}_1(\cdot,\cdot))$ and let $u \in \mathcal{F}$.

(a) $0 \le \mathrm{Cap}_u(G) \le \mathcal{E}_1[u]$. Moreover $\mathrm{Cap}_u(\mathcal{G}) = 0$ if and only if $u \in \mathcal{G}^\perp$ and $\mathrm{Cap}_u(\mathcal{G}) = \mathcal{E}_1[u]$ if and only if $u \in \mathcal{G}$. Notice that "$\perp$" is meant with respect to $\mathcal{E}_1(\cdot,\cdot)$ rather than $\langle\cdot,\cdot\rangle$.
(b) $\mathrm{Cap}_u(\mathcal{G}) + \mathrm{Cap}_u(\mathcal{G}^\perp) = \mathcal{E}_1[u]$.
(c) $\mathrm{Cap}_u(\mathcal{G}_1) \le \mathrm{Cap}_u(\mathcal{G}_2)$ if $\mathcal{G}_1$ and $\mathcal{G}_2$ are closed subspaces of $(\mathcal{F}, \mathcal{E}_1(\cdot,\cdot))$ such that $\mathcal{G}_1 \subset \mathcal{G}_2$.
(d) If u is a normalized eigenvector of H with eigenvalue $\lambda = \inf \sigma(H)$, then $\mathrm{Cap}_u(\mathcal{G}) \in [0,1]$. Moreover
$$\mathrm{Cap}_u(\mathcal{G}) = \langle u, P_{\mathcal{G}} u\rangle$$
in this case.
(e) $\mathrm{Cap}_{\alpha u}(\mathcal{G}) = |\alpha|^2 \mathrm{Cap}_u(\mathcal{G})$ for each scalar α.

Next we examine what $\mathrm{Cap}_u(\mathcal{G})$ looks like if $\mathcal{H}$ is an L^2-space and $\mathcal{G}$ consists of functions that vanish on a prescribed set. As we shall see in Proposition 4.5 (a), the capacity of Definition 4.2 admits a similar description as in the special case of a regular Dirichlet form.

Let X be a Hausdorff space which is equipped with a strictly positive Borel measure m (i.e., each open set has positive measure) and let A be an arbitrary subset of X. Define $\mathcal{F}_A$ as the closure of the space

$$\overset{\circ}{\mathcal{F}}_A := \{v \in \mathcal{F} : v = 0 \ m\text{-a.e. on some neighborhood of } A\}, \tag{4.11}$$

where the closure is taken with respect to the topology induced by the scalar product $\mathcal{E}_1(\cdot,\cdot)$. A result similar to Proposition 4.5 (a) below was proved in [Nol97b].

Proposition 4.5. *For any $u \in \mathcal{F}$ the following equations hold.*

(a) $\mathrm{Cap}_u(\mathcal{F}_A^\perp) = \inf\{\mathcal{E}_1[v] : v \in \mathcal{F}, v = u \ m\text{-}a.e. \text{ on some nhd. of } A\}$,
(b) $\mathrm{Cap}_u(\mathcal{F}_A^\perp) = \inf\{\mathrm{Cap}_u(\mathcal{F}_U^\perp) : U \supset A, U \text{ open}\}$.

Proof. (a) First observe that

$$\begin{aligned}
&\inf\{\mathcal{E}_1[v] : v \in \mathcal{F}, v = u \ m\text{-a.e. on some neighborhood of } A\}\\
&= \inf\{\mathcal{E}_1[u-v] : v \in \overset{\circ}{\mathcal{F}}_A\}\\
&= \inf\{\mathcal{E}_1[u-v] : v \in \mathcal{F}_A\}.
\end{aligned} \tag{4.12}$$

Let $P_{\mathcal{F}_A}$ be the orthogonal projection onto $\mathcal{F}_A$ in $(\mathcal{F}, \mathcal{E}_1(\cdot,\cdot))$. From equation (4.12) we see that

$$\begin{aligned}\mathrm{Cap}_u(\mathcal{F}_A^\perp) &= \mathcal{E}_1[u - P_{\mathcal{F}_A}u] \\ &\geq \inf\{\mathcal{E}_1[v] : v \in \mathcal{F}, v = u \ \ m\text{-a.e. on some neighborhood of } A\}.\end{aligned}$$

To get the converse inequality let v be an arbitrary function in $\mathcal{F}_A$. Then, by the Cauchy–Schwarz inequality,

$$\begin{aligned}\mathrm{Cap}_u(\mathcal{F}_A^\perp)^2 &= \mathcal{E}_1((1-P_{\mathcal{F}_A})u, u)^2 \\ &= \mathcal{E}_1((1-P_{\mathcal{F}_A})u, u-v)^2 \\ &\leq \mathcal{E}_1[(1-P_{\mathcal{F}_A})u]\mathcal{E}_1[u-v] \\ &= \mathrm{Cap}_u(\mathcal{F}_A^\perp)\mathcal{E}_1[u-v].\end{aligned}$$

Hence $\mathrm{Cap}_u(\mathcal{F}_A^\perp) \leq \mathcal{E}_1[u-v]$. Since v was arbitrary in $\mathcal{F}_A$, this inequality implies together with (4.12)

$$\begin{aligned}\mathrm{Cap}_u(\mathcal{F}_A^\perp) &\leq \inf\{\mathcal{E}_1[u-v] : v \in \mathcal{F}_A\} \\ &= \inf\{\mathcal{E}_1[v] : v \in \mathcal{F}, v = u \ \ m\text{-a.e. on some neighborhood of } A\}.\end{aligned}$$

The proof of (a) is complete.

(b) Let $U \supset A$ be an open set. Then clearly $\mathcal{F}_U^\perp \supset \mathcal{F}_A^\perp$ and hence, by Remark 4.4 (c),

$$\mathrm{Cap}_u(\mathcal{F}_U^\perp) \geq \mathrm{Cap}_u(\mathcal{F}_A^\perp),$$

implying

$$\mathrm{Cap}_u(\mathcal{F}_A^\perp) \leq \inf\{\mathrm{Cap}_u(\mathcal{F}_U^\perp) : U \supset A, U \text{ open}\}.$$

For the other inequality let $\varepsilon > 0$ be arbitrary. From (a) we see that there is some $v \in \mathcal{F}$ with $v = u$ m-a.e. on some open neighborhood V of A and such that

$$\mathcal{E}_1[v] \leq \mathrm{Cap}_u(\mathcal{F}_A^\perp) + \varepsilon.$$

We also see from (a) that $\mathcal{E}_1[v] \geq \mathrm{Cap}_u(\mathcal{F}_V^\perp)$. Thus

$$\inf\{\mathrm{Cap}_u(\mathcal{F}_U^\perp) : U \supset A, U \text{ open}\} \leq \mathrm{Cap}_u(\mathcal{F}_V^\perp) \leq \mathrm{Cap}_u(\mathcal{F}_A^\perp) + \varepsilon,$$

and the proposition is proved. □

We remark that Proposition 4.5 can be used for an alternative proof of Proposition 4.3 which avoids the use of the machinery of Hunt processes. The proof is carried out in [Nol99].

4.4. Other capacities

After having discussed the capacity of a regular Dirichlet form and its generalization to the general Hilbert space setting we now want to give a brief overview on other notions of capacity. First we define the capacity of a non-regular and non-symmetric Dirichlet form. In Section 4.4.2 we treat capacities on $\mathbb{R}^d$ that are not induced by a Dirichlet form, but by higher-order differential operators.

4.4.1. The capacity of a non-regular, non-symmetric Dirichlet form
Dirichlet forms on $L^2(X, m)$ which are not necessarily symmetric are considered in [MR92]. Furthermore the topological assumptions on X are weakened considerably: X is merely assumed to be a Hausdorff space. Also the local compactness is no longer needed.

Definition 4.6. Let $\mathcal{F}$ be a dense subspace of $L^2(X, m)$. A (not necessarily symmetric) non-negative bilinear form $\mathcal{E} : \mathcal{F} \times \mathcal{F} \to \mathbb{R}$ is called a *Dirichlet form on* $L^2(X, m)$ if the following conditions are satisfied.

(i) The symmetric part

$$\tilde{\mathcal{E}_1}(u, v) := \frac{1}{2}(\mathcal{E}_1(u, v) + \mathcal{E}_1(v, u))$$

is a closed form on $L^2(X, m)$,

(ii) $\mathcal{E}$ fulfills the *weak sector condition* with some continuity constant $K > 0$, i.e.,

$$|\mathcal{E}_1(u, v)| \le K\, \mathcal{E}_1(u, u)^{1/2} \mathcal{E}_1(v, v)^{1/2}$$

for each $u, v \in \mathcal{F}$,

(iii) $u \in \mathcal{F}$ implies $v := u^+ \wedge 1 \in \mathcal{F}$ and

$$\mathcal{E}(u + v, u - v) \ge 0, \quad \mathcal{E}(u - v, u + v) \ge 0.$$

It is easy to see that the condition (iii) is equivalent to the Markov property (C.2) in the case of a symmetric form. It is shown in [MR92] that if $(\mathcal{E}, \mathcal{F})$ is a Dirichlet form, there is a unique operator H (not necessarily self-adjoint) corresponding to $(\mathcal{E}, \mathcal{F})$ in the sense that

$$\mathcal{E}(u, v) = \langle Hu, v\rangle,$$

valid for $v \in \mathcal{F}$ and u in the domain of H. It follows from the Hille–Yosida theorem that there is a strongly continuous semigroup $(S_t)_{t>0}$ on $L^2(X, m)$ associated to H and hence to $(\mathcal{E}, \mathcal{F})$. The adjoint semigroup $(S_t^*)_{t>0}$ is also strongly continuous and is called the *cosemigroup associated to* $(\mathcal{E}, \mathcal{F})$. Analogously $\left((H^* + \alpha)^{-1}\right)_{\alpha>0}$ is called the *coresolvent of* H.

The definition of capacity uses the concept of balayaged functions which will be described briefly in the following proposition.

Proposition 4.7. *Let U be an open subset of X and let $h : X \to \mathbb{R}$ be an arbitrary function. Assume that*

$$\mathcal{L}_{h,U} := \{u \in \mathcal{F} : u \ge h \quad m\text{-a.e. on } U\} \ne \emptyset.$$

Then there is a unique function $h_U \in \mathcal{L}_{h,U}$, called the balayaged function of h on U *such that $\mathcal{E}_1(h_U, u) \ge \mathcal{E}_1[h_U]$ for all $u \in \mathcal{L}_{h,U}$. Analogously there is a unique function $\hat{h_U} \in \mathcal{L}_{h,U}$, called the* cobalayaged function of h on U *such that $\mathcal{E}_1(u, \hat{h_U}) \ge \mathcal{E}_1[\hat{h_U}]$ for all $u \in \mathcal{L}_{h,U}$.*

A proof can easily be deduced from Stampaccia's theorem, see [MR92] Theorem 2.6. Using balayaged functions the capacity $\mathrm{Cap}_{h,g}(\cdot)$ can be defined for functions h, g such that one of the following two conditions hold.

(i) $h, g \in \mathcal{F}$, h is 1-excessive (i.e., $S_t h \le e^t h$) and g is 1-coexcessive (i.e., $S_t^* g \le e^t g$),

(ii) $g = h$ and $h_U \wedge h$, $\hat{h_U} \wedge h$ are 1-excessive resp. 1-coexcessive functions in $\mathcal{F}$ for all open $U \subset X$ with $\mathcal{L}_{h,U} \neq \emptyset$.

Given h, g with (i) or (ii) define for open $U \subset X$,

$$\mathrm{Cap}_{h,g}(U) := \begin{cases} \mathcal{E}_1(h_U, \hat{g_U}) & \text{if } \mathcal{L}_{h,U} \neq \emptyset \text{ and } \mathcal{L}_{g,U} \neq \emptyset, \\ \infty & \text{else} \end{cases} \tag{4.13}$$

and extend this definition to arbitrary $A \subset X$ by

$$\mathrm{Cap}_{h,g}(A) := \inf\{\mathrm{Cap}_{h,g}(U) : U \supset A, U \text{ open}\}.$$

It is easy to see that $h = g = 1$ fulfills condition (ii). If $(\mathcal{E}, \mathcal{F})$ is a symmetric regular Dirichlet form, then $\mathrm{Cap}_{1,1}(\cdot)$ coincides with the capacity of Definition C.14.

Now suppose that

$$h = (H+1)^{-1}\phi, \quad g = (H^*+1)^{-1}\phi, \tag{4.14}$$

where $\phi \in L^2(X, m)$ with $0 \le \phi \le 1$ m-a.e. Then h is 1-excessive and g is 1-coexcessive, so $\mathrm{Cap}_{h,g}(\cdot)$ is properly defined. It is remarkable that the sets of zero capacity do not depend on h, g in this case. This leads to the concept of $\mathcal{E}$-exceptional sets: A subset $A \subset X$ is called *$\mathcal{E}$-exceptional* if $\mathrm{Cap}_{h,g}(A) = 0$ for some/any choice of ϕ in (4.14). Hence quasi-continuous and quasi-everywhere can be defined in the obvious way. Since $(\mathcal{E}, \mathcal{F})$ was not assumed to be a regular Dirichlet form, one cannot expect that each $u \in \mathcal{F}$ admits quasi-continuous m-versions, as is true in the case of a regular symmetric Dirichlet form on a locally compact metric space. Hence an analog of parts (d)–(f) of Theorem C.17 cannot be true in general, but *if* each $u \in \mathcal{F}$ admits a quasi-continuous m-version $\tilde{u}$ and X fulfills some mild topological assumption, it is indeed true that for arbitrary $A \subset X$

$$\mathrm{Cap}_{h,g}(A) = \mathcal{E}_1(h_A, \hat{g_A}),$$

where h_A is the unique element (which turns out to exist) of

$$\mathcal{L}_{h,A} := \{u \in \mathcal{F} : \tilde{u} \ge \tilde{h} \text{ quasi-everywhere on } A\}$$

which satisfies

$$\mathcal{E}_1(h_A, u) \ge \mathcal{E}_1[h_A]$$

for all $u \in \mathcal{L}_{h,A}$; $\hat{g_A}$ is defined similarly.

4.4.2. CAPACITIES ON $\mathbb{R}^d$ In this subsection we suppose that $X = \mathbb{R}^d$. There are many different definitions of capacity available concerning more general operators than partial differential operators of order smaller than 2. The various variants are conditioned by the various aspects of applications. For an overview of the different notions of capacity see Schulze, Wildenhain [SW77] IX.7 and IX.8. A comprehensive treatment can be found in the books of Maz'ja [Maz85], Landkof [Lan72] or, more recently, Adams and Hedberg [AH95]. We confine ourselves to the capacity in the sense of Maz'ja.

Definition 4.8. Fix $p > 1$ and $\ell \in \mathbb{N}$. For $\Omega \subset \mathbb{R}^d$ open and $K \subset \Omega$ compact the *(p, ℓ)-capacity of K in Ω* is defined as

$$(p,\ell)\text{-Cap}(K,\Omega) := \inf\left\{\int_\Omega \sum_{|\alpha|=\ell} |D^\alpha \phi|^p dx : \begin{array}{c}\phi \in C_c^\infty(\Omega), 0 \le \phi \le 1, \phi = 1\\ \text{on some neighborhood of } K\end{array}\right\}. \tag{4.15}$$

From the spectral theoretical viewpoint the case $p = 2$ is the most important one. The $(2, l)$-capacity can be used to give necessary and sufficient criteria for the operator $(-\Delta)^\ell$ to have discrete spectrum. It is also possible to give capacitary criteria for the semi-boundedness of Schrödinger operators, see [Maz85], Section 12.5 for details. Let us remark that these considerations are not restricted to powers of the Laplacian but can be generalized to elliptic operators of arbitrary order defined via quadratic forms.

Suppose that $p = 2$ and $\ell = 1$. Then (4.15) looks similar to the Definition C.14 (d) of the zero-order capacity in the case of the regular Dirichlet form which corresponds to the Dirichlet Laplacian in $L^2(\Omega)$, see Section 1.1. But there are three differences: Firstly in (4.15) only smooth functions with compact support are used whereas in Definition C.14 (d) all functions in the form domain $\mathcal{F} = H_0^1(\Omega)$ were considered. Secondly in (4.15) the infimum is taken only over those functions that are equal to 1 on K in contrast to Definition C.14 (d) where the functions were allowed to be greater than or equal to 1. Finally in (4.15) the functions must take values between 0 and 1, unlike in Definition C.14 (d). But it turns out that all these differences do not affect the value of the infimum.

Proposition 4.9. *Let $K \subset \Omega$ be compact. Then*

$$(2,1)\text{-Cap}(K,\Omega) = \text{Cap}_0(K),$$

where $\text{Cap}_0(\cdot)$ *is the zero-order capacity of the regular Dirichlet form which corresponds to the Dirichlet Laplacian in $L^2(\Omega)$, see Definition* C.14 (d).

The proof is carried out e.g., in [DMN97].

5. Capacitary estimates for domain perturbations

This section contains the main results of the thesis in hand. Although a major part is already contained in [Nol97b], [DMN97], [Nol97a], [Nol00] or [Nol99], some results are generalized or the proofs are simplified. Section 5.1 contains two upper bounds for the bottom eigenvalue, the first of which is a generalization of McGillivray's work [McG96], whereas the second eigenvalue estimate works in the general Hilbert space setting of Section 4.3.

Proving lower bounds for the bottom eigenvalue turns out to be a harder problem than proving upper bounds, even in the case of the Laplacian. This is due to the fact that one cannot simply guess a good trial function u_0 for the perturbed ground-state and then estimate $\mathcal{E}[u_0]$. Nevertheless, in Section 5.2 we shall obtain a lower bound for the shift of the bottom eigenvalue which is of the same magnitude as the upper bounds of Section 5.1. By using a simpler method we obtain another lower bound which generalizes Thirring's inequality.

Assuming that the underlying self-adjoint operator has purely discrete spectrum, we prove in Section 5.3 a capacitary upper bound for the second eigenvalue. It is not possible to obtain capacitary lower bounds for higher eigenvalues by using the classical versions of capacity as is seen from the example of the Dirichlet Laplacian on a square in $\mathbb{R}^2$.

In Section 5.4 we treat additive perturbations together with domain perturbations and prove an upper bound for the first eigenvalue under the assumption that the additive perturbation is infinitesimally form bounded.

Finally, Section 5.5 contains a capacitary characterization for the spectral bound being unaffected by a given domain perturbation.

5.1. Upper bounds for the bottom eigenvalue

During the last decades many mathematicians studied the problem of analyzing the connection between domain perturbations of a given self-adjoint operator and changes of its spectrum. Among others Rauch and Taylor [Rau75], [Tay76], [Tay79], [RT75a], [RT75b] worked on that subject and pointed out the relevance of this kind of problem for various areas of mathematical physics. The following result due to Taylor [Tay76] shows that the capacity can be used to give quantitative upper and lower bounds for the bottom eigenvalue of the Laplacian.

Theorem 5.1. *If $\Omega \subset \mathbb{R}^d$ is a bounded domain, λ is the lowest eigenvalue for $-\Delta$ on $\Omega \setminus K$ with Dirichlet boundary conditions on ∂K, Neumann boundary conditions on $\partial\Omega$, $K \subset \Omega$ compact, then there are constants $c_1, c_2 > 0$ such that for small* $\mathrm{Cap}_0(K)$,

$$c_1 \mathrm{Cap}_0(K) \leq \lambda \leq c_2 \mathrm{Cap}_0(K),$$

where $\mathrm{Cap}_0(\cdot)$ *is the classical zero-order capacity in $\mathbb{R}^d$, i.e., the capacity of Definition* C.14 (d) *in the case of the Laplacian in $L^2(\mathbb{R}^d)$.*

Taylor's result has been generalized in many directions. For example Gesztesy and Zhao proved in [GZ94] that the bottom eigenvalue of certain Schrödinger operators remains unaffected by domain perturbations if and only if the perturbed domain differs from the unperturbed one only by a set of zero capacity. The proof is based on Brownian motion and on the Feynman–Kac formula. One year later

Arendt and Monniaux gave an analytical proof of this result which used a domination argument for semigroups as its main ingredient [AM95]. Their result allowed the potential to vary as well. We refer to the survey article [DMN97] for precise statements and the proofs as well as for further literature on this subject. Let us also remark that similar estimates for the bottom eigenvalue also hold for the Laplace–Beltrami operator on a Riemannian manifold, see [CF78], [CF88], [Cou95], [Oza82], [Oza83]. Another generalization of Taylor's Theorem was given by McGillivray [McG96] who proved the same estimate in the context of regular Dirichlet forms. We next state this result precisely and show that one can improve the estimate by introducing the so-called *ground-state transformation.*

Let $(\mathcal{E}, \mathcal{F})$ be a regular Dirichlet form in $L^2(X,m)$ and let H be the associated self-adjoint and non-negative operator, see Appendix C.1. Assume that the spectrum of H is purely discrete. Fix a compact subset K of X and put $Y := X \setminus K$. According to Theorem C.20 (a) the restriction of $\mathcal{E}$ to the space

$$\mathcal{F}_K := \{u \in \mathcal{F} : \tilde{u} = 0 \text{ q.e. on } K\}$$

defines a regular Dirichlet form in $L^2(Y,m)$ with associated self-adjoint and non-negative operator H^Y. The spectrum of H^Y remains discrete by the minimax principle, see Appendix A. Denote the bottom eigenvalues of H and H^Y by λ and λ^Y respectively. Recall that a Borel subset E of X is called *invariant* for the semigroup $(e^{-tH})_{t>0}$ if $f = f\chi_E$ implies $e^{-tH}f = \chi_E(e^{-tH}f)$ for all $t > 0$. The semigroup $(e^{-tH})_{t>0}$ is called *irreducible* if X and $\emptyset$ are the only invariant sets, modulo null sets. According to Proposition 1.4.3 in [Dav89] irreducibility has the following important consequences for the bottom eigenvalue λ of H:

(i) λ is a simple eigenvalue,
(ii) the eigenfunction ϕ corresponding to the ground-state energy λ may be chosen such that $\phi(x) > 0$ m-a.e. on X.

The semigroup is said to be *ultracontractive* if for each $t > 0$ there is a constant $c_t \geq 0$ such that

$$\|e^{-tH}f\|_{L^\infty(X,m)} \leq c_t \, \|f\|_{L^2(X,m)}$$

for all $f \in L^2(X,m)$. Note that ultracontractivity implies that each eigenfunction of H is bounded which was the main reason for assuming ultracontractivity in Theorem 5.2 below.

Theorem 5.2. (*I. McGillivray* 1996) *Let $(\mathcal{E}, \mathcal{F})$ be a regular Dirichlet form in $L^2(X,m)$ such that its associated self-adjoint operator H has purely discrete spectrum. If $(e^{-tH})_{t>0}$ is irreducible and ultracontractive, then there are constants $c_1, c_2 \in (0,\infty)$ such that*

$$\lambda^Y - \lambda \leq c_1 \mathrm{Cap}(K)$$

for all compact subsets K of X with capacity smaller than c_2.

5.1.1. THE GROUND-STATE TRANSFORMATION One drawback of Theorems 5.1 and 5.2 is that these results are only valid for sets of sufficiently small capacity. If in the situation of Theorem 5.2 the obstacle region K comes close to the boundary of X one must expect that the capacity increases so McGillivray's result cannot be applied.

We next show that this problem can be circumvented by introducing the ground-state transformation by which we mean the unitary map

$$U : L^2(X,m) \to L^2(X,\phi^2 m), \quad f \mapsto \phi^{-1} f.$$

Here ϕ is the unique function in the eigenspace $\ker(H-\lambda)$ which is normalized and strictly positive m-a.e. Define the self-adjoint operator H' in $L^2(X,\phi^2 m)$ by

$$\operatorname{dom}(H') := \{Uu : u \in \operatorname{dom}(H)\}, \quad H'u := UHU^{-1}u.$$

Then $\sigma(H') = \sigma(H)$ and the constant functions are eigenfunctions of H' with eigenvalue λ. In particular these eigenfunctions are automatically bounded. This observation allows us to drop the ultracontractivity condition in Theorem 5.2, see Theorem 5.4 below. Let $(\mathcal{E}',\mathcal{F}')$ be the closed form associated to H'. It is easy to see that $(\mathcal{E}',\mathcal{F}')$ is again a Dirichlet form with

$$\mathcal{F}' = \{\phi^{-1}u : u \in \mathcal{F}\} \quad \text{and} \quad \mathcal{E}'(\phi^{-1}u,\phi^{-1}v) = \mathcal{E}(u,v)$$

for $u,v \in \mathcal{F}$. Denote by $\operatorname{Cap}'(U)$ the $\mathcal{E}'$-capacity of some open set $U \subset X$. If U is such that $\phi(x) \le 1$ m-a.e. on U, then

$$\begin{aligned}
\operatorname{Cap}'(U) &= \inf\{\mathcal{E}_1'(u',u') : u' \in \mathcal{F}', u' \ge 1 \ m\text{-a.e. on } U\} \\
&= \inf\{\mathcal{E}_1'(\phi^{-1}u,\phi^{-1}u) : u \in \mathcal{F} : u \ge \phi \ m\text{-a.e. on } U\} \\
&= \inf\{\mathcal{E}_1(u,u) : u \in \mathcal{F} : u \ge \phi \ m\text{-a.e. on } U\} \\
&\le \operatorname{Cap}(U).
\end{aligned}$$

These calculations carry over to arbitrary $A \subset X$. For instance, if $X = \Omega$ is a sufficiently regular domain and H is the Dirichlet Laplacian in $L^2(\Omega)$, then $\phi(x) \to 0$ as $x \to \partial\Omega$. Therefore we can hope that the new capacity $\operatorname{Cap}'(\cdot)$ yields a better estimate for the ground-state shift $\lambda^Y - \lambda$. Before we can prove the generalization of Theorem 5.2 we have to guarantee that $(\mathcal{E}',\mathcal{F}')$ is regular. This turns out to be true if the ground-state ϕ is a continuous function which clearly is not a serious restriction. For example, if H is a (second-order) elliptic differential operator the eigenfunctions are even C^∞ because of the elliptic regularity theorem, see Appendix B.

The following Lemma is taken from [Nol97a].

Lemma 5.3. *Let $(\mathcal{E},\mathcal{F})$ be a regular Dirichlet form in $L^2(X,m)$ possessing the following properties:*

(i) *The associated self-adjoint operator H has purely discrete spectrum,*
(ii) *the semigroup $(e^{-tH})_{t>0}$ is irreducible.*
(iii) *the ground-state ϕ is continuous on X.*

Then $(\mathcal{E}',\mathcal{F}')$ is regular and $(e^{-tH'})_{t>0}$ is irreducible.

Proof. The proof of the irreducibility is straightforward, so let us concentrate on the regularity of $(\mathcal{E}', \mathcal{F}')$. We have to show the denseness of $C_0(X) \cap \mathcal{F}'$ in $\mathcal{F}'$ as well as in $(C_0(X), \|\cdot\|_\infty)$. The first statement is again straightforward, so we will only prove that $C_0(X) \cap \mathcal{F}'$ is dense in $C_0(X)$ with respect to uniform norm. Let $g \in C_0(X)$ arbitrary. Because of the continuity of ϕ we conclude $h := \phi g \in C_0(X)$, and the regularity of $\mathcal{E}$ guarantees the existence of a sequence $h_n \in \mathcal{F} \cap C_0(X)$ such that

$$h_n \to h \text{ in } C_0(X). \tag{5.1}$$

Urysohn's theorem (cf. [Rud66], Theorem 2.1.2) gives us some $\eta \in C_0(X)$ with

$$\eta \equiv 1 \text{ on } \operatorname{supp}(h). \tag{5.2}$$

Now we use again regularity of $\mathcal{E}$ to find for arbitrary $\varepsilon > 0$ a function $\eta' \in C_0(X) \cap \mathcal{F}$ satisfying

$$\|\eta - \eta'\|_\infty < \varepsilon. \tag{5.3}$$

Put

$$g_n := \eta' \phi^{-1} h_n.$$

The proof is completed if we show

$$g_n \in \mathcal{F}' \cap C_0(X) \tag{5.4}$$

and

$$g_n \to g \text{ in } C_0(X). \tag{5.5}$$

Proof of (5.4)*:* Obviously we have $g_n \in C_0(X)$. Because of

$$\mathcal{F}' = \{\phi^{-1} u : u \in \mathcal{F}\}$$

it suffices to show that $\eta' h_n \in \mathcal{F}$. This is a consequence of $\eta', h_n \in \mathcal{F} \cap L^\infty(X, m)$ and Theorem C.13 (a). It remains to prove (5.5), but this is easy:

$$\begin{aligned} \|g_n - g\|_\infty &= \|\eta' \phi^{-1} h_n - \phi^{-1} h\|_\infty \\ &= \|\eta' \phi^{-1} h_n - \eta \phi^{-1} h\|_\infty \\ &\leq \|\eta' \phi^{-1} h_n - \eta' \phi^{-1} h\|_\infty + \|\eta' \phi^{-1} h - \eta \phi^{-1} h\|_\infty \\ &\leq \|\eta' \phi^{-1}\|_\infty \|h_n - h\|_\infty + \|\phi^{-1} h\|_\infty \|\eta' - \eta\|_\infty \\ &\leq 2\varepsilon \|\phi^{-1} h\|_\infty \end{aligned}$$

if n is sufficiently large because of (5.3) and (5.1) (note that $\eta', h \in C_0(X)$ implying $\|\phi^{-1} h\|_\infty < \infty$). □

With the aid of Lemma 5.3 we can prove a generalization of Theorem 5.2. Theorem 5.4 below is a slight improvement of Theorem 6 in [Nol97a], where a similar estimate has been proved with larger constant c_δ.

Theorem 5.4. *Under the hypothesis of Lemma* 5.3 *the following is true: For each* $0 \leq \delta < \frac{1}{2}(1+\lambda)$ *there is some constant* $c_\delta > 0$ *such that*

$$\lambda^Y - \lambda \leq c_\delta \mathrm{Cap}'(K),$$

whenever K *is a compact subset* K *of* X *whose capacity is smaller than* δ. *The constant* c_δ *is given explicitly by* $c_\delta := \frac{2(1+\lambda)}{1+\lambda-2\delta}$.

Proof. From Lemma 5.3 we conclude that $(\mathcal{E}', \mathcal{F}')$ is again a regular Dirichlet form. Let $(\Omega', \mathcal{M}', \mathcal{M}'_t, X'_t, \mathbb{P}'_x)$ be the Hunt process associated to $(\mathcal{E}', \mathcal{F}')$ and define the set S'_0 of measures of finite energy integral and their associated 1-potentials $U'_1\mu$ in the obvious way. Further let $\nu := \mu^1_K \in S'_0$ be the unique measure corresponding to K and 1 in the sense of Theorem C.23 (a). Because of

$$\|U'_1\nu\|_{L^\infty(X,m)} = \|\mathbb{E}'_{(\cdot)}\{e^{-\tau'_K}\}\|_{L^\infty(X,m)} \leq 1$$

and

$$\nu(X) = \int_X 1 \mathrm{d}\nu = \mathcal{E}'_1(1, U'_1\nu) < \infty$$

we see from Theorem C.24 that

$$\nu(X) = \nu(K) \leq \mathrm{Cap}'(K).$$

From Dynkin's formula, see Theorem C.23 (b), we obtain

$$\begin{aligned} U'_1\nu &= U'_1\mu_K^{(H'+1)^{-1}(H'+1)1} \\ &= (H'+1)^{-1}(H'+1)1 - (H'^Y+1)^{-1}((H'+1)1)_{|Y} \\ &= 1 - (1+\lambda)(H'^Y+1)^{-1}1_{|Y}, \end{aligned}$$

i.e.,

$$(H'^Y+1)^{-1}1_{|Y} = (1+\lambda)^{-1}1 - (1+\lambda)^{-1}U'_1\nu,$$

and

$$\mathcal{E}'_1((H'^Y+1)^{-1}1_{|Y}, (1+\lambda)^{-1}U'_1\nu) = 0.$$

Putting everything together we get

$$\begin{aligned} 1+\lambda^Y &= \inf_{u\in\mathcal{F}'_K} \frac{\mathcal{E}'_1[u]}{\|u\|^2_{L^2(X,\phi^2 m)}} \\ &\leq \frac{\mathcal{E}'_1[(H'^Y+1)^{-1}1_{|Y}]}{\|(H'^Y+1)^{-1}1_{|Y}\|^2_{L^2(X,\phi^2 m)}} \end{aligned}$$

$$
\begin{aligned}
&\le \frac{\mathcal{E}_1'[(1+\lambda)^{-1}1]}{\|(1+\lambda)^{-1}1-(1+\lambda)^{-1}U_1'\nu\|^2_{L^2(X,\phi^2m)}}\\
&\le \frac{(1+\lambda)^{-1}}{(1+\lambda)^{-2}-2(1+\lambda)^{-2}\langle 1,U_1'\nu\rangle_{L^2(X,\phi^2m)}}\\
&= \frac{1+\lambda}{1-2(1+\lambda)^{-1}\mathcal{E}_1'(1,U_1'\nu)}\\
&= \frac{(1+\lambda)^2}{1+\lambda-2\nu(K)}\\
&= \frac{(1+\lambda)^2}{1+\lambda-2\mathrm{Cap}'(K)}\\
&= 1+\lambda+\frac{2(1+\lambda)\mathrm{Cap}'(K)}{1+\lambda-2\mathrm{Cap}'(K)}\\
&\le 1+\lambda+c_\delta\mathrm{Cap}'(K),
\end{aligned}
$$

finishing the proof of the theorem. □

5.1.2. An upper bound in the general Hilbert space setting The results of Section 5.1.1 are restricted to operators which correspond to regular Dirichlet forms. It is a consequence of the Beurling–Deny formulae (4.10) that the scope of regular Dirichlet forms in $L^2(\Omega)$, $\Omega \subset \mathbb{R}^d$ an open set, is limited to (pseudo-) differential operators of order at most 2. In particular the clamped plate cannot be handled as it may be identified with the biharmonic operator Δ^2 subject to Dirichlet boundary conditions. Another class of operators which are not included are semi-bounded operators $H \ge \lambda$ having discrete spectrum in some interval $[\lambda,\mu)$ and essential spectrum in $[\mu,\infty)$. Examples of such kind of operators are one- and many-particle Schrödinger operators in quantum mechanics. In particular the Schrödinger operator of the hydrogen atom is of this kind, see [BS91], [Wei95a], [Wei95b].

We shall overcome these restrictions by proving abstract eigenvalue estimates which use the capacity of Definition 4.2. To this aim we assume from now on unless otherwise stated, that $H^{\mathcal{G}}$ is the domain perturbation of some self-adjoint and semi-bounded operator H with respect to the subspace $\mathcal{G}$. This was explained in detail in Section 3.2. In what follows we use the notation of this section without further notification. Our first result gives an upper bound for the spectral bound $\lambda^{\mathcal{G}} = \inf\sigma(H^{\mathcal{G}})$ of $H^{\mathcal{G}}$ in terms of the ϕ_n-capacities of $\mathcal{G}^\perp$ where the ϕ_n's are normalized eigenfunctions of H.

Theorem 5.5. *Let $\{\lambda_n\}$ be the (finite or infinite) set of eigenvalues for H with corresponding normalized eigenfunctions ϕ_n. Then*

$$
\lambda^{\mathcal{G}} \le \inf_n \left(\lambda_n + \frac{\mathrm{Cap}_{\phi_n}(\mathcal{G}^\perp)}{1-\frac{\mathrm{Cap}_{\phi_n}(\mathcal{G}^\perp)}{1+\lambda_n-\lambda}}\right).
$$

Proof. Let P be the orthogonal projection onto $\mathcal{G}^\perp$, put $\kappa_n := \mathrm{Cap}_{\phi_n}(\mathcal{G}^\perp)$ and $\mu_n := \lambda_n - \lambda$. Then

$$\langle \phi_n, P\phi_n \rangle = \frac{1}{\mu_n + 1}\mathcal{E}_1(\phi_n, P\phi_n) = \frac{1}{\mu_n + 1}\mathcal{E}_1[P\phi_n] = \frac{\kappa_n}{\mu_n + 1}.$$

Moreover, by the Cauchy–Schwarz inequality

$$\|P\phi_n\| \geq |\langle P\phi_n, \phi_n \rangle| = \frac{\kappa_n}{\mu_n + 1}.$$

Hence

$$\begin{aligned}
1 + \lambda^{\mathcal{G}} - \lambda &= \inf\left\{\frac{\mathcal{E}_1[u]}{\|u\|^2} : u \in \mathcal{G} \setminus \{0\}\right\} \\
&\leq \frac{\mathcal{E}_1[(1-P)\phi_n]}{\|(1-P)\phi_n\|^2} \\
&= \frac{1 + \mu_n - \kappa_n}{1 - 2\mathrm{Re}\langle \phi_n, P\phi_n \rangle + \|P\phi_n\|^2} \\
&\leq \frac{1 + \mu_n - \kappa_n}{1 - 2\frac{\kappa_n}{\mu_n + 1} + \frac{\kappa_n^2}{(\mu_n+1)^2}} \\
&= \frac{(1+\mu_n)^2}{1 + \mu_n - \kappa_n} \\
&= 1 + \mu_n + \frac{\kappa_n + \mu_n \kappa_n}{1 + \mu_n - \kappa_n} \\
&= 1 + \lambda_n - \lambda + \frac{\kappa_n}{1 - \frac{\kappa_n}{1 + \lambda_n - \lambda}},
\end{aligned}$$

and the desired estimate follows. □

As an immediate consequence we note the following corollary.

Corollary 5.6.

(a) *Suppose that λ is an eigenvalue of H with normalized eigenfunction $\phi \notin \mathcal{G}^\perp$. Then*

$$\lambda^{\mathcal{G}} - \lambda \leq \frac{\mathrm{Cap}_\phi(\mathcal{G}^\perp)}{1 - \mathrm{Cap}_\phi(\mathcal{G}^\perp)}. \tag{5.6}$$

(b) *Fix $\delta \in [0, 1)$. For each closed subspace $\mathcal{G}$ of $(\mathcal{F}, \mathcal{E}_1(\cdot, \cdot))$ with $\mathrm{Cap}_\phi(\mathcal{G}^\perp) \leq \delta$ we have*

$$\lambda^{\mathcal{G}} - \lambda \leq \frac{1}{1 - \delta}\mathrm{Cap}_\phi(\mathcal{G}^\perp).$$

Remark 5.7.

(a) Note that we did not assume that λ is a simple eigenvalue of H. If λ is degenerate we may apply Corollary 5.6 (a) for any normalized ϕ in the eigenspace $\ker(H - \lambda)$ and obtain

$$\lambda^{\mathcal{G}} - \lambda \leq \inf\left\{\frac{\mathrm{Cap}_\phi(\mathcal{G}^\perp)}{1 - \mathrm{Cap}_\phi(\mathcal{G}^\perp)} : \phi \in \ker(H - \lambda), \|\phi\| = 1\right\}.$$

(b) The condition $\phi \notin \mathcal{G}^{\perp}$ is necessary to rule out the possibility $\mathrm{Cap}_{\phi}(\mathcal{G}^{\perp}) = 1$ (see Remark 4.4) because the denominator of the right-hand side in (5.6) vanishes if $\phi \in \mathcal{G}^{\perp}$. Intuitively speaking $\phi \in \mathcal{G}^{\perp}$ means that ϕ was "completely thrown away" by restricting $\mathcal{E}$ to the subspace $\mathcal{G}$. From this point of view it becomes clear that (5.6) will only be a good estimate if $\mathrm{Cap}_{\phi}(\mathcal{G}^{\perp})$ is sufficiently small, see also Example 5.2. Roughly speaking, $\mathrm{Cap}_{\phi}(\mathcal{G}^{\perp})$ measures how much of ϕ will be lost if the form $\mathcal{E}$ is restricted to the smaller subspace $\mathcal{G}$, and $\mathrm{Cap}_{\phi}(\mathcal{G}^{\perp})/(1 - \mathrm{Cap}_{\phi}(\mathcal{G}^{\perp}))$ gives a bound for the energy of the projected function $P_{\mathcal{G}}\phi$.

The following simple example shows that $\phi \in \mathcal{G}^{\perp}$ might happen if $\mathcal{H} = L^2(X, m)$ with a non-connected set X.

Example. Let $(\mathcal{E}, H_0^1((0,\pi) \cup (4,5)))$ be the quadratic form of the Dirichlet Laplacian on $L^2((0,\pi) \cup (4,5))$. The ground-state of H is given by

$$\phi(x) = \begin{cases} \sin x, & x \in (0,\pi), \\ 0, & \text{else}, \end{cases}$$

and we see that $\phi \in \mathcal{G}^{\perp}$ if $\mathcal{G} \subset H_0^1(4,5)$.

We next show that in the case of a Dirichlet Schrödinger operator on a *domain*, i.e., an open and connected subset of $\mathbb{R}^d$, we always have $\phi \notin \mathcal{G}^{\perp}$. Let us first describe the definition of a Schrödinger operator via sesquilinear forms. Let Ω be an open subset of $\mathbb{R}^d$ and let $q \in L^1_{loc}(\Omega)$ be a real-valued function such that multiplication by q is form-bounded with respect to the form

$$(u,v) \mapsto \int_{\Omega} \nabla\overline{u}\, \nabla v \,\mathrm{d}x, \quad u,v \in \mathcal{F} := H_0^1(\Omega).$$

It is well known (see e.g., [RS75], Section X.2) that if the relative bound is less than 1, the form

$$(u,v) \mapsto \int_{\Omega} \nabla\overline{u}\, \nabla v \,\mathrm{d}x + \int_{\Omega} q\overline{u}v \,\mathrm{d}x, \quad u,v \in \mathcal{F},$$

is closed and semi-bounded from below. Denote the corresponding self-adjoint operator by H and let $\lambda := \inf \sigma(H)$ be its spectral bound. Let $(\mathcal{E}, \mathcal{F})$ be the non-negative form associated to $H - \lambda$. For the definition of the domain perturbation let Λ be an open subset of Ω. In order to avoid trivialities we assume $\Lambda \neq \Omega$ and $\Lambda \neq \emptyset$. Put $\mathcal{G} := H_0^1(\Lambda)$. Then $\mathcal{G} \subset \mathcal{F}$ and hence $(\mathcal{E}, \mathcal{G})$ defines a self-adjoint operator $H^{\mathcal{G}}$ in $\mathcal{H}^{\mathcal{G}} := L^2(\Lambda)$.

Theorem 5.8. *In the situation just described suppose that Ω is a domain and that λ is an eigenvalue of H with normalized eigenfunction ϕ. Then $\mathrm{Cap}_{\phi}(\mathcal{G}^{\perp}) < 1$.*

Proof. Connectedness of Ω implies that the semigroup $(e^{-tH})_{t>0}$ is irreducible. This together with the positivity preserving property of $(e^{-tH})_{t>0}$ implies that ϕ is strictly positive almost everywhere on Ω (for details concerning this type of argumentation see e.g., [Dav89] or [Goe77]). Then for any $u \in \mathcal{G} \setminus \{0\}$ with $u \geq 0$ m-a.e.

$$\mathcal{E}_1(\phi, u) = \langle \phi, u \rangle = \int_\Lambda \phi u \, \mathrm{d}x > 0,$$

hence $\phi \notin \mathcal{G}^\perp$, i.e. $\mathrm{Cap}_\phi(\mathcal{G}^\perp) < 1$. □

Corollary 5.6 (a) can be used to prove a capacitary criterion for the perturbed operator to have a bottom eigenvalue.

Corollary 5.9. *If the spectral bound λ is an isolated eigenvalue of H of finite multiplicity and some normalized ϕ in the eigenspace* $\ker(H - \lambda)$ *satisfies*

$$\lambda + \frac{\mathrm{Cap}_\phi(\mathcal{G}^\perp)}{1 - \mathrm{Cap}_\phi(\mathcal{G}^\perp)} < \inf(\sigma(H) \setminus \{\lambda\}),$$

then $\lambda^{\mathcal{G}}$ is an isolated eigenvalue of $H^{\mathcal{G}}$ of finite multiplicity.

Proof. Let k be the dimension of the eigenspace $\ker(H - \lambda)$. By Corollary 5.6 (a) and the minimax principle, see Appendix A, we have

$$\begin{aligned}
\lambda^{\mathcal{G}} &\leq \lambda + \frac{\mathrm{Cap}_\phi(\mathcal{G}^\perp)}{1 - \mathrm{Cap}_\phi(\mathcal{G}^\perp)} \\
&< \inf(\sigma(H) \setminus \{\lambda\}) \\
&= \inf\left\{\sup\left\{\frac{(\mathcal{E} + \lambda)[u]}{\|u\|^2} : u \in L \setminus \{0\}\right\} : L \subset \mathcal{F},\, \dim L = k+1\right\} \\
&\leq \inf\left\{\sup\left\{\frac{(\mathcal{E} + \lambda)[u]}{\|u\|^2} : u \in L \setminus \{0\}\right\} : L \subset \mathcal{G},\, \dim L = k+1\right\}.
\end{aligned}$$

Therefore $\lambda^{\mathcal{G}}$ is not the least upper bound of the sequence

$$\left(\inf\left\{\sup\left\{\frac{(\mathcal{E} + \lambda)[u]}{\|u\|^2} : u \in L \setminus \{0\}\right\} : L \subset \mathcal{G},\, \dim L = n\right\}\right)_{n \in \mathbb{N}},$$

and the assertion follows again by the minimax principle. □

5.2. Lower bounds for the bottom eigenvalue

McGillivray proved in the context of regular Dirichlet forms that the shift of the spectral bound can be estimated from below by some constant times the capacity [McG96], but his result needs strong additional assumptions and does not carry over to more general operators. As we shall see in this section it is possible to obtain capacitary lower bounds for the shift $\lambda^{\mathcal{G}} - \lambda$ similar to those of Corollary 5.6 if λ is a simple and isolated eigenvalue of H.

Theorem 5.10. *Suppose that λ is a simple eigenvalue of H with normalized eigenfunction ϕ. Let $\mu \in [0, \infty)$ be the spectral gap, i.e.,*

$$\mu := \inf\{\sigma(H) \setminus \{\lambda\}\} - \lambda. \tag{5.7}$$

Then

$$\lambda^{\mathcal{G}} - \lambda \geq \frac{\mu \mathrm{Cap}_\phi(\mathcal{G}^\perp)}{1+\mu(1-\mathrm{Cap}_\phi(\mathcal{G}^\perp))} \geq \frac{\mu}{\mu+1}\mathrm{Cap}_\phi(\mathcal{G}^\perp).$$

Proof. The second inequality is obvious, so let us prove the first one. Let $\mathcal{K}$ be the eigenspace of H and λ. Denote by $P_\mathcal{K}$ and $\mathcal{K}^\perp$ the orthogonal projection onto $\mathcal{K}$ and the orthogonal complement of $\mathcal{K}$ respectively, where orthogonality is meant with respect to the scalar product $\mathcal{E}_1(\cdot,\cdot)$ on $\mathcal{F}$. Put $P_{\mathcal{K}^\perp} := 1 - P_\mathcal{K}$. Then for any $u \in \mathcal{F}$,

$$\mathrm{Cap}_u(\mathcal{K}^\perp) = \mathcal{E}_1[P_{\mathcal{K}^\perp}u] \geq (1+\mu)\|P_{\mathcal{K}^\perp}u\|^2.$$

Observe that $P_\mathcal{K}u$ and $P_{\mathcal{K}^\perp}u$ are not only orthogonal with respect to $\mathcal{E}_1(\cdot,\cdot)$ but are also orthogonal with respect to $\langle\cdot,\cdot\rangle$. This gives us

$$\begin{aligned}\mathrm{Cap}_u(\mathcal{K}^\perp) &\geq (\mu+1)(\|u\|^2 - \|P_\mathcal{K}u\|^2)\\ &= (\mu+1)(\|u\|^2 - \mathcal{E}_1[P_\mathcal{K}u])\\ &= (\mu+1)(\|u\|^2 - \mathrm{Cap}_u(\mathcal{K})).\end{aligned} \tag{5.8}$$

Let $P_\mathcal{G}$ be the orthogonal projection onto $\mathcal{G}$. Since $\mathcal{K}$ is spanned by ϕ and $\mathcal{E}_1[\phi] = 1$ we obtain for $u \in \mathcal{G}$

$$\begin{aligned}\mathrm{Cap}_u(\mathcal{K}) &= \mathcal{E}_1[P_\mathcal{K}u] = \mathcal{E}_1[\mathcal{E}_1(\phi,u)\phi] = |\mathcal{E}_1(\phi,u)|^2 = |\mathcal{E}_1(\phi,P_\mathcal{G}u)|^2\\ &= |\mathcal{E}_1(P_\mathcal{G}\phi,u)|^2 \leq \mathcal{E}_1[P_\mathcal{G}\phi]\mathcal{E}_1[u] = \mathcal{E}_1[u]\mathrm{Cap}_\phi(\mathcal{G}).\end{aligned} \tag{5.9}$$

Now let $u \in \mathcal{G}$ with $\|u\| = 1$. Then, by Remark 4.4 (b), (5.8) and (5.9)

$$\begin{aligned}(1+\mu\mathrm{Cap}_\phi(\mathcal{G}))\mathcal{E}_1[u] &= \mu\mathrm{Cap}_\phi(\mathcal{G})\mathcal{E}_1[u] + \mathrm{Cap}_u(\mathcal{K}) + \mathrm{Cap}_u(\mathcal{K}^\perp)\\ &\geq \mu\mathrm{Cap}_u(\mathcal{K}) + \mathrm{Cap}_u(\mathcal{K}) + (\mu+1)(1-\mathrm{Cap}_u(\mathcal{K}))\\ &= \mu+1.\end{aligned}$$

Consequently

$$\begin{aligned}1+\lambda^{\mathcal{G}}-\lambda &= \inf\{\mathcal{E}_1[u] : u\in\mathcal{G}, \|u\|=1\}\\ &\geq \frac{\mu+1}{1+\mu\mathrm{Cap}_\phi(\mathcal{G})}\\ &= 1+\frac{\mu-\mu\mathrm{Cap}_\phi(\mathcal{G})}{1+\mu\mathrm{Cap}_\phi(\mathcal{G})}\\ &= 1+\frac{\mu\mathrm{Cap}_\phi(\mathcal{G}^\perp)}{1+\mu(1-\mathrm{Cap}_\phi(\mathcal{G}^\perp))},\end{aligned}$$

which completes the proof. □

From Corollary 5.6 (a) and Theorem 5.10 we immediately obtain the following characterization for the shift $\lambda^{\mathcal{G}} - \lambda$ to be positive.

Corollary 5.11. *Suppose that λ is a simple and isolated eigenvalue of H with eigenfunction ϕ. Then $\lambda^{\mathcal{G}} > \lambda$ if and only if $\mathrm{Cap}_\phi(\mathcal{G}^\perp) > 0$.*

In practice it seems to be difficult to compute the capacities involved in Theorems 5.5 and 5.10 since both $\mathcal{G}$ and $\mathcal{G}^\perp$ are typically infinite-dimensional, e.g., if H is an operator in $L^2(\mathbb{R}^d)$ and $H^{\mathcal{G}}$ arises from H by imposing boundary conditions on a set with non-empty interior. It is possible, however, to prove another lower bound for the shift of the spectral bound which involves the u-capacity of the eigenspace $\mathcal{K}$ for $u \in \mathcal{G}$ rather than the ϕ-capacity of $\mathcal{G}$ for $\phi \in \mathcal{K}$ as in Theorems 5.5 and 5.10. Since $\mathcal{K}$ happens to be one-dimensional in many interesting cases, e.g., if H is the Dirichlet Laplacian on some bounded domain in $\mathbb{R}^d$, it is possible to compute the u-capacity of $\mathcal{K}$ explicitly in these cases.

Theorem 5.12. *Suppose that λ is an eigenvalue of H. Denote the corresponding eigenspace by $\mathcal{K}$ and let*

$$\delta := \sup\{\mathrm{Cap}_u(\mathcal{K}) : u \in \mathcal{G}, \|u\| = 1\}.$$

Then $\lambda^{\mathcal{G}} - \lambda \geq \mu(1-\delta)$, where μ is as in (5.7) the spectral gap.

Proof. Let $P_{\mathcal{K}}$ and $\mathcal{K}^\perp$ be the orthogonal projection onto $\mathcal{K}$ and the orthogonal complement of $\mathcal{K}$ respectively and put $P_{\mathcal{K}^\perp} := 1 - P_{\mathcal{K}}$. Then for $u \in \mathcal{G}$ such that $\|u\| = 1$:

$$\begin{aligned}
\mathcal{E}_1[u] &= \mathcal{E}_1[P_{\mathcal{K}}u] + \mathcal{E}_1[P_{\mathcal{K}^\perp}u] = \|P_{\mathcal{K}}u\|^2 + \mathcal{E}_1[P_{\mathcal{K}^\perp}u] \\
&\geq \|P_{\mathcal{K}}u\|^2 + (\mu+1)\|P_{\mathcal{K}^\perp}u\|^2 = 1 + \mu\|P_{\mathcal{K}^\perp}u\|^2 \\
&= 1 + \mu(1 - \|P_{\mathcal{K}}u\|^2) = 1 + \mu(1 - \mathcal{E}_1[P_{\mathcal{K}}u]) \\
&= 1 + \mu(1 - \mathrm{Cap}_u(\mathcal{K})) \\
&\geq 1 + \mu(1-\delta),
\end{aligned}$$

and the theorem follows by

$$\lambda^{\mathcal{G}} - \lambda + 1 = \inf\{\mathcal{E}_1[u] : u \in \mathcal{G}, \|u\| = 1\}.$$

□

As an application of Theorem 5.12 we will prove a generalization of Thirring's inequality which reads as follows: Let $\Omega, \Lambda \subset \mathbb{R}^d$ be bounded domains with $\Lambda \subset \Omega$. Denote by $(\lambda_j(\Omega))_{j\geq 1}$ and $(\lambda_j(\Lambda))_{j\geq 1}$ the eigenvalues of the Dirichlet Laplacian in $L^2(\Omega)$ and in $L^2(\Lambda)$ respectively. Then

$$\lambda_1(\Lambda) \geq \lambda_1(\Omega) \int_\Lambda |\phi|^2 \mathrm{d}x + \lambda_2(\Omega) \int_{\Omega\setminus\Lambda} |\phi|^2 \mathrm{d}x.$$

A proof of this result can be found in [Szn98]. For the generalization of Thirring's inequality we use the previous notation and assume additionally that (X, m) is a measure space, $\mathcal{H} = L^2(X, m)$ and $\mathcal{G}$ is a subspace of $\mathcal{F}$ such that $\mathcal{H}^{\mathcal{G}} = L^2(Y, m)$ with some measurable set $Y \subset X$.

Corollary 5.13. *Suppose that λ is a simple eigenvalue of H with normalized eigenfunction ϕ. Then*

$$\lambda^{\mathcal{G}} \geq \lambda \int_Y |\phi|^2 \mathrm{d}m + (\lambda + \mu) \int_{X \setminus Y} |\phi|^2 \mathrm{d}m,$$

where μ is as in (5.7) the spectral gap.

Proof. Since λ is simple the orthogonal projection $P_{\mathcal{K}}$ onto the eigenspace $\mathcal{K} = \ker(H - \lambda)$ is given by

$$P_{\mathcal{K}} u = \mathcal{E}_1(\phi, u)\phi = \langle \phi, u\rangle \phi, \quad u \in \mathcal{F}.$$

Hence if $u \in \mathcal{G}$ such that $\|u\| = 1$:

$$\operatorname{Cap}_u(\mathcal{K}) = \mathcal{E}_1[P_{\mathcal{K}} u] = |\langle \phi, u\rangle|^2 \mathcal{E}_1[\phi] = \left| \int_Y \overline{\phi} u \, \mathrm{d}m \right|^2 \leq \int_Y |\phi|^2 \mathrm{d}m,$$

and we see that

$$\delta := \sup\{\operatorname{Cap}_u(\mathcal{K}) : u \in \mathcal{G}, \|u\| = 1\} \leq \int_Y |\phi|^2 \mathrm{d}m.$$

Thus by Theorem 5.12

$$\begin{aligned} \lambda^{\mathcal{G}} &\geq \mu(1 - \delta) + \lambda \geq \mu \int_{X \setminus Y} |\phi|^2 \mathrm{d}m + \lambda \\ &= (\mu + \lambda) \int_{X \setminus Y} |\phi|^2 \mathrm{d}m + \lambda \int_Y |\phi|^2 \mathrm{d}m. \end{aligned}$$

□

Note that we did not assume that $\lambda + \mu$ is an eigenvalue of H.

Remark 5.14. Let us look at Corollary 5.13 in the special case of domain perturbations on small balls in $\mathbb{R}^d$. More precisely, assume that the measure space X is some open subset Ω of $\mathbb{R}^d$, $\mathrm{d}m = \mathrm{d}x$ is the Lebesgue measure and $Y = \Omega \setminus B_r(x_0)$, where $B_r(x_0)$ is the closed ball of radius r centered at x_0. If ϕ is continuous at x_0 and $\phi(x_0) \neq 0$ we have by Corollary 5.13

$$\lambda^{\mathcal{G}} - \lambda \geq \lambda \int_{\Omega \setminus B_r(x_0)} |\phi|^2 \mathrm{d}x + (\lambda + \mu) \int_{B_r(x_0)} |\phi|^2 \mathrm{d}x - \lambda = \mu \int_{B_r(x_0)} |\phi|^2 \mathrm{d}x \geq c r^d$$

for some constant $c > 0$ and sufficiently small r.

Example. Let Ω be an open and bounded subset of $\mathbb{R}^d$ and let

$$H = \sum_{|\alpha|, |\beta| \leq m} (-1)^\alpha D^\alpha (a_{\alpha\beta} D^\beta u), \quad a_{\alpha\beta} = \overline{a_{\beta\alpha}} \in \mathbb{C}$$

be an elliptic constant coefficient operator of order $2m$, defined as the closure of the form

$$(u, v) \mapsto \sum_{|\alpha|, |\beta| \leq m} \int_\Omega a_{\alpha\beta} D^\alpha \overline{u} D^\beta v \, \mathrm{d}x,$$

initially defined on $C_c^\infty(\Omega)$, the space of smooth functions with compact support. Then the spectrum of H is discrete by Theorem 14.6 of [Agm65] and all eigenfunctions are analytic (see e.g., [Joh49], [Hör83a], Section 8 and [Hör83b], Section 11). Assume that the bottom eigenvalue λ is simple which happens to be true e.g., in the case of second-order differential operators on a bounded domain. Fix a point $x_0 \in \Omega$ such that the first eigenfunction ϕ of H does not vanish in x_0. Define the self-adjoint operator H_{r,x_0} in $L^2(\Omega \setminus B_r(x_0))$ in the same way as H, but with initial form domain $C_c^\infty(\Omega \setminus B_r(x_0))$. Denote its bottom eigenvalue by λ_{r,x_0}. Then Remark 5.14 gives us

$$\lambda_{r,x_0} - \lambda \geq cr^d$$

for some constant $c > 0$ and sufficiently small r. This estimate is far from optimal since we shall see in Section 6, Theorem 6.2, that for $d > 2m$ one even has

$$\lambda_{x,x_0} - \lambda \geq \operatorname{const} r^{d-2m}.$$

The proof of this result is based on Theorem 5.10.

Let us finally illustrate the upper and lower bounds by an almost trivial example. Further examples are given in Section 6.

Example. Let $\mathcal{F} = \mathcal{H} = \mathbb{R}^2$, $H = \begin{pmatrix} 0 & 0 \\ 0 & \mu \end{pmatrix}$ with some $\mu > 0$ and let $\mathcal{G}$ be the linear span of $e_\alpha := (\cos\alpha, \sin\alpha)$ where $\alpha \in [0, \pi)$. Since $\mathcal{E}[e_\alpha] = \mu \sin^2 \alpha$ we have $H^{\mathcal{G}} = \mu \sin^2 \alpha$ and $\lambda^{\mathcal{G}} = \mu \sin^2 \alpha$. Now let us compute the ϕ-capacity of $\mathcal{G}^\perp$ where $\phi = (1, 0)$ is the ground-state of H. Since $\mathcal{E}_1[e_\alpha]^{1/2} = (1 + \mu \sin^2 \alpha)^{1/2} =: c_\alpha$ we get

$$\operatorname{Cap}_\phi(\mathcal{G}) = \mathcal{E}_1[\mathcal{E}_1(\phi, c_\alpha^{-1} e_\alpha) c_\alpha^{-1} e_\alpha] = c_\alpha^{-4} \mathcal{E}_1[\cos\alpha e_\alpha] = \frac{\cos^2 \alpha}{1 + \mu \sin^2 \alpha}$$

which gives us

$$\operatorname{Cap}_\phi(\mathcal{G}^\perp) = 1 - \operatorname{Cap}_\phi(\mathcal{G}) = \frac{(1 + \mu)\sin^2 \alpha}{1 + \mu \sin^2 \alpha}.$$

Hence, by Theorem 5.6

$$\lambda^{\mathcal{G}} \leq \frac{\operatorname{Cap}_\phi(\mathcal{G}^\perp)}{1 - \operatorname{Cap}_\phi(\mathcal{G}^\perp)} = (1 + \mu) \tan^2 \alpha,$$

and by Theorem 5.10

$$\lambda^{\mathcal{G}} \geq \frac{\mu \operatorname{Cap}_\phi(\mathcal{G}^\perp)}{1 + \mu(1 - \operatorname{Cap}_\phi(\mathcal{G}^\perp))} = \frac{\mu(1 + \mu)\sin^2 \alpha}{1 + \mu \sin^2 \alpha} \frac{1}{1 + \mu \frac{\cos^2 \alpha}{1 + \mu \sin^2 \alpha}} = \mu \sin^2 \alpha,$$

and we see that the upper bound is a good estimate only if $\operatorname{Cap}_\phi(\mathcal{G}^\perp)$ is sufficiently small, whereas the lower bound always gives the correct value.

5.3. Higher eigenvalues

There is an enormous literature on estimates for eigenvalues of the Laplacian or other special operators as well as relations between these eigenvalues such as the ratio between the first and the second one. We do not try to list these results but remark instead that there are only few results which can treat general classes of operators. One of those is the previously mentioned work of McGillivray [McG96] who proved in the context of regular Dirichlet forms that under an ultracontractivity condition the shift of the higher eigenvalues may be estimated from above by some constant times the capacity of the set on which the domain perturbation takes place. We shall obtain a similar estimate for the second eigenvalue in our general Hilbert space setting.

To keep things simple we assume from now on that H has purely discrete spectrum, i.e., there is an orthonormal basis $(\phi_j)_{j\in\mathbb{N}}$ of $\mathcal{H}$ and a non-decreasing sequence $(\lambda_j)_{j\in\mathbb{N}}$ of real numbers such that

$$\operatorname{dom}(H) = \{u \in \mathcal{H} : (\lambda_j\langle\phi_j, u\rangle)_{j\in\mathbb{N}} \in \ell^2\}$$

and

$$Hu = \sum_{j\in\mathbb{N}} \lambda_j\langle\phi_j, u\rangle\phi_j$$

for all $u \in \operatorname{dom}(H)$.

Since we are interested mainly in the shift of the eigenvalues we may assume without loss of generality that the bottom eigenvalue λ_1 is equal to zero. Define $\mathcal{E}, \mathcal{F}, \mathcal{G}, \mathcal{H}^{\mathcal{G}}, H^{\mathcal{G}}$ as in the previous sections. It is a consequence of the minimax principle that the spectrum of $H^{\mathcal{G}}$ is again discrete and that each eigenvalue, not only the lowest, gets shifted to the right. Hence there is an orthonormal basis $(\psi_j)_{j\in\mathbb{N}}$ of $\mathcal{H}^{\mathcal{G}}$ and a sequence $(\lambda_j^{\mathcal{G}})_{j\in\mathbb{N}}$ satisfying $\lambda_j^{\mathcal{G}} \geq \lambda_j$ such that we have the following representation.

$$H^{\mathcal{G}}v = \sum_{j\in\mathbb{N}} \lambda_j^{\mathcal{G}}\langle\psi_j, v\rangle\psi_j$$

for all

$$v \in \operatorname{dom}(H^{\mathcal{G}}) = \{w \in \mathcal{G} : (\lambda_j^{\mathcal{G}}\langle\psi_j, w\rangle)_{j\in\mathbb{N}} \in \ell^2\}.$$

Now the main result on the shift of the second eigenvalue reads as follows:

Theorem 5.15. *Suppose that*

$$\operatorname{Cap}_{\phi_1}(\mathcal{G}^\perp) < \lambda_2 \frac{(1+\lambda_2-\operatorname{Cap}_{\phi_2}(\mathcal{G}^\perp))^2}{(1+\lambda_2)^4 + (1+\lambda_2-\operatorname{Cap}_{\phi_2}(\mathcal{G}^\perp))^2\lambda_2}. \tag{5.10}$$

Then

$$\begin{aligned}\lambda_2^{\mathcal{G}} &- \lambda_2 \\ &\leq \lambda_2 \frac{(1-\operatorname{Cap}_{\phi_1}(\mathcal{G}^\perp))(1+\lambda_2)^2\operatorname{Cap}_{\phi_2}(\mathcal{G}^\perp) + (1+\lambda_2)^4\operatorname{Cap}_{\phi_1}(\mathcal{G}^\perp)}{(1-\operatorname{Cap}_{\phi_1}(\mathcal{G}^\perp))(1+\lambda_2-\operatorname{Cap}_{\phi_2}(\mathcal{G}^\perp))^2\lambda_2 - \operatorname{Cap}_{\phi_1}(\mathcal{G}^\perp)(1+\lambda_2)^4}.\end{aligned}$$

As we shall see in Example 6.1, inequality (5.10) is fulfilled in great generality for elliptic differential operators if the domain perturbation takes place on a sufficiently small ball. In this context we prove an inequality which links the two capacities $\mathrm{Cap}_{\phi_1}(\mathcal{G}^\perp)$ and $\mathrm{Cap}_{\phi_2}(\mathcal{G}^\perp)$ by

$$\mathrm{Cap}_{\phi_1}(\mathcal{G}^\perp) \leq c\mathrm{Cap}_{\phi_2}(\mathcal{G}^\perp) \tag{5.11}$$

for some constant $c > 0$ depending on the eigenfunctions ϕ_1 and ϕ_2 but independent of the class of $\mathcal{G}'s$ in question. If (5.11) holds we can prove that the shift of the second eigenvalue can at most grow linearly with respect to the capacity $\mathrm{Cap}_{\phi_2}(\mathcal{G}^\perp)$. Let us first state and prove this corollary before turning to the proof of Theorem 5.15.

Corollary 5.16. *Let λ_1 be a simple eigenvalue of H. Then, for each $c > 0$ there are constants $c_1, \varepsilon > 0$ such that for all closed subspaces $\mathcal{G}$ of $\mathcal{F}$ with $\mathrm{Cap}_{\phi_2}(\mathcal{G}^\perp) < \varepsilon$ and satisfying inequality* (5.11) *we have*

$$\lambda_2^{\mathcal{G}} - \lambda_2 \leq c_1 \mathrm{Cap}_{\phi_2}(\mathcal{G}^\perp).$$

The constant c_1 can be given explicitly:

$$c_1 = 2(1+\lambda_2)^2(1+2c(1+\lambda_2)^2).$$

Proof. Since λ_1 is simple we have $\lambda_2 > 0$. In view of inequality (5.11) there is a constant $\gamma_0 > 0$ such that the hypothesis of Theorem 5.15 is satisfied whenever $\mathrm{Cap}_{\phi_2}(\mathcal{G}^\perp) \leq \gamma_0$. Similarly we can achieve that

$$\mathrm{Cap}_{\phi_2}(\mathcal{G}^\perp) \leq \frac{\lambda_2}{4c} \frac{(1+\lambda_2 - \mathrm{Cap}_{\phi_2}(\mathcal{G}^\perp))^2}{(1+\lambda_2)^4} \tag{5.12}$$

whenever $\mathrm{Cap}_{\phi_2}(\mathcal{G}^\perp) \leq \gamma_1$ for some other constant $\gamma_1 > 0$. Put

$$\gamma := \min\{\gamma_0, \gamma_1, \lambda_2, \frac{1}{2c}\} > 0.$$

Hence, if $\mathrm{Cap}_{\phi_2}(\mathcal{G}^\perp) \leq \gamma$, Theorem 5.15 gives us

$$\begin{aligned} &\lambda_2^{\mathcal{G}} - \lambda_2 \\ &\leq \frac{(1+\lambda_2)^2(1-\mathrm{Cap}_{\phi_1}(\mathcal{G}^\perp))\mathrm{Cap}_{\phi_2}(\mathcal{G}^\perp) + (1+\lambda_2)^4\mathrm{Cap}_{\phi_1}(\mathcal{G}^\perp)}{\lambda_2(1+\lambda_2-\mathrm{Cap}_{\phi_2}(\mathcal{G}^\perp))^2(1-\mathrm{Cap}_{\phi_1}(\mathcal{G}^\perp)) - (1+\lambda_2)^4\mathrm{Cap}_{\phi_1}(\mathcal{G}^\perp)}\lambda_2, \\ &= \frac{(1+\lambda_2)^2\mathrm{Cap}_{\phi_2}(\mathcal{G}^\perp) + (1+\lambda_2)^4\frac{\mathrm{Cap}_{\phi_1}(\mathcal{G}^\perp)}{1-\mathrm{Cap}_{\phi_1}(\mathcal{G}^\perp)}}{(1+\lambda_2-\mathrm{Cap}_{\phi_2}(\mathcal{G}^\perp))^2 - \frac{(1+\lambda_2)^4\mathrm{Cap}_{\phi_1}(\mathcal{G}^\perp)}{\lambda_2(1-\mathrm{Cap}_{\phi_1}(\mathcal{G}^\perp))}}. \end{aligned}$$

Since $\mathrm{Cap}_{\phi_2}(\mathcal{G}^\perp) \leq \frac{1}{2c}$ we have $\mathrm{Cap}_{\phi_1}(\mathcal{G}^\perp) \leq 1/2$. Hence

$$\frac{\mathrm{Cap}_{\phi_1}(\mathcal{G}^\perp)}{1-\mathrm{Cap}_{\phi_1}(\mathcal{G}^\perp)} \leq 2\mathrm{Cap}_{\phi_1}(\mathcal{G}^\perp).$$

From this inequality and (5.11) we conclude

$$\begin{aligned}\lambda_2^{\mathcal{G}} - \lambda_2 &\le \frac{(1+\lambda_2)^2\mathrm{Cap}_{\phi_2}(\mathcal{G}^\perp) + 2\mathrm{Cap}_{\phi_1}(\mathcal{G}^\perp)(1+\lambda_2)^4}{(1+\lambda_2-\mathrm{Cap}_{\phi_2}(\mathcal{G}^\perp))^2 - 2\mathrm{Cap}_{\phi_1}(\mathcal{G}^\perp)(1+\lambda_2)^4/\lambda_2}\\ &\le \frac{(1+\lambda_2)^2\mathrm{Cap}_{\phi_2}(\mathcal{G}^\perp) + 2c(1+\lambda_2)^4\mathrm{Cap}_{\phi_2}(\mathcal{G}^\perp)}{(1+\lambda_2-\mathrm{Cap}_{\phi_2}(\mathcal{G}^\perp))^2 - 2c\mathrm{Cap}_{\phi_2}(\mathcal{G}^\perp)(1+\lambda_2)^4/\lambda_2}.\end{aligned}$$

Because of inequality (5.12) the calculation continues and we obtain

$$\begin{aligned}\lambda_2^{\mathcal{G}} - \lambda_2 &\le 2\frac{(1+\lambda_2)^2 + 2c(1+\lambda_2)^4}{(1+\lambda_2-\mathrm{Cap}_{\phi_2}(\mathcal{G}^\perp))^2}\mathrm{Cap}_{\phi_2}(\mathcal{G}^\perp)\\ &\le 2(1+\lambda_2)^2(1+2c(1+\lambda_2)^2)\mathrm{Cap}_{\phi_2}(\mathcal{G}^\perp),\end{aligned}$$

where in the last step we used that $\mathrm{Cap}_{\phi_2}(\mathcal{G}^\perp) \le \lambda_2$. □

Before we can prove Theorem 5.15 we have to prepare some lemmas, the first of which already gives an upper bound for the second eigenvalue but involves the scalar product $|\langle P_{\mathcal{G}}\phi_2, \psi_1\rangle|$ which shall be estimated later on.

Lemma 5.17. *Let $P_{\mathcal{G}}$ be the orthogonal projection onto $\mathcal{G}$ in the Hilbert space $(\mathcal{F}, \mathcal{E}_1(\cdot,\cdot))$. Suppose that*

$$|\langle P_{\mathcal{G}}\phi_2, \psi_1\rangle| < \frac{1+\lambda_2-\mathrm{Cap}_{\phi_2}(\mathcal{G}^\perp)}{1+\lambda_2}.$$

Then

$$1+\lambda_2^{\mathcal{G}} \le \frac{1+\lambda_2-\mathrm{Cap}_{\phi_2}(\mathcal{G}^\perp) - |\langle P_{\mathcal{G}}\phi_2, \psi_1\rangle|^2(1+\lambda_1^{\mathcal{G}})}{(1+\lambda_2-\mathrm{Cap}_{\phi_2}(\mathcal{G}^\perp))^2 - |\langle P_{\mathcal{G}}\phi_2, \psi_2\rangle|^2(1+\lambda_2)^2}(1+\lambda_2)^2.$$

Proof. Put

$$\alpha := -\langle P_{\mathcal{G}}\phi_2, \psi_1\rangle.$$

By multiplying ϕ_2 with some constant of modulus 1 we can achieve that α is real without changing the value of $\mathrm{Cap}_{\phi_2}(\mathcal{G}^\perp)$. Since

$$\langle P_{\mathcal{G}}\phi_2 + \alpha\psi_1, \psi_1\rangle = 0,$$

we get

$$\begin{aligned}1+\lambda_2^{\mathcal{G}} &= \inf\left\{\frac{\mathcal{E}_1[u]}{\|u\|^2} : u \in \mathcal{G}\setminus\{0\}, \langle u, \psi_1\rangle = 0\right\}\\ &\le \frac{\mathcal{E}_1[P_{\mathcal{G}}\phi_2 + \alpha\psi_1]}{\|P_{\mathcal{G}}\phi_2 + \alpha\psi_1\|^2}.\end{aligned}$$

We now estimate numerator and denominator separately. Let $P_{\mathcal{G}^\perp} = 1 - P_{\mathcal{G}}$ be the projection onto $\mathcal{G}^\perp$. The numerator can be calculated exactly:

$$\begin{aligned}
&\mathcal{E}_1[P_{\mathcal{G}}\phi_2 + \alpha\psi_1] = \mathcal{E}_1[P_{\mathcal{G}}(\phi_2 + \alpha\psi_1)] \\
&\quad = \mathcal{E}_1[\phi_2 + \alpha\psi_1] - \mathcal{E}_1[P_{\mathcal{G}^\perp}(\phi_2 + \alpha\psi_1)] \\
&\quad = \mathcal{E}_1[\phi_2 + \alpha\psi_1] - \mathcal{E}_1[P_{\mathcal{G}^\perp}\phi_2] \\
&\quad = \lambda_2 + 1 + 2\alpha\mathrm{Re}(\mathcal{E}_1(\phi_2, \psi_1)) + \alpha^2\mathcal{E}_1[\psi_1] - \mathrm{Cap}_{\phi_2}(\mathcal{G}^\perp) \\
&\quad = \lambda_2 + 1 + 2\alpha(\lambda_1^{\mathcal{G}} + 1)\mathrm{Re}(\langle P_{\mathcal{G}}\phi_2, \psi_1\rangle) + \alpha^2(\lambda_1^{\mathcal{G}} + 1) - \mathrm{Cap}_{\phi_2}(\mathcal{G}^\perp) \\
&\quad = \lambda_2 + 1 - \mathrm{Cap}_{\phi_2}(\mathcal{G}^\perp) - \alpha^2(\lambda_1^{\mathcal{G}} + 1).
\end{aligned}$$

For the denominator we have the following estimate:

$$\begin{aligned}
&\|P_{\mathcal{G}}\phi_2 + \alpha\psi_1\|^2 = \|P_{\mathcal{G}}\phi_2\|^2 + 2\alpha\mathrm{Re}(\langle P_{\mathcal{G}}\phi_2, \psi_1\rangle) + \alpha^2 \\
&\quad = \|(1 - P_{\mathcal{G}^\perp})\phi_2\|^2 - \alpha^2 \\
&\quad = 1 - 2\mathrm{Re}(\langle \phi_2, P_{\mathcal{G}^\perp}\phi_2\rangle) + \|P_{\mathcal{G}^\perp}\phi_2\|^2 - \alpha^2 \\
&\quad \geq 1 - 2\mathrm{Re}(\langle \phi_2, P_{\mathcal{G}^\perp}\phi_2\rangle) + |\langle \phi_2, P_{\mathcal{G}^\perp}\phi_2\rangle|^2 - \alpha^2 \\
&\quad = 1 - \frac{2}{1+\lambda_2}\mathrm{Re}(\mathcal{E}_1(\phi_2, P_{\mathcal{G}^\perp}\phi_2)) + \frac{1}{(1+\lambda_2)^2}|\mathcal{E}_1(\phi_2, P_{\mathcal{G}^\perp}\phi_2)|^2 - \alpha^2 \\
&\quad = \left(1 - \frac{\mathrm{Cap}_{\phi_2}(\mathcal{G}^\perp)}{1+\lambda_2}\right)^2 - \alpha^2 \\
&\quad = \frac{(1 + \lambda_2 - \mathrm{Cap}_{\phi_2}(\mathcal{G}^\perp))^2 - \alpha^2(1+\lambda_2)^2}{(1+\lambda_2)^2},
\end{aligned}$$

and the assertion follows immediately. □

In view of this lemma it is clear that an upper bound for $|\langle P_{\mathcal{G}}\phi_2, \psi_1\rangle|$ is useful. The next lemma reduces the problem of finding upper bounds for $|\langle P_{\mathcal{G}}\phi_2, \psi_1\rangle|$ to finding lower bounds for $|\langle \phi_1, \psi_1\rangle|$.

Lemma 5.18. *The following inequality holds.*

$$|\langle P_{\mathcal{G}}\phi_2, \psi_1\rangle|^2 \leq (\lambda_2 + 1)^2(1 - |\langle \phi_1, \psi_1\rangle|^2).$$

Proof. First we get rid of the projection $P_{\mathcal{G}}$ by observing that

$$\begin{aligned}
\langle P_{\mathcal{G}}\phi_2, \psi_1\rangle &= \frac{1}{\lambda_1^{\mathcal{G}} + 1}\mathcal{E}_1(P_{\mathcal{G}}\phi_2, \psi_1) \\
&= \frac{1}{\lambda_1^{\mathcal{G}} + 1}\mathcal{E}_1(\phi_2, \psi_1) \\
&= \frac{\lambda_2 + 1}{\lambda_1^{\mathcal{G}} + 1}\langle \phi_2, \psi_1\rangle \\
&= \frac{\lambda_2 + 1}{\lambda_1^{\mathcal{G}} + 1}\langle \phi_2, \psi_1 - \langle \phi_1, \psi_1\rangle\phi_1\rangle.
\end{aligned}$$

Then the Cauchy–Schwarz inequality gives us

$$\begin{aligned}|\langle P_{\mathcal{G}}\phi_2,\psi_1\rangle|^2 &\le \left(\frac{\lambda_2+1}{\lambda_1^{\mathcal{G}}+1}\right)^2 \|\psi_1-\langle\phi_1,\psi_1\rangle\phi_1\|^2\\ &= \left(\frac{\lambda_2+1}{\lambda_1^{\mathcal{G}}+1}\right)^2 (1-|\langle\phi_1,\psi_1\rangle|^2)\\ &\le (\lambda_2+1)^2(1-|\langle\phi_1,\psi_1\rangle|^2).\end{aligned}$$

□

Finally we give a lower bound for $|\langle\phi_1,\psi_1\rangle|$.

Lemma 5.19. *If λ_1 is simple, i.e., if $\lambda_2 > 0$, the following inequality holds.*

$$|\langle\phi_1,\psi_1\rangle|^2 \ge \frac{\lambda_2-\lambda_1^{\mathcal{G}}}{\lambda_2}.$$

Proof. This is seen by the following calculation.

$$\begin{aligned}1+\lambda_1^{\mathcal{G}} &= \mathcal{E}_1[\psi_1]\\ &= \mathcal{E}_1\left[\sum_{j=1}^{\infty}\langle\phi_j,\psi_1\rangle\phi_j\right]\\ &= \sum_{j=1}^{\infty}|\langle\phi_j,\psi_1\rangle|^2\mathcal{E}_1[\phi_j]\\ &= |\langle\phi_1,\psi_1\rangle|^2+\sum_{j=2}^{\infty}(\lambda_j+1)|\langle\phi_j,\psi_1\rangle|^2\\ &\ge |\langle\phi_1,\psi_1\rangle|^2+(1+\lambda_2)\sum_{j=2}^{\infty}|\langle\phi_j,\psi_1\rangle|^2\\ &= |\langle\phi_1,\psi_1\rangle|^2+(1+\lambda_2)(1-|\langle\phi_1,\psi_1\rangle|^2)\\ &= 1+\lambda_2-\lambda_2|\langle\phi_1,\psi_1\rangle|^2.\end{aligned}$$

□

Since $\lambda_1^{\mathcal{G}}$ can be estimated from above in terms of the ϕ_1-capacity of $\mathcal{G}^\perp$, we are now in the position to prove a capacitary upper bound for $|\langle P_{\mathcal{G}}\phi_2,\psi_1\rangle|$ which involves only λ_2 and $\mathrm{Cap}_{\phi_1}(\mathcal{G}^\perp)$.

Lemma 5.20. *If λ_1 is simple and $\mathrm{Cap}_{\phi_1}(\mathcal{G}^\perp) < 1$, then*

$$|\langle P_{\mathcal{G}}\phi_2,\psi_1\rangle|^2 \le \frac{(\lambda_2+1)^2}{\lambda_2}\frac{\mathrm{Cap}_{\phi_1}(\mathcal{G}^\perp)}{1-\mathrm{Cap}_{\phi_1}(\mathcal{G}^\perp)}.$$

Proof. In view of Lemma 5.18 and Lemma 5.19 we have

$$\begin{aligned}
|\langle P_{\mathcal{G}}\phi_2, \psi_1\rangle|^2 &\leq (\lambda_2+1)^2(1-|\langle\phi_1,\psi_1\rangle|^2) \\
&\leq (\lambda_2+1)^2\left(1-\frac{\lambda_2-\lambda_1^{\mathcal{G}}}{\lambda_2}\right) \\
&= \frac{(\lambda_2+1)^2}{\lambda_2}\lambda_1^{\mathcal{G}} \\
&\leq \frac{(\lambda_2+1)^2}{\lambda_2}\frac{\mathrm{Cap}_{\phi_1}(\mathcal{G}^\perp)}{1-\mathrm{Cap}_{\phi_1}(\mathcal{G}^\perp)},
\end{aligned}$$

where in the last inequality we used the capacitary upper bound of Corollary 5.6 (a) for the first eigenvalue. □

Putting together what we have achieved so far, we can now easily prove Theorem 5.15.

Proof. (of Theorem 5.15)

Put $\alpha := -\langle P_{\mathcal{G}}\phi_2, \psi_1\rangle$. Observe that (5.10) implies that λ_1 is simple and that $\mathrm{Cap}_{\phi_1}(\mathcal{G}^\perp) < 1$. Hence Lemma 5.20 and inequality (5.10) yield

$$\alpha^2 \leq \frac{(\lambda_2+1)^2}{\lambda_2}\frac{\mathrm{Cap}_{\phi_1}(\mathcal{G}^\perp)}{1-\mathrm{Cap}_{\phi_1}(\mathcal{G}^\perp)} < \left(\frac{1+\lambda_2-\mathrm{Cap}_{\phi_2}(\mathcal{G}^\perp)}{1+\lambda_2}\right)^2. \tag{5.13}$$

Therefore the hypothesis of Lemma 5.17 is verified and we get

$$\begin{aligned}
&\lambda_2^{\mathcal{G}}-\lambda_2 = 1+\lambda_2^{\mathcal{G}}-(1+\lambda_2) \\
&\leq \frac{1+\lambda_2-\mathrm{Cap}_{\phi_2}(\mathcal{G}^\perp)-\alpha^2(1+\lambda_1^{\mathcal{G}})}{(1+\lambda_2-\mathrm{Cap}_{\phi_2}(\mathcal{G}^\perp))^2-\alpha^2(1+\lambda_2)^2}(1+\lambda_2)^2-(1+\lambda_2) \\
&= \frac{(1+\lambda_2)(1+\lambda_2-\mathrm{Cap}_{\phi_2}(\mathcal{G}^\perp))\mathrm{Cap}_{\phi_2}(\mathcal{G}^\perp)+\alpha^2(1+\lambda_2)^2(\lambda_2-\lambda_1^{\mathcal{G}})}{(1+\lambda_2-\mathrm{Cap}_{\phi_2}(\mathcal{G}^\perp))^2-\alpha^2(1+\lambda_2)^2} \\
&\leq \frac{(1+\lambda_2)^2\mathrm{Cap}_{\phi_2}(\mathcal{G}^\perp)+\alpha^2(1+\lambda_2)^2\lambda_2}{(1+\lambda_2-\mathrm{Cap}_{\phi_2}(\mathcal{G}^\perp))^2-\alpha^2(1+\lambda_2)^2}.
\end{aligned}$$

Now the first inequality of (5.13) gives us

$$\begin{aligned}
\lambda_2^{\mathcal{G}}-\lambda_2 &\leq \frac{(1+\lambda_2)^2\mathrm{Cap}_{\phi_2}(\mathcal{G}^\perp)+(1+\lambda_2)^4\frac{\mathrm{Cap}_{\phi_1}(\mathcal{G}^\perp)}{1-\mathrm{Cap}_{\phi_1}(\mathcal{G}^\perp)}}{(1+\lambda_2-\mathrm{Cap}_{\phi_2}(\mathcal{G}^\perp))^2-\frac{(1+\lambda_2)^4\mathrm{Cap}_{\phi_1}(\mathcal{G}^\perp)}{\lambda_2(1-\mathrm{Cap}_{\phi_1}(\mathcal{G}^\perp))}} \\
&= \frac{(1+\lambda_2)^2(1-\mathrm{Cap}_{\phi_1}(\mathcal{G}^\perp))\mathrm{Cap}_{\phi_2}(\mathcal{G}^\perp)+(1+\lambda_2)^4\mathrm{Cap}_{\phi_1}(\mathcal{G}^\perp)}{\lambda_2(1+\lambda_2-\mathrm{Cap}_{\phi_2}(\mathcal{G}^\perp))^2(1-\mathrm{Cap}_{\phi_1}(\mathcal{G}^\perp))-(1+\lambda_2)^4\mathrm{Cap}_{\phi_1}(\mathcal{G}^\perp)}\lambda_2,
\end{aligned}$$

and Theorem 5.15 is proved. □

Applications of this theorem are given in Section 6.

Remark 5.21. Let H be a self-adjoint operator which corresponds to a regular Dirichlet form in $L^2(X,m)$. If Y is an open subset of X such that $K := X \setminus Y$ is compact and H^Y arises from H by imposing Dirichlet boundary conditions on K, one can *not* prove a capacitary lower bound of the form

$$\lambda_j^Y - \lambda_j \geq \text{const}\,\text{Cap}(K) \tag{5.14}$$

for the second eigenvalue or for higher eigenvalues (here $\text{Cap}(\cdot)$ denotes the capacity of the regular Dirichlet form associated to H). This is seen from the following example (cf. Figure 3 below). Let $\Omega = (0,\pi)^2 \subset \mathbb{R}^2$, $H = -\Delta$ with Dirichlet

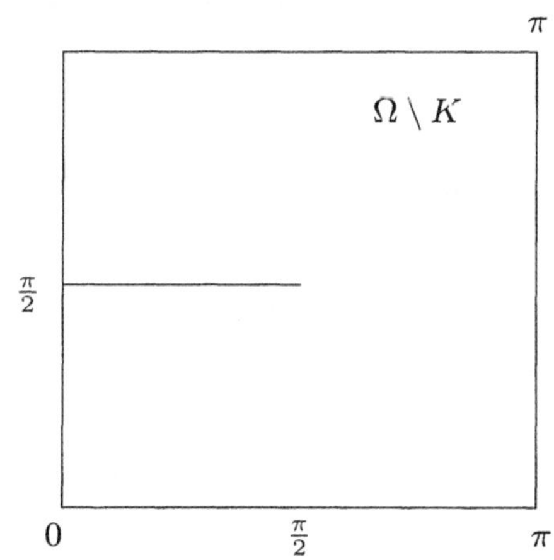

FIGURE 3. A domain with $\lambda_2 = \lambda_2^Y$

boundary conditions on $\partial\Omega$ and let H^Y be the Dirichlet Laplacian in $L^2(\Omega \setminus K)$, where

$$K = \{(x,y) \in \Omega : x \leq \pi/2, y = \pi/2\}.$$

One can easily prove that the second eigenvalues λ_2 and λ_2^Y of H and H^Y coincide although the capacity of K is positive, see [GZ94] or [DMN97].

The crucial point with this example is that the second eigenfunction ϕ_2 of H given by

$$\phi_2(x,y) = \sin(x)\sin(2y)$$

already belongs to the form domain $\mathcal{G} = H_0^1(\Omega \setminus K)$ of H^Y. Let us ask what this means for the general capacity of Definition 4.2: Since $\phi_2 \in \mathcal{G}$ the projection of ϕ_2 onto $\mathcal{G}^\perp$ vanishes, i.e., we have $\text{Cap}_{\phi_2}(\mathcal{G}^\perp) = 0$, unlike as in the classical case. Hence the inequality $\lambda_2^Y - \lambda_2 \geq \text{const}\,\text{Cap}_{\phi_2}(\mathcal{G}^\perp)$ indeed holds for this particular example but we cannot prove it in general.

5.4. Additive perturbations

It is also possible to include additive perturbations together with the domain perturbation treated in the previous sections. In what follows we ask for an estimate for the spectral bound of operators $H^{\mathcal{G},V}$ of the form

$$H^{\mathcal{G},V} = H^{\mathcal{G}} + V,$$

where $H^{\mathcal{G}}$ is as in Definition 3.1 the domain perturbation with respect to the closed subspace $\mathcal{G}$ of $(\mathcal{F}, \mathcal{E}_1(\cdot,\cdot))$ and V is a suitable perturbation in $\mathcal{H}^{\mathcal{G}} = \overline{\mathcal{G}}^{\mathcal{H}}$.

Recall that a symmetric operator V in $\mathcal{H}^{\mathcal{G}}$ is called *infinitesimally form bounded* with respect to $H^{\mathcal{G}}$ if the domain of V contains the domain of $H^{\mathcal{G}}$ and for each $\varepsilon > 0$ there is some $c'_\varepsilon \geq 0$ such that for all $u \in \mathrm{dom}(H^{\mathcal{G}})$ the following inequality holds:

$$|\langle Vu, u\rangle| \leq \varepsilon |\langle H^{\mathcal{G}} u, u\rangle| + c'_\varepsilon \|u\|^2.$$

Observe that this is the case if and only if for each $\varepsilon > 0$ there is some $c_\varepsilon \geq 0$ such that

$$|\langle Vu, u\rangle| \leq \varepsilon \mathcal{E}_1[u] + c_\varepsilon \|u\|^2, \quad u \in \mathrm{dom}(H^{\mathcal{G}}). \tag{5.15}$$

The latter version is more convenient for our purpose. We shall always assume that c_ε is chosen as small as possible. Then $\varepsilon \mapsto c_\varepsilon$ is a monotonically decreasing function on $(0,\infty)$. It is well known, see Theorem VI.1.33 of [Kat80], that if V is infinitesimally form bounded with respect to $H^{\mathcal{G}}$, then the form

$$(u,v) \mapsto \langle H^{\mathcal{G}} u, v\rangle + \langle Vu, v\rangle, \quad u, v \in \mathrm{dom}(H^{\mathcal{G}})$$

is closable and semi-bounded from below (actually it would be enough to assume form boundedness with some bound strictly less than 1, see [Kat80], Section VI.6). Hence its closure $\mathcal{E} + \lambda + V$ determines a unique self-adjoint operator $H^{\mathcal{G},V}$ in $\mathcal{H}^{\mathcal{G}}$. Moreover the form domain of $H^{\mathcal{G},V}$ equals $\mathcal{G}$ and one has

$$(\mathcal{E}_1 + V)[u] \leq (1+\varepsilon)\mathcal{E}_1[u] + c_\varepsilon \|u\|^2$$

for all $u \in \mathcal{G}$. In the next theorem we prove an upper bound for the spectral bound $\lambda^{\mathcal{G},V} = \inf \sigma(H^{\mathcal{G},V})$ of $H^{\mathcal{G},V}$.

Theorem 5.22. *Let λ be an eigenvalue of H with normalized eigenfunction $\phi \notin \mathcal{G}$ and let V be a symmetric operator in $\mathcal{H}^{\mathcal{G}}$ which is infinitesimally form bounded with respect to $H^{\mathcal{G}}$. Then*

$$\lambda^{\mathcal{G},V} - \lambda \leq \inf_{\varepsilon>0} \left\{ c_\varepsilon + \frac{\varepsilon + \mathrm{Cap}_\phi(\mathcal{G}^\perp)}{1 - \mathrm{Cap}_\phi(\mathcal{G}^\perp)} \right\},$$

where c_ε is the constant in inequality (5.15).

Proof. Let $P_{\mathcal{G}^\perp}$ be the orthogonal projection (in $(\mathcal{F}, \mathcal{E}_1(\cdot,\cdot))$ onto $\mathcal{G}^\perp$. Then, by (5.15)

$$\begin{aligned} 1+\lambda^{\mathcal{G},V}-\lambda &= \inf\left\{\frac{(\mathcal{E}_1+V)[u]}{\|u\|^2} : u \in \mathcal{G}\setminus\{0\}\right\} \\ &\leq \frac{(\mathcal{E}_1+V)[(1-P_{\mathcal{G}^\perp})\phi]}{\|(1-P_{\mathcal{G}^\perp})\phi\|^2} \\ &\leq (1+\varepsilon)\frac{\mathcal{E}_1[(1-P_{\mathcal{G}^\perp})\phi]}{\|(1-P_{\mathcal{G}^\perp})\phi\|^2} + c_\varepsilon. \end{aligned}$$

From the proof of Theorem 5.5 we see that

$$\frac{\mathcal{E}_1[(1-P_{\mathcal{G}^\perp})\phi]}{\|(1-P_{\mathcal{G}^\perp})\phi\|^2} \leq \frac{1}{1-\mathrm{Cap}_\phi(\mathcal{G}^\perp)},$$

and the result follows. □

If V is a symmetric operator in $\mathcal{H}$ which is infinitesimally form bounded with respect to H we can define $H^{\mathcal{G},V}$ for any closed subspace of $\mathcal{G}$ by restricting the form $\mathcal{E}+\lambda+V$ to $\mathcal{G}$. In this case we can say more.

Corollary 5.23. *Fix $0<\delta_1<\delta_2<1$. Let λ be an eigenvalue of H with normalized eigenfunction ϕ and let V be a symmetric operator in $\mathcal{H}$ which is infinitesimally form bounded with respect to H. Then for every closed subspace $\mathcal{G}$ of $\mathcal{F}$ with $\mathrm{Cap}_\phi(\mathcal{G}^\perp) \in [\delta_1, \delta_2]$ the following inequality holds.*

$$\lambda^{\mathcal{G},V} - \lambda \leq c_{\delta_1} + \frac{2}{1-\delta_2}\mathrm{Cap}_\phi(\mathcal{G}^\perp),$$

where c_{δ_1} is the constant in inequality (5.15) *with $\varepsilon = \delta_1$*

Proof. Since $\mathrm{Cap}_\phi(\mathcal{G}^\perp) \leq \delta_2 < 1$ the hypothesis of Theorem 5.22 on ϕ is fulfilled. Thus Theorem 5.22 gives us

$$\begin{aligned} \lambda^{\mathcal{G},V}-\lambda &\leq \left\{\inf_{\varepsilon>0} c_\varepsilon + \frac{\varepsilon+\mathrm{Cap}_\phi(\mathcal{G}^\perp)}{1-\mathrm{Cap}_\phi(\mathcal{G}^\perp)}\right\} \\ &\leq c_{\mathrm{Cap}_\phi(\mathcal{G}^\perp)} + \frac{2\mathrm{Cap}_\phi(\mathcal{G}^\perp)}{1-\mathrm{Cap}_\phi(\mathcal{G}^\perp)} \\ &\leq c_{\delta_1} + \frac{2}{1-\delta_2}\mathrm{Cap}_\phi(\mathcal{G}^\perp). \end{aligned}$$

□

For applications of this result, see Section 6.2.

5.5. Operators with spectral bound of arbitrary type

In Section 5 the spectral bound $\lambda := \inf \sigma(H)$ was always supposed to be an eigenvalue. We proved capacitary estimates on the magnitude of the shift of the spectral bound when introducing a domain perturbation. These estimates measured the shift in terms of the capacity with respect to a normalized eigenfunction to λ. In this section we no longer assume that λ is an eigenvalue. Theorem 5.24 below gives a capacitary characterization for the spectral bound being unaffected under a domain perturbation.

Assume as in Section 3.2 that $(\mathcal{H}, \langle\cdot,\cdot\rangle)$ is an arbitrary real or complex Hilbert space and H is a self-adjoint operator in $\mathcal{H}$ which is semi-bounded from below with spectral bound $\lambda := \inf \sigma(H)$. Define $(\mathcal{E}, \mathcal{F})$ as the non-negative closed quadratic form which corresponds to $H - \lambda$. Given a closed subspace $\mathcal{G}$ of $(\mathcal{F}, \mathcal{E}_1(\cdot,\cdot))$, let $H^{\mathcal{G}}$ be the domain perturbation of H with respect to $\mathcal{G}$ in the sense of Definition 3.1 and put $\lambda^{\mathcal{G}} := \inf \sigma(H^{\mathcal{G}})$. The next theorem states that there is no shift of the spectral bound if and only if one can find normalized $\phi \in \mathcal{F}$ of arbitrarily small energy $\mathcal{E}[\phi]$ such that also the capacities $\mathrm{Cap}_\phi(\mathcal{G}^\perp)$ remain arbitrary small.

Theorem 5.24. *The following assertions are equivalent:*

(i) $\lambda^{\mathcal{G}} = \lambda$.

(ii) *For each* $\varepsilon > 0$ *there is a* $\phi_\varepsilon \in \mathcal{F}$ *such that* $\|\phi_\varepsilon\| = 1$, $\mathcal{E}[\phi_\varepsilon] < \varepsilon$ *and* $\mathrm{Cap}_{\phi_\varepsilon}(\mathcal{G}^\perp) < \varepsilon$.

Proof. The proof of (i) $\Rightarrow$ (ii) is easy. Since

$$\inf\{\mathcal{E}[u] : u \in \mathcal{G} : \|u\| = 1\} = \inf \sigma(H^{\mathcal{G}}) - \lambda = 0$$

we can find for each $\varepsilon > 0$ some $\phi_\varepsilon \in \mathcal{G}$ with $\mathcal{E}[\phi_\varepsilon] < \varepsilon$ and $\|\phi_\varepsilon\| = 1$. But $\phi_\varepsilon \in \mathcal{G}$ implies $\mathrm{Cap}_{\phi_\varepsilon}(\mathcal{G}^\perp) = 0$, see Remark 4.4 (a). For the other implication let $P_{\mathcal{G}^\perp}$ be the orthogonal projection onto $\mathcal{G}^\perp$. Then

$$\|P_{\mathcal{G}^\perp}\phi_\varepsilon\|^2 \leq \mathrm{Cap}_{\phi_\varepsilon}(\mathcal{G}^\perp) < \varepsilon$$

which gives us

$$\begin{aligned} 1 + \lambda^{\mathcal{G}} - \lambda &\leq \frac{\mathcal{E}_1[(1 - P_{\mathcal{G}^\perp})\phi_\varepsilon]}{\|(1 - P_{\mathcal{G}^\perp})\phi_\varepsilon)\|^2} \\ &\leq \frac{\mathcal{E}_1[\phi_\varepsilon]}{1 - 2\mathrm{Re}\langle\phi_\varepsilon, P_{\mathcal{G}^\perp}\phi_\varepsilon\rangle + \|P_{\mathcal{G}^\perp}\phi_\varepsilon\|^2} \\ &\leq \frac{1 + \varepsilon}{1 - 2|\langle\phi_\varepsilon, P_{\mathcal{G}^\perp}\phi_\varepsilon\rangle|} \\ &\leq \frac{1 + \varepsilon}{1 - 2\|P_{\mathcal{G}^\perp}\phi_\varepsilon\|} \\ &\leq \frac{1 + \varepsilon}{1 - 2\sqrt{\varepsilon}}. \end{aligned}$$

The result follows by letting ε tend to zero. □

Example. Let $H = -\frac{d^2}{dx^2}$ in $\mathcal{H} = L^2(\mathbb{R})$. The corresponding form is then given by

$$\mathcal{E}(u,v) = \int_{\mathbb{R}} u'v'\mathrm{d}x, \quad (u, v \in \mathcal{F} = H^1(\mathbb{R})).$$

Obviously the spectrum of H is purely absolutely continuous and is equal to the half-axis $[0,\infty)$. Let $\mathcal{G} = H_0^1((0,\infty))$. Then $H^{\mathcal{G}}$ is the Dirichlet Laplacian in $\mathcal{H}^{\mathcal{G}} = L^2((0,\infty))$. To derive from Theorem 5.24 that $\lambda^{\mathcal{G}} = 0$ (which is of course obvious for other reasons too) it is sufficient to consider for instance the sequence $\phi_n(x) = c_n\chi_{[0,n]}(x)\sin(\pi x/n)$, where c_n is such that $\|\phi_n\| = 1$.

6. Applications to differential operators of arbitrary order

The general results of Section 5 are applied to the special case of differential operators on open subsets of $\mathbb{R}^d$. We shall treat operators of arbitrary even order $2m$ with minimal assumptions on the coefficients. In Section 6.1 we show that the eigenvalue estimates of Section 5 yield estimates of the form

$$c_1 r^{d-2m} \leq \lambda^{\mathcal{G}} - \lambda \leq c_2 r^{d-2m}$$

and

$$\lambda_2^{\mathcal{G}} - \lambda_2 \leq c_3 r^{d-2m}$$

if $d > 2m$ and the domain perturbation takes place on a ball of sufficiently small radius r. The results of Section 5.4 are applied in Section 6.2 where we look at perturbations consisting of the above domain perturbation together with an additive perturbation belonging to the Stummel class of suitable order.

For further relevant literature which is not restricted to second-order operators we refer to [Maz85], [Wei84] and [Fre82].

6.1. Capacitary eigenvalue estimates for domain perturbations

Let $\Omega \subset \mathbb{R}^d$ be an open set. We want to treat operators in $L^2(\Omega)$ which are formally given by

$$(Hu)(x) = \sum_{|\alpha|,|\beta|\leq m} (-1)^{|\alpha|} D^\alpha(a_{\alpha\beta}(x)D^\beta u(x)) \tag{6.1}$$

with coefficient functions $a_{\alpha\beta} = \overline{a_{\beta\alpha}} \in L^1_{loc}(\Omega)$. To give a rigorous definition of H we consider the quadratic form

$$a(u,v) := \sum_{|\alpha|,|\beta|\leq m} \int_\Omega a_{\alpha\beta}(x) D^\alpha u D^\beta \overline{v}\,\mathrm{d}x, \tag{6.2}$$

initially defined on $C_c^\infty(\Omega)$, the space of smooth functions with compact support. In order to obtain a self-adjoint operator we need to know that a is closable. The requirement of closability is an implicit assumption on the coefficients $a_{\alpha\beta}$ and there are many criteria known on the coefficients $a_{\alpha\beta}$ implying closability of a, see e.g., Davies [Dav95], [Dav97] or Agmon [Agm65], Section 7. We will always assume that for some constants $b \in \mathbb{R}$, $c > 0$, and all $u \in C_c^\infty(\Omega)$ the following inequality holds.

$$c^{-1}\|u\|_m^2 \le a(u,u) + b\|u\|^2 \le c\|u\|_m^2, \tag{6.3}$$

where $\|\cdot\|_m$ is the Sobolev norm

$$\|u\|_m := \left(\sum_{|\alpha|\le m} \int_\Omega |D^\alpha u|^2 \mathrm{d}x \right)^{1/2}.$$

Observe that a quadratic form satisfying inequality (6.3) is always closable and that (6.3) is still valid for $u \in H_0^m(\Omega)$, the Sobolev space of order m, because $H_0^m(\Omega)$ is by definition the completion of $C_c^\infty(\Omega)$ with respect to the norm $\|\cdot\|_m$. This also makes clear that the closure $\overline{a}$ of a has domain $\mathcal{F} = H_0^m(\Omega)$. Let H be the self-adjoint operator corresponding to $(\overline{a}, \mathcal{F})$, put $\lambda = \inf \sigma(H)$ and let $\mathcal{E} := \overline{a} - \lambda$ be the non-negative form associated to $H - \lambda$.

Although there is in general no explicit description of the domain $\mathrm{dom}(H)$ of H we call H an operator with Dirichlet boundary conditions since H is the Friedrichs extension of the operator (6.1), initially defined on $C_c^\infty(\Omega)$, provided that the coefficients $a_{\alpha\beta}$ are sufficiently smooth.

We now want to investigate the u-capacity (see Definition 4.2) of subspaces of $\mathcal{F}$ consisting of functions that vanish on some set $A \subset \Omega$. Therefore we define $\mathcal{F}_A$ as in (4.11). If $A = \overline{B}_r(x_0) := \{x \in \Omega : |x - x_0| \le r\}$ we write $\mathcal{F}_{r,x_0}$ for short. Lemma 6.1 below compares the u-capacity of $\mathcal{F}_A$ with the $(m,2)$-capacity $\mathrm{Cap}_{m,2}(A)$ treated in the book of Adams and Hedberg [AH95]. Recall that this capacity is defined by

$$\mathrm{Cap}_{m,2}(A) := \inf\{\|u\|_m^2 : u \in C_c^\infty(\mathbb{R}^d), u \ge 1 \text{ on some nhd. of } A\},$$

see Section 2.2 of [AH95] for details.

Lemma 6.1. *Let $a_{\alpha\beta} = \overline{a_{\beta\alpha}} \in L^1_{loc}(\Omega)$ such that inequality (6.3) holds and let $u_0 \in H_0^m(\Omega)$.*

(a) *For each $\varepsilon > 0$ there is a constant $c_\varepsilon > 0$ such that for all $x_0 \in \Omega$ and all $r \in (0,1)$ with $B_{r(1+\varepsilon)}(x_0) \subset \Omega$ the following inequality holds.*

$$\mathrm{Cap}_{u_0}(\mathcal{F}_{r,x_0}^\perp) \le c_\varepsilon r^{-2m} \|u_0\|^2_{H_0^m(B_{r(1+\varepsilon)}(x_0))}.$$

(b) *If U is an open subset of Ω satisfying*

$$\operatorname*{essinf}_{x\in U} \mathrm{Re}(e^{i\theta} u_0(x)) > 0$$

for some $\theta \in [0, 2\pi)$, then there is some constant $c_U > 0$ such that for all $A \subset U$,

$$\mathrm{Cap}_{u_0}(\mathcal{F}_A^\perp) \ge c_U \,\mathrm{Cap}_{m,2}(A).$$

Proof. (a) Choose $\psi \in C_c^\infty(\mathbb{R}^d)$ such that $\psi(x) = 1$ if $|x| \le 1$ and $\psi(x) = 0$ if $|x| \ge 1 + \varepsilon$. Put

$$\Psi_r : \mathbb{R}^d \to \mathbb{R}, \quad \Psi_r(x) := \psi\left(\frac{x - x_0}{r}\right).$$

Then $\Psi_r u_0 \in H_0^m(\Omega)$, $\Psi_r u_0 = u_0$ on $B_r(x_0)$ and $\operatorname{supp}(\Psi_r) \subset B_{r(1+\varepsilon)}(x_0)$. In view of Proposition 4.5 (a) and inequality (6.3) we have

$$\operatorname{Cap}_{u_0}(\mathcal{F}_{r,x_0}^{\perp}) \leq \mathcal{E}_1[\Psi_r u_0] \leq c\|\Psi_r u_0\|_m^2 = c\|\Psi_r u_0\|_{H_0^m(B_{r(1+\varepsilon)}(x_0))}^2.$$

In what follows the symbol "const" stands for a generic constant, possibly changing its value from line to line, which is independent of r. By Leibniz' formula and the Cauchy–Schwarz inequality we obtain for $|\alpha| \leq m$,

$$\begin{aligned}
|D^\alpha \Psi_r u_0|^2 &\leq \text{const} \sum_{\beta\leq\alpha} \left|(D^\beta \Psi_r)\,(D^{\alpha-\beta} u_0)\right|^2 \\
&\leq \text{const} \sum_{\beta\leq\alpha} r^{-2|\beta|} \left|D^{\alpha-\beta} u_0\right|^2 \\
&\leq \text{const}\, r^{-2m} \sum_{\beta\leq\alpha} \left|D^{\alpha-\beta} u_0\right|^2 .
\end{aligned}$$

Hence the result follows by

$$\begin{aligned}
\operatorname{Cap}_{u_0}(\mathcal{F}_{r,x_0}^{\perp}) &\leq c \sum_{|\alpha|\leq m} \int_{B_{r(1+\varepsilon)}(x_0)} |D^\alpha \Psi_r u_0|^2 \,\mathrm{d}x \\
&\leq \text{const}\, r^{-2m} \sum_{|\alpha|\leq m} \int_{B_{r(1+\varepsilon)}(x_0)} \sum_{\beta\leq\alpha} \left|D^{\alpha-\beta} u_0\right|^2 \mathrm{d}x \\
&\leq \text{const}\, r^{-2m} \|u_0\|_{H_0^m(B_{r(1+\varepsilon)}(x_0))}^2 .
\end{aligned}$$

(b) Put

$$\gamma := \operatorname*{essinf}_{x\in U} \operatorname{Re}(e^{i\theta} u_0(x)).$$

Then by Proposition 4.5 (a) and inequality (6.3)

$$\begin{aligned}
&\operatorname{Cap}_{u_0}(\mathcal{F}_A^{\perp}) \\
&\quad = \inf\{\mathcal{E}_1[u] : u \in H_0^m(\Omega), u = u_0 \ m\text{-a.e. on some nhd. of } A\} \\
&\quad = \gamma^2 \inf\{\mathcal{E}_1[u] : u \in C_c^\infty(\Omega), u = e^{i\theta}\gamma^{-1} u_0 \ m\text{-a.e. on some nhd. of } A\} \\
&\quad \geq c^{-1}\gamma^2 \inf\{\|u\|_m^2, u \in C_c^\infty(\Omega), u = e^{i\theta}\gamma^{-1} u_0 \ m\text{-a.e. on some nhd. of } A\} \\
&\quad \geq c^{-1}\gamma^2 \inf\{\|u\|_m^2, u \in C_c^\infty(\Omega), u \geq 1 \ m\text{-a.e. on some nhd. of } A\}.
\end{aligned}$$

This proves the desired estimate with $c_U = c^{-1}\gamma^2$. □

Next we combine the inequalities just proved with the capacitary eigenvalue estimates that have been proved in Section 5. In the following theorems H_{r,x_0} denotes the self-adjoint operator in $L^2(\Omega\setminus\overline{B}_r(x_0))$ which is associated to $(\overline{a}, \mathcal{F}_{r,x_0})$ and $\lambda_{r,x_0} := \inf\sigma(H_{r,x_0})$ its spectral bound.

Theorem 6.2. *Let $d > 2m$ and let $a_{\alpha\beta} = \overline{a_{\beta\alpha}} \in L^1_{loc}(\Omega)$ such that inequality (6.3) holds. Moreover suppose that $\lambda = \inf\sigma(H)$ is a simple and isolated eigenvalue with* *eigenfunction* $\phi \notin \mathcal{F}_{r,x_0}^{\perp}$. *If* $x_0 \in \Omega$ *is such that the conditions of Lemma* 6.1 *hold with* $u_0 = \phi$, *then there are constants* $c_1, c_2 > 0$ *such that for small* $r > 0$,

$$c_1 r^{d-2m} \leq \lambda_{r,x_0} - \lambda \leq c_2 r^{d-2m}. \tag{6.4}$$

Proof. The proof follows easily from Corollary 5.6 (b), Theorem 5.10 and Lemma 6.1. □

Remark 6.3. Inequality (6.4) is already contained in [Maz85], Chapter 10 in the case of operators being defined on a sqare in $\mathbb{R}^d$. See also the original work [Maz73].

If the coefficients $a_{\alpha\beta}$ are (real-)analytic and Ω is a domain, we can prove the same lower bound even if $\phi(x_0) = 0$.

Theorem 6.4. *Let* $d > 2m$ *and let* $a_{\alpha\beta} = \overline{a_{\beta\alpha}}$ *be (real-)analytic functions on a domain* Ω *such that* H *is elliptic (i.e.,* $\sum_{|\alpha|=m} \sum_{\beta+\beta'=\alpha} a_{\beta\beta'}(x)\xi^\alpha \neq 0$ *for all* $(x, \xi) \in \Omega \times \mathbb{R}^d \setminus \{0\}$*) and inequality* (6.3) *holds. If* λ *is a simple and isolated eigenvalue of* H, *then for any* $x_0 \in \Omega$ *there is a constant* $c > 0$ *such that for small* $r > 0$

$$\lambda_{r,x_0} - \lambda \geq c r^{d-2m}.$$

Proof. It is known (see e.g., [Joh49], [Hör83a], Section 8 and [Hör83b], Section 11) that the eigenfunction ϕ of H and λ is analytic, i.e., for each unit vector $v \in \mathbb{R}^d$ the mapping

$$f_v : (-r, r) \to \mathbb{C}, \quad t \mapsto \phi(x_0 + tv)$$

can be extended to an analytic function (here $r > 0$ is such that $\{x \in \mathbb{R}^d : |x - x_0| < r \subset \Omega)$. It follows that the case $\phi(x_0) \neq 0$ is already covered by Theorem 6.2, so let us assume that $\phi(x_0) = 0$. Since $\phi \not\equiv 0$ and Ω is a domain there is some unit vector v such that $f_v \not\equiv 0$. Hence the set of zeros for f_v must be discrete. By decreasing r we may assume that $t = 0$ is the only zero of f_v in $(-r, r)$. Put $y_0 := x_0 + \frac{r}{2}v$. Then $\phi(y_0) \neq 0$ and $B_{\frac{r}{2}}(y_0) \subset B_r(x_0) \subset \Omega$. By Theorem 6.2 there is some $c' > 0$ such that

$$\lambda_{r,x_0} - \lambda \geq \lambda_{\frac{r}{2},y_0} - \lambda \geq c' \left(\frac{r}{2}\right)^{d-2m} = c r^{d-2m},$$

with $c = 2^{2m-d}c'$. This completes the proof. □

Next we treat the second eigenvalue.

Example. Let H and H_{r,x_0} be the operators defined in Example 5.2. Suppose in addition that $d > 2m$ and that the second eigenfunction ϕ_2 of H does not vanish in x_0. Then there is a constant $c > 0$ such that for small enough r,

$$\lambda_{2,r,x_0} - \lambda_2 \leq c r^{d-2m},$$

where λ_2 and λ_{2,r,x_0} are the second eigenvalues of H and H_{r,x_0} respectively.

Proof. This follows immediately from Corollary 5.16 and Lemma 6.1. □

6.2. Stummel class perturbations

Let H_0 be a symmetric constant-coefficient partial differential operator in $L^2(\mathbb{R}^d)$ of order $2m$, i.e.,

$$H_0 u = \sum_{|\alpha|\leq 2m} (-i)^{|\alpha|} a_\alpha \frac{\partial^\alpha u}{\partial x^\alpha}, \quad u \in C_c^\infty(\mathbb{R}^d), a_\alpha \in \mathbb{R}.$$

Assume that H_0 is elliptic, i.e.,

$$\sum_{|\alpha|=2m} a_\alpha \xi^\alpha \neq 0 \text{ for all } \xi \in \mathbb{R}^d \setminus \{0\}.$$

Then H_0 is essentially self-adjoint and the unique self-adjoint extension of H_0, which we again denote by H_0, has domain $H^{2m}(\mathbb{R}^d)$. Now suppose that Ω is an open and bounded subset of $\mathbb{R}^d$ and that for $|\alpha| < 2m$ we are given measurable functions $q_\alpha : \mathbb{R}^d \to \mathbb{R}$ with $\operatorname{supp}(q_\alpha) \subset \Omega$. Moreover assume that each q_α belongs to the *Stummel class* of order ρ_α with ρ_α satisfying

$$\rho_\alpha < 4m - 2\alpha. \tag{6.5}$$

Recall that this means that the function

$$x \mapsto \begin{cases} \int_{|x-y|\leq 1} |q_\alpha(y)|^2 |x-y|^{\rho_\alpha - d}\, dy, & \text{if } d > \rho_\alpha, \\ \int_{|x-y|\leq 1} |q_\alpha(y)|^2\, dy, & \text{if } d \leq \rho_\alpha \end{cases}$$

remains bounded. It is known, see [Wei76] Theorems 10.17 and 10.18, that then the operator

$$Vu := \sum_{|\alpha|<2m} (-i)^{|\alpha|} q_\alpha \frac{\partial^\alpha u}{\partial x^\alpha}$$

maps $H^{2m}(\mathbb{R}^d)$ into $L^2(\mathbb{R}^d)$ and that V is *infinitesimally small* with respect to H_0, i.e., for each $\varepsilon > 0$ there is some $\tilde{c_\varepsilon} \geq 0$ such that

$$\|Vu\|^2 \leq \varepsilon \|H_0 u\|^2 + \tilde{c_\varepsilon} \|u\|^2$$

for all $u \in H^{2m}(\mathbb{R}^d)$. In particular V is infinitesimally form bounded with respect to H_0 since this is implied by infinitesimal smallness, see [Kat80], Theorem VI.1.38.

In order to obtain discreteness of the spectrum we localize to Ω. Let H be the Friedrichs extension in $L^2(\Omega)$ of the restriction of H_0 to $C_c^\infty(\Omega)$ and put $\lambda = \inf \sigma(H)$. Then the quadratic form $\mathcal{E}$ which corresponds to $H - \lambda$ has domain $\mathcal{F} = H_0^m(\Omega)$ and the restriction of V to $C_c^\infty(\Omega)$ remains infinitesimally form bounded with respect to H. Hence for each $\varepsilon > 0$ there is a constant $c_\varepsilon \geq 0$ such that

$$|\langle Vu, u\rangle| \leq \varepsilon \mathcal{E}_1[u] + c_\varepsilon \|u\|^2.$$

Since Ω is bounded the spectrum of H is discrete, see [Agm65], Theorem 14.6. Let ϕ be a normalized function in the eigenspace $\ker(H - \lambda)$. As an immediate consequence of Corollary 5.23 we obtain the following result.

Theorem 6.5. *Fix $0 < \delta_1 \leq \delta_2 < 1$. For any closed subspace $\mathcal{G}$ of the form domain $H_0^m(\Omega)$ with $\mathrm{Cap}_\phi(\mathcal{G}^\perp) \in [\delta_1, \delta_2]$ we have*

$$\lambda^{\mathcal{G},V} - \lambda \leq c_{\delta_1} + \frac{2}{1-\delta_2}\mathrm{Cap}_\phi(\mathcal{G}^\perp),$$

where $\lambda^{\mathcal{G},V}$ is the spectral bound of $H^{\mathcal{G},V}$ defined as the self-adjoint operator which corresponds to the form

$$(u,v) \mapsto \mathcal{E}(u,v) + \sum_{|\alpha|<2m} \langle q_\alpha u, v\rangle, \quad u, v \in \mathcal{G}.$$

Example. Let $d \in \{5,6,7\}$, $\Omega = (0,1)^d$, $H = \Delta^2$ in $L^2(\Omega)$, defined as the closure of the quadratic form

$$u \mapsto \int_\Omega |\Delta u|^2 \mathrm{d}x, \quad u \in C_c^\infty(\Omega).$$

Further let $V \in L^2_{loc,unif}(\mathbb{R}^d)$ with $\mathrm{supp}(V) \subset \Omega$. Define

$$\Lambda := \Omega \setminus \{x \in \Omega : |x - (1/2, \ldots, 1/2)| \leq r\}$$

and let $H^{\Lambda,V} := \Delta^2 + V$ in $L^2(\Lambda)$. Then there are constants $c_1, c_2 \geq 0$ such that for small enough r,

$$\lambda^{\Lambda,V} - \lambda \leq c_1 + c_2 r^{d-4},$$

where $\lambda^{\Lambda,V} = \inf \sigma(H)$ is the spectral bound of $H^{\Lambda,V}$.

Proof. Since $V \in L^2_{loc,unif}(\mathbb{R}^d)$, V belongs to the Stummel class of order d. Because $d < 8$ the hypothesis (6.5) is fulfilled. Hence

$$\lambda^{\Lambda,V} - \lambda \leq c + \frac{2}{1-\delta}\mathrm{Cap}_\phi(H_0^2(\Lambda)^\perp)$$

with some $c \geq 0$. Moreover $d > 4$ implies $\mathrm{Cap}_\phi(H_0^m(\Lambda)^\perp) \leq c' r^{d-4}$, see Lemma 6.1 and Theorem 6.2. □

7. Scaling methods

This final section is dedicated to scaling properties of the zero-order capacities. In Section 7.1 we derive the scaling property

$$\mathrm{Cap}_0(rA) = r^{d-2}\mathrm{Cap}_0(A)$$

for the classical zero-order capacity in $\mathbb{R}^d$, $d \geq 3$. After introducing a zero-order version of the general capacity of Section 4.3 we prove in Section 7.2 zero-order counterparts of some estimates of Section 5. In Section 7.3 we specialize to differential operators with homogeneous symbol and prove an appropriate scaling property for the zero-order capacity of Section 7.2 for this class of operators. After introducing a general method for optimizing eigenvalue estimates by scaling we finally show that the results of Section 7.3 cannot be improved by this method.

7.1. The scaling property of the classical zero-order capacity in $\mathbb{R}^d$

Let U be an open subset of $\mathbb{R}^d$ where $d \geq 3$. The classical (zero-order) capacity in $\mathbb{R}^d$ can be defined by

$$\mathrm{Cap}_0(U) := \begin{cases} \inf\{\int_{\mathbb{R}^d} |\nabla u|^2 \mathrm{d}x : u \in C_c^\infty(\mathbb{R}^d), u \geq 1 \text{ on } U\}, \\ \infty, \text{ if there is no such } u. \end{cases} \tag{7.1}$$

The capacity can be extended to arbitrary subsets of $\mathbb{R}^d$ by

$$\mathrm{Cap}_0(A) := \inf\{\mathrm{Cap}_0(U) : U \supset A, U \text{ open}\}.$$

See Section 4.1 for a physical motivation of this definition in the context of electrostatics. Let $\mathcal{E}$ be the quadratic form of the Laplacian in $L^2(\mathbb{R}^d)$ and let $\mathcal{F} := H^1(\mathbb{R}^d)$ be its form domain. Then

$$\mathcal{E}[u] = \int_{\mathbb{R}^d} |\nabla u|^2 \mathrm{d}x, \quad u \in \mathcal{F}.$$

Therefore we may write (7.1) as

$$\mathrm{Cap}_0(U) = \begin{cases} \inf\{\mathcal{E}[u] : u \in \mathcal{F}, u \geq 1 \quad \text{a.e. on } U\}, \\ \infty, \text{ if there is no such } u. \end{cases}$$

This formula makes plain the name "zero-order capacity" since $\mathcal{E}$ is used in the definition rather than $\mathcal{E}_1$ as in Definition 4.2. In this context we also call the capacity of Definition 4.2 the "first-order capacity".

Clearly the symbol

$$\sum_{j=1}^{d} |\xi_j|^2$$

of the Laplacian is a homogeneous function of degree 2. This allows us to compute the zero-order capacity of a ball in $\mathbb{R}^d$. For simplicity we consider only the case $d \geq 3$. Because of the translation invariance of the Laplacian we may restrict ourselves to balls which are centered at the origin. Let

$$B_r := \{x \in \mathbb{R}^d : |x| < r\}$$

be the open ball of radius r and let $\psi_r(x) := r^{-1}x$, $x \in \mathbb{R}^d$. Then

$$\begin{aligned} \mathrm{Cap}_0(B_r) &= \inf\left\{\int_{\mathbb{R}^d} |\nabla u|^2 \mathrm{d}x : u \in C_c^\infty(\mathbb{R}^d), u \geq 1 \text{ on } B_r\right\} \\ &= \inf\left\{\int_{\mathbb{R}^d} |\nabla(u \circ \psi_r)|^2 \mathrm{d}x : u \in C_c^\infty(\mathbb{R}^d), u \geq 1 \text{ on } B_1\right\} \\ &= r^{-2} \inf\left\{\int_{\mathbb{R}^d} |(\nabla u)|^2 \circ \psi_r \mathrm{d}x : u \in C_c^\infty(\mathbb{R}^d), u \geq 1 \text{ on } B_1\right\} \\ &= r^{d-2} \inf\left\{\int_{\mathbb{R}^d} |(\nabla u)|^2 \circ \psi_r \, |\det D\psi_r| \, \mathrm{d}x : \begin{array}{l} u \in C_c^\infty(\mathbb{R}^d), \\ u \geq 1 \text{ on } B_1 \end{array}\right\} \\ &= r^{d-2} \mathrm{Cap}_0(B_1). \end{aligned} \tag{7.2}$$

Therefore it is enough to know the capacity of the unit ball which is given by

$$\mathrm{Cap}_0(B_1) = \frac{4\pi^{d/2}}{\Gamma(\frac{d}{2}-1)}. \tag{7.3}$$

This formula can be proved stochastically with the aid of Brownian motion (see [PS78], Section 3.1) or analytically as it is possible to explicitly write down the zero-order equilibrium potential e_{B_1}, i.e., the unique minimizer of the infimum in (7.1). It is given by

$$e_{B_1}(x) = \begin{cases} 1 & \text{if } x \in B_1, \\ |x|^{2-d} & \text{if } x \in \mathbb{R}^d \setminus B_1. \end{cases}$$

For $|x| > 1$ we have $|\nabla e_{B_1}|^2 = (2-d)^2|x|^{2-2d}$. Therefore we get

$$\mathrm{Cap}_0(B_1) = \int_{\mathbb{R}^d} |\nabla e_{B_1}|^2 \mathrm{d}x = (2-d)^2\omega_d \int_1^\infty r^{2-2d} r^{d-1} \mathrm{d}r = (d-2)\omega_d,$$

where

$$\omega_d := \frac{2\pi^{d/2}}{\Gamma(d/2)}$$

is the surface area of the unit ball in $\mathbb{R}^d$. Now equation (7.3) follows immediately from the functional equation of the gamma function.

With the same method one proves that the scaling property (7.2) of the zero-order capacity holds not only for balls but is true for any subset A of $\mathbb{R}^d$, i.e., we have

$$\mathrm{Cap}_0(rA) = r^{d-2}\mathrm{Cap}_0(A), \tag{7.4}$$

where

$$rA := \{rx : x \in A\}$$

denotes the set A, scaled by r.

7.2. The zero-order capacity in the general Hilbert space setting

If H is the Laplacian in $\mathcal{H} := L^2(\mathbb{R}^d)$ there is no scaling property for the first-order capacity of Definition 4.2. This is due to the fact that in this case $\lambda = \inf \sigma(H) = 0$, i.e., the form $\mathcal{E}_1(\cdot,\cdot)$ corresponds to $-\Delta + 1$ whose symbol is

$$\sum_{j=1}^{d} |\xi_j|^2 + 1$$

and the constant term destroys the homogeneity. Therefore we introduce a zero-order capacity in the general Hilbert space setting which can be defined in a straightforward way if H is a strictly positive operator. As we shall see in Theorem 7.4, it is possible to prove zero-order counterparts of the results of Section 5. This leads to bounds which behave appropriately under scaling. Since the scaling property (7.4) can be used to optimize estimates for eigenvalues, this method is of general interest.

Remark 7.1. There are also good reasons to prefer the first-order capacity: Firstly the operator associated to $\mathcal{E}_1$ is, per definition, $H - \lambda + 1$, where $\lambda = \inf \sigma(H)$ is the spectral bound of H. If λ is an eigenvalue of H with normalized eigenfunction ϕ this implies

$$\mathcal{E}_1[\phi] = \langle (H - \lambda + 1)\phi, \phi\rangle = 1 \text{ and } 0 \leq \mathrm{Cap}_\phi(\mathcal{G}) \leq 1,$$

for any closed subspace $\mathcal{G}$ of $(\mathcal{F}, \mathcal{E}_1(\cdot,\cdot))$. Therefore our notation simplifies most of the formulas and makes the proofs more transparent and easier to read.

Moreover one can only hope for a homogeneous capacity in the case of non-negative differential operators with homogeneous symbol. The theory presented in Section 5 is much more general as it allows arbitrary self-adjoint and semi-bounded operators in general Hilbert spaces, i.e., the underlying Hilbert space need not necessarily be an L^2-space. In this setting there is no meaning of scaling anyway.

The definition of the abstract zero-order capacity reads as follows.

Definition 7.2. Let H be a self-adjoint operator in $\mathcal{H}$ with positive spectral bound λ. Let $\mathcal{G}$ be a closed subspace of $(\mathrm{dom}(H^{1/2}), \langle H^{1/2}\cdot, H^{1/2}\cdot\rangle)$. The *zero-order capacity of $\mathcal{G}$ with respect to u* is defined by

$$\mathrm{Cap}_u^0(\mathcal{G}) := \langle H^{1/2}P_{\mathcal{G}}u, H^{1/2}P_{\mathcal{G}}u\rangle = \|H^{1/2}P_{\mathcal{G}}u\|^2,$$

where $P_{\mathcal{G}}$ is the orthogonal projection onto the space $\mathcal{G}$ in the Hilbert space $(\mathrm{dom}(H^{1/2}), \langle H^{1/2}\cdot, H^{1/2}\cdot\rangle)$.

Remark 7.3.

(a) Clearly $\langle H^{1/2}\cdot, H^{1/2}\cdot\rangle$ and $\mathcal{E}_1(\cdot,\cdot)$ (see Definition 4.2) are equivalent scalar products on $\mathcal{F} = \mathrm{dom}(H^{1/2})$. Hence the classes of perturbations one can handle are the same.

(b) If $\lambda = 1$, then the zero-order capacity of Definition 7.2 and the first-order capacity of Definition 4.2 coincide.

It is not difficult to prove zero-order versions of Corollary 5.6 (a) and Theorem 5.10.

Theorem 7.4. *Let λ be an eigenvalue of H with normalized eigenfunction ϕ.*

(a) *If $\phi \notin \mathcal{G}$, then*

$$\lambda^{\mathcal{G}} - \lambda \leq \frac{\lambda \mathrm{Cap}_\phi^0(\mathcal{G}^\perp)}{\lambda - \mathrm{Cap}_\phi^0(\mathcal{G}^\perp)}. \tag{7.5}$$

(b) *Suppose that λ is a simple eigenvalue of H. Let $\mu := \inf\{\sigma(H) \setminus \{\lambda\}\} - \lambda$ be the spectral gap. Then*

$$\lambda^{\mathcal{G}} - \lambda \geq \lambda\mu \frac{\mathrm{Cap}_\phi^0(\mathcal{G}^\perp)}{\lambda^2 + \mu\lambda - \mu \mathrm{Cap}_\phi^0(\mathcal{G}^\perp))}. \tag{7.6}$$

Proof. (a) Let $P_{\mathcal{G}^\perp}$ be the orthogonal projection onto $\mathcal{G}^\perp$ in the Hilbert space $(\mathrm{dom}(H^{1/2}), \langle H^{1/2}\cdot, H^{1/2}\cdot\rangle)$. Then

$$\begin{aligned}
\lambda^{\mathcal{G}} &= \inf\left\{\frac{\|H^{1/2}u\|^2}{\|u\|^2} : u \in \mathcal{G}\right\}\\
&\le \frac{\|H^{1/2}(1-P_{\mathcal{G}^\perp})\phi\|^2}{\|(1-P_{\mathcal{G}^\perp})\phi\|^2}\\
&= \frac{\|H^{1/2}\phi\|^2 - \|H^{1/2}P_{\mathcal{G}^\perp}\phi\|^2}{1-2\mathrm{Re}\langle\phi, P_{\mathcal{G}^\perp}\phi\rangle + \|P_{\mathcal{G}^\perp}\phi\|^2}\\
&= \frac{\lambda - \mathrm{Cap}^0_\phi(\mathcal{G}^\perp)}{1-\frac{2}{\lambda}\mathrm{Re}\langle H\phi, P_{\mathcal{G}^\perp}\phi\rangle + \|P_{\mathcal{G}^\perp}\phi\|^2}\\
&\le \frac{\lambda - \mathrm{Cap}^0_\phi(\mathcal{G}^\perp)}{1-\frac{2}{\lambda}\mathrm{Re}\langle H^{1/2}\phi, H^{1/2}P_{\mathcal{G}^\perp}\phi\rangle + |\langle P_{\mathcal{G}^\perp}\phi, \phi\rangle|^2}\\
&= \frac{\lambda - \mathrm{Cap}^0_\phi(\mathcal{G}^\perp)}{1-\frac{2}{\lambda}\mathrm{Cap}^0_\phi(\mathcal{G}^\perp) + \frac{1}{\lambda^2}\mathrm{Cap}^0_\phi(\mathcal{G}^\perp)^2}\\
&= \frac{\lambda}{1-\frac{\mathrm{Cap}^0_\phi(\mathcal{G}^\perp)}{\lambda}}\\
&= \lambda + \frac{\lambda\mathrm{Cap}^0_\phi(\mathcal{G}^\perp)}{\lambda - \mathrm{Cap}^0_\phi(\mathcal{G}^\perp)}.
\end{aligned}$$

(b) Let $\mathcal{K}$ be the one-dimensional eigenspace of H and λ. Denote by $P_\mathcal{K}$ and $\mathcal{K}^\perp$ the orthogonal projection onto $\mathcal{K}$ and the orthogonal complement of $\mathcal{K}$ respectively, where orthogonality is meant with respect to the scalar product $\langle H^{1/2}\cdot, H^{1/2}\cdot\rangle$ on $\mathrm{dom}(H^{1/2})$. Put $P_{\mathcal{K}^\perp} := 1 - P_\mathcal{K}$. Since $\mathcal{K}$ is spanned by ϕ and $\|H^{1/2}\phi\| = \sqrt{\lambda}$ we get for any $u \in \mathcal{F}$,

$$P_\mathcal{K}u = \langle H^{1/2}\frac{1}{\sqrt{\lambda}}\phi, H^{1/2}u\rangle\frac{1}{\sqrt{\lambda}}\phi = \langle\phi, u\rangle\phi.$$

Hence

$$\begin{aligned}
\langle P_{\mathcal{K}^\perp}u, P_\mathcal{K}u\rangle &= \langle\phi, u\rangle\langle P_{K^\perp}u, \phi\rangle\\
&= \frac{1}{\lambda}\langle\phi, u\rangle\langle P_{\mathcal{K}^\perp}u, H\phi\rangle\\
&= \frac{1}{\lambda}\langle\phi, u\rangle\langle H^{1/2}P_{\mathcal{K}^\perp}u, H^{1/2}\phi\rangle.\\
&= 0.
\end{aligned}$$

Consequently

$$\mathrm{Cap}^0_u(\mathcal{K}^\perp) = \|H^{1/2}P_{\mathcal{K}^\perp}u\|^2 \ge (\lambda+\mu)\|P_{\mathcal{K}^\perp}u\|^2$$

and

$$\|u\|^2 = \|P_\mathcal{K}u\|^2 + \|P_{\mathcal{K}^\perp}u\|^2.$$

Putting all these facts together we obtain for $u \in \mathcal{G}$ with $\|u\| = 1$,

$$\begin{aligned}
&\Big(1 + \frac{\mu}{\lambda^2}\mathrm{Cap}_\phi^0(\mathcal{G})\Big)\, \|H^{1/2}u\|^2 \\
&= \mathrm{Cap}_u^0(\mathcal{K}) + \mathrm{Cap}_u^0(\mathcal{K}^\perp) + \frac{\mu}{\lambda^2}\|H^{1/2}P_\mathcal{G}\phi\|^2\|H^{1/2}u\|^2 \\
&\geq \mathrm{Cap}_u^0(\mathcal{K}) + (\lambda+\mu)\|P_{\mathcal{K}^\perp}u\|^2 + \frac{\mu}{\lambda^2}\left|\langle H^{1/2}P_\mathcal{G}\phi, H^{1/2}u\rangle\right|^2 \\
&= \|H^{1/2}P_\mathcal{K}u\|^2 + (\lambda+\mu)(1 - \|P_\mathcal{K}u\|^2) + \frac{\mu}{\lambda^2}\left|\langle H^{1/2}\phi, H^{1/2}u\rangle\right|^2 \\
&= \lambda|\langle u,\phi\rangle|^2 + (\lambda+\mu)(1 - |\langle u,\phi\rangle|^2) + \mu|\langle u,\phi\rangle|^2 \\
&= \lambda + \mu.
\end{aligned}$$

This inequality gives us the following estimate for the perturbed bottom eigenvalue:

$$\begin{aligned}
\lambda^\mathcal{G} &= \inf\{\|H^{1/2}u\|^2 : u \in \mathcal{G}, \|u\| = 1\} \\
&\geq \frac{\lambda^2(\lambda+\mu)}{\lambda^2 + \mu\mathrm{Cap}_\phi^0(\mathcal{G})} \\
&= \lambda + \lambda\mu\frac{\lambda - \mathrm{Cap}_\phi^0(\mathcal{G})}{\lambda^2 + \mu\mathrm{Cap}_\phi^0(\mathcal{G})} \\
&= \lambda + \lambda\mu\frac{\mathrm{Cap}_\phi^0(\mathcal{G}^\perp)}{\lambda^2 + \lambda\mu - \mu\mathrm{Cap}_\phi^0(\mathcal{G}^\perp)}.
\end{aligned}$$

The proof is complete. □

Remark 7.5.

(a) If $\lambda = 1$ we recover the estimates of Corollary 5.6 (a) and Theorem 5.10 respectively.

(b) From part (b) of Theorem 7.4 we can immediately conclude the following lower bound which is slightly worse but more handy:

$$\lambda^\mathcal{G} - \lambda \geq \frac{\mu}{\lambda+\mu}\mathrm{Cap}_\phi^0(\mathcal{G}^\perp).$$

7.3. Scaling invariance of eigenvalue estimates

We first describe the type of operators we are interested in. Let Ω be an open subset of $\mathbb{R}^d$ and let $a_{\alpha\beta} = \overline{a_{\beta\alpha}} \in L^1_{loc}(\Omega)$. Consider the homogeneous bilinear form

$$a_0(u,v) := \sum_{|\alpha|=|\beta|=m} \int_\Omega a_{\alpha\beta}\overline{D^\alpha u}D^\beta v\,\mathrm{d}x, \tag{7.7}$$

initially defined for $u, v \in C_c^\infty(\Omega)$. Assume that a_0 is semi-bounded from below and closable. Let a be any closed extension of a_0 (not necessarily the closure of a_0) and denote the form domain of a by $\mathcal{F}$. Let H be the unique self-adjoint operator which corresponds to a in the sense of Theorem 2.2, put $\lambda := \inf\sigma(H)$ and $\mathcal{E}_1 := a - \lambda + 1$.

Observe that a_0 is closable if

$$c^{-1}\|u\|_m^2 \le a_0(u,u) + b\|u\|^2 \le c\|u\|_m^2,$$

for some $b \in \mathbb{R}, c > 0$ and all $u \in C_c^\infty(\Omega)$, where $\|\cdot\|_m$ denotes the Sobolev norm of order m. In this case the above inequality carries over to all $u \in H_0^m(\Omega)$ and the closure $\overline{a_0}$ of a_0 has domain $H_0^m(\Omega)$. See also the discussion at the beginning of Section 6.

Define for each $r > 0$ the open set

$$\Omega_r := \{rx : x \in \Omega\}$$

and the scaling transformation

$$\psi_r : \Omega_r \to \Omega, \quad \psi_r(x) = r^{-1}x.$$

Further define the self-adjoint operator H_r in $\mathcal{H}_r := L^2(\Omega_r)$ as the operator which corresponds to the closed form

$$a_r(u,v) := \sum_{|\alpha|=|\beta|=m} \int_{\Omega_r} (a_{\alpha\beta} \circ \psi_r)\overline{D^\alpha u} D^\beta v \,\mathrm{d}x, \tag{7.8}$$

which is defined for

$$u, v \in \mathcal{F}_r := \{w \circ \psi_r : w \in \mathcal{F}\}.$$

Then

$$\mathrm{dom}(H_r) = \{u \circ \psi_r : u \in \mathrm{dom}(H)\}, \text{ and } H_r(u \circ \psi_r) = r^{-\ell}(Hu) \circ \psi_r,$$

where $\ell := 2m$ is the order of the differential operator H. For instance, if H is the Dirichlet Laplacian in $L^2(\Omega)$, then $\ell = 2$ and H_r is the Dirichlet Laplacian in $L^2(\Omega_r)$. Now suppose that it is given a closed subspace $\mathcal{G}$ of $(\mathcal{F}, \mathcal{E}_1(\cdot,\cdot))$. Define the operator $H^\mathcal{G}$ in $\mathcal{H}^\mathcal{G} := \overline{\mathcal{G}}^\mathcal{H}$ via the construction described in detail in Section 3.2. In the same way we obtain an operator $H_r^\mathcal{G}$ acting in an associated Hilbert space $\mathcal{H}_r^\mathcal{G}$. Denote the spectral bounds by $\lambda^\mathcal{G}$ and $\lambda_r^\mathcal{G}$ respectively. Because of the homogeneity we have $\lambda_r = r^{-\ell}\lambda$ and $\lambda_r^\mathcal{G} = r^{-\ell}\lambda^\mathcal{G}$, i.e.,

$$\lambda^\mathcal{G} - \lambda = r^\ell(\lambda_r^\mathcal{G} - \lambda_r). \tag{7.9}$$

Therefore an estimate for $\lambda_r^\mathcal{G} - \lambda_r$ also yields an estimate for the shift of the spectral bound of the operator H. Once one has proved a general estimate for the shift of the spectral bound one may apply this result for each pair $H_r, H_r^\mathcal{G}$, $r > 0$. Hence one can try to improve the estimate for $\lambda^\mathcal{G} - \lambda$ by optimizing the right-hand side of equation (7.9) over r.

We shall make this idea concrete in the case of the estimates of Theorem 7.4. For this we use the zero-order capacity of Definition 7.2. In order to use it we have to assume that H is a strictly positive operator. First we show that the zero-order capacity has appropriate scaling properties.

Proposition 7.6. *Let $\mathcal{G}$ be a closed subspace of $(\mathcal{F}, \mathcal{E}_1(\cdot,\cdot))$ and let $u \in \mathcal{F}$ arbitrary. For $r > 0$ define*

$$\mathcal{G}_r := \{v \circ \psi_r : v \in \mathcal{G}\}, \quad u_r := u \circ \psi_r.$$

Then the following is true:

(a) $\mathrm{Cap}^0_{u_r}(\mathcal{G}_r) = r^{d-\ell}\mathrm{Cap}^0_u(\mathcal{G})$,

(b) *For any subset A of Ω put*

$$\begin{aligned}\mathcal{F}_A &:= \{v \in \mathcal{F} : v = 0 \ \textit{m-a.e. on some neighborhood of } A\},\\ \mathcal{F}_{rA} &:= \{v \in \mathcal{F}_r : v = 0 \ \textit{m-a.e. on some neighborhood of } rA\}.\end{aligned}$$

Then

$$\mathrm{Cap}^0_{u_r}(\mathcal{F}_{rA}^\perp) = r^{d-\ell}\mathrm{Cap}^0_u(\mathcal{F}_A^\perp).$$

Proof. (a) We first prove the identity

$$P_{\mathcal{G}_r}(u \circ \psi_r) = (P_{\mathcal{G}}u) \circ \psi_r. \tag{7.10}$$

To this aim, note that for any $v, w \in \mathcal{F}$ we have

$$a_r(v \circ \psi_r, w \circ \psi_r) = r^{d-\ell}a(v, w),$$

implying

$$G_r^\perp := \{v \in \mathcal{G}_r : a_r(v, w) = 0 \text{ for all } w \in \mathcal{G}_r\} = \{w \circ \psi_r : w \in \mathcal{G}^\perp\}.$$

Write $u = v + w$ with $v \in \mathcal{G}$ and $w \in \mathcal{G}^\perp$. Then

$$P_{\mathcal{G}_r}(u \circ \psi_r) = P_{\mathcal{G}_r}(v \circ \psi_r) + P_{\mathcal{G}_r}(w \circ \psi_r) = v \circ \psi_r = (P_{\mathcal{G}}v) \circ \psi_r = (P_{\mathcal{G}}u) \circ \psi_r,$$

which proves (7.10). It is enough to prove the assertion (a) for all $u \in \mathrm{dom}(H)$. Using identity (7.10) we obtain

$$\begin{aligned}\mathrm{Cap}^0_{u_r}(\mathcal{G}_r) &= \|H_r^{1/2}P_{\mathcal{G}_r}u_r\|^2 = \langle H_r^{1/2}P_{\mathcal{G}_r}u_r, H_r^{1/2}u_r\rangle = \langle P_{\mathcal{G}_r}u_r, H_r u_r\rangle\\ &= r^{-\ell}\langle P_{\mathcal{G}_r}(u \circ \psi_r), (Hu) \circ \psi_r\rangle\\ &= r^{-\ell}\langle (P_{\mathcal{G}}u) \circ \psi_r, (Hu) \circ \psi_r\rangle\\ &= r^{d-\ell}\langle P_{\mathcal{G}}u, Hu\rangle\\ &= r^{d-\ell}\langle H^{1/2}P_{\mathcal{G}}u, H^{1/2}u\rangle\\ &= r^{d-\ell}\mathrm{Cap}^0_u(\mathcal{G}).\end{aligned}$$

For the proof of (b) note that obviously $\mathcal{F}_{rA} = (\mathcal{F}_A)_r$. From the proof of the first part of this proposition we see that $((\mathcal{F}_A)_r)^\perp = ((\mathcal{F}_A)^\perp)_r$. Consequently

$$\mathrm{Cap}^0_{u_r}((\mathcal{F}_{rA})^\perp) = \mathrm{Cap}^0_{u_r}(((\mathcal{F}_A)_r)^\perp) = \mathrm{Cap}^0_{u_r}(((\mathcal{F}_A)^\perp)_r) = r^{d-\ell}\mathrm{Cap}^0_u(\mathcal{F}_A^\perp).$$

□

As an application of Proposition 7.6 we show that the eigenvalue estimates of Theorem 7.4 cannot be improved by optimizing over r. If ϕ is a normalized eigenfunction of H with eigenvalue λ, then clearly

$$\tilde{\phi_r} := r^{-d/2}\,\phi \circ \psi_r$$

is a normalized eigenfunction of H_r with eigenvalue $\lambda_r = r^{-\ell}\lambda$. Hence Theorem 7.4 (a) together with identity (7.9) yields

$$\lambda^{\mathcal{G}} - \lambda \le r^{\ell} \frac{\lambda_r \mathrm{Cap}^0_{\tilde{\phi}_r}(\mathcal{G}_r^{\perp})}{\lambda_r - \mathrm{Cap}^0_{\tilde{\phi}_r}(\mathcal{G}_r^{\perp})}, \tag{7.11}$$

and from part (b) we get

$$\lambda^{\mathcal{G}} - \lambda \ge r^{\ell} \lambda_r \mu_r \frac{\mathrm{Cap}^0_{\tilde{\phi}_r}(\mathcal{G}_r^{\perp})}{\lambda_r^2 + \mu_r\lambda_r - \mu_r \mathrm{Cap}^0_{\tilde{\phi}_r}(\mathcal{G}_r^{\perp}))}. \tag{7.12}$$

Theorem 7.7. *The numbers on the right-hand sides of inequalities* (7.11) *and* (7.12) *coincide with the right-hand sides of* (7.5) *and* (7.6) *respectively. In particular they are independent of* r.

Proof. By Remark 4.4 and Proposition 7.6 (a) we have

$$\begin{aligned}
\mathrm{Cap}^0_{\tilde{\phi}_r}((\mathcal{G}_r)^{\perp}) &= \mathrm{Cap}^0_{r^{-d/2}\phi\circ\psi_r}((\mathcal{G}_r)^{\perp}) \\
&= r^{-d}\mathrm{Cap}^0_{\phi\circ\psi_r}((\mathcal{G}_r)^{\perp}) \\
&= r^{-d}\mathrm{Cap}^0_{\phi\circ\psi_r}((\mathcal{G}^{\perp})_r) \\
&= r^{-\ell}\mathrm{Cap}^0_{\phi}(\mathcal{G}^{\perp}).
\end{aligned}$$

This together with the identity $\lambda_r = r^{-\ell}\lambda$ gives us

$$\begin{aligned}
r^{\ell}\frac{\lambda_r \mathrm{Cap}^0_{\tilde{\phi}_r}(\mathcal{G}_r^{\perp})}{\lambda_r - \mathrm{Cap}^0_{\tilde{\phi}_r}(\mathcal{G}_r^{\perp})} &= \frac{\lambda_r \mathrm{Cap}^0_{\phi}(\mathcal{G}^{\perp})}{\lambda_r - r^{-\ell}\mathrm{Cap}^0_{\phi}(\mathcal{G}^{\perp})} \\
&= \frac{r^{-\ell}\lambda \mathrm{Cap}^0_{\phi}(\mathcal{G}^{\perp})}{r^{-\ell}(\lambda - \mathrm{Cap}^0_{\phi}(\mathcal{G}^{\perp}))} \\
&= \frac{\lambda \mathrm{Cap}^0_{\phi}(\mathcal{G}^{\perp})}{\lambda - \mathrm{Cap}^0_{\phi}(\mathcal{G}^{\perp})}.
\end{aligned}$$

Similarly we obtain

$$\begin{aligned}
r^{\ell}\lambda_r\mu_r \frac{\mathrm{Cap}^0_{\tilde{\phi}_r}(\mathcal{G}_r^{\perp})}{\lambda_r^2 + \mu_r\lambda_r - \mu_r\mathrm{Cap}^0_{\tilde{\phi}_r}(\mathcal{G}_r^{\perp})} &= \frac{\lambda r^{-\ell}\mu r^{-\ell}\mathrm{Cap}^0_{\phi}(\mathcal{G}^{\perp})}{r^{-2\ell}\lambda^2 + r^{-2\ell}\lambda\mu + r^{-\ell}\mu r^{-\ell}\mathrm{Cap}^0_{\phi}(\mathcal{G}^{\perp})} \\
&= \lambda\mu \frac{\mathrm{Cap}^0_{\phi}(\mathcal{G}^{\perp})}{\lambda^2 + \mu\lambda - \mu\mathrm{Cap}^0_{\phi}(\mathcal{G}^{\perp})}.
\end{aligned}$$

□

8. Open problems

This concluding section describes some open problems which seem to be worthwhile for further research.

(1) Treatment of higher eigenvalues

(a) *Upper bounds.* In the case of regular Dirichlet forms it is known that all eigenvalues of the perturbed operator admit capacitary upper bounds when the original operator is subjected to a domain perturbation on a compact set [McG96].

In Section 5.3 it has been proved that this is true in the general Hilbert space setting for the second eigenvalue, but the proof of this result was much more difficult than the corresponding proof for the bottom eigenvalue. Clearly, it would be desirable to have upper bounds for each eigenvalue of the perturbed operator. Since the minimax principle gives an expression for all eigenvalues, it should be possible to prove an analog of Theorem 5.15 and Corollary 5.16 for all eigenvalues. The idea is the following: By the minimax principle we have

$$\lambda_j^{\mathcal{G}} = \inf\{\mathcal{E}\,[u] : u \in \mathcal{G}, \|u\| = 1, u \perp \psi_1, \ldots, \psi_{k-1}\},$$

where $\lambda_k^{\mathcal{G}}$ are the eigenvalues of the perturbed operator $H^{\mathcal{G}}$ with corresponding eigenvectors ψ_k. Therefore

$$\lambda_k^{\mathcal{G}} \leq \mathcal{E}[P_{\mathcal{G}}\phi_k + \sum_{j=1}^{k-1} \alpha_j \psi_j],$$

where the α_j's have to be chosen appropriately.

(b) *Lower bounds.* In Remark 5.21 we have seen that in general one cannot expect capacitary lower bounds for higher eigenvalues by using the classical capacity. We also saw that inequality (5.14) becomes true when the classical capacity is replaced by the general capacity. Therefore one can conjecture that this did not happen by coincidence, but capacitary lower bounds are indeed true when using the general capacity. Unfortunately, the proof of the lower bound for the first eigenvalue (Theorem 5.10) does not carry over to the case of higher eigenvalues, so a more general method is needed to understand the behavior of higher eigenvalues in general.

(2) Capacity and scattering theory

Given $H^{\mathcal{G}}$, the domain perturbation of H with respect to the subspace $\mathcal{G}$ (according to Definition 3.1), one can ask whether the pair $(H, H^{\mathcal{G}})$ is a complete scattering system with respect to the identification operator $\mathrm{id} : \mathcal{H}^G \hookrightarrow \mathcal{H}$, i.e., the question is whether the two-space wave operators exist and are complete. In [DC98] it has been proved that this is the case for the Dirichlet Laplacians $(-\Delta_\Omega, -\Delta_\Lambda)$, Λ, Ω open subsets of $\mathbb{R}^d$, if the symmetric difference $\Omega \bigtriangleup \Lambda$ has finite capacity. One year later this result was generalized to Kato–Feller operators [DGCZ95]. In particular, completeness of the scattering system implies equality of the absolutely continuous spectra of H and $H^{\mathcal{G}}$. So proving completeness is not only of interest for operators which are relevant for quantum mechanics. Since the proof of the above-mentioned results uses stochastic methods, it is not possible to generalize it to a larger class of operators, e.g., higher-order differential operators. On the other hand, in [Ede99] there is a very general approach for proving completeness which is based on purely functional analytic properties. With the aid of these results it seems to be possible to prove completeness for the pair $(H, H^{\mathcal{G}})$ in the general Hilbert space setting.

(3) Numerical analysis

The results of Section 5 gave estimates on eigenvalues in terms of the capacity. For concrete applications one needs to compute these capacities, but this turns out to be difficult, even in the case of the classical capacity. On the other hand, it is quite easy to give estimates for these capacities by choosing good trial functions. At this point one could use the powerful tools of numerical analysis to find such trial functions. Besides giving quantitative bounds in concrete cases, numerical considerations can also be used to illustrate the abstract results.

(4) Non-selfadjointness

Although the results of Section 6 on differential operators made no restrictions on the order of the operators in question, there remain important problems which cannot be treated with the methods described in this thesis because of our assumption on the self-adjointness. For instance in [BEGK99], a connection between domain perturbations and the Wentzell–Freidlin theory for non-self-adjoint generators was discovered. Therefore it seems to be of interest to analyze the behavior of the spectrum by considering domain perturbations of non-self-adjoint operators. In particular, operators whose spectrum is a discrete set which is contained in some sector of the complex plane could be treated by passing from symmetric forms to sectorial forms which are well studied, [Kat80].

In concrete cases the infimum λ of the real part of the spectrum is often an isolated eigenvalue and there is no other eigenvalue with real part λ. It is an open problem to examine the behavior of λ under domain perturbations. To this aim the theory of capacity for self-adjoint operators would have to be translated to sectorial operators.

Appendix A. The minimax principle

Let H be a non-negative self-adjoint operator in the Hilbert space $\mathcal{H}$. Then $\sigma(H) \subset \mathbb{R}$, and the bottom of the spectrum of H is given by

$$\begin{aligned} \lambda &= \inf\{\langle Hu, u\rangle : u \in \operatorname{dom}(H), \|u\| = 1\} \\ &= \inf\{\|H^{\frac{1}{2}}u\|^2 : u \in \operatorname{dom}(H^{\frac{1}{2}}), \|u\| = 1\}. \end{aligned}$$

The number $\alpha \in \sigma(H)$ belongs, by definition, to the *discrete spectrum of* H if and only if α is an isolated eigenvalue of finite multiplicity. Otherwise λ is said to be in the *essential spectrum*. For any finite-dimensional subspace $\mathcal{G}$ of $\operatorname{dom}(H^{\frac{1}{2}})$ and

$n \in \mathbb{N}$ define

$$\begin{aligned} \lambda_H(\mathcal{G}) &= \sup\{\|H^{\frac{1}{2}}u\|^2 : u \in \mathcal{G}, \|u\| = 1\}, \\ \lambda_n &= \inf\{\lambda_H(\mathcal{G}) : \dim(\mathcal{G}) = n\}. \end{aligned}$$

Then $\lambda_1 = \lambda$ and $(\lambda_n)_{n\in\mathbb{N}}$ is an increasing sequence of real numbers. The assertion of the minimax principle is the following: The least upper bound of $(\lambda_n)_{n\in\mathbb{N}}$ equals the bottom λ_∞ of the essential spectrum of H. The sequence $(\lambda_n)_{n\in\mathbb{N}}$, omitting all values equal to λ_∞ (if there are such) coincides with the discrete spectrum in $[\lambda, \lambda_\infty)$ and each eigenvalue is repeated according to its multiplicity.

By an obvious modification one can also include self-adjoint operators which are merely semi-bounded.

Appendix B. Sobolev spaces of fractional order and the elliptic regularity theorem

For any $s \in \mathbb{R}$ define $k_s : \mathbb{R}^d \to \mathbb{R}$ by $k_s(x) := (1 + |x|^2)^s$. Further define positive measures μ_s on $\mathbb{R}^d$ by $\mathrm{d}\mu_s := k_s \mathrm{d}x$, $\mathrm{d}x$ being the Lebesgue measure in $\mathbb{R}^d$. Each $f \in L^2(\mathbb{R}^d, \mu_s)$ may be regarded as a tempered distribution in $\mathbb{R}^d$. To see this, let $S(\mathbb{R}^d)$ be the topological vector space of all Schwartz functions and let $(\phi_n)_{n\in\mathbb{N}}$ be a sequence in $S(\mathbb{R}^d)$ that converges to zero with respect to the topology of $S(\mathbb{R}^d)$. Then, for each $t \in \mathbb{R}$,

$$\sup_{x\in\mathbb{R}^d} k_t(x)|\phi_n(x)| \to 0$$

as $n \to \infty$. Choose $t \in \mathbb{R}$ such that $k_{-s-2t} \in L^1(\mathbb{R}^d)$. Then, by the Cauchy–Schwarz inequality

$$\begin{aligned} \left|\int_{\mathbb{R}^d} f\phi_n \mathrm{d}x\right|^2 &\le \int_{\mathbb{R}^d} |f|^2 k_{s/2}^2 \mathrm{d}x \int_{\mathbb{R}^d} |\phi_n|^2 k_{-s/2}^2 \mathrm{d}x \\ &= \|f\|^2_{L^2(\mathbb{R}^d,\mu_s)} \int_{\mathbb{R}^d} |\phi_n|^2 k_t^2 k_{-s-2t} \mathrm{d}x \\ &\le \|f\|^2_{L^2(\mathbb{R}^d,\mu_s)} \left(\sup_{x\in\mathbb{R}^d} |\phi_n(x)| k_t(x)\right)^2 \|k_{-s-2t}\|_{L^1(\mathbb{R}^d)}. \end{aligned}$$

This shows, that $\int_{\mathbb{R}^d} f\phi_n \mathrm{d}x \to 0$ as n tends to infinity. Therefore the map

$$T_f : S(\mathbb{R}^d) \to \mathbb{C}, \quad T_f(\phi) = \int_{\mathbb{R}^d} f\phi \, \mathrm{d}x$$

is a continuous linear functional on $S(\mathbb{R}^d)$, i.e., T_f belongs to the dual space $S'(\mathbb{R}^d)$ of $S(\mathbb{R}^d)$ which is called the space of *tempered distributions*. Denote by F the Fourier transform in $S'(\mathbb{R}^d)$ and recall that F is a continuous and bijective linear map in $S'(\mathbb{R}^d)$ whose inverse is also continuous. In terms of F the *Sobolev space of order* $s \in \mathbb{R}$ is defined by

$$H^s(\mathbb{R}^d) := F^{-1}(L^2(\mathbb{R}^d, \mu_s)) = \{F^{-1}f : f \in L^2(\mathbb{R}^d, \mu_s)\}.$$

$H^s(\mathbb{R}^d)$ becomes a Hilbert space with respect to the scalar product

$$\langle F^{-1}f, F^{-1}g\rangle_{H^s(\mathbb{R}^d)} := \langle f, g\rangle_{L^2(\mathbb{R}^d,\mu_s)}.$$

Let Ω be an open subset of $\mathbb{R}^d$ and denote the space of distributions in Ω by $D'(\Omega)$. A distribution $T \in D'(\Omega)$ is said to be *locally* H^s if for any $x \in \Omega$ there is a neighborhood Λ of x and a distribution $\tilde{T} \in H^s(\mathbb{R}^d)$ such that $T(\phi) = \tilde{T}(\phi)$ for all $\phi \in C_c^\infty(\Lambda)$.

Let $N \in \mathbb{N}$ be fixed and suppose that $f_\alpha \in C^\infty(\Omega)$ for each multi-index α with $|\alpha| \leq N$. Assume that there is at least one f_α with $|\alpha| = N$ which is not identically zero. These data determine a differential expression L by which we mean the map

$$L : D'(\Omega) \to D'(\Omega), \quad L(T) := \sum_{|\alpha|\leq N} f_\alpha D^\alpha T.$$

The number N is called the *order of* L. L is said to be *elliptic* if

$$\sum_{|\alpha|=N} f_\alpha(x)y^\alpha \neq 0$$

for all $x \in \Omega$ and $y \in \mathbb{R}^d \setminus \{0\}$. Observe that ellipticity only depends on those f_α with $|\alpha| = N$. The lower order terms play no role.

Now the theorem on elliptic regularity reads as follows.

Theorem B.1. *In the situation just described suppose that L is elliptic and $T, \tilde{T}$ are distributions in Ω such that*

(i) $L(\tilde{T}) = T$.
(ii) *T is locally H^s for some $s \in \mathbb{R}$.*

Then $\tilde{T}$ is locally H^{s+N}.

Appendix C. Markov processes and Hunt processes

In this appendix we summarize the necessary background information on Hunt processes and introduce the notion of equivalence for Hunt processes. Almost all of the material can be found in [FOT94].

Throughout this appendix X will denote a locally compact separable metric space and m will be a strictly positive Radon measure on X of full support, i.e., m is a measure defined on the Borel sets $\mathcal{B}(X)$ of X which is finite on compact sets, strictly positive on non-empty open sets and $\operatorname{supp}(m) = X$. Let X_Δ be the one-point compactification of X. When X is already compact, Δ is adjoint as an isolated point. A σ-algebra $\mathcal{B}_\Delta$ on X_Δ is given by

$$\mathcal{B}_\Delta = \mathcal{B}(X) \cup \{B \cup \{\Delta\} : B \in \mathcal{B}(X)\}.$$

Definition C.1.

(a) A *stochastic process with state space* $(X_\Delta, \mathcal{B}_\Delta)$ is a quadruple $(\Omega, \mathcal{M}, (X_t)_{t\geq 0}, \mathbb{P})$ such that

   (i) $(\Omega, \mathcal{M}, \mathbb{P})$ is a probability space,
   (ii) X_t is a measurable map from Ω to X_Δ for each $t \geq 0$.

(b) A filtration $(\mathcal{M}_t)_{t\geq 0}$, i.e., an increasing family of sub-σ-algebras of $\mathcal{M}$, is called *admissible* with respect to $(\Omega, \mathcal{M}, (X_t)_{t\geq 0}, \mathbb{P})$ if each X_t is measurable from $(\Omega, \mathcal{M}_t)$ to $(X_\Delta, \mathcal{B}_\Delta)$. It is called *right continuous*, if for each $t \geq 0$,

$$\mathcal{M}_t = \bigcap_{s>t} \mathcal{M}_s.$$

(c) Given an admissible filtration $(\mathcal{M}_t)_{t\geq 0}$, a $[0,\infty]$-valued function τ on Ω is called an $(\mathcal{M}_t)$-*stopping time* if $\{\tau \leq t\} \in \mathcal{M}_t$ for each $t \geq 0$. The σ-algebra $\mathcal{M}_\tau$ is then defined by

$$\mathcal{M}_\tau = \{A \in \mathcal{M} : A \cap \{\tau \leq t\} \in \mathcal{M}_t \text{ for each } t \geq 0\}.$$

Having defined these standard notions from probability theory we come to the definition of a Markov process as a stochastic process with additional properties.

Definition C.2.

(a) A quadruple $M = (\Omega, \mathcal{M}, (X_t)_{t\geq 0}, (\mathbb{P}_x)_{x\in X_\Delta})$ is called a Markov process on $(X, \mathcal{B}(X))$ if
   (i) For each $x \in X_\Delta$, $(\Omega, \mathcal{M}, (X_t)_{t\geq 0}, \mathbb{P}_x)$ is a stochastic process with state space $(X_\Delta, \mathcal{B}_\Delta)$.
   (ii) The function

$$x \mapsto \mathbb{P}_x\{X_t \in E\}$$

   is $\mathcal{B}$-measurable for each $t \geq 0$ and each $E \in \mathcal{B}_\Delta$.
   (iii) There is an admissible filtration $(\mathcal{M}_t)_{t\geq 0}$ such that

$$\mathbb{P}_x(\{X_{t+s} \in E\}|\mathcal{M}_t) = \mathbb{P}_{X_t}(\{X_s \in E\})$$

   $\mathbb{P}_x$-a.e. for any $x \in X$, $t, s \geq 0$ and $E \in \mathcal{B}$.
   (iv) $\mathbb{P}_\Delta(\{X_t = \Delta\}) = 1$ for each $t \geq 0$.

(b) A Markov process is called *normal* if

$$\mathbb{P}_x(\{X_0 = x\}) = 1 \text{ for each } x \in X.$$

(c) A Markov process is said to be a *strong Markov process with respect to the admissible filtration* $(\mathcal{M}_t)_{t\geq 0}$ if $(\mathcal{M}_t)_{t\geq 0}$ is right continuous and the equality

$$\mathbb{P}_\mu(\{X_{\tau+s} \in E\}|\mathcal{M}_\tau) = \mathbb{P}_{X_\tau}(\{X_s \in E\})$$

holds whenever μ is a probability measure on X_Δ, $E \in \mathcal{B}_\Delta$, $s \geq 0$ and τ is an $(\mathcal{M}_t)$-stopping time. Here the probability measure $\mathbb{P}_\mu$ on $(X_\Delta, \mathcal{B}_\Delta)$ is defined by

$$\mathbb{P}_\mu(E) := \int_{X_\Delta} \mathbb{P}_x(E)\, d\mu(x). \tag{C.1}$$

(d) A Markov process is called m-symmetric if its *transition function*

$$p_t(x, E) := \mathbb{P}_x(\{X_t \in E\}), \quad x \in X, t \geq 0, E \in \mathcal{B}_\Delta$$

is m-symmetric in the sense that

$$\int_{X_\Delta} u(x) \left(\int_{X_\Delta} v(y) p_t(x, dy) \right) dx = \int_{X_\Delta} v(x) \left(\int_{X_\Delta} u(y) p_t(x, dy) \right) dx$$

for any non-negative measurable functions u and v.

All these definitions were necessary to introduce the main object of this appendix, the Hunt process which will be defined now.

Definition C.3. A normal Markov process $M = (\Omega, \mathcal{M}, X_t, \mathbb{P}_x)$ on $(X, \mathcal{B})$ is called a Hunt process if

(i) $X_\infty(\omega) = \Delta$ for each $\omega \in \Omega$.
(ii) $X_t(\omega) = \Delta$ for each $t \geq \zeta(\omega) := \inf\{t \geq 0 : X_t(\omega) = \Delta\}$. The function ζ is called the *life time* of M.
(iii) For each $t \in [0, \infty]$ there is a map $\Theta_t : \Omega \to \Omega$ such that $X_s \circ \Theta_t = X_{t+s}$, $s \geq 0$.
(iv) For each $\omega \in \Omega$ the path $t \mapsto X_t(\omega)$ is right continuous on $[0, \infty)$ and has a left limit on $(0, \infty)$ (such a path is usually called *cadlag*).
(v) There is an admissible filtration $(\mathcal{M}_t)_{t \geq 0}$ such that M is strong Markov with respect to $(\mathcal{M}_t)_{t \geq 0}$.
(vi) For any $(\mathcal{M}_t)$-stopping time τ_n increasing to τ and for any probability measure μ on X_Δ the following equality holds

$$\mathbb{P}_\mu(\{\lim_{n \to \infty} X_{\tau_n} = X_\tau, \tau < \infty\}) = \mathbb{P}_\mu(\{\tau < \infty\}),$$

where $\mathbb{P}_\mu$ is defined in (C.1).

The connection between regular Dirichlet forms and Hunt processes consists in the deep fact that there is a one-to-one correspondence between these two objects, see Theorem C.18. But the Dirichlet form determines the associated Hunt process only up to equivalence, a phrase which we now want to make clear. First we have to introduce nearly Borel sets.

Definition C.4. Let M be a Hunt process.

(a) A set $B \subset X$ is called *nearly Borel* if for each probability measure μ on X_Δ there are $B_1, B_2 \in \mathcal{B}_\Delta$ such that $B_1 \subset B \subset B_2$ and

$$\mathbb{P}_\mu(\{\exists t \geq 0 : X_t \in B_2 \setminus B_1\}) = 0.$$

Here $\mathcal{B}_\Delta$ is the σ-algebra on the one-point compactification B_Δ of B, which may be constructed from B as X_Δ was constructed from X in the beginning of this section.
(b) A nearly Borel set $B \subset X$ is called M-invariant if for each $x \in B$ we have

$$\mathbb{P}_x(\{X_t \in B_\Delta, X_{t-} \in B_\Delta \text{ for all } t \geq 0\}) = 1,$$

where $X_{t-} = \lim_{s \uparrow t} X_s$.

Now equivalence between Hunt processes is defined as follows.

Definition C.5. Two Hunt processes M_1, M_2 are said to be *equivalent* if there is some nearly Borel set $N \subset X$ such that

(i) $X \setminus N$ is an invariant set for M_1 and M_2.
(ii) $m(N) = 0$.
(iii) The transition functions of M_1 and M_2 coincide outside N.

Finally we need an optional sampling theorem for supermartingales, which is used in the proof of the stochastic representation of the equilibrium potential. First we have to define a suitable class of supermartingales for which this theorem holds.

Definition C.6. Let $(\Omega, \mathcal{M}, (Y_t)_{t\geq 0}, \mathbb{P})$ be a stochastic process with state space $\mathbb{R}$ (equipped with the σ-algebra of all Borel sets) and let $(\mathcal{M}_t)_{t\geq 0}$ be an admissible filtration.

(a) $(\Omega, \mathcal{M}, (\mathcal{M}_t)_{t\geq 0}, (Y_t)_{t\geq 0}, \mathbb{P})$ is said to be a supermartingale if for each $s > t \geq 0$:

$$\mathbb{E}\{|Y_t|\} < \infty \quad \text{and} \quad \mathbb{E}\{Y_s|\mathcal{M}_t\} \leq Y_t.$$

(b) A supermartingale $(\Omega, \mathcal{M}, (\mathcal{M}_t)_{t\geq 0}, (Y_t)_{t\geq 0}, \mathbb{P})$ is said to be *of class (DL)* if for each $t \geq 0$,

$$\lim_{n\to\infty} \sup_{\tau\in\mathcal{A}_t} \mathbb{E}\{|Y_\tau| : |Y_\tau| \geq n\} = 0,$$

where $\mathcal{A}_t$ denotes the family of all $(\mathcal{M}_t)$-stopping times τ with $\tau \leq t$.

Using this definition the optional sampling theorem of Doob reads as follows.

Theorem C.7. (*Doob's optional sampling*). *Let $(\Omega, \mathcal{M}, (\mathcal{M}_t)_{t\geq 0}, (Y_t)_{t\geq 0}, \mathbb{P})$ be a supermartingale of class* (DL) *which is right continuous. Further let τ be a $(\mathcal{M}_t)$-stopping time and $0 \leq t \leq s$. Then*

$$\mathbb{E}\{Y_{\min(s,\tau)}|\mathcal{M}_t\} \leq Y_{\min(t,\tau)}.$$

A proof can be found in [DC], Appendix C.

C.1. Basics on regular Dirichlet forms

In this appendix we state without proof the necessary facts on regular Dirichlet forms which are needed to understand the previous sections. This is by no means a complete treatment of the subject, it is not even a complete survey. The reader should consult [FOT94], [Fuk80] or [MR92] for all the details and the proofs of the theorems. Most of this appendix is taken from [FOT94].

We start with describing the setup and fixing notation. Let X be a locally compact separable metric space and m be a strictly positive Radon measure on X of full support, i.e., m is a measure defined on the Borel sets $\mathcal{B}(X)$ of X which is finite on compact sets, strictly positive on non-empty open sets and $\operatorname{supp}(m) = X$.

We use the notation $L^2(X,m)$ for the Hilbert space of all square integrable (equivalence classes of) real-valued functions on X equipped with the scalar product

$$\langle u, v\rangle := \int_X uv\, dm.$$

We write $L^\infty(X,m)$ for the Banach space of all m-essentially bounded real-valued functions on X, the space of continuous functions on X will be denoted by $C(X)$, its subspace consisting of continuous functions with compact support by $C_c(X)$. The pointwise maximum and minimum of two functions u, v is abbreviated by $u \vee v$ and $u \wedge v$ respectively. Hence $u^+ := u \vee 0$ and $u^- := -(u \wedge 0)$ denote the positive and the negative part of u respectively.

Definition C.8.

(a) A non-negative symmetric closed form $(\mathcal{E}, \mathcal{F})$ on $L^2(X,m)$ is called a *Dirichlet form* if the following *Markov property* is satisfied.

$$u \in \mathcal{F} \text{ implies } u^+ \wedge 1 \in \mathcal{F} \text{ and } \mathcal{E}[u^+ \wedge 1] \le \mathcal{E}[u]. \tag{C.2}$$

(b) A Dirichlet form $(\mathcal{E}, \mathcal{F})$ is called *regular* if $\mathcal{F} \cap C_c(X)$ is not only dense in $(\mathcal{F}, \mathcal{E}_1(\cdot,\cdot))$ but also dense in $C_c(X)$ with respect to the uniform norm $\|\cdot\|_\infty$.

Sometimes it is useful to replace the Markov property (C.2) by a smooth version, explained in the next lemma.

Lemma C.9. *Let $(\mathcal{E}, \mathcal{F})$ be a non-negative symmetric closed form. Then $(\mathcal{E}, \mathcal{F})$ fulfills the Markov property if and only if for each $\varepsilon > 0$ there is a real function $\phi_\varepsilon : \mathbb{R} \to \mathbb{R}$ such that*

(i) *$\phi_\varepsilon(t) = t$ for $t \in [0,1]$.*
(ii) *$-\varepsilon \le \phi_\varepsilon(t) \le 1 + \varepsilon$ for every $t \in \mathbb{R}$.*
(iii) *$t_1 < t_2$ implies that $0 \le \phi_\varepsilon(t_2) - \phi_\varepsilon(t_1) \le t_2 - t_1$.*
(iv) *$u \in \mathcal{F}$ implies $\phi_\varepsilon \circ u \in \mathcal{F}$ and $\mathcal{E}[\phi_\varepsilon \circ u] \le \mathcal{E}[u]$.*

Remark C.10. Let H be the unique self-adjoint operator which corresponds to $(\mathcal{E}, \mathcal{F})$ in the sense of Theorem 2.2. Since $H \ge 0$ we can construct the semigroup $(e^{-tH})_{t>0}$ of H and the resolvent $(H+\alpha)^{-1}$, $\alpha > 0$. It is natural to ask about characterizations of the Markov property (C.2) in terms of the semigroup and the resolvent. Before we can give the answer we have to introduce some more notation.

Definition C.11. A bounded linear operator T on $L^2(X,m)$ is said to have the *Markov property* if $u \in L^2(X,m)$, $0 \le u \le 1$ m-almost everywhere implies that $0 \le Tu \le 1$ m-almost everywhere.

Theorem C.12. *Let $(\mathcal{E}, \mathcal{F})$ be a non-negative symmetric closed form on $L^2(X,m)$ and let H be the self-adjoint operator which is associated to $(\mathcal{E}, \mathcal{F})$. The following are equivalent.*

(i) *$(\mathcal{E}, \mathcal{F})$ fulfills the Markov property* (C.2).
(ii) *For each $t > 0$ the operator e^{-tH} fulfills the Markov property of Definition* C.11.

(iii) *For each* $\alpha > 0$ *the operator* $\alpha(H+\alpha)^{-1}$ *fulfills the Markov property of Definition* C.11.

Theorem C.13. *A regular Dirichlet form* $(\mathcal{E}, \mathcal{F})$ *possesses the following properties.*

(a) *If* $u, v \in \mathcal{F} \cap L^\infty(X, m)$, *then* $uv \in \mathcal{F}$ *and*

$$\mathcal{E}[uv] \le \|u\|_\infty^2 \mathcal{E}[v] + \|v\|_\infty^2 \mathcal{E}[u].$$

(b) *If* $u \in L^2(X, m)$, $v \in \mathcal{F}$ *are* 1-excessive, *i.e.*,

$$e^{-tH}u \le e^t u, \quad e^{-tH}v \le e^t v, \quad (t > 0),$$

and $u \le v$, *then* $u \in \mathcal{F}$ *and* $\mathcal{E}_1[u] \le \mathcal{E}_1[v]$.

Every Dirichlet form admits a set function, the *capacity*, which is defined for arbitrary subsets of X.

Definition C.14.

(a) For $U \subset X$ open define the *capacity of* U by

$$\operatorname{Cap}(U) := \inf\{\mathcal{E}_1[u] : u \in \mathcal{F}, u \ge 1 \quad m\text{-a.e. on } U\}.$$

If there is no such u the convention $\operatorname{Cap}(U) = \infty$ applies. The capacity is extended to arbitrary $A \subset X$ by

$$\operatorname{Cap}(A) := \inf\{\operatorname{Cap}(U) : U \supset A, U \text{ open}\}.$$

(b) A statement is said to hold *quasi-everywhere* (*q.e.*) if it holds outside of some set of zero capacity.
(c) A function defined on X is said to be *quasi-continuous* (*q.c.*) if for each $\varepsilon > 0$ there is an open set $U \subset X$ with $\operatorname{Cap}(U) < \varepsilon$ such that $u|U^c$ is continuous.
(d) The *zero-order capacity* $\operatorname{Cap}_0(\cdot)$ is defined in the same way as in (a) but with $\mathcal{E}_1[u]$ replaced by $\mathcal{E}[u]$.

The following result is an important consequence of the regularity of $(\mathcal{E}, \mathcal{F})$.

Theorem C.15. *Suppose that* $(\mathcal{E}, \mathcal{F})$ *is a regular Dirichlet form. Then each* $u \in \mathcal{F}$ *admits a quasi-continuous* m*-version* $\tilde{u}$, *i.e., there is a quasi-continuous function* $\tilde{u}$ *which coincides* m*-almost everywhere with* u. *Two quasi-continuous* m*-versions of* u *coincide quasi-everywhere.*

The notation $u = \tilde{v}$ will always indicate that u is a quasi-continuous m-version of v.

Next we present a lemma which is often used to prove that certain statements which are assumed to hold m-a.e. are indeed valid quasi-everywhere. Part (b) can be proved with the aid of the Banach–Alaoglu and the Banach–Saks theorems.

Lemma C.16. *Let* $(\mathcal{E}, \mathcal{F})$ *be a regular Dirichlet form on* $L^2(X, m)$.

(a) *If* $u_n \in \mathcal{F}$ *is a sequence such that* $u_n \to u$ *in* $(\mathcal{F}, \mathcal{E}_1(\cdot,\cdot))$, *then there is a subsequence that converges to* $\tilde{u}$ *quasi-everywhere.*

(b) *Let* $u_n \in \mathcal{F}$ *such that* $\sup_{n\in\mathbb{N}} \mathcal{E}_1[u_n] < \infty$. *If* $u \in \mathcal{F}$ *such that* $u_n \to u$ *in* $L^2(X,m)$, *then there is a subsequence* u_{n_j} *such that its Cesaro means* $\frac{1}{k}(u_{n_1} + \cdots + u_{n_k})$ *converge to* u *in* $(\mathcal{F}, \mathcal{E}_1(\cdot,\cdot))$.

The most important properties of capacity are given in the following theorem.

Theorem C.17. *Let* $(\mathcal{E}, \mathcal{F})$ *be a regular Dirichlet form on* $L^2(X,m)$.

(a) *Let* $A_1, A_2 \subset X$ *arbitrary. If* $A_1 \subset A_2$, *then* $\mathrm{Cap}(A_1) \le \mathrm{Cap}(A_2)$.

(b) *If* $B \subset X$ *is a Borel set, then*

$$\mathrm{Cap}(B) = \sup\{\mathrm{Cap}(K) : K \subset B, K \textit{ compact }\}.$$

(c) *If* A_n, $n \in \mathbb{N}$, *is an arbitrary sequence of subsets of* X, *then*

$$\mathrm{Cap}\left(\bigcup_{n=1}^{\infty} A_n\right) \le \sum_{n=1}^{\infty} \mathrm{Cap}(A_n).$$

(d) *For arbitrary* $A \subset X$ *the capacity of* A *may be computed as*

$$\mathrm{Cap}(A) = \inf\{\mathcal{E}_1[u] : u \in \mathcal{F}, \tilde{u} \ge 1 \textit{ q.e. on } A\}. \tag{C.3}$$

(e) *The infimum in* (C.3) *is attained for a unique element* $e_A \in \mathcal{F}$, *called the* equilibrium potential of A. *Moreover* $0 \le e_A \le 1$ *m-a.e. and* $\tilde{e_A} = 1$ *quasi-everywhere on* A; e_A *is a 1-excessive function, i.e., for each* $t > 0$ *the following inequality holds.*

$$e^{-tH} e_A \le e^t e_A,$$

where H *is the self-adjoint operator corresponding to* $(\mathcal{E}, \mathcal{F})$ *in the sense of Remark* C.10.

(f) *For arbitrary* $A \subset X$ *there is a unique measure* ν_A *on* X, *called the* equilibrium measure of A, *satisfying the following properties.*
  (i) $\mathrm{supp}(\nu_A) \subset \overline{A}$.
  (ii) ν_A *charges no set of zero capacity.*
  (iii) $\mathcal{E}_1(e_A, u) = \int_X u \, d\nu_A$ *for each* $u \in \mathcal{F} \cap C_c(X)$.

As an immediate consequence of (e) and (f) we see that for a closed set $F \subset X$ the following equalities hold true:

$$\mathrm{Cap}(F) = \mathcal{E}_1[e_F] = \nu_F(F) = \nu_F(X). \tag{C.4}$$

We now want to describe the connections between Hunt processes and regular Dirichlet forms. Given an m-symmetric Hunt process $M = (\Omega, \mathcal{M}, X_t, \mathbb{P}_x)$ on $(X, \mathcal{B}(X))$, it is easy to construct a regular Dirichlet form: The *transition function*

$$p_t(x, E) := \mathbb{P}_x(\{X_t \in E\}), \qquad (t > 0, x \in X, E \in \mathcal{B}(X))$$

of M determines a semigroup $(S_t)_{t>0}$ via

$$S_t f(x) := \int_X f(y) p_t(x, dy) = \mathbb{E}_x(f(X_t)).$$

In this context the semigroup property is often given via the *Chapman–Kolmogorov equation* The above semigroup clearly fulfills the Markov property of Definition (C.11). Since it turns out to be strongly continuous as well, there is a regular Dirichlet form on $L^2(X,m)$ which corresponds to $(S_t)_{t>0}$. This Dirichlet form is called *the Dirichlet form of* M. It is not so easy to see that the converse of these considerations is also true:

Theorem C.18. *Given a regular Dirichlet form* $(\mathcal{E},\mathcal{F})$ *on* $L^2(X,m)$ *there is an* m*-symmetric Hunt process* M *whose Dirichlet form is the given one* $(\mathcal{E},\mathcal{F})$. M *is unique up to equivalence.*

For a proof see [FOT94] Theorems 7.2.1 and 4.2.7. For the construction of the process it plays an important role that $(\mathcal{E},\mathcal{F})$ is regular. According to Theorem C.15 every $u \in \mathcal{F}$ admits a quasi-continuous m-version. This allows one to ignore successively sets of capacity zero.

In view of Theorem C.18 one can ask for a stochastic interpretation of analytical objects. In Theorem C.19 below we give a stochastic representation of the equilibrium potential e_B where $B \subset X$ is a Borel set. This theorem is identical to Theorem 4.1 where the proof is carried out.

Theorem C.19. *Let* $(\mathcal{E},\mathcal{F})$ *be a regular Dirichlet form on* $L^2(X,m)$ *with the corresponding Hunt process* $M = (\Omega,\mathcal{M},\mathcal{M}_t,X_t,\mathbb{P}_x)$ *and let* $B \subset X$ *be a Borel set of finite capacity. Then*

$$e_B(x) = \mathbb{E}_x\{e^{-\tau_B}\},$$

where $\tau_B := \inf\{t > 0 : X_t \in B\}$ *denotes the first hitting time of* B.

A major part of this thesis is devoted to the analysis of the spectral consequences of domain perturbations of an operator or its associated form, and it is a natural question whether the restriction of a regular Dirichlet form $(\mathcal{E},\mathcal{F})$ to a smaller domain $\mathcal{F}_B$ consisting of functions which vanish on B (in a sense to be made precise), is again a regular Dirichlet form. This turns out to be rather difficult and depends on several properties of the set $B \subset X$. Fortunately everything works fine if B is closed. This is the content of part (a) of the following theorem, which is taken from Theorem 4.4.3 in [FOT94].

Theorem C.20. *For any Borel set* $B \subset X$ *define a closed subspace of* $(\mathcal{F},\mathcal{E}_1(\cdot,\cdot))$ *by*

$$\mathcal{F}_B := \{u \in \mathcal{F} : \tilde{u} = 0 \text{ q.e. on } B\}.$$

(a) *If* B *is closed, then the restriction of* $\mathcal{E}$ *to the form domain* $\mathcal{F}_B$ *is a regular Dirichlet form in* $L^2(X \setminus B, m)$.

(b) *Let* $P_{\mathcal{F}_B^\perp}$ *be the orthogonal projection onto* $\mathcal{F}_B^\perp$ *in* $(\mathcal{F},\mathcal{E}_1(\cdot,\cdot))$. *Then*

$$(P_{\mathcal{F}_B^\perp}u)(\cdot) = \mathbb{E}_{(\cdot)}\{e^{-\tau_B}\tilde{u}(X_{\tau_B})\}.$$

For any open subset Y of X put $B := X\setminus Y$. By Theorem C.20 (a) the restriction of $\mathcal{E}$ to $\mathcal{F}_B$ corresponds to a unique self-adjoint and non-negative operator H^Y in $L^2(Y,m)$. The Dynkin formula establishes a connection between the resolvents of H and H^Y. Before we can state this result we have to introduce measures of finite energy integral and their associated potentials.

Definition C.21. A positive Radon measure μ on X is said to be *of finite energy integral*, if there is a constant $\gamma \geq 0$ with

$$\int_X |\tilde{u}|\, d\mu \leq \gamma \sqrt{\mathcal{E}_1(u,u)} \tag{C.5}$$

for all $u \in \mathcal{F} \cap C_c(X)$. The collection of such measures is denoted S_0.

For any measure μ of finite energy integral the map $u \mapsto \int_X |\tilde{u}| d\mu$ defines a continuous linear functional on $(\mathcal{F}, \mathcal{E}_1(\cdot,\cdot))$. The Riesz representation theorem entails that there exists a function $U_1\mu \in \mathcal{F}$ such that

$$\mathcal{E}_1(U_1\mu, u) = \int_X \tilde{u}\, d\mu$$

for all $u \in \mathcal{F} \cap C_c(X)$.

Definition C.22. The function $U_1\mu \in \mathcal{F}$ is called the *1-potential of* μ .

Example. Let $f \in L^2(X,m)$ be a non-negative function. Then $\mu := f dm$ is a measure of finite energy integral and its 1-potential is given by $U_1\mu = (H+1)^{-1}f$.

Now Dynkin's formula reads as follows.

Theorem C.23. *Let $(\mathcal{E}, \mathcal{F})$ be a regular Dirichlet form in $L^2(X,m)$.*

(a) *For any compact $K \subset X$ and any 1-potential u there is a unique measure $\mu_K^u \in S_0$ supported in K such that*

$$(\widetilde{U_1\mu_K^u})(x) = \mathbb{E}_x\{e^{-\tau(K)}\tilde{u}(X_{\tau_K})\}.$$

(b) *(Dynkin's formula): Let $f \in L^2(X,m)$ with $f \geq 0$ m-a.e. and denote the restriction of f to Y by $f_{|Y}$. Then*

$$(H+1)^{-1}f = (H^Y+1)^{-1}(f_{|Y}) + U_1\mu_K^{(H+1)^{-1}f}$$

and

$$\mathcal{E}_1\left((H^Y+1)^{-1}(f_{|Y}), U_1\mu_K^{(H+1)^{-1}f}\right) = 0.$$

The measures of finite energy integral can also be used to compute the capacity of compact subsets of X.

Theorem C.24. *Let K be a compact subset of X. Then*

$$\mathrm{Cap}(K) = \sup\left\{\mu(K) : \begin{array}{c} \mu \in S_0, \mathrm{supp}(\mu) \subset K, \\ \mu(X) < \infty, \|\widetilde{U_1\mu}\|_\infty \leq 1 \end{array}\right\}.$$

Finally we need a highly non-trivial result on the set of irregular points. A point $x \in X$ is called *regular for the Borel set* $B \subset X$, if $\mathbb{P}_x\{\tau_B = 0\} = 1$, where $\tau_B = \inf\{t > 0 : X_t \in B\}$ is the first hitting time of B; x is called *irregular for* B if $\mathbb{P}_x\{\tau_B = 0\} = 0$. It is a consequence of Blumenthal's 0-1 law that x is irregular if and only if it is not regular.

Theorem C.25. *For any Borel set B the set of irregular points in B is of zero capacity.*

For the proof see [FOT94] Theorems A.2.6 (i), 4.1.3, and 4.2.1 (ii).

Acknowledgments

I would like to thank my supervisor, Prof. Dr. Michael Demuth, for his support and his many mathematical suggestions. This work has also benefited from fruitful discussions with the members of the "Stochastic Spectral Theory" group in Clausthal: Sven Eder, Walter Renger, and Eckhard Giere. Moreover, I would like to thank Bruce Saint Hilaire for helping me to improve the English in several parts of the manuscript.

References

[Ada78] R.A. Adams. *Sobolev Spaces.* New York: Academic Press, 1978.

[AGHKH88] S. Albeverio, F. Gesztesy, R. Hoegh-Krohn, and H. Holden. *Solvable Models in Quantum Mechanics.* Texts and Monographs in Physics. Springer-Verlag, 1988.

[Agm65] S. Agmon. *Elliptic Boundary Value Problems.* Van Nostrand Mathematical Studies, 1965.

[Agr97] M.S. Agranovich. Elliptic boundary problems. In *Partial differential equations IX*, volume 79 of *Encyclopaedia of Mathematical Sciences*, pages 1–144. Springer-Verlag, 1997.

[AH95] D.R. Adams and L.I. Hedberg. *Function Spaces and Potential Theory*, volume 314 of *Grundlehren der Mathematischen Wissenschaften.* Springer Verlag, 1995.

[AM95] W. Arendt and S. Monniaux. Domain perturbation for the first eigenvalue of the Dirichlet Schrödinger operator. *Oper. Theory Adv. Appl.*, 78:9–19, 1995.

[BEGK99] A. Bovier, M. Eckhoff, V. Gayrard, and M. Klein. Metastability in stochastic dynamics of disordered mean-field models. Preprint, 1999.

[BH86] J. Bliedtner and W. Hansen. *Potential theory. An Analytic and Probabilistic Approach to Balayage.* Springer-Verlag, 1986.

[BH91] N. Bouleau and F. Hirsch. *Dirichlet forms and analysis on Wiener space.* Walter de Gruyter, 1991.

[Bra94] J.F. Brasche. Dirichletformen. Manuscript, 1994.

[BS91] F.A. Berezin and M.A. Shubin. *The Schrödinger equation*, volume 66 of *Mathematics and its Applications.* Kluwer Academic Publishers Group, 1991. Translated from the 1983 Russian edition.

[CD92] C.V. Coffman and R.J. Duffin. On the fundamental eigenfunctions of a clamped punctured disk. *Adv. Appl. Math.*, 13:142–151, 1992.

[CF78] I. Chavel and E.A. Feldman. Spectra of domains in compact manifolds. *J. Funct. Anal.*, 30:198–222, 1978.

[CF88] I. Chavel and E.A. Feldman. Spectra of manifolds less a small domain. *Duke Math. J.*, 56:399–414, 1988.

[Cou95] G. Courtois. Spectrum of manifolds with holes. *J. Funct. Anal.*, 134:194–221, 1995.

[Dav89] E.B. Davies. *Heat Kernels and Spectral Theory.* Cambridge University Press, 1989.

[Dav95] E.B. Davies. Uniformly elliptic operators with measurable coefficients. *J. Funct. Anal*, 132:141–169, 1995.

[Dav97] E.B. Davies. L^p spectral theory of higher order elliptic differential operators. *Bull. Lond. Math. Soc.*, 29:513–546, 1997.

[DC] M. Demuth and J. van Casteren. *Stochastic Spectral Theory for Selfadjoint Feller Operators.* Birkhäuser Verlag, Basel, 2000.

[DC98] M. Demuth and J. van Casteren. Completeness of scattering systems with obstacles of finite capacity. In I. Gohberg, editor, *Differential and Integral Operators*, volume I, pages 39–50. Birkhäuser Verlag, 1998.

[DGCZ95] M. Demuth, F. Gesztesy, J. van Casteren, and Z. Zhao. Finite capacities in spectral theory. In M. Demuth and B.-W. Schulze, editors, *Partial Differential Operators and Mathematical Physics*, volume 78 of *Operator Theory, Advances and Applications*, pages 89–97. Birkhäuser Verlag, 1995.

[DMN97] M. Demuth, I. McGillivray, and A. Noll. Capacity and spectral theory. In M. Demuth, E. Schrohe, B.-W. Schulze, and J. Sjöstrand, editors, *Spectral Theory, Microlocal Analysis, Singular Manifolds*, volume 14 of *Advances in Partial Differential Equations*, pages 12–77. Akademie Verlag, Berlin, 1997.

[Ede99] S. Eder. *Factorization Theorems for compact operators and semigroup differences.* PhD thesis, Clausthal, Mathematisch-Naturwissenschaftliche Fakultät, 1999.

[FOT94] M. Fukushima, Y. Oshima, and M. Takeda. *Dirichlet Forms and Symmetric Markov Processes*, volume 19 of *Studies in Mathematics.* Walter de Gruyter Co, Berlin, 1994.

[Fre82] J. Frehse. Capacitary methods in the theory of partial differential equations. *Jahresber. d. Dt. Math.-Verein*, 84:1–44, 1982.

[Fuk80] M. Fukushima. *Dirichlet Forms and Markov Processes.* North Holland, Amsterdam, 1980.

[Goe77] H.-W. Goelden. On the non-degeneracy of the ground state of Schrödinger operators. *Math. Z.*, 155:239–247, 1977.

[GZ94] F. Gesztesy and Z. Zhao. Domain perturbations, Brownian motion and ground states of Dirichlet Schrödinger operators. *Math. Z.*, 215:143–150, 1994.

[Hör83a] L. Hörmander. *The Analysis of Linear Partial Differential Operators I.* Grundlehren der mathematischen Wissenschaften. Springer Verlag, Berlin-New York, 1983.

[Hör83b] L. Hörmander. *The Analysis of Linear Partial Differential Operators II.* Grundlehren der mathematischen Wissenschaften. Springer Verlag, Berlin-New York, 1983.

[HS78] P.-R. Halmos and V.-S. Sunder. *Bounded integral operators on L^2 spaces.* Ergebnisse der Mathematik und ihrer Grenzgebiete. Springer-Verlag, 1978.

[Jac62] J.D. Jackson. *Classical Electrodynamics.* John Wiley & Sons, Inc., 1962.

[Joh49] F. John. On linear partial differential equations with analytic coefficients. *Commun. Pure Appl. Math.*, 2:209–253, 1949.

[Joh71] F. John. *Partial Differential Equations.* Springer-Verlag, 3rd edition, 1971.

[Kat80] T. Kato. *Perturbation Theory for Linear Operators.* Springer Verlag, Berlin-Heidelberg-New York, second edition, 1980.

[Kor65] V.B. Korotkov. Integral operators with Carleman kernels. *Sov. Math. Dokl.*, 6:1496–1499, 1965. Translation from Dokl. Akad. Nauk SSSR 165, 748-751 (1965).

[Lan72] N.S. Landkof. *Foundations of Modern Potential Theory.* Grundlehren der Mathematischen Wissenschaften. Springer-Verlag, 1972.

[Maz73] V.G. Maz'ja. On (p, l)-capacity, embedding theorems and the spectrum of a selfadjoint elliptic operator. *Math. USSR-Ivz.*, 7:357–387, 1973. English translation.

[Maz85] V.G. Maz'ja. *Sobolev Spaces.* Springer Series in Soviet Mathematics. Springer Verlag Berlin, New York, 1985.

[McG96] I. McGillivray. Capacitary estimates for Dirichlet eigenvalues. *J. Funct. Anal.*, 139:244–259, 1996.

[MR84] P.J. McKenna and M. Rao. Lower bounds for the first eigenvalue of the Laplacian with Dirichlet boundary conditions and a theorem of Hayman. *Appl. Anal.*, 18:55–66, 1984.

[MR92] Z.M. Ma and M. Röckner. *Introduction to the Theory of (Non-Symmetric) Dirichlet Forms.* Springer Verlag, 1992.

[Nol97a] A. Noll. Capacitary estimates for the ground-state shift. *Universitatis Iagellonicae Acta Mathematica*, 34:81–88, 1997.

[Nol97b] A. Noll. A generalization of Dynkin's formula and capacitary estimates for semibounded operators. In M. Demuth and B.-W. Schulze, editors, *Differential Equations, Asymptotic Analysis and Mathematical Physics*, volume 100 of *Mathematical Research*, pages 252–259. Akademie Verlag, Berlin, 1997.

[Nol99] A. Noll. Capacity in abstract Hilbert spaces and applications to higher order differential operators. *Comm. P. D. E.*, 24:759–775, 1999.

[Nol00] A. Noll. Domain perturbations, shift of eigenvalues and capacity. *J. Funct. Anal.*, 170:246–263, 2000.

[Oza81] S. Ozawa. Singular variation of domains and eigenvalues of the Laplacian. *Duke Math. J.*, 48(4):767–778, 1981.

[Oza82] S. Ozawa. The first eigenvalue of the Laplacian on two-dimensional manifolds. *Tohoku Math. J.*, 34:7–14, 1982.

[Oza83] S. Ozawa. Electrostatic capacity and eigenvalues of the Laplacian. *J. Fac. Sci. Univ. Tokyo*, 30:53–62, 1983.

[PS78] S. Port and C. Stone. *Brownian Motion and Classical Potential Theory.* Probability and Mathematical Statistics. Academic Press, 1978.

[Rau75] J. Rauch. The mathematical theory of crushed ice. In *Partial Differential Equations and Related Topics*, volume 446 of *Lect. Notes in Math.*, pages 370–379. Springer, Berlin, 1975.

[RS75] M. Reed and B. Simon. *Methods of Modern Mathematical Physics II: Fourier Analysis, Selfadjointness.* Academic Press, New York, 1975.

[RS78] M. Reed and B. Simon. *Methods of Modern Mathematical Physics IV: Analysis of operators.* Academic Press, New York, 1978.

[RS80] M. Reed and B. Simon. *Methods of Modern Mathematical Physics I: Functional Analysis.* Academic Press, New York, 2nd edition, 1980.

[RSS94] G.V. Rozenblum, M.A. Shubin, and M.Z. Solomyak. Spectral theory of differential operators. In *Partial differential equations VII*, volume 66 of *Encyclopaedia of Mathematical Sciences.* Springer-Verlag, 1994.

[RT75a] J. Rauch and M. Taylor. Electrostatic screening. *J. Math. Phys.*, 16:284–288, 1975.

[RT75b] J. Rauch and M. Taylor. Potential and scattering theory on wildly perturbed domains. *J. Funct. Anal.*, 18:27–59, 1975.

[Rud66] W. Rudin. *Real and Complex Analysis.* McGraw-Hill, Series in higher mathematics, 1966.

[Rud73] W. Rudin. *Functional Analysis.* McGraw-Hill, Series in higher mathematics, 1973.

[Sim79] B. Simon. *Functional Integration and Quantum Physics.* Academic Press, New York, 1979.

[Sim82] B. Simon. Schrödinger semigroups. *Bull. Amer. Math. Soc.*, 7:447–526, 1982.

[SW77] B.-W. Schulze and G. Wildenhain. *Methoden der Potentialtheorie für elliptische Differentialgleichungen beliebiger Ordnung*, volume 46 of *Mathematische Lehrbücher und Monographien.* Akademie Verlag, 1977.

[Szn98] A.-S. Sznitman. *Brownian motion, obstacles and random media.* Springer Monographs in Mathematics. Berlin. Springer-Verlag, 1998.

[Tay76] M. Taylor. Scattering length and perturbations of $-\Delta$ by positive potentials. *J. Math. Anal. Appl.*, 53:291–312, 1976.

[Tay79] M. Taylor. Estimate on the fundamental frequency of a drum. *Duke Math. J.*, 46:447–453, 1979.

[Tre67] F. Treves. *Topological vector spaces, distributions and kernels.* Academic Press, 1967.

[Wei76] J. Weidmann. *Lineare Operatoren in Hilberträumen.* B. G. Teubner, Stuttgart, 1976.

[Wei84] J. Weidmann. Stetige Abhängigkeit der Eigenwerte und Eigenfunktionen elliptischer Differentialoperatoren vom Gebiet. *Math. Scand.*, 54:51–69, 1984.

[Wei95a] J. Weidmann. Mathematische Grundlagen der Quantenmechanik I. Manuscript, University of Frankfurt, 1995.

[Wei95b] J. Weidmann. Mathematische Grundlagen der Quantenmechanik II. Manuscript, University of Frankfurt, 1995.

André Noll
Technical University of Darmstadt
Schloßgartenstraße 7
D-64289 Darmstadt, Germany
e-mail: `noll@mathematik.tu-darmstadt.de`

Operator Theory:
Advances and Applications, Vol. 145, 363–394

An Interpolation Family between Gabor and Wavelet Transformations: Application to Differential Calculus and Construction of Anisotropic Banach Spaces

Bruno Nazaret and Matthias Holschneider

Abstract. In this paper we define an interpolation family of transformations, whose extremes are the Gabor and wavelet transformations, in order to extend the Cordoba–Fefferman results (see [CF]) and to define a differential calculus at the first order. This interpolation family is based on the representation through translated and modulated versions of an analyzing function, with the additional property that this family is naturally localized in paraboloids. This will allow us at the end of the paper to construct anisotropic Banach spaces of functions by pullback techniques.

1. Introduction

In signal treatment, there are mostly two transformations which are used. The first, the Gabor transformation, is based on the representation of the Weyl–Heisenberg group over $L^2(\mathbb{R}^n)$ (see [AAGM]). More precisely, $g \in L^2(\mathbb{R}^n)$ being given, we define for all $b, w \in \mathbb{R}^n$,

$$g_{b,\omega}(x) = e^{i\omega(x-b)}g(x-b)$$

and, for all $s \in L^2(\mathbb{R}^n)$, the Gabor transform of s with respect to g by

$$\mathcal{W}_g^{(0)}s(b,\omega) = \langle g_{b,\omega}, s\rangle .$$

One can prove that $\mathcal{W}_g^{(0)}$ is unitary and then invertible. Up to a constant, its inverse transformation is given by, for all $r \in L^2(T^*\mathbb{R}^n)$,

$$\mathcal{M}_g^{(0)}r(x) = \int_{T^*\mathbb{R}^n} dbd\omega g_{b,\omega}(x)r(b,w).$$

The second, the wavelet transformation, defined by Grossman and Morlet in [GM1], [GM2] at the beginning of 1980s for the analysis of seismic signals and developed by Meyer [CM] and his collaborators [Mal], [Dau1], [Dau2], [Jaf], is based on the representation of the affine group over $L^2(\mathbb{R}^n)$ (see [AAGM]). It replaces the frequency by a scale notion. Here, we define the wavelet transform of s with respect to g by: for all $b \in \mathbb{R}^n$, $a > 0$,

$$\mathcal{W}[g,s](b,a) = \mathcal{W}_g s(b,a) = \langle g_{b,a}, s\rangle$$

where

$$g_{b,a}(x) = \frac{1}{a^n} g\left(\frac{x-b}{a}\right)$$

is a translated and dilated version of the analyzing function g. In the case of wavelets, the invertibility is more difficult to handle. The natural candidate to be the inverse is, as in the Gabor transformation, its adjoint, defined by:

$$\mathcal{M}_g s(x) = \int_{\mathbb{R}^n}\int_0^\infty \frac{dadb}{a} g_{b,a}(x) r(b,a)$$

for $r \in L^2\left(\mathbb{R}^n \times \mathbb{R}^+, \frac{dadb}{a}\right)$. The result is the following; defining

$$m_{g,h}(\xi) = \int_0^\infty \frac{da}{a}\overline{\widehat{g}(a\xi)}\widehat{h}(a\xi)$$

we will say that g is admissible (condition A1) if there exists a constant $c > 0$ such that for all $\xi \neq 0$,

$$0 < c^{-1} \leq m_{g,g}(\xi) \leq c < \infty$$

and that g is strictly admissible if, for all $\xi \neq 0$, $m_{g,g}(\xi) \equiv 1$. Then, if g is admissible, there exists a reconstruction wavelet h (that is such that $\mathcal{M}_h\mathcal{W}_g = Id$). In the case where g is strictly admissible, g is its own reconstruction wavelet and $\mathcal{W}_g$ is unitary.

These two transformations do not present the same interest. The wavelet transform acts as a mathematical microscope. Indeed, suppose that the analyzing function is localized around a given point x_0 with variance 1. Then, the function $g_{b,a}$ is localized around $(x_0 - b)$ with variance a. This property allows the wavelet transformation to be used for analyzing pointwise and microlocal regularity (see [Hol1], [Hol2], [Jaf], [HT]). Furthermore, the Fourier representation of the wavelet transform, obtained by the Plancherel theorem,

$$\mathcal{W}_g s(b,a) = \frac{1}{(2\pi)^n}\int_{\mathbb{R}^n} d\xi e^{ib\xi}\overline{\widehat{g}(a\xi)}\widehat{s}(\xi),$$

shows that the wavelet can be interpreted as a time-frequency analysis. Its main disadvantage is that it possesses a bad angular resolution since, a going from 0 to ∞, the support of $\widehat{g}(a\xi)$ is translated in a conical set in the Fourier space.

On the contrary, the Gabor transform does not admit the pointwise regularity of functions since the width of the window $(g, \widehat{g})$ is constant. But, considering the Fourier representation of this transform,

$$\mathcal{W}_g^{(0)} s(b,\omega) = \frac{1}{(2\pi)^n}\int_{\mathbb{R}^n} d\xi e^{ib(\xi-\omega)}\overline{\widehat{g}(\xi-\omega)}\widehat{s}(\xi),$$

one can see that it isolates the central frequency direction.

In order to avoid these difficulties, the first idea would be to perform all these operations on g (*i.e.*, translations, dilations, and modulations), but it does not work because the obtained transformations are not L^2-continuous. So, we have to make some restrictions. (This idea was already studied by Folland in [Fol].)

Let $\alpha \in (0,1)$ be given. We define the analysis of order α of s with respect to g by the following: $\forall (b,\omega) \in T^*\mathbb{R}^n$,

$$\mathcal{W}_g^{(\alpha)} s(b,\omega) = \left\langle g_{b,\omega}^{(\alpha)}, s \right\rangle \tag{1}$$

where

$$g_{b,\omega}^{(\alpha)}(x) = \langle\omega\rangle^{\frac{n\alpha}{2}} e^{i\omega(x-b)} g\left(\langle\omega\rangle^{\alpha}(x-b)\right). \tag{2}$$

Here, the C^∞ function $\langle\omega\rangle$ is given by:

$$\begin{cases} \langle\omega\rangle = |\omega| \text{ for } |\omega| \geq 1, \\ 1/2 \leq \langle\omega\rangle \leq 1 \text{ for } |\omega| \leq 1 \end{cases} \tag{3}$$

($|\omega|$ stands for the euclidean norm of ω). In the same manner, we define the synthesis of order α of the function r defined on the phase space with respect to h: $\forall x \in \mathbb{R}^n$,

$$\mathcal{M}_h^{(\alpha)} r(x) = \int_{T^*\mathbb{R}^n} dbd\omega h_{b,\omega}^{(\alpha)}(x) r(b,w). \tag{4}$$

Here, $\mathcal{W}_g^{(\alpha)} s$ analyzes s around b at the scale $|\omega|^{-\alpha}$. Using the Plancherel formula, we obtain:

$$\mathcal{W}_g^{(\alpha)} s(b,\omega) = \frac{1}{(2\pi)^n} \int_{\mathbb{R}^n} d\xi \langle\omega\rangle^{-\frac{n\alpha}{2}} e^{ib\xi} \overline{\widehat{g}\left(\langle\omega\rangle^{-\alpha}(\xi-\omega)\right)} \widehat{s}(\xi). \tag{5}$$

Then, in Fourier space, the analyzing function family moves into paraboloids with opening α. Unfortunately, with this representation, we lost the inversion formula, though we will see later that there exists an approximate one. Now, let us remark that, for $\alpha = 0, 1$, we obtain respectively the Gabor and a wavelet type transformations. For $\alpha = 1/2$, we have the FBI (Fourier Bros Iagolnizer) transformation introduced by Bros–Iagolnizer ([BI]) and Sjöstrand ([Sjo]) for studying analytic microlocal classes and used by Cordoba–Fefferman ([CF]) for representing pseudo-differential operators as multipliers on the phase space at the first order.

In the first section of this paper, we study the transformations $\mathcal{W}^{(\alpha)}$, $\mathcal{M}^{(\alpha)}$ and their continuity on the Schwartz space, and prove that "inversion" formulas hold modulo regularizing operators of any order, given *a priori*. Then, we define $\mathcal{W}_g^{(\alpha)} s$ for s being a tempered distribution by duality.

In the second section, we shall study the natural operator algebras generated by the transformation $\mathcal{W}^{(\alpha)}$ and its adjoint $\mathcal{M}^{(\alpha)}$, and look at the properties of multiplier operators on the phase space through $\mathcal{W}^{(\alpha)}$. Here, we generalise the Cordoba–Fefferman results for every exponent $\alpha \in (0,1)$. We give also the action of $\mathcal{W}^{(\alpha)}$ on diffeomorphisms and local change of coordinates.

Finally, in the last section, we use $\mathcal{W}^{(\alpha)}$ to characterize Sobolev spaces $H^m(\mathbb{R}^n)$and construct anisotropic Banach spaces of functions. We finish with the study of an example, based on an L^∞-localization over the space.

2. The interpolation family

2.1. Analysis and reconstruction in $\mathcal{S}(\mathbb{R}^n)$

We study here the continuity of the transformations defined above on the Schwartz space. For that aim, let us introduce the localization functions, which are used in the definition of semi-norms in the Fréchet space $\mathcal{S}(\mathbb{R}^n)$. For $k \in \mathbb{R}$ and $x \in \mathbb{R}^n$, we set

$$\kappa_k(x) = (1 + |x|)^{-k}.$$

Then, the following classical technical lemma holds.

Lemma 2.1. $\forall k \in \mathbb{R}$, $\forall x, x' \in \mathbb{R}^n$,

$$\kappa_k(x + x') \leq \mathcal{O}(1)\kappa_k(x)\kappa_{-|k|}(x').$$

Remark. Throughout this article, $\mathcal{O}(1)$ denotes a numerical constant, which may vary from one line to another.

Proof. In the case where $k \leq 0$, one has that $1 + |x + x'| \leq (1 + |x|)(1 + |x'|)$, which concludes the argument.

Suppose now that $k > 0$. Obviously, one can assume that $|x|, |x'| \geq 1$. It is sufficient to show that the following quantity is bounded:

$$\frac{|x'|^k}{|x|^k}\kappa_k(x + x'),$$

and is equal to

$$\left(\frac{|x'|}{|x|} + \left||x'|e_x + \frac{|x'|}{|x|}x'\right|\right)^{-k}$$

where e_x denotes the unit vector with direction x. Let us proceed by reductio ad absurdum and then suppose that there exist subsequences (x_p), (x'_p) such that

$$\lim_{p \longrightarrow 0}\left(\frac{|x'_p|}{|x_p|} + |x'_p|\left|e_{x_p} + \frac{x'_p}{|x_p|}\right|\right) = 0,$$

hence $|x'_p|/|x_p| \rightarrow_{p\rightarrow\infty} 0$ and $\left|e_{x_p} + |x_p|^{-1}x'_p\right| \rightarrow_{p\rightarrow\infty} 0$, since $|x'_p|$ is bounded below. In addition, $(e_{x_p})_{p\in\mathbb{N}}$ being bounded, one can extract from it a subsequence which converges to a unit vector e. Then, $|x'_p|/|x_p| \rightarrow 1$, a contradiction. □

We now give another technical lemma, which is fundamental in localization questions for the transformations $\mathcal{W}^{(\alpha)}$, and whose proof may be found in [Hol2].

Lemma 2.2. *Let $k > n$, $b \in \mathbb{R}^n$, and $a > 0$. Then, we have*

$$|\mathcal{W}[\kappa_k, \kappa_k](b, a)| \leq \mathcal{O}(1)\frac{1}{(1+a)^n}\kappa_k\left(\frac{b}{1+a}\right).$$

We can now prove the main result of the section, concerning the continuity of the transformations $\mathcal{W}^{(\alpha)}$ and $\mathcal{M}^{(\alpha)}$.

Theorem 2.1. $\mathcal{W}^{(\alpha)} : \mathcal{S}(\mathbb{R}^n) \times \mathcal{S}(\mathbb{R}^n) \rightarrow \mathcal{S}(T^*\mathbb{R}^n)$ *and* $\mathcal{M}^{(\alpha)} : \mathcal{S}(\mathbb{R}^n) \times \mathcal{S}(T^*\mathbb{R}^n) \rightarrow \mathcal{S}(\mathbb{R}^n)$ *are continuous.*

Proof. We begin by showing a result about the properties of rapidly decreasing functions.

Lemma 2.3. *$\varphi : \mathbb{R}^{2n} \to \mathbb{R}$ is rapidly decreasing if and only if φ is rapidly decreasing with respect to x uniformly in y, and with respect to y uniformly in x.*

Proof. The first assertion is obvious. Suppose now that φ verifies the two uniform growth conditions. Then, $\forall k, l \in \mathbb{N}$, $\exists C_k, C_l$;

$$\begin{aligned} (|x| \geq p, y \in \mathbb{R}^n) &\implies |\varphi(x,y)| \leq C_k p^{-2k} \\ \text{and } (|y| \geq q, x \in \mathbb{R}^n) &\implies |\varphi(x,y)| \leq C_l q^{-2k} \\ \text{then } (|x| \geq p, |y| \geq q) &\implies |\varphi(x,y)| \leq \min\left(C_k p^{-2k}, C_l q^{-2k}\right). \end{aligned}$$

One can assume that $C_k p^{-2k} \leq C_l q^{-2k}$. Then, one gets

$$|\varphi(x,y)| \leq C_k p^{-2k} \leq \sqrt{C_k C_l}\, p^{-k} q^{-k}$$

and the conclusion. □

Now, we prove the continuity of $\mathcal{W}^{(\alpha)}$. In order to use Lemma 2.3, we show the two following points.

- Localization in b:

$$\left(\mathcal{W}_g^{(\alpha)} s\right)(b,\omega) = \int_{\mathbb{R}^n} dx \langle\omega\rangle^{\frac{n\alpha}{2}} e^{-i\omega(x-b)} \overline{g\left(\langle\omega\rangle^\alpha (x-b)\right)} s(x).$$

Let $k \in \mathbb{N}$ be given.

$$\begin{aligned} \left|\left(\mathcal{W}_g^{(\alpha)} s\right)(b,\omega)\right| &\leq \int_{\mathbb{R}^n} dx \langle\omega\rangle^{\frac{n\alpha}{2}} \kappa_k\left(\langle\omega\rangle^\alpha (x-b)\right) \kappa_k(x) \\ &\leq \langle\omega\rangle^{-\frac{n\alpha}{2}} \mathcal{W}\left[\kappa_k, \kappa_k\right]\left(b, \langle\omega\rangle^{-\alpha}\right) \\ &\leq \mathcal{O}(1) \kappa_k\left(\frac{b}{1+\langle\omega\rangle^{-\alpha}}\right) \end{aligned}$$

using Lemma 2.2. Moreover, $1 + \langle\omega\rangle^{-\alpha} \geq 1$, which implies the desired localization.

- Localization in ω:

Here we use the frequency representation of the transformation $\mathcal{W}^{(\alpha)}$.

$$\left(\mathcal{W}_g^{(\alpha)} s\right)(b,\omega) = \int_{\mathbb{R}^n} d\xi \langle\omega\rangle^{-\frac{n\alpha}{2}} e^{ib\xi} \overline{\widehat{g}\left(\langle\omega\rangle^{-\alpha}(\xi-\omega)\right)} \widehat{s}(\xi).$$

hence

$$\begin{aligned} \left|\left(\mathcal{W}_g^{(\alpha)} s\right)(b,\omega)\right| &\leq \langle\omega\rangle^{-\frac{n\alpha}{2}} \int_{\mathbb{R}^n} d\xi \kappa_k\left(\langle\omega\rangle^{-\alpha}(\xi-\omega)\right) \kappa_k(\xi) \\ &\leq \langle\omega\rangle^{\frac{n\alpha}{2}} \mathcal{W}\left[\kappa_k, \kappa_k\right]\left(b, \langle\omega\rangle^{\alpha}\right) \\ &\leq \mathcal{O}(1) \left(\frac{\langle\omega\rangle^{\frac{\alpha}{2}}}{1+\langle\omega\rangle^\alpha}\right)^n \kappa_k\left(\frac{\omega}{1+\langle\omega\rangle^\alpha}\right) \end{aligned}$$

by the same lemma. In addition, since $\alpha < 1$, one has that

$$\kappa_k\left(\frac{\omega}{1+\langle\omega\rangle^\alpha}\right) \leq \mathcal{O}(1)\kappa_k\left(|\omega|^{-\alpha}\omega\right)$$

which implies the conclusion.

Now, application of Lemma 2.3 gives us the continuity of $\mathcal{W}^{(\alpha)}$ as a functional with values in the space of rapidly decreasing functions. The continuity as a functional with values in the Schwartz space comes from the two following identities: for $|\omega| > 1$ and $1 \leq j \leq n$,

$$\begin{aligned} \partial_{b_j}\mathcal{W}_g^{(\alpha)}s &= i\omega_j\mathcal{W}_g^{(\alpha)}s - |\omega|^\alpha\mathcal{W}_{\partial_j g}^{(\alpha)}s, \\ \partial_{\omega_j}\mathcal{W}_g^{(\alpha)}s &= -i\mathcal{W}_g^{(\alpha)}(x_j s) - \alpha|\omega|^{-2-\alpha}\omega_j\mathcal{W}_{\partial_j g}^{(\alpha)}(x_j s). \end{aligned}$$

This completes the proof of the continuity of $\mathcal{W}^{(\alpha)}$. In order to show the same result for $\mathcal{M}^{(\alpha)}$, we use the same arguments:

$$\left(\mathcal{M}_h^{(\alpha)}r\right)(x) = \int_{T^*\mathbb{R}^n} dbd\omega\langle\omega\rangle^{\frac{n\alpha}{2}}e^{i\omega(x-b)}h\left(\langle\omega\rangle^\alpha(x-b)\right)r(b,\omega);$$

then

$$\begin{aligned} \left|\left(\mathcal{M}_h^{(\alpha)}r\right)(x)\right| &\leq \int_{T^*\mathbb{R}^n} dbd\omega\langle\omega\rangle^{\frac{n\alpha}{2}}|r(b,\omega)|\left|h\left(\langle\omega\rangle^\alpha(x-b)\right)\right| \\ &\leq \int_{\mathbb{R}^n} d\omega\langle\omega\rangle^{-\frac{n\alpha}{2}}\kappa_k(\omega)\int_{\mathbb{R}^n} db\langle\omega\rangle^{n\alpha}\kappa_k\left(\langle\omega\rangle^\alpha(b-x)\right)\kappa_k(b) \\ &\leq \mathcal{O}(1)\int_{\mathbb{R}^n} d\omega\left(\frac{1}{\langle\omega\rangle^{\frac{n\alpha}{2}}+\langle\omega\rangle^{-\frac{n\alpha}{2}}}\right)^n\kappa_k(\omega)\kappa_k\left(\frac{x}{1+\langle\omega\rangle^{-\alpha}}\right). \end{aligned}$$

Now, if $|\omega| \leq 1$, one has that $\frac{|x|}{1+\langle\omega\rangle^{-\alpha}} \geq \frac{2}{3}|x|$, hence

$$\kappa_k\left(\frac{x}{1+\langle\omega\rangle^{-\alpha}}\right) \leq \mathcal{O}(1)\kappa_k(x).$$

Conversely, if $|\omega| \geq 1$, one has that

$$\kappa_k\left(\frac{x}{1+\langle\omega\rangle^{-\alpha}}\right) \leq \mathcal{O}(1)\kappa_k(x).$$

Then, actually,

$$\left|\left(\mathcal{M}_h^{(\alpha)}r\right)(x)\right| \leq \mathcal{O}(1)\left(\int_{\mathbb{R}^n} d\omega\kappa_k(\omega)\right)\kappa_k(x)$$

gives the continuity as a functional with values in the space of rapidly decreasing functions. In the same manner as for the analysis, the identity

$$\left(\partial_j\mathcal{M}_h^{(\alpha)}r\right) = \mathcal{M}_h^{(\alpha)}(\omega_j r) + \mathcal{M}_{\partial_j h}^{(\alpha)}(|\omega|r),$$

for $1 \leq j \leq n$, completes the proof of the theorem. □

2.2. "Inversion" formula

In what follows, we shall assume that $\alpha \in (0,1)$. We have defined the transformations $\mathcal{W}^{(\alpha)} : \mathcal{S}(\mathbb{R}^n) \times \mathcal{S}(\mathbb{R}^n) \to \mathcal{S}(T^*\mathbb{R}^n)$ and $\mathcal{M}^{(\alpha)} : \mathcal{S}(\mathbb{R}^n) \times \mathcal{S}(T^*\mathbb{R}^n) \to \mathcal{S}(\mathbb{R}^n)$, which are continuous. In this section, we shall study the operator $\mathcal{M}_h^{(\alpha)}\mathcal{W}_g^{(\alpha)}$, for $g, h \in \mathcal{S}(\mathbb{R}^n)$. In the sequel, we will note, for a function m, $M(m)$ the multiplication operator by m.

Proposition 2.2. *If $s \in \mathcal{S}(\mathbb{R}^n)$, then $\forall x \in \mathbb{R}^n$,*

$$\begin{aligned} (\mathcal{M}_h^{(\alpha)}\mathcal{W}_g^{(\alpha)}s)(x) &= \mathcal{F}^{-1}\left(M\left(m_{g,h}^{(\alpha)}\right)\mathcal{F}s\right)(x) \\ \text{where } m_{g,h}^{(\alpha)}(\xi) &= \int_{\mathbb{R}^n} d\omega \langle\omega\rangle^{-n\alpha}\overline{\widehat{g}}\left(\langle\omega\rangle^{-\alpha}(\xi-\omega)\right)\widehat{h}\left(\langle\omega\rangle^{-\alpha}(\xi-\omega)\right). \end{aligned}$$

Proof. Let us recall that

$$(\mathcal{W}_g^{(\alpha)}s)(b,\omega) = \int dx \langle\omega\rangle^{\frac{n\alpha}{2}} e^{-i\omega(x-b)}\overline{g}\left(\langle\omega\rangle^{\alpha}(x-b)\right)s(x)$$

which gives us the following Fourier representation:

$$(\mathcal{W}_g^{(\alpha)}s)(b,\omega) = \frac{1}{(2\pi)^n}\int d\xi \langle\omega\rangle^{\frac{-n\alpha}{2}} e^{ib\xi}\overline{\widehat{g}}\left(\langle\omega\rangle^{-\alpha}(\xi-\omega)\right)\widehat{s}(\xi).$$

It follows that

$$\begin{aligned} (\mathcal{M}_h^{(\alpha)}\mathcal{W}_g^{(\alpha)}s)(x) &= \frac{1}{(2\pi)^n}\iint dbd\omega \langle\omega\rangle^{\frac{n\alpha}{2}} e^{i\omega(x-b)}h\left(\langle\omega\rangle^{\alpha}(x-b)\right) \\ &\quad \times \int d\xi \langle\omega\rangle^{\frac{-n\alpha}{2}} e^{ib\xi}\overline{\widehat{g}}\left(\langle\omega\rangle^{-\alpha}(\xi-\omega)\right)\widehat{s}(\xi). \end{aligned}$$

Thus, by the Fubini theorem, one gets

$$\begin{aligned} (\mathcal{M}_h^{(\alpha)}\mathcal{W}_g^{(\alpha)}s)(x) &= \frac{1}{(2\pi)^n}\int d\xi \widehat{s}(\xi)e^{ix\xi} \\ &\quad \times \int d\omega \langle\omega\rangle^{-n\alpha}\overline{\widehat{g}}\left(\langle\omega\rangle^{-\alpha}(\xi-\omega)\right)\widehat{h}\left(\langle\omega\rangle^{-\alpha}(\xi-\omega)\right) \end{aligned}$$

and the result. □

Studying the Fourier multiplier $m_{g,h}^{(\alpha)}$, we obtain two approximate inversion formulas. The first presents the advantage of holding for all analyzing and reconstruction functions, and the second is more precise.

Theorem 2.3. *Let $g, h \in \mathcal{S}(\mathbb{R}^n)$. Then, we have*

$$\mathcal{M}_h^{(\alpha)}\mathcal{W}_g^{(\alpha)} = (2\pi)^{-n}\langle g, h\rangle_{\mathbb{R}^n} Id + T$$

where T is a smoothing Fourier multiplier in the Hörmander class $OPS_{\alpha,0}^{1-\alpha}(\mathbb{R}^n)$ (see [Hor1, Hor2] *for the definition of such classes).*

Proof. First, let us introduce some notation:

- e_ξ denotes the unit vector with direction ξ.
- For $\omega \in \mathbb{R}^n \backslash \{0\}$, we set $\Phi_e(\omega) = \langle\omega\rangle^{-\alpha}(e-\omega)$ where $e \in \mathbb{S}^{n-1}$.
- $f = \overline{\mathcal{F}g}\mathcal{F}h$.

In addition, we take ξ given, such that $|\xi| \geq 2$.

1st step Study of $\Phi = \Phi_{e_1}$ where $e_1 = (1, 0, \ldots, 0)$.
First, one has that $\Phi(e_1) = 0$. Moreover, a simple computation gives

$$\forall |\omega| \geq 1, \forall y \in \mathbb{R}^n, D\Phi(\omega) \cdot y = |\omega|^{-\alpha-2}(\omega, y)(e_1 - \omega) - |\omega|^{-\alpha} y$$

thus, $D\Phi(e_1) = -Id$ is invertible. We derive from this, by a local inversion theorem, that there exists $r > 0$ such that $\Phi : B(e_1, r) \to \Phi(B(e_1, r)) := V_r(e_1)$ is a diffeomorphism. Let us remark that one can choose r small enough such that $\overline{B(0,1)} \cap \overline{B(\xi, r|\xi|)} = \emptyset$.

Now, one makes a last remark; let R be a rotation such that $R(e_1) = e$. Then, Φ and Φ_e are conjugate by R (*i.e.*, $\Phi = R^{-1}\Phi_e R$) which implies that the radius r can be chosen independently from $e \in \S^{n-1}$.

2nd step Here, we prove Lemma 2.4 below.
Let us split $\mathbb{R}^n$ into two parts $\Omega_1 = \{|\omega - \xi| \geq r|\xi|\}$ and $\Omega_2 = \mathbb{R}^n \backslash \Omega_1$. Then, we set $m_{g,h}^{(\alpha)} = m_2 + m_1$ where

$$m_j(\xi) = \int_{\Omega_j} d\omega \langle\omega\rangle^{-n\alpha} f\left(\langle\omega\rangle^{-\alpha}(\xi - \omega)\right) \quad j = 1, 2.$$

Lemma 2.4. *m_1 is the symbol of an infinitely smoothing operator. In other words,*

$$m_1(\xi) = \mathcal{O}(|\xi|^{-\infty}) \qquad (|\xi| \to \infty).$$

Proof of the lemma.

$$|m_1(\xi)| \leq \int_{\Omega_1} d\omega \langle\omega\rangle^{-n\alpha} f^*\left(\langle\omega\rangle^{-\alpha}(\xi - \omega)\right)$$

where f^* is rapidly decreasing, non-increasing with respect to $|x|$, and such that $|f(x)| \leq f^*(x)$. Let us split Ω_1 again into two parts, for ω near or far from 0. Then, let $\Omega_1^0 = \Omega_1 \cap \{|\omega| \leq 1\}$ and $\Omega_1^\infty = \{|\omega| \geq 1\}$. The integral over Ω_1^0 clearly define a bounded function of ξ. In order to estimate the integral over Ω_1^∞, one writes that

$$|\omega|^{-\alpha} \geq (1 + 1/r)^{-\alpha} |\xi - \omega|^{-\alpha},$$

hence

$$\left| \int_{\Omega_1^\infty} d\omega \langle\omega\rangle^{-n\alpha} f^*\left(\langle\omega\rangle^{-\alpha}(\xi - \omega)\right) \right|$$
$$\leq \mathcal{O}(1) \int_{|\omega - \xi| \geq r|\xi|} d\omega |\xi - \omega|^{-n\alpha} f^*\left(|\xi - \omega|^{1-\alpha}\right)$$

and then, since f^* is rapidly decreasing, one gets the result. □

3rd step Asymptotic expansion of m_2.

$$m_2(\xi) = \int_{\Omega_2} d\omega |\omega|^{-n\alpha} f(|\omega|^{-\alpha}(\xi - \omega)).$$

Setting in the integral $\omega' = \omega/|\xi|$, one gets that

$$m_2(\xi) = |\xi|^{n(1-\alpha)} \int_{B(e_\xi, r)} d\omega |\omega|^{-n\alpha} f(|\xi|^{1-\alpha}|\omega|^{-\alpha}(e_\xi - \omega)).$$

Now, one can set $u = \Phi_{e_\xi}(\omega)$:

$$m_2(\xi) = \int_{V_r(\xi)} \frac{du}{|1 + \alpha|\omega|^{-2}(e_\xi - \omega, \omega)|} |\xi|^{n(1-\alpha)} f(|\xi|^{1-\alpha} u).$$

The point here is to expand the function $k(u) = \frac{1}{|1+\alpha|\omega|^{-2}(e_\xi - \omega, \omega)|}$ around 0. First, let us note that, for $|u|$ small, one can remove the absolute value. In addition, by the definition of Φ_{e_ξ}, one has that

$$1 + \alpha|\omega|^{-2}(e_\xi - \omega, \omega) = 1 + \alpha|\omega|^{\alpha-1}|u|(e_u, e_\omega),$$

thus

$$\begin{aligned} k(u) &= 1 - \alpha|\omega|^{\alpha-1}(e_u, e_\omega)|u| + \mathcal{O}(|u|^2) \\ &= 1 - \alpha(e_u, e_\omega)|u| + \mathcal{O}(|u|^2) \\ &= 1 - \alpha(e_u, e_\xi)|u| + |u|^2 \epsilon(u) \end{aligned}$$

where $\epsilon(u)$ is uniformly bounded in ξ.

Note that this estimate only holds in a neighborhood of 0, that is $V_r(\xi)$. Let us define the principal part in the entire space $\mathbb{R}^n$. Since f is rapidly decreasing, this is equivalent to adding a rapidly decreasing function in ξ. (The proof would be the same as the proof of Lemma 2.4.) Then, one may easily prove that:

$$m_2(\xi) = \int_{\mathbb{R}^n} du f(u) - \alpha \left(\int_{\mathbb{R}^n} du (e_u, e_\xi) f(u) \right) |\xi|^{\alpha-1} + \mathcal{O}\left(|\xi|^{2(\alpha-1)}\right)$$

for $|\xi|$ large enough.

4th step Conclusion of the proof.

One has obtained:

$$m_{g,h}^{(\alpha)}(\xi) = \int_{\mathbb{R}^n} du f(u) + r_{g,h}^{(\alpha)}(\xi)$$

where $r_{g,h}^{(\alpha)}(\xi) = \mathcal{O}\left(|\xi|^{1-\alpha}\right)$ for $|\xi| \geq R$. Now, let χ be a cut-off function with support in $\overline{B(0,R)}$, and consider the following decomposition:

$$r_{g,h}^{(\alpha)} = \chi r_{g,h}^{(\alpha)} + (1-\chi) r_{g,h}^{(\alpha)}.$$

The first part $\chi r_{g,h}^{(\alpha)}$ is smooth and compactly supported, and so is the symbol of an infinitely smoothing operator. By the study above, the second part $(1-\chi) r_{g,h}^{(\alpha)}$ is the symbol of a smoothing operator of order $1 - \alpha$. To complete the proof, it remains to show that the derivatives verify the Hörmander class estimates. For that aim, let us write that, for all $\beta \in I\!N^n$,

$$\partial^\beta m_{g,h}^{(\alpha)}(\xi) = \int_{\mathbb{R}^n} d\xi \langle\omega\rangle^{-(n+\beta)\alpha} f\left(\langle\omega\rangle^{-\alpha}(\xi-\omega)\right).$$

Applying exactly the four steps of the demonstration above, one may obtain the desired estimate. □

Now, we state a more precise result, which needs an admissibility condition on the analyzing function g. (Another condition will appear later for the L^2-theory.) This result proves the existence of a reconstruction function modulo a smoothing operator of arbitrary order, this order being given *a priori*.

Theorem 2.4. *Let $g \in \mathcal{S}(\mathbb{R}^n)$ be such that $\widehat{g}(0) \neq 0$, and let $N \in I\!N$. Then, there exists a function $h_N \in \mathcal{S}(\mathbb{R}^n)$ such that*

$$\mathcal{M}_{h_N}^{(\alpha)} \mathcal{W}_g^{(\alpha)} = Id + R_N$$

where $R_N \in OPS_{\alpha,0}^{N(1-\alpha)}(\mathbb{R}^n)$.

Proof. In the proof of Theorem 2.3, one has seen that

$$m_{g,h}^{(\alpha)}(\xi) = m_2(\xi) + m_1(\xi)$$

with $m_1(\xi) = \mathcal{O}\left(|\xi|^{-\infty}\right)$, $(|\xi| \longrightarrow \infty)$ and that

$$m_2(\xi) = \int_{V_r(\xi)} du k(u)(\widehat{\overline{g}h})(|\xi|^{1-\alpha}u)|\xi|^{n(1-\alpha)}.$$

For the result above, one expanded k to the first order. Let us extend this expansion to further orders. One then obtains terms such as

$$\int_{V_r(\xi)} du P_{e_\xi}(u)\widehat{\overline{g}h}(u).$$

Let us set $f = \widehat{\overline{g}h}$ and let us choose f with compact support in $V_r(\xi)$ such that $\widehat{f}(0) = 1$, and $\forall \beta \in I\!N^n$, verifying $N > |\beta| \geq 1$, one has $\partial_\beta \widehat{f}(0) = 0$. Under this assumption, choosing r small enough such that $\widehat{g}$ does not vanish on $V_r(\xi)$, one can set $\widehat{h_N}(u) = \frac{f(u)}{\widehat{\overline{g}}(u)}$. Actually, all terms of order at most N in the asymptotic expansion vanish, and h_N is the desired reconstruction function. Actually, the estimate on the derivatives is obtained in the same way as in Theorem 2.3. □

2.3. Analysis of distribution

This section is devoted to the definition of the transformation $\mathcal{W}^{(\alpha)}$ for tempered distributions, in order to construct later anisotropic microlocal classes. For that aim, we proceed by duality, using the continuity of the analysis and the reconstruction. Then, let g be an analyzing function in the Schwartz class and η be a tempered distribution. We set, for $r \in \mathcal{S}(T^*\mathbb{R}^n)$,

$$\left\langle r, \mathcal{W}_g^{(\alpha)}\eta\right\rangle = \left\langle \mathcal{M}_g^{(\alpha)} r, \eta\right\rangle. \tag{6}$$

Then, the continuity of $\mathcal{W}^{(\alpha)}$ implies that $\mathcal{W}_g^{(\alpha)}\eta$ is a tempered distribution on the phase space $T^*\mathbb{R}^n$. Furthermore, the continuity of the functional $\mathcal{W}_g^{(\alpha)}$ is a mere consequence of equation (6). In fact, we obtain a more precise result for the distribution $\mathcal{W}_g^{(\alpha)}\eta$.

Proposition 2.5. *Let $g \in \mathcal{S}(\mathbb{R}^n)$ and $\eta \in \mathcal{S}'(\mathbb{R}^n)$. Then, in the sense of distributions, $\mathcal{W}_g^{(\alpha)}\eta$ is equal to the C^∞-function defined on $T^*\mathbb{R}^n$ by:*

$$\Omega(b,\omega) = \langle g_{b,\omega}, \eta\rangle \, .$$

The proof follows immediately from the definitions.

This proposition shows that even if the distribution is very irregular, its transform is smooth. Here, the regularity properties of distributions are closely related to the localization of the coefficients through the analysis by $\mathcal{W}_g^{(\alpha)}$.

In the same manner, we define also, for $h \in \mathcal{S}(\mathbb{R}^n)$, the continuous functional $\mathcal{M}_h^{(\alpha)}$ from $\mathcal{S}'(T^*\mathbb{R}^n)$ to $\mathcal{S}'(\mathbb{R}^n)$ by: $\forall\Omega \in \mathcal{S}'(T^*\mathbb{R}^n)$, $\forall s \in \mathcal{S}(\mathbb{R}^n)$,

$$\left\langle s, \mathcal{M}_h^{(\alpha)}\Omega\right\rangle = \left\langle \mathcal{W}_h^{(\alpha)}s, \Omega\right\rangle . \tag{7}$$

Remark that, using the inversion formula in Theorem 2.3, $\mathcal{M}_h^{(\alpha)}\Omega$ can be very irregular.

3. $\mathcal{W}^{(\alpha)}$ transformations and operator algebras

Here, we generalize the Cordoba–Fefferman results ([CF]). A pseudo-differential operator T being given, we want to write it in the form $T = \mathcal{M}_h^{(\alpha)}\widetilde{T}\mathcal{W}_g^{(\alpha)}$ up to a smoothing operator and at arbitrary order, where $\widetilde{T}$ acts on the phase space and has an asymptotic expansion in the different levels of a graded algebra, which we shall present and study here. Our first point is to define the "identity" part of this algebra.

3.1. A graded kernels operator algebra

Definition 3.1. *Let $\mathcal{K} = Op(K)$ be an operator acting on functions of $T^*\mathbb{R}^n$. We will say that $\mathcal{K}$ belongs to $OPS_0^{(\alpha)}(T^*\mathbb{R}^n)$ iff*

$$|K(b,\omega;b',\omega')| = \varphi\left(\langle\omega\rangle^\alpha|b-b'|, \langle\omega\rangle^{-\alpha}|\omega-\omega'|\right)$$

where $\varphi : \mathbb{R}^2 \to \mathbb{R}$ is rapidly decreasing.

We begin by giving the equivalent following definition.

Proposition 3.2. *In both dilation terms of Definition* 3.1, *we can replace ω by ω'.*

In order to prove 3.2, let us give a technical lemma.

Lemma 3.1. *A function f is localised in $\langle\omega\rangle^{-\alpha}(\omega-\omega')$ if and only if it is localised in $\langle\omega'\rangle^{-\alpha}(\omega-\omega')$.*

Proof. It is sufficient to prove one of the assertions. Then, let us suppose that f is localised in $\langle\omega\rangle^{-\alpha}(\omega-\omega')$. One has that

$$\langle\omega'\rangle^{-\alpha}(\omega-\omega')=\left(\frac{\langle\omega\rangle}{\langle\omega'\rangle}\right)^{\alpha}\langle\omega\rangle^{-\alpha}(\omega-\omega').$$

Here, four cases can happen.
1^{rst} case $|\omega|,|\omega'|\geq 1$. Then, one writes that

$$\begin{aligned}\frac{\langle\omega'\rangle}{\langle\omega\rangle} &= \frac{|\omega'|}{|\omega|}\leq 1+\frac{|\omega-\omega'|}{|\omega|}\\ &\leq 1+|\omega|^{-\alpha}|\omega-\omega'|\\ &\leq 2|\omega|^{-\alpha}|\omega-\omega'|\end{aligned}$$

since one can suppose without loss of generality that $\langle\omega\rangle^{-\alpha}(\omega-\omega')\geq 1$. Consequently, one obtains

$$\langle\omega'\rangle^{-\alpha}|\omega-\omega'|\geq 2^{-\alpha}\left(\langle\omega\rangle^{-\alpha}|\omega-\omega'|\right)^{1-\alpha}$$

which gives the result.
2nd case $|\omega|,|\omega'|\leq 1$. It is immediate because of the boundedness of the functions.
3rd case $|\omega|\leq 1$, $|\omega'|\geq 1$. In that case, $\langle\omega'\rangle$ is equivalent to $|\omega-\omega'|$ since $\alpha<1$, which implies the desired estimate.
4th case The proof is analogous to the third one. □

Proof of Proposition 3.2. By Lemma 2.3, we can look at the localizations (in b and ω) separately. In addition, Lemma 3.1 implies that the localization in the ω variable is solved. Let us suppose having the localization in $\langle\omega\rangle^{\alpha}(b-b')$ and in $\langle\omega\rangle^{\alpha}(\omega-\omega')$ (and so equivalently in $\langle\omega'\rangle^{\alpha}(\omega-\omega')$). One has that

$$\langle\omega'\rangle^{\alpha}(b-b')=\left(\frac{\langle\omega'\rangle}{\langle\omega\rangle}\right)^{\alpha}\langle\omega\rangle^{\alpha}(b-b').$$

Furthermore, in the proof of Lemma 3.1, one proves that

$$\frac{\langle\omega'\rangle}{\langle\omega\rangle}\sim\langle\omega\rangle^{-\alpha}(\omega-\omega')$$

which ends the proof. □

Proposition 3.3. *$OPS_0^{(\alpha)}(T^*\mathbb{R}^n)$ is an algebra.*

Proof. Let $\mathcal{K}=OP(K)$ and $\mathcal{K}'=Op(K')$ in $OPS_0^{(\alpha)}(T^*\mathbb{R}^n)$. Then, $\mathcal{K}\circ\mathcal{K}'=Op(K'')$ with

$$K''(b,\omega;b',\omega')=\iint_{T^*\mathbb{R}^n}db''d\omega''K(b,\omega;b'',\omega'')K'(b'',\omega'';b',\omega'). \tag{8}$$

Now, let k, l be some integers;

$$\begin{aligned}|K''(b,\omega;b',\omega')| &\leq \iint_{T^*\mathbb{R}^n} db''d\omega''\kappa_k\left(\langle\omega\rangle^\alpha|b''-b|\right)\kappa_l\left(\langle\omega\rangle^{-\alpha}|\omega''-\omega|\right)\\ &\qquad\times\kappa_k\left(\langle\omega'\rangle^\alpha|b''-b'|\right)\kappa_l\left(\langle\omega'\rangle^{-\alpha}|\omega''-\omega'|\right)\\ &\leq \int_{\mathbb{R}^n} db''\kappa_k\left(\langle\omega\rangle^\alpha|b''-b|\right)\kappa_k\left(\langle\omega'\rangle^\alpha|b''-b'|\right)\\ &\qquad\times\int_{\mathbb{R}^n} d\omega''\kappa_l\left(\langle\omega\rangle^{-\alpha}|\omega''-\omega|\right)\kappa_l\left(\langle\omega'\rangle^{-\alpha}|\omega''-\omega'|\right).\end{aligned}$$

We study each part separately. First, one has that

$$\begin{aligned}&\int d\omega''\kappa_l(\langle\omega\rangle^{-\alpha}|\omega''-\omega|)\kappa_l(\langle\omega'\rangle^{-\alpha}|\omega''-\omega'|)\\ &= \langle\omega\rangle^{n\alpha}\mathcal{W}[\kappa_l,\kappa_l]\left(\frac{\omega-\omega'}{\langle\omega'\rangle^\alpha},\frac{\langle\omega\rangle^\alpha}{\langle\omega'\rangle^\alpha}\right)\\ &\leq \mathcal{O}(1)\langle\omega\rangle^{n\alpha}\kappa_l\left(\frac{|\omega-\omega'|}{\langle\omega'\rangle^\alpha+\langle\omega\rangle^\alpha}\right).\end{aligned}$$

On the other hand, the integral in b'' gives

$$\begin{aligned}&\int d\omega''\kappa_k\left(\langle\omega\rangle^\alpha|b''-b|\right)\kappa_k\left(\langle\omega'\rangle^\alpha|b''-b'|\right)\\ &\leq \mathcal{O}(1)\langle\omega\rangle^{-n\alpha}\kappa_k\left(\tfrac{|b-b'|}{\langle\omega'\rangle^{-\alpha}+\langle\omega\rangle^{-\alpha}}\right),\end{aligned}$$

hence one obtains the following estimate:

$$|K''(b,\omega;b',\omega')| \leq \mathcal{O}(1)\kappa_l\left(\frac{|\omega-\omega'|}{\langle\omega'\rangle^\alpha+\langle\omega\rangle^\alpha}\right)\kappa_k\left(\frac{|b'-b|}{\langle\omega'\rangle^{-\alpha}+\langle\omega\rangle^{-\alpha}}\right). \tag{9}$$

Let us remark first that if one of the ω, ω' variables has its norm less than or equal to 1, then the argument is concluded. It remains to show that equation (9) implies the desired localization on high frequencies. This shall be done in the following lemma, which ends the proof of the proposition. □

Lemma 3.2. *Let $K : (T^*\mathbb{R}^n)^2 \longrightarrow \mathbb{R}$ be such that:*

$$K(b,\omega;b',\omega') = \varphi\left(\frac{|\omega-\omega'|}{|\omega|^\alpha+|\omega'|^\alpha},\frac{|b-b'|}{|\omega|^{-\alpha}+|\omega'|^{-\alpha}}\right)$$

where φ is rapidly decreasing. Then, K verifies

$$K(b,b';\omega,\omega') = \psi\left(|\omega'|^\alpha|b-b'|,|\omega'|^{-\alpha}|\omega-\omega'|\right)$$

with ψ rapidly decreasing.

Proof. We first study the $|\omega|^{-\alpha}|\omega-\omega'|$ localization. The estimate

$$\frac{|\omega-\omega'|}{|\omega|^\alpha+|\omega'|^\alpha}=|\omega|^{-\alpha}|\omega-\omega'|\frac{1}{1+\frac{|\omega'|^\alpha}{|\omega|^\alpha}}\geq 3^\alpha(|\omega|^{-\alpha}|\omega-\omega'|)^{1-\alpha}$$

holds, hence

$$|\omega|^{-\alpha}|\omega-\omega'|\leq 3^{\frac{\alpha}{1-\alpha}}\left(\frac{|\omega-\omega'|}{|\omega|^\alpha+|\omega'|^\alpha}\right)^{\frac{1}{1-\alpha}}$$

which implies the $|\omega|^{-\alpha}|\omega-\omega'|$ localization (and also the $|\omega'|^{-\alpha}|\omega-\omega'|$ localization). Now, let us look at the $|\omega|^\alpha|b-b'|$ localization. In the same manner as above, one has that

$$\frac{|b-b'|}{|\omega|^{-\alpha}+|\omega'|^{-\alpha}}=|\omega'|^\alpha|b-b'|\frac{1}{1+\frac{|\omega'|^\alpha}{|\omega|^\alpha}}\geq 3^{-\alpha}|\omega'|^\alpha|b-b'|(|\omega|^\alpha|\omega-\omega'|)^{-\alpha},$$

hence

$$|\omega'|^\alpha|b-b'|\leq 3^\alpha(|\omega|^\alpha|\omega-\omega'|)^\alpha\frac{|b-b'|}{|\omega|^{-\alpha}}|\omega'|^{-\alpha}$$

and one obtains the result by the $|\omega|^{-\alpha}|\omega-\omega'|$ localization. □

This algebra is a basic component in the sense that, as the following result shows, the typical elements of $OPS^{(\alpha)}(T^*\mathbb{R}^n)$ are the cross kernel operators $\mathcal{W}_g^{(\alpha)}\mathcal{M}_h^{(\alpha)}$.

Proposition 3.4. *Let g, h be in $\mathcal{S}(\mathbb{R}^n)$. Then,*

$$\mathcal{W}_g^{(\alpha)}\mathcal{M}_h^{(\alpha)}=Op[p_{g,h}^{(\alpha)}]\in OPS_0^{(\alpha)}(T^*\mathbb{R}^n)$$

where $p_{g,h}^{(\alpha)}(b,\omega;b',\omega')=\langle g_{b,\omega},h_{b',\omega'}\rangle$.

Proof. From Fubini's theorem, one immediately obtains that $\mathcal{W}_g^{(\alpha)}\mathcal{M}_h^{(\alpha)}=Op\left[\langle g_{b,\omega},h_{b',\omega'}\rangle\right]$. Let us now look at the localization of this kernel. First, one treats the $|\omega-\omega'|$ localization. One has the following Fourier representation:

$$\begin{aligned}&p_{g,h}^{(\alpha)}(b,\omega;b',\omega')\\&\quad=\frac{(\langle\omega\rangle\langle\omega'\rangle)^{-\frac{n\alpha}{2}}}{(2\pi)^n}\int_{\mathbb{R}^n}d\xi e^{i\xi(b-b')}\overline{\widehat{g}}\left(\langle\omega\rangle^{-\alpha}(\xi-\omega)\right)\widehat{h}\left(\langle\omega'\rangle^{-\alpha}(\xi-\omega')\right),\end{aligned}$$

hence

$$\begin{aligned}&\left|p_{g,h}^{(\alpha)}(b,\omega;b',\omega')\right|\\&\quad\leq\frac{(\langle\omega\rangle\langle\omega'\rangle)^{-\frac{n\alpha}{2}}}{(2\pi)^n}\int_{\mathbb{R}^n}d\xi\left|\widehat{g}\left(\langle\omega\rangle^{-\alpha}(\xi-\omega)\right)\right|\left|\widehat{h}\left(\langle\omega'\rangle^{-\alpha}(\xi-\omega')\right)\right|.\end{aligned}$$

Now, let $k \in I\!N$ be given,

$$
\begin{aligned}
&\left|p_{g,h}^{(\alpha)}(b,\omega;b',\omega')\right| \\
&\leq \mathcal{O}(1)\left(\langle\omega\rangle\langle\omega'\rangle\right)^{-\frac{n\alpha}{2}}\int_{\mathbb{R}^n} d\xi\kappa_k\big(\langle\omega\rangle^{-\alpha}(\xi-\omega)\big)\kappa_k\big(\langle\omega'\rangle^{-\alpha}(\xi-\omega')\big) \\
&\leq \mathcal{O}(1)\left(\frac{\langle\omega\rangle}{\langle\omega'\rangle}\right)^{\frac{n\alpha}{2}} \mathcal{W}[\kappa_k,\kappa_k]\left(\langle\omega'\rangle^{\alpha}(b-b'),\left(\frac{\langle\omega\rangle}{\langle\omega'\rangle}\right)^{\alpha}\right) \\
&\leq \mathcal{O}(1)\left(\frac{\langle\omega\rangle}{\langle\omega'\rangle}\right)^{\frac{n\alpha}{2}} \kappa_k\left(\frac{\omega-\omega'}{\langle\omega\rangle^{\alpha}+\langle\omega'\rangle^{\alpha}}\right).
\end{aligned}
$$

On the other hand, for the $|b-b'|$ localization, the same computation gives the following estimate: $\forall k \in I\!N$,

$$
p_{g,h}^{(\alpha)}(b,\omega;b',\omega') = \mathcal{O}(1)\left(\frac{\langle\omega'\rangle}{\langle\omega\rangle}\right)^{\frac{n\alpha}{2}} \kappa_k\left(\frac{b-b'}{\langle\omega\rangle^{-\alpha}+\langle\omega'\rangle^{-\alpha}}\right).
$$

Finally, the result follows from Lemmas 3.1 and 3.2. □

We also give another result, which follows obviously from the proposition above, but has a great interest. It is the equivalent of the reproducing kernel equation for the wavelet transform.

Proposition 3.5. *Let g, h be in $\mathcal{S}(\mathbb{R}^n)$ and let N be an integer. We assume that $\widetilde{g}(0) \neq 0$. Then,*

$$
\exists \mathcal{K}_N \in OPS_0^{(\alpha)}(T^*\mathbb{R}^n); \quad \mathcal{W}_h^{(\alpha)} = \mathcal{K}_N \mathcal{W}_g^{(\alpha)} + \mathcal{W}_h^{(\alpha)} R_N
$$

where R_N is a smoothing Fourier multiplier belonging to $OPS_{\alpha,0}^{N(1-\alpha)}(\mathbb{R}^n)$.

Proof. It suffices to take $\mathcal{K}_N = \mathcal{W}_h^{(\alpha)}\mathcal{M}_{h_N}^{(\alpha)}$, where h_N is a reconstruction function of g at order N. □

Now, we define the other components of the graded algebra. First, we introduce weighted functions. Let us consider a function $r : T^*\mathbb{R}^n \longrightarrow \mathbf{C}$, for which we suppose that it verifies the following growth condition:

$$
\left|\frac{r(b',\omega')}{r(b,\omega)}\right| \leq \mathcal{O}(1)\left(\langle\omega\rangle^{\alpha}|b-b'| + \langle\omega\rangle^{-\alpha}|\omega-\omega'|\right)^M. \tag{10}
$$

Now, we define the set of kernels on the phase space $S_r^{(\alpha)}(T^*\mathbb{R}^n)$ by:

$$
s \in S_r^{(\alpha)}(T^*\mathbb{R}^n) \Longleftrightarrow, |s(b,\omega;b',\omega')| \leq r(b',\omega')F(b,\omega;b',\omega')
$$

where $F \in S_0^{(\alpha)}(T^*\mathbb{R}^n)$, $F \geq 0$. In the case where $r(b,\omega) = \langle\omega\rangle^m$ (which verifies the condition (10) by Lemma 3.1), one notes this set $S_m^{(\alpha)}(T^*\mathbb{R}^n)$. Condition (10) immediately implies the following lemma, concerning dependence on the variables.

Lemma 3.3. *We assume that the function r verifies the condition* (10). *If $s \in S_r^{(\alpha)}(T^*\mathbb{R}^n)$, then there exists $F \in S_0^{(\alpha)}(T^*\mathbb{R}^n)$, $F \geq 0$ such that:*

$$|s(b,\omega;b',\omega')| \leq r(b,\omega)F(b,\omega;b',\omega').$$

Now, for $s \in S_r^{(\alpha)}(T^*\mathbb{R}^n)$, we define the operator

$$Op(s) \cdot u(b,\omega) = \int_{T^*\mathbb{R}^n} db'd\omega' s(b,\omega;b',\omega')u(b',\omega')$$

and we denote the set $\left\{Op(s), s \in S_r^{(\alpha)}(T^*\mathbb{R}^n)\right\}$ by $OPS_r^{(\alpha)}(T^*\mathbb{R}^n)$. Then, the following composition result holds.

Proposition 3.6. *Suppose that the functions r_1, r_2 both verify condition* (10). *Then,*

$$OPS_{r_1}^{(\alpha)}(T^*\mathbb{R}^n) \circ OPS_{r_2}^{(\alpha)}(T^*\mathbb{R}^n) \subset OPS_{r_1r_2}^{(\alpha)}(T^*\mathbb{R}^n).$$

Proof. Let $s_1 \in S_{r_1}^{(\alpha)}(T^*\mathbb{R}^n)$ and $s_2 \in S_{r_2}^{(\alpha)}(T^*\mathbb{R}^n)$. Then, $Op(s_1) \circ Op(s_2) = Op(s_3)$ with

$$s_3(b,\omega;b'',\omega'') = \int_{T^*\mathbb{R}^n} db'd\omega' s_1(b,\omega;b',\omega')s_2(b',\omega';b'',\omega'').$$

One must prove that $s_3 \in S_{r_1r_2}^{(\alpha)}(T^*\mathbb{R}^n)$. By definition, there exist $F_1, F_2 \in S_0^{(\alpha)}(T^*\mathbb{R}^n)$ such

$$|s_3(b,\omega;b'',\omega'')| \leq r_1(b,\omega)\int_{T^*\mathbb{R}^n} db'd\omega' r_2(b',\omega')F_1(b,\omega;b',\omega')F_2(b',\omega';b'',\omega'').$$

By Lemma 3.3, one may write

$$|s_3(b,\omega;b'',\omega'')| \leq (r_1r_2)(b,\omega)F_3(b,\omega;b'',\omega'')$$

where F_3 is defined by $Op(F_3) = Op(F_1) \circ Op(F_2')$, with $F_1, F_2' \in S_0^{(\alpha)}(T^*\mathbb{R}^n)$. Proposition 3.3 completes the proof. □

Remark: Lemma 3.3 above means that the elements of $OPS_r^{(\alpha)}(T^*\mathbb{R}^n)$ are products of multipliers on the phase space which have some growth by elements of $OPS_0^{(\alpha)}(T^*\mathbb{R}^n)$. (We shall see the importance of this remark in the next section, dealing with decomposition of pseudo-differential operators through $\mathcal{W}^{(\alpha)}$ transformations.)

We finish the section by defining the graded algebra we promised at the beginning. Let us set

$$OPS^{(\alpha)}(T^*\mathbb{R}^n) = \bigcup_{m\in\mathbb{R}} OPS_m^{(\alpha)}(T^*\mathbb{R}^n).$$

Proposition 3.7. *$OPS^{(\alpha)}(T^*\mathbb{R}^n)$ is a graded operator algebra.*

3.2. Representation of PDO through $\mathcal{W}^{(\alpha)}$ transformation

Let $T \in OPS^m_{\gamma,\delta}(\mathbb{R}^n)$ be a pseudo-differential operator with symbol σ_T in a Hörmander class (see [Hor1, Hor2] for the definition of these classes). Then, setting $\widetilde{T_P} = \mathcal{W}_g^{(\alpha)} T \mathcal{M}_{h_P}^{(\alpha)}$, where $P \in I\!N$, and g, h_P verify Theorem 2.4, we get

$$T = \mathcal{M}_h^{(\alpha)} \widetilde{T_P} \mathcal{W}_g^{(\alpha)} + R_P$$

where R_P is a smoothing operator of order $P(1-\alpha)$. We shall study in this section the operator $\widetilde{T}$. For that aim, we introduce, for μ, ν two multi-indices and for a real m, the quantity

$$m(\mu, \nu) = m - (|\mu|(\gamma - \alpha) + |\nu|(\alpha - \delta)) .$$

Theorem 3.8. *We assume that* $0 \leq \delta < \alpha < \gamma \leq 1$. *Then, for all* $N \in I\!N$, *we have the decomposition*

$$\widetilde{T} = \sum_{k=0}^{N} \sum_{|\mu|+|\nu|=k} \mathcal{K}_{\mu,\nu}^{(N,k)}$$

where $\mathcal{K}_{\mu.\nu}^{(N.k)} \in OPS^{(\alpha)}_{m(\mu.\nu)}(T^*\mathbb{R}^n)$.

Proof. It follows from a direct computation that $\widetilde{T} = Op\left[\langle g_{b,\omega}, Th_{b',\omega'}\rangle\right]$ with

$$Th_{b',\omega'}(x) = \frac{1}{(2\pi)^n} \int_{\mathbb{R}^n} d\xi \langle\omega'\rangle^{-\frac{n\alpha}{2}} e^{i\xi(x-b')} \sigma_T(x,\xi) \widehat{h}\left(\langle\omega'\rangle^{-\alpha}(\xi - \omega')\right) .$$

Now, let us make a Taylor expansion of the symbol σ_T around (b', ω').

$$\begin{aligned} \sigma_T(x,\xi) &= \sum_{k=0}^{N-1} \sum_{|\mu|+|\nu|=k} C_{\mu,\nu} \partial_x^\mu \partial_\xi^\nu \sigma_T(b',\omega')(x-b')^\mu (\xi-\omega')^\nu \\ &\quad + \sum_{|\mu|+|\nu|=N} C_{\mu,\nu} \rho(x,b';\xi,\omega')(x-b')^\mu(\xi-\omega')^\nu \end{aligned}$$

where

$$\rho(x,b';\xi,\omega') = \int_0^1 dt(1-t)^{N-1} \partial_x^\mu \partial_\xi^\nu \sigma_T\left(b' + t(x-b'), \omega' + t(\xi - \omega')\right) .$$

First, we treat each term of the main part in the expansion:

$$C_{\mu.\nu} \partial_x^\mu \partial_\xi^\nu \sigma_T(b',\omega') \int_{\mathbb{R}^n} d\xi \langle\omega'\rangle^{-\frac{n\alpha}{2}} e^{i\xi(x-b')} (x-b')^\mu (\xi-\omega')^\nu \widehat{h}\left(\langle\omega'\rangle^{-\alpha}(\xi-\omega')\right) .$$

Let us set $\widehat{h_2}(\omega) = \langle\omega\rangle^\nu \widehat{h}(\omega)$. One gets

$$C_{\mu,\nu} \partial_x^\mu \partial_\xi^\nu \int_{\mathbb{R}^n} d\xi \langle\omega'\rangle^{-\frac{n\alpha}{2}+\alpha|\nu|} e^{i\xi(x-b')} (x-b')^\mu \widehat{h_2}\left(\langle\omega'\rangle^{-\alpha}(\xi-\omega')\right) .$$

Now, one sets $\widehat{\widehat{h}}(\omega) = (-i\partial_\omega^\mu)\, \widehat{h_2}(\omega)$, which gives

$$C_{\mu,\nu} \langle\omega'\rangle^{\alpha(|\nu|-|\mu|)} \partial_x^\mu \partial_\xi^\nu \sigma_T(b',\omega') \int_{\mathbb{R}^n} d\xi \langle\omega'\rangle^{-\frac{n\alpha}{2}} e^{i\xi(x-b')} \widehat{\widehat{h}}\left(\langle\omega'\rangle^{-\alpha}(\xi-\omega')\right) .$$

Then, the matrix elements are written as

$$C_{\mu,\nu}\langle\omega'\rangle^{\alpha(|\nu|-|\mu|)}\partial_x^\mu\partial_\xi^\nu\sigma_T(b',\omega')\left\langle g_{b,\omega},\widetilde{h}_{b',\omega'}\right\rangle$$

and we obtain the desired result by estimates on the symbol σ_T.

Now, one treats the remainder. The expression one needs to estimate is the following:

$$\sum_{|\mu|+|\nu|=N} C_{\mu,\nu}\int_{\mathbb{R}^n} d\xi\langle\omega'\rangle^{-\frac{n\alpha}{2}}e^{i\xi(x-b')}(x-b')^\mu(\xi-\omega')^\nu\rho(x,b';\xi,\omega')$$
$$\times\widehat{h}\left(\langle\omega'\rangle^{-\alpha}(\xi-\omega')\right).$$

This estimate is a sum of type $\sum R_{\mu,\nu}h_{b',\omega'}$ where $R_{\mu,\nu}$ is a pseudo-differential operator. We will obtain the conclusion if we prove that, for each term of the sum, there exists some real L such that $\langle\omega'\rangle^{-L}R_{\mu,\nu}h_{b',\omega'}$ and its Fourier transform are rapidly decreasing, in particular if the family

$$\{\langle\omega'\rangle^{\frac{n\alpha}{2}}D_{\langle\omega'\rangle^\alpha}E_{-\omega'}T_{-b'}\langle\omega'\rangle^{-L}R_{\mu,\nu}h_{b',\omega'},b',\omega'\in\mathbb{R}^n\}$$

is bounded in $\mathcal{S}(\mathbb{R}^n)$. Setting $\widehat{\widehat{h}}(\omega)=\langle\omega\rangle^\nu\widehat{h}(\omega)$, one gets the expression

$$\sum_{|\mu|+|\nu|=N} C_{\mu,\nu}\int_{\mathbb{R}^n} d\xi\langle\omega'\rangle^{-\frac{n\alpha}{2}+\alpha|\nu|}e^{i\xi(x-b')}(x-b')^\mu\rho(x,b';\xi,\omega')$$
$$\times\widehat{\widehat{h}}\left(\langle\omega'\rangle^{-\alpha}(\xi-\omega')\right)$$

which gives, by applying the inverse transformations,

$$\sum_{|\mu|+|\nu|=N} C_{\mu,\nu}\langle\omega'\rangle^{-\frac{n\alpha}{2}+\alpha(|\nu|-|\mu|)}$$
$$\times\int_{\mathbb{R}^n} d\xi e^{ix\langle\omega'\rangle^{-\alpha}(\xi-\omega')}x^\mu\rho\left(\langle\omega'\rangle^{-\alpha}x+b',b';\xi,\omega'\right)\widehat{\widehat{h}}\left(\langle\omega'\rangle^{-\alpha}(\xi-\omega')\right).$$

Changing the integration variable ξ by $\langle\omega'\rangle^{-\alpha}(\xi-\omega')$, one obtains

$$\sum_{|\mu|+|\nu|=N} C_{\mu,\nu}\langle\omega'\rangle^{\alpha(|\nu|-|\mu|)}$$
$$\times\int_{\mathbb{R}^n} d\xi e^{ix\xi}x^\mu\widehat{\widehat{h}}(\xi)\rho\left(\langle\omega'\rangle^{-\alpha}x+b',b';\langle\omega'\rangle^\alpha\xi+\omega',\omega'\right).$$

In order to simplify the expressions, we set

$$\begin{aligned}
&f_{b',\omega'}(x,\xi)\\
&\quad=\langle\omega\rangle^{\alpha(|\nu|-[\mu|)}\rho\left(\langle\omega'\rangle^{-\alpha}x+b',b';\langle\omega'\rangle^\alpha\xi+\omega',\omega'\right)\\
&\quad=\langle\omega\rangle^{\alpha(|\nu|-|\mu|)}\int_0^1 dt(1-t)^{N-1}\partial_x^\mu\partial_\xi^\nu\sigma_T\left(b'+t\langle\omega'\rangle^{-\alpha}x,\omega'+t\langle\omega'\rangle^\alpha\xi\right).
\end{aligned}$$

Now, let us consider an integral with parameters

$$K_{b',\omega'}(x) = \int d\xi e^{ix\xi} k_{b',\omega'}(x,\xi).$$

Then, the following lemma holds

Lemma 3.4. *We assume that $\langle\omega'\rangle^{-L} k_{b',\omega'}$ is rapidly decreasing with respect to ξ, and all its partial derivatives in ξ too, uniformly in x, b', ω'. Then,*

$$\{\langle\omega'\rangle^{-L} K_{b',\omega'}\}$$

is a bounded family in $\mathcal{S}(\mathbb{R}^n)$.

Proof. Let $(s_\lambda)_{\lambda\in\Lambda}$ be a family in $\mathcal{S}(\mathbb{R}^n)$, and set $S_\lambda = \mathcal{F}^{-1}(s_\lambda)$. By the continuity of the inverse Fourier transform, if $(s_\lambda)_{\lambda\in\Lambda}$ is bounded in $\mathcal{S}(\mathbb{R}^n)$, then $(S_\lambda)_{\lambda\in\Lambda}$ too, and in particular, in the space of rapidly decreasing functions. Let us set $s_{b',\omega',\lambda}(\xi) = \langle\omega'\rangle^{-L} k_{b',\omega'}(\lambda,\xi)$. Consequently, $S_{b',\omega',\lambda}$ is rapidly decreasing with respect to x, uniformly in b', ω', λ. One obtains the conclusion by specializing $\lambda = x$. □

Let us come back to the proof of the theorem. First, one remarks that Lemma 3.4 cannot be applied to the derivatives of $K_{b',\omega'}$. Indeed, the derivatives in x does not allow the specialization of the parameter λ. We need some additional work before. One has that

$$\partial_x^\beta K_{b',\omega'}(x) = \sum_{0\le|\epsilon|\le|\beta|} C'_{\epsilon,\beta} \int_{\mathbb{R}^n} d\xi e^{ix\xi} \xi^{\beta-\epsilon} \partial_x^\epsilon k_{b',\omega'}(x,\xi).$$

Then, it is clear that the right assumption on $k_{b',\omega'}$ is:
$\forall\beta,\epsilon \in I\!N^n$, $\forall k \in I\!N$, $\exists C_{k,b',\omega'}$;

$$\forall x,\xi,b',\omega', \quad \left|\partial_x^\beta \partial_\xi^\epsilon k_{b',\omega'}(x,\xi)\right| \le C_{k,b',\omega'} \langle\omega'\rangle^L \kappa_k(\xi).$$

Since $\widetilde{h}$ is in the Schwartz space, it is sufficient that $f_{b',\omega'}$ verifies the following assumption:
$\exists L \in \mathbb{R}$; $\forall\beta,\epsilon \in I\!N^n$, $\exists L' \in \mathbb{R}$, $\exists C_{\beta,\epsilon} \in \mathbb{R}$;

$$\forall x,\xi,b',\omega', \quad \left|\partial_x^\beta \partial_\xi^\epsilon f_{b',\omega'}(x,\xi)\right| \le C_{\beta,\epsilon} \kappa_{L'}(\xi) \langle\omega'\rangle^L.$$

Then, let $\beta,\epsilon \in I\!N^n$ be given. With the estimates on the symbol σ_T, one gets that

$$\begin{aligned} \partial_x^\beta \partial_\xi^\epsilon f_{b',\omega'}(x,\xi) &= \langle\omega'\rangle^{\alpha(|\epsilon|+|\nu|-|\beta|-|\mu|)} \int_0^1 dt (1-t)^{N-1} t^{|\beta|+|\epsilon|} \\ &\quad \times \partial_x^{\mu+\beta} \partial_\xi^{\nu+\epsilon} \sigma_T\left(b' + t\langle\omega'\rangle^{-\alpha} x, \omega' + t\langle\omega'\rangle^\alpha \xi\right), \\ \left|\partial_x^\beta \partial_\xi^\epsilon f_{b',\omega'}(x,\xi)\right| &\le C\langle\omega'\rangle^{\alpha(|\epsilon|+|\nu|-|\beta|-|\mu|)} \int_0^1 dt (1-t)^{N-1} \\ &\quad \times \kappa_{-m+\gamma(|\nu|+|\epsilon|)-\delta(|\mu|+|\beta|)}\left(\omega' + t\langle\omega'\rangle^\alpha \xi\right). \end{aligned}$$

We suppose that $|\xi|, |\omega'| \geq 1$ (low frequencies do not modify the result). Let us set $l = m - \gamma(|\nu| + |\epsilon|) + \delta(|\mu| + |\beta|)$ and distinguish two cases.
– $l > 0$: then, writing that

$$\begin{aligned}\kappa_{-l}\left(\omega' + \langle\omega'\rangle^{\alpha}\xi\right) &\leq \mathcal{O}(1)\left(|\omega'| + |\omega'|^{\alpha}\xi\right)^{l}\\ &\leq \mathcal{O}(1)|\omega'|^{\alpha l}\left(|\omega'|^{1-\alpha} + \xi\right)^{l}\\ &\leq \mathcal{O}(1)|\omega'|^{\alpha l}|\omega'|^{(1-\alpha)l}|\xi|^{l}\end{aligned}$$

one obtains that

$$\begin{aligned}\left|\partial_x^{\beta}\partial_{\xi}^{\epsilon} f_{b',\omega'}(x,\xi)\right| &\leq \mathcal{O}(1)\kappa_l(\xi)\langle\omega\rangle^{m(\mu,\nu)-(|\epsilon|(\gamma-\alpha)+|\beta|(\alpha-\delta))}\\ &\leq \mathcal{O}(1)\kappa_l(\xi)\langle\omega\rangle^{m(\mu,\nu)}\end{aligned}$$

with the assumptions on α, γ, δ, and the argument is concluded.
– $l < 0$: this case is more difficult to treat. One has to show that there exists some real L' such that

$$|\omega'|^{-l}|\xi|^{L'l}\kappa_l\left(\omega' + t|\omega'|^{\alpha}\xi\right)$$

is uniformly bounded with respect to t, ω', ξ. This quantity equals

$$\frac{1}{\left(|\omega'|^{-1}|\xi|^{L'} + |\xi|^{L'}\,|e_{\omega'} + t|\omega'|^{\alpha-1}\xi|\right)^{-l}}. \tag{11}$$

One shall proceed by reductio ad absurdum. Let us suppose that there exist sequences (t_p), (ω'_p), (ξ_p) such that (11) tends to $+\infty$ as p goes to ∞. Then, one has that

$$\lim_{p\to\infty} |\xi_p|^{L'}|\omega'_p|^{-1} = 0$$

and that

$$\lim_{p\to\infty}\left|e_{\omega'_p} + t_p|\omega'_p|^{\alpha-1}\xi_p\right| = 0.$$

Since the sequences (t_p) and $(e_{\omega'_p})$ are bounded, one can assume that they converge respectively to $t_0 \in [0,1]$ and $e \in \mathbb{S}^{n-1}$. It follows that

$$\lim_{p\to\infty} |\xi_p||\omega'_p|^{\alpha-1} \geq t_0^{-1}$$

(here, we adopt the convention $1/0 = \infty$). Choosing $L' > \frac{1}{1-\alpha}$ in

$$|\xi_p|^{L'}|\omega'_p|^{-1} = \left(|\xi_p||\omega'_p|^{-1}\right)^{L'}|\omega'_p|^{(1-\alpha)L'-1}$$

one obtains a contradiction. In the same manner as in the case $l > 0$, one gets the desired estimate. The theorem is proved. □

Now, we give a result which makes precise the form of the main part and the order of the remainder terms. Its demonstration obviously follows from the proof of 3.8.

Theorem 3.9. *Under the same assumption as in Theorem* 3.8, $\widetilde{T}$ *expands as follows:*

$$\widetilde{T} = \sum_{|\mu|+|\nu|<N,\ m(\mu,\nu)>m-N\min(\gamma-\alpha,\alpha-\delta)} \mathcal{K}_{\mu,\nu}^{(N)}$$

with $\mathcal{K}_{\mu,\nu}^{(N)} \in OPS_{m(\mu,\nu)}^{(\alpha)}(T^*\mathbb{R}^n)$*. Furthermore, the kernels are written as*

$$K_{\mu,\nu}^{(N)}(b,\omega,b',\omega') = k_{\mu,\nu}^{(N)}(b',\omega')\left\langle g_{b,\omega'}, \widetilde{h}_{b',\omega'}\right\rangle$$

where $\mathcal{F}\widetilde{h} = \left(-i\partial_\xi^\mu\right)(\xi^\nu \mathcal{F}h)$.

Then, one obtains a decomposition result of the symbol σ_T through the $\mathcal{W}^{(\alpha)}$ at the first order, by using Proposition 3.5. In addition, we can remark that we need not a precise reconstruction function (Theorem 2.3 is then sufficient). For higher orders, the result does not hold anymore. Indeed, some kernels appear which, though they are localised, are not absorbed in the analysis with respect to g.

3.3. Multipliers on phase space

We study here the most simple operators which are generated by the $\mathcal{W}^{(\alpha)}$ transforms, that is the multipliers through these transformations. Let $m : T^*\mathbb{R}^n \longrightarrow \mathbf{C}$ be a function defined on $T^*\mathbb{R}^n$. For $\alpha \in (0,1)$, and g, h in the Schwartz class verifying the hypothesis of Theorem 2.3, we set

$$Op_{g,h}^{(\alpha)}(m) = \mathcal{M}_h^{(\alpha)} M(m) \mathcal{W}_g^{(\alpha)}.$$

These operators have a great importance in our first-order differential calculus. Indeed, the result of the last section can be rewritten (at the first order) as follows:

Proposition 3.10. *We adopt the notation of Section* 3.2*. Then,* $T \in OPS_{\gamma,\delta}^m(\mathbb{R}^n)$ *being given and assuming that* $\gamma > \alpha > \delta$*, we can write that:*

$$T = Op_{g,h}^{(\alpha)}(\sigma_T) + \mathcal{M}_h^{(\alpha)} \mathcal{K}_{rest} \mathcal{W}_g^{(\alpha)}$$

where $\mathcal{K}_{rest} \in OPS_{m-\min(\gamma-\alpha,\alpha-\delta,1-\alpha)}^{(\alpha)}$.

Remark: We shall see in the section dealing with function spaces that the rest is really smoothing, since the multiplication by a term such as $\langle\omega\rangle^{-k}$ with $k \geq 0$ sends the Sobolev spaces $H^m(\mathbb{R}^n)$ to $H^{m+k}(\mathbb{R}^n)$ after transformation by $\mathcal{W}^{(\alpha)}$.

Our aim here is to obtain a result, which would allow us to write that, m, m' being given functions defined on $T^*\mathbb{R}^n$, we have the identity

$$Op_{g,h}^{(\alpha)}(m) \circ Op_{g,h}^{(\alpha)}(m') = Op_{g,h}^{(\alpha)}(m \cdot m') + T_{rest}$$

where T_{rest} is smoothing in some sense. We have already seen in Proposition 3.10 that if m' is in some Hörmander class of symbol, this holds, using the first order composition theorem of pseudo-differential operators. Now, we give weaker conditions on the multiplier m' for this identity to hold.

Theorem 3.11. *Let* $\alpha \in (0,1)$*, and* g*,* h *be in the Schwartz class. We assume that the multiplier* $m : T^*\mathbb{R}^n \longrightarrow \mathbf{C}$ *verifies: if* $|\beta| + |\gamma| \leq 1$,

$$\left|\partial_{b,\omega}^{\beta,\gamma} m\left(b+b',\omega+\omega'\right)\right| \leq r(b,\omega)\left(\langle\omega\rangle^\alpha |b| + \langle\omega\rangle^{-\alpha}|\omega'|\right)^K \tag{12}$$

where $K \in \mathbb{R}$ is given and $r(b,\omega) = o(m(b,\omega))$ as $|b| + |\omega|$ goes to ∞ and verifies condition (10)*. Then, the following identity holds:*

$$\mathcal{W}_g^{(\alpha)} \mathcal{M}_h^{(\alpha)} M(m) = M(m) \mathcal{W}_g^{(\alpha)} \mathcal{M}_h^{(\alpha)} + \mathcal{R}$$

with $\mathcal{R} \in OPS_r^{(\alpha)}(T^*\mathbb{R}^n)$.

Proof. The Taylor formula, applied to m, gives:

$$m(b',\omega') = m(b,\omega) + \int_0^1 dt Dm\left(b + t(b'-b), \omega + t(\omega' - \omega)\right) \cdot (b' - b, \omega' - \omega).$$

Let us note the remainder Q. By hypothesis (12), one gets the estimate for Q,

$$|Q(b,\omega;b',\omega')| \le r(b,\omega) \left(\langle\omega\rangle^{\alpha}|b'-b| + \langle\omega\rangle^{-\alpha}|\omega' - \omega|\right)^{K+1} \int_0^1 dt\, t^K.$$

Consequently, one obtains that

$$\left[\mathcal{W}_g^{(\alpha)} \mathcal{M}_h^{(\alpha)}, M(m)\right] = \mathcal{R}$$

where the kernel R of $\mathcal{R}$ verifies

$$|R(b,\omega;b',\omega')| \le \mathcal{O}(1) r(b,\omega) \left(\langle\omega\rangle^{\alpha}|b-b'| + \langle\omega\rangle^{-\alpha}|\omega' - \omega|\right)^{K+1} \langle g_{b,\omega}, h_{b',\omega'}\rangle .$$

This completes the proof. □

Theorem 3.12. *Let us suppose that m, m' satisfy condition* (10) *and that m' verifies* (12)*. Let us note $R \in OPS_{\alpha,0}^{1-\alpha}(\mathbb{R}^n)$ the pseudo-differential operator of inversion Theorem 2.3. Then, for all $N \in I\!N$, there exists r verifying condition* (10) *with the additional property that $r(b,\omega) = o(mm'(b,\omega))$ as $|b| + |\omega|$ goes to ∞ and there exists $\mathcal{R}_N \in OPS_{|\omega|^{\alpha-1}mm'}^{(\alpha)}(T^*\mathbb{R}^n) \cup OPS_r^{(\alpha)}(T^*\mathbb{R}^n)$ such that:*

$$Op_{g,h}^{(\alpha)}(m) \circ Op_{g,h}^{(\alpha)}(m') = Op_{g,h}^{(\alpha)}(mm') + \mathcal{M}_h^{(\alpha)} \mathcal{R}_N \mathcal{W}_g^{(\alpha)} + Op(mm')R^N.$$

The proof is based on the following lemma

Lemma 3.5. *We take the notation of Theorem* 3.12*. Then, for all $N \in I\!N$, there exists $\widetilde{\mathcal{R}}_N \in OPS_{\alpha-1}^{(\alpha)}$, such that*

$$\mathcal{W}_h^{(\alpha)} = \widetilde{\mathcal{R}}_N \mathcal{W}_g^{(\alpha)} + \mathcal{W}_h^{(\alpha)} R^N.$$

Proof. Let us set $\widetilde{R} = \mathcal{W}_g^{(\alpha)} R \mathcal{M}_h^{(\alpha)}$. Through the pseudo-differential operator R belonging to a limit class with respect to Theorem 3.8, one may show, following the proof of 3.8, that $\widetilde{R} \in OPS_{\alpha-1}^{(\alpha)}(T^*\mathbb{R}^n)$ (details are left to the reader). Then, a recurrence using inversion Theorem 2.3 ends the proof. □

Proof of Theorem 3.12. It follows immediately from Theorem 3.11 and Lemma 3.5. □

3.4. $\mathcal{W}^{(\alpha)}$ transformations and diffeomorphisms

In this section, we shall study the action of diffeomorphisms on the coefficient space. For that aim, we want to find an operator $\Psi[\varphi]$ acting on functions of the phase space such that, up to a smoothing operator:

$$\mathcal{W}_g^{(\alpha)}\Phi[\varphi] = \Psi[\varphi]\mathcal{W}_g^{(\alpha)} \tag{13}$$

where $g \in \mathcal{S}(\mathbb{R}^n)$ is an analysing function, $\varphi : \mathbb{R}^n \longrightarrow \mathbb{R}^n$ a diffeomorphism bounded with all its derivatives, and $\Phi[\varphi]$ is defined by:

$$\Phi[\varphi] : s(y) \mapsto \frac{1}{J\left(\varphi^{-1}(y)\right)} s\left(\varphi^{-1}(y)\right).$$

Here, J denotes the Jacobian of φ. The adjoint of $\Phi[\varphi]$ is $\Phi^*[\varphi] : r(x) \mapsto r\left(\varphi(x)\right)$, in the following sense: for all $r, s : \mathbb{R}^n \longrightarrow \mathbb{R}^n$,

$$\int_{\mathbb{R}^n} dx \overline{(\Phi[\varphi]s)\,(x)} r(x) = \int_{\mathbb{R}^n} dx \overline{s(x)}\,(\Phi^*[\varphi|r)\,(x).$$

Here, we need to change slightly the definition of the $\mathcal{W}^{(\alpha)}$ transformation, in order to compensate the loss of covariance in the transformation. We introduce in the definition a matrix parameter, which gives a certain latitude in the relative direction between the frequency modulation and dilation. The analysing family then becomes, for $g \in \mathcal{S}(\mathbb{R}^n)$, $(b,\omega) \in T^*\mathbb{R}^n$, $C \in GL_n(\mathbb{R})$,

$$g_{b,\omega,C} = \langle C\rangle^{\frac{n\alpha}{2}} e^{i\omega(x-b)} g\left(\langle C\rangle^{\alpha}(x-b)\right).$$

The reader may be easily convinced that it does not change any of the results we give before, but offers us the possibility to do linear deformations on the tangent space.

Let us now go back to the diffeomorphism action on a $\mathcal{W}^{(\alpha)}$ transform. In order to find $\Psi[\varphi]$, we make in the following theorem an asymptotic expansion of the matrix elements $K^{(\alpha)}[g,\varphi]$ of the operator $\mathcal{W}_g^{(\alpha)}\Phi[\varphi]\mathcal{W}_g^{(\alpha)}$, for $g \in \mathcal{S}(\mathbb{R}^n)$.

We begin by setting some notation:

- $\partial\Phi_x[\varphi] : s(y) \mapsto \frac{1}{J(\varphi^{-1}(x))} s\left(\partial\varphi_x^{-1} y\right)$ (it describes the action of φ on the tangent space).
- For N_1, N_2, N_3 integers, we define $\mathcal{N}(N_1,N_2,N_3)$ being the following set:

$$\left\{(k_1,k_2,k_3) \in I\!N^3, 0 \le k_1 < N_1, 2 \le k_2 < N_2, 0 \le k_3 < N_3, k_1+k_3 \ge 1\right\}.$$

- For a function F defined on $(T^*\mathbb{R}^n)^2$, we will say that F satisfies the $(*)$ condition if:

$$(*) \quad |F(b,\omega;b',\omega')| \le \rho\left(\langle\omega\rangle^{\alpha}(b-\varphi(b')), \langle\omega\rangle^{-\alpha}\left(\omega - \partial\varphi_{\varphi(b')}\omega'\right)\right)$$

where the function ρ is rapidly decreasing.

Then, the result is the following:

Theorem 3.13. *We assume that $\alpha > 1/2$. Then, the kernel $K^{(\alpha)}[g,\varphi]$ is written as follows: for all integers N_1, N_2, N_3,*

$$\begin{aligned} K^{(\alpha)}[g,\varphi] &= K_0^{(\alpha)}[g,\varphi] \\ &+ \sum_{(k_1,k_2,k_3)\in\mathcal{N}(N_1,N_2,N_3)} K_{k_1,k_2,k_3}^{(\alpha)}[g,\varphi]\langle C'\omega'\rangle^{k_3(1-k_2\alpha)-k_1\alpha} \\ &+ R_{N_1,N_2,N_3}^{(\alpha)}\langle C'\omega'\rangle^{N_3(1-N_2\alpha)-N_1\alpha} \end{aligned}$$

where the kernels $K^{(\alpha)}$ and the remainder $R^{(\alpha)}$ satisfy the $(*)$ condition. Furthermore,

$$K_0^{(\alpha)}[g,\varphi](b,\omega;b',\omega') = p_{g,\partial\Phi_{\varphi(b)}[\varphi]g}^{(\alpha)}\left(b,\omega,C;\varphi(b'),{}^t\,\partial\varphi_{\varphi(b')}^{-1}\omega',C'^t\partial\varphi_{\varphi(b')}\right).$$

Remark: If we only expand at the first order, we find exactly the adjoint of the searched operator $\Psi[\varphi]$:

$$\Psi^*[\varphi] : r(b,\omega,C) \mapsto \frac{1}{J(b)} r\left(\varphi^{-1}(b),{}^t\,\partial\varphi_{\varphi(b)}\omega, C^t\partial\varphi_{\varphi(b)}^{-1}\right).$$

Remark that the fact that the functions really depend on the matrix parameter is not important for the estimates because of the boundedness of $\partial\varphi$. In addition, if φ is affine, the relation (13) exactly holds (*i.e.*, without a smoothing term).

Proof. The matrix elements of $\mathcal{W}_g^{(\alpha)}\Phi[\varphi]\mathcal{M}_g^{(\alpha)}$ is

$$K^{(\alpha)}[g,\varphi](b,\omega,C;b',\omega',C') = \langle g_{b,\omega,C}, \Phi[\varphi]g_{b',\omega',C'}\rangle .$$

Let us set

$$\Phi_{b,\omega,C}[\varphi]g = D^*_{\langle C\rangle^{-\alpha}}E^*_{\partial\varphi^{-1}(\varphi(b))}T^*_{\varphi(b)}\Phi[\varphi]g_{b,\omega,C}.$$

One has that

$$\Phi_{b,\omega,C}[\varphi]g(x) = \frac{g\left(\langle C\rangle^{\alpha}\left(\varphi^{-1}\left(\varphi(b)+\langle C\rangle^{-\alpha}x\right)-b\right)\right)}{J\left(\varphi^{-1}\left(\varphi(b)+\langle C\rangle^{-\alpha}x\right)\right)} \times e^{i\omega\left(\varphi^{-1}\left(\varphi(b)+\langle C\rangle^{-\alpha}x\right)-b-\partial\varphi^{-1}(\varphi(b))\langle C\rangle^{-\alpha}x\right)}. \tag{14}$$

The first order in the asymptotic expansion is given by the limit, as $|\omega|$ tends to infinity, of the quantity (14). Let us remark that, since $\alpha > 1/2$,

$$\lim_{|\omega|\to\infty}\Phi_{b,\omega,C}g = \partial\Phi_{\varphi(b)}[\varphi]g \tag{15}$$

in the C^∞-topology.

Now, one treats the rest of the expansion. The non-oscillating term expands without any difficulties as follows:

$$\left(\partial\Phi_{\varphi(b)}[\varphi]g\right)(x) + \sum_{k>1} A_k(b)g(x)\langle C\rangle^{-k\alpha} \tag{16}$$

where $A_k(b)$ are some differential operators (uniformly bounded in b). One writes for the oscillating term

$$\begin{aligned} & e^{i\omega\left(\varphi^{-1}(\varphi(b)+\langle C\rangle^{-\alpha}x)-b-\partial\varphi^{-1}_{\varphi(b)}\langle C\rangle^{-\alpha}x\right)} \\ = \; & e^{i\omega\sum_{|\beta|\leq N}\left(\partial\varphi^{-1}\right)^{\beta}_{\varphi(b)}x^{\beta}\langle C\rangle^{-|\beta|\alpha}} + r_N(x,b,\omega) \end{aligned}$$

where $r_N(b,x,\omega)=\mathcal{O}(1)|x|^{N+1}\langle C\rangle^{-(N+1)\alpha}$. Moreover,

$$\begin{aligned} & e^{i\omega\sum_{|\beta|\leq N}\left(\partial\varphi^{-1}\right)^{\beta}_{\varphi(b)}x^{\beta}\langle C\rangle^{-|\beta|\alpha}} \\ = \; & \sum_{k\leq N'}\frac{i^k}{k!}\left(\omega,\sum_{|\beta|\leq N}\left(\partial\varphi^{-1}\right)^{\beta}_{\varphi(b)}x^{\beta}\langle C\rangle^{-|\beta|\alpha}\right)^k \\ & +\widetilde{r}_{N,N'}(b,x,\omega). \end{aligned}$$

One gets for the oscillating part

$$\sum_{k\leq N'}\frac{i^k}{k!}\sum_{|\beta|\leq N,|\gamma|\leq k}\left(e_\omega,\left(\partial\varphi^{-1}\right)^{\beta}x^{\beta}\right)^{\gamma}\langle C\rangle^{|\gamma|(1-|\beta|\alpha)}+\widetilde{r}_{N,N'}(b,x,\omega).$$

As a conclusion, this term is equal to

$$\sum_{k\leq N,k'\leq N'}F_{k,k'}(b,x,\omega)\langle C\rangle^{k'(1-k\alpha)}+\widetilde{r}_{N,N'}(b,x,\omega) \tag{17}$$

where

$$F_{k,k'}(b,x,\omega)=\sum_{\gamma,\beta}C_{\gamma,\beta}\left(e_\omega,\left(\partial\varphi^{-1}\right)^{\beta}_{\varphi(b)}x^{\beta}\right)^{\gamma}$$

is polynomial growing in x and uniformly bounded in b, ω, since φ is bounded with all its derivatives. Furthermore, the remainder satisfies the estimate

$$|\widetilde{r}_{N,N'}(b,x,\omega)|\leq\mathcal{O}(1)\langle C\rangle^{(N'+1)(1-(N+1)\alpha)}|x|^{(N'+1)(N+1)}.$$

Putting (16) and (17) together, one gets that

$$\begin{aligned} \Phi_{b,\omega,C}[\varphi]g \; = \; & \partial\Phi_{\varphi(b)}[\varphi]g+\sum_{\mathcal{N}}F_{k_2,k_3}(b,x,\omega)A_{k_1}(b)g\langle C\rangle^{k_3(1-k_2\alpha)-k_1\alpha} \\ & +R_{N_1,N_2,N_3}(b,x,\omega) \end{aligned}$$

with

$$R_{N_1,N_2,N_3}(b,x,\omega)=\mathcal{O}(1)|x|^{N_1}|\omega|^{N_3(1-N_2\alpha)-N^1\alpha}.$$

Now, writing that

$$K^{(\alpha)}[g,\varphi](b,\omega;b',\omega')=\langle\widetilde{g}_{b,\omega,C;b',\omega',C'},\Phi_{b',\omega',C'}[\varphi]g\rangle$$

where

$$\widetilde{g}_{b,\omega,C;b',\omega',C'}=D_{\langle C\rangle^\alpha}E_{\partial\varphi^{-1}_{\varphi(b')}\omega'}T_{-\varphi(b')}g_{b,\omega,C},$$

one gets the announced expansion. The estimate on the kernels $K^{(\alpha)}$ follows from the boundedness in $\mathcal{S}(\mathbb{R}^n)$ of the family

$$\left\{ D^*_{\langle C\rangle^{-\alpha}} E^*_{\partial\varphi^{-1}_{\varphi(b)}\omega} T^*_{\varphi(b)} F_{k_2,k_3}(b,\omega) A_{k_1}(b) g \right\}.$$

Actually, it remains to compute the first-order term in the expansion. By (15), one has

$$\begin{aligned} K_0^{(\alpha)}(b,\omega;b',\omega') &= \langle \widetilde{g}_{b,\omega,C;b',\omega',C'}, \partial\Phi_{\varphi(b')}[\varphi]g\rangle \\ &= \left\langle g_{b,\omega,C}, \left(\partial\Phi_{\varphi(b')}[\varphi]g\right)_{\varphi(b'),{}^t\partial\varphi^{-1}_{\varphi(b')}\omega',\mathbf{C}'^t\partial\varphi_{\varphi(\mathbf{b}')}} \right\rangle \end{aligned}$$

which gives the conclusion. □

4. Application to construction of directional regularity Banach spaces

4.1. The H^m-theory

We begin by giving the characterization of Hilbert–Sobolev spaces through the $\mathcal{W}^{(\alpha)}$ transformation coefficients. Those spaces need another admissibility condition on the analyzing function.

Definition 4.1. *We will say that a function $g \in \mathcal{S}(\mathbb{R}^n)$ satisfies the* (A2) *condition if and only if*

$$\exists c > 0; \forall \xi \in \mathbb{R}^n, \quad 0 < c^{-1} \leq m_{g,g}^{(\alpha)}(\xi) \leq c < \infty. \tag{A2}$$

Remark: By Theorem 2.3, the boundedness is not the problem. But, it may happen that the symbol vanishes somewhere. Basically, the best examples of functions satisfying (A2) is the Gaussians.

Before studying Sobolev spaces, we define the $\mathcal{W}^{(\alpha)}$ transform of functions in $L^2(\mathbb{R}^n)$.

Proposition 4.2. *Let $g, s \in \mathcal{S}(\mathbb{R}^n)$. Then,*

$$\|\mathcal{W}_g^{(\alpha)} s\|^2_{L^2(\mathbb{R}^n)} = \left\langle \widehat{s}, m_{g,g}^{(\alpha)} \widehat{s} \right\rangle_{L^2(\mathbb{R}^n)}.$$

Proof. It immediately follows from the fact that the adjoint of $\mathcal{W}_g^{(\alpha)}$ is $\mathcal{M}_g^{(\alpha)}$. □

By the way, $\mathcal{W}_g^{(\alpha)}$ is continuous from $\mathcal{S}(\mathbb{R}^n)$ to $L^2(\mathbb{R}^n)$ for the L^2 norm. From the density of the Schwartz space in $L^2(\mathbb{R}^n)$, it follows that we can extend $\mathcal{W}_g^{(\alpha)}$ to $L^2(\mathbb{R}^n)$ by continuity. Moreover, if g satisfies (A2), the fact that the symbol $m_{g,g}^{(\alpha)}$ is bounded below gives us the following characterization result:

Theorem 4.3. *Let $g \in \mathcal{S}(\mathbb{R}^n)$ verify* (A2). *Then,*

$$s \in L^2(\mathbb{R}^n) \Longleftrightarrow \mathcal{W}_g^{(\alpha)} s \in L^2(T^*\mathbb{R}^n).$$

The same manner of characterization holds for Sobolev spaces.

Theorem 4.4. *Let $g \in \mathcal{S}(\mathbb{R}^n)$ verify* (A2). *Then,*

$$s \in H^m(\mathbb{R}^n) \Longleftrightarrow \mathcal{W}_g^{(\alpha)} s \in L^2(T^*\mathbb{R}^n, dbd\omega\langle\omega\rangle^{2m}) = L^2_m(T^*\mathbb{R}^n).$$

Proof. In the same way as in the proof of Proposition 4.2, one has

$$\forall g, s \in \mathcal{S}(\mathbb{R}^n), \quad \|\mathcal{W}_g^{(\alpha)} s\|_{L^2(T^*\mathbb{R}^n, dnd\omega\langle\omega\rangle^m)} = \left\langle \widehat{s}, \widetilde{m}_{g,g}^{(\alpha)} \widehat{s} \right\rangle_{L^2(\mathbb{R}^n, |\xi|^{2m} d\xi)}$$

where

$$\widetilde{m}_{g,g}^{(\alpha)}(\xi) = |\xi|^{-2m} \int_{\mathbb{R}^n} d\omega \langle\omega\rangle^{n\alpha+2m} \left|\widehat{g}\left(\langle\omega\rangle^{-\alpha}(\xi-\omega)\right)\right|^2.$$

With the same arguments as in Theorem 2.3, one may prove that the asymptotic expansion is modified by the addition of a term as $C_m|\xi|^{1-\alpha}$ and then that this multiplier is bounded. It follows from a density argument that, if $s \in H^m(\mathbb{R}^n)$, then $\mathcal{W}_g^{(\alpha)} s \in L^2_m(T^*\mathbb{R}^n)$.

Let us show the converse assertion. It leads us to show that the admissibility condition is not modified by the addition of $\langle\omega\rangle^{2m}$. If $m \geq 0$, then $\widetilde{m}_{g,g}^{(\alpha)}(\xi) \geq m_{g,g}^{(\alpha)}(\xi)$, which gives the result, since g verifies (A2). Now, assume that $m < 0$ and proceed by reductio ad absurdum. Suppose that $\widetilde{m}_{g,g}^{(\alpha)}$ is not bounded by below. In that case, there would exist a sequence $(\xi_j)_{j\in I\!N}$ such that $\widetilde{m}_{g,g}^{(\alpha)}(\xi_j) \to 0$. In addition, the expansion of $\widetilde{m}_{g,g}^{(\alpha)}$ shows that (ξ_j) is bounded. Then, up to a sequence, it converges to $\xi \in \mathbb{R}^n$ verifying, by continuity of $\widetilde{m}_{g,g}^{(\alpha)}$, $\widetilde{m}_{g,g}^{(\alpha)}(\xi) = 0$, which implies that, for every ω, $g\left(\langle\omega\rangle^{-\alpha}(\xi-\omega)\right) = 0$, hence $m_{g,g}^{(\alpha)}(\xi) = 0$ and is a contradiction. □

4.2. Admissible Banach spaces

The H^m-theory being built, we are now able to define anisotropic Banach spaces with the $\mathcal{W}^{(\alpha)}$ transformations by pullback technics. Indeed, these function spaces shall be included in the Sobolev spaces scale. For that aim, we consider a Banach space $B(T^*\mathbb{R}^n)$ of locally integrable functions on the phase space $T^*\mathbb{R}^n$. Moreover, we suppose that:

$$\mathcal{S}(T^*\mathbb{R}^n) \subset B(T^*\mathbb{R}^n) \subset \mathcal{S}'(T^*\mathbb{R}^n).$$

Let us now introduce the following definition.

Definition 4.5. *Let $B(T^*\mathbb{R}^n)$ be a Banach space defined as above, and $\alpha \in (0,1)$. $B(T^*\mathbb{R}^n)$ is said to be α-admissible if and only if it satisfies the following two conditions:*

$$\exists m, m' \in \mathbb{R};\, L^2_m(T^*\mathbb{R}^n) \subset B(T^*\mathbb{R}^n) \subset L^2_{m'}(T^*\mathbb{R}^n). \tag{B1}$$

$$\forall K \in S_0^{(\alpha)}(T^*\mathbb{R}^n), Op[K] : B(T^*\mathbb{R}^n) \longrightarrow B(T^*\mathbb{R}^n) \text{ is continuous.} \tag{B2}$$

Now, in order to define anisotropic Banach spaces, we need to study the dependence with respect to the analyzing function.

Proposition 4.6. *Let $\alpha \in (0,1)$. Consider an α-admissible Banach space $B(T^*\mathbb{R}^n)$, $g \in \mathcal{S}(\mathbb{R}^n)$ an admissible analyzing function and $s \in \mathcal{S}'(\mathbb{R}^n)$ a tempered distribution such that $\mathcal{W}_g^{(\alpha)} s \in B(T^*\mathbb{R}^n)$. Then, for all $h \in \mathcal{S}(\mathbb{R}^n)$, $\mathcal{W}_h^{(\alpha)} s \in B(T^*\mathbb{R}^n)$.*

Proof. Since g is admissible, by Proposition 3.5, one gets for all integer N the existence of $K_N \in S_0^{(\alpha)}(T^*\mathbb{R}^n)$, and $R_N \in S_{\alpha,0}^{N(1-\alpha)}(\mathbb{R}^n)$ verifying that

$$\mathcal{W}_h^{(\alpha)} = Op[K_N]\mathcal{W}_g^{(\alpha)} + \mathcal{W}_h^{(\alpha)} R_N.$$

Then, by condition (B2), one has that $Op[K_N]\mathcal{W}_g^{(\alpha)} s \in B(T^*\mathbb{R}^n)$. On the other hand, since $\mathcal{W}_g^{(\alpha)} s \in B(T^*\mathbb{R}^n) \subset L^2_{m'}(\mathbb{R}^n)$, one knows by condition (B1) that the distribution s is in the Sobolev space $H^{m'}(\mathbb{R}^n)$. As a consequence, for N large enough, $R_N s \in H^m(\mathbb{R}^n)$, and then $\mathcal{W}_h^{(\alpha)} s \in L^2_m(\mathbb{R}^n) \subset B(T^*\mathbb{R}^n)$, which completes the proof. □

This result proves that the following definition is well posed, that is does not depend on the analyzing function.

Definition 4.7. *Let $\alpha \in (0,1)$. We consider an α-admissible Banach space and g an analyzing function in the Schwartz space. We define the Banach space $B(\mathbb{R}^n)$ of distributions on $\mathbb{R}^n$ as follows: for a tempered distribution s,*

$$s \in B(\mathbb{R}^n) \iff \mathcal{W}_g^{(\alpha)} s \in B(T^*\mathbb{R}^n).$$

We define the norm in $B(\mathbb{R}^n)$ by:

$$\|s\|_{B(\mathbb{R}^n)} = \|\mathcal{W}_g^{(\alpha)} s\|_{B(T^*\mathbb{R}^n)}.$$

Remark Proposition 4.6 and Theorem 4.4 show that if we take another admissible analyzing function h in the definition of $B(\mathbb{R}^n)$, then the norm induced by h is equivalent.

The next step in building these Banach spaces is to give more developed conditions on $B(T^*\mathbb{R}^n)$, for this space to be admissible. There are four conditions:

$$\mathcal{S}(T^*\mathbb{R}^n) \subset B(T^*\mathbb{R}^n) \subset \mathcal{S}'(T^*\mathbb{R}^n). \tag{C1}$$

$$\||s|\|_{B(T^*\mathbb{R}^n)} = \|s\|_{B(T^*\mathbb{R}^n)}. \tag{C2}$$

Condition (C2) means that $B(T^*\mathbb{R}^n)$ is a Banach lattice.

We can estimate for $s_{b',\omega'}(b,\omega) = s(b + \langle\omega\rangle^{-\alpha} b', \omega + wa^\alpha\omega')$,
$\|s_{b',\omega'}\|_{B(T^*\mathbb{R}^n)} \leq \rho(b',\omega')\|s\|_{B(T^*\mathbb{R}^n)}$ (C3)
where ρ is at most polynomial growing.

$$\begin{aligned}&\mu \in L^\infty(T^*\mathbb{R}^n), s \in B(T^*\mathbb{R}^n) \Longrightarrow \mu s \in B(T^*\mathbb{R}^n),\\ &\|M[\mu]\|_{B(T^*\mathbb{R}^n)\to B(T^*\mathbb{R}^n)} = \|\mu\|_{L^\infty(T^*\mathbb{R}^n)}.\end{aligned} \tag{C4}$$

This condition says that the multiplier algebra of $B(T^*\mathbb{R}^n)$ contains the bounded functions.

These conditions are sufficient to imply the invariance condition (B2), as it is shown in the next lemma.

Lemma 4.1. *Suppose that* $B(T^*\mathbb{R}^n)$ *verify condition* (C1)–(C4). *Then, condition* (B2) *holds for* $B(T^*\mathbb{R}^n)$.

Proof. Let $s \in B(T^*\mathbb{R}^n)$, and $\mathcal{K} = Op(K) \in S_0^{(\alpha)}(T^*\mathbb{R}^n)$. Then, by a change of variables, one may write

$$\mathcal{K}s = \int_{T^*\mathbb{R}^n} db'd\omega'(Ks)\left(b + \langle\omega\rangle^{-\alpha}b', \omega + \langle\omega\rangle^{\alpha}\omega'\right).$$

Now, passing to the absolute value, then to the norm on both sides, and using hypothesis (C3), one obtains

$$\|\mathcal{K}s\|_{B(T^*\mathbb{R}^n)} \leq \|s\|_{B(T^*\mathbb{R}^n)} \int_{T^*\mathbb{R}^n} db'd\omega' K\left(b + \langle\omega\rangle^{-\alpha}b', \omega + \langle\omega\rangle^{\alpha}\omega'\right)\rho(b',\omega').$$

Since $K \in S_0^{(\alpha)}(T^*\mathbb{R}^n)$, the last integral is finite, and the lemma is proved □

The end of the section is devoted to a short study of the multiplier operators in admissible Banach spaces. Let $B(T^*\mathbb{R}^n)$ verify conditions (C1)–(C4), and let r be a weighted function. We assume that r satisfies both (10) and (12). Then, we can define the function space

$$rB(T^*\mathbb{R}^n) = \{rs, s \in B(T^*\mathbb{R}^n)\}$$

on which we put the norm $\|s\|_{rB} = \|s/r\|_B$.

Proposition 4.8. *With the assumptions above,* $rB(T^*\mathbb{R}^n)$ *is* α*-admissible.*

Proof. Let $\mathcal{K} \in OPS_0^{(\alpha)}(T^*\mathbb{R}^n)$ and let $s \in B(T^*\mathbb{R}^n)$. Then,

$$\frac{\mathcal{K}rs(b,\omega)}{r(b,\omega)} = \mathcal{K}s + \frac{\mathcal{K}_{rest}s(b,\omega)}{r(b,\omega)}$$

with $\mathcal{K}_{rest} \in OPS_{r'}^{(\alpha)}(T^*\mathbb{R}^n)$, and with $r' = o(r)$, by Theorem 3.11. It remains to show that the remainder term is in $B(T^*\mathbb{R}^n)$, which is obvious by the estimate in $OPS_{r'}^{(\alpha)}(T^*\mathbb{R}^n)$. □

As a consequence, we can define the space $B_r(\mathbb{R}^n)$ by pullback from $rB(T^*\mathbb{R}^n)$. Then, we obtain the following continuity result.

Proposition 4.9. *Let* r *verify* (10), (12), *and let* $B(T^*\mathbb{R}^n)$ *be an admissible space. Then,* $Op_{g,h}^{(\alpha)}(r) : B(\mathbb{R}^n) \longrightarrow B_r(\mathbb{R}^n)$ *is continuous.*

Proof. Immediate. □

We finish by giving the analogue of Theorem 3.12 for admissible Banach spaces.

Theorem 4.10. *Let* m, m' *verify* (10), (12), *and let* $B(T^*\mathbb{R}^n)$ *be* α*-admissible. Then, there exists* $s = o(mm')$ *and* R *continuous from* $B(\mathbb{R}^n)$ *to* $B_s(\mathbb{R}^n)$ *such that:*

$$Op_{g,h}^{(\alpha)}(m) \circ Op_{g,h}^{(\alpha)}(m') = Op_{g,h}^{(\alpha)}(mm') + R.$$

Proof. Adopting the notation of Theorem 3.12, it suffices to take s equal to $\max(r, |\omega|^{\alpha-1}mm')$. □

4.3. An example

We finish this paper with the study of an example based on the L^∞-spaces. We define the following function space on $T^*\mathbb{R}^2$ (for simplicity, we only consider the case $n=2$): for $m \geq 1$,

$$r \in B_m\left(T^*\mathbb{R}^2\right) \Longleftrightarrow |r(b,\omega)| \leq \mathcal{O}(1)\varphi_m(\omega) \text{ uniformly in } b$$

where

$$\varphi_m(\omega) = \left(1 + |\omega_1|^{2m} + |\omega_2|^2\right)^{1/2}$$

and we put on $B_m\left(T^*\mathbb{R}^2\right)$ the norm $\|s\|_m = \|s/\varphi_m\|_{L^\infty}$. Then, we have the following result

Theorem 4.11. *$B_m\left(T^*\mathbb{R}^2\right)$ is an α-admissible Banach space for $m < 1/\alpha$.*

Proof. First, $B_m\left(T^*\mathbb{R}^2\right)$ obviously verifies condition (B1). It remains to verify condition (B2) for $B_m(T^*\mathbb{R}^2)$. Then, let $r \in B_m(T^*\mathbb{R}^2)$ and $\mathcal{K} = OP(K) \in OPS_0^{(\alpha)}(T^*\mathbb{R}^2)$ be given.

$$(\mathcal{K}r)(b,\omega) = \int_{T^*\mathbb{R}^2} db'd\omega' K(b,\omega : b',\omega')r(b',\omega').$$

By definition of $OPS_0^{(\alpha)}(T^*\mathbb{R}^2)$, one has that, for all $k, l \in \mathbb{N}$,

$$(\mathcal{K}r)(b,\omega) \leq \int_{T^*\mathbb{R}^2} db'd\omega' \kappa_k\left(\langle\omega\rangle^\alpha(b-b')\right)\kappa_l\left(\langle\omega\rangle^{-\alpha}(\omega-\omega')\right)r(b',\omega').$$

Let us treat the integral in the b' variable.

$$\int_{\mathbb{R}^2} db'\kappa_k\left(\langle\omega\rangle^\alpha(b-b')\right) = \int_{\mathbb{R}^2} db'\kappa_k\left(\langle\omega\rangle^\alpha b'\right) = \langle\omega\rangle^{-2\alpha}\int_{\mathbb{R}^2}\kappa_k(b') = \mathcal{O}(1)\langle\omega\rangle^{-2\alpha}.$$

It remains to estimate the ω' integral. First, one remarks that one can take $|\omega| \geq 1$ without loss of generality. Then, the inequality becomes

$$(\mathcal{K}r)(b,\omega) \leq \mathcal{O}(1)|\omega|^{-2\alpha}\int_{\mathbb{R}^2} d\omega'\varphi_m(\omega')\kappa_l\left(|\omega|^{-\alpha}(\omega-\omega')\right).$$

Let us set for simplicity

$$f_m(\omega) = \int_{\mathbb{R}^2} d\omega'\varphi_m(\omega')\kappa_l\left(|\omega|^{-\alpha}(\omega-\omega')\right).$$

Here, we split $\mathbb{R}^2$ into two parts, $\Omega_S^\epsilon = \{\langle\omega\rangle^{-\alpha}|\omega-\omega'| \leq \langle\omega\rangle^\epsilon\}$ and $\Omega_M^\epsilon = (\Omega_M^\epsilon)^c$, ϵ being a given positive real. Then, one has that

$$\begin{aligned} f_{m,S}^\epsilon &= \int_{\Omega_S^\epsilon} d\omega'\kappa_l\left(|\omega|^{-\alpha}(\omega-\omega')\right)\varphi_m(\omega') \\ &\leq \kappa_{l_1}(\omega)\int_{|\omega''|\geq|\omega|^\epsilon} d\omega''\kappa_{l_2}(\omega'')\varphi_m\left(\omega+|\omega|^\alpha\omega''\right) \end{aligned}$$

for l large enough. Then, one can see that $f^{\epsilon}_{m,S}$ is rapidly decreasing for all $\epsilon > 0$. Now, we estimate the second part of $f_m(\omega)$. One writes that

$$\begin{aligned} f^{\epsilon}_{m,M}(\omega) &= |\omega|^{2\alpha} \int_{|\omega''|\le|\omega|^{\epsilon}} d\omega'' \kappa_l(\omega'')\varphi_m\left(\omega + |\omega|^{\alpha}\omega''\right) \\ &\le |\omega|^{2\alpha} \left(\int_{\mathbb{R}^2} d\omega''\kappa_l(\omega'')\right)\left(\sup_{|\omega''|\le|\omega|^{\epsilon}} \varphi_m\left(\omega + |\omega|^{\alpha}\omega''\right)\right) \\ &\le \mathcal{O}(1)|\omega|^{2\alpha}\left(\left||\omega|^{\alpha+\epsilon} + |\omega_1|\right|^{2m} + \left||\omega|^{\alpha+\epsilon} + |\omega_2|\right|^2 + 1\right)^{1/2}. \end{aligned}$$

Hence, one needs to prove that

$$f^{\epsilon}_{m,M}(\omega) \le \mathcal{O}(1)|\omega|^{2\alpha}\varphi_m(\omega).$$

Since $|\omega| \ge 1$, one has to treat three cases.

1st case: $|\omega_1|, |\omega_2| > 1$ and $|\omega_1| \ge |\omega_2|$. In that case, one has that $|\omega|^2 \le 2|\omega_1|^2$ and, for ϵ small enough that $\alpha + \epsilon < 1$. Then, one obtains that

$$\begin{aligned} f^{\epsilon}_{m,M}(\omega) &\le \mathcal{O}(1)|\omega|^{2\alpha}\left(\left||\omega| + |\omega_1|\right|^{2m} + \left||\omega| + |\omega_2|\right|^2 + 1\right)^{1/2} \\ &\le \mathcal{O}(1)|\omega|^{2\alpha}\left(|\omega_1|^{2m} + |\omega|^2 + |\omega_2|^2 + 1\right)^{1/2} \\ &\le \mathcal{O}(1)|\omega|^{2\alpha}\varphi_m(\omega) \end{aligned}$$

and one gets the result for the first case.

2nd case: $|\omega_1|, |\omega_2| > 1$, $|\omega_1| < |\omega_2|$. Here, the estimate becomes

$$\begin{aligned} f^{\epsilon}_{m,M}(\omega) &\le \mathcal{O}(1)|\omega|^{2\alpha}\left(|\omega_1|^{2m} + |\omega|^{2m(\alpha+\epsilon)} + |\omega_2|^2 + 1\right)^{1/2} \\ &\le \mathcal{O}(1)|\omega|^{2\alpha}\varphi_m(\omega) \end{aligned}$$

since $\alpha m < 1$, and the result follows

3rd case: one of $|\omega_1|, |\omega_2|$ is lower than 1. Here, the conclusion is immediate, since $\alpha < 1$.

The theorem is proved. □

References

[AAGM] S.T. Ali, J.P. Antoine, J.P. Gazeau, U.A. Mueller. Coherent states and their generalizations: A mathematical overview. *Rev. Math. Phys.*, 7:1013–1104, 1995.

[BI] J. Bros, D. Iagolnitzer. Support essentiel et structure analytique des distributions. In *Séminaire Goulaouic-Lions-Schwartz*, exp no 18, 1975.

[CF] C. Fefferman A. Cordoba. Wave packets and Fourier integral operators. *Comm. Partial Diff. Eq.*, 3:979–1005, 1978.

[CM] R. Coiffman, Y. Meyer. *Ondelettes et opérateurs* I, II & III. Hermann, 1990.

[CZ] A.P. Calderon, A. Zygmund. Singular integral operator and differential equations. *Amer. J. Math.*, 79:901–921, 1957.

[Dau1] I. Daubechies. Wavelets, time-frequency localization and signal analysis. *IEEE Trans. Inf. Th.*, 36:961–1005, 1990.

[Dau2] I. Daubechies. *Ten lectures on wavelets.* SIAM, 1992.

[Fol] G.B. Folland. *Harmonic analysis on phase space.* Number 122 in Annals of Mathematics Studies. Princeton University Press, 1989.

[GM1] A. Grossmann, J. Morlet, T. Paul. Integral transforms associated to square integrable representations I: General results. *J. Math. Phys.*, 26:2473–2479, 1985.

[GM2] A. Grossmann, J. Morlet, T. Paul. Integral transforms associated to square integrable representations II: Examples. *Ann. IHP, Phys. Th.*, 45:293–309, 1986.

[Hol1] M. Holschneider. Analyse d'objets fractals par transformation en ondelettes. Thèse de doctorat, Marseille 1989.

[Hol2] M. Holschneider. Wavelet analysis of partial differential oprators. In Demuth, Schrohe, Schulze, Sjöstrand (eds), *Schrödinger Operators, Markov Semigroups, Wavelet Analysis, Operator Algebras.* Advances in Partial Differential Equations. Akademie Verlag Berlin, 1996.

[HT] M. Holschneider, Ph. Tchamitchian. Pointwise regularity of Riemann's nowhere differentiable function. *Inventiones Mathematicae*, 105:157–175, 1991.

[Hor1] L. Hörmander. *The analysis of partial differential operators 1.* Springer-Verlag, 1982.

[Hor2] L. Hörmander. *The analysis of partial differential operators 3.* Springer-Verlag, 1985.

[Jaf] S. Jaffard. Estimations höldériennes ponctuelles des fonctions au moyen de leurs coefficients d'ondelettes. *C.R. Acad. Sc. Paris, Sér. I*, 308, 1989.

[Mal] S. Mallat. A theory for multiresolution signal decomposition : the wavelet representation. *IEEE Trans. Pattern Anal. and Mach. Int.*, 11:674–693, 1989.

[Sjo] J. Sjöstrand. Singularités analytiques microlocales. *Astérisque*, 95, 1982.

Bruno Nazaret
Department of Mathematics
University of Cergy-Pontoise
2, Avenue Adolphe Chauvin
F-95302 Cergy-Pontoise, France

Matthias Holschneider
Department of Mathematics
University of Potsdam
PF 60 15 53
D-14415 Potsdam, Germany

Operator Theory:
Advances and Applications, Vol. 145, 395–418

Formes de torsion analytique et fibrations singulières

Xiaonan Ma

Abstract. In this paper, we generalize to a relative situation a result of Bismut [**B2**] on the Quillen metric associated to a singular fibration. To establish our results, we make essential use of the immersion theorem of Bismut for analytic torsion forms [**B3**].

Résumé. L'objet de cet article est d'étudier des formes de torsion analytique au voisinage de fibres dégénérant en diviseurs à croisements normaux. On calcule la partie finie de la renormalisation à l'aide des formes de torsion analytique sur la normalisée de ces fibres. Ce résultat étend une formule de Bismut [**B2**] à une situation en famille.

Introduction

Les formes de torsion analytique de Bismut-Köhler sont l'extension de la torsion analytique de Ray-Singer pour une submersion holomorphe. Ces formes sont contenues dans la définition de l'image directe de Gillet et Soulé [**GS**] en géométrie d'Arakelov.

Soit $\pi : X \to S$ un morphisme de variétés compactes complexes qui est une submersion sur le complémentaire d'une sous variété Σ de codimension 2, où π a des singularités ordinaires. Soit $\Delta = \pi(\Sigma)$. Soit ξ un fibré holomorphe sur X. On pose $\lambda(j^*\xi) = \det(R^\bullet\pi\xi)^{-1}$. Dans [**B2**], Bismut a calculé le comportement de la métrique de Quillen sur $\lambda(j^*\xi)$ près de Δ. Après extraction d'une divergence logarithmique, il a décrit la métrique limite en fonction de la métrique de Quillen sur la normalisation des fibres singulières.

Le but de cet article est de généraliser un résultat de Bismut [**B2**, Théorème 0.2] à une situation en famille. On le donne comme le Théorème 2.1. Pour montrer le Théorème 2.1, on procède comme en [**B2**]. C'est à dire, on utilise la technique de la déformation du cône normal [**BaFM**], et le théorème d'immersion de [**B3**, Théorème 0.1]. Comme tous les calculs sont presque les mêmes que dans [**B2**], on omit toujours les détails des calculs. On se concentre souvent de vérifier que les cohomologies intermédiaires soient localement libres, et les relations des classes de Bott-Chern correpondantes.

Cet article est organisé de la manière suivante. Dans la Section 1, en utilisant le résultat de [**B3**], on calcule le comportement des formes de torsion analytique près des fibres singulières. Dans la Section 2, on énonce le Théorème 2.1 qui, à l'aide des formes de torsion analytique sur la normalisée de ces fibres, calcule sa renormalisation. Dans la Section 3, on montre le Théorème 2.1 dans le cas $Z = Z^1 \cup_\Sigma Z^2$. Dans la Section 4, on montre le Théorème 2.1 dans le cas général. Dans la Section 5, on vérifie la compatibilité du Théorème 2.1 et [**BerB**, Théorème 3.1].

Remerciements

Je tiens à remercier le Professeur J.M. Bismut pour ses encouragements et pour d'utiles discussions. Je voudrais remercier aussi ICTP, Trieste, spécialement Professeur M.S. Narasimhan pour hospitalité.

1. La divergence logarithmique des formes de torsion analytique près des fibres singulières

Cette section est organisée de la façon suivante. Dans a), on définit une fibration singulière. Dans b), on donne des hypothèses et des notations. Dans c), on applique le Théorème d'immersion de [**B3**] à notre situation. Dans d), on calcule la partie divergente des formes de torsion analytique près des fibres singulières. Dans e), on indique les formules d'anomalie pour la renormalisation des formes de torsion analytique.

a) Une fibration singulière

Soit V, B deux variétés complexes compactes Kählériennes. Soit S une variété complexe de dimension 1. Soit Σ une sous variété complexe de V de codimension 2, soit Δ un ensemble de points isolés dans S. Soit $\pi : V \to B \times S$ une application holomorphe de fibre compacte Y. Soit $p_1 : B \times S \to B$ et $p_2 : B \times S \to S$ les projections naturelles. On note $\pi_1 = p_1 \circ \pi$ et $\pi_2 = p_2 \circ \pi$.

On suppose que $\pi : V \setminus \Sigma \to B \times S$ (resp. $\pi : \Sigma \to B \times \Delta$) est une submersion et que si $x \in \Sigma$, il existe un système de coordonnées holomorphes $(x^1, \dots, x^n)$ près de x, une coordonnée (s^1) près de $\pi_2(x)$, un système de coordonnées holomorphes $(b^1, \dots, b^m)$ près de $\pi_1(x)$ tels que localement

$$\begin{aligned} &\Sigma = (x^1 = 0, x^2 = 0), \\ &\Delta = (s^1 = 0), \\ &\pi_1(x^1, \dots, x^n) = (x^3, \dots, x^{m+2}), \\ &\pi_2(x^1, \dots, x^n) = x^1 x^2. \end{aligned} \tag{1.1}$$

Pour $b \in B, s \in S$, soit $X_b = \pi_1^{-1}(b)$, $Z_s = \pi_2^{-1}(s)$. Alors $\pi_1 : V \to B$ est une submersion holomorphe de fibre compacte X.

On note aussi $Z = \pi_2^{-1}(\Delta)$, on considère les immersions $i : \Sigma \to V, j_s : Z_s \to V, j : Z \to V$. Soit V_Σ l'éclaté de V le long de Σ, et soit $\overline{p} : V_\Sigma \to V$ la projection canonique. Soit $\rho : \widehat{Z} \to Z$ la normalisée de Z. Alors $\widehat{Z}$ est un diviseur lisse de V_Σ. On pose $\widehat{\Sigma} = \rho^{-1}(\Sigma)$. Alors $\rho : \widehat{\Sigma} \to \Sigma$ est un revêtement double. Soit κ le

$\mathbf{Z}_2$ -fibré sur Σ, $\kappa = \det \rho_* \mathcal{O}_{\widehat{\Sigma}}$. Alors Z, $\widehat{Z}$, $\widehat{\Sigma}$ et Σ se fibrent sur $B \times \Delta$ de fibres compactes Z_B, $\widehat{Z}_B, \widehat{\Sigma}_B$ et Σ_B. Pour $b \in B, s \in \Delta$, si on note $\widehat{Y}_{(b,s)}$ la normalisée de $Y_{(b,s)}$, on a $\widehat{Z}_B = \widehat{Y}_{(b,s)}$.

b) Hypothèses et notations

Soit ξ un fibré vectoriel holomorphe sur V.

Dans la suite, on suppose que les images directes $R^\bullet \pi_* \xi$, $R^\bullet(\pi j \rho)_*(j\rho)^*\xi$, $R^\bullet(\pi i)_* i^* \xi$, $R^\bullet(\pi i \rho)_*(i\rho)^*\xi$ et $R^\bullet(\pi i)_*(i^*\xi \otimes \kappa)$ sont localement libres sur $B \times S$, $B \times \Delta$, $B \times \Delta$, $B \times \Delta$ et $B \times \Delta$. Soit $H(Y, \xi_{|Y})$ la cohomologie de $\xi_{|Y}$ le long des fibres Y. Alors ils sont des fibrés holomorphes $\mathbf{Z}$-gradués. Plus exactement $R^\bullet \pi_* \xi = H(Y, \xi_{|Y})$.

Soit g^{TV} une métrique Kählérienne sur TV, soient g^{TX}, g^{TY} et g^{TZ_s} les métriques induites par g^{TV} sur TX, TY et TZ_s sur V, $V \setminus \Sigma$ et $V \setminus Z$. Soit h^ξ une métrique hermitienne sur ξ. On munit la métrique triviale sur κ. De plus, les variétés $\widehat{Z}$ et Σ sont lisses, la métrique g^{TV} induite donc des métriques $g^{T\widehat{Z}_B}, g^{T\Sigma_B}$ sur $T\widehat{Z}/B \times \Delta, T\Sigma/B \times \Delta$.

Pour $b \in B \times (S \setminus \Delta)$, soit $\Omega(Y_b, \xi_{|Y_b})$ le complexe de Dolbeault des formes $\mathcal{C}^\infty$ sur la fibre Y_b à valeur dans ξ. On munit $\Omega(Y, \xi_{|Y})$ de la métrique hermitienne L_2 associée à g^{TY}, h^ξ [**B3**, §2.6]. On identifie $H(Y, \xi_{|Y})$ aux formes harmoniques dans le complexe de Dolbeault relatif $\Omega(Y, \xi_{|Y})$. Soit $h^{R\pi_*\xi}$ la métrique L_2 associée sur $R^\bullet \pi_* \xi$ sur $B \times (S \setminus \Delta)$ [**B3**, §2.6], [**BerB**, §1c)]. De même, on note $h^{R(\pi i)_*(i^*\xi \otimes \kappa)}$ et $h^{R(\pi j\rho)_*(j\rho)^*\xi}$ les métriques L_2 sur $R^\bullet(\pi i)_*(i^*\xi \otimes \kappa)$ et $R^\bullet(\pi j \rho)_*(j\rho)^*\xi$ sur $B \times \Delta$ induites par $g^{T\widehat{Z}_B}, g^{T\Sigma_B}$ et h^ξ.

Si K est un fibré vectoriel holomorphe avec une métrique g^K sur B, si Q est un polynôme caractéristique, on désigne par $Q(K, g^K) \in P^B$ la forme de Chern-Weil associée à la courbure de la connexion holomorphe hermitienne sur K.

Soit P^B l'espace des formes réelles $\mathcal{C}^\infty$ sur B qui sont la somme de formes de type (p,p). Soit $P^{B,0}$ l'espace de $\alpha \in P^B$ qui s'écrivent $\alpha = \partial\beta + \bar\partial\gamma$ où β, γ sont $\mathcal{C}^\infty$ sur B.

Comme B est compacte Kählérienne, $P^{B,0}$ est un sous espace vectoriel fermé de P^B pour la convergence uniforme [**B3**, §6.5].

Définition 1.1. *Soit* $\|.\|_k (k \in \mathbf{N})$ *la norme sur* $P^B/P^{B,0}$ *définie de la manière suivante: pour* $\alpha \in P^B$, *on pose*

$$\|\alpha\|_k = \inf_{\beta \in P^{B,0}} |\alpha - \beta|_{\mathcal{C}^k} = \inf_{\beta \in P^{B,0}} \sup_{x \in B, |a| \le k} |\frac{\partial^a}{\partial x^a}(\alpha - \beta)(x)|. \tag{1.2}$$

Pour $s \in S \setminus \Delta$, soient ω^{Z_s}, ω^V, $\omega^{\widehat{Z}}$ et ω^Σ les $(1,1)$-formes réelles fermées sur Z_s V, $\widehat{Z}$ et Σ associées aux métriques g^{TZ_s}, g^{TV}, $g^{T\widehat{Z}}$ et $g^{T\Sigma}$.

Soit $T(\omega^{Z_s}, h^\xi) \in P^B$, $T(\omega^V, h^\xi) \in P^{B \times (S\setminus\Delta)}$, $T(\omega^{\widehat{Z}}, h^\xi) \in P^{B \times \Delta}$ et $T(\omega^\Sigma, h^{i^*\xi\otimes\kappa}) \in P^{B\times\Delta}$ les formes de torsion analytique construites dans Bismut-Köhler [**BK**] sur B, $B\times(S\setminus\Delta)$, $B\times\Delta$ et $B\times\Delta$ associées à $(\pi j_s, \omega^{Z_s}, h^\xi)$, (π, ω^V, h^ξ),

$(\pi j\rho, \omega^{\widehat{Z}}, h^\xi)$ et $(\pi i, \omega^\Sigma, h^{i^*\xi\otimes\kappa})$. Les formes $T(\omega^V, h^\xi)$ vérifient l'équation

$$\frac{\overline{\partial}\partial}{2i\pi} T(\omega^V, h^\xi) = \mathrm{ch}(R^\bullet\pi_*\xi, h^{R\pi_*\xi}) - \int_Y \mathrm{Td}(TY, g^{TY})\mathrm{ch}(\xi, g^\xi). \tag{1.3}$$

et les formes $T(\omega^{Z_s}, h^\xi)$, etc, vérifient des équations analogues.

Pour $s \in S$, soit $\tau_s : B \to B \times S$ l'application de $y \in B$ à $(y, s) \in B \times S$. Alors pour $s \in S \setminus \Delta$, on a

$$\tau_s^* T(\omega^V, h^\xi) = T(\omega^{Z_s}, h^\xi) \quad \text{dans} \quad P^B/P^{B,0}. \tag{1.4}$$

c) Théorème d'immersion pour les formes de torsion analytique

Soit $\mathrm{Gr}(\pi_2)$ le graphe de π_2 dans $V \times S$. Alors $\mathrm{Gr}(\pi_2)$ est un diviseur lisse dans $V \times S$. Soit $[\mathrm{Gr}(\pi_2)]$ le fibré en droite sur $V \times S$ associé à $\mathrm{Gr}(\pi_2)$. Soit l l'identification $x \in V \to l(x) = (x, \pi_2(x)) \in \mathrm{Gr}(\pi_2)$. Soit m l'immersion $\mathrm{Gr}(\pi_2) \to V \times S$. Soit $n : V \times S \to V, n' : V \times S \to S$ les projections canoniques. On déduit facilement que

$$[\mathrm{Gr}(\pi_2)]_{|\mathrm{Gr}(\pi_2)} = n'^* TS_{|\mathrm{Gr}(\pi_2)}, \quad N_{\mathrm{Gr}(\pi_2)/V\times S} = n'^* TS_{|\mathrm{Gr}(\pi_2)}. \tag{1.5}$$

Soit j_s l'immersion $Z_s \to V$. Soit $[Z_s]$ le fibré en droite sur V associé à Z_s. Soient $\sigma_{[\mathrm{Gr}(\pi_2)]}$, $\sigma_{[Z_s]}$ les sections canoniques des fibrés en droite $[\mathrm{Gr}(\pi_2)]$, $[Z_s]$. Si $s \in S$, alors $[\mathrm{Gr}(\pi_2)]_{V\times\{s\}} = [Z_s]$ et $\sigma_{[\mathrm{Gr}(\pi_2)]_{V\times\{s\}}} = \sigma_{[Z_s]}$. Sur $V \times S$, on a une suite exacte de faisceaux

$$\begin{aligned} 0 \to & \mathcal{O}_{V\times S}(n^*\xi \otimes [-\mathrm{Gr}(\pi_2)]) \\ & \overset{\sigma_{[\mathrm{Gr}(\pi_2)]}}{\longrightarrow} \mathcal{O}_{V\times S}(n^*\xi) \to m_*\mathcal{O}_{\mathrm{Gr}(\pi_2)}((nm)^*\xi) \to 0. \end{aligned} \tag{1.6}$$

Pour $s \in S$, la restriction de (1.6) à $V \times \{s\}$ est exactement

$$0 \to \mathcal{O}_V(\xi \otimes [-Z_s]) \overset{\sigma_{[Z_s]}}{\to} \mathcal{O}_V(\xi) \to j_{s*}\mathcal{O}_{Z_s}(j_s^*\xi) \to 0. \tag{1.7}$$

On note (η, v) le complexe sur $V \times S$,

$$(\eta, v) : 0 \to n^*\xi \otimes [-\mathrm{Gr}(\pi_2)] \overset{\sigma_{[\mathrm{Gr}(\pi_2)]}}{\longrightarrow} n^*\xi \to 0. \tag{1.8}$$

Rappelle que l'application $(\pi_1 \circ n, n') : V \times S \to B \times S$ est une submersion de fibre compacte X. Soit $H(X, \eta_{|X})$ l'hypercohomologie de $(\mathcal{O}_{X_b\times\{s\}}(\eta_{|X_b\times\{s\}}), v)$. Alors par (1.6), $H(X, \eta_{|X})$ est isomorphe à $H(Y, \xi_{|Y}) = R^\bullet\pi_*\xi$, et $H(X, \eta_{|X})$ est un fibre vectoriel sur $B \times S$.

Soit g^{TS} une métrique hermitienne sur TS. Soit $g^{[\mathrm{Gr}(\pi_2)]}$ une métrique hermitienne sur $[\mathrm{Gr}(\pi_2)]$. Soit $g^{-[\mathrm{Gr}(\pi_2)]}$ la métrique hemitienne sur $[-\mathrm{Gr}(\pi_2)]$ induite par $g^{[\mathrm{Gr}(\pi_2)]}$. Soit $h^\eta = h^\xi \otimes g^{[-\mathrm{Gr}(\pi_2)]} \oplus h^\xi$ la métrique sur $\eta = n^*\xi \otimes [-\mathrm{Gr}(\pi_2)] \oplus n^*\xi$.

Soit $h^{H(X,\eta_{|X})}$ la métrique L_2 sur $H(X, \eta_{|X})$ associée aux métriques g^{TX}, h^η définie par [**B3**, §3.2] et $\widetilde{\mathrm{ch}}(H(X, \eta_{|X}), h^{R\pi_*\xi}, h^{H(X,\eta_{|X})}) \in P^{B\times(S\setminus\Delta)}/P^{B\times(S\setminus\Delta),0}$

la classe de Bott-Chern associée à [**BGS1**, §1f)]. Par [**B3**, §3.2], on peut construire des formes de torsion analytique $T(n^*\omega^V, h^\eta) \in P^{B\times S}/P^{B\times S,0}$, telle que

$$\frac{\overline{\partial}\partial}{2i\pi}T(n^*\omega^V, h^\eta) = \mathrm{ch}(H(X,\eta_{|X}), h^{H(X,\eta_{|X})}) - \int_{X\times\{s\}} \mathrm{Td}(TX, g^{TX})\mathrm{ch}(\eta, h^\eta). \tag{1.9}$$

Sur $V\setminus\Sigma$, on a une suite exacte de fibrés vectoriels holomorphes

$$0 \to TY \to TX \to \pi_2^* TS \to 0.$$

Soit $\widetilde{\mathrm{Td}}(TX, g^{TX}, \pi_2^* g^{TS})$ la classe de Bott-Chern associée à la suite exacte ci-dessus, on a

$$\frac{\overline{\partial}\partial}{2i\pi}\widetilde{\mathrm{Td}}(TX, g^{TX}, \pi_2^* g^{TS}) = \mathrm{Td}(TX, g^{TX}) - \mathrm{Td}(TY, g^{TY})\pi_2^*\mathrm{Td}(TS, g^{TS}).$$

Pour $s\in S\setminus\Delta$, on pose

$$I_s = \tau_s^*\Big[T(n^*\omega^V, h^\eta) - T(\omega^V, h^\xi) + \widetilde{\mathrm{ch}}(H(X,\eta_{|X}), h^{H(X,\eta_{|X})}, h^{R\pi_*\xi})\Big]. \tag{1.10}$$

On a l'analogue de [**B2**, Théorème 5.6], **Théorème 1.2**. *Pour $s\in S\setminus\Delta$, l'identité suivante est vraie dans $P^B/P^{B,0}$:*

$$\begin{aligned} I_s = &\int_{X\times\{s\}} \frac{\mathrm{Td}(TX, g^{TX})\mathrm{ch}(\xi, h^\xi)}{\mathrm{Td}([\mathrm{Gr}(\pi_2)], g^{[\mathrm{Gr}(\pi_2)]})}\log(\|\sigma_{[\mathrm{Gr}(\pi_2)]}\|^2) \\ &- \int_{Y_s} \frac{\widetilde{\mathrm{Td}}(TX, g^{TX}, \pi_2^* g^{TS})}{\mathrm{Td}([\mathrm{Gr}(\pi_2)], g^{[\mathrm{Gr}(\pi_2)]})}\mathrm{ch}(\xi, h^\xi). \end{aligned} \tag{1.11}$$

Preuve. En appliquant la formule de [**B3**, Théorème 0.1] à l'immersion $j_s : Z_s \to V$ ($s\in S\setminus\Delta$) qui est exactement l'immersion $\mathrm{Gr}(\pi_2)\cap V\times\{s\} \to V\times\{s\}$, et à la suite exacte de faisceaux (1.6) sur $V\times\{s\}$. En procédant comme en [**B2**, Théorème 5.6], on a (1.11). □

d) La divergence logarithmique des formes de torsion analytique près des fibres singulières

On pose

$$E(x) = \frac{x-\sinh(x)}{2x(1-\cosh(x))}. \tag{1.12}$$

Alors si $\zeta(s)$ est la fonction zêta de Riemann,

$$E(x) = -\sum_{n\geq 0, n \text{ pair}} \frac{(n+2)\zeta(-n-1)x^n}{(n+1)!}, \quad E(0) = \frac{1}{6}. \tag{1.13}$$

On identifie E au genre additif correspondant.

Soit $h'^{R\pi_*\xi}$ une métrique hermitienne $\mathcal{C}^\infty$ sur $R^\bullet\pi_*\xi$ sur $B\times S$. Soit $\widetilde{\mathrm{ch}}(R^\bullet\pi_*\xi, h^{R\pi_*\xi}, h'^{R\pi_*\xi}) \in P^{B\times(S\setminus\Delta)}/P^{B\times(S\setminus\Delta),0}$ la classe de Bott-Chern de [**BGS1**, §1f)] telle que

$$\frac{\overline{\partial}\partial}{2i\pi}\widetilde{\mathrm{ch}}(R^\bullet\pi_*\xi, h^{R\pi_*\xi}, h'^{R\pi_*\xi}) = \mathrm{ch}(R^\bullet\pi_*\xi, h'^{R\pi_*\xi}) - \mathrm{ch}(R^\bullet\pi_*\xi, h^{R\pi_*\xi}). \tag{1.14}$$

Soit $\sigma_{[\Delta]}$ la section canonique du diviseur $[\Delta]$. Soit $g^{[\Delta]}$ une métrique sur $[\Delta]$.

Théorème 1.3. *Pour $s_0 \in \Delta$, la limite*

$$\lim_{S\setminus\Delta\ni s\to s_0}\Big\{\tau_s^*\Big[T(\omega^V, h^\xi) + \widetilde{\mathrm{ch}}(R^\bullet\pi_*\xi, h^{R\pi_*\xi}, h'^{R\pi_*\xi})\Big] \tag{1.15}$$
$$+\frac{1}{2}\log(\|\sigma_{[\Delta]}\|_s^2)\int_{\Sigma_B}\mathrm{Td}(T\Sigma_B)E(N_{\Sigma/V})\mathrm{ch}(\xi)\Big\}.$$

existe dans $P^B/P^{B,0}$. On note cette limite comme $T(\omega^Z, h^\xi, h'^{R\pi_*\xi}) \in P^B/P^{B,0}$.

On considère $T(\omega^Z, h^\xi, h'^{R\pi_*\xi}) \in P^B/P^{B,0}$ comme les formes de torsion analytique de $Z_{s_0} \to B$.

Avant démontrer le Théorème 1.3, on vérifie que ce Théorème est compatible à [**B2**, Théorème 5.9], [**BerB**, Théorème 3.1].

Soit $\lambda(\xi)$ et $\lambda(R^\bullet\pi_*\xi)$ les fibrés en droite sur S qui sont les inverses des déterminants de $H(Z,\xi_{|Z})$ et $H(B, R^\bullet\pi_*\xi_{|B})$. Alors par [**KM**], le fibré en droite $\lambda(\xi)\otimes\lambda^{-1}(R^\bullet\pi_*\xi)$ a une section canonique non nulle σ sur S. Soit $\|\ \|_{\lambda(\xi)}$ et $\|\ \|_{\lambda(R^\bullet\pi_*\xi)}$ les métriques de Quillen sur $\lambda(\xi)$ et $\lambda(R^\bullet\pi_*\xi)$ sur $S\setminus\Delta$ associées à g^{TZ_s}, h^ξ et $g^{TB}, h^{R\pi_*\xi}$. Soit $\|\ \|_{\lambda(\xi)\otimes\lambda^{-1}(R^\bullet\pi_*\xi)}$ la métrique correspondante sur $\lambda(\xi)\otimes\lambda^{-1}(R^\bullet\pi_*\xi)$ sur $S\setminus\Delta$. Alors pour $s\in S\setminus\Delta$, en appliquant [**BerB**, Théorème 3.1], on a

$$\begin{aligned}\log\|\sigma\|^2_{\lambda(\xi)\otimes\lambda^{-1}(R^\bullet\pi_*\xi)} &= -\textstyle\int_{B\times\{s\}}\mathrm{Td}(TB, g^{TB})T(\omega^V, h^\xi)\\ &\quad+\textstyle\int_{Z_s}\widetilde{\mathrm{Td}}(TZ, TB, g^{TZ}, g^{TB})\mathrm{ch}(\xi, h^\xi).\end{aligned} \tag{1.16}$$

Ici, comme en (1.11), la classe de Bott-Chern $\widetilde{\mathrm{Td}}(\cdot,\cdot)$ correspond à la suite exacte

$$0\to TY\to TZ\to\pi_1^*TB\to 0.$$

En utilisant [**B2**, Théorèmes 2.1 et 5.9] et [**BGS2**, Théorème 1.23] (se réferer aussi à §5), quand $s\in S\setminus\Delta\to s_0\in\Delta$,

$$-\int_{B\times\{s\}}\mathrm{Td}(TB, g^{TB})\Big[T(\omega^V, h^\xi) + \widetilde{\mathrm{ch}}(R^\bullet\pi_*\xi, h^{R\pi_*\xi}, h'^{R\pi_*\xi})\Big]$$
$$-\frac{1}{2}\log(\|\sigma_{[\Delta]}\|_s^2)\int_\Sigma\mathrm{Td}(T\Sigma)E(N_{\Sigma/V})\mathrm{ch}(\xi)$$

a une limite. C'est à dire, le Théorème 1.3 est compatible à [**BerB**], [**B2**]. □

Preuve du Théorème 1.3. Dans notre cas, par une modification simple de l'argument de [**B2**], on sait que les résultats analogues de [**B2**, Théorèmes 2.2 et 3.3] sont aussi vrais.

En procédant comme en [**B2**, Théorème 5.7], quand $s \in S \setminus \Delta \to s_0 \in \Delta$, l'asymptotique du terme à droite de (1.11) dans $P^B/P^{B,0}$ est exactement la même formule que celle de [**B2**, (5.20)].

Comme $T(n^*\omega^V, h^\eta) \in P^{B\times S}$, on a montré le Théorème 1.3. □

Par le Théorème 1.3, dans $P^B/P^{B,0}$, on pose

$$I_\lambda = \lim_{s\to s_0} \Big[I_s - \frac{1}{2}\log(\|\sigma_{[\Delta]}\|_s^2) \int_{\Sigma_B} \mathrm{Td}(T\Sigma_B)E(N_{\Sigma/V})\mathrm{ch}(\xi)\Big]. \tag{1.17}$$

e) Formules d'anomalie pour des formes de torsion analytique

Soit $g'^{TV}, h'^\xi, h''^{R\pi_*\xi}$ des métriques sur $TV, \xi, R^\bullet\pi_*\xi$. Pour $s_0 \in \Delta$ fixé, on peut construire aussi $T(\omega'^Z, h'^\xi, h''^{R\pi_*\xi}) \in P^B/P^{B,0}$ comme dans le Théorème 1.3.

Par [**B2**, Théorème 1.1], le fibré vectoriel TY sur $V_\Sigma \setminus P(N_{\Sigma/V}) \simeq V\setminus\Sigma$ s'étend en un sous-fibré $v(TV/B\times S)$ de p^*TV sur V_Σ. Alors sur $\widehat{Z}$, on a $T\widehat{Y} = v(TV/B\times S)$. Soit $g^{v(TV/B\times S)}$, $g'^{v(TV/B\times S)}$ les métriques induites par g^{TV}, g'^{TV} sur $v(TV/B\times S)$. Soit $g'^{T\widehat{Y}}$ la métrique sur $T\widehat{Y}$ induite par g'^{TV}. Soit

$$\begin{aligned}
&\widetilde{\mathrm{Td}}(v(TV/B\times S), g^{v(TV/B\times S)}, g'^{v(TV/B\times S)}) \in P^{V_\Sigma}/P^{V_\Sigma,0},\\
&\widetilde{\mathrm{ch}}(\xi, h^\xi, h'^\xi) \in P^V/P^{V,0}, \quad \widetilde{\mathrm{Td}}(T\widehat{Y}, g^{T\widehat{Y}}, g'^{T\widehat{Y}}) \in P^{\widehat{Z}}/P^{\widehat{Z},0},\\
&\widetilde{\mathrm{ch}}(R^\bullet\pi_*\xi, h'^{R\pi_*\xi}, h''^{R\pi_*\xi}) \in P^{B\times S}/P^{B\times S,0}
\end{aligned}$$

les classes de Bott-Chern [**BGS1**]. On a

$$\frac{\bar\partial\partial}{2i\pi}\widetilde{\mathrm{ch}}(\xi, h^\xi, h'^\xi) = \mathrm{ch}(\xi, h'^\xi) - \mathrm{ch}(\xi, h^\xi). \tag{1.18}$$

et les autres formes ci-dessus vérifient des équations analogues.

Alors par (1.15), [**B2**, Théorème 2.1], [**BK**, Théorème 3.10], on a

Théorème 1.4. *On a l'identité suivante dans $P^B/P^{B,0}$*

$$\begin{aligned}
T(\omega'^Z, h'^\xi, h''^{R\pi_*\xi}) - T(\omega^Z, h^\xi, h'^{R\pi_*\xi}) &= \widetilde{\mathrm{ch}}(R^\bullet\pi_*\xi, h'^{R\pi_*\xi}, h''^{R\pi_*\xi})\\
&- \int_{\widehat{Y}} \Big(\widetilde{\mathrm{Td}}(T\widehat{Y}, g^{T\widehat{Y}}, g'^{T\widehat{Y}})\mathrm{ch}(\xi, h^\xi) + \mathrm{Td}(T\widehat{Y}, g'^{T\widehat{Y}})\widetilde{\mathrm{ch}}(\xi, h^\xi, h'^\xi)\Big)\\
&-2\int_{\Sigma_B}\Big\{\mathrm{ch}(\xi, h^\xi)\int_{P(N_{\Sigma/V})}\widetilde{\mathrm{Td}}(v(TV/B\times S), g^{v(TV/B\times S)}, g'^{v(TV/B\times S)})\\
&+\widetilde{\mathrm{ch}}(\xi, h^\xi, h'^\xi)\int_{P(N_{\Sigma/V})}\mathrm{Td}\Big(v(TV/B\times S), g'^{v(TV/B\times S)}\Big)\Big\}.
\end{aligned} \tag{1.19}$$

2. Comparaison des formes de torsion analytique

Pour $s \in \Delta$ fixé, soit $\widehat{Z}_s$ la normalisation de Z_s et $\rho : \widehat{Z}_s \to Z_s$ est la projection canonique. On note $k : \Sigma_s \to Z_s$ l'immersion triviale. Dans la suite, on omit l'indice s.

Alors $\rho_* \mathcal{O}_{\widehat{Z}}((j\rho)^*\xi)$ est un $\mathcal{O}_Z$-faisceau sur Z, et il existe un morphisme naturel $\mathcal{O}_Z(j^*\xi) \to \rho_*\mathcal{O}_{\widehat{Z}}((j\rho)^*\xi)$ et un morphisme de différence $\rho_*\mathcal{O}_{\widehat{Z}}((j\rho)^*\xi) \to k_*(\mathcal{O}_\Sigma(i^*\xi \otimes \kappa))$. Sur Z, on a une suite exacte de faisceaux [**B2**, (5.53)]

$$(L, v) : 0 \to \mathcal{O}_Z(j^*\xi) \to \rho_*\mathcal{O}_{\widehat{Z}}((j\rho)^*\xi) \to k_*(\mathcal{O}_\Sigma(i^*\xi \otimes \kappa)) \to 0. \tag{2.1}$$

D'après (2.1), on a une suite exacte de fibrés vectoriels holomorphes sur B

$$(R^\bullet \pi_* L, v) : \cdots \to R^{i-1}(\pi i)_*(i^*\xi \otimes \kappa) \to R^i(\pi j)_*\xi \to R^i(\pi j \rho)_*(j\rho)^*\xi \to \cdots . \tag{2.2}$$

On munit $R^\bullet\pi_* L$ de la métrique $h^{R\pi_* L} = h'^{R\pi_*\xi} \oplus h^{R(\pi j\rho)_*(j\rho)^*\xi} \oplus h^{R(\pi i)_*(i^*\xi\otimes\kappa)}$. Soit $\widetilde{\mathrm{ch}}(R^\bullet\pi_* L,\, h^{R\pi_* L}) \in P^B/P^{B,0}$ la classe de Bott-Chern [**BGS1**] telle que

$$\begin{aligned}\frac{\overline{\partial}\partial}{2i\pi}\widetilde{\mathrm{ch}}(R^\bullet\pi_* L, h^{R\pi_* L}) &= \mathrm{ch}\Big(R(\pi j\rho)_*(j\rho)^*\xi, h^{R(\pi j\rho)_*(j\rho)^*\xi}\Big) \\ &\quad -\mathrm{ch}(R^\bullet\pi_*\xi, h'^{R\pi_*\xi}) - \mathrm{ch}\Big(R(\pi i)_*(i^*\xi\otimes\kappa), h^{R(\pi i)_*(i^*\xi\otimes\kappa)}\Big).\end{aligned} \tag{2.3}$$

On rappelle que $T(\omega^Z, h^\xi, h'^{R\pi_*\xi}) \in P^B/P^{B,0}$ est définie par le Théorème 1.3, et que $T(\omega^{\widehat{Z}}, h^\xi)$ et $T(\omega^\Sigma, h^{i^*\xi\otimes\kappa})$ sont définies à §1b). On pose

$$T'(\omega^Z, h^\xi, h'^{R\pi_*\xi}) = T(\omega^{\widehat{Z}}, h^\xi) - T(\omega^\Sigma, h^{i^*\xi\otimes\kappa}) - \widetilde{\mathrm{ch}}(R^\bullet\pi_* L, h^{R\pi_* L}). \tag{2.4}$$

L'objet de cet article est de donner une formule pour

$$T'(\omega^Z, h^\xi, h'^{R\pi_*\xi}) - T(\omega^Z, h^\xi, h'^{R\pi_*\xi}) \in P^B/P^{B,0}. \tag{2.5}$$

Soit $\zeta(s)$ la fonction zêta de Riemann. Soit $R(x)$ la série de Gillet-Soulé [**GS**]

$$R(x) = \sum_{n\geq 0,\, n\,\mathrm{impair}} \Big(2\frac{\zeta'(-n)}{\zeta(-n)} + \sum_{j=1}^n \frac{1}{j}\Big)\zeta(-n)\frac{x^n}{n!}. \tag{2.6}$$

Si $f(x)$ est une série entière, on pose $f(x) = \Sigma_{k\geq 0} f^{[k]}x^k$. On définit $S(x), T(x)$ par

$$\begin{aligned} S(x) &= \sum_{n\geq 1,\, n\,\mathrm{impair}}\sum_{j=1}^n \frac{(-1)^j C_n^j}{j}\zeta(-n)\frac{x^n}{n!}, \\ T(x) &= \sum_{n\geq 2,\, n\,\mathrm{pair}} \frac{1}{(n+3)!}\Big[\frac{(1+x)^{n+1}\log(1+x)}{1+2x}\Big]^{[n]} x^n.\end{aligned} \tag{2.7}$$

On identifie $R(x), S(x), T(x)$ aux genres additifs correspondants. On pose

$$\begin{aligned} Q(x^1, x^2) &= \frac{1}{x^1 - x^2} \\ &\quad\times \int_{x^1}^{x^2} \frac{\mathrm{Td}(-x^1)\mathrm{Td}(-x^2) - \mathrm{Td}(-s)\mathrm{Td}(-(x^1+x^2-s))}{\mathrm{Td}(-s)(s-x^1)(s-x^2)} ds.\end{aligned} \tag{2.8}$$

Alors $Q(x^1, x^2)$ est une fonction symétrique de (x^1, x^2). On peut considérer Q comme une fonction ad-invariante des matrices 2×2. On identifie Q au genre correspondant. De plus $Q(x, -x)$ et $\mathrm{Td}(x)Q(x, 0)$ sont des fonctions paires.

On rappelle que TY sur $V_\Sigma \setminus P(N_{\Sigma/V}) \simeq V \setminus \Sigma$ s'étend en un sous-fibré $v(TV/B \times S)$ de $\overline{p}^*TV$ sur V_Σ [**B3**, Théorème 1.1]. Soit $g^{v(TV/B\times S)}$ la métrique induite par g^{TV} sur $v(TV/B \times S)$. Soit U le fibré en droite universel sur $P(N_{\Sigma/V})$. Soit $p_\Sigma : P(N_{\Sigma/V}) \to \Sigma$ la projection canonique. Sur $P(N_{\Sigma/V})$, on a la suite exacte [**B2**, (0.10)]

$$0 \to p_\Sigma^* T\Sigma/B \times \Delta \to v(TV/B \times S) \to U \otimes p_\Sigma^*\kappa \to 0. \tag{2.9}$$

Soit $g^{N_{\Sigma/V}}$ une métrique sur $N_{\Sigma/V}$ telle que le scindement local $N_{\Sigma/V} = N_{\Sigma/Z^1} + N_{\Sigma/Z^2}$ (où N_{Σ/Z^i} représente le fibré normal dans l'une des deux branches locales de Y) soit orthogonal, et que l'isomorphisme canonique $\Lambda^2 N_{\Sigma/V} \otimes p_\Sigma^*\kappa \simeq \pi_2^*[\Delta]$ soit une isométrie [**B2**, §6.3 et §9.4]. Par [**BGS1**, §1f)], on peut associer une classe de Bott-Chern

$$\widetilde{\mathrm{Td}}(v(TV/B \times S), g^{v(TV/B\times S)}, g^{U\otimes p_\Sigma^*\kappa}) \in P^{P(N_{\Sigma/V})}/P^{P(N_{\Sigma/V}),0}$$

à la suite exacte de fibrés holomorphes hermitiens (2.9) telle que

$$\begin{aligned}\frac{\overline{\partial}\partial}{2i\pi}&\widetilde{\mathrm{Td}}(v(TV/B \times S), g^{v(TV/B\times S)}, g^{U\otimes p_\Sigma^*\kappa}) \\ &= \mathrm{Td}(v(TV/B \times S), g^{v(TV/B\times S)}) \\ &\quad - p_\Sigma^*\mathrm{Td}(T\Sigma/B \times \Delta, g^{T\Sigma/B\times\Delta})\mathrm{Td}(U \otimes p_\Sigma^*\kappa, g^{U\otimes p_\Sigma^*\kappa}).\end{aligned} \tag{2.10}$$

Théorème 2.1. *On a la formule suivante dans* $P^B/P^{B,0}$,

$$\begin{aligned}T'(\omega^Z, h^\xi, h'^{R\pi_*\xi}) &- T(\omega^Z, h^\xi, h'^{R\pi_*\xi}) \\ = 2\int_{\Sigma_B} &\mathrm{ch}(\xi, g^\xi)\int_{P(N_{\Sigma/V})} \widetilde{\mathrm{Td}}\Big(v(TV/B \times S), g^{v(TV/B\times S)}, g^{U\otimes p_\Sigma^*\kappa}\Big) \\ - \int_{\Sigma_B} &\mathrm{Td}(T\Sigma_B)\Big(\mathrm{Td}(N_{\Sigma/V})T(N_{\Sigma/V}) + 2Q(N_{\Sigma/V})\Big)\mathrm{ch}(\xi) \\ + \int_{\hat{\Sigma}_B} &\mathrm{Td}(T\widehat{\Sigma}_B)\Big(\mathrm{Td}(x)\Big(\frac{R+S}{x} + Q(x,0)\Big)\Big)(N_{\widehat{\Sigma}/\widehat{Z}})\mathrm{ch}(\xi).\end{aligned} \tag{2.11}$$

Remarque 2.2.

i) Si B est un point, le Théorème 2.1 est [**B2**, Théorème 0.2].

ii) Dans le Théorème 2.1, si V est projective, par l'argument de résolution acyclique, on peut supposer seulement que

$$R^\bullet\pi_*\xi,\ R^\bullet(\pi j\rho)_*(j\rho)^*\xi \qquad \text{et} \qquad R^\bullet(\pi i)_*(i^*\xi \otimes \kappa)$$

sont localement libres sur $B \times S$, $B \times \Delta$ et $B \times \Delta$.

3. Preuve du Théorème 2.1: le cas où la fibre exceptionnelle est scindée

Cette section est organisée de la façon suivante. Dans a), on introduit notre hypothèse sur Z. Dans b), on compare des formes de torsion analytique. Dans c), on montre le Théorème 2.1 sous notre hypothèse sur Z.

a) Une hypothèse simple sur Z

Dans la suite, on fixe toujours $s_0 \in \Delta$. Soit $Z = Z_{s_0}$.

On suppose qu'il existe deux diviseurs lisses Z^1 et Z^2 dans V intersectés transversalement le long de Σ tels que

$$Z = Z^1 \cup_\Sigma Z^2. \tag{3.1}$$

Alors le fibré κ sur Σ est trivial. Soit $i : \Sigma \to V$, $j^k : Z^k \to V, j : Z \to V$ les immersions naturelles. Pour $k = 1, 2$, soit $\sigma_{[Z^k]}$ la section canonique de $[Z^k]$. Alors la section canonique $\sigma_{[Z]}$ de $[Z]$ est donnée par

$$\sigma_{[Z]} = \sigma_{[Z^1]} \otimes \sigma_{[Z^2]}. \tag{3.2}$$

Par (3.2), on identifie $\Lambda^2([-Z^1] \oplus [-Z^2])$ à $[-Z]$.

Les complexes de Koszul $(\Lambda([-Z^k], \sigma_{[Z^k]})$, $(\Lambda[-Z], \sigma_{[Z]})$, $(\Lambda([-Z^1] \oplus [-Z^2])$, $\sigma_{[Z^1]} \oplus \sigma_{[Z^2]})$ sont des résolutions des $\mathcal{O}_V$-faisceaux $j^k_* \mathbf{C}, j_* \mathbf{C}, i_* \mathbf{C}$ sur V. On a le complexe de faisceaux sur V

$$\begin{array}{ccccccccc}
 & & 0 & & 0 & & 0 & & \\
 & & \uparrow & & \uparrow & & \uparrow & & \\
0 & \to & i_*\mathbf{C} & \longrightarrow & 0 & \to & 0 & \to & 0 \\
 & & \uparrow & & \uparrow & & \uparrow & & \\
0 & \to & \mathbf{C} & \longrightarrow & \mathbf{C} & \to & 0 & \to & 0 \\
 & & \uparrow{\scriptstyle i_{\sigma_{[Z^1]}} \oplus (-i_{\sigma_{[Z^2]}})} & & \uparrow{\scriptstyle e} & & \uparrow & & \\
0 & \to & [-Z^1] \oplus [-Z^2] & \xrightarrow{\sigma_{[Z^1]} \oplus \sigma_{[Z^2]}} & \mathbf{C} \oplus \mathbf{C} & \to & j^1_*\mathbf{C} \oplus j^2_*\mathbf{C} & \to & 0 \\
 & & \uparrow{\scriptstyle i_{\sigma_{[Z^1]}} \oplus (-i_{\sigma_{[Z^2]}})} & & \uparrow{\scriptstyle e} & & \uparrow{\scriptstyle e} & & \\
0 & \to & [-Z] & \xrightarrow{\sigma_{[Z]}} & \mathbf{C} & \to & j_*\mathbf{C} & \to & 0 \\
 & & \uparrow & & \uparrow & & \uparrow & & \\
 & & 0 & & 0 & & 0 & &
\end{array} \tag{3.3}$$

Dans (3.3), $e : \mathbf{C} \to \mathbf{C} \oplus \mathbf{C}$ est le complexe diagonal $f \to (f, f)$ et $e : \mathbf{C} \oplus \mathbf{C} \to \mathbf{C}$ est donné par $(f, g) \to (f - g)$, et $j^1_* \mathbf{C} \oplus j^2_* \mathbf{C} = (j\rho)_* \mathbf{C}$, $e : j_* \mathbf{C} \to j^1_* \mathbf{C} \oplus j^2_* \mathbf{C} = (j\rho)_* \mathbf{C}$ est induit par le morphisme de $\mathcal{O}_Z$-faisceaux sur Z, $\mathbf{C} \to \rho_* \mathbf{C}$ considéré dans (2.1).

Dans (3.3), les trois premières lignes (à partir du bas) sont des suites exactes de faisceaux. Les deux premières lignes correspondent aux complexes de Koszul pour les diviseurs Z, Z^1 et Z^2 dans V. Les deux premières colonnes dans (3.3) sont aussi des suites exactes de faisceaux. La première colonne correspond au complexe de Koszul associée à $\Sigma = Z^1 \cap Z^2$.

Par un argument de suite spectrale, la cohomologie totale du bicomplexe (3.3) est exactement $i_* \mathbf{C}$.

Bien sûr, tous les complexes de $\mathcal{O}_Z$-faisceaux peuvent être tensorisés par ξ. Donc on a un bicomplexe de fibrés holomorphes sur V,

$$\begin{array}{ccccc} & 0 & & 0 & \\ & \uparrow & & \uparrow & \\ 0\to & \xi \longrightarrow & & \xi\to & 0\\ & \mu\uparrow & & \mu\uparrow & \\ 0\to & ([-Z^1]\oplus[-Z^2])\otimes\xi \xrightarrow{v} & & \xi\oplus\xi\to & 0\\ & \mu\uparrow & & \mu\uparrow & \\ 0\to & [-Z]\otimes\xi\xrightarrow{v} & & \xi\to & 0\\ & \uparrow & & \uparrow & \\ & 0 & & 0 & \end{array} \tag{3.4}$$

Dans (3.4), on numérote les lignes de haut en bas par $(\xi_\bullet^i)(i=0,1,2)$ et les colonnes de droite à gauche par $(\xi_i^\bullet)(i=0,1)$. Par (1.8), on sait que

$$(\xi_\bullet^2, v) = (\eta, v)_{|V\times\{s_0\}}. \tag{3.5}$$

b) Comparaison des formes de torsion analytique

Pour $k=1,2$, soit $g^{[Z^k]}$ une métrique hermitienne sur $[Z^k]$. On munit $[Z]$ de la métrique $g^{[Z]}=g^{[Z^1]}\otimes g^{[Z^2]}$. On munit aussi $[Z^1]\oplus[Z^2]$ de la métrique $g^{[Z^1]\oplus[Z^2]}= g^{[Z^1]}\oplus g^{[Z^2]}$. Alors $\Lambda([-Z^k]),\Lambda([-Z]),\Lambda([-Z^1]\oplus[-Z^2])$ sont munis naturels des métriques hermitiennes $g^{\Lambda([-Z^k])},g^{\Lambda([-Z])},g^{\Lambda([-Z^1]\oplus[-Z^2])}$.

Soit $H(X,\xi^\bullet_{\bullet|X})$ l'hypercohomologie de $(\mathcal{O}_X(\xi^\bullet_{\bullet|X}),\mu+v)$. Soient $H(X,\xi^i_{\bullet|X})$ et $H(X,\xi^\bullet_{i|X})$ les hypercohomologies de $(\mathcal{O}_X(\xi^i_{\bullet|X}),v)$ et $(\mathcal{O}_X(\xi^\bullet_{i|X}),\mu)$. Comme dans [**B3**, §3.2], par la théorie de Hodge, on peut équiper des métriques L_2, $h^{H(X,\xi^\bullet_{\bullet|X})}$, $h^{H(X,\xi^i_{\bullet|X})}$ et $h^{H(X,\xi^\bullet_{i|X})}$ $(i=0,1,2)$ sur $H(X,\xi^\bullet_{\bullet|X})$, $H(X,\xi^i_{\bullet|X})$ et $H(X,\xi^\bullet_{i|X})$ sur B associées à $g^{\Lambda([-Z^k])},g^{\Lambda([-Z])},g^{\Lambda([-Z^1]\oplus[-Z^2])}$, h^ξ et g^{TX}. D'après (3.3), sur B, on a

$$\begin{aligned} &H(X,\xi^0_{\bullet|X})=H(X,\xi^\bullet_{0|X})=0,\\ &H^p(X,\xi^1_{\bullet|X})=R^p(\pi j\rho)_*(j\rho)^*\xi),\\ &H^p(X,\xi^2_{\bullet|X})=H^p(X,\eta_{|X})=R^p\pi_*\xi,\\ &H^{p-1}(X,\xi^\bullet_{\bullet|X})=H^p(X,\xi^\bullet_{1|X})=R^p(\pi i)_*i^*\xi,\\ &h^{H(X,\xi^\bullet_{1|X})}=h^{H(X,\xi^\bullet_{\bullet|X})},\qquad h^{H(X,\xi^2_{\bullet|X})}=h^{H(X,\eta_{|X})}. \end{aligned} \tag{3.6}$$

Par [**B3**, §3.2], on peut construire des formes de torsion analytique $T(\omega^V,h^{\xi^\bullet_\bullet})$, $T(\omega^V,h^{\xi^\bullet_i})$ et $T(\omega^V,h^{\xi^i_\bullet})\in P^B$, telles que

$$\frac{\overline{\partial}\partial}{2i\pi}T(\omega^V,h^{\xi^\bullet_\bullet})=\mathrm{ch}(H(X,\xi^\bullet_{\bullet|X}),h^{H(X,\xi^\bullet_{\bullet|X})})-\int_X\mathrm{Td}(TX,g^{TX})\mathrm{ch}(\xi^\bullet_\bullet,h^{\xi^\bullet_\bullet}). \tag{3.7}$$

et les formes $T(\omega^V,h^{\xi^\bullet_i}),T(\omega^V,h^{\xi^i_\bullet})$ vérifient des équations analogues. Remarque aussi $\mathrm{ch}(\xi^i_\bullet,h^{\xi^i_\bullet})=\sum(-1)^j\mathrm{ch}(\xi^i_j,h^{\xi^i_j})$. Alors, par (1.9), on a

$$i^*_{s_0}T(n^*\omega^V,h^\eta)=T(\omega^V,h^{\xi^2_\bullet}). \tag{3.8}$$

On rappelle que le complexe L est défini en (2.1). Soit $h'^{R\pi_*L} = h^{H(X,\xi^2_{\bullet|X})} \oplus h^{H(X,\xi^1_{\bullet|X})} \oplus h^{H(X,\xi^\bullet_{\bullet|X})}$ la métrique sur $R^\bullet\pi_* L = R\pi_*\xi \oplus R(\pi j\rho)_*(j\rho)^*\xi \oplus R(\pi i)_* i^*\xi$. Comme en (2.3), soit $\widetilde{\mathrm{ch}}(R^\bullet\pi_* L, h'^{R\pi_* L}) \in P^B/P^{B,0}$ la classe de Bott-Chern pour le complexe (2.2).

Théorème 3.1. *On a l'identité suivante dans* $P^B/P^{B,0}$,

$$-T(\omega^V, h^{\xi^\bullet_1}) = \sum_{i=1}^{2} (-1)^i T(\omega^V, h^{\xi^i_\bullet}) + \widetilde{\mathrm{ch}}(R^\bullet\pi_* L, h'^{R\pi_* L}). \tag{3.9}$$

Preuve. Pour le bicomplexe $E = (\Omega(X, \xi^\bullet_\bullet), \overline{\partial}^X + v, \mu)$, si on le filtre par μ, alors le premier terme de la suite spectrale associée est $E_1^{p,i} = H^p(X, \xi^i_{\bullet|X})$. D'où le suite spectrale (E_r, d_r) dégénére à E_2. D'après l'argument de [**Ma1**, Théorème 11.1], [**B3**, Théorème 0.2], pour le complexe (3.4), dans $P^B/P^{B,0}$, on a

$$\begin{aligned} T(\omega^V, h^{\xi^\bullet_\bullet}) &= \textstyle\sum_{i=0}^{2} (-1)^i T(\omega^V, h^{\xi^i_\bullet}) + \widetilde{\mathrm{ch}}(R^\bullet\pi_* L, h'^{R\pi_* L}), \\ T(\omega^V, h^{\xi^\bullet_\bullet}) &= \textstyle\sum_{i=0}^{1} (-1)^i T(\omega^V, h^{\xi^\bullet_i}). \end{aligned} \tag{3.10}$$

On vérifie facilement qu'on a

$$T(\omega^V, h^{\xi^\bullet_0}) = 0, \qquad T(\omega^V, h^{\xi^0_\bullet}) = 0. \tag{3.11}$$

Par (3.10), (3.11), on a (3.9). □

c) Preuve du Théorème 2.1

Soit

$$\begin{aligned} I_{\lambda'} &= T(\omega^V, h^{\xi^\bullet_1}) - T(\omega^\Sigma, h^\xi) + \widetilde{\mathrm{ch}}\Big(H(X, \xi^\bullet_{1|X}), h^{H(X,\xi^\bullet_{1|X})}, h^{R(\pi i)_*(i^*\xi)}\Big), \\ I_{\lambda''} &= T(\omega^V, h^{\xi^1_\bullet}) - T(\omega^{\widehat{Z}}, h^\xi) + \widetilde{\mathrm{ch}}\Big(H(X, \xi^1_{\bullet|X}), h^{H(X,\xi^1_{\bullet|X})}, h^{R(\pi j\rho)_*(j\rho)^*\xi}\Big). \end{aligned} \tag{3.12}$$

Alors en utilisant [**B3**, Théorème 0.1] et en procédant comme en [**B2**, Théorème 6.1], on sait que dans $P^B/P^{B,0}$, $I_{\lambda'}, I_{\lambda''}$ sont égaux formellement aux mêmes formules que les termes à droite de [**B2**, (6.11)].

On rappelle que I_λ est défini en (1.17). Par [**BGS1**, Théorème 1.20], [**Ma1**, (10.7)], dans $P^B/P^{B,0}$, on a

$$\begin{aligned} &\widetilde{\mathrm{ch}}(R^\bullet\pi_* L, h'^{R\pi_* L}) - \widetilde{\mathrm{ch}}(R^\bullet\pi_* L, h^{R\pi_* L}) \\ &= \widetilde{\mathrm{ch}}\Big(R(\pi i)_* i^*\xi, h^{H(X,\xi^\bullet_{1|X})}, h^{R(\pi i)_*(i^*\xi)}\Big) \\ &-\widetilde{\mathrm{ch}}\Big(R(\pi j\rho)_*(j\rho)^*\xi, h^{H(X,\xi^1_{\bullet|X})}, h^{R(\pi j\rho)_*(j\rho)^*\xi}\Big) \\ &+\widetilde{\mathrm{ch}}(R^\bullet\pi_*\xi, h^{H(X,\xi^2_{\bullet|X})}, h'^{R\pi_*\xi}). \end{aligned} \tag{3.13}$$

Par (1.10), (1.15), (3.7), (3.9) et (3.13), dans $P^B/P^{B,0}$, on a

$$T'(\omega^Z, h^\xi, h'^{R\pi_*\xi}) - T(\omega^Z, h^\xi, h'^{R\pi_*\xi}) = I_\lambda + I_{\lambda'} - I_{\lambda''}. \tag{3.14}$$

Par (3.14), en procédant comme dans [**B2**, §6 et §7], on a le Théorème 2.1. □

4. Preuve du Théorème 2.1: le cas général

Dans cette section, on montre le Théorème 2.1 dans le cas général. On ne suppose donc plus que $Z = Z^1 \cup_\Sigma Z^2$. L'idée principale de la preuve est d'utiliser la technique de la déformation du cône normal [**BaFM**], [**BGS4**]. Elle réduit notre problème à un problème analytique naturellement localisé sur $P(N_{\Sigma/V} \oplus 1)$, auquel la technique de [**B2**, §4 et §6] peut être appliquée.

Cette section est organisée de la façon suivante. Dans a), soit W l'éclaté de $V \times \mathbf{P}^1$ le long de $\Sigma \times \{\infty\}$, on relève le diviseur Z à un diviseur D dans W. Dans b), on calcule l'asymptotique des formes de torsion analytique sur $B \times \mathbf{P}^1$. Dans c), en utilisant les résultats de la Section 3, on sait que le problème peut être localisé sur $P(N_{\Sigma/V} \oplus 1)$. Et on établit un résultat analogue de [**B2**, Théorème 9.16]. Dans d), on montre le Théorème 2.1 dans le cas général.

Pour simplifier les notations, on suppose toujours que $\Delta = \{s_0\}$. On utilise les même notations qu'aux Sections 1 et 2.

a) Diviseur D

Soit W l'éclaté de $V \times \mathbf{P}^1$ le long de $\Sigma \times \{\infty\}$. Soit p, q les projections $p : W \to V$, $q : W \to \mathbf{P}^1$. Pour $z \in \mathbf{P}^1$, on pose

$$W_z = q^{-1}(z). \tag{4.1}$$

Si $z \neq \infty$, l'application $p : W_z \to V$ est une identification de variétés complexes. Alors $p^* Z$ est un diviseur sur W.

Définition 4.1. *Soit D le diviseur sur W*

$$D = p^* Z - 2P(N_{\Sigma/V} \oplus 1). \tag{4.2}$$

Par [**B3**, Proposition 9.2], D est un diviseur effectif dans W, transversaux avec W_∞, et à croisements normaux le long de $\Sigma \times \mathbf{P}^1$. On vérifie facilement que D est exactement l'éclaté de $Z \times \mathbf{P}^1$ le long de $\Sigma \times \{\infty\}$. En particulier, $W_\infty \cap D$ est l'union de la normalisation $\widehat{Z} = V_\Sigma \cap D$ de Z et de $P(N_{\Sigma/V} \oplus 1) \cap D$ qui est un diviseur aux croisements normaux $P(N_{\widehat{\Sigma}/\widehat{Z}} \oplus 1)$ dans $P(N_{\Sigma/V} \oplus 1)$, et ils sont transversaux le long de $\widehat{\Sigma}$.

Soit i_D l'immersion $D \to W$. Soit p_D, q_D les restrictions de p, q à D. Alors $p_D(D) \subset Z$. On note $\overline{q} = (\pi_1 \circ p, q) : W \to B \times \mathbf{P}^1$, $\overline{q}_D = \overline{q}_{|D}$. Pour $s \in \mathbf{P}^1$, on note aussi $D_z = p_D^{-1}(z)$. Soit D_B la fibre de la fibration $\pi \circ p : D \to B \times \{s_0\}$. Soit $H(Z_B, j^*\xi_{|Z_B})$ la cohomologie de ξ le long des fibres Z_B, alors $H(Z_B, j^*\xi_{|Z_B}) = R\pi_* \xi_{|B \times \{s_0\}}$. Par [**B2**, Proposition 9.3], on a

Proposition 4.2. *L'application linéaire $\alpha \in H(Z_B, j^*\xi_{|Z_B}) \to p_D^* \alpha \in H(D_B, (jp_D)^*\xi_{|D_B})$ induit une trivialisation de $R^\bullet \overline{q}_{D*}(jp_D)^*\xi$ sur $\mathbf{P}^1$.*

b) L'asymptotique des formes de torsion analytique sur $\mathbf{P}^1$

Soit g^{TS} une métrique sur TS. Soit $\sigma_{[\infty]}$ la section canonique de $[\infty]$ sur $\mathbf{P}^1$, soit $g^{[\infty]}$ la métrique canonique sur $[\infty] = \mathcal{O}(-1)$.

On rappelle que par (1.5), $[Z]_{|Z} = \pi_2^* TS$. Soit $g^{[Z]}$ une métrique hermitienne sur $[Z]$ telle que

$$([Z], g^{[Z]})_{|Z} = \pi_2^*(TS, g^{TS}). \tag{4.3}$$

Évidemment

$$[P(N_{\Sigma/V} \oplus 1)]_{|\Sigma \times P^1} = q^*[\infty]_{|\Sigma \times P^1}. \tag{4.4}$$

Soit $g^{[P(N_{\Sigma/V}\oplus 1)]}$ la métrique sur $[P(N_{\Sigma/V} \oplus 1)]$ sur W telle que

$$\Big([P(N_{\Sigma/V} \oplus 1)], g^{[P(N_{\Sigma/V}\oplus 1)]}\Big)_{|\Sigma \times P^1} = q^*([\infty], g^{[\infty]}). \tag{4.5}$$

Par (4.2), on a

$$[D] = q^*[Z] \otimes [P(N_{\Sigma/V} \oplus 1)]^{-2}. \tag{4.6}$$

Soit $g^{[D]}$ la métrique sur $[D]$ induite par $g^{[Z]}$ et $g^{[P(N_{\Sigma/V}\oplus 1)]}$. Par (4.2)–(4.6), on a

$$([D], g^{[D]})_{|\Sigma \times P^1} = \Big((\pi_2 p)^* TS \otimes q^*[-2\infty], (\pi_2 p)^* g^{TS} \otimes q^* g^{[-2\infty]}\Big)_{|\Sigma \times P^1}. \tag{4.7}$$

On rappelle aussi que le complexe

$$(\xi_D, v) : 0 \to \mathcal{O}_W(p^*\xi \otimes [-D]) \overset{\sigma_{[D]}}{\to} \mathcal{O}_W(p^*\xi) \to 0. \tag{4.8}$$

est une résolution canonique de $i_{D*}\mathcal{O}_D((jp_D)^*\xi)$. Soit $H(X, \xi_{D|X})$ l'hypercohomologie de $(\mathcal{O}_X(\xi_D), v)$, alors par (4.8), $H(X, \xi_{D|X}) \simeq R\overline{q}_{D*}(jp_D)^*\xi$.

Comme V est Kählérienne, W est aussi Kählérienne. Soit g^{TW} une métrique Kählérienne sur TW. Alors g^{TW} induit une métrique Kählérienne g^{TX} sur TX, et on munit les métriques hermitiennes associées sur les sous-variétés de W. Soit h^ξ une métrique hermitienne sur ξ.

Soit $\widetilde{\pi} : W \to B \times S \times \mathbf{P}^1$ (resp. $\overline{\pi} : W \to B \times \mathbf{P}^1$) définie par

$$\widetilde{\pi} = (\pi_1 p, \pi_2 p, q) \quad (\text{resp. } \overline{\pi} = (\pi_1 p, q)). \tag{4.9}$$

Plus précisément, $\widetilde{\pi} : W \setminus (W_\infty \cup Z \times \mathbf{P}^1) \to B \times (S \setminus \Delta) \times (\mathbf{P}^1 \setminus \{\infty\})$ est une submersion de fibre compacte Y. Soit $p_{B\times S} : B \times S^1 \times \mathbf{P}^1 \to B \times S$ la projection naturelle. Par la Proposition 4.2, $R^\bullet\widetilde{\pi}_*\xi$ est localement libre sur $B \times S \times \mathbf{P}^1$, et $R^\bullet\widetilde{\pi}_*\xi = p_{B\times S}^* R\pi_*\xi$. Soit $h^{R\widetilde{\pi}_*\xi}$ la métrique L_2 sur $R\widetilde{\pi}_*\xi$ sur $B\times(S\setminus\Delta)\times(\mathbf{P}^1\setminus\{\infty\})$ associée à h^ξ, g^{TW}. Soit $h'^{R\widetilde{\pi}_*\xi}$ une métrique hermitienne sur $R\widetilde{\pi}_*\xi$ sur $B \times S \times \mathbf{P}^1$.

Soit $h^{H(X,\xi_{D|X})}$ la métrique L_2 sur $H(X, \xi_{D|X})$ associée aux métriques h^ξ, $h^{[D]}$, g^{TX} définie par [**B3**, §3.2] sur $B \times (\mathbf{P}^1 \setminus \{\infty\})$. En remplaçant formellement V par W, π par $\widetilde{\pi}$, ξ par $p^*\xi$ à la Section 1, soient

$$\begin{aligned} T(\omega^W, h^\xi) &\in P^{B\times(S\setminus\Delta)\times(\mathbf{P}^1\setminus\{\infty\})}/P^{B\times(S\setminus\Delta)\times(\mathbf{P}^1\setminus\{\infty\}),0}, \\ T(\omega^W, h^{\xi_D}) &\in P^{B\times(\mathbf{P}^1\setminus\{\infty\})}/P^{B\times(\mathbf{P}^1\setminus\{\infty\}),0} \end{aligned}$$

les formes de torsion analytique associées à $(\widetilde{\pi}, \omega^W, h^\xi)$, $(\overline{\pi}, \omega^W, h^{\xi_D})$ pour (4.8) [**BK**], [**B3**, §3.2], en particulier,

$$\frac{\overline{\partial}\partial}{2i\pi} T(\omega^W, h^{\xi_D}) = \mathrm{ch}(H(X, \xi_{D|X}), h^{H(X,\xi_{D|X})}) - \int_X \mathrm{Td}(TX, g^{TX}) \mathrm{ch}(\xi_D, h^{\xi_D}).$$

et

$$\mathrm{ch}(\xi_D, h^{\xi_D}) = \mathrm{ch}(p^*\xi, h^{p^*\xi}) - \mathrm{ch}(p^*\xi \otimes [-D], h^{p^*\xi\otimes[-D]}).$$

Pour $z \in \mathbf{P}^1, s \in S$, soient $\tau_{s,z} : B \to B \times S \times \mathbf{P}^1$ et $\tau_z : B \to B \times \mathbf{P}^1$ les applications de $y \in B$ à $(y, s, z) \in B \times S \times \mathbf{P}^1$ et $(y, z) \in B \times \mathbf{P}^1$.

Définition 4.3. *Pour $z \in \mathbf{P}^1 \setminus \{\infty\}$, $s_0 \in \Delta$, dans $P^B/P^{B,0}$, on pose*

$$\begin{aligned} T_z(\omega^D, h^\xi, h'^{R\widetilde{\pi}_*\xi}) = \lim_{S\setminus\Delta\ni s\to s_0} \Big\{ \tau_{s,z}^* \Big[T(\omega^W, h^\xi) + \widetilde{\mathrm{ch}}(R^\bullet\widetilde{\pi}_*\xi, h^{R\widetilde{\pi}_*\xi}, h'^{R\widetilde{\pi}_*\xi}) \Big] \\ + \frac{1}{2}\log(\|\sigma_{[\Delta]}\|_s^2 \|\sigma_{[\infty]}\|_z^{-4}) \int_{\Sigma_B} \mathrm{Td}(T\Sigma_B) E(N_{\Sigma/V}) \mathrm{ch}(\xi) \Big\}. \end{aligned} \tag{4.10}$$

En effet, pour $z \in \mathbf{P}^1 \setminus \{\infty\}$, la Définition 4.3 correspond exactement au Théorème 1.3, dans lequel V est remplacé par W_z. Comme en (1.17), $I_{\lambda,z}$ est bien défini, et dans $P^B/P^{B,0}$, on a

$$I_{\lambda,z} = \tau_z^* T(\omega^W, h^{\xi_D}) - T_z(\omega^D, h^\xi, h'^{R\widetilde{\pi}_*\xi}) - \tau_z^* \widetilde{\mathrm{ch}}(R\widetilde{\pi}_*\xi, h'^{R\widetilde{\pi}_*\xi}, h^{H(X,\xi_{D|X})}). \tag{4.11}$$

En procédant comme dans [**B2**, Théorème 9.9], on a

Théorème 4.4. *La fonction $z \in \mathbf{P}^1 \setminus \{\infty\} \to I_{\lambda,z} \in P^B/P^{B,0}$ se prolonge en une fonction continue sur $\mathbf{P}^1$.*

Remarque 4.5. Si on note $I_{\lambda,\infty} = \lim_{z\to\infty} I_{\lambda,z} \in P^B/P^{B,0}$, alors $I_{\lambda,\infty}$ a une formule analogue que le terme à droite de [**B2**, (9.37)].

On rappelle que U est le fibré en droite universel sur $P(N_{\Sigma/V})$.

Théorème 4.6. *Quand $z \to \infty$*

$$\lim_{z\to\infty} \Big\{ T_z(\omega^D, h^\xi, h'^{R\widetilde{\pi}_*\xi}) - \log|z|^2 [\int_{\widehat{\Sigma}_B} \mathrm{Td}(T\widehat{\Sigma}_B) E(U) \mathrm{ch}(\xi)] \Big\} \tag{4.12}$$

a une limite dans $P^B/P^{B,0}$. On note cette limite comme $T_\infty(\omega^D, h^\xi, h'^{R\widetilde{\pi}_*\xi}) \in P^B/P^{B,0}$.

Preuve. On rappelle que la fibration $W \to B \times \mathbf{P}^1$ vérifie les hypothèses de la Section 1. Comme $H(X, \xi_{D|X})$ est localement libre sur $B \times \mathbf{P}^1$, par les Théorèmes 1.2 et 1.3 et [**B2**, (9.47)], pour la fibration $\overline{q} : W \to B \times \mathbf{P}^1$, on a

$$\begin{aligned} \lim_{z\to\infty} \Big\{ \tau_z^* T(\omega^W, h^{\xi_D}) - \tau_z^* \widetilde{\mathrm{ch}}(R\widetilde{\pi}_*\xi, h'^{R\widetilde{\pi}_*\xi}, h^{H(X,\xi_{D|X})}) \\ - \log|z|^2 [\int_{\widehat{\Sigma}_B} \mathrm{Td}(T\widehat{\Sigma}_B) E(U) \mathrm{ch}(\xi)] \Big\} \end{aligned}$$

existe dans $P^B/P^{B,0}$. Par le Théorème 4.4, on a le Théorème 4.6. □

Pour la fibration $\widetilde{\pi} : W \setminus W_\infty \to B \times S \times (\mathbf{P}^1 \setminus \{\infty\})$, $Z \times (\mathbf{P}^1 \setminus \{\infty\})$ sont les fibres singulières, comme en (2.4), sur $B \times \mathbf{P}^1 \setminus \{\infty\}$, on peut définir

$$T'(\omega^D, h^\xi, h'^{R\widetilde{\pi}_*\xi}) \in P^{B\times(P^1\setminus\{\infty\})}/P^{B\times(P^1\setminus\{\infty\}),0}.$$

Théorème 4.7. *Quand* $z \to \infty$

$$\lim_{z\to\infty} \Big\{\tau_z^* T'(\omega^D, h^\xi, h'^{R\widetilde{\pi}_*\xi}) - \log|z|^2 [\int_{\widehat{\Sigma}_B} \mathrm{Td}(T\widehat{\Sigma}_B) E(U) \mathrm{ch}(\xi)]\Big\} \tag{4.13}$$

a une limite dans $P^B/P^{B,0}$, *qu'on le note comme* $T'_\infty(\omega^D, h^\xi, h'^{R\widetilde{\pi}_*\xi}) \in P^B/P^{B,0}$.

Preuve. Comme en [**B2**, (9.40)],

$$N_{\widehat{\Sigma}_\infty/D} = N_{P(N_{\Sigma/V})/W}|_{\widehat{\Sigma}_\infty} = (U \oplus U^{-1})|_{\widehat{\Sigma}}.$$

Par les Théorèmes 1.2 et 1.3, on a le Théorème 4.7. □

Remarque 4.8. Pour $(\tau_z^* T' - T_z)(\omega^D, h^\xi, h'^{R\widetilde{\pi}_*\xi})$, on a aussi un résultat analogue que [**B2**, Théorème 9.12].

c) Localisation du calcul à ∞ sur $P(N_{\Sigma/V} \oplus 1)$

On a $P(N_{\Sigma/V}) \subset P(N_{\Sigma/V} \oplus 1) \subset W_\infty$, et $\widehat{\Sigma}_\infty = D \cap P(N_{\Sigma/V})$. Soit

$$D' = D \cap P(N_{\Sigma/V} \oplus 1). \tag{4.14}$$

Alors D' est un diviseur à croisements normaux le long de Σ dans $P(N_{\Sigma/V} \oplus 1)$, et $\Sigma \subset D'$ est l'ensemble des points singuliers dans D'. Soit j' l'immersion $D' \to P(N_{\Sigma/V} \oplus 1)$. Soit $j'_1 : D' \to D_\infty = W_\infty \cap D$ l'immersion naturelle.

Par §4a), $\widehat{\Sigma}_\infty$ s'identifie à $\widehat{\Sigma}$ dans §1a), et $\rho : \widehat{\Sigma}_\infty \to \Sigma$ est un revêtement double et $\kappa = \det \rho_! \mathcal{O}_{\widehat{\Sigma}}$. Soit $p_\Sigma : P(N_{\Sigma/V} \oplus 1) \to \Sigma$ la projection canonique. Sur Σ, on a une suite exacte de faisceaux

$$(K, v) : \quad 0 \to \mathcal{O}_\Sigma \xrightarrow{e} \rho_* \mathcal{O}_{\widehat{\Sigma}} \xrightarrow{f} \mathcal{O}_\Sigma(\kappa) \to 0. \tag{4.15}$$

En (4.15), e est l'application diagonale $x \to (x, x)$, f est l'application de différence $(x, y) \to x - y$. Rappelle que $i : \Sigma \to V$ est l'immersion naturelle. Par (4.15), on a un complexe acyclique de fibrés holomorphes sur B

$$(R(\pi i)_* K, v) : \cdots \to H^j(\widehat{\Sigma}_B, (\pi i \rho)^* \xi) \to H^j(\Sigma_B, i^*\xi \otimes \kappa) \to H^{j+1}(\Sigma_B, i^*\xi) \to \cdots. \tag{4.16}$$

Soit k l'immersion $\Sigma \to D'$, soit $\widehat{D}'$ la normalisation de D'. Alors $\widehat{D}'$ est l'éclaté de D' le long de Σ. Soit $\rho' : \widehat{D}' \to D'$ la projection canonique. Alors on a une suite exacte de faisceaux

$$0 \to \mathcal{O}_{D'} \to \rho'_* \mathcal{O}_{\widehat{D}'} \to k_* \mathcal{O}_\Sigma(\kappa) \to 0. \tag{4.17}$$

Sur $P(N_{\Sigma/V} \oplus 1) \subset W_\infty$, le complexe

$$(\xi_{D'}, v) : 0 \to \mathcal{O}_{P(N_{\Sigma/V}\oplus 1)}([-D'] \otimes (ip)^*\xi) \xrightarrow{\sigma_{[D']}} \mathcal{O}_{P(N_{\Sigma/V}\oplus 1)}((ip)^*\xi) \to 0. \tag{4.18}$$

est une résolution canonique de $j'_* \mathcal{O}_{D'}((ipj')^*\xi)$ sur $P(N_{\Sigma/V} \oplus 1)$, et $(\xi_{D'}, v)$ est exactement la restriction du complexe (ξ_D, v) en (4.8) à $P(N_{\Sigma/V} \oplus 1)$.

On note $D'_B, \widehat{D}'_B, P(N_{\Sigma/V} \oplus 1)_B, P(N_{\Sigma/V})_B, V_{\Sigma|B}$ les fibres des fibrations de $D', \widehat{D}'$, $P(N_{\Sigma/V} \oplus 1)$, $P(N_{\Sigma/V})$, V_Σ à B. Par un argument de suite spectrale, on sait que [**B3**, §9e)]

$$\begin{aligned} &H(D'_B, (ipj')^*\xi) = H(\Sigma_B, i^*\xi),\\ &H(\widehat{D}'_B, (ipj'\rho')^*\xi) = H(\widehat{\Sigma}_B, (i\rho')^*\xi),\\ &H(P(N_{\Sigma/V} \oplus 1)_B, [-D'] \otimes (ip)^*\xi) = 0,\\ &H(P(N_{\Sigma/V} \oplus 1)_B, (ip)^*\xi) = H(\Sigma_B, i^*\xi). \end{aligned} \tag{4.19}$$

Plus précisément, on vérifie facilement que, sous l'isomorphisme canonique de (4.19), les suites exactes longues de cohomologie associées à (4.15), (4.17) sont (4.16).

On rappelle que Σ et $P(N_{\Sigma/V}\oplus 1)$ ont été équipés naturellement des métriques induites par g^{TW} comme dans §4a). La métrique g^{TW} induit aussi une métrique sur $T\widehat{D}'$. De même, tous les fibrés holomorphes considérés $(ip)^*\xi, [-D'], \ldots$ sont déjà équipés des métriques hermitiennes induites, par exemple, $g^{TP(N_{\Sigma/V}\oplus 1)_B}, g^{T\widehat{D}'_B}$ sont les métriquess induites sur les fibrés tangents relatifs $TP(N_{\Sigma/V} \oplus 1)_B$, $T\widehat{D}'_B$.

Soit $H(P(N_{\Sigma/V} \oplus 1)_B, \xi_{D'})$ l'hypercohomologie de $(\mathcal{O}_{P(N_{\Sigma/V}\oplus 1)_B}(\xi_{D'}), v)$. Par (4.18), on sait que

$$H(P(N_{\Sigma/V} \oplus 1)_B, \xi_{D'}) = H(D'_B, (ipj')^*\xi) = H(\Sigma_B, i^*\xi). \tag{4.20}$$

Soit $h^{H(P(N_{\Sigma/V}\oplus 1)_B, \xi_{D'})}$, $h^{H(\widehat{D}'_B, (ipj'\rho')^*\xi)}$ les métriques L_2 sur $H(\Sigma_B, i^*\xi)$, $H(\widehat{\Sigma}_B, (i\rho')^*\xi)$ sur B induites par $(g^{P(N_{\Sigma/V}\oplus 1)_B}, h^{\xi_D})$, $(g^{T\widehat{D}'_B}, h^\xi)$ [**B3**, §3.2]. Rappelle que $h^{R(\pi i)_*(i^*\xi\otimes\kappa)}$ est la métrique L_2 sur $R(\pi i)_*(i^*\xi \otimes \kappa)$ définie à §1b). On munit $R^\bullet(\pi i)_*K$ de la métrique $h^{R(\pi i)_*K} = h^{H(P(N_{\Sigma/V}\oplus 1)_B, \xi_{D'})} \oplus h^{H(\widehat{D}'_B, (ipj'\rho')^*\xi)} \oplus h^{R(\pi i)_*(i^*\xi\otimes\kappa)}$. Alors par (4.16), il existe une classe de Bott-Chern $\widetilde{\mathrm{ch}}(R^\bullet(\pi i)_*K, h^{R(\pi i)_*K}) \in P^B/P^{B,0}$ [**BGS1**, §1f)] telle que

$$\begin{aligned} \frac{\bar\partial\partial}{2i\pi}\widetilde{\mathrm{ch}}(R^\bullet(\pi i)_*K, h^{R(\pi i)_*K}) =& \mathrm{ch}(H(\widehat{\Sigma}_B, (i\rho')^*\xi), h^{H(\widehat{D}'_B, (ipj'\rho')^*\xi)})\\ &- \mathrm{ch}(H(\Sigma_B, i^*\xi), h^{H(P(N_{\Sigma/V}\oplus 1)_B, \xi_{D'})})\\ &- \mathrm{ch}(R^\bullet(\pi i)_*(i^*\zeta \otimes \kappa), h^{R(\pi i)_*(i^*\xi\otimes\kappa)}). \end{aligned} \tag{4.21}$$

Soit $T(\omega^{P(N_{\Sigma/V}\oplus 1)}, h^{\xi_{D'}}) \in P^B/P^{B,0}$ les formes de torsion analytique associées au complexe (4.18) et $(\pi i p_\Sigma, h^{\xi_D})$ [**B3**, §3.2]. Soit $T(\omega^{\widehat{D}'}, h^\xi) \in P^B/P^{B,0}$ les formes de torsion analytique associées à $(ipj'\rho' : \widehat{D}' \to B \times \{s_0\}, \omega^{\widehat{D}'}, h^\xi)$. Soit

$$\begin{aligned} I_{\lambda(ipj')} = &-T(\omega^{P(N_{\Sigma/V}\oplus 1)}, h^{\xi_{D'}}) + T(\omega^{\widehat{D}'}, h^\xi)\\ &-T(\omega^\Sigma, h^{i^*\xi\otimes\kappa}) - \widetilde{\mathrm{ch}}(R^\bullet(\pi i)_*K, h^{R(\pi i)_*K}). \end{aligned} \tag{4.22}$$

Rappelle que V_Σ est l'éclaté de V le long de Σ. Par la Section 4a) et (4.8), le complexe

$$(\xi_{D|V_\Sigma}, v) : 0 \to (p^*\xi \otimes [-D])_{|V_\Sigma} \overset{\sigma_{[D]}}{\to} p^*\xi_{|V_\Sigma} \to 0. \tag{4.23}$$

est une résolution canonique de $\mathcal{O}_{\widehat{Z}}(p^*\xi)$ sur V_Σ. Le complexe

$$(\xi_{D|P(N_{\Sigma/V})}, v) : 0 \to (p^*\xi \otimes [-D])_{|P(N_{\Sigma/V})} \overset{\sigma_{[D]}}{\to} p^*\xi_{|P(N_{\Sigma/V})} \to 0. \tag{4.24}$$

est une résolution canonique de $\mathcal{O}_{\widehat{\Sigma}}(p^*\xi)$ sur $P(N_{\Sigma/V})$.

Soient $H(V_{\Sigma|B}, \xi_D)$ et $H(P(N_{\Sigma/V})_B, \xi_D)$ les hypercohomologies de

$$(\mathcal{O}_{V_{\Sigma|B}}(\xi_{D|V_\Sigma}), v), \ (\mathcal{O}_{P(N_{\Sigma/V})_B}(\xi_{D|P(N_{\Sigma/V})}), v).$$

Par (4.23) et (4.24), on a

$$\begin{aligned} H(V_{\Sigma|B}, \xi_D) &= H(\widehat{Z}_B, (j\rho)^*\xi) = R(\pi j \rho)_*(j\rho)^*\xi, \\ H(P(N_{\Sigma/V})_B, \xi_D) &= H(\widehat{\Sigma}_B, (i\rho)^*\xi). \end{aligned} \tag{4.25}$$

Soient $h^{H(V_{\Sigma B}, \xi_D)}$, $h^{H(P(N_{\Sigma/V})_B, \xi_D)}$ les métriques L_2 associées sur $H(\widehat{Z}_B, (j\rho)^*\xi)$, $H(\widehat{\Sigma}_B, (i\rho)^*\xi)$ sur B [**B3**, §3.2].

On rappelle que le complexe $(R^\bullet\pi_*L, v)$ est défini en (2.2). On munit $R^\bullet\pi_*L$ de la métrique

$$h'^{R\pi_*L} = \oplus h'^{R\widetilde{\pi}_*\xi} \oplus h^{H(V_{\Sigma|B}, \xi_D)} \oplus h^{R(\pi i)_*(i^*\xi \otimes \kappa)}.$$

Soit $\widetilde{\mathrm{ch}}(R^\bullet\pi_*L, h'^{R\pi_*L}) \in P^B/P^{B,0}$ la classe de Bott-Chern [**BGS1**] telle qu'elle vérifie une équation analogue de (2.3).

On munit $R^\bullet(\pi i)_*K$ de la métrique

$$h'^{R(\pi i)_*K} = \oplus h^{H(P(N_{\Sigma/V} \oplus 1)_B, \xi_{D'})} \oplus h^{H(V_{\Sigma|B}, \xi_D)} \oplus h^{R(\pi i)_*(i^*\xi \otimes \kappa)}.$$

Comme en (4.21), on l'associe une classe de Bott-Chern $\widetilde{\mathrm{ch}}(R^\bullet(\pi i)_*K, h'^{R(\pi i)_*K}) \in P^B/P^{B,0}$. On a aussi [**B2**, (9.35)]

$$N_{P(N_{\Sigma/V})/W} = U \oplus U^{-1}. \tag{4.26}$$

Soit $W_{P(N_{\Sigma/V})}$ l'éclaté de W le long de $P(N_{\Sigma/V}) \subset W_\infty$ et $p_{P(N_{\Sigma/V})} : W_{P(N_{\Sigma/V})} \to W$ la projection correspondante. Alors, $\widehat{W}_\infty$ la normalisation de W_∞ est un diviseur lisse dans $W_{P(N_{\Sigma/V})}$, qui sont transversaux le long de $P(U \oplus U^{-1})$.

Soient $\widehat{D}, \widehat{D}_\infty$ les normalisations de D, D_∞. Soit $\rho_D : \widehat{D} \to D$, $\widehat{\rho} : \widehat{D}_\infty \to D_\infty$ les projections naturelles. Soit $i_1 : \Sigma \to D_\infty$ l'immersion naturelle.

Sur D_∞, on a un bicomplexe de faisceaux

$$\begin{array}{ccccccc} & & 0 & & 0 & & 0 \\ & & \downarrow & & \downarrow & & \downarrow \\ 0 \to & & \mathcal{O}_{W_\infty|D_\infty} & \xrightarrow{v_1} & \rho_{D*}\mathcal{O}_{\widehat{D}|D_\infty} & \xrightarrow{v_2} & i_{1*}\rho_{D*}\mathcal{O}_{\widehat{\Sigma}} \to 0 \\ & & \mu_1\downarrow & & \mu_1\downarrow & & \mu_1\downarrow \\ 0 \to & & p_{P(N_{\Sigma/V})*}\mathcal{O}_{\widehat{W}_\infty|D_\infty} & \xrightarrow{v_1} & \widehat{\rho}_*\mathcal{O}_{\widehat{D}_\infty} & \xrightarrow{v_2} & i_{1*}\rho_{D*}\mathcal{O}_{\widehat{\Sigma}} \to 0 \\ & & \mu_2\downarrow & & \mu_2\downarrow & & \downarrow \\ 0 \to & & i_{1*}\rho_*\mathcal{O}_{\widehat{\Sigma}_\infty} & \xrightarrow{v_1} & i_{1*}\rho_*\mathcal{O}_{\widehat{\Sigma}_\infty} & \to & 0 \to 0 \\ & & \downarrow & & \downarrow & & \downarrow \\ & & 0 & & 0 & & 0 \end{array} \tag{4.27}$$

Dans (4.27), on a

$$\begin{aligned} p_{P(N_{\Sigma/V})*}\mathcal{O}_{\widehat{W}_\infty|D_\infty} &= \mathcal{O}_{D_\infty \cap V_\Sigma} \oplus \mathcal{O}_{D'}, \\ \widehat{\rho}_*\mathcal{O}_{\widehat{D}_\infty} &= \mathcal{O}_{D_\infty \cap V_\Sigma} \oplus \rho'_*\mathcal{O}_{\widehat{D}'}. \end{aligned} \tag{4.28}$$

Les v_1, μ_1 sont des morphismes naturelles, v_2, μ_2 sont des morphismes de différence. De plus, comme $W_\infty = V_\Sigma \cup_{P(N_{\Sigma/V})} P(N_{\Sigma/V} \oplus 1)$, les v_1, μ_1 des troisièmes lignes et colonnes sont des identités.

Après avoir tensorisé (4.27) par $p^*\xi$, on note les lignes de haut en bas par $L^i_\bullet$ $(i = 0, 1, 2)$, et les colonnes de gauche à droite par $L^\bullet_i$ $(i = 0, 1, 2)$. On note aussi $R\pi_* L^i_\bullet$, $R\pi_* L^\bullet_i$ $(i = 0, 1, 2)$ les suites exactes longues de la cohomologie de $L^i_{\bullet|D_\infty|B}$, $L^\bullet_{i|D_\infty|B}$. Par (4.27), on a un bicomplexe de fibrés vectoriels sur B

$$\begin{array}{cccccccc}
 & & 0 & & 0 & & 0 & \\
 & & \downarrow & & \downarrow & & \downarrow & \\
\cdots \to & & H^i(D_{\infty|B}, p^*\xi) & \xrightarrow{v_1} & X_i & \xrightarrow{v_2} & H^i(\Sigma_{|B}, i^*\xi \otimes \kappa) & \to \cdots \\
 & & \mu_1 \downarrow & & \mu_1 \downarrow & & \mu_1 \downarrow & \\
\cdots \to & & H^i(D_{\infty|B}, p_{P(N_{\Sigma/V})*}\mathcal{O}_{\widehat{W}_\infty} \otimes p^*\xi) & \xrightarrow{v_1} & Y_i & \xrightarrow{v_2} & H^i(\Sigma_{|B}, i^*\xi \otimes \kappa) & \to \cdots \\
 & & \mu_2 \downarrow & & \mu_2 \downarrow & & \mu_2 \downarrow & \\
\cdots \to & & H^i(\Sigma_{|B}, i^*\xi \otimes \kappa) & \xrightarrow{v_1} & Z_i & \xrightarrow{v_2} & 0 & \to \cdots
\end{array} \tag{4.29}$$

(avec $X_i = H^i(D_{\infty|B}, \rho_{D*}\mathcal{O}_{\widehat{D}} \otimes p^*\xi)$, $Y_i = H^i(D_{\infty|B}, \widehat{\rho}_*\mathcal{O}_{\widehat{D}_\infty} \otimes p^*\xi)$, $Z_i = H^i(\Sigma_{|B}, i^*\xi \otimes \kappa))$ dont les lignes sont exactes.

Soit $h'^{R\pi_* L^i_j}$ $(i, j = 0, 1, 2)$ une métrique hermitienne sur $R^\bullet\pi_* L^i_j$ sur B. Alors elle induit des métriques $h'^{R\pi_* L^i_\bullet}$ et $h'^{R\pi_* L^\bullet_j}$ sur $R^\bullet\pi_* L^i_\bullet$ et $R^\bullet\pi_* L^\bullet_j$. Soit $\widetilde{\mathrm{ch}}(R^\bullet\pi_* L^i_\bullet, h'^{R\pi_* L^i_\bullet})$ et $\widetilde{\mathrm{ch}}(R^\bullet\pi_* L^\bullet_j, h'^{R\pi_* L^\bullet_j}) \in P^B/P^{B,0}$ les classes de Bott-Chern associées [**BGS1**, §1f)] qui vérifient des équations analogues à (4.21). Par (4.27), (4.29), d'après un argument de suite spectrale, en utilisant [**Ma1**, Théorème 10.7], dans $P^B/P^{B,0}$, on a

$$\sum_{i=0}^{2} (-1)^i \widetilde{\mathrm{ch}}(R^\bullet\pi_* L^i_\bullet, h'^{R\pi_* L^i_\bullet}) = \sum_{j=0}^{2} (-1)^j \widetilde{\mathrm{ch}}(R^\bullet\pi_* L^\bullet_j, h'^{R\pi_* L^\bullet_j}). \tag{4.30}$$

Rappelle que $p_\Sigma : P(N_{\Sigma/V} \oplus 1) \to \Sigma$ est la projection naturelle. Soient $T(\omega^{V_\Sigma}, h^{\xi_D})$, $T(\omega^{P(N_{\Sigma/V})}, h^{\xi_D}) \in P^B/P^{B,0}$ les formes de torsion analytique de (4.23), (4.24) associées à $(\pi_1 p, h^{\xi_D})$, $(\pi i p_{\Sigma|P(N_{\Sigma/V})}, h^{\xi_D})$ [**B3**, §3.2]. On pose

$$\begin{aligned}
I_{\lambda_1} &= T(\omega^{V_\Sigma}, h^{\xi_D}) - T(\omega^{\widehat{Z}}, h^\xi) \\
&\quad - \widetilde{\mathrm{ch}}\Big(H(\widehat{Z}_B, (j\rho)^*\xi), h^{R(\pi j\rho)_*(j\rho)^*\xi}, h^{H(X_{\Sigma B}, \xi_D)}\Big), \\
I_{\lambda_2} &= T(\omega^{P(N_{\Sigma/V})}, h^{\xi_D}) - T(\omega^\Sigma, h^{i^*\xi\otimes\kappa}) \\
&\quad - \widetilde{\mathrm{ch}}\Big(H(\widehat{\Sigma}_B, (i\rho)^*\xi), h^{R(\pi i)_*(i^*\xi\otimes\kappa)}, h^{H(P(N_{\Sigma/V})_B, \xi_D)}\Big).
\end{aligned} \tag{4.31}$$

Pour les formes de torsion analytique $T(\omega^W, h^{\xi_D})$ associées à $\overline{q} : W \to B \times \mathbf{P}^1$ et le complexe (ξ_D, v) (cf. (4.8)), comme $\overline{q}$ est singulière sur $B \times \{\infty\}$, en procédant comme dans les Sections 1 et 2, on définit $T(\omega^{W_\infty}, h^{\xi_D}, h'^{R\widetilde{\pi}_*\xi})$ et $T'(\omega^{W_\infty}, h^{\xi_D}, h'^{R\widetilde{\pi}_*\xi}) \in P^B/P^{B,0}$. Comme on a fait après (2.2), on définit

$$h^{R\pi_* L^\bullet_0} = h'^{R\pi_*\xi} \oplus h^{H(D_{\infty|B}, p_{P(N_{\Sigma/V})*}\mathcal{O}_{\widehat{W}_\infty} \otimes p^*\xi)} \oplus h^{H(\Sigma_{|B}, \rho_*\mathcal{O}_{\Sigma_\infty} \otimes i^*\xi)}$$

la métrique L^2 sur $R^\bullet \pi_* L_0^\bullet$ induite par h^{ξ_D}, g^{TW}. Comme $W_\infty = V_\Sigma \cup_{P(N_{\Sigma/V})} P(N_{\Sigma/V} \oplus 1)$, par (2.4), on a

$$\begin{aligned} T'(\omega^{W_\infty}, h^{\xi_D}, h'^{R\overline{\pi}_* \xi}) = T(\omega^{V_\Sigma}, h^{\xi_D}) + T(\omega^{P(N_{\Sigma/V} \oplus 1)}, h^{\xi_D}) \\ -T(\omega^{P(N_{\Sigma/V})}, h^{\xi_D}) - \widetilde{\mathrm{ch}}(R^\bullet \pi_* L_0^\bullet, h^{R\pi_* L_0^\bullet}). \end{aligned} \tag{4.32}$$

On rappelle aussi que $\widehat{D}$ est la normalisation de D et $\rho' : \widehat{D} \to D$ est la projection canonique. Par la Proposition 4.2, $R(\overline{q}_D \rho')_* (jp_D \rho')^* \xi$ est localement libre sur $B \times \mathbf{P}^1$. Comme dans les Sections 1 et 2, on peut définir $T(\omega^{\widehat{D}_\infty}, h^\xi, h'^{R\overline{\pi}_* \xi})$ et $T'(\omega^{\widehat{D}_\infty}, h^\xi, h'^{R\overline{\pi}_* \xi}) \in P^B / P^{B,0}$ sur B pour la fibration $p_D \rho' : \widehat{D} \to B\{s_0\} \times \mathbf{P}^1$.

On a une équation analogue que [**B2**, (9.92)].

Proposition 4.9. *On a l'équation suivante dans* $P^B / P^{B,0}$,

$$\begin{aligned} I_{\lambda,\infty} - I_{\lambda_1} + I_{\lambda_2} - (T' - T)(\omega^{\widehat{D}_\infty}, h^\xi, h'^{R\overline{\pi}_* \xi}) \\ +(T' - T)(\omega^{W_\infty}, h^{\xi_D}, h'^{R\overline{\pi}_* \xi}) \\ = (T'_\infty - T_\infty)(\omega^D, h^\xi, h'^{R\overline{\pi}_* \xi}) - I_{\lambda(ipj')}. \end{aligned} \tag{4.33}$$

Preuve. Par [**Ma1**, (10.7)], Remarque 4.5, (4.11)–(4.13), (4.22), (4.30)–(4.32), on a (4.33). □

Théorème 4.10. *Le terme* $(T'_\infty - T_\infty)(\omega^D, h^\xi, h'^{R\overline{\pi}_* \xi}) - I_{\lambda(ipj')} \in P^B / P^{B,0}$ *est égal exactement le même terme à droite de* [**B2**, (9.69)].

Preuve. En effet, le terme $\log(\frac{\| \ \|_{\lambda((jp_D)^*\xi)}}{\| \ \|'_{\lambda((jp_D)^*\xi)}})^2(\infty)$ dans [**B2**, (9.69)] correspend à $(T'_\infty - T_\infty)$, le terme $\log(\frac{\| \ \|_{\widetilde{\lambda}((ipj')^*\xi)}}{\| \ \|'_{\lambda((ipj')^*\xi)}})^2$ dans [**B2**, (9.69)] correspend à $I_{\lambda(ipj')}$. Par (4.33), et en procédant comme en [**B2**, Théorème 9.16], on a le Théorème. □

d) Preuve du Théorème 2.1

En procédant comme en [**B2**, §9f) et §9g)], et en remplaçant les métriques de Quillen par les formes de torsion analytique, on a le Théorème 2.1. □

5. Fonctorialité de la renormalisation de formes de torsion analytique

Dans cette section, on explique la compatibilité du Théorème 2.1 et [**BerB**, Théorème 3.1]. Comme ça concerne des fibrés singulières, la vérification n'est pas toujours triviale.

Cette section est organisée de la manière suivante. Dans a), on étudie des classes de Bott-Chern près des fibres singulières. Dans b), on donne une version de [**BerB**, Théorème 3.1] aux fibres singulières.

Dans cette section, on fait les mêmes hypothèses et utilise toujours les mêmes notations qu'aux Sections 1 et 2.

a) Classes de Bott-Chern près des fibres singulières

On considère la suite exacte de fibrés holomorphes hermitiens sur $V \setminus \Sigma$

$$0 \to TY \to TZ \to \pi_1^* TB \to 0. \tag{5.1}$$

Soit $\widetilde{\mathrm{Td}}(TZ, TB, g^{TZ}, g^{TB}) \in P^{V\setminus\Sigma}/P^{V\setminus\Sigma,0}$ la classe de Bott-Chern de [**BGS1**] telle que

$$\begin{aligned}&\frac{\overline{\partial}\partial}{2i\pi}\widetilde{\mathrm{Td}}(TZ, TB, g^{TZ}, g^{TY}) \\ &\quad = \mathrm{Td}(TZ, g^{TZ}) - \pi_1^*(\mathrm{Td}(TB, g^{TB}))\mathrm{Td}(TY, g^{TY}).\end{aligned} \tag{5.2}$$

Par [**B2**, Théorème 1.1], le fibré verctoriel TZ sur $V_\Sigma \setminus P(N_{\Sigma/V}) \simeq V \setminus \Sigma$ s'étend en un sous-fibré $v(TV/S)$ de $\overline{p}^* TV$ sur V_Σ, de plus, le fibré $v(TV/B \times S)$ défini à §1e) est un sous-fibré de $v(TV/S)$. Sur $\widehat{Z}$, on a

$$T\widehat{Z} = v(TV/S). \tag{5.3}$$

Soit $g^{T\widehat{Z}}, g^{v(TV/S)}$ les métriques sur $T\widehat{Z}, v(TV/S)$ induites par g^{TV}. Par (2.9), on a un bicomplexe de fibrés vectoriels holomorphes sur $P(N_{\Sigma/V})$,

$$\begin{array}{ccccccc}
 & 0 & & 0 & & 0 & \\
 & \downarrow & & \downarrow & & \downarrow & \\
0 \to & \pi_\Sigma^* T\Sigma/B \times \Delta & \to & v(TV/B \times S) & \to & U \otimes p_\Sigma^*\kappa & \to 0 \\
 & \downarrow & & \downarrow & & \downarrow & \\
0 \to & \pi_\Sigma^* T\Sigma/S & \to & v(TV/S) & \to & U \otimes p_\Sigma^*\kappa & \to 0 \\
 & \downarrow & & \downarrow & & \downarrow & \\
0 \to & \pi_\Sigma^* TB & \to & \pi_\Sigma^* TB & \to & 0 & \to 0 \\
 & \downarrow & & \downarrow & & \downarrow & \\
 & 0 & & 0 & & 0 &
\end{array} \tag{5.4}$$

Soient

$$\begin{aligned}&\widetilde{\mathrm{Td}}(v(TV/S), g^{v(TV/S)}, g^{U\otimes p_\Sigma^*\kappa}), \widetilde{\mathrm{Td}}(v(TV/S), TB, g^{v(TV/S)}, g^{TB}) \\ &\qquad\qquad \in P^{P(N_{\Sigma/V})}/P^{P(N_{\Sigma/V}),0}\end{aligned}$$

les classes de Bott-Chern associées aux deuxièmes ligne et colonne de (5.4). Soit $\widetilde{\mathrm{Td}}(T\Sigma/S, TB, g^{T\Sigma/S}, g^{TB}) \in P^\Sigma/P^{\Sigma,0}$ la classe de Bott-Chern associée à la première colonne de (5.4). Par (5.4), et l'unicité de la classe de Bott-Chern [**BGS1**, Théorème 1.29], dans $P^{P(N_{\Sigma/V})}/P^{P(N_{\Sigma/V}),0}$, on a ([**B2**, (9.72)])

$$\begin{aligned}&\widetilde{\mathrm{Td}}(v(TV/S), g^{v(TV/S)}, g^{U\otimes p_\Sigma^*\kappa}) \\ &\quad - \widetilde{\mathrm{Td}}(v(TV/B \times S), g^{v(TV/B\times S)}, g^{U\otimes p_\Sigma^*\kappa})\mathrm{Td}(TB, g^{TB}) \\ &\quad = \widetilde{\mathrm{Td}}(v(TV/S), TB, g^{v(TV/S)}, g^{TB}) \\ &\quad - \widetilde{\mathrm{Td}}(T\Sigma/S, TB, g^{T\Sigma/S}, g^{TB})\mathrm{Td}(U \otimes p_\Sigma^*\kappa, g^{U\otimes p_\Sigma^*\kappa}).\end{aligned} \tag{5.5}$$

Proposition 5.1. *Pour $\mu \in P^V$ fixé, quand $s \in S \setminus \Delta \to s_0 \in \Delta$, on a*

$$\int_{Z_s} \mu \widetilde{\mathrm{Td}}(TZ, TB, g^{TZ}, g^{TB}) = \int_{\widehat{Z}} \mu \widetilde{\mathrm{Td}}(T\widehat{Z}, TB, g^{T\widehat{Z}}, g^{TB}) \tag{5.6}$$
$$+2 \int_{P(N_{\Sigma/V})} \mu \widetilde{\mathrm{Td}}(v(TV/S), TB, g^{v(TV/S)}, g^{TB}) + O(s - s_0).$$

Preuve. Soit $\mu_1 \in P^{V_\Sigma}$, par l'argument dans [**B2**, Théorème 2.1], quand $s \in S \backslash \Delta \to s_0 \in \Delta$, on a

$$\int_{Z_s} \mu_1 = \int_{\widehat{Z}} \mu_1 + 2 \int_{P(N_{\Sigma/V})} \mu_1 + O(s - s_0). \tag{5.7}$$

On a aussi

$$\widetilde{\mathrm{Td}}(TZ, TB, g^{TZ}, g^{TB})_{B \times \{s\}} = \widetilde{\mathrm{Td}}(v(TV/S), TB, g^{v(TV/S)}, g^{TB})_{B \times \{s\}}. \tag{5.8}$$

Par (5.7) et (5.8), on a la Proposition. □

b) Fonctorialité des formes de torsion analytique

On rappelle que $h'^{R\pi_*\xi}$ est une métrique $\mathcal{C}^\infty$ sur $R^\bullet\pi_*\xi$ sur $B \times S$.

Soit $\lambda(\xi)$ et $\lambda(R^\bullet\pi_*\xi)$ les fibrés en droite sur S qui sont les inverses des déterminants de $H(Z, \xi_{|Z})$ et $H(B, R^\bullet\pi_*\xi_{|B})$. Alors par [**KM**], le fibré en droite $\lambda(\xi) \otimes \lambda^{-1}(R^\bullet\pi_*\xi)$ a une section canonique non nulle σ_1 sur S. Soit $\| \ \|'_{\lambda(R^\bullet\pi_*\xi)}$ la métrique de Quillen sur $\lambda(R^\bullet\pi_*\xi)$ associées aux métriques $g^{TB}, h'^{R\pi_*\xi}$ sur TB, $R\pi_*\xi$. Soit $\| \ \|_{\lambda(\xi)_{|\Delta}}$ la métrique renormalisée définie dans [**B2**, (0.7)]. Soit $\| \ \|_{\lambda(\xi)_{|\Delta} \otimes \lambda^{-1}(R^\bullet\pi_*\xi)}$ la métrique sur $\lambda(\xi)_{|\Delta} \otimes \lambda^{-1}(R^\bullet\pi_*\xi)$ induite par $\| \ \|_{\lambda(\xi)_{|\Delta}}$ et $\| \ \|'_{\lambda(R^\bullet\pi_*\xi)}$.

Théorème 5.2. *On a l'identité*

$$\log(\|\sigma_1\|^2_{\lambda(\xi)_{|\Delta} \otimes \lambda^{-1}(R^\bullet\pi_*\xi)}) \tag{5.9}$$
$$= \int_B \mathrm{Td}(TB, g^{TB}) T(\omega^Z, h^\xi, h'^{R\pi_*\xi})$$
$$- \int_{\widehat{Z}} \widetilde{Td}(T\widehat{Z}, TB, g^{T\widehat{Z}}, g^{TB}) \mathrm{ch}(\xi, h^\xi)$$
$$-2 \int_{P(N_{\Sigma/V})} \widetilde{\mathrm{Td}}(v(TV/S), TB, g^{v(TV/S)}, g^{TB}) \mathrm{ch}(\xi, h^\xi).$$

Preuve. Par (1.16), Théorème 1.3, Proposition 5.1, on a le Théorème 5.2. □

Soient σ_2 et σ_3 les sections canoniques non-nulles de

$$\lambda((j\rho)^*\xi) \otimes \lambda^{-1}(R^\bullet(\pi j\rho)_*(j\rho)^*\xi), \quad \lambda(i^*\xi) \otimes \lambda^{-1}(R^\bullet(\pi i)_* i^*\xi).$$

Soient τ_1 et τ_2 les sections canoniques non-nulles des inverses du déterminant de cohomologie associées à (2.1) et (2.2). Pour simplifier, on note $\|\tau_i\|$, $\|\sigma_i\|$ les métriques de Quillen associées. Par un argument de suite spectrale [**KM**], on a

$$\sigma_1 \otimes \sigma_2^{-1} \otimes \sigma_3 = \tau_1 \otimes \tau_2^{-1}. \tag{5.10}$$

Par [**BGS2**, Théorème 1.23], on a

$$\log \|\tau_2\|^2 = \int_B \mathrm{Td}(TB, g^{TB})\widetilde{ch}(R^\bullet \pi_* L, h^{R\pi_* L}). \tag{5.11}$$

Si on applique [**BerB**, Théorème 3.1] à σ_2, σ_3, [**B2**, Théorème 0.2] (ou Théorème 2.1) à la section τ_1, par le Théorème 2.1, (5.5), (5.9) et (5.11), à partir de l'équation des métriques de Quillen pou (5.10), on ramène finalement l'équation suivante

$$\int_\Sigma \widetilde{\mathrm{Td}}(T\Sigma/S, TB, g^{T\Sigma/S}, g^{TB})\Big[\int_{P(N_{\Sigma/V})_\Sigma} \mathrm{Td}(U\otimes p_\Sigma^*\kappa, g^{U\otimes p_\Sigma^*\kappa}) + \frac{1}{2}\Big] = 0. \tag{5.12}$$

Mais par [**B2**, (6.64)], on a

$$\int_{P(N_{\Sigma/V})_\Sigma} \mathrm{Td}(U\otimes p_\Sigma^*\kappa, g^{U\otimes p_\Sigma^*\kappa}) = -\frac{1}{2}. \tag{5.13}$$

Donc le Théorème 2.1 et [**BerB**, Théorème 3.1] sont compatibles.

On peut interpréter le Théorème 2.1 et la compatibilité de Théorème 2.1 et [**BerB**, Théorème 3.1] comme "la renormalisation commute à la limite adiabatique pour la métrique de Quillen".

References

[BaFM] Baum P., Fulton W., MacPherson R., Riemann-Roch for singular varieties, *Publ. Math. IHES* 45 (1975), 106–146.

[BerB] Berthomieu A., Bismut J.-M., Quillen metric and higher analytic torsion forms, *J. Reine Angew. Math* 457, 1994, 85–184.

[BBos] Bismut J.-M., Bost J.B., Fibrés déterminants, métriques de Quillen et dégénérecence des courbes, Acta Math. 165 (1990), 1–103.

[B2] Bismut J.-M., Quillen metrics and singular fibres in arbitrary relative dimension, *J. of. Alg. Geom.* 6 (1997), 19–149.

[B3] Bismut J.-M., *Families of immersions, and higher analytic torsion*, Astérisque 244, 1997.

[BGS1] Bismut J.-M., Gillet H., Soulé C., Analytic torsion and holomorphic determinant bundles.I, *Comm. Math. Phys.* 115 (1988), 49–78.

[BGS2] Bismut J.-M., Gillet H., Soulé C., Analytic torsion and holomorphic determinant bundles.III, *Comm. Math. Phys.* 115 (1988), 301–351.

[BGS4] Bismut J.-M., Gillet H., Soulé C., Complex immersions and Arakelov geometry, *Progress in Math., no. 86, Birkhäuser, Boston* 1990, 249–331.

[BK] Bismut J.-M., Köhler K., Higher analytic torsion forms for direct images and anomaly formulas, *J. of Alg. Geom.* 1 (1992), 647–684.

[Bo] Bott R., Vector fields and characteristic numbers, *Michigan Math. J.* 14 (1967), 231–244.

[GrH] Griffiths P., Harris J., *Principles of Algebraic Geometry*, New York, Wiley 1978.

[GS] Gillet H., Soulé C., Analytic torsion and the arithmetic Todd genus, *Topology* 30 (1991), 21–54.

[KM] Knudsen P.F., Mumford D., The projectivity of the moduli space of stable curves, I, Préliminaires on "det" and "div", *Math. Scand.* 39 (1976), 19–55.

[Ma1] Ma Xiaonan., Formes de torsion analytique et familles de submersions. I, *Bull. Soc. Math. France* 127 (1999), 541–621.

[Ma2] Ma Xiaonan., Formes de torsion analytique et familles de submersions. II, *Asian J. of Math.*, 4 (2000), 633–668.

[RS] Ray D.B., Singer I.M., Analytic torsion for complex manifolds, *Ann. of Math.* 98 (1973), 154–177.

Xiaonan Ma
Humboldt-Universität zu Berlin
Institut für Mathematik
Rudower Chaussee 25
D-12489 Berlin, Germany
e-mail: `xiaonan@mathematik.hu-berlin.de`

Operator Theory:
Advances and Applications, Vol. 145, 419–437

Regularisation of Secondary Characteristic Classes and Unusual Index Formulas for Operator-valued Symbols

Grigori Rozenblum

Abstract. The usual expression for the secondary characteristic classes is regularized so that it fits for the case of Hilbert bundles. This gives index formulas for pseudodifferential operators with operator-valued symbols.

1. Introduction

The present article continues the paper [14] where analytical index formulas for pseudodifferential operators in the Euclidean space with operator-valued symbols were obtained. The special feature of these formulas was that they look different compared with usual index formulas, well known for the case of scalar or matrix symbols: the latter simply do not make sense in the operator-valued situation since they contain the trace of non-trace-class operators. New formulas, involving certain regularisation based on the suspension homomorphism in cyclic co-homologies, can be naturally called 'unusual index formulas'. In [14], the operators in the Euclidean space were considered. Now we turn our attention to operators acting on arbitrary compact manifolds and obtain unusual formulas for this case. Again, as in [14], there are two main ingredients in our work. First, is the analysis of a special class of operator-valued symbols, possessing the property of improvement under differentiation in co-variables, this improvement expressed in the terms of Shatten classes. The second ingredient is the regularisation of classical Chern–Weyl formulas for the characteristic classes involved in the index formulas, by means of cyclic co-homology methods.

The co-homological form of the classical Atiyah–Singer index formula can be expressed in two versions. The more usual one involves the Chern character, the even co-homological class on the co-ball bundle of the manifold constructed from the symbol of the operator and connections on the start and target vector bundles.

2000 *Mathematics Subject Classification.* Primary 35P15; Secondary 47A10, 47B38, 47B65, 47D06.

Key words and phrases. Pseudodifferential operators, index, characteristic classes, cyclic co-homology.

This work was completed with the support of grant from the Swedish Royal Academy of Sciences.

Another version, applicable only for the case when these latter bundles are isomorphic, involves the Chern–Simons character, an odd-dimensional co-homology class constructed from the symbol of the operator.

Consider the most simple case of an elliptic zero-order operator in $\mathbb{R}^m$ with symbol $a(x,\xi)$, invertible outside some ball B_R (with boundary S_R) in $\mathbb{R}^{2m}$. The index of the pseudodifferential operator $Ops(a)$ depends only on the values of the symbol a on the sphere S_R, or, to be more exact, on the class of this symbol in the group $K^1(S_R)$. Here, the two types of formulas mentioned above are

$$\text{ind } Ops(a) = \int\limits_{B_R} ch_{2m}(a), \tag{1.1}$$

and

$$\text{ind } Ops(a) = \int\limits_{S_R} cs_{2m-1}(a). \tag{1.2}$$

The analytical expression for the Chern character, or, to be more precise, its highest degree component $ch_{2m}(a)$ (which is the only one involved in (1.1)) is given by

$$ch_{2m}(a) = c_m \operatorname{tr}(((dr + r(da)r)da)^m), \tag{1.3}$$

where r is a regularizer for a, i.e., symbol coinciding with a^{-1} outside the ball B_R, and the $2m-1$-degree component of the Chern–Simons character $cs_{2m-1}(a)$ equals

$$cs_{2m-1}(a) = c'_m \operatorname{tr}((a^{-1}da)^{2m-1}). \tag{1.4}$$

Note, that (1.2), (1.4) involve only those values of the symbol on which the index actually depends, while (1.1), (1.3) contain values of the symbol inside the ball, although the index does not depend on these values. This circumstance favors odd index formulas, admitting a more natural interpretation for the case of more general symbol algebras, not being algebras of (matrix- or operator-valued) functions on the co-spheric bundle (see [10],[3]).

When we pass to the operator-valued case (to emphasize this, we boldface operator symbols), the analogy of (1.1), (1.3) requires the symbol $\mathbf{r}$ to be a *very* good regularizer of $\mathbf{a}$: the operators $\mathbf{ar}-\mathbf{1}, \mathbf{ra}-\mathbf{1}$ must belong to the trace class. Thus, although the index (for a proper class of symbols) still depends only on the values of the symbol on the sphere, a special continuation of the symbol inside the ball is required, admitting this nice regularizer. On the other hand, (1.2), (1.4) still express the index in the terms of the 'essential' values of the symbol $\mathbf{a}$, however, a new condition arises: the operator under the trace sign in (1.4) must belong to the trace class – but, in typical applications, it does not. This situation is dealt with in [14], where the expression (1.4) for the Chern–Simons character was modified, or, in other words, regularized, by algebraic methods, so that it makes sense for wider (and, in fact, arising in applications) classes of symbols and, substituted into (1.2), gives a correct expression for the index of the pseudodifferential operator.

The approach in [14] originates in [13], where a similar regularisation procedure was used for Toeplitz operators on the line, with operator-valued symbols.

On a general manifold X, co-homological index formulas have a more complicated structure; they contain the full Chern (Chern–Simons) character as well as the Todd class reflecting the topology of the manifold. The analytical description of these formulas can be found, for example, in [6]. Analogies of (1.1)–(1.4) are

$$\text{ind } Ops(\mathbf{a}) = \int_{T^*X} \mathcal{T}(X) ch(\mathbf{a}), \tag{1.5}$$

$$\text{ind } Ops(\mathbf{a}) = \int_{S^*X} \mathcal{T}(X) sc(\mathbf{a}), \tag{1.6}$$

where[1]

$$ch(\mathbf{a}) = \text{Tr}\,(\exp((d\mathbf{r} + \mathbf{r}d\mathbf{a}\mathbf{r})d\mathbf{a}) - \mathbf{1}), \tag{1.7}$$

$$cs(\mathbf{a}) = -\frac{1}{\pi i}\int_0^1 \text{Tr}\,(\mathbf{a}^{-1}d\mathbf{a}\exp(-\frac{(\mathbf{a}^{-1}d\mathbf{a})^2 t(1-t)}{2\pi i}))dt, \tag{1.8}$$

and $\mathcal{T}(X) = \sum \tau_{4j}$ is the complete Todd class of the manifold X. Again, in the operator-valued case, the operators under the trace sign in (1.5), (1.7) belong to the trace class provided a proper continuation of the symbol $\mathbf{a}$ from the co-spheric bundle S^*X to the co-ball bundle B^*X is constructed, admitting an up-to-trace-class-error regularizer $\mathbf{r}$. Formulas (1.6), (1.8), on the other hand, do not make sense unless the operator-valued form $\mathbf{a}^{-1}d\mathbf{a}$ belongs to some Shatten ideal, depending on the dimension and topology of the manifold (generally, the need for trace class property may arise). It is not reasonable to restrict the consideration to such operator symbols. In fact, as it is well known, pseudodifferential operators acting in the space of functions with values in infinite-dimensional spaces arise in a natural way in the analysis on singular manifolds (see, e.g., [4], [15]). The properties of the symbols depend, to begin with, on the dimension of the manifold and on the form of the singularity, therefore, they can't be postulated arbitrarily. Thus, some regularisation of odd index formulas (1.6), (1.8), and, in particular, of the Chern–Simons character, is needed. An additional feature of our problem is that, unlike the matrix case, homogeneous symbols are not the interesting ones. The symbols arising in applications are quite far from being homogeneous, and therefore even the notion of the principal symbol does not make immediate sense.

In the present paper we extend analysis of classes of operator-valued symbols, introduced in [14], to the case of manifolds. We explain what is the natural notion of the principal symbol for this class, and it is in the terms of this principal symbol that the index of the operator must be expressed. Similar to [14], we define an index as a homomorphism from the $\mathbf{K}_1$-group of an algebra of symbols to $\mathbb{Z}$. To find the

[1] Compared with the usual formula, the unit operator is subtracted in the expression (1.7). This does not effect the index since $\mathcal{T}(X)$ contains only dx differentials, so the result of integration in (1.5) does not change. On the other hand, in the infinite-dimensional case, it is necessary to subtract the identity operator, otherwise trace never makes sense.

analytical description of this homomorphism, we construct regularisations of the Chern–Simons character (1.8) related to these symbols. As a result, we obtain new index formulas for elliptic pseudodifferential operators with operator-valued symbols. Note that, unlike the finite-dimensional case, we do not lose in generality when passing to the odd-dimensional index formulas: infinite-dimensional Hilbert bundles are always trivial.

To make the exposition more independent of [14], we describe here the general scheme of regularisation.

Let $\mathfrak{S}$ be a Banach $*$-algebra with unit, $M(\mathfrak{S})$ be the set of matrices over $\mathfrak{S}$. The groups $\mathbf{K}_j(\mathfrak{S}), j=0,1$ are the usual K-groups in the theory of Banach $*$-algebras, in particular, $\mathbf{K}_1(\mathfrak{S})$ consists of equivalence classes of invertible matrices in $M(\mathfrak{S})$, i.e., elements in $GL(\mathfrak{S})$. If $\mathfrak{S}$ does not have a unit, one attaches it and thus replaces $M(\mathfrak{S})$ by $M(\mathfrak{S})^+$ in the latter definition. (We use boldface $\mathbf{K}$ in order to distinguish operator algebras K-groups from topological ones.) The notion of these $\mathbf{K}$-groups carries over to local $*$-sub-algebras of $\mathfrak{S}$, i.e., sub-algebras closed with respect to the holomorphic functional calculus in $\mathfrak{S}$. Important here is the fact that the $\mathbf{K}$-groups of a dense local sub-algebra in $\mathfrak{S}$ are isomorphic to the ones of $\mathfrak{S}$ (see, e.g., [1]).

For a normed $*$-algebra $\mathfrak{S}$, the group $C^k_\lambda(\mathfrak{S})$ of cyclic cochains consists of $(k+1)$- linear continuous functionals $\varphi(a_0,a_1,\dots,a_k)$, which are cyclic in the sense $\varphi(a_0,a_1,\dots,a_k)=(-1)^k\varphi(a_1,a_2,\dots,a_0)$. The Hochschild co-boundary operator $b: C^k_\lambda(\mathfrak{S})\to C^{k+1}_\lambda(\mathfrak{S})$ generates, in a usual way, co-homology groups $HC^k_\lambda(\mathfrak{S})$.

There is a coupling of $HC^{2\nu-1}_\lambda(\mathfrak{S})$ and $K_1(\mathfrak{S})$ (see [3], Ch. III.3):

$$[\varphi]\times_\nu[a]=\gamma_\nu(\varphi\otimes\mathrm{tr})(a^{-1}-1,a-1,a^{-1}-1,a-1,\dots,a^{-1}-1,a-1), \quad (1.9)$$

where tr is the matrix trace and γ_ν is the normalisation constant, chosen in [2] to be equal to $(2i)^{-1/2}2^{-2\nu+1}\Gamma(\nu+\frac{1}{2})$, for functoriality reasons. The sense of constructing index formulas is to find a proper cyclic cocycle φ, for which the coupling (1.9) gives the index of an operator with symbol a under a certain 'quantisation', i.e., the rule associating an operator to the symbol.

The suspension (or periodicity) homomorphism in cyclic co-homologies is $S: HC^{2\nu-1}_\lambda(\mathfrak{S})\to HC^{2\nu+1}_\lambda(\mathfrak{S})$. This operation is *not*, in general, an isomorphism. In fact, it is a monomorphism, with range isomorphic to the kernel of the homomorphism $\mathbf{I}$ associating to every cyclic cocycle representing a class in $HC^{2\nu+1}_\lambda(\mathfrak{S})$, the class of the same cocycle in the Hochschild co-homology group $H^{2\nu+1}(\mathfrak{S})$ (see [3], III.1.γ). The suspension homomorphism is consistent with the coupling (1.9):

$$[\varphi]\times_\nu[a]=S[\varphi]\times_{\nu+1}[a],\ [a]\in\mathbf{K}_1(\mathfrak{S}),[\varphi]\in H^{2\nu-1}_\lambda(\mathfrak{S}). \quad (1.10)$$

Based upon (1.10), the regularisation goes in the following manner.

Suppose that for a certain 'nice' sub-algebra $\mathfrak{S}_0$ in $\mathfrak{S}$, we have a functional on $GL(\mathfrak{S}_0)$ being a sum of expressions of the form (1.9)

$$\Phi(a)=\sum_j[\varphi_j]\times_{\nu_j}[a], \quad (1.11)$$

where φ_j are some cyclic cocycles of degree $2\nu_j - 1$. For a wider $*$-algebra $\mathfrak{S}$, the cocycles φ_j may be not extendible by continuity. However, when we apply sufficiently many times the suspension S to functionals φ_j in (1.11), it may (and in fact, it often does) turn out that the suspended functionals admit such extension by continuity. In particular, when the original functional Φ describes the index of the operator with symbol a under a certain sanitation, the suspension produces unusual index formulas on $\mathfrak{S}$.

In [14], this strategy was implemented for pseudodifferential operators in the Euclidean space, where the original expression (1.11) for the index involves only one term. For general manifolds, (1.6)–(1.8) contains, generally, the sum of several such terms, see (1.6). So we take as Φ the integral of the product of the Chern–Simons character of the symbol $\mathbf{a}$ with an arbitrary scalar closed form ψ. After applying the iterated suspension, we arrive at an expression of the functional, being the integral of ψ multiplied by a new form, which is natural to consider as the regularized Chern–Simons form, since it defines the same co-homology class. Substituting this latter form into (1.6) we get unusual index formulas.

The research was partly supported by grant from the Royal Swedish Academy of Science. The author is grateful to Professors B. Fedosov, B.-W. Schulze and B. Sternin for helpful discussions.

2. Operator-valued symbols

The calculus of pseudodifferential operators with operator-valued symbols, acting on functions defined in a domain of the Euclidean space, was developed in [14]. The main feature of this calculus was the choice of the notion of *negligible* operator. In the classical calculus, one usually considers infinitely smoothing operators as negligible, and it is in the terms of this notion, that the main facts of calculus are expressed, such as pseudo locality, composition formula, change of variables etc. In the operator-valued case, infinitely smoothing operators are not sufficiently nice. In [14], we proposed to consider the trace class operators as negligible for the needs of calculus.

In the present section we establish further properties of this operator-valued calculus, having in mind the study of operators on compact manifolds. We start with the definition of the classes $\mathcal{S}_q^\gamma$ consisting of symbols, being bounded operators in some separable Hilbert space $\mathfrak{K}$. These classes involve a certain sufficiently large integer N, concrete choice of which depends on particular applications. By $|\cdot|_p$ we denote the norm (quasi-norm for $p < 1$) in the Shatten ideal $\mathfrak{s}_p$.

Our symbols satisfy the following conditions.

Definition 2.1. Let $\gamma \le 0, q > 0$. The class $\mathcal{S}_q^\gamma = \mathcal{S}_q^\gamma(\mathbb{R}^m \times \mathbb{R}^{m'})$ consists of functions[2] $\mathbf{a}(y, \eta)$, $(y, \eta) \in \mathbb{R}^m \times \mathbb{R}^{m'}$, such that for any (y, η), $\mathbf{a}(y, \eta)$ is a bounded

[2]Not following the unwritten tradition, we denote the variables by (y, η), and not by (x, ξ). This somehow reflects the fact that in the most usual situation, when considering pseudodifferential

operator in $\mathfrak{K}$ and, moreover,

$$\|D_\eta^\alpha D_y^\beta \mathbf{a}(y,\eta)\| \le C_{\alpha,\beta}(1+|\eta|)^{-|\alpha|+\gamma}, \tag{2.1}$$

$$|D_\eta^\alpha D_y^\beta \mathbf{a}(y,\eta)|_{\frac{q}{-\gamma+|\alpha|}} \le C_{\alpha,\beta}, \quad |\alpha| \le \gamma+q, \tag{2.2}$$

$$|D_\eta^\alpha D_y^\beta \mathbf{a}(y,\eta)|_1 \le C_{\alpha,\beta}(1+|\eta|)^{q+\gamma-|\alpha|}, \quad |\alpha| > \gamma+q \tag{2.3}$$

for $|\alpha|, |\beta| \le N$ (where all derivatives are understood in the strong sense).

Remark 1. In [14] the estimate (2.2) was required for all $|\alpha|, |\beta| \le N$. Our definition here introduces a somewhat larger class, which follows from the obvious inequality $|\mathbf{a}|_1 \le |\mathbf{a}|_p \|\mathbf{a}\|^{1-p}$, $p < 1$. In fact, it is (2.3) that is actually needed in the proofs of all the facts in the pseudodifferential calculus (this is mentioned in [14] and one can even see this in the formulation of the key Proposition 2.2 below), while it is easier to check (2.3) in concrete situations.

Having a symbol $\mathbf{a}(y,y',\eta) \in \mathcal{S}_q^\gamma(\mathbb{R}^{2m}\times\mathbb{R}^m)$, we define the pseudodifferential operator with this symbol as

$$(OPS(\mathbf{a})u)(y) = (\mathbf{a}(y,y',D_y)u)(y) = (2\pi)^{-m}\iint e^{i(y-y')\eta}\mathbf{a}(y,y',\eta)u(y')d\eta dy', \tag{2.4}$$

where $u(y)$ is a function on $\mathbb{R}^m$ with values in $\mathfrak{K}$. In particular, if $\mathbf{a}$ does not depend on y', this is the usual formula involving the Fourier transform:

$$\mathbf{a}(y,D_y)u = OPS(\mathbf{a}) = \mathcal{F}^{-1}\mathbf{a}(y,\eta)\mathcal{F}u.$$

The following proposition proved in [14] gives a sufficient condition for a pseudodifferential operator to belong to trace class.

Proposition 2.2. *Let the operator-valued symbol* $\mathbf{a}(y,y',\eta)$ *in* $\mathbb{R}^{2m}\times\mathbb{R}^m$ *be smooth with respect to* y, y', *let all* y, y'*-derivatives* $D_y^\beta D_{y'}^{\beta'}\mathbf{a}$ *up to some (sufficiently high) order* N *be trace class operators with trace class norm bounded uniformly in* y, y'. *Suppose that* $g(y), h(y) = O((1+|y|)^{-2m})$. *Then the operator* $h\mathbf{a}(y,y',D_y)g$ *belongs to* $\mathfrak{s}_1(L_2(\mathbb{R}^m;\mathfrak{K}))$.

All usual properties of the pseudodifferential calculus hold for our operator-valued version, with trace class operators acting as negligible. In particular, the composition rule takes the following form (see Theorem 3.6 in [14]).

Proposition 2.3. *If* $\mathbf{a} \in \mathcal{S}_q^{\gamma_1}, \mathbf{b} \in \mathcal{S}_q^{\gamma_2}, \gamma_1, \gamma_2 \le 0$, *then, for* N *large enough,*

$$h(y)Ops(\mathbf{a})Ops(\mathbf{b}) - h(y)Ops(\mathbf{a}\circ_N \mathbf{b}) \in \mathfrak{s}_1,$$

where

$$\mathbf{a}\circ_N\mathbf{b} = \sum_{|\alpha|\le N}(\alpha!)^{-1}D^\alpha\mathbf{a}\partial^\beta\mathbf{b} \tag{2.5}$$

and $h(y)$ *is a function decaying as* $(1+|y|)^{-m-1}$.

---

operators on a manifold X with singularities lying on a submanifold Y, it is the variables on T^*Y which the operator symbol depend on.

The property of pseudo-locality established in [14] (Theorem 3.5), has the following form.

Proposition 2.4. *Let the symbol* $\mathbf{a}(y, y', \eta)$ *belong to* $S_q^\gamma(\mathbb{R}^{2n} \times \mathbb{R}^n)$ *for some* $q > 0, \gamma \leq 0$, *let* φ, ψ *be bounded functions with disjoint supports, at least one of them being compactly supported. Then (for* N *large enough) the operator* $\varphi\mathbf{a}(y, y', D)\psi$ *belongs to* $\mathfrak{s}_1(L_2(\mathbb{R}^m; \mathfrak{K}))$, *moreover, the* $\mathfrak{s}_1$*-norm of this operator is controlled by* L_∞*-norms of* φ, ψ, *the distance between their supports, and constants* $C_{\alpha,\beta}$ *in* (2.1), (2.2), (2.3).

In order to consider an operator on manifolds, we need this property in a somewhat stronger form.

Theorem 2.5. *Let the symbol* $\mathbf{a}(y, y', \eta)$ *belong to* $S_q^\gamma(\mathbb{R}^{2m} \times \mathbb{R}^m)$ *for some* $q > 0, \gamma \leq 0$, *have compact support in one of variables* y, y' *and, moreover,* $\mathbf{a}(y, y', \eta) = \mathbf{b}(y, y', \eta)|y - y'|^{2N}$ *for* $|y - y'| \leq \delta$, *for some fixed positive* $\delta > 0$, *large enough* N *and a symbol* $\mathbf{b} \in S_q^\gamma$. *Then the operator* $\mathbf{a}(y, y', D)$ *belongs to the trace class, with trace class norm estimated to be the constants* $C_{\alpha,\beta}$ *in* (2.1), (2.2), (2.3) *and similar constants for* $\mathbf{b}$.

Proof. The proof follows the lines of the reasoning in [14] leading to Proposition 2.4. The first-order partial differential operator $L = L(D_\eta) = -i|y - y'|^{-2}(y - y')D_\eta$ has the property $Le^{i\eta(y-y')} = e^{i\eta(y-y')}$, and since $|y - y'|^{-N}\mathbf{a}$ does not have singularities, we can insert L^N into the expression for any N. After integration by parts (first formal, but then justified in the usual way), we obtain that (2.4) equals

$$(2\pi)^{-m} \int e^{i\eta(y-y')}\varphi'(y)|y-y'|^{-2N}((y-y')D_\eta)^N\mathbf{a}(y, y', \eta)\psi'(y')u(y')dy'd\eta. \quad (2.6)$$

Now, provided N is large enough, we arrive at the conditions of Proposition 2.2. □

We note an important particular case of Theorem 2.5.

Corollary 2.6. *If the symbol* $\mathbf{a}(y, y', \eta)$ *in* S_q^γ *vanishes in a neighborhood of the diagonal* $y = y'$, *then the operator* (2.4) *belongs to the trace class.*

Now we are able to establish for our classes the analogies of two important results of the scalar calculus (see, e.g., [8]).

Theorem 2.7. *An operator with the symbol* $\mathbf{a}(y, y', \eta) \in S_q^\gamma$ *which has compact support in one of the variables* y, y' *can, up to a trace class error, be represented as an operator with some other symbol* $\mathbf{a}'$, *not depending on* y', *moreover*

$$\mathbf{a}'(y, \eta) = \sum_{|\alpha| \leq N} (\alpha!)^{-1} D_{y'}^\alpha \partial_\eta^\alpha \mathbf{a}(y, y', \eta)|_{y'=y}. \quad (2.7)$$

The *proof* literally follows the (rather long one) in [8]. One considers the starting section of the Taylor expansion of the symbol $\mathbf{a}$ near the diagonal $y' = y$ in $y - y'$. These terms give the expression in (2.7). The remainder term has zero of high order on the diagonal, and therefore gives a trace class operator, according to Theorem 2.5. □

The second important fact, formulated but not proved in [14], concerns the behavior of pseudodifferential operators under the change of variables.

Theorem 2.8. *Let the symbol* $\mathbf{a}(y,\eta)$ *belong to* $\mathcal{S}_q^\gamma$ *and have compact support* Ω *in* y. *Let* $z=\kappa(y)$ *be a diffeomorphism defined on a neighborhood of* Ω. *Then, up to a trace class error, the operator*

$$B=\kappa\circ Ops(\mathbf{a})\circ\kappa^{-1} \tag{2.8}$$

is a pseudodifferential operator with the usual expression for the symbol $\mathbf{b}(z,\zeta)$ *of* B, *in particular,*

$$\mathbf{b}(z,\zeta)-\mathbf{a}(\kappa^{-1}(z),{}^TD\kappa\zeta)\in\mathcal{S}_q^{\gamma-1}. \tag{2.9}$$

Again, the *proof* follows the pattern in [8]. First, cut off the complement to a neighborhood of the diagonal $y=y'$, which, according to Corollary 2.6 gives a trace class perturbation to the operator. Then, after the change of variables, the operator $B=\kappa\circ Ops(\mathbf{a})\circ\kappa^{-1}$ takes the form

$$Bv(z)=\int\int e^{i(\kappa^{-1}(z)-\kappa^{-1}(z'))\eta}\mathbf{a}(\kappa^{-1}(z),\eta)u(\kappa^{-1}(z))|\partial y/\partial z|dzd\eta.$$

Now, the Kuranishi trick, the linear change of variables $\eta=G(y,y')\zeta$ near the diagonal $y'=y$, transforms B to the usual pseudodifferential form, with the phase function $(z-z')\zeta$ and with a certain symbol depending on z,z',ζ, which, by means of Theorem 2.7 and Corollary 2.6 reduces to a symbol not depending on z'. □

This latter result enables one to define pseudodifferential operators on compact manifolds.

Let Y be a compact manifold of dimension m. We consider the space $L_2(Y)=L_2(Y,\mathfrak{K})$, where $\mathfrak{K}$ is an infinite-dimensional Hilbert space. (There is no need to consider Hilbert bundles over Y, since all such bundles, unlike the finite-dimensional case, are trivial.)

Definition 2.9. An operator A in $L_2(Y)$ is called pseudodifferential in the class $\mathcal{L}_q^\gamma(Y)$ if for some co-ordinate covering Y by open sets U_ι with co-ordinate mappings κ_ι to domains in $\mathbb{R}^m$, a partition of unity φ_ι subordinated to U_ι and another system ψ_ι of cut-off functions with supports in U_ι such that $\psi_\iota\varphi_\iota=\varphi_\iota$, the operators

$$\kappa_\iota\psi_\iota A\varphi_\iota\kappa_\iota^{-1} \tag{2.10}$$

are pseudodifferential operators with symbols in $\mathcal{S}_q^\gamma$, plus some trace class operator.

According to Theorem 2.8, the property described by this definition does not depend on the choice of the covering, cut-offs and mappings. Moreover, the symbols at a point $y\in Y$ obtained by different choices are related by the formula of the type (2.9).

Definition 2.10. A pseudodifferential operator $A \in \mathcal{L}_q^0$ is called elliptic if for any point $y \in U_\iota$ and cut-off functions $\varphi_\iota, \psi_\iota$, such that $\varphi_\iota(y) \neq 0$, the symbol $\mathbf{b}_\iota(z,\zeta)$ of operator (2.10) is invertible for z in a neighborhood of $\kappa_\iota(y)$ and for sufficiently large $|\zeta|$, with norm-bounded inverse.

Theorem 2.8 implies that the property of the operator to be elliptic does not depend on the choices made in the definition.

In the usual, finite-dimensional, pseudodifferential calculus, one easily associates canonically to a pseudodifferential operator on a manifold its leading symbol, being a matrix defined on the co-spheric bundle on the manifold; there, the crucial role is played by the homogeneity property of symbols. In the operator-valued context, homogeneous symbols are not the ones which are interesting for applications (see, e.g., [4], [15], [14]). In what follows we are going to associate an invertible operator family on the co-spheric bundle to an *elliptic operator*. This association, although non-canonical, generates a homomorphism from the group of elliptic operators to the $\mathbf{K}_1$-group of a proper algebra of operator-valued functions. We follow the pattern of the construction in [9]. For the given elliptic operator A, fix some co-ordinate covering, cut-off functions and co-ordinate mappings, as in Definition 2.9. Fix also some Riemannian metric on Y. For each neighborhood U_ι, let $\mathbf{b}_\iota(z,\zeta)$ be the symbol of the corresponding to U_ι pseudodifferential operator in a domain in the Euclidean space. Define then the operator function on T^*U_ι as $\mathbf{a}_\iota(y,\eta) = \mathbf{b}_\iota(\kappa_\iota(y), {}^T D\kappa_\iota(y)^{-1}\eta)$. This symbol is invertible for $|\eta|$ large enough (the length is understood in the sense of the Riemannian metric). Symbols obtained in this way over the same point in the intersection of co-ordinate neighborhoods differ by a term in $\mathcal{S}_q^{-1}$; this implies that by gluing such symbols together by means of a partition of unity, we get an operator function $\mathbf{a}$ on T^*Y, invertible for large $|\eta|$. Let R be so large, that this $\mathbf{a}$ is invertible for $|\eta| \geq R$. We associate to our operator the invertible operator-function $\mathbf{c}(y,\eta) = \mathbf{a}(y, R\eta)$, for $|\eta| = 1$. One easily checks that this operator function is differentiable in y, and it improves its properties under η-differentiation: if T_ν are vector fields on S^*Y which are tangent to $S^*Y|_y$ at every point $y \in Y$ then

$$T_{\nu_1} T_{\nu_2} \dots T_{\nu_k} \mathbf{c}(y,\eta) \in \mathfrak{s}_{\frac{q}{k}}(\mathfrak{K}),\ k \leq q. \tag{2.11}$$

Denote the class of such symbols, satisfying (2.11) for $k = 1$ by $\mathfrak{S}_q(S^*Y)$. Then, as it follows from our construction, different choices of coverings, cut-offs etc. lead to homotopic elements in $\mathrm{GL}\,(\mathfrak{S}_q(S^*Y))$. Thus, we have associated to any elliptic operator A a class $[\sigma(A)]$ in the group $\mathbf{K}_1(\mathfrak{S}_q(S^*Y))$. It is clear that this is a homomorphism.

As is usual in the elliptic theory, the index homomorphism factors through the above symbol homomorphism.

Theorem 2.11. *Elliptic operators in $\mathcal{L}_q^0(Y)$ are Fredholm, and the index depends only on its class $[\sigma(A)]$ in $\mathbf{K}_1(\mathfrak{S}_q(S^*Y))$.*

Proof. The reasoning goes along the standard lines, following the proof in the matrix case, say, as in [5],[14]. We explain the main issues, emphasizing the operator specifics. First, the Fredholm property is proved by constructing the regularizer R for the elliptic operator A. In our case, it is important that it regularizes A up to a trace class term:

$$AR-1\in\mathfrak{s}_1(L_2(Y;\mathfrak{K})), RA-1\in\mathfrak{s}_1(L_2(Y;\mathfrak{K})). \tag{2.12}$$

This is performed by the usual construction, first locally, then gluing together local representations. One takes a symbol $\mathbf{r}_0\in\mathcal{S}^0_q$ which equals $\mathbf{a}^{-1}$ for large $|\eta|$, as the initial approximation for $\mathbf{r}$. Then one adds 'lower order terms' to $\mathbf{r}_0$, to compensate the extra terms in the composition formula (2.5), and as a result, the operator version of the composition rule gives (2.12). For large $|\eta|$, the terms in $\mathbf{r}\circ_N\mathbf{a},\mathbf{a}\circ_N\mathbf{r}$ contain products of $\mathbf{a},\mathbf{a}^{-1}$ and their derivatives. Let $\mathbf{r}\circ_N\mathbf{a}|_M$ denote the sum of those terms in $\mathbf{r}\circ_N\mathbf{a}$ where the sum of orders of derivatives does not exceed $2M$. This symbol, for $N\ge M$, does not depend on N, and, for M large enough,

$$Ops(\mathbf{r}\circ_N\mathbf{a}|_M)-1, Ops(\mathbf{a}\circ_N\mathbf{r}|_M)-1\in\mathfrak{s}_1. \tag{2.13}$$

Moreover, the symbols $\mathbf{r}\circ_N\mathbf{a}|_M-\mathbf{1},\mathbf{a}\circ_N\mathbf{r}|_M-\mathbf{1}$ have compact support and belong to the trace class.

The next step is to obtain a rough analytical index formula. It has almost the same form as the analytical index formula (0.6) in [5] (the additional parameter was present there, for bookkeeping the orders of derivatives):

$$\operatorname{ind} OPS(\mathbf{a})=(2\pi)^{-m}\int\limits_{T^*Y}\operatorname{Tr}(\mathbf{a}\circ_N\mathbf{r}|_M-\mathbf{r}\circ_N\mathbf{a}|_M)dyd\eta. \tag{2.14}$$

One interprets (2.14) in the following way. Of course, the concrete expression for the integrand depends on the co-ordinate system and is not a density on T^*Y. However, as it is shown in [5], after integration in η, the integrand in (2.14) becomes a density on Y, thus it can be integrated over Y. The proof of (2.14) is the same as for the analogous formula for the Euclidean space in [14] (see Proposition 4.1 there).

In fact, the particular form of the expression in (2.14) does not matter. The only thing which is important, is that (2.14) involves integration of some expression containing the symbol and its derivatives over a compact set in T^*Y. This enables us to repeat the reasoning in [14]. First, it follows from (2.14) that if two elliptic symbols $\mathbf{a}_1$ and $\mathbf{a}_2$, invertible for $|\eta|\ge R$, coincide in the co-ball bundle $B^*_RY=\{|\eta|\le R\}$, then the indices of operators coincide: in fact, the integrands in (2.14) for $\mathbf{a}_1$ and $\mathbf{a}_2$ coincide inside B^*_RY and are zero outside it. Further, if two such symbols are equal only on the co-spheric bundle $S^*_RY=\partial B^*_RY$, then their indices still coincide since they can be made equal inside B^*_RY by means of a homotopy of elliptic symbols (in fact, the homotopy acting inside B^*_RY only). Finally, if the elliptic symbols $\mathbf{a}_1,\mathbf{a}_2$ are invertible on and outside $\mathfrak{S}^*_{R_1}Y,\mathfrak{S}^*_{R_2}Y$ and their values on these co-spherical bundles are homotopic in the class of invertible symbols, they can be deformed by a homotopy in the class of elliptic symbols to the previous situation. We refer to Section 4 in [14] for the details of the construction. □

Thus it makes sense, for a given invertible element $\mathbf{a} \in \mathfrak{S}_q$, to denote by ind ($[\mathbf{a}]$) the index of any elliptic pseudodifferential operator A on Y, for which $\sigma(A) = [\mathbf{a}]$.

Remark 2. A result similar to Theorem 2.11 was formulated in [12]; the proof there was given for homogeneous symbols, while for general, 'hypo-elliptic' symbols it was only declared that the approach in [9] can be used. This may be correct, but not automatically. In the approach by Hörmander, the case of non-homogeneous symbols is reduced to the case of homogeneous ones by means of an approximation procedure, where the given operator and its adjoint are approximated by operators with homogeneous symbols in the strong sense. This works fine for the usual matrix-valued symbols. However, in the infinite-dimensional case, strong convergence behaves badly under the operation of taking adjoint, and this surely would require some extra conditions to be imposed on the symbols, as well as cause some other complications.

3. Regularisation

Now we will follow the general scheme of [14] in finding the regularisation of the Chern–Simons character of an automorphism of a Hilbert bundle.

Let M be a compact manifold of dimension n and $\mathfrak{K}$ be an infinite-dimensional Hilbert space. By $\mathfrak{S}_0 = \mathfrak{S}_0(M, \mathfrak{K})$ we denote the algebra of functions on M with values being finite rank operators in $\mathfrak{K}$. Thus $GL(\mathfrak{S}_0)$ is the set of invertible operator-valued functions on M having form $\mathbf{1}+\mathbf{k}$ with $\mathbf{k} \in \mathfrak{S}_0$. For any $q \in (0, \infty)$, the algebra $\mathfrak{S}_q^0$ consists of functions having values in $\mathfrak{s}_q(\mathfrak{K})$. This algebra produces $GL(\mathfrak{S}_q^0)$. We always suppose that differentiation preserves the Shatten class properties of the operator-functions in question. In natural topologies, $GL(\mathfrak{S}_q^0)$ is dense in $GL(\mathfrak{S}_0)$.

For $\mathbf{a} \in GL(\mathfrak{S}_0)$, the Chern–Simons character is given by the formula

$$cs(\mathbf{a}) = -\frac{1}{2\pi i}\int_0^1 \operatorname{tr}(\mathbf{a}^{-1}d\mathbf{a}\exp(\frac{1}{-2\pi i}t(1-t)(\mathbf{a}^{-1}d\mathbf{a})^2))dt$$

$$= \sum_{\nu=1} \beta_\nu \operatorname{tr}((\mathbf{a}^{-1}d\mathbf{a})^{2\nu-1}), \tag{3.1}$$

where

$$\beta_\nu = -(\frac{1}{2\pi i})^\nu \frac{(\nu-1)!}{(2\nu-1)!}. \tag{3.2}$$

This expression makes sense for $\mathbf{a} \in GL(\mathfrak{S}_0)$, and, moreover, for $\mathbf{a} \in GL(\mathfrak{S}_1^0)$, since the expression under the trace sign contains at least one trace class operator. Our task is to find some other expression, valid for $\mathbf{a} \in GL(\mathfrak{S}_q^0)$, for a given q, and defining the same co-homology class as (3.1).

We consider a single term in (3.1), with a fixed ν. Take an arbitrary closed (scalar-valued) form ψ of degree $n-(2\nu-1)$ and consider

$$\psi[\mathbf{a}] = \langle \psi, cs(\mathbf{a})\rangle = \int_M \psi cs(\mathbf{a}) \tag{3.3}$$

for $\mathbf{a} \in GL(\mathfrak{S}_1^0)$, so (3.3) makes sense. We can rewrite (3.3) as

$$\psi[\mathbf{a}] = \beta_\nu \int_M \psi \operatorname{tr}((\mathbf{a}^{-1}d\mathbf{a})^{2\nu-1})$$

$$= (-1)^{\nu-1}\beta_\nu \int_M \psi \operatorname{tr}((\mathbf{a}^{-1}-\mathbf{1})d(\mathbf{a}-\mathbf{1})d(\mathbf{a}^{-1}-\mathbf{1})\dots d(\mathbf{a}-\mathbf{1})). \tag{3.4}$$

Consider the multilinear functional, cochain, on $\mathfrak{S}_1^0$, corresponding to (3.4):

$$\varphi_\psi(\mathbf{a}_0,\mathbf{a}_1,\dots,\mathbf{a}_k) = (-1)^{\nu-1}\beta_\nu \int_M \psi \operatorname{tr}(\mathbf{a}_0 d\mathbf{a}_1\dots d\mathbf{a}_k); \mathbf{a}_0,\dots,\mathbf{a}_k \in \mathfrak{S}_1, k=2\nu-1.$$

Due to the cyclicity property of the trace and anti-commutativity of the multiplication of the forms, this cochain is cyclic:

$$\varphi_\psi(\mathbf{a}_1,\mathbf{a}_2,\dots,\mathbf{a}_k,\mathbf{a}_0) = (-1)^{\nu-1}\beta_\nu \int_M \psi \operatorname{tr}(\mathbf{a}_1 d\mathbf{a}_2\dots d\mathbf{a}_0)$$

$$= (-1)^{\nu-1}\beta_\nu \int_M \operatorname{tr}(d\mathbf{a}_0\psi\mathbf{a}_1\dots d\mathbf{a}_k) = (-1)^{\nu-1}\beta_\nu \int_M \psi \operatorname{tr}(d(\mathbf{a}_0\mathbf{a}_1)\dots d\mathbf{a}_k)$$

$$-(-1)^{\nu-1}\beta_\nu \int_M \psi \operatorname{tr}(\mathbf{a}_0 d\mathbf{a}_1\dots d\mathbf{a}_k) = -\varphi_\psi(\mathbf{a}_0,\dots,\mathbf{a}_k).$$

Here we used the fact that the scalar form $\operatorname{Tr}(d(\mathbf{a}_0\mathbf{a}_1)\dots d\mathbf{a}_k))$ is exact. A similar calculation shows that the cochain φ_ψ is closed with respect to the Hochschild co-boundary operation,

$$(b\varphi_\psi)(\mathbf{a}_0,\dots,\mathbf{a}_{k+1}) = \varphi_\psi(\mathbf{a}_0\mathbf{a}_1,\dots,\mathbf{a}_{k+1}) - \varphi_\psi(\mathbf{a}_0,\mathbf{a}_1\mathbf{a}_2,\dots,\mathbf{a}_{k+1}) + \dots$$
$$+\varphi_\psi(\mathbf{a}_{k+1}\mathbf{a}_0,\dots,\mathbf{a}_k) = 0.$$

Thus φ_ψ is a cyclic cocycle and it defines a class in the cyclic co-homology group $HC^k(\mathfrak{S}_1^0)$. Now we apply the iterated suspension homomorphism S to the cocycle φ_ψ. In the calculations below we follow the reasoning in [14], with modifications needed to deal with our new situation. We use the representation of S in the terms of the universal differential graded algebra for $\mathfrak{S}_1^0$ as in [3], [10].

In general, for a given algebra $\mathfrak{S}$, denote by $\tilde{\mathfrak{S}}$ the algebra obtained by adjoining the unit $\mathbf{1}$. For each $k \in \mathbb{N}$, let Ω^k be the linear space

$$\Omega^k = \Omega^k(\mathfrak{S}) = \tilde{\mathfrak{S}} \otimes_{\mathfrak{S}} \mathfrak{S}^{\otimes k}; \quad \Omega = \oplus\Omega^k.$$

The differential $d: \Omega^\mu \to \Omega^{\mu+1}$ is given by

$$d((\mathbf{a}_0+\lambda\mathbf{1})\otimes\mathbf{a}_1\otimes\dots\otimes\mathbf{a}_k) = \lambda\mathbf{1}\otimes\mathbf{a}_0\otimes\dots\otimes\mathbf{a}_k \in \Omega^{k+1}.$$

By construction, one has $d^2 = 0$. One defines a right $\mathfrak{S}$-module structure on Ω^μ by setting

$$(\tilde{\mathbf{a}}_0 \otimes \mathbf{a}_1 \otimes \cdots \otimes \mathbf{a}_k)\mathbf{a} = \sum_{i=0}^{k} (-1)^{k-i} \tilde{\mathbf{a}}_0 \otimes \cdots \otimes \mathbf{a}_i \mathbf{a}_{i+1} \otimes \cdots \otimes \mathbf{a}.$$

This right action of $\mathfrak{S}$ extends in a natural way to the product in Ω:

$$\omega(\tilde{\mathbf{b}}_0 \otimes \mathbf{b}_1 \otimes \cdots \otimes \mathbf{b}_k) = (\omega\tilde{\mathbf{b}}_0) \otimes \mathbf{b}_1 \otimes \cdots \otimes \mathbf{b}_k, \; \omega \in \Omega^{k'}.$$

The product satisfies

$$\tilde{\mathbf{a}}_0 d\mathbf{a}_1 \dots d\mathbf{a}_k = \tilde{\mathbf{a}}_0 \otimes \mathbf{a}_1 \otimes \cdots \otimes \mathbf{a}_k, \; \mathbf{a}_i \in \mathfrak{S}$$

and gives Ω the structure of a graded differential algebra. This algebra is universal in the sense that any homomorphism ρ of $\mathfrak{S}$ into some differential graded algebra (Ω', d') extends to a homomorphism ρ^* of (Ω, d) to (Ω', d') respecting the product of differentials:

$$\rho^*(\tilde{\mathbf{a}}_0 d\mathbf{a}_1 \dots d\mathbf{a}_k) = \rho(\mathbf{a}_0) d(\rho(\mathbf{a}_1)) \dots d(\rho(\mathbf{a}_k)).$$

The particular situation where we use this universality property is for $\mathfrak{S} = \mathfrak{S}_1^0$ and for the differential algebra Ω' being the algebra of operator-valued differential forms on M,

$$\omega' = \tilde{\mathbf{a}}_0 d'\mathbf{a}_1 \dots d'\mathbf{a}_k, \; \mathbf{a}_i \in \mathfrak{S}_1^0$$

where in the product, for the terms of the form

$$d'\mathbf{a} = \sum \mathbf{a}^i dx_i,$$

the usual product of operators and the exterior product of differentials is used. Thus in the degree zero, two differential algebras coincide, and we take identity as the homomorphism ρ involved in the definition of the universality property. We will omit the prime symbol in the sequel.

According to Proposition 4, Ch. III.1, in [3], any cyclic cocycle $\varphi \in C_\lambda^k(\mathfrak{S})$ of dimension k can be represented as

$$\varphi(\mathbf{a}_0, \dots, \mathbf{a}_k) = \hat{\varphi}(\mathbf{a}_0 d\mathbf{a}_1 \dots d\mathbf{a}_k),$$

where $\hat{\varphi}$ is a closed graded trace of dimension k on $\Omega(\mathfrak{S})$. In our particular case, for the cocycle φ_ψ of degree $k = 2\nu - 1$, this representation is generated by

$$\hat{\varphi}_\psi(\tilde{\mathbf{a}}_0 d\mathbf{a}_1 \dots d\mathbf{a}_k) = (-1)^{\nu-1}\beta_\nu \int_M \psi \operatorname{tr}(\tilde{\mathbf{a}}_0 d\mathbf{a}_1 \dots d\mathbf{a}_k),$$

where b_ν are given in (3.2).

In the terms of the above model, on the cocycle level, S admits the following description. For the algebra of complex numbers $\mathbb{C}$, we consider the graded differential algebra $\Omega(\mathfrak{S}) \otimes \Omega(\mathbb{C})$, with elements having the form

$$(\mathbf{a}_0 \otimes w_0) d(\mathbf{a}_1 \otimes w_1) \dots d(\mathbf{a}_k \otimes w_k),$$

where $w_0, \dots, w_k \in \tilde{\mathbb{C}}$, with differential

$$d(\mathbf{a} \otimes w) = (d\mathbf{a}) \otimes w + \mathbf{a} \otimes dw.$$

For a cyclic cocycle $\varphi \in C^k_\lambda(\mathfrak{S})$ and cyclic cocycle $\sigma \in C^p_\lambda(\mathbb{C})$, following [3], we define the cup product $\varphi\sharp\sigma \in C^{k+p}_\lambda(\mathfrak{S} \otimes \mathbb{C}) = C^{k+p}_\lambda(\mathfrak{S})$ by setting

$$\varphi\sharp\sigma(\mathbf{a}_0, \dots, \mathbf{a}_{k+p}) = (\hat{\varphi} \otimes \hat{\sigma})((\mathbf{a}_0 \otimes e)d(\mathbf{a}_1 \otimes e)\dots d(\mathbf{a}_{k+p} \otimes e)). \tag{3.5}$$

Here, e is the unit in $\mathbb{C}$, i.e., the element $1 + 01 \in \tilde{\mathbb{C}}$, $\hat{\varphi}, \hat{\sigma}$ are graded closed traces of degree, respectively, k and p in $\Omega(\mathfrak{S}), \Omega(\mathbb{C})$ representing φ, σ and thus only terms of bidegree (k, p) survive in (3.5). It is shown in [3] that $\varphi\sharp\sigma$ is a cyclic cocycle. In particular, take $\psi = \sigma_1 \in C^2_\lambda(\mathbb{C})$, $\sigma_1(e, e, e) = 1$. Cup product with σ_1 generates the homomorphism S in cyclic co-homologies.

Now we consider iterations of S. For an even integer $p = 2l$, we set $\sigma_l = \sigma_1^{\sharp l}$ where $\sharp l$ denotes taking to the power l in the sense of $\sharp$ operation. According to Corollary 13, Ch. III.1 in [3], $\sigma_l(e, e, \dots, e) = l!$. To the cocycle σ_l, there corresponds the graded closed trace $\hat{\sigma}_l$ of degree p on $\Omega^*(\mathbb{C})$, moreover $\hat{\sigma}_l(ede\dots de) = l!$. Cup multiplication with the cocycle σ_l generates the iterated homomorphism S^l in cyclic cohomologies of the algebra $\mathfrak{S}$. We will study the structure of $S^l\varphi \in C^{k+p}_\lambda(\mathfrak{S})$, for $\varphi \in C^k_\lambda(\mathfrak{S})$. According to (3.5),

$$S^l\varphi(\mathbf{a}_0, \mathbf{a}_1 \dots \mathbf{a}_{k+p}) = \widehat{S^l\varphi}((\mathbf{a}_0 \otimes e)d(\mathbf{a}_1 \otimes e)\dots d(\mathbf{a}_{k+p} \otimes e))$$

$$= \hat{\varphi} \otimes \hat{\sigma}((\mathbf{a}_0 \otimes e)(d\mathbf{a}_1 \otimes e + \mathbf{a}_1 \otimes de)\dots(d\mathbf{a}_{k+p} \otimes e + \mathbf{a}_{k+p} \otimes de)). \tag{3.6}$$

Since $\hat{\varphi}$ is a graded trace of degree k and $\hat{\sigma}_l$ is a graded trace of degree $p = 2l$, only the terms of bidegree (k, p) contribute to (3.6). There are a lot of such terms, and each of them involves the value of $\hat{\varphi}$ on a certain ordered product of $\mathbf{a}_i$ and $d\mathbf{a}_i$, where exactly $k + 1$ factors are of the form $\mathbf{a}_i$, including $\mathbf{a}_0$, as well as the value of $\hat{\sigma}_l$ on the corresponding ordered product of the elements e and de, with $k + 1$ factors e, including the first one, and p factors de. Quite a lot of these terms vanish. In fact, since e is an idempotent, $e = e^2$, we have

$$de = e(de) + (de)e, \quad e(de)e = 0, \quad e(de)(de) = (de)(de)e. \tag{3.7}$$

Therefore, if some term in the expansion of (3.6) contains the product of an odd number of consecutive de surrounded by e, it vanishes. Thus only those terms survive where each group of consecutive de in the product contains an even number of de. For any such product, using (3.7), we can rearrange the factors e, de and arrive at the expression $\hat{\sigma}_l(ede\dots de)$ which equals $l!$. This leaves us with the contribution to (3.6) involving $\mathbf{a}_i$ and $d\mathbf{a}_i$. In these terms, the variables $\mathbf{a}_i$ enter in a very special way. Since, for $i \neq 0$, $\mathbf{a}_i$ stand on the places where we had de in (3.6), and $d\mathbf{a}_i$ on the places where we had e, only those terms survive in (3.6), where an even number of variables $\mathbf{a}_i$ stand in succession, not counting $\mathbf{a}_0$. This gives us the following characterisation of $S^l\varphi$.

Proposition 3.1. *The image $S^l\varphi$ in C_λ^{k+p} of a cyclic cocycle $\varphi \in C_\lambda^k$ under the iterated homomorphism S^l equals*

$$S^l\varphi(\mathbf{a}_0, \dots, \mathbf{a}_{k+p}) = l! \sum_{u_i, v_i} \hat{\varphi}(\mathbf{a}_0 \prod_i A_i B_i), \tag{3.8}$$

where the summation is performed over collections of u_i, v_i such that $1 = v_0 \le u_1 < v_1 < \cdots < u_s \le v_s = k+p+1$, $u_i - v_i$ are even, $A_i = \mathbf{a}_{v_{i-1}} \dots \mathbf{a}_{u_i - 1}$, $B_i = d\mathbf{a}_{u_i} \dots d\mathbf{a}_{v_i - 1}, \sum(v_i - u_i) = k$.

Getting an explicit analytical description of (3.8) for a general collection of $\mathbf{a}_i$ might be quite troublesome. We, however, are interested only in the value of $S^l\varphi$ on a very special collection of these variables. In fact, according to (3.2), (3.4), we are interested in evaluating $\varphi_\psi(\mathbf{a}_0, \dots, \mathbf{a}_k)$, for $\mathbf{a}_{2i} = \mathbf{a}^{-1} - 1$, $\mathbf{a}_{2i+1} = (\mathbf{a} - 1)$, $\mathbf{a} \subset GL(\mathfrak{S}_1)$, so that $\mathbf{a}_i \in \mathfrak{S}_1$. This enables us to simplify the expression (3.8).

Proposition 3.2. *For $\mathbf{a} \in GL(\mathfrak{S}_1^0)$,*

$$\begin{aligned} &S^l\varphi_\psi(\mathbf{a}^{-1} - \mathbf{1}, \mathbf{a} - \mathbf{1}, \dots, \mathbf{a}^{-1} - \mathbf{1}, \mathbf{a} - \mathbf{1}) \\ &= (l!)^{-1}\beta_\nu \int\limits_M \psi \operatorname{tr} \left[(\frac{d}{ds})^l (\mathbf{a}^{-1}(1 - s\mathbf{c})^{-1} d\mathbf{a})^k |_{s=0} \right], \end{aligned} \tag{3.9}$$

$$\begin{aligned} &S^l\varphi_\psi(\mathbf{a}^{-1} - \mathbf{1}, \mathbf{a} - \mathbf{1}, \dots, \mathbf{a}^{-1} - \mathbf{1}, \mathbf{a} - \mathbf{1}) \\ &= \beta_\nu \operatorname{tr} \int\limits_M \psi(\mathbf{c} + \mathbf{a}^{-1} d\mathbf{a})^{k+l}, \ \mathbf{c} = (\mathbf{a} - \mathbf{1})(\mathbf{a}^{-1} - \mathbf{1}). \end{aligned} \tag{3.10}$$

Proof. We have to show that for our specific choice of variables, the expression (3.8) takes the form (3.9), (3.10). We can see that in (3.8), each term A_i equals $\mathbf{c}^{\frac{(u_i - v_{i-1})}{2}}$. This means that all terms in (3.8) can be obtained in the following way. Write down the expression $\mathbf{a}^{-1} d\mathbf{a} d(\mathbf{a}^{-1}) \dots d\mathbf{a}$, with ν factors $d\mathbf{a}$ and $\nu - 1$ factors $d(\mathbf{a}^{-1})$. Before each $d\mathbf{a}$, $d(\mathbf{a}^{-1})$ insert several factors $\mathbf{c}$, so that there are l of them altogether. Summing all such products and taking into account that $d(\mathbf{a}^{-1}) = -\mathbf{a}^{-1} d\mathbf{a}\mathbf{a}^{-1}$ and that $\mathbf{c}$ and $\mathbf{a}$ commute, we come to the formula

$$S^l\varphi_\psi(\mathbf{a}^{-1} - 1, \dots, \mathbf{a} - 1) = (-1)^{\nu-1}\beta_\nu \int_M \psi \sum_{\sum \kappa_i = l} \prod_i (\mathbf{c}^{\kappa_i} \mathbf{a}^{-1} d\mathbf{a}). \tag{3.11}$$

To describe (3.11) more explicitly, introduce an extra variable s and consider the expression depending on s:

$$\Psi(s, \mathbf{a}) = (-1)^{\nu-1}\beta_\nu \int_M \psi \operatorname{tr} (((1 - s\mathbf{c})^{-1} \mathbf{a}^{-1} d\mathbf{a})^k). \tag{3.12}$$

For s small enough, $1 - s\mathbf{c}$ is invertible, and therefore (3.12) can be rewritten as

$$\Psi(s, \mathbf{a}) = (-1)^{\nu-1}\beta_\nu \int_M \psi((1 + s\mathbf{c} + s^2\mathbf{c}^2 + \cdots)\mathbf{a}^{-1} d\mathbf{a})^k. \tag{3.13}$$

Now we can see that (3.11) equals the coefficient at s^l in (3.13), and this gives us (3.9). As for (3.10), it is clear that the term of degree $k = 2\nu - 1$ in the integrand, the only one that has correct degree and thus survives under the integration, exactly equals the integrand in (3.11). □

Now we are able to find the regularized expression for the functional $\psi[\mathbf{a}] = \int_M \psi cs(\mathbf{a})$.

Proposition 3.3. *For* $\mathbf{a} \in GL(\mathfrak{S}_1^0)$ *and a closed* $n - (2\nu - 1)$ *form* ψ,

$$\psi[\mathbf{a}] = \int_M \psi cs_{2\nu-1}^{(l)} = \int_M (l!)^{-1}\beta_{\nu,l}\psi \operatorname{tr}\left[(\frac{d}{ds})^l (\mathbf{a}^{-1}(1 - s\mathbf{c})^{-1} d\mathbf{a})^k|_{s=0} \right], \quad (3.14)$$

where

$$\beta_{\nu,l} = \beta_\nu \gamma_\nu^{-1} \gamma_{\nu+l} = \frac{l!(\nu + l - 1)!}{(2\nu + 2l - 1)!},$$

γ_ν *are constants in* (1.9). *The functional in* (3.13) *admits extension by continuity to* $\mathbf{a} \in GL(\mathfrak{S}_q^0)$ *for* $q < 2\nu + 2l - 1$.

Proof. The expression (3.14) for the functional follows immediately from (3.9) and the periodicity relation (1.9). To prove the extension property, note that each summand in (3.9) contains $2\nu - 1$ factors $d\mathbf{a}$ and l factors $\mathbf{c}$; since for $\mathbf{a} \in GL(\mathfrak{S}_q^0)$, we have $d\mathbf{a} \in \mathfrak{s}_q$, $\mathbf{c} \in \mathfrak{s}_{\frac{q}{2}}$, and therefore the whole product belongs to $\mathfrak{s}_1$. □

Since the form ψ can be chosen arbitrarily, the relation (3.14) can be expressed in the following way.

Theorem 3.4. *For* $\mathbf{a} \in GL(\mathfrak{S}_1)$, *the forms* $cs_{2\nu-1}(\mathbf{a})$ *and* $cs_{2\nu-1}^{(l)}(\mathbf{a})$ *define the same co-homology class in* $H^{2\nu-1}(M;\mathbb{C})$. *The latter form can be extended to* $\mathbf{a} \in GL(\mathfrak{S}_q^0)$, $q < 2\nu - 1 + 2l$.

4. The full Chern–Simons character and unusual index formulas

In order to use the expression for the components of the Chern–Simons character found in Section 3 for calculating the index, we consider the full Chern–Simons character

$$cs^{(l)}(\mathbf{a}) = \sum_\nu sc_{2\nu-1}^{(l)}(\mathbf{a}). \quad (4.1)$$

We are going to find for $cs^{(l)}(\mathbf{a})$ an expression similar to (3.1), therefore we perform a certain summation procedure for the terms in (3.14).

Consider the formal sum

$$h(\omega) = \sum_{\nu=1}^{\infty} \frac{(\nu + l - 1)!}{(2\nu + 2l - 1)!}\omega^{2\nu-1}.$$

We can transform it as

$$\omega^{2l-1} \sum_{\nu=1}^{\infty} \frac{1}{(\nu + l - 1)!} \int_0^1 (t(1 - t)\omega^2)^{\nu+l-1} dt. \quad (4.2)$$

Setting $\sigma = t(1-t)\omega^2$, we can write

$$h(\omega) = \omega^{2l-1} \int_0^1 e_l(\sigma)dt,$$

where $e_l(\zeta) = \exp(\zeta) - 1 - \zeta - \zeta^2/2! - \cdots - \zeta^{l-1}/(l-1)!$. Simple calculations give

$$e_l(\zeta) = (l!)^{-1} \int_0^\zeta (\zeta - \zeta')^{l-1} \exp(\zeta')d\zeta',$$

thus $\exp_l(z) = \zeta^{-l} e_l(z)$ is the Riesz mean of order l of the exponential function. In this notation, (4.2) takes the form

$$h(\omega) = \int_0^1 (t(1-t))^{l-1}\omega \exp_l(t(1-t)\omega^2)dt.$$

We apply these formal calculations for $\omega = (-2\pi i)^{-1/2}(1 - s\mathbf{c})\mathbf{a}^{-1}d\mathbf{a}$. The formally infinite series are in fact finite, since they cannot contain forms of order higher than n, and this gives us the regularisation for (4.1).

Theorem 4.1. *Let* $\mathbf{a} \in GL(\mathfrak{S}_1^0)$. *Define*

$$cs^{(l)}(\mathbf{a}) = \frac{1}{-2\pi i}$$

$$\times \operatorname{Tr}(\frac{d}{ds})^l \left[\int_0^1 ((\mathbf{1} - s\mathbf{c})\mathbf{a}^{-1}d\mathbf{a})(t(1-t))^{l-1} \exp_l(t(1-t)\frac{((1-s\mathbf{c})\mathbf{a}^{-1}d\mathbf{a})^2}{2\pi i})dt \right] |_{s=0}. \tag{4.3}$$

Then $cs^{(l)}(\mathbf{a})$ *and* $cs(\mathbf{a})$ *define the same co-homology class in* $H^{odd}(M)$, *moreover* $cs^{(l)}(\mathbf{a})$ *extends by continuity to* $\mathbf{a} \in GL(\mathfrak{S}_q^0)$, $q < 2l+1$.

Now we can apply our regularized Chern–Simons character for calculating the index of pseudodifferential operators with operator-valued symbols.

For a given m-dimensional manifold Y, we set $M = S^*Y$, $n = 2m-1$. We consider an elliptic pseudodifferential operator A having the symbol in $\mathcal{S}_q$. As it is shown in Section 2, the index of the operator A depends only on the stable homotopy class of its principal symbol $\mathbf{a}$ in $\mathbf{K}_1(\mathfrak{S}_q(S^*Y))$.

Suppose first that $\mathbf{a} \in GL(\mathfrak{S}_q^0(S^*Y))$. Recall that this means that the operator-valued function $\mathbf{a}$ is invertible and differs from the identity operator by an operator in the Shatten class $\mathfrak{s}_q$.

Theorem 4.2. *If* $\mathbf{a} \in GL(\mathfrak{S}_q^0(S^*Y))$, *then*

$$\operatorname{ind} A = \int\limits_{S^*Y} \mathcal{T}(Y) cs^{(l)}(\mathbf{a}), \tag{4.4}$$

where $cs^{(l)}(\mathbf{a})$ *is defined in* (4.3).

Proof. For $\mathbf{a} \in GL(\mathfrak{S}_1^0(S^*Y))$, the index formula (4.4) follows from the usual one and Theorem 4.1. Further, according to Theorem 2.11 and again Theorem 4.1, both parts in (4.4) are continuous in the topology of $GL(\mathfrak{S}_q^0(S^Y))$. □

Finally, we extend the index formula to principal symbols in $\mathfrak{S}_q(S^*Y)$. Recall that these symbols, defined in Section 2, satisfy (2.11). Here the regularisation of the Chern–Simons classes given by (4.3) is not sufficient because for a symbol $\mathbf{a} \in GL(\mathfrak{S}_q)$, the symbol $\mathbf{c} = (\mathbf{a} - \mathbf{1})(\mathbf{a}^{-1} - \mathbf{1})$ is, generally, only bounded and multiplication by it does not improve trace ideals properties. Therefore one more regularisation is needed.

Proposition 4.3. *Let $\mathbf{a} \in \mathfrak{S}_q$ be an invertible symbol on S^*Y. Then there exists an invertible operator-valued function $\mathbf{a}_0(y)$ acting in the space $\mathfrak{K} \oplus \mathbb{C}^d$, for some finite d, differentiable on Y so that $\mathbf{a}_0(y) - (\mathbf{a}(y,\eta) \oplus \mathbf{1}_d) \in \mathfrak{s}_q$.*

Remark 3. In fact, this operator $\mathbf{a}_0$ can be explicitly constructed, given $\mathbf{a}$, but the particular choice is of no importance here.

Proof. Everything is very easy in the case of the manifold Y having vanishing Euler characteristic. For such manifolds, there exists a smooth section $\eta(y)$ of the co-spheric bundle of Y, and one can set $\mathbf{a}_0(y) = \mathbf{a}(y, \eta(y))$. For a manifold with a nonzero Euler characteristic, the construction is more complicated. Set $\mathbf{a}(y,0) = \int_{S_y^*Y} \mathbf{a}(y,\eta)d\eta$. The definition of the class $\mathfrak{S}_q$ implies that for any $\eta \in S_y^*(Y)$, $\mathbf{a}(y,0)$ differs from $\mathbf{a}(y,\eta)$, by an operator in $\mathfrak{s}_q(\mathfrak{K})$. Since $\mathbf{a}(y,\eta)$ is invertible, this implies that $\mathbf{a}(y,0)$ is a Fredholm operator in $\mathfrak{K}$ with zero index. Having such a Fredholm family, one can, in a standard way, include $\mathbf{a}(y,0)$ as the upper left corner in a matrix $\mathbf{a}_0(y) = \begin{pmatrix} \mathbf{a}(y,0) & \mathbf{g}(y) \\ \mathbf{h}(y) & \mathbf{k}(y) \end{pmatrix}$, with operator functions $\mathbf{g} : \mathbb{C}^d \to \mathfrak{K}$, $\mathbf{h} : \mathfrak{K} \to \mathbb{C}^d$ and $\mathbf{k} : \mathbb{C}^d \to \mathbb{C}^d$, so that this block-matrix is invertible (see e.g., in [7] the details of the construction.) Since the operator-functions $\mathbf{g}, \mathbf{h}, \mathbf{k}$ have finite rank, the difference $(\mathbf{a}(y,\eta) \oplus \mathbf{1}_d) - \mathbf{a}_0(y)$ belongs to $\mathfrak{s}_q$. □

Using this operator function $\mathbf{a}_0(y)$, we can now establish the index formula. First, note that the symbols $\mathbf{a}$ and $\tilde{\mathbf{a}} = \mathbf{a} \oplus \mathbf{1}_d$ represent the same class in the group $\mathbf{K}_1(\mathfrak{S}_q)$, therefore they have the same index. The index of the symbol $\mathbf{a}_0(y)$ equals zero, since the corresponding operator, multiplication by the operator-valued function $\mathbf{a}_0(y)$, is an isomorphism. Therefore, the index of the symbol

$$\mathbf{b}(y,\eta) = \mathbf{a}_0(y)^{-1}\tilde{\mathbf{a}}(y,\eta) \tag{4.5}$$

equals the index of the symbol $\mathbf{a}$. But now, the symbol $\mathbf{b}$ belongs to the class $\mathfrak{S}_q^{0+}$, and Theorem 4.2 can be applied. This gives our final result.

Theorem 4.4. *For an invertible symbol $\mathbf{a} \in GL(\mathfrak{S}_q^0)$, define the symbol $\mathbf{b}$ as in (4.5). Then*

$$\operatorname{ind}([\mathbf{a}]) = \int_{S^*Y} \mathcal{T}(Y) cs^{(l)}(\mathbf{b}),$$

where the regularized Chern–Simons character is given by (4.3).

References

[1] B. Blackadar, *K-theory for Operator Algebras*. Springer-Verlag, New York et al. 1986.

[2] A. Connes, *Noncommutative Differential geometry*. Publ. Math. I.H.E.S **62** (1985), 257–360.

[3] A. Connes, *Noncommutative Geometry*. Academic Press, NY, 1994.

[4] C. Dorschfeldt, B.-W. Schulze, *Pseudo-differential operators with operator-valued symbols in the Mellin-edge-approach*. Ann. Global Anal. Geom. **12** (1994), 135–171.

[5] B. Fedosov, *Analytic formulas for the index of elliptic operators*. Trans. Moscow Math. Soc. **30** (1974), 159–240.

[6] B. Fedosov, *Index theorems*. Encyclop. Math. Sci. Vol. 65 (Partial Differential Equations, VIII) 155–251, Springer, Berlin, 1996.

[7] B. Fedosov, B.-W. Schulze, N. Tarkhanov, *On the index of elliptic operators on a wedge*. J. Funct. Anal. **157** (1998), 164–209.

[8] A. Grigis, J. Sjöstrand, *Microlocal analysis for differential operators. An Introduction*. Cambridge Univ. Press, Cambridge, 1994.

[9] L. Hörmander, *On the index of pseudodifferential operators*. Elliptische Differentialgleichungen, Band II, Akademie-Verlag, 127–146, 1969.

[10] M. Karoubi, *Homologie Cyclique et K-théorie*. Astérisque, **149**, 1987.

[11] M. Karoubi, *Théorie générale des classes caractéristiques secondaires*. K-theory, **4** (1990), 55–88.

[12] G. Luke, *Pseudodifferential operators on Hilbert bundles*. Journal Diff. Equat. **12** (1972), 566–589.

[13] B. Plamenevsky, G. Rozenblum, *Pseudodifferential operators with discontinuous symbols: K-theory and index formulas*. (Russian) Funktsional. Anal. i Prilozhen. **26** (1992), 4, 45–56, (Translation in: Functional Anal. Appl. **26** (1993), 266–275).

[14] G. Rozenblum, *On some analytical index formulas related to operator-valued symbols*. Electr. J. Diff. Equat. (2002), 17, 31 pp.

[15] B.-W. Schulze, B. Sternin, V.Shatalov, *Differential equations on singular manifolds: semiclassical theory and operator algebra*. Wiley, Berlin, 1998.

Grigori Rozenblum
Department of Mathematics
Chalmers Univ. of Technology
S-41296 Göteborg, Sweden
e-mail: `grigori@math.chalmers.se`

Subseries
Advances in Partial Differential Equations

Subseries editors:

Bert-Wolfgang Schulze
Institut für Mathematik
Universität Potsdam
14415 Potsdam
Germany

Michael Demuth
Institut für Mathematik
Technische Universität Clausthal
38678 Clausthal-Zellerfeld
Germany

Sergio Albeverio
Institut für Angewandte Mathematik
Universität Bonn
53115 Bonn
Germany

Elmar Schrohe
Institut für Mathematik
Universität Potsdam
14415 Potsdam
Germany

The subseries *Advances in Partial Differential Equations* is intended to report on recent developments in all areas of partial differential equations, in particular, on those belonging to the tradition defined by the series *Operator Theory: Advances and Applications* and new areas. The *Advances* volumes will consist primarily of expository research articles, presenting both an overview of the state of a field and new results, in such areas as microlocal analysis and its applications to asymptotic phenomena, mathematical physics, spectral theory, as well as symplectic geometry, the analysis of singularities of solutions, and geometric analysis. Selected articles of this kind may also appear as a separate monographic volume. A certain number of short communications will also be published.

Proposals for manuscripts may be sent to B.-W. Schulze or to one of the other subseries editors.

www.ingramcontent.com/pod-product-compliance
Lightning Source LLC
Chambersburg PA
CBHW071410230926
51846CB00001B/43

9783034894296